AF411629

Integrated Analytical Systems

Series Editor

Radislav A. Potyrailo, GE Global Research, Niskayuna, NY, USA

For further volumes:
http://www.springer.com/series/7427

Series Preface

> In my career I've found that "thinking outside the box" works better if I know what's "inside the box."
>
> Dave Grusin, composer and jazz musician

> Different people think in different time frames: scientists think in decades, engineers thinkin years, and investors think in quarters.
>
> Stan Williams, Director of Quantum Science Research, Hewlett Packard Laboratories

> Everything can be made smaller, never mind physics;
> Everything can be made more efficient, never mind thermodynamics;
> Everything will be more expensive, never mind common sense.
>
> Tomas Hirschfeld, pioneer of industrial spectroscopy

Integrated Analytical Systems

Series Editor: Dr. Radislav A. Potyrailo, GE Global Research, Niskayuna, NY

The field of analytical instrumentation systems is one of the most rapidly progressing areas of science and technology. This rapid development is facilitated by (1) the advances in numerous areas of research that collectively provide the impact on the design features and performance capabilities of new analytical instrumentation systems and by (2) the technological and market demands to solve practical measurement problems.

The book series *Integrated Analytical Systems* reflects the most recent advances in all key aspects of development and applications of modern instrumentation for chemical and biological analysis. These key development aspects include: (1) innovations in sample introduction through micro-and nano-fluidic designs, (2) new types and methods of fabrication of physical transducers and ion detectors, (3) materials for sensors that became available due to the breakthroughs in biology, combinatorial materials science and nanotechnology, (4) innovative data processing and mining methodologies that provide dramatically reduced rates of false alarms, and (5) new scenarios of applications of the developed systems.

A multidisciplinary effort is required to design and build instruments with previously unavailable capabilities for demanding new applications. Instruments with more sensitivity are required today to analyze ultra-trace levels of environmental pollutants, pathogens in water, and low vapor pressure energetic materials in air. Sensor systems with faster response times are desired to monitor transient in-vivo events and bedside patients. More selective instruments are sought to analyze specific proteins in vitro and analyze ambient urban or battlefield air. Distributed sensors for multiparameter measurements (often including not only chemical and biological but also physical measurements) are needed for surveillance over large terrestrial areas or for personal health monitoring as wearable sensor networks. For these and many other applications, new analytical instrumentation is urgently needed. This book series is intended to be a primary source on both fundamental and practical information of where analytical instrumentation technologies are now and where they are headed in the future.

September 2012   Radislav A. Potyrailo

Michael A. Carpenter · Sanjay Mathur
Andrei Kolmakov

Editors

Metal Oxide Nanomaterials for Chemical Sensors

 Springer

Editors
Michael A. Carpenter
College of Nanoscale Science
    and Engineering
State University of New York
Albany, NY
USA

Andrei Kolmakov
Department of Physics
Southern Illinois University
Carbondale, IL
USA

Sanjay Mathur
Institute of Inorganic Chemistry
University of Cologne
Cologne
Germany

ISBN 978-1-4614-5394-9   ISBN 978-1-4614-5395-6   (eBook)
DOI 10.1007/978-1-4614-5395-6
Springer New York Heidelberg Dordrecht London

Library of Congress Control Number: 2012950395

Preface

Metal oxide materials due to their unique combination of redox chemistry, optical, electrical and semiconductor properties, have for many years played a key role in the successful implementation of chemical sensor technology. Given the intrinsic advantages of confinement effects and fundamentally new material properties of nanoscopic materials, there is a strong drive to exploit the potential of nanosized metal oxide materials and their new morphologies for chemical sensing applications. However, the multi-dimensional interplay among interfacial interactions, chemical composition, preparation method, and end-use conditions of metal oxide nanomaterials strongly affects the sensor functionality, which often makes device integration and development for real-world applications very challenging.

Research efforts to improve the performances of present metal oxide sensor technology through the variation of both surface chemistry, morphology and microstructure via combinatorial and chemically directed design and novel synthetic methods has begun to yield libraries of materials for use and development as chemical sensing materials. Interestingly, while the general reaction mechanism for both oxidizing and reducing gases on metal oxides is thought to be understood, there are many details within the reaction mechanism which induces the subsequent sensing signal that are not definitively characterized and points out the need for further development.

The contents of this book present a state-of-the-art collection and critical survey of recent developments in the implementation of metal oxide nanomaterial research methodologies for the discovery and optimization of new sensor materials, methods and sensing systems. The book should be of interest to a diverse and broad readership belonging to both academia and industrial research units as it provides a detailed description and analysis of (i) metal oxide nanomaterial sensing principles (ii) advances in metal oxide nanomaterial synthesis/deposition methods, including liquid and vapor processing techniques (iii) advances, challenges and insights gained from the in situ/ex situ analysis of reaction mechanisms and (iv) technical development and integration challenges in the fabrication of sensing arrays and devices.

Chapter 1 describes the generally accepted reaction mechanism of both oxidizing and reducing gases with metal oxides. In many ways this reaction mechanism is thought to be well understood. However, a direct measurement of the surface species produced via the myriad of interfacial reactions taking place, both during active gas exposure and while sensing signal measurement is taking place, i.e. operando conditions, has been difficult and in many cases inconclusive with respect to the proposed reaction mechanism. This first chapter provides a very detailed review of this topic and its associated measurement challenges.

As a follow-on to the first chapter's discussion of reaction mechanisms, Chap. 2 details many of the classic methods used in studying the oxidation or reduction reactions on metal oxides. These ultra-high vacuum surface science experiments are able to produce and characterize perfectly clean model metal oxide surfaces for study. Surface science methods that are working on closing the gap between model UHV and real sensor exposure conditions are described with respect to the methods and their limitations. Lastly, unique TiO_2 metal oxide surfaces prepared using grazing incidence low energy ion sputtering are described, which provide new insight for some unique sensor applications. These illustrations are also an example for expanding the use of surface science techniques to include unique surface preparation as well as its characterization and study.

The design and synthesis of metal oxides for sensing applications can benefit from a chemical principles approach as detailed in Chap. 3. Specifically, the redox reactions of metal oxide materials with a variety of gases is examined from a determination of the acidity and basicity of the metal oxide surface. Current concepts of interrelationships between metal oxide chemical composition, crystal and surface structure and its activity in the reaction with gas phase components are considered. Details are provided on how these calculations are made and applied towards the doping of SnO_2 nanomaterials with a range of dopants with varying acidic/basic character. The variation in response of these materials with respect to the selectivity and the overall sensor response to both oxidizing and reducing target gases is detailed.

Chapter 4 begins with an introduction into the use of metal oxides for chemical sensing applications both from a materials development standpoint as well as a description of the generally accepted reaction mechanism for oxidizing and reducing gases. The chemistry of metal oxides is rich with possibilities, even more so when one considers the range of dopants that can be added to a given metal oxide material. Such a variation modifies the reactive properties of the metal oxide towards the target gases as well as modifying its corresponding thermal dependence. To develop libraries of materials, Chap. 4 includes a description of a combinatorial synthesis and testing methodology for a wide variety of metal oxide nanomaterials. A series of experimental approaches are described with a range of sensing examples provided.

The characteristic sensing dependence as a function of control of the crystalline character is developed in Chap. 5 with studies pertaining to selected microstructures. The synthesis and characterization of these selected microstructures are described for TiO_2 (anatase) and dopant-stabilized ε-WO_3. While anatase was shown to have unique activities over rutile, ε-WO$_3$ shows a high sensitivity and unique selectivity to polar gas molecules. Such unique dependencies can be particularly useful in the development of both sensitive and selective sensing arrays.

The need of developing materials with an increasing level of control is continued in Chap. 6. Molecular beam epitaxial (MBE) growth of metal oxide thin films is a method enabling the development of nanomaterials with a higher level of crystalline ordering than that achievable by physical vapor deposition methods. Such a high level of ordering has been found to be beneficial, for instance, in increasing the oxygen ion conductivity of metal oxide films and can provide interesting characteristics from a chemical sensors perspective. The MBE deposition method and its associated materials characterization techniques are described and examples of the use of MBE grown metal oxide materials for sensing applications are provided.

Chapter 7 outlines a methodology to provide atomic level control over the chemistry of the metal oxide as well as the ability to coat geometries and structures with angstrom levels of film thickness control. Atomic layer deposition (ALD) methods have been well developed for a variety of needs related to the integrated circuit (IC) industry. However, many of the properties of this technique which are so attractive for the IC industry are also of interest for the development of metal oxide nanomaterials for chemical sensors. The characteristics of the ALD method are outlined and applications for coating both thin films on flat and very porous substrate materials are described in the context of a series of sensing applications.

While the chemistry available for producing a variety of metal oxides is rich with possibilities there are also a number of new methods, beyond the well known colloidal wet chemistry or vapor phase processing methods that are available for production of unique metal oxide nanomaterials. Chapter 8 details one of the more recent efforts which utilizes microwave irradiation processing methods for production of a rich array of metal oxides and composites as well as microstructures. The experimental processes used to achieve this library of materials is outlined and sensing applications of a subset of the materials is provided.

Part II of the book is concentrated on describing novel morphologies and the signal transduction principles in metal oxide-based sensors. Chapter 9 provides a detailed introduction into the synthesis and characterization of metal oxide nanowires as well as their currently accepted general reaction/sensing mechanism. Both conductometric and field effect device structures are introduced. The benefits of using nanowires with diameters on the size scale of the Debye screening length are discussed with respect to both the enhanced sensitivity as well as their reactive properties.

Chapter 10 continues the discussion of metal oxide nanowire-based sensors with a focus on the most commonly used ZnO and SnO$_2$ nanowires. These types of sensors are discussed with respect to both gas phase and biochemical sensor device development and applications. Furthermore, the possibility of using ZnO-based nanowire materials for optical detection schemes as well as integration into wireless structures are detailed as well, which provides strong evidence for the ubiquity of metal oxide nanomaterials in sensing devices.

In many cases the improvement of a sensing material can be realized by increasing its surface area. While for nanowire-based devices the benefit for a reduction in the nanowire diameter is realized by the creation of a depletion layer that envelops the entire wire and is significantly modulated upon reaction with the target gases, thus creating a more sensitive sensor. Furthermore, increasing the surface area of the nanowire through the creation of complex morphologies such as a fish-bone type structure as well as many other 3-D structures can lead to unique adsorption sites for reaction and transduction. Chapter 11 describes the synthesis and characterization of a variety of metal oxide chemistries with complex morphologies. The benefits with respect to both sensitivity and selectivity, operation temperature reduction, enhanced response times and stability are described.

While the use of the optical properties of ZnO for sensing applications was briefly introduced in Chap. 9, a detailed description of the optical properties of metal oxides is provided in Chap. 12. An optical transduction method can be advantageous given that it can be considered a wireless technique and is thus compatible with harsh environment conditions. Furthermore, in the development of multi-transduction sensing array platforms, combining both electronic and optical techniques may offer unique selectivity measurement opportunities. In this chapter the intrinsic and extrinsic photoluminescence of metal oxides and their dependence on target gas exposures are shown to be used for the detection of oxidizing and reducing gases. The size dependence of photoluminescence with regards to both quantum effects as well as changes in surface dominated processes is discussed with respect to sensing applications. While photoluminescence has proven to be useful for the detection of target gases, changes in the absorption properties of noble metals (Cu, Ag and Au) embedded in metal oxides have also proven to be optical beacons for the development of harsh environment compatible chemical sensors.

Part III of the book is focused on new device architectures and integration challenges of metal oxides into sensing device structures. Chapter 13 begins by outlining the unique possibilities that metal oxides with hetero-contacts and phase boundaries offer as a design platform for sensing applications. These details are highlighted with examples of engineered nanostructures of various compositions (pure, doped, composites, heterostructures) and forms (particles, tubes, wires, films). In addition the system architecture can be further enhanced through surface functionalization and the addition of a pre-concentrator system to promote enhanced transduction.

While both changes in the metal oxide chemistry as well as morphology can have pronounced effects on the sensing properties of a particular nanomaterial, the reaction temperature is also a dominating factor in the sensing characteristics. Chapter 14 provides a detailed description of a sensing device structure which uses temperature not only to affect the reaction properties of the target species, but through the collection of the sensing signals as a function of both temperature and time, a sensitive and selective sensing device can be achieved. Interpretation of these multi-parameter data sets using statistical algorithms provides both a characterization of the sensitivity as well as the selectivity of these sensing arrays.

Sensor arrays based on metal oxide nanowires for the so called "electronic nose" applications are the focus of Chap. 15. Details with respect to nanowire growth and integration onto sensor array platforms are described. The benefits and implications of such nanowire sensor arrays for a range of sensing applications is provided. Finally, the interpretation of sensing array data using pattern recognition algorithms to provide the necessary sensitivity and selectivity performance factors for electronic nose applications is detailed.

Chapter 16 begins this discussion with integration of metal oxide nanomaterials onto MEMS device structures such as microhotplates. These challenges include functionalization of the microhotplate with metal oxides formed using both liquid and vapor phase methods. A series of examples are provided for acquisition of sensing data as a function of temperature, sensor array integration as well as multiparametric data acquisition and interpretation.

To summarize, the last decade was famous due to the appearance of new paradigms for the development of metal oxide nanomaterials-based chemical sensors leading to new principles in receptor and transduction principles. This book reviews only a beginning of this exciting journey as it is clear that there will be many years of exciting discoveries ahead.

Michael A. Carpenter
Sanjay Mathur
Andrei Kolmakov

Contents

**Part I   Understanding, Characterization and Synthesis
of Modern Metal Oxide Nanomaterials**

1   Insights into the Mechanism of Gas Sensor Operation   3
Aleksander Gurlo

**2   Surface Science Studies of Metal Oxide
Gas Sensing Materials**   35
Junguang Tao and Matthias Batzill

**3   Design, Synthesis and Application of Metal Oxide-Based
Sensing Elements: A Chemical Principles Approach**   69
Valery Krivetskiy, Marina Rumyantseva and Alexander Gaskov

**4   Combinatorial Approaches for Synthesis of Metal Oxides:
Processing and Sensing Application**   117
Clemens J. Belle and Ulrich Simon

5   Selective Crystal Structure Synthesis and Sensing Dependencies . . .   167
Lisheng Wang and Perena Gouma

**6   Synthesis of Metal Oxide Nanomaterials for Chemical
Sensors by Molecular Beam Epitaxy**   189
Manjula I. Nandasiri, Satyanarayana V. N. T. Kuchibhatla
and Suntharampillai Thevuthasan

7   Atomic Layer Deposition for Metal Oxide Nanomaterials   225
Xiaohua Du

8   Microwave Synthesis of Metal Oxide Nanoparticles   245
Natalie P. Herring, Asit B. Panda, Khaled AbouZeid,
Serial H. Almahoudi, Chelsea R. Olson, A. Patel and M. S. El-Shall

**Part II   Novel Morphologies and Signal Transduction Principles
            in Metal Oxide-Based Sensors**

**9   Metal Oxide Nanowires: Fundamentals
      and Sensor Applications**   287
Zhiyong Fan and Jia G. Lu

10   ZnO Nanowires for Gas and Bio-Chemical Sensing   321
Stephen J. Pearton, David P. Norton and Fan Ren

**11   Metal Oxide Nanowire Sensors with Complex
      Morphologies and Compositions**   345
Qiuhong Li, Lin Mei, Ming Zhuo, Ming Zhang and Taihong Wang

12   Optical Sensing Methods for Metal Oxide Nanomaterials.   365
Nicholas A. Joy and Michael A. Carpenter

Part III   New Device Architectures and Integration Challenges

**13   Metal Oxide Nano-architectures and Heterostructures
      for Chemical Sensors**   397
Thomas Fischer, Aadesh P. Singh, Trilok Singh,
Francisco Hernández-Ramírez, Daniel Prades and Sanjay Mathur

**14   Evaluation of Metal Oxide Nanowire Materials
      With Temperature-Controlled Microsensor Substrates.**   439
Kurt D. Benkstein, Baranidharan Raman, David L. Lahr
and Steven Semancik

**15   Multisensor Micro-Arrays Based on Metal Oxide Nanowires
      for Electronic Nose Applications**   465
Victor V. Sysoev, Evgheni Strelcov and Andrei Kolmakov

**16   Microhotplates and Integration with Metal-Oxide
      Nanomaterials**   503
Emanuele Barborini

Concluding Remarks and Outlook   539

Index   543

Introduction

Modern chemical sensors can be defined as a transducer, which is comprised of, or coated with, a chemically responsive layer. The transducer converts forms of energy such as electrical, optical, or mechanical into a measurable signal. The chemically responsive layer, however, is what makes chemical sensors unique from physical sensors. This layer is needed as there are nearly an infinite number of detectable chemicals that are important for a variety of applications including automotive, mining, pharmaceutical and many other industries as well as personal safety and homeland security. In order for the chemical sensor to respond to the target chemical of interest, the chemically selective layer needs to preferentially interact with the target chemical. This interaction causes a change in the transducer's properties, which produces a change in the sensor signal. If these target chemicals could be detected using chemical sensors with the required detection limit, sufficient selectivity as well as an acceptable cost factor, the industrial and personal applications of chemical sensors would penetrate a variety of markets. These basic chemical sensing concepts are not new, and no doubt were envisioned with some of the very first chemical sensor demonstrations. The earliest of which may be likened to nineteenth and early twentieth century miners who were certainly grateful for their "hand held" air quality monitor, the canary. Once their canaries stopped singing (*acoustical transduction*) and fell off their perch (*observed signal*), the miners would immediately evacuate the mine as this was an indication of a buildup of combustible gases (methane) or carbon monoxide. Either of which could cause a fatal condition for the miners and unfortunately for the canary, it was nearly always a fatal condition.

With the passage of time and the industrial revolution a tremendous amount of development work has been completed over the years in the field of chemical sensors and thankfully it appears that the industry has moved beyond the use of canaries. While air quality monitoring is one application of chemical sensors, monitoring of our other natural resources, water and soil quality are also of vital importance. Chemical sensors, or their need, can also be found in a variety of technological areas including: combustion, agricultural, industrial processing, food industry, medical, public safety, among many others. As the technology and processes in each of these individual areas mature the sensor required typically becomes more challenging to develop with regards to their reliability, detection limits, sensitivity, selectivity and cost.

Recently there have been several new publications that have described the history and fundamental operating principles of a variety of chemical sensors [1–3] and these details need not be included in this book. For the current work we will focus on the use of metal oxide nanomaterials for chemical sensors. One of the first landmark demonstrations, which in hindsight is a precursor to modern metal oxide-based chemical sensors was the Nernst lamp developed in the late nineteenth century [4]. Nernst's studies on the electrolytic conduction of metal oxides showed that while they were non-conductors at room temperature, at high temperatures, upwards of 700 K, these metal oxides became conductors and in being resistively heated they also produced a brilliant white light. Furthermore, they could be reliably operated in the presence of air and did not need to be encapsulated in a non-oxidizing environment, as did its incandescent predecessors. Thus in 1904 the Nernst lamp replaced those comprised of carbon filaments, but this was short-lived as the Nernst lamp was eventually replaced by the tungsten lamp [5]. Interestingly though, the Nernst lamp used yttria-stabilized zirconia (YSZ) as its incandescent source and as such is closely related to Lambda oxygen sensors which are used in automobiles even today. YSZ has over time been shown to be stable in oxidizing atmospheres at temperatures even as high as 1,200 K and is an excellent oxygen ion conductor, leading to its widespread use as an electrolyte in solid oxide sensors and fuel cells. Lambda sensors typically have a hollow cylinder of YSZ, with the outer wall exposed to the exhaust gas and the inner wall exposed to the ambient atmosphere. At elevated temperatures (700 K) oxygen is reduced on the YSZ electrolyte and the gradient of oxygen ions at elevated temperatures induces an ionic potential across the electrodes placed on the inner and outer walls of the YSZ cylinder.

The Lambda oxygen sensor developed during the late 1960s by the Robert Bosch GmbH Company used a very similar design as described above. This device was first used in cars as part of an emissions control system in 1976. The use of this device allowed for optimization of the fuel to air mixing ratio for more efficient combustion of fuel and in doing so reduces the concentrations of unburnt fuel and oxides of nitrogen (NOx) that are emitted into the atmosphere. The reduction in these emissions have been critical in the improvement of tropospheric air quality over the last three decades, making the integration of O_2 sensors in cars a key enabling technology for the improvement of vehicular emissions.

The Taguchi gas sensor, sold by Figaro Engineering, is another example of a very successful chemical sensor and these devices have been mass-produced since 1968. This sensor was developed out of research conducted by Mr. Naoyoshi Taguchi and uses the chemiresistive properties of semiconductor metal oxides as the transduction mechanism. Tin oxide is the most popular metal oxide used in these sensors and by functionalizing them with various catalytically active materials (Pd, Pt, Au, etc.) an entire product line of sensors which detect common combustible and reducing gases such as H_2, CO, CH_4 as well as gases of interest to various industrial applications can be purchased.

What kind of properties makes metal oxides so attractive for their use in gas sensors? Metal oxides in general have proven to be resistant to high temperatures, often are catalytically active, are optically transparent in the visible wavelengths, and are wide band gap materials that are often used as electrical insulators or semiconductors in a variety of electrical devices [6]. The electrical and optical properties and the chemical reactive characteristics of metal oxides are the key attributes for chemical sensing applications. Metal oxides such as YSZ, SnO_2, TiO_2, ZrO_2, CeO_2, ZnO, CuO and many others commonly undergo oxidation and reduction cycles under appropriate conditions. Metal oxides exposed to oxidizing gases, such as O_2 or NO_2 as an example, will typically result in reactions of the following general form:

(1) $O_2 + 2e^- \rightarrow 2O^-$
(2) $2O^- + 2e^- \rightarrow 2O^{2-}$
(3) $NO_2 \rightarrow NO + O$
(4) $O + e^- \rightarrow O^-$
(5) $O^- + e^- \rightarrow O^{2-}$

While this is a greatly simplified reaction scheme, as it does not show all of the possible intermediates, the common theme for oxidizing gases is that they tend to form O^- and/or O^{2-} ions. The formation of oxygen ionic species at the surface leads to buildup of the depletion (accumulation) region in n- (p-) type of semiconducting metal oxides, which effectively modulates their conductance. On the other hand, in order for the doubly ionized oxygen anion to be stabilized it needs to diffuse into an oxygen vacancy site formed at a lattice defect. These modifications of the defect sites will in turn affect the optical (luminescent, color) properties of oxide.

Reducing gases such as H_2 or CO are thought to react readily with the aforementioned oxygen anions along the following simplified reaction paths:

(6) $H_2 + O^- \rightarrow H_2O + e^-$ **or** $H_2 + O^{2-} \rightarrow H_2O + 2e^-$
(7) $CO + O^- \rightarrow CO_2 + e^-$ **or** $CO + O^{2-} \rightarrow CO_2 + 2e^{2-}$

A few logical questions can be raised at this point: how are these electrons donated from/to the sensing material, how does this produce a measurable affect on the transduction mechanism of choice and how does size as well as morphology of the sensing element influence the transduction function? For chemical sensors that rely on the change in resistance of an n-type (p-type) metal oxide semiconductor as the transduction mechanism, a decrease in valence band electrons used for production of oxygen anions with the accompanying formation of the depletion (accumulation) regions in the sensing material will cause an increase (decrease) in resistance. Likewise upon exposure to reducing type gases, there will be a reduction in the number of ionosorbed oxygen species and the electrons will be donated back to the metal oxide semiconductor causing a decrease (increase) in resistance. For systems that rely on the modulation of optical properties as a transduction mechanism the change in oxygen anions upon reaction with the target gases produces a strong signal as well. This has been most commonly observed with ZnO and TiO$_2$ [7, 8]. In both of these cases there are oxygen defect states that are a source of photoluminescence upon excitation with a UV photon. If the defect states (or oxygen vacancy) are not populated with oxygen anions, this channel is available for photoluminescence, and likewise it turns off when oxygen anions occupy the defect state. Therefore in the presence of an oxidizing gas, photoluminescence is reduced, while in the presence of a reducing gas, photoluminescence increases. The interaction mechanism is slightly different for Lambda type sensors. In this case oxygen still follows the simplified reactions noted in schemes (1) and (2) above, however, the transduction mechanism is dictated by the diffusion of the incorporated O^{2-} bulk ions. This is driven by a concentration gradient of O^{2-} ions between the ambient and emission side of the YSZ component, and the quantity measured is the change in ionic conductance. It is interesting to note that the details of the above reaction mechanisms are strongly material and materials processing specific and in many cases are an active area of research in the optimization of these materials for enhanced sensitivity and selectivity characteristics.

While the use of metal oxides has provided many useful chemical sensing results, it is quite common to add either dopants or catalytically active materials to the metal oxide to modify its reactive/sensing properties. As an example: the addition of yttria (Y$_2$O$_3$) dopants to zirconia creates the commonly used yttria-stabilized zirconia metal oxide that is used in the Lambda sensors as well as in solid oxide fuel cells. Catalytically active materials such as Pt, Pd, Au, Fe$_2$O$_3$ among many others are also added to metal oxides to modify the reaction paths and rates with the aim of increasing the detection limits as well as enhancing the selectivity towards detection of specific target gases.

For over 40 years the most visible application of chemical sensors was employment of O$_2$ sensors in automobiles and CO detectors in households. Currently there exists a strong demand in chemical sensors for bio-medical applications, homeland security, environmental control as well as in appliances within our homes, or in our portable electronics. To meet these needs improvement of selectivity, sensitivity, power consumption along with reduction in cost and size for the chemical sensing systems has to be realized. It is here where over the last 10–15 years, advances in the areas of nanoscale science and engineering of metal oxide composite materials have shown promise for their incorporation and potentially disruptive advancement of chemical sensor technologies. Advantages that are realized by nanoscale materials include a number of new functional opportunities as well as financial aspects. In the simplest case, the surface area to volume ratio of a spherical particle increases by a factor of $1/r$ with a reduction in particle size. Particles with diameters below 10 nm contain a significant fraction of their material in the surface layer of the particle. This is important as the catalytic reactions responsible for sensing action take place at the surfaces of these metal oxide materials. The reduction in size will not only reduce costs but may also provide novel opportunities in realization of new receptor/transduction principles in chemical sensing. Specifically, the reduction in size of metal oxides often contributes to emerging catalytic reaction characteristics, which can be taken advantage of during the design of an optimized material for detection of the target gases. As the catalytic properties of metal oxides also typically are tailored by decorating the metal oxide with metallic particles, this reduction in size is also beneficial for modifying the catalytic properties of nanocomposite materials comprised of both metal and metal oxide nanoparticles [9]. Further yet, the metal oxide nanocomposites can be processed to include 3-D assemblies of either particles or nanowires. The porosity of these assemblies can be highly beneficial for enhanced adsorption of the target gases, and with this 3-D heterostructures can be developed with individual components within the heterostructure designed for particular interactions. However, a balance here clearly needs to be struck, as highly porous materials can have enhanced adsorption properties, but for use in a chemical sensor this property does not want to be enhanced to the point where it acts like a chemical sieve or trap for the target chemicals, as an optimal chemical sensing material must have both a fast response and recovery time.

While optimizing the chemistry, size and microstructure of metal oxide and metal doped metal oxide composite materials for catalytic chemical reactions is at the forefront of the design of materials for chemical sensors, the reduction in size to the nanoscale also allows for enhanced transduction properties. As noted above chemiresistor-based sensing devices operate under the principle that upon adsorption and or reaction of gases the resistance across the conducting channel changes in a characteristic fashion. When the metal oxide composite material which coats the electrodes is comprised of either a bed of percolating nanoparticles or nanowires the resistance change can be dictated both by the individual contact resistances between particles or wires, but also through the formation of a depletion layer inside the individual nanoparticle (nanowire) [10]. As electrons are donated from the conduction band to enable chemical reactions an electron depletion layer is formed, and resistance across an n-type material will increase. As this layer approaches the respective materials Debye length the conductance changes can be quite significant. This is more clearly relevant when the Debye length, which is typically on the order of several nanometers, is on the order of the size of the nanoparticle or nanowire. In these cases the depletion layer envelops the whole nanoparticle or nanowire and resistivity changes should be the largest and most beneficial for sensitive analyte detection. In metal oxides composed of nanosized grains almost all the carriers are trapped in surface states and only a few thermally activated carriers are available for conduction. In this configuration the transition from activated to strongly de-activated carrier density, produced by reactions with the target gas species, has a huge effect on sensor conductance. Thus, the technological challenge has moved to the fabrication of metal oxide-based sensors with crystallite sizes as small as possible which maintain their stability over long-term operation at high temperature.

To decouple and better understand the contribution of the interconnects and individual nanoparticles (nanowires) in the sensing process, recent work has been focused on using single particles or single nanowires between two electrodes as model sensing devices. This is easiest to visualize for a single chemiresistor pinned between two electrodes [10]. In this approach, the nanowires are usually placed between two electrodes, and biased at constant voltage, reading the change in resistance (RES). This approach works on the same principle with thin film conductometric gas sensors. Sensors based on single crystalline nanowires may be advantageous over their planar counterparts as almost all of the adsorbed species are active in producing a surface depletion layer compared with a thin surface. If the nanowire diameter is on the order of its Debye length, then reactions which deplete electrons from the conduction band will form a depletion layer that spans the entire core of the nanowire, resulting in a dramatic increase in resistivity for this nanowire. These features together with transport of charge carriers in one-dimensional channels are responsible for enhanced sensitivity and thus ultra low detection limits. There are other benefits as well, in that one can not only measure the change in resistance of the nanowire, but by applying a constant voltage to the wire, the wire itself is resistively heated [11]. This is an important point because most metal oxide-based materials need to be operated at elevated temperatures in order for them to function as a chemical sensing material. Typically these temperatures range between 100 and 500 °C. By integrating the heater into the very design of the nanowire-based chemiresistor, a component required for integration into a chemical sensing platform has been removed and therefore simplifies its production and minimizes the sensor power consumption to record low levels in the micro-watt range. While the use of individual nanowires is proving to enhance the sensing properties of these materials, it is still challenging and expensive to produce this kind of sensor, as the nanowire is still most often "picked up" and "pinned" in place on electrodes to ensure good electrical contact. However, this single nanowire approach is clearly not a scalable method yet. While work is in progress to develop methods to grow or align the individual wires of a given size and composition between electrical contacts, this is still an area of research with much work to be done to ensure that this is a scalable method which can be compatible with modern device integration paradigms.

In order to find new principles of analyte detection and to enhance the selectivity as well as sensitivity limits by miniaturization of active sensing elements, heterostructured sensor devices utilizing multiple transduction techniques are currently being explored. An example of this type of device structure using an integrated RES and surface ionization (SI) approach has recently been proposed. In this device a metal oxide is electrically contacted directly opposite a counter electrode. Resistive measurements are performed between the two electrodes making contact with the nanowire, whereas SI recordings are carried out between the nanowire and the counter electrode. Surface ionization is an alternative method of gas detection, which was studied in the past by several groups mostly for noble and refractory metals, whereby ionosorbed analyte ion species are extracted into free space by a counter electrode positioned at a short distance above the hot emitter surface [12–14]. In contrast to earlier reports on metals, the recent developments have demonstrated that the SI principle can be applied to SnO_2 nanowires as the emitter surface which is the most widely employed material in the fabrication of RES metal oxide gas sensors [15].

In closing, while metal oxide nanomaterials have many promising attributes that are pushing the state of the art in chemical sensor development, system integration is a major challenge. For metal oxide-based chemical sensors these include incorporation of a variety of ancillary components including heaters, gas sampling mechanisms (active or passive), power sources, and the associated electronics for signal processing and data analysis. While the use of nanomaterials allows for reduced requirements for power consumption, this nanomaterial still needs to be incorporated into a bulk device for all current applications of chemical sensors. Again, this is a common integration problem for many types of devices and if the market requires bulk quantities of units, a process which uses parallel manufacturing paradigms is inevitably a requirement in order to drive down production costs. A solution to this problem would be integration onto a platform such as silicon, which allows one to take advantage of the parallel processing methods, which have been developed over the past decades for the MEMS and integrated circuit (IC) industries. While not every process step utilizing metal oxide nanomaterials is compatible with the strict material and processing protocols used in the IC industry, there are components, which can likely be coupled post-processing. Therefore it is not surprising that modern device fabrication methods using a combination of MEMS and silicon IC processing techniques are currently being explored to probe this next step in the development of chemical sensors for pervasive use across a variety of applications. The development of novel metal oxide nanomaterials and new sensing paradigms in combination with integration strategies for production of scalable chemical sensors promises to lead to a variety of new research directions as well as sensors which will satisfy the needs of a range of technological fields for many years to come.

Michael A. Carpenter
Sanjay Mathur
Andrei Kolmakov

References

1. Janata J (2009) Principles of chemical sensors, 2nd edn. Springer, New York
2. Korotcentkov G (ed) (2010) Chemical sensors, vols 1–6. Momentum Press, New York
3. Aswal DK, Gupta SK (eds) (2007) The science and technology of chemiresistor gas sensors. Nova Science Publishers, New York
4. Monmouth Smith H (1898) The Nernst Lamp. Science 8:689
5. Hans-Georg B, Scholz G, Scholz F (1983) The Nernst Lamp and its inventor. Zeitschrift fuer Chemie 23(8):277–283
6. Barsoum MW (2003) Fundamentals of ceramics. Taylor and Francis, Bristol

7. Valerini D, Creti A, Caricato AP, Lomascolo M, Rella R, Martino M (2010) Optical gas sensing through nanostructured ZnO films with different morphologies. Sens Act B 145:167
8. Setaro A, Lettieri S, Diamare D, Maddalena P, Malagu C, Carotta MC, Martinelli G (2008) Nanograined anatase titania-based optochemical gas detection. New J Phys 10:053030
9. Haruta M (1997) Size- and support-dependency in the catalysis of gold. Catal Today 36(1):53–166
10. Hernandez-Ramirez F, Prades JD, Tarancon A, Barth S, Casals O, Jimenez-Diaz R, Pellicer E, Rodríguez J, Juli MA, Romano-Rodriguez A, Morante JR, Mathur S, Helwig A, Spannhake J, Mueller G (2007) Portable microsensors based on individual SnO$_2$ nanowires. Nanotechnology 18:495501
11. Strelcov E, Dmitriev S, Button B, Cothren J, Sysoev V, Kolmakov A (2008) Evidence of the self heating effect on surface reactivity and gas sensing of metal oxide nanowires. Nanotechnology 19:355502/1–355502/5
12. Rasulev UK, Zandberg EY (1988) Surface ionization of organic compounds and its applications. Prog Surf Sci 28:181–412
13. Sears WM, Moen VA, Miremadi BK, Frindt RF, Morrison SR (1987) Positive ion emission from a platinum hot wire gas sensor. Sens Actuators 11:209–220
14. Fujii T, Katai T (1986) Surface ionization mass spectrometry of organic compounds. I. nitrogen containing aliphatic organic compounds. Int J Mass Spec Ion Process 71:129–140
15. Hernandez-Ramirez F, Prades JD, Hackner A, Fischer T, Mueller G, Mathur S, Morante JR (2011) Miniaturized ionization gas sensors from single metal oxide nanowires. Nanoscale 3(2):630–634

Part I
Understanding, Characterization and Synthesis of Modern Metal Oxide Nanomaterials

Chapter 1
Insights into the Mechanism of Gas Sensor Operation

Aleksander Gurlo

Abstract Since the development of the first models of gas detection on metal-oxide-based sensors much effort has been made to describe the mechanism responsible for gas sensing. Despite progress in recent years, a number of key issues remain the subject of controversy; for example, the disagreement between the results of electrophysical and spectroscopic characterization, as well as the lack of proven mechanistic description of surface reactions involved in gas sensing. In the present chapter the basics as well as the main problems and unresolved issues associated with the chemical aspects of gas sensing mechanism in chemiresistors based on semiconducting metal oxides are addressed.

> *"Sensors have a 'life cycle' consisting of preparation,*
> *activation, operation with deactivation and, possible,*
> *regeneration. Thus understanding the performance in terms of*
> *reaction and conductance mechanisms is only a part of the*
> *total understanding of a sensor."*
> Dieter Kohl, *Sensors and Actuators* 1989, 18, 71.

A. Gurlo (✉)
Fachbereich Material- und Geowissenschaften, Technische Universitaet Darmstadt,
Darmstadt, Germany
e-mail: gurlo@materials.tu-darmstadt.de

M. A. Carpenter et al. (eds.), *Metal Oxide Nanomaterials for Chemical Sensors*,
Integrated Analytical Systems, DOI: 10.1007/978-1-4614-5395-6_1,
© Springer Science+Business Media New York 2013

1.1 Chemiresistors: From Semiconductor Surfaces to Gas Detectors

Since the early 1920s numerous investigations have demonstrated the influence of the gas atmosphere on conductivity, free carrier mobility, surface potential, and work function on a number of semiconductors (see summary of early works in [1–13]). This led to the understanding that the surface of semiconductors is highly sensitive to chemical reactions and chemisorptive processes [3, 14–20] and resulted finally in the "theory of surface traps" (Brattain and Bardeen [21]), "boundary layer theory of chemisorption" [10, 22, 23] (Engell, Hauffe and Schottky) and "electron theory of chemisorption and catalysis on semiconductors" (Wolkenstein [5–7, 24]). They laid also the theoretical foundations for the subsequent development of metal-oxide-based gas sensors.

Although from this understanding to the use of semiconductors as gas sensors "was, in principle, a small step" [25], the idea of using the changes in conductivity of a semiconducting metal oxide for gas detection was not conceived until the middle of the 1950s. The earliest written evidence came in 1956, in the Diploma Thesis performed in Erlangen under supervision of Mollwo and Heiland and entitled "Oxygen detection in gases changes in the conductivity of a semiconductor (ZnO)" [26], the results discussed later in [1, 27]: "If one exposes a zinc oxide layer which has been given a previous heating at 500 K in a high vacuum to oxygen at a constant pressure, the conductivity falls very rapidly initially and more slowly later. If one then increases the oxygen pressure suddenly, the current of the conductivity exhibits a kink when plotted as a function of the time. In this change the slopes immediately before and immediately after the kink point are proportional to the partial pressure of oxygen. One can use this effect to relate a known and an unknown concentration of oxygen often even under conditions in which one has a mixture of gases..." (cited from Ref. [1]). In 1957, Heiland showed that the "well-conducting surface layer on zinc oxide crystals provides a new, very sensitive test for atomic hydrogen" [28] and Myasnikov demonstrated that ZnO films can be used as a highly-sensitive oxygen-analyzer [29]. Later he developed this "to the method of semicondutor probes", which allows for "studying free radical processes" and for detecting "free active particles and to measure their concentration under stationary and non-stationary conditions in gases and liquids" [30]. However, the conditions under which ZnO was able to operate as a "sensing device" were far from the real ambient conditions (and, accordingly, from a practical application); the "sensitive" effects were observed: (i) in vacuum conditions, exposed to oxygen or hydrogen, (ii) after "activation" or "sensitization" of the surface by heating in H_2 and in UHV.

The practical use of metal-oxide-based gas sensors in normal ambient conditions was not considered until 1962, when Seiyama et al. reported that a ZnO film can be used as a detector of inflammable gases in air [31] (see also [32]), and Taguchi claimed that a sintered SnO_2 block can also work in the same way [33] (for the history of TGS (Taguchi Gas Sensor) sensors, see [34]). The latter approach became very successful, leading to the foundation of the first sensor company (Figaro Engineering Inc.), which established mass production and started selling the TGS sensors in 1968.

Since then, many different metal oxides have been investigated as sensing materials (see, for example, Ref. [35] for a comprehensive review), however, tin dioxide (SnO_2)—alone or "activated" with small quantities of noble metals/their oxides (Pd, Pt, Au)—has remained the most commonly used and the best-understood prototype material in commercial gas sensors [36] as well as in the basic studies of the gas sensing mechanism [35–43].

1.2 Characterization Methodology: From Prototype Surfaces to Operating Sensors

The detailed characterization of metal oxide sensors requires the "simultaneous measurement of the gas response and the determination of molecular adsorption properties for a better understanding of gas sensing mechanisms" [44]. This measurement can be done either on clean and well-defined surfaces in ultrahigh vacuum (UHV) conditions or at temperatures and pressures that mimic real sensor operating conditions ("in situ" [45]). Continuous progress has been made during the past few years for the latter strategy, i.e. toward the use of in situ and operando spectroscopic techniques.

The "crossing of interests" [46] and "bridges of physics and of chemistry across the semiconductor surface" [47] determined experimental methodology applied for the gas-semiconductor studies in general and gas sensing studies in particular in the course of the last 50 years.

The first systematic methodological approach ("design concept for chemical sensors") in gas sensing-studies was explicitly formulated in 1985, in a series of papers entitled "Development of chemical sensors: empirical art or systematic research?" ([48–50], see also [51]]).

The underlying concept was that by "studying the surface of single crystals under well-defined conditions, one might try to achieve a better separation of parameters influencing the properties of gas sensors" [52]. The reactions were addressed by surface spectroscopic methods under ultra-high-vacuum (UHV) conditions on well-defined "prototype" structures while the sensor performance was tested under realistic measuring conditions on the structures of practical importance ("sensors").

This "comparative approach" advanced the basic understanding of surface reactions and the corresponding conduction mechanism responsible for gas sensing. However, it showed also the limits of surface science in gas-sensing studies and led to the understanding that if spectroscopic and electrical data are not obtained simultaneously, they must be obtained (i) under the same conditions and (ii) on identical samples. A comprehensive description of surface reactions on SnO_2 published in 1989 resulted from simultaneous thermal desorption spectroscopy (TDS; i.e. reactive scattering of a molecular beam) and conductance measurements [52]. These measurements were applied to SnO_2 single crystals and thin evaporated films exposed to a certain dose of CH_3COOH, CO or CH_4 in UHV conditions while at the sensor operating temperature.

As an alternative to sensing studies on single crystals or thin films, sensing characterization studies have focused on a combination of electrical measurements with spectroscopic investigations of catalysis reactions on *polycrystalline, high surface-area materials* with the aim to "link semiconductor studies with catalytic studies" [9]. However, most of the studies were performed under conditions far from the real working conditions of sensors (for the summary of numerous studies on semiconducting metal oxides, see references [4, 13, 53]). Besides spectroscopic and catalytic (kinetic) investigations (SnO_2: kinetic studies of CO oxidation [54], IR spectroscopic studies of water, CO_2 and CO adsorption [55], (summarized in Ref. [56]), EPR investigations of oxygen adsorption, [57], (reviewed in references [58, 59])), the improvements were concentrated on devising systems and in situ cells for combined (i.e., performed under the same conditions on "identical" samples) and simultaneous electrical, catalytic and spectral investigations.

These activities, however, were overlooked by the sensor community at that time, as in situ electrical characterization of realistic ("polycrystalline") samples, namely, the Hall effect measurements (1982 [60]), changes in work function (CPD) by the Kelvin method (1983 [61]), ac impedance spectroscopy (1991 [62, 63]), simultaneous work function change and conductance measurements (1991 [64]) were preferred for studying the mechanism of operating sensors [99].

Later, this approach was followed systematically in the number of works (reviewed in references [38, 65], recent works in references [66–70] and references therein) to elucidate a mechanism of gas detection on SnO_2-based sensors. Local electronic properties (e.g., the density of states in the region near the band gap) of a sensing material were determined by scanning tunnelling microscopy and spectroscopy (STM-STS) in vacuum conditions [71–73] or under N_2, CO and NO_2 (at room temperature) [74].

By the end of the 1990s, the spectroscopic techniques for gas-sensing studies were differentiated according to conditions under which they can be applied: those that may be applied "under in situ real operation conditions of the sensors" and those that may be applied "under ideal conditions far away from real practical world" [75]. This differentiation subsequently resulted in the systematic combination of phenomenological and spectroscopic measurement techniques under working conditions of sensors [38], and thus lead to the in situ and operando methodology.

Continuous progress has been made during the past few years for the latter strategy, that is, the use of in situ and operando spectroscopic techniques (see [76, 77]):

- In situ spectroscopy: *spectroscopic characterization of sensing materials under operation conditions or conditions relevant to operation conditions; herein, the sensing performance of this material may be not characterized or may be characterized in a separate experiment,*

- Operando spectroscopy: *spectroscopic characterization of an active sensing element in real time and under operating conditions with the simultaneous read-out of the sensor activity and simultaneous monitoring of gas composition.*

These definitions determine the boundary conditions under which an "operando" experiment is performed:

1. on a *sensing element,* which itself is a complex device and consists of several parts: in solid-state devices with an electrical response, for example, the sensing layer is deposited onto a substrate to which electrodes for an electrical read-out are attached ("transducer"); therefore the assessment of their interfaces is of paramount importance for understanding the overall sensing mechanism;
2. *in real time:* a sensor is devised to respond to the changes in the gas atmosphere as fast as possible; accordingly, it demands a fast spectroscopic response;
3. *under operating conditions:* these can vary from ambient conditions (RT and atmospheric pressure) to high temperatures and pressures;
4. *with simultaneous read-out of sensor activity:* the gas concentration to be measured is transduced by the sensor into an electrical or other convenient output, depending on the modus operandi of sensor (optical, mechanical, thermal, magnetic, electronic, or electrochemical) and the transducer technology;
5. *with simultaneous monitoring of gas composition*; on-line gas analysis in gas sensing plays a twofold role: (i) the output compositions and concentrations provide data about reaction products and possible reaction paths and (ii) the input concentration verifies the sensor input data (concentration of the component to be detected).

The operando methodology couples electrical ("phenomenological") and spectroscopic techniques and, aims to correlate the sensor activity with the spectroscopic data obtained under the same conditions on the same sample (Fig. 1.1). In an ideal case, one would obtain four types of information: (i) gas-phase changes (and reaction products) from on-line gas analysis, (ii) species adsorbed on the surface, (iii) changes in the oxide surface and lattice, and (iv) sensor activity.

1.3 Mechanism of Gas Detection: Never Ending Story About Oxygen

Epigraph

> Due to the electron affinity of oxygen, the electron can be transferred to the chemisorbed oxygen and, consequently, there will be no chemisorbed oxygen atoms, but ions, in the surface

K. Hauffe, *Adv. Catal.* **1955**, *7*, 213– 257.

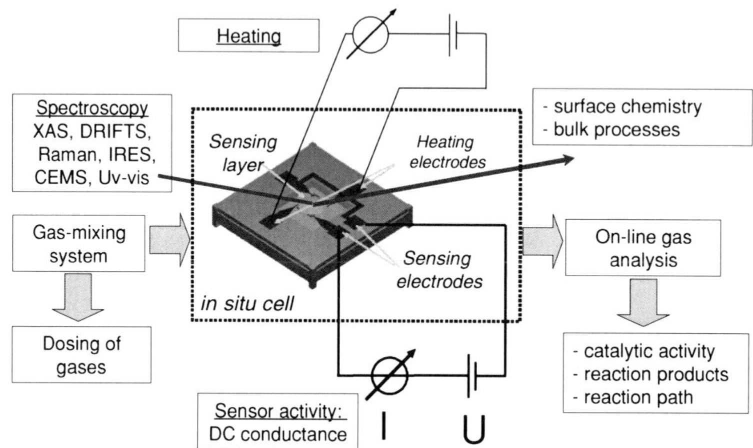

Fig. 1.1 Methodological approach for simultaneous spectroscopic and electrical ("phenomenological") characterisation of metal-oxide-based gas sensors. Modified from ref. [76]

Since the development of the first models of gas detection on metal-oxide-based sensors [78, 79] much effort has been made to describe the mechanism responsible for gas sensing (see, for example, [80–82]). Despite progress in recent years, a number of key issues remain the subject of controversy; for example, the disagreement between electrophysical and spectroscopic investigations, as well as the lack of a proven mechanistic description of surface reactions involved in gas sensing.

Nowadays, the influence of the gas atmosphere on the electrical transport properties of semiconductors and, accordingly, the operation of metal-oxide-based gas sensors is currently described by the combination of two different models; they are the ionosorption and the reduction-reoxidation mechanisms (Table 1.1). The *ionosorption model* considers only the space-charge effects/changes of the electric surface potential that results from the "ionosorption" of gaseous molecules. The *reduction-reoxidation model* explains the sensing effects by changes in the oxygen stoichiometry, that is, by the variation of the amount of the (sub-) surface oxygen vacancies and their ionization. The latter involves explicitly the diffusion of oxygen (or oxygen vacancies) from/in the bulk of the sensing material.

1.4 Oxygen Ionosorption

The electrical conductivity and work function can be described as collective physical properties of semiconductors which are changed by an ionosorption process and are accessible to measurement. *The key in the mechanistic description of gas sensing is "oxygen ionosorption" and reaction of reducing gases with ionosorbed oxygen ions.*

Table 1.1 Gas sensing mechanism on SnO_2 according to ionosorption and oxygen vacancy models

Gas detection	Ionosorption model	Oxygen vacancy model
Oxygen	$O_2(ads) + e^-(CB) \leftrightarrow O_2^-(ads)$ $O_2^-(ads) + e^-(CB) \leftrightarrow O_2^{2-}(ads) \leftrightarrow 2O^-(ads)$	$2V_O^\bullet + O_2(gas) + 2e^-(CB) \leftrightarrow 2O_O^\times$
CO/presence of oxygen	$CO(gas) + O^-(ads) \leftrightarrow CO_2(gas) + e^-(CB)$	$CO(gas) + O_O^\times \leftrightarrow CO_2(gas) + V_O^\times$ $V_O^\times \leftrightarrow V_O^\bullet + e^-(CB)$
CO/absence of oxygen	$CO(gas) \leftrightarrow CO^+(ads) + e^-(CB)$	$V_O^\bullet \leftrightarrow V_O^{\bullet\bullet} + e^-(CB)$
NO_2	$NO_2(gas) + e^-(CB) \leftrightarrow NO_2^-(ads)$	$NO_2(gas) + V_O^\bullet \leftrightarrow NO_2^-(ads) + V_O^{\bullet\bullet}$ $2NO_2(gas) + O_2^-(ads) + V_O^\bullet \leftrightarrow 2NO_3^-(ads) + V_O^{\bullet\bullet}$
Water vapour	$H_2O(gas) + O^-(ads) + 2Sn_{Sn}^\times \leftrightarrow 2(Sn_{Sn}^\times - OH) + e^-(CB)$ $H_2O(gas) + Sn_{Sn}^\times + O_O^\times \leftrightarrow (Sn_{Sn}^\times - OH) + OH_O^\bullet + e^-(CB)$	$H_2O(gas) + 2Sn_{Sn}^\times + O_O^\times \leftrightarrow 2(Sn_{Sn}^\times - OH) + V_O^\bullet + e^-(CB)$

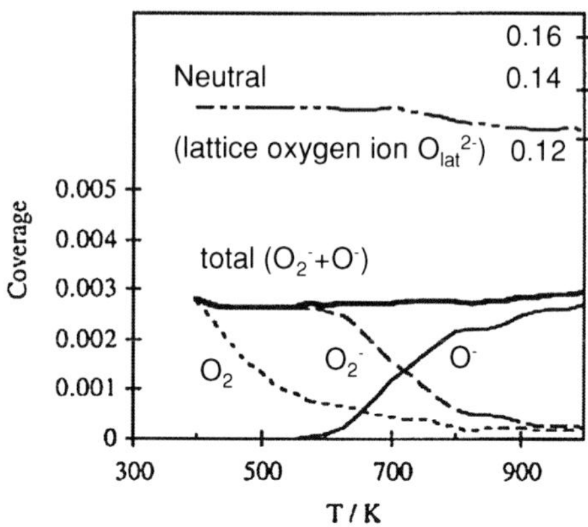

Fig. 1.2 The simulated equilibrium coverages of the oxygen species. The transition from O_2^- to O^- is calculated to be around 700 K (intersection at 427 °C). Copyright Elsevier, reproduced with permission from Ref. [100]

The oxygen influence on the electrical conductivity and work function is very well documented. For SnO_2, for example, exposure of single crystals (Ref. [83] and refs therein), polycrystalline samples (porous films [84], powders [57], pressed bars[85]) as well as one dimensional nanostructures [86, 87] to oxygen leads to the (i) decrease in the electrical conductivity and in the concentration of conduction electron density (Hall effect measurements [57]), (ii) increase in the work function observed in UHV conditions (XPS/UPS [88]) and under atmospheric pressure (simultaneous Contact Potential Difference, CPD, and conductance measurements [84]). Similar effects have been also observed on TiO_2 and ZnO (see early publications on TiO_2 [89, 90] [91] and on ZnO [3, 92–95]).

The magnitude of the changes depends strongly on the oxide temperature (see for example [85] and [84]), particle size and pre-treatment (history). On high-surface area and reduced samples the changes are much higher in comparison to single crystals and oxidised samples. The reduced samples show activity at temperatures as low as room temperature (r.t.), for oxidised samples higher temperatures (>100 °C) are needed. This difference between oxidised and reduced samples is usually ignored by the "ionosorption theory".

Because the detailed mechanism of oxygen adsorption cannot be derived directly from electrophysical investigations[96], the chemistry of adsorbed surface oxygen on SnO_2 was adapted from the "ionosorption model" [97–100]. It was assumed that the thermally stimulated processes of oxygen adsorption, dissociation and charge transfer involve only conduction electrons [4, 81]:

$O_2(gas) \leftrightarrow O_2(ads)$	*Physisorption*
$O_2(ads) + e^-(CB) \leftrightarrow O_2^-(ads)$	*Ionosorption*
$O_2^-(ads) + e^-(CB) \leftrightarrow O_2^{2-}(ads) \leftrightarrow 2O^-(ads)$	*Ionosorption*
$O^-(ads) + e^-(CB) \leftrightarrow O^{2-}(ads)$	*Ionosorption*
$O^{2-}(ads) \leftrightarrow O^{2-}(1st\ bulk\ layer)$	*Diffusion*

The nature of the ionised oxygen species is assumed to depend on the adsorption temperature (Fig. 1.2). At low temperatures (150–200 °C) oxygen adsorbs on SnO_2 non-dissociatively in its molecular form (as charged O_{2ads}^- ions). At high temperatures (between 200 and 400 °C or even higher) it dissociates to atomic oxygen (as charged O_{ads}^- or O_{ads}^{2-} ions) [4, 37, 75, 80, 81, 98, 99, 101]. Neutral oxygen species such as physisorbed oxygen, $O_{2, phys}$, are assumed not to play any role in gas sensing. The same holds for the lattice oxygen ions, O_{lat}^{2-}, in bulk materials at temperatures not high enough for fast oxygen exchange reactions (see detailed discussion below).

At this point, a problem of semantics starts to bring additional confusion, especially in the operational use of the terms "charged" species and the "charge transfer" at the surface. In semiconductor physics, the charge transfer implies by definition the transfer of free charge carriers, that is, conduction electrons or holes. Accordingly, the species that influence the electrical conductivity are regarded as "charged" or "ionized". They are represented by free oxygen ions. The species that do not influence the conductivity are regarded as "neutral". They are represented by physisorbed oxygen molecules.

The phenomenological model describes the oxygen ionosorption on an n-type semiconductor as follows:

- ionosorbed oxygen species are formed due to the transfer of conduction electrons from the semiconductor;
- they can be regarded as free oxygen ions which are electrostatically stabilized in the vicinity of the surface;
- there are no other adsorbed oxygen species besides physisorbed oxygen and oxygen ions;
- physisorbed oxygen is electrically neutral and oxygen ions are electrically active ("charged") species.

The simplified picture showing the influence of adsorption on surface conductivity and work function is as follows. An oxygen molecule becomes physisorbed at the surface. In the next step, an electron from the oxide's conduction band is trapped at the adsorbed oxygen molecule. The adsorbed oxygen molecule and surface itself become negatively charged. The flow of electrons from the semiconductor into the chemisorbed layer, without any diffusion of ionic species at the same time, induces a space charge between the interior of the semiconductor and its surface. The negative surface charge is compensated by a positive charge and a space-charge layer forms below it. This positive space-charge layer has reduced electron densities as compared to the bulk and is called an "electron-depleted layer or a charge depleted layer". As a result the energy band, pertaining to the surface, bends upwards with respect to the Fermi level. This causes the creation of barriers on the surface, ($q\Delta V_S > 0$), due to the increasing work function, ($q\Delta V_S > 0$), and decreasing surface conductance ($G = Gexp(-q\Delta V_S/kT)$) (Fig. 1.3). The process of charge transfer continues until equilibrium is reached and a steady state is achieved. To prevent very high double-layer potentials, the total amount of the "charged" species is limited to 10^{-5}–10^{-3} monolayer which corresponds approximately to 1 V of the surface potential V_S (this is the so-called

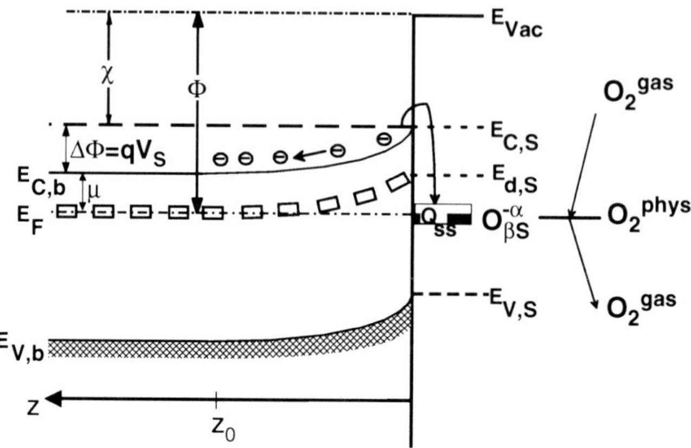

Fig. 1.3 Band bending on an n-type semiconductor after ionosorption of oxygen. Work functions Φ of semiconductors contain three contributions; e.g. the energy difference between the Fermi level and conduction band in the bulk $(E_C-E_F)_b$, band bending qV_S (q denotes elementary charge) and electron affinity χ: $\Phi = (E_C - E_F)_b + qV_S + \chi$ (due to the definition, $V_S = E_{C,S}-E_{C,B}$). For ionosorption the work function follows only the change in band bending $(\Delta\Phi = q\Delta V_S)$. The z_0 denotes the depth of the depletion region; μ—the electrochemical potential; $E_{V,B}$ and $E_{V,S}$—valence band edge in the bulk and at *the surface, respectively*; $E_{d,S}$—*donor level at the surface*; $E_{C,B}$ *and* $E_{C,S}$—*conductance band* edge in the bulk and at the surface, respectively; E_F—Fermi level; $O_{2,gas}$ is an oxygen molecule in the ambient atmosphere; $O_{2,phys}$—a physisorbed oxygen species; $O_{\beta S}^{-\alpha}$—a chemisorbed oxygen species ($\alpha = 1$ and $\alpha = 2$ for singly and doubly ionised forms, respectively; $\beta = 1$ and $\beta = 2$ for atomic and molecular forms, respectively)

Weisz limitation, see original [18] and discussion in [4]). Within the framework of this concept, the operation of SnO_2-based sensors is described as follows: oxygen adsorbs in a delocalized manner, trapping electrons from the conduction band and forming ions—"charged" molecular ($O_2^{-\ ads}$) and atomic (O^-ads, O_2^-ads) species—electrostatically stabilized at the surface in the vicinity of metal cations. This happens under real working conditions of sensors, between 100 and 450 °C, at atmospheric pressure, at 20.5 vol. % background oxygen.

Reducing gases, like CO, react with the oxygen ions (by either Eley–Rideal or Langmuir–Hinshelwood mechanism) freeing electrons that return to the conduction band.

The ionosorption theory explains also the increase in the sensing performance with decreasing crystal size. Firstly, the reactivity of nanomaterials is mainly determined by the so-called "smoothly scalable" size-dependent properties which are related to the fraction of atoms at the surface [102]. As the crystal size decreases, the surface-to-volume ratio increases proportionally with the inverse of the crystal size. The increase in the total surface-to-volume ratio with respect to the size decrease generates more "reactivity" due to a dominant surface-like behavior caused by an increased fraction of atoms at the surface [102]. Thus, all properties which depend on the surface-to-volume ratio change continuously and

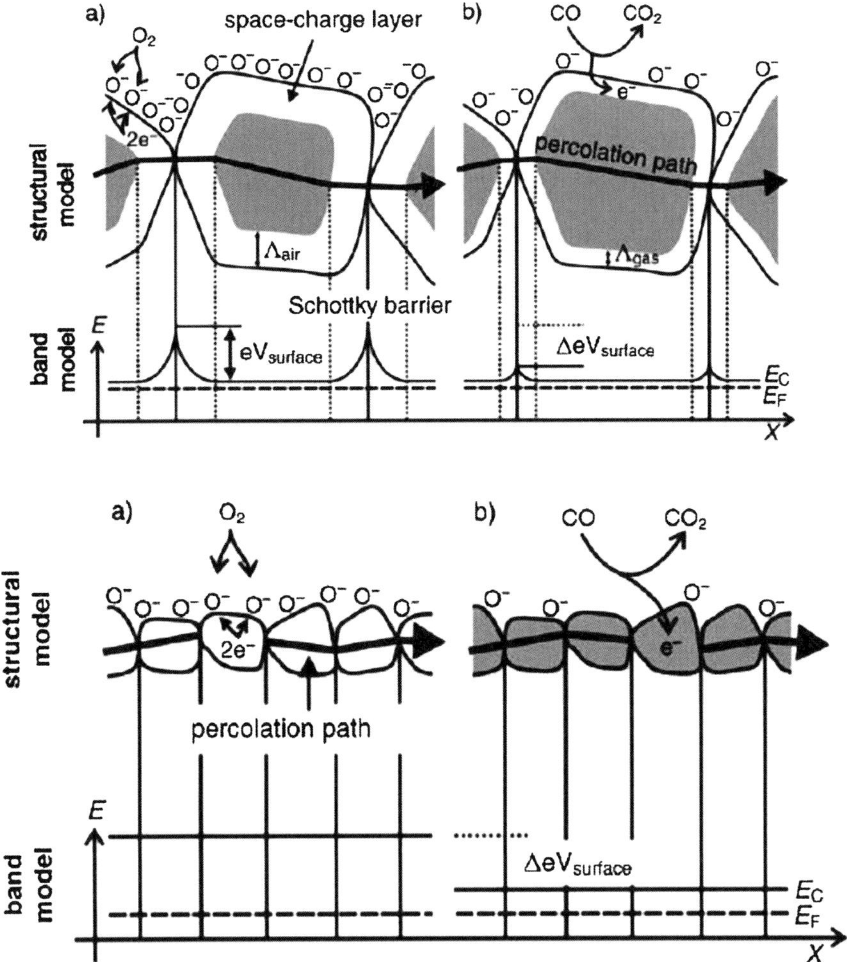

Fig. 1.4 The mechanism controlling the conductivity change and its magnitude depends on the ratio between grain size (D) and Debye screening length (Λ). If $D > 2\Lambda$, the depletion of the surface between the grain boundaries controls the conductivity. In this case low response to the analyte is expected as only a small part of the semiconductor is affected by interaction with analyte. If $D \leq 2\Lambda$, the whole grain depleted and changes in the surface oxygen concentration affects the whole semiconductor resulting in high response. Copyright Wiley–VCH, reproduced with permission from Ref. [43]

extrapolate rapidly at very low crystal sizes. As a consequence, nanoparticles with increased surface-to-volume ratio are expected to be more reactive and accordingly, more gas sensitive.

With decreasing crystal size there is also a transition from a partly to a completely charge depleted particle that can be observed, depending on the ratio between the crystal and the Debye screening length L_D (Λ in Fig. 1.4) (for calculation, see for

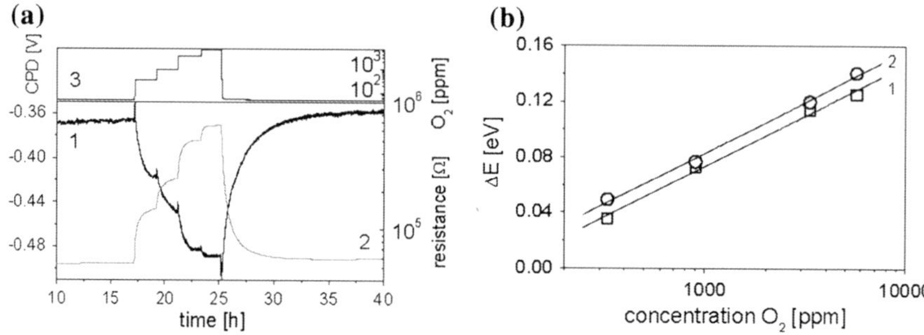

Fig. 1.5 **a** The contact potential difference (CPD $= -\Delta\Phi$) *1* and the resistance *2* have been recorded at different O$_2$ concentrations *3* on the nanocrystalline SnO$_2$ at 400 °C in dry nitrogen at atmospheric pressure (adapted from Ref. [84]). **b** Calculated from **a** work function change ($\Delta\Phi$) *1* and band bending ($q\Delta V_S = kT\ln(G_0/G)$) *2* changes. Copyright Wiley–VCH, reproduced with permission [145]

example [37, 103]). For partly depleted particles, when surface reactions do not influence the conduction in the entire layer, the conduction process takes place in the bulk region. Formally, two resistances occur in parallel, one influenced by surface reactions and the other not; the conduction is parallel to the surface, and this explains the limited sensitivity [37, 39]. Fully depleted particles possess higher sensitivity as the charge depletion layer fully impacts the conduction channel within the nano-particle, thus achieving better performance in gas exposure experiments [43].

Summarizing, the atomic charged oxygen ion (O$^-$ads) is assumed to be of particular importance in gas sensing because "the O$^-$ ion appears to be more reactive of the two possibilities and thus more sensitive to the presence of organic vapours or reducing agents..." [81]. Accordingly, "there are two important questions to resolve here: First, under what conditions does O$^-$ dominate over O$_2^-$? Second, what is the total surface charge as a function of ... temperature and partial oxygen pressure?" [81] As a consequence, ambitious efforts have been made (i) to calculate the surface coverage by different types of ionosorbed oxygen [6, 104–106] (Fig. 1.2) and (ii) to correlate the overall conductance of the sensors with the chemical state of charged oxygen species at the surface [37, 107].

The contradiction arises when connecting the main statements of the iono-sorption model to common chemical sense and spectroscopic findings.

A first example of this is to note that oxygen ionosorption should be reflected in *equal* changes in the work function and band bending, $kT\ln(G_0/G) = q\Delta V_S = \Delta\Phi$ (see also Fig. 1.3). These values can be independently obtained, for example, in the simultaneous CPD (here $\Delta V_{CPD} = -\Delta\Phi = q\Delta V_S$) and conductance mea-surements (here $q\Delta V_S = kT\ln(G_0/G)$, see an example in [38].

However, even if one can measure *formal* evidence for the pure oxygen ion-osorption ($kT\ln(G_0/G) = \Delta\Phi$, Fig. 1.5, the transient changes observed ($q\Delta V_s = 0$, $\Delta\Phi = 0$ after 50 h) reflects very slow surface processes. These slow changes are in sharp contrast to the fast charging expected at the oxide surface (see Ref. [108], charging takes less than 5 ms even at 250 K) and this discrepancy is not explained by "ionosorption theory".

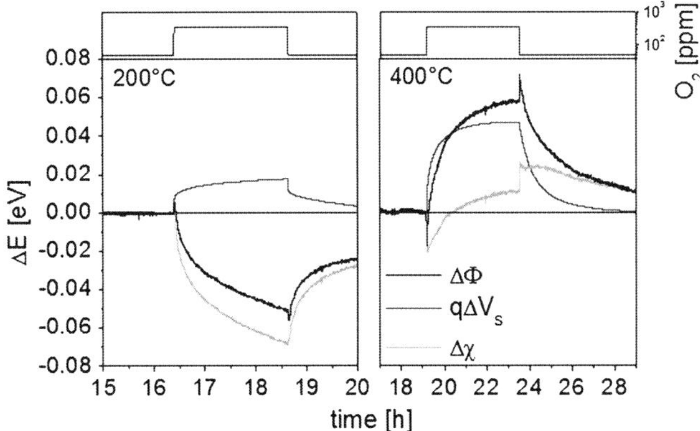

Fig. 1.6 Changes in work function (*black*), band bending (*dark grey*) and electron affinity (*light grey*) due to 300 ppm O_2 (——-) at 200 °C (*left*) and at 400 °C (*right*). Copyright Elsevier, reproduced with permission from Ref. [67]

The results shown in Fig. 1.5 strongly suggest that at 400 °C all other species, besides ionic ones, can be regarded as being of secondary importance. However, at 200 °C a completely different behavior of the changes of the work function appear. This is illustrated by Fig. 1.6 where changes in work function, band bending and electronic affinity due to a pulse of 300 ppm oxygen are displayed for 200–400 °C, respectively. The most important difference is the strong decrease in electronic affinity at 200 °C. Such effects did not appear at 400 °C. As shown in Ref. [38], the changes in electronic affinity are connected with the formation or loss of dipolar species between adsorbate and adsorbent accompanied by localized bonding. Therefore, in order to get an explanation of the experimental results we have to allow for the possibility of dipole formation arising from the adsorption of neutral molecular oxygen species (Fig. 1.7). These species are neglected in all mechanistic description of gas sensing on semiconducting metal oxides.

A critical look at the available experimental data shows that the concept of oxygen ionosorption is based exclusively on phenomenological measurements. Despite trying for a long time, there has not been any convincing spectroscopic evidence for "ionosorption". Neither superoxide ion O_2^-, nor charged atomic oxygen O^-, nor peroxide ions O_2^{2-}, nor CO^+ have been observed under real working conditions of sensors (see a recent review [109]).

With regard to the two main forms of charged oxygen species on the surface (superoxide ion O_2^- and charged atomic oxygen O^-) and widely used in the mechanistic description of gas sensing properties and modelling of oxide conduction mechanism, it appears that:

1. The superoxide ion (O_2^-) has been observed only after low-temperature adsorption <150 °C on reduced SnO_2;
2. There has not been any spectroscopic evidence for the formation of charged atomic oxygen (O^-) on SnO_2.

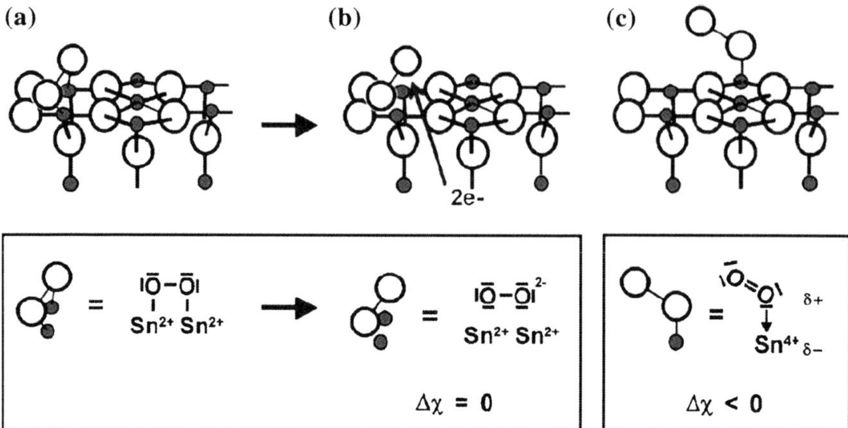

Fig. 1.7 Adsorption of O_2 on a reduced SnO_2 (110) surface. There is a stable state for a twisted conformation on a fourfold Sn^{2+} site (**a**) It is expected to accept negative charge under building of ionosorbed species and thus without influence on electron affinity (**b**) A stable conformation tilted from the normal, where a Lewis acid/base interaction leads to a local dipole with a negative partial charge on the tin and thus to a decrease in χ, is also reported (**c**) Copyright Elsevier, reproduced with permission from Ref. [67]

Moreover, several findings, such as (i) the formation of superoxide ion ($O_2{}^-$)only at low adsorption temperatures(<150 °C) on reduced SnO_2, (ii) absence of a high-temperature oxygen desorption (peak at 400–550 °C, attributed to adsorbed oxygen) if the superoxide-ion is present at the surface, (iii) decrease in oxygen intensity with increasing evacuation temperature; herewith the amount of O_2 desorbed is equal to the number of superoxide ion centres, (iv) correlation between TPD, EPR, IR and electrophysical studies on reduced SnO_2, allows us to conclude that the *superoxide ion does not undergo transformations into charged atomic oxygen at the surface and represents a dead-end form of low-temperature oxygen adsorption on reduced metal oxide.*

As known, the superoxide ion can undergo the following chemical changes on the surface: (i) lose an electron (to the CB) and leave as gaseous O_2 and (ii) gain an additional electron (becoming a peroxide ion $O_2{}^{2-}$), followed by cleaving to form atomic oxygen and the lattice oxygen anion (O^{2-}). According to the spectroscopic data (there is no evidence either for peroxide ion or for charged atomic oxygen) the transformation from superoxide ion to atomic oxygen does not happen on SnO_2. This indicates two competing channels for oxygen adsorption—molecular and dissociative. A similar mechanism has been recently postulated for TiO_2 [110] and Ag [111]. On TiO_2, only η^2-coordinated dioxygen decomposes to oxygen adatom and a filled oxygen vacancy (in contrast, the η^1-coordinated dioxygen desorbs at 410 K [110]). On Ag, upon heating, physisorbed oxygen transforms into molecular chemisorbed I α-O_2 ("end-form") which does not dissociate into the atomic form due to the high conversion barrier; only molecular chemisorbed II β-O_2 ("transformable form") accessible only through a direct interaction from the gas phase (and not accessible from physisorbed form) dissociates into atomic oxygen [111].

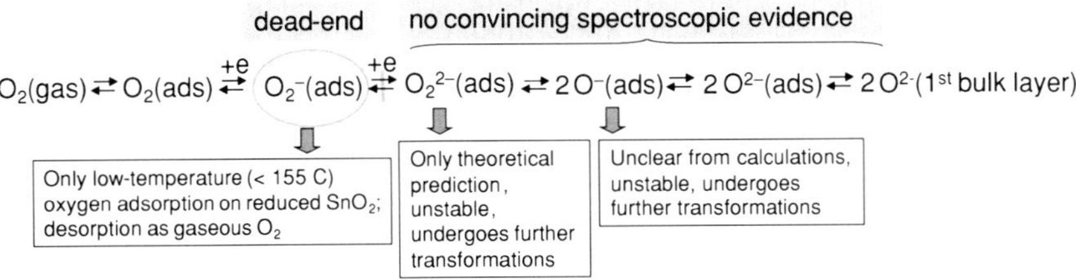

Fig. 1.8 Scheme of oxygen interaction with metal oxides showing the superoxide ion as a dead-end form. Modified from Ref. [145]

The long sought efforts to quench "high-temperature" oxygen species (claimed to represent charged atomic oxygen—O^-) has not yielded any measureable success. No such paramagnetic species have been observed on high-temperature oxygen treated oxides (TiO_2, SnO_2, ZnO) [95, 112, 113]. Moreover, the EPR evidence of the surface O^- species formed due to oxygen adsorption is very contradictory. Furthermore, from a review of the literature there isn't convincing evidence of their formation on n-type semiconducting oxides due to their direct interaction with dioxygen. Likewise, it is not possible to connect the high-temperature peak in the TPD spectra and the change in the conductivity with the formation of surface O^- species. Consequently, the conclusions on O^- formation on SnO_2 are not supported by any spectroscopic data. Accordingly, the picture of oxygen adsorption on SnO_2 has to be modified in the following way (Fig. 1.8).

1.5 Oxygen-Vacancy Model (Reduction-Reoxidation Mechanism)

This model focuses on oxygen vacancies at the surface, which are considered to be "the determining factor in the chemiresistive behavior" [114]. Tin dioxide, the most extensively investigated sensing material, is oxygen-deficient and, therefore, an n-type semiconductor, whose oxygen vacancies act as electron donors. Alternate reduction and reoxidation of the surface by gaseous oxygen (Mars—van Krevelen mechanism)control the surface conductivity and therefore the overall sensing behaviour. In this model, the mechanism of CO detection is represented as follows: (i) CO removes oxygen from the surface of the lattice to give CO_2, thereby producing an oxygen vacancy; (ii) the vacancy becomes ionized, thereby introducing electrons into the conduction band and increasing the conductivity; (iii) if oxygen is present, it fills the vacancy; in this process one or more electrons are taken from the conduction band, which results in a decrease in conductivity.

Numerous experimental and theoretical works have evaluated the reduction-reoxidation mechanism (see, for example, references [52, 114–119]); this mechanism still dominates in almost all spectroscopic studies (see, for example, references [120–125], Table 1.2). For example, it was found that oxygen promotes

Table 1.2 Some examples of main finding from in situ and operando case studies of gas sensing on semiconducting metal oxides

Method	Oxides	Gases	Main findings
IRES	WO_3, $AlVO_4$ and Co_3O_4 [124]	O_2 in N_2, C_3H_6 and acetone in air	Ionisation of oxygen vacancies
DRIFTS	CdGeON [147, 148]	O_2 in N_2	Filling of oxygen vacancies; change of the Ge coordination number
XAS	SnO_2 and Pd/SnO_2 [136, 149]	CO and H_2 in N_2	Sn^{4+} and Pd^{2+} reduction as secondary processes, CO and H_2 oxidation by ionosorbed oxygen
XAS	Pt/SnO_2 [150, 151]	CO in N_2, H_2S	Variation in Pt oxidation state in reducing and oxidising atmospheres
XAS	SnO_2; Pt/SnO_2 [152, 153]	Air, CO/air, CO/N_2, O_2	Variation in Pt oxidation state in reducing and oxidising atmospheres
CEMS	Bi_2O_3-SnO_2 [152, 153]	He, CO/He, CH_4/He	Oxidation by lattice oxygen atoms, formation of oxygen vacancies
FTIR	TiO_2, SnO_2, In_2O_3, WO_3 [154–159]	CO_2, CO, O_2, O_3, NOx	Variations of the free carriers density
FTIR	SnO_2 [123, 128]	O_2/N_2, CO/air, He/air	Photoionisation of ionised oxygen vacancy with increasing oxygen content
	SnO_2, MoOx-SnO_2, Pd/SnO_2, WOx-SnO_2 [129, 160–163]	O_2, CO, NO, NO_2	Formation of oxygen vacancies and their ionisation
DR UV-vis	SnO_2 [164]	O_2, hydrazine	Formation of oxygen vacancies and their ionisation
CEMS	SnO_2 and Ru, Pt, Pt/SnO_2 [165, 166]	NO/argon, air	Formation of oxygen vacancies and their ionisation
	SnO_2 and Pd/SnO_2 [121]	CO/N_2, air	Formation of Sn (II) as indicator of oxygen vacancy formation
EPR	SnO_2; Ru, Pt, Pt/SnO_2 [120, 165–168]	Dry and humid air, CO in air and N_2, NO/Ar, H_2	Formation of oxygen vacancies and their ionisation

water vapour dissociation on SnO_2 at 330–400 °C [126]: an increase in the concentration of hydroxyl groups (peaks at 3640 cm^{-1}) was observed for low oxygen (2000 ppm) and water vapour (3 ppm) concentrations and evolved towards saturation. This effect was explained by the reaction:

$$H_2O \ (gas) + O^-(ads) + 2Sn_{Sn}{}^{\times} \leftrightarrow 2(Sn_{Sn}{}^{\times} - OH) + e^-(CB)$$

At first sight this seems to provide evidence for the ionosorption model. However, this effect (i.e. the increase in the concentration of hydroxyl groups during oxygen exposure) can be explained just as well, by completely different processes within the framework of the oxygen vacancy model. For example, an EPR signal of single ionised oxygen vacancies ($V_O^{\bullet}$) at 1.89 was observed after wet

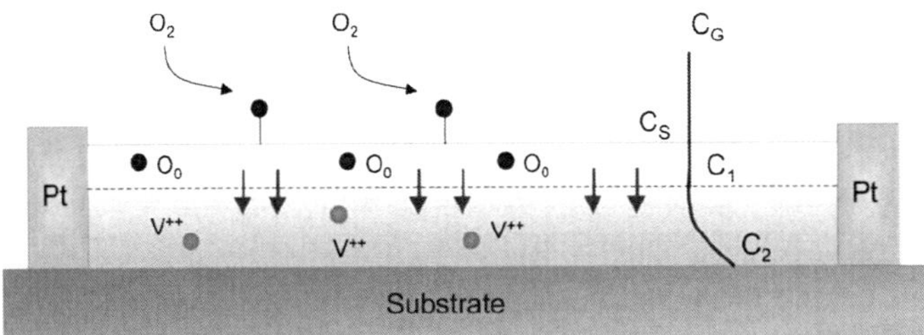

Fig. 1.9 One-dimensional model of oxygen diffusion in nanowires. According to the ionosorption model, adsorbed oxygen creates a depletion region close to the surface (dashed line) and then a fast change of R_{NW} is observed. The new equilibrium between oxygen in the environment, C_G, and the concentration of oxygen at both the nanowire surface, C_S, and its external shell, C_1, creates a gradient with the inside C_2 favoring ion migration into the bulk. This diffusion is associated with long-term drifts of R_{NW}. Copyright Wiley–VCH, reproduced with permission from Ref. [134]

air treatment of SnO_2 at 200 °C [127]. Accordingly, the observed influence of water and oxygen can be described by the two reactions:

$$38; H_2O \text{ (gas)} + 2Sn_{Sn}{}^\times + O_O{}^\times \leftrightarrow 2(Sn_{Sn}{}^\times - OH) + V_O{}^\bullet + e^-(CB)$$

$$38; 2V_O{}^\bullet + O_2 \text{ (gas)} + 2e^-(CB) \leftrightarrow 2O_O{}^\times$$

However, the problems associated with oxygen adsorption and detection of reducing gases in an oxygen-free atmosphere are questioning the validity of the reduction-reoxidation model (see detailed discussion below).

As we mentioned above, ionosorbed oxygen has never been observed in operando and in situ studies on metal oxide sensors under working conditions [123, 124]. By contrast, operando and in situ spectroscopy provides very strong evidence of the reaction and ionization of oxygen vacancies under operating conditions of sensors [120–124].

The in situ FT-IR studies [123, 128] of SnO_2 under working conditions (at 375–450 °C) showed an increase of the intensity in the broad band in the region of 2300–800 cm^{-1} (so-called X-band) with increasing oxygen content. The proximity of the absorption edge with respect to the ionization energy of the second level of oxygen vacancies (1400–1500 cm^{-1} $\sim$ 170–180 meV) is indicative of the electronic transition from this level to the conduction band (i.e. photoionisation of $V_O^\bullet$ to $V_O^{\bullet\bullet}$)[129]. Accordingly, this band can serve as an indicator of the electron concentration in the neighborhood of oxygen vacancies in the SnO_2. Similar effects were observed on Ga_2O_3, $AlVO_4$, WO_3 [124]. However, this interpretation was considered to be in contrast to the early electrophysical measurements on SnO_2 [130] which showed that the donor levels in SnO_2 are located at around 30–150 meV below the conduction band, and will be completely ionized at the sensor operating temperatures [131, 132]. The next problem is related to (i) the ionization of oxygen vacancies and consequently, and (ii) to the diffusion processes in the oxide lattice. For SnO_2, for example, it is assumed that the surface defects do not

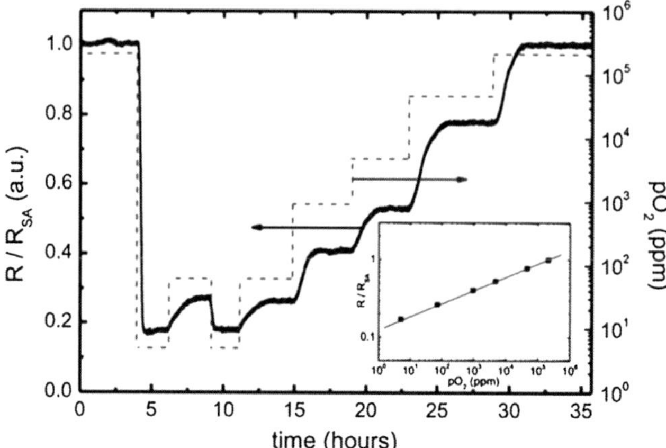

Fig. 1.10 Response of one SnO_2 nanowire with r = 20 nm to increasing oxygen partial pressure at room temperature (T = 298 K). Resistance is normalized to the experimental value in synthetic air environment. (Inset) log–log plot of resistance as function of oxygen partial pressure. A linear behavior of slope n = 1/6 is observed. Copyright Wiley–VCH, reproduced with permission from Ref. [134]

act as electron donors; they have to migrate a small distance into the bulk to become ionized [52]. The diffusion coefficients for this process are low, and, accordingly, the defects are immobilized at the operating temperatures [116]. Nevertheless, diffusion at grain boundaries and at the surface can be much faster than bulk diffusion [12].

A different situation seems to appear in quasi-one-dimensional SnO_2 structures (nanowires) [133, 134], where the oxygen adsorption involves two steps (i) healing of oxygen vacancies by adsorbed oxygen (fast) and (ii) diffusion of as-formed oxygen ions in the bulk healing (annealing) oxygen vacancies (slow) (Fig. 1.9).

Surprisingly, even at room temperature an exponent of 1/6 was found in the power law for the oxygen partial pressure dependence of the sensor signal (i.e. conductivity) indicating an intrinsic case in the defect chemistry of SnO_2 (Fig. 1.10) [116]:

$$38; \frac{1}{2}O_2 + V_O^{\bullet\bullet} + 2e' \leftrightarrow O_O$$

$$38; \left[e'\right] = 2^{1/3}K_O^{-1/3}p_O^{-1/6}, \text{ considering the electroneutrality condition}$$

$$38; \left[e'\right] = 2[V_O^{\bullet\bullet}]$$

These findings contrast with bulk film studies which have shown that (i) the surface exchange reaction, i.e. the incorporation of adsorbed oxygen together with the fast electron transfer, is the rate-determining step [118] and (ii) significant oxygen exchange is observed on SnO_2 only at temperatures above 400 C [135].

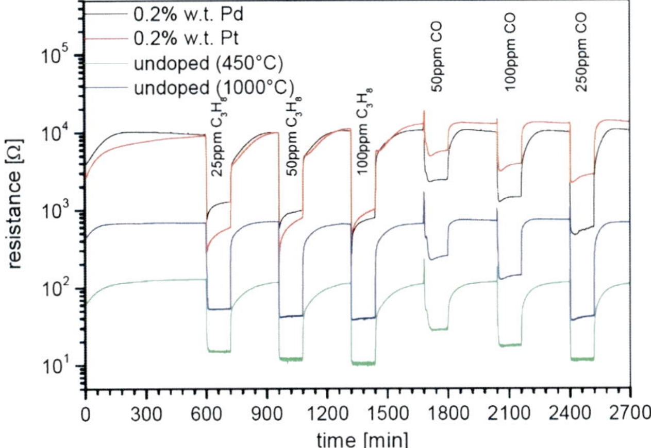

Fig. 1.11 Transient resistance change of undoped, Pd and Pt doped SnO_2 sensors in oxygen free atmosphere (N_2 balance) [146]. Note, that the sensor resistance recovers its initial value after removing the target gas from the surrounding atmosphere

1.6 Reduction as a Secondary Process: an Open Issue of Detection of Reduction Gases in Oxygen-Free Conditions

The mechanism of detection of reducing gases in oxygen-free atmospheres requires consideration of the following four experimentally confirmed observations:

1. Recovery of sensor resistance to its initial value in an inert gas (N_2, Ar, He) after removing a reducing gas (CO, CH_4, H_2) from the test atmosphere.
2. Missing correlation between the degree of oxide reduction and the magnitude of gas sensing response.
3. Missing correlation between the gas combustion (oxidation) and the magnitude of gas sensing response.
4. Decrease of sensor signals (relative resistance change) with increasing oxygen concentration.

We have to note that a lack of experiments does not allow addressing properly these issues; some important points are discussed below.

Let us take as an example of CO detection in the oxygen-free conditions (alternating CO/N_2 and N_2 flows): What happens when CO is removed from the surrounding atmosphere? From electrical measurements one knows that the sensor resistance (or conductance) recovers its initial value (Fig. 1.11). However, within the framework of the reduction-reoxidation mechanism, gaseous oxygen is required for the reverse process ("vacancy refilling"). Unfortunately, the consideration of this problem has been avoided in spectroscopic studies by alternating

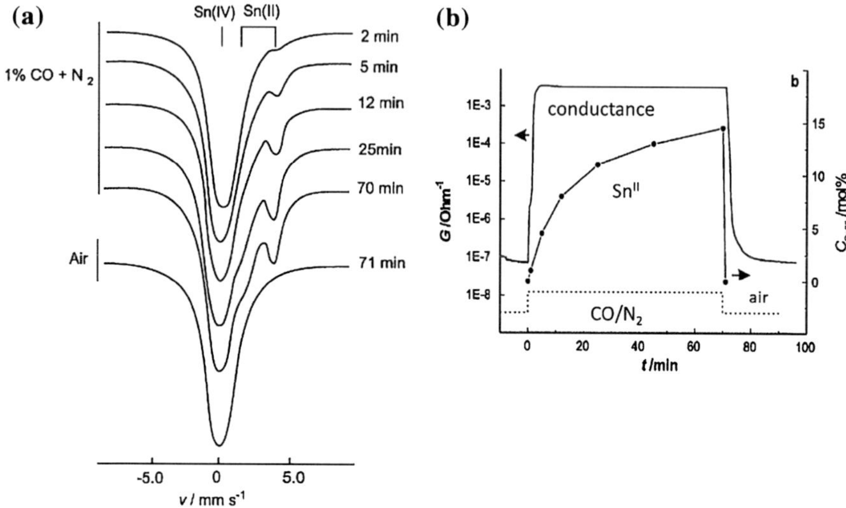

Fig. 1.12 **a** Mössbauer spectra of SnO$_2$ in a 1 % CO/N$_2$ atmosphere and then in dry air at 380 °C; **b** Electrical response of SnO$_2$ and change of Sn(II) concentration in 1 % CO/N$_2$ and dry air at 380 °C under 4 l/h gas flow rate. Copyright the Royal Chemical Society, reproduced with permission from Ref. [121]

CO/N$_2$ (or Ar) and O$_2$/N$_2$ (or Ar) flows, whereas "realistic" conditions require alternating CO/N$_2$ (or Ar) and N$_2$ (or Ar) flows.

The oxide reduction as well as reduction/reoxidation of catalytic additives like Pd and Pt seems to be a secondary process that is not connected with the overall sensor response. Figure 1.12 a and b shows the electrical response combined with a change of Sn(II) concentration revealed by in situ Mössbauer spectra for nanocrystalline SnO$_2$ in the presence of CO and dry air at 380 °C. It was shown that the conductance changes simultaneously with the change of the tin oxidation state (which in turn indicates the formation of oxygen vacancies). A rapid and pronounced increase in Sn(II) spectral contribution was observed just after CO admission into the reactor. The Sn(II) component disappeared 1 min after air admission. It was also noted that (i) a very low Sn(II) content (1 mol %) was sufficient for the conductance to change by 1000 times and (ii) a further increase of Sn(II) concentration up to 14 mol % under exposure to CO did not significantly change the conductance. Similar findings have been also reported for H$_2$ detection with Pd-promoted SnO$_2$ sensors. Figure 1.13 [136] show the correlation between the electrical conductance and the oxidation states of Pd and Sn during the cycling of Pd–SnO$_2$ film in H$_2$ and O$_2$ gas mixtures. At 373 K, the conductance changes without any variation of the Pd and Sn oxidation states. At higher temperatures, the oxidation state of Pd varies considerably depending on the atmospheric composition. However, there is no direct correlation between the conductance and the oxidation states of Pd and Sn, i.e. even at 573 K the conductance changes by several orders of magnitude without any measurable variation of the oxidation

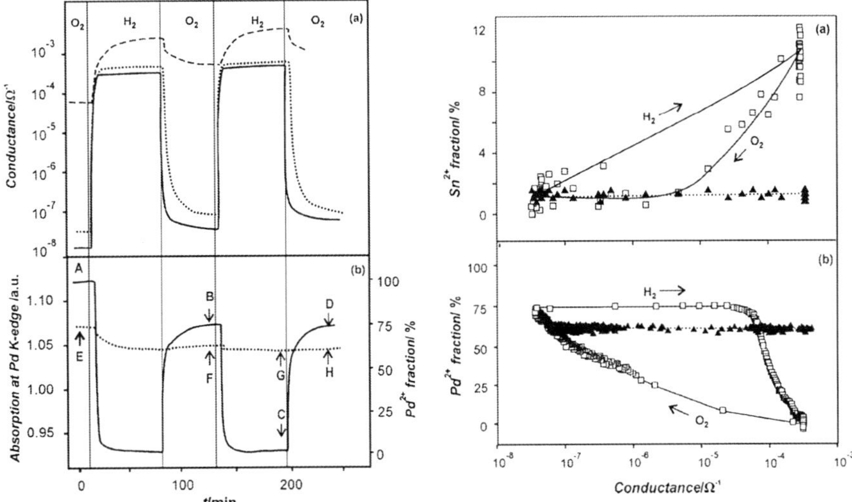

Fig. 1.13 (On the *left*) Variation of the electrical conductance **a** normalized absorption at Pd K-edge (b, *left scale*) and Pd^{2+} fraction in the Pd^{2+}/Pd^0 mixture (b, *right scale*) for Pd–SnO$_2$ film at 573 K (*solid lines*) and 373 K (*dotted lines*) during the alternative exposure to 20 % O$_2$ in He and 1000 ppm H$_2$ in He. The broken line corresponds to pure SnO$_2$ at 573 K. (On the right) Operando XAS and conductance studies. The correlation between the conductance of Pd–SnO$_2$ film and the oxidation states of tin **a** and palladium **b** at 573 K (*white squares, solid lines*) and 373 K (*black triangles, dotted lines*). Fraction of Pd^{2+} is the concentration of Pd^{2+} in the Pd^{2+}/Pd^0 mixtures; the fraction of Sn^{2+} is the concentration of Sn^{2+} in the Sn^{2+}/Sn^{4+}mixture. The arrows indicate the direction in which the system changes during exposure to H$_2$ and O$_2$ Copyright the Royal Chemical Society, reproduced with permission from Ref. [136]

states of both metals. These results indicate that oxidation and reduction of Pd nanoparticles and SnO$_2$ matrix are the secondary processes, which are not responsible for the sensitivity to H$_2$.

No direct correlation exists between sensor activity (i.e. sensor signal) and consumption of target gases obtained from the on-line gas analysis. As demonstrated in Fig. 1.14 herein at 200 °C the sensor shows relatively high activity in C$_3$H$_8$ detection (sensor signal is about 6–400 ppm C$_3$H$_8$) the combustion, however, is almost negligible [137]. The same also holds for higher temperatures and other analytes. Several recent works have also demonstrated that there has been no direct correlation between sensor (SnO$_2$, TiO$_2$...) response to CO and the CO$_2$ production ("catalytic activity") [138–140].

The strong argument is coming from the observation that in the "oxygen-free" atmosphere the sensor response (i.e. relative change in the conductance) is even higher than in "oxygen" containing atmosphere (Fig. 1.15). To explain these findings, an assumption about the formation of ionosorbed donor-like CO$^+$ species is made (see, for example [141] and Fig. 1.16). However, like in the case of ionosorbed oxygen species we face here a problem of common chemical understanding and missing spectroscopic evidence.

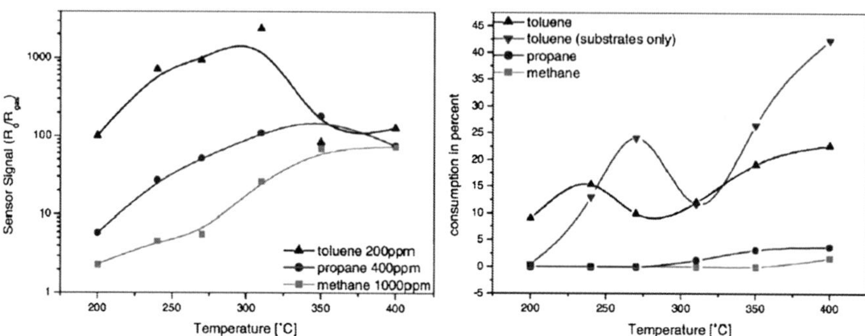

Fig. 1.14 (*Left*) Sensor signal of SnO$_2$ sensors exposed to different analytes in dry air dependent on operating temperature of sensors and (*right*) overall gas combustion measured by on-line PAS Copyright Elsevier, reproduced with permission from Ref. [137]

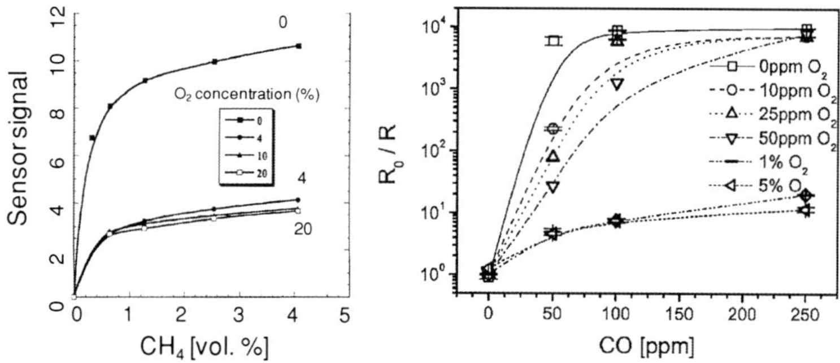

Fig. 1.15 (*left*) Relative change in the conductance (sensor signal, defined as $(G_{CH4}-G_0)/G_0$, where G_{CH4} and G_0 are the conductance values measured at 450 C under CH$_4$ supporting gas, respectively.) for undoped SnO$_2$ as a function of CH$_4$ concentrations for different O$_2$ concentrations (N$_2$ balance). Copyright Elsevier, reproduced with permission from Ref. [85]. (*Right*) Sensor signal of a 0.2 wt % Pt doped SnO2 thick film sensor at 300 C as a function of CO concentration for different O$_2$ concentrations (0, 10, 25, 50 ppm, 1 and 5 %). Copyright Elsevier, reproduced with permission from Ref. [141]

1.7 Summary and Outlook

Since the development of the first models of gas detection on metal-oxide-based sensors much effort has been made to describe the mechanism responsible for gas sensing. Despite progress in recent years, a number of key issues remain the subject of controversy; for example, the disagreement between electrophysical and spectroscopic investigations, as well as the lack of proven mechanistic description of surface reactions involved in gas sensing. The state-of-the-art description and understanding of gas sensing on metal oxides cannot explain all effects observed on *operating* metal oxide sensors.

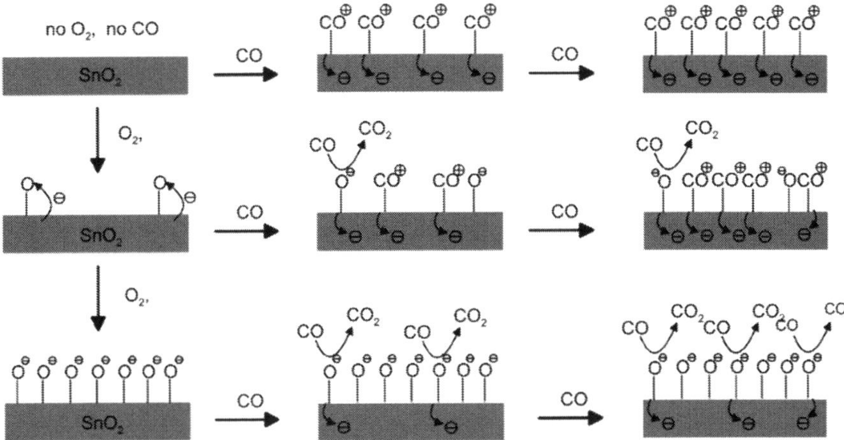

Fig. 1.16 Schematic representation CO interaction with SnO_2 as function of the O_2 concentration: in the absence of O_2, CO adsorbs as electron donor and leads to a decrease of the sensor resistance. In the presence of O_2, CO reacts mainly with adsorbed O_2 species. The contribution of each mechanism depends on the O_2 concentration. Copyright Elsevier, reproduced with permission from Ref. [141]

Obviously, our ability for understanding fundamental physicochemical phenomena is limited by dominant physicochemical paradigms. Accordingly, the interpretation of spectroscopic data depends on the model a priori chosen for the mechanistic description.

Besides general applicability for describing the mechanism of gas sensing on semiconducting metal oxides, the ionosorpion model works very well in explaining also quite unobvious results under a priori made unproven assumptions; typical examples are:

- higher response to reducing gases such as CO in absence of oxygen compared with that in the presence of oxygen; here the assumption is made about the formation of ionosorbed donor-like CO^+ species (see, for example [141]);
- large response for smaller crystals; obviously this interpretation does not involve and does not require the explicit formation of any ions at the surface. In case of the interaction with gaseous oxygen, the key is only the decrease of electron density (depletion) in the surface regions of a metal oxide; this effect is more pronounced in materials with higher surface/volume ratio (see, for example, Ref. [103]);
- reversible p- to n- and to p-type transition on semiconducting metal oxides induced by gas adsorption [142, 143].

The ionosorption theory fails to explain several important issues, among them:

1. the missing spectroscopic evidence as well as theoretical confirmation of ionosorbed oxygen species and
2. electron affinity changes due to the oxygen adsorption.

Excluding the electron transfer to adsorbed species by assuming the localization of electrons in solid material in close vicinity to adsorbed/gaseous species eliminates immediately the issue of missing spectroscopic evidence of oxygen ions. The polarizability effects can also be involved for explaining the competition between e.g. CO and oxygen: in the presence of CO, adsorbed O_2 molecules will also attract the electrons from the highly polarizable CO molecules thus reducing the O_2 strength to attract the conduction electrons [144]. This assumption can also explain large response for smaller crystals (see above). However, it fails to explain the higher response—meaning the increase in the conductance—to reducing gases such as CO in absence of oxygen compared with that in the presence of oxygen.

Even if numerous experimental and theoretical works have evaluated the reduction-reoxidation mechanism and this mechanism still dominates in almost all spectroscopic studies, the reduction-reoxidation model fails to explain

1. the missing correlation between the conductance of sensing materials and the degree of reduction *under operating conditions* (from operando studies);
2. kinetics of oxygen exchange;
3. recovery of the sensor resistance to its initial value after exposure to reducing gases in oxygen-free conditions where—according to the reduction-reoxidation mechanism, gaseous oxygen is required for the reverse process ("vacancy refilling");
4. lack of gaseous products of the oxidation of reducing gases (e.g. CO_2) from simultaneous on-line gas analysis.

References

1. Heiland G, Mollwo E, Stockmann F (1959) Electronic processes in zinc oxide. Solid State Phys 8:191–323
2. Many A, Goldstein Y, Grover NB (1971) Semiconductor surfaces. North-Holland Publishing Company, Amsterdam, 496
3. Morrison SR (1955) Surface-barrier effects in adsorption, illustrated by zinc oxide. Adv Catal 7:259–301
4. Morrison SR (1977) The chemical physics of surfaces, Plenum Press, New York, 415
5. Wolkenstein T (1960) Electron theory of catalysis on semiconductors. Adv Catal 12:189–264
6. Wolkenstein T (1987) Electronic processes on the surface of semiconductors during chemisorption, Consult Bur New York, 431
7. Wolkenstein T (1964) Elektronentheorie der Katalyse an Halbleitern, Verlin: VEB
8. Hauffe K (1955) Application of the theory of semiconductors to problems of heterogeneous catalysis. Adv Catal 7:213–57
9. Hauffe K (1955) Application of the semiconductor theory to problems of heterogeneous catalysis. Angewandte Chemie 67:189–207
10. Engell HJ, Hauffe K (1953) The boundary-film theory of chemisorption. Interpreting the reaction on the solid-gas interface (Die Randschichttheorie der Chemisorption . Ein Beitrag zur Deutung von Vorgängen an der Grenzfläche Festkörper/Gas). Zeitschrift fuer Elektrochemie und Angewandte Physikalische Chemie 57:762–73

11. Hauffe K (1955) Reaktionen in und an Festen Stoffen (Erste Auflage). Springer, Berlin, 696
12. Hauffe K (1966) Reaktionen in und an Festen Stoffen (Zweite Auflage). Springer, Berlin, 696
13. Kiselev VF, Krylov OV (1987) Electronic phenomena in adsorption and catalysis on semiconductors and dielectrics. Springer series in surface sciences, Springer, Berlin, 279
14. Roginskii SZ (1949) Principles of catalyst theory. Problemy Kinetiki i Kataliza 6:9–53
15. Wagner C (1950) The mechanism of the decomposition of nitrous oxide on zinc oxide as catalyst. J Chem Phys 18:69–71
16. Bevan DJ, Anderson MJS (1950) electronic conductivityconductivity and surface equilibria of zinc oxide. Discussions of the Faraday society 8:238–246
17. Boudart M (1952) Electronic chemical potential in chemisorption and catalysis. J Am Chem Soc 74:1531–5
18. Weisz PB (1953) Effects of electronic-charge transfer between adsorbate and solid on chemisorption and catalysis. J Chem Phys 21:1531–8
19. Morrison SR (1953) Changes of surface conductivity of germanium with ambient. J Phys Chem 57:860–3
20. Garrett CGB (1960) Quantitative considerations concerning catalysis at a semiconductor surface. J Chem Phys 33(4):966–979
21. Brattain WH, Bardeen J (1953) Surface properties of germanium. Bell Sys Tech J 32:1–41
22. Engell HJ (1954) Randschichteffekte an der Grenzfläche Hableiter/Vakuum und Halbleiter/Gasraum. Halbleiterprobleme 1:249–272
23. Hauffe K (1956) Gas reactions on semiconducting surfaces and space charge boundary layers. In: Kingston RH (ed) Semiconductor surface physics, University of Pennsylvania Press, Philadelphia 259–282
24. Vol'kenshtein FF (1949) Electronic theory of promotion and poisoning of ionic catalysts. Problemy Kinetiki i Kataliza 6(Geterogennyi Kataliz):66–82
25. Morrison SR (1982) Semiconductor gas sensors. Sens Actuators 2(4):329–341
26. Kefeli A (1956) Sauerstoffnachweis in Gasen durch Leitfähigkeitsänderung eines Halbleiters(zno). Diploma Thesis, Institut für Angewandte Physik, Universität Erlangen, Erlangen
27. Heiland G (1982) Homogeneous semiconducting gas sensors. Sens Actuators 2(4):343–61
28. Heiland G (1957) Effect of hydrogen on the electrical conductivity on the surface of zinc oxide crystals (Zum Einfluss von Wasserstoff auf die elektrische Leitfähigkeit an der Oberfläche von Zinkoxydkristallen). Zeitschrift fuer Physik 148:15–27
29. Myasnikov IA (1957) The relation between the electric conductance and the adsorptive and sensitizing properties of zinc oxide. I. Electron phenomena in zinc oxide during adsorption of oxygen. Zhurnal Fizicheskoi Khimii 31:1721–30
30. Kupriyanov LY (1996) Semiconductor sensors in physico-chemical studies. In: Middelhoek S (ed) Handbook of sensors and actuators, vol 4. Elsevier, Amsterdam p 412
31. Seiyama T, Kato A, Fujiishi K, Nagatani M (1962) A new detector for gaseous components using semiconductive thin films. Anal Chem 34:1502–1503
32. Seiyama T, Kagawa S (1966) Detector for gaseous components with semiconductive thin films. Anal Chem 38(8):1069–73
33. Taguchi N (1962) Gas-detecting device. Jpn Pat 45–38200
34. Chiba A (1992) Development of the TGS gas sensor. In: Yamauchi S (ed) Chemical sensor technology, Elsevier, Amsterdam pp 1–18
35. Eranna G, Joshi BC, Runthala DP, Gupta RP (2004) Oxide materials for development of integrated gas sensors—a comprehensive review. Crit Rev Solid State Mater Sci 29(3–4):111–188
36. Ihokura K, Watson J (1994) Stannic oxide gas sensors, principles and applications. CRC Press, Boca Raton p 187
37. Barsan N, Weimar U (2001) Conduction model of metal oxide gas sensors. J Electroceram 7(3):143–167
38. Barsan N, Weimar U (2003) Understanding the fundamental principles of metal oxide based gas sensorsgas sensors; the example of CO sensing with SnO_2 sensors in the presence of humidity. J Phys-Condens Matter 15(20):813–839

39. Gurlo A, Barsan N, Weimar U (2006) Gas sensors based on semiconducting metal oxidesmetal oxides. In: Fierro JLG (ed) Metal oxides: chemistry and applications, CRS Press, Boca Raton, pp 683–738

40. Ahlers S, Müller G, Doll T (2005) A rate equation approach to the gas sensitivity of thin film metal oxide materials. Sens Actuators, B: Chem 107(2):587–599

41. Park CO, Akbar SA (2003) Ceramics for chemical sensing. J Mater Sci 38(23):4611–4637

42. Korotcenkov G (2005) Gas response control through structural and chemical modification of metal oxide films: state of the art and approaches. Sens Actuators, B: Chem 107(1):209–232

43. Franke ME, Koplin TJ, Simon U (2006) Metal and metal oxide nanoparticles in chemiresistors: does the nanoscale matter? Small 2(1):36–50

44. Batzill M, Diebold U (2006) Characterizing solid state gas responses using surface charging in photoemission: water adsorption on SnO2(101). J Phys-Condens Matter 18(8):129–134

45. Bell NA, Brooks JS, Forder SD, Robinson JK, Thorpe SC (2002) Backscatter Fe-57 Mossbauer studies of iron(II) phthalocyanine. Polyhedron 21(1):115–118

46. Wolkenstein T (1969) Introduction. In: Hauffe K and Wolkenstein T (eds) Symposium on electronic phenomena in chemisorption and catalysis on semiconductors, Walter de Gruyter & Co, Berlin, pp 67–82

47. Weisz PB (1956) Bridges of physics and chemistry across the semiconductor surface. In: Kingston RH (ed) Semiconductor surface physics, University of Pennsylvania Press, Philadelphia, pp 247–258

48. Goepel W (1985) Entwicklung chemischer Sensoren: empirische Kunst oder systematischeForschung? Teil 2 (Development of chemical sensors: empirical art or systematic research? Part 2). Technisches Messen 52(3):92–105

49. Goepel W (1985) Entwicklung chemischer sensoren: empirische Kunst oder systematischeForschung? Teil 3 (Development of chemical sensors: empirical art or systematic research? Part 3). Technisches Messen 52(2):175–182

50. Goepel W (1985) Entwicklung chemischer sensoren: empirische Kunst oder systematischeForschung? (Development of chemical sensors: empirical art or systematic research?). Technisches Messen 52(5):175–82

51. Goepel W (1985) Chemisorption and charge transfer at ionic semiconductor surfaces: implications in designing gas sensors. Prog Surf Sci 20(1):9–103

52. Kohl D (1989) Surface processes in the detection of reducing gases with tin dioxide-based devices. Sens Actuators 18(1):71–113

53. Kiselev VF, Krylov OV (1985) Adsorption processes on semiconductor and dielectric surfaces I. Springer series in chemical physics, 287

54. Fuller MJ, Warwick ME (1973) Catalytic oxidation of carbon monoxide on tin(IV) oxide. J Catal 29(3):441–50

55. Thornton EW, Harrison PG (1975) Tin oxide surfaces. I Surface hydroxyl groups and the chemisorption of carbon dioxide and carbon monoxidecarbon monoxide on tin(IV) oxide. J Chem Soc, Faraday Trans 1: Phys Chem Condens Phases 71(3):461–72

56. Harrison PG (1989) Tin(IV) Oxide: surface chemistry, catalysis and gas sensing. In: Harrison PG (ed) Chemistry of Tin, Blackie, Glasgow

57. Mizokawa Y, Nakamura S (1975) ESR and electric conductance studies of the fine powdered tin dioxide. Jpn J Appl Phys 14(6):779–88

58. Che M, Tench AJ (1982) Characterization and reactivity of mononuclear oxygen species on oxide surfaces. Adv Catal 31:77–133

59. Che M, Tench AJ (1983) Characterization and reactivity of molecular oxygen species on oxide surfaces. Adv Catal 32:1–148

60. Ogawa H, Nishikawa M, Abe A (1982) Hall measurement studies and an electrical conduction model of tin oxide ultrafine particle films. J Appl Phys 53(6):4448–4455

61. Mizsei J, Harsanyi J (1983) Resistivity and work function measurements on Pd-doped SnO_2 sensor surface. Sens Actuators 4:397–402

62. Schierbaum KD, Weimar U, Goepel W, Kowalkowski R (1991) Conductance, work function and catalytic activity of SnO_2-based gas sensors. Sens Actuators, B: Chem B3(3):205–214
63. Gutierrez J, Ares L, Horillo MC, Sayago I, Agapito J, Lopez L (1991) Use of complex impedance spectroscopy in chemical sensor characterization. Sens Actuators B: Chem 4(3–4):359–363
64. Mizsei J, Lantto V (1991) Simultaneous response of work function and resistivity of some SnO_2-based samples to H_2 and H_2S. Sens Actuators B: Chem 4(1–2):163–168
65. Kappler J (2001) Characterization of high-performance SnO_2 gas sensorsgas sensors for CO detection by in-situin-situ techniques (Ph.D. Thesis, University of Tübingen) Aachen: Shaker Verlag
66. Oprea A, Moretton E, Barsan N, Becker WJ, Wollenstein J, Weimar U (2006) Conduction model of SnO_2 thin films based on conductance and Hall effect measurements. J Appl Phys 100(3):033716
67. Sahm T, Gurlo A, Barsan N, Weimar U (2006) Basics of oxygen and SnO_2 interaction; work function change and conductivity measurements. Sens Actuators, B: Chem 118(1–2):78–83
68. Sahm T, Gurlo A, Barsan N, Weimar U, Madler L (2005) Fundamental studies on SnO_2 by means of simultaneous work function change and conduction measurements. Thin Solid Films 490(1):43–47
69. Gurlo A, Barsan N, Oprea A, Sahm M, Sahm T, Weimar U (2004) An n- to p-type conductivity transition induced by oxygen adsorption on alpha-Fe_2O_3. Appl Phys Let 85(12):2280–2282
70. Gurlo A, Sahm M, Oprea A, Barsan N, Weimar U (2004) A p- to n-transition on α-Fe_2O_3-based thick film sensors studied by conductance and work function change measurements. Sens Actuators B-Chem 102(2):291–298
71. Dunstan PR, Maffeis TGG, Ackland MP, Owen GT, Wilks SP (2003) The correlation of electronic properties with nanoscale morphological variations measured by SPM on semiconductor devices. J Phys Condens Matter 15(42):S3095–S3112
72. Maffeis TGG, Owen GT, Malagu C, Martinelli G, Kennedy MK, Kruis FE, Wilks SP (2004) Direct evidence of the dependence of surface state density on the size of SnO_2 nanoparticles observed by scanning tunnelling spectroscopy. Surf Sci 550(1–3):21–25
73. Maffeis TGG, enny MP, Teng KS, Wilks SP, Ferkel HS, Owen GT (2004) Macroscopic and microscopic investigations of the effect of gas exposure on nanocrystalline SnO_2 at elevated temperature. Appl Surf Sci 234(1–4):82–85
74. Arbiol J, Gorostiza P, Cirera A, Cornet A, Morante JR (2001) In situ analysis of the conductance of SnO_2 crystalline nanoparticles in the presence of oxidizing or reducing atmosphere by scanning tunneling microscopy. Sens Actuators B-Chem 78(1–3):57–63
75. Barsan N, Schweizer-Berberich M, Gopel W (1999) Fundamental and practical aspects in the design of nanoscaled SnO_2 gas sensors: a status report. Fresenius J Anal Chem 365(4):287–304
76. Gurlo A, Riedel R (2007) In situ and operando spectroscopy for assessing mechanisms of gas sensing. Angewandte Chemie—Int Edn 46(21):3826–3848
77. Barsan N, Koziej D, Weimar U (2007) Metal oxide-based gas sensor research: how to? Sens Actuators B-Cheml 121(1):18–35
78. Windischmann H, Mark P (1979) A model for the operation of a thin-film tin oxide (SnOx) conductance-modulation carbon monoxide sensor. J Electrochem Soc 126(4):627–33
79. Schulz M, Bohn E, Heiland G (1979) Messung von Fremdgasen in der Luft mit Halbleitersensoren (Measurement of extraneous gases in air by means of semiconducting sensors). Tech Messen 46(11):405–14
80. Williams DE (1987) Conduction and gas response of semiconductor gas sensors. In: Moseley PT and Totfield BC (eds) Solid state gas sensors, Adam Hilger, Philadelphia, pp 71–123
81. Madou MJ, Morrison SR (1989) Chemical sensing with solid state devices. Academic Press, San Diego, p 556

82. Gas Sensors. Sberveglieri G (ed) 1992, Kluwer, Dordrecht, p 409
83. Jarzebski ZM, Marton JP (1976) Physical properties of tin(IV) oxide materials. II. Electrical properties. J Electrochem Soc 123(9):299–310
84. Sahm T, Gurlo A, Barsan N, Weimar U (2005) Basics of oxygen and SnO_2 interaction; work function change and conductivity measurements. In Eurosensors XIX, European conference on solid-state transducers, Barcelona
85. Tournier G, Pijolat C (1999) Influence of oxygen concentration in the carrier gas on the response of tin dioxide sensor under hydrogen and methane. Sens Actuators, B: Chem B61(1–3):43–50
86. Arnold MS, Avouris P, Pan ZW, Wang ZL (2003) Field-effect transistors based on single semiconducting oxide nanobelts. J Phys Chem B 107(3):659–663
87. Kalinin SV, Shin J, Jesse S, Geohegan D, Baddorf AP, Lilach Y, Moskovits M, Kolmakov A (2005) Electronic transport imaging in a multiwire SnO_2 chemical field-effect transistor device. J Appl Phys 98(4)
88. Szuber J, Czempik G, Larciprete R, Adamowicz B (2000) The comparative XPS and PYS studies of SnO_2 thin films prepared by L-CVD technique and exposed to oxygen and hydrogen. Sens Actuators, B: Chem B70(1–3):177–181
89. Figurovskaya EN, Kiselev VF, Vol'kenshtein FF (1965) Influence of chemisorption of oxygen on the work function and electrical conductivity of TiO_2. Doklady Akademii Nauk SSSR 161(5):1142–1145
90. Kiselev VF (1967) Borderline between physical and chemical adsorption. Zeitschrift fuer Chemie 7(10):369–378
91. Gopel W, Rocker G, Feierabend R (1983) Intrinsic defects of TiO_2(110)—interaction with chemisorbed O_2, H_2, CO, and CO_2. Phy Rev B 28(6):3427–3438
92. Heiland G (1954) Zum Einfluss von Wasserstoff auf die elektrische Leitfähigkeit von ZnO-Kristallen. Zeitschrift der Physik 138:459–464
93. Goepel W (1978) Reactions of oxygen with zinc oxide-(1010) surfaces. J Vac Sci Technol 15(4):1298–310
94. Fujitsu S, Koumoto K, Yanagida H, Watanabe Y, Kawazoe H (1999) Change in the oxidation state of the adsorbed oxygen equilibrated at 25 °C on ZnO surface during room temperature annealing after rapid quenching. Jpn J Appl Phys Part 1: Regular papers, Short notes and review papers 38(3A):1534–1538
95. Na BK, Walters AB, Vannice MA (1993) Studies of gas adsorptiongas adsorption on zinc oxide using ESR, FTIR spectroscopy, and MHE (microwave Hall effect) measurements. J Catal 140(2):585–600
96. Kiselev VF, Krylov OV (1987) Springer series in surface sciences, electronic phenomena in adsorption and catalysis on semiconductors and dielectrics. 7:279
97. McAleer JF, Moseley PT, Norris JOW, Williams DE (1987) Tin dioxide gas sensors. Part 1. Aspects of the surface chemistry revealed by electrical conductance variations. J Chem Soc, Faraday Trans 1: Phys Chem Condens Phases 83(4):1323–1346
98. Harrison PG, Willett MJ (1989) Tin oxide surfaces. 20. Electrical properties of tin(IV) oxide gel: nature of the surface conductance in air as a function of temperature. J Chem Soc, Faraday Trans 1: Phys Chem Condens Phases 85(8):1921–1932
99. Willett MJ (1991) Spectroscopy of surface reactions In: Moseley PT, Norris JO and Williams DE (eds) Techiques and mecahnism in gas sensing, Adam Hilger, Bristol, pp 61–107
100. Pulkkinen U, Rantala TT, Rantala TS, Lantto V (2001) Kinetic Monte Carlo simulation of oxygen exchange of SnO_2 surface. J Mol Catal A: Chem 166(1):15–21
101. Lantto V, Romppainen P (1987) Electrical studies on the reactions of carbon monoxide with different oxygen species on tin dioxide surfaces. Surf Sci 192(1):243–264
102. Roduner E (2006) Size matters: why nanomaterials are different. Chem Soc Rev 35(7):583–592
103. Oprea A, Gurlo A, Barsan N, Weimar U (2009) Transport and gas sensing properties of In_2O_3 nanocrystalline thick films: a hall effect based approach. Sens Actuators B-Chem 139(2):322–328

104. Geistlinger H (1993) Electron theory of thin-film gas sensors Sens Actuators, B: Chem 17(1):47–60
105. Geistlinger H, Eisele I, Flietner B, Winter R (1996) Dipole- and charge transfer contributions to the work function change of semiconducting thin films: experiment and theory. Sens Actuators, B: Chem B34(1–3):499–505
106. Rothschild A, Komem Y (2003) Numerical computation of chemisorption isotherms for device modeling of semiconductor gas sensors. Sens Actuators B-Chem 93(1–3):362–369
107. Gurlo A, Barsan N, Ivanovskaya MùWeimar U, Gopel W (1998) In$_2$O$_3$ and MoO$_3$-In$_2$O$_3$ thin film semiconductor sensors: interaction with NO$_2$ and O$_3$. Sens Actuators B-Chem 47(1–3):92–99
108. Wahlstrom E, Vestergaard EK, Schaub R, Ronnau A, Vestergaard M, Laegsgaard E, Stensgaard I, Besenbacher F (2004) Electron transfer-induced dynamics of oxygen molecules on the TiO$_2$(110) surface. Science 303(5657):511–513
109. Gurlo A (2006) Interplay between O$_2$ and SnO$_2$: oxygen ionosorption and spectroscopic evidence of adsorbed oxygen. ChemPhysChem 7:2041–2052
110. Henderson MA, Epling WS, Perkins CL, Peden CHF, Diebold U (1999) Interaction of molecular oxygen with the vacuum-annealed TiO$_2$(110) surface: molecular and dissociative channels. J Phys Chem B 103(25):5328–5337
111. Bartolucci F, Franchy R, Barnard JC, Palmer RE (1998) Two chemisorbed species of O$_2$ on Ag(110). Phys Rev Lett 80(23):5224–5227
112. Iwamoto M, Yoda Y, Yamazoe N, Seiyama T (1978) Study of metal oxide catalysts by temperature programmed desorption. 4. Oxygen adsorption on various metal oxides. J Phys Chem 82(24):2564–2570
113. Tanaka K, Blyholder G (1972) Adsorbed oxygen species on zinc oxide in the dark and under illumination. J Phys Chem 76(22):3184–7
114. Zemel JN (1988) Theoretical description of gas-film interaction on tin oxide (SnOx). Thin Solid Films 163:189–202
115. Kolmakov A, Moskovits M (2004) Chemical sensing and catalysiscatalysis by one-dimensional metal-oxide nanostructures. Ann Rev Mater Res 34:151–180
116. Maier J, Gopel W (1998) Investigations of the bulk defect chemistry of polycrystalline tin(IV) oxide. J Solid State Chem 72(2):293–302
117. Gopel W, Schierbaum K, Wiemhofer HD, Maier J (1989) Defect chemistry of tin(IV)-oxide in bulk and boundary-layers. Solid State Ionics 32(3):440–443
118. Kamp B, Merkle R, Lauck R, Maier J (2005) Chemical diffusion of oxygen in tin dioxide: Effects of dopantsdopants and oxygen partial pressure. J Solid State Chem 178(10): 3027–3039
119. Kamp B, Merkle R, Maier J (2001) Chemical diffusion of oxygen in tin dioxide. Sens Actuators, B: Chem B77(1–2):534–542
120. Armelao L, Barreca D, Bontempi E, Canevali C, Depero LE, Mari CM, Ruffo R, Scotti R, Tondello E, Morazzoni F (2002) Can electron paramagnetic resonance measurements predict the electrical sensitivity of SnO$_2$-based film? Appl Magn Reson 22(1):89–100
121. Safonova O, Bezverkhy I, Fabrichnyi P, Rumyantseva M, Gaskov A (2002) Mechanism of sensing CO in nitrogen by nanocrystalline SnO$_2$ and SnO$_2$(Pd) studied by Mossbauer spectroscopy and conductance measurements. J Mater Chem 12(4):1174–1178
122. Morandi S, Ghiotti G, Chiorino A, Comini E (2005) FT-IR and UVUV-Vis-NIR characterisation of pure and mixed MoO$_3$ and WO$_3$ thin films. Thin Solid Films 490(1):74–80
123. Lenaerts S, Roggen J, Maes G (1995) FT-IR characterization of tin dioxide gas sensor materials under working conditions. Spectrochim Acta Part A Mol Biomol Spectrosc 51A(5):883–894
124. Pohle R, Fleischer M, Meixner H (2001) Infrared emission spectroscopic study of the adsorption of oxygen on gas sensors based on polycrystalline metal oxide films. Sens Actuators B-Chem 78(1–3):133–137
125. Sergent N, Gelin P, Perier-Camby L, Praliaud H, Thomas G (2003) Study of the interactions between carbon monoxide and high specific surface area tin dioxide: Thermogravimetric analysis and FTIR spectroscopy. J Therm Anal Calorim 72(3):1117–1126

126. Koziej D, Barsan N, Weimar U, Szuber J, Shimanoe K, Yamazoe N (2005) Water-oxygen interplay on tin dioxide surface: implication on gas sensing. Chem Phys Lett 410(4–6): 321–323

127. Di Nola P, Morazzoni F, Scotti R, Narducci D (1993) Paramagnetic point defects in tin dioxide and their reactivity with surrounding gases. Part 1—interaction of oxygen lattice centers with vapor-phase water, air, inert and combustible gases, as revealed by electron paramagnetic resonance spectroscopy. J Chem Soc, Faraday Trans 89(20):3711–3713

128. Lenaerts S, Honore M, Huyberechts G, Roggen J, Maes G (1994) In situ infrared and electrical characterization of tin dioxide gas sensors in nitrogen/oxygen mixtures at temperatures up to 720 K. Sens Actuators, B: Chem 19(1–3):478–482

129. Ghiotti G, Chiorino A, Boccuzzi F (1989) Infrared study of surface chemistry and electronic effects of different atmospheres on tin dioxide. Sens Actuators 19(2):151–7

130. Harbeck S (2005) Characterisation and functionality of SnO_2 gas sensors using vibrational spectroscopy. Ph.D. thesis, Faculty of Chemistry, Universität tübingen, http://w210.ub.uni-tuebingen.de/dbt/volltexte/2005/1693/, Tuebingen

131. Fonstad CG, Rediker RH (1971) Electrical properties of high-quality stannic oxide crystals. J Appl Phys 42(7):2911–2918

132. Samson S, Fonstad CG (1973) Defect structure and electronic donordonor levels in stannic oxide crystals. J Appl Phys 44(10):4618–4621

133. Lopez N, Prades JD, Hernandez-Ramirez F, Morante JR, Pan J, Mathur S (2010) Bidimensional versus tridimensional oxygen vacancy diffusion in SnO_{2-x} under different gas environments. Phys Chem Chem Phys 12(10):2401–2406

134. Hernandez-Ramirez F, Prades JD, Tarancon AùBarth S, Casals O, Jimenez-Diaz R, Pellicer EùRodriguez J, Morante JR, Juli MA, Mathur S, Romano-Rodriguez A (2008) Insight into the role of oxygen diffusion in the sensing mechanisms of SnO_2 nanowires. Adv Funct Mater 18(19):2990–2994

135. Boreskov GK (1964) The catalysis of isotopic exchange in molecular oxygen. Adv Catal 15:285–339

136. Safonova OV, Neisius T, Ryzhikov A, Chenevier B, Gaskov AM, Labeau M (2005) Characterization of the H_2 sensing mechanism of Pd-promoted SnO_2 by XAS in operando conditions. Chem Commun 41:5202–5204

137. Schmid W, Barsan N, Weimar U (2003) Sensing of hydrocarbons with tin oxide sensors: possible reaction path as revealed by consumption measurements. Sens Actuators B-Chem 89(3):232–236

138. Delabie L, Honore M, Lenaerts S, Huyberechts G, Roggen J, Maes G (1997) The effect of sintering and Pd-doping on the conversion of CO to CO_2 on SnO_2 gas sensor materials. Sens Actuators B-Chem 44(1–3):446–451

139. Dutta PK, De Lucia MF (2006) Correlation of catalytic activity and sensor response in TiO_2 high temperature gas sensors. Sens Actuators, B: Chem 115(1):1–3

140. Schmid W, Barsan N, Weimar U (2004) Sensing of hydrocarbons and CO in low oxygen conditions with tin dioxide sensors: possible conversion paths. Sens Actuators B-Chem 103(1–2):362–368

141. Hahn SH, Barsan N, Weimar U, Ejakov SG, Visser JH, Soltis RE (2003) CO sensing with SnO_2 thick film sensors: role of oxygen and water vapour. Thin Solid Films 436(1):17–24

142. Gurlo A, Sahm M, Oprea A, Barsan N, Weimar U (2004) A p- to n-transition on alpha-Fe_2O_3-based thick film sensors studied by conductance and work function change measurements. Sens Actuators B-Chem 102(2):291–298

143. Gurlo A, Barsan N, Oprea A, Sahm M, Sahm T, Weimar U (2004) An n- to p-type conductivity transition induced by oxygen adsorption on alpha-Fe_2O_3. Appl Phys Lett 85(12):2280–2282

144. Arulsamy AD, Elersic K, Modic M, Cvelbar U, Mozetic M (2010) Reversible carrier-type transitions in gas-sensing oxides and nanostructures. Chem Phys Chem 11(17):3704–3712

145. Gurlo A (2006) Interplay between O_2 and SnO_2: oxygen ionosorption and spectroscopic evidence for adsorbed oxygen. Chem Phys Chem 7(10):2041–2052

146. Hahn S (2002) SnO$_2$ thick film sensors at ultimate limits: performance at low O$_2$ and H$_2$O concentrations. Size reduction by CMOS technology, Ph.D. thesis, Faculty of Chemistry, Universität Tübingen, Tuebingen
147. Benitez JJ, Centeno MA, Merdrignac OMùGuyader J, Laurent YJ, Odriozola A (1995) DRIFTS chamber for in situ and simultaneous study of infrared and electrical response of sensors. Appl Spectrosc 49(8):1094–6
148. Benitez JJ, Centeno MA, dit Picard CL, Merdrignac O, Laurent Y, Odriozola JA (1996) In situ diffuse reflectance infrared spectroscopy (DRIFTS) study of the reversibility of CdGeON sensors towards oxygen. Sens Actuators, B: Chem B31(3):197–202
149. Safonova OV, Neisius T, Chenevier B, Matko I, Labeau M, Gaskov A (2004) In situ XAS studies on the effect of Pd and Pt clusters on the mechanism of SnO$_2$ based gas sensors. In 13th international congress on catalysis, Paris, France, July 2004
150. Gaidi M, Chenevier B, Labeau M, Hazemann JL In situ simultaneous XAS and electrical characterizations of Pt-doped tin oxide thin film deposited by pyrosol method for gas sensors application. Sens Actuators, B: Chem B120(1):313–315
151. Gaidi M, Labeau M, Chenevier B, Hazemann JL (1998) In situ EXAFS analysis of the local environment of Pt particles incorporated in thin films of SnO$_2$ semi-conductor oxide used as gas-sensors. Sens Actuators B-Chem 48(1–3):277–284
152. Gaidi M, Hazemann JL, Matko I, Chenevier B, Rumyantseva M, Gaskov A, Labeau M (2000) Role of Pt aggregates in Pt/SnO$_2$ thin films used as gas sensors—investigations of the catalytic effect. J Electrochem Soc 147(8):3131–3138
153. Sharma SS, Nomura K, Ujihira Y (1992) Moessbauer studies on tin-bismuth oxide carbon monoxide selective gas sensor. J Appl Phys 71(4):2000–5
154. Nomura K, Sharma SS, Ujihira Y (1993) Characterization of tin oxide films gas sensor by in situ conversion electron Moessbauer spectrometry (CEMS). Nucl Instrum Methods Phys Res, Sect B B76(1–4):357–9
155. Baraton M.-I, Merhari L (2005) Investigation of the gas detection mechanism in semiconductor chemical sensors by FTIR spectroscopy. Synth React Inorg, Met-Org, Nano-Met Chem 35(10):733–742
156. Baraton M-I, Merhari L (2004) Nanoparticles-based chemical gas sensors for outdoor air quality monitoring. Nano-Micro Interface 227–238
157. Baraton M-I (1996) FT-IR surface study of nanosized ceramic materials used as gas sensors. Sens Actuators, B: Chem B31(1–2):33–8
158. Baraton MI (1994) Infrared and Raman characterization of nanophase ceramic materials. High Temp Chem Processes 3:545–554
159. Baraton MI, Merhari L, Ferkel H, Castagnet JF (2002) Comparison of the gas sensing properties of tin, indium and tungsten oxides nanopowders: carbon monoxide and oxygen detection. Mater Sci Eng C-Biomimetic Supramolecular Syst 19(1–2):315–321
160. Baraton MI, Merhari L, Keller P, Zweiacker K, Meyer JU (1999) Novel electronic conductance CO$_2$ sensors based on nanocrystalline semiconductors. Materials research society symposium proceedings, (Microcrystalline and Nanocrystalline Semiconductors–1998) 536:341–346
161. Chiorino A, Ghiotti G, Prinetto F, Carotta MCùMalagu C, Martinelli G (2001) Preparation and characterization of SnO$_2$ and WOx-SnO$_2$ nanosized powders and thick films for gas sensing. Sens Actuators, B: Chem B78(1–3):89–97
162. Chiorino A, Ghiotti G, Prinetto F, Carotta MC, Gallana M, Martinelli G (1999) Characterization of materials for gas sensors. Surface chemistry of SnO$_2$ and MoOx-SnO$_2$ nano-sized powders and electrical responses of the related thick films. Sens Actuators B-Chem 59(2–3):203–209
163. Chiorino A, Ghiotti G, Carotta MC, Martinelli G (1998) Electrical and spectroscopic characterization of SnO$_2$ and Pd-SnO$_2$ thick films studied as CO gas sensors. Sens Actuators, B: Chem B47(1–3):205–212

164. Chiorino A, Ghiotti G, Prinetto F, Carotta MC, Martinelli G, Merli M (1997) Characterization of SnO_2-based gas sensors a spectroscopic and electrical study of thick films from commercial and laboratory-prepared samples. Sens Actuators, B Chem B 44(1–3):474–482
165. Popescu DA, Herrmann JM, Ensuque A, Bozon-Verduraz F (2001) Nanosized tin dioxide: spectroscopic (UV-VIS, NIR, EPR) and electrical conductivity studies Phys Chem Chem Phys 3(12):2522–2530
166. Canevali C, Mari CM, Mattoni M, Morazzoni F, Ruffo R, Scotti R, Russo U, Nodari L (2004) Mechanism of sensing NO in argon by nanocrystalline SnO_2: electron paramagnetic resonance, Mossbauer and electrical study. Sens Actuators, B: Chem B100(1–2):228–235
167. Canevali C, Mari CM, Mattoni M, Morazzoni F, Nodari L, Ruffo R, Russo U, Scotti R (2005) Interaction of NO with nanosized Ru-, Pd-, and Pt-doped SnO_2: electron paramagnetic resonance, Mossbauer, and electrical investigation. J Phys Chem B 109(15):7195–7202
168. Morazzoni F, Canevali C, Chiodini N, Mari C, Ruffo R, Scotti R, Armelao L, Tondello E, Depero LE, Bontempi E (2001) Nanostructured Pt-doped tin oxide films: sol-gel preparation, spectroscopic and electrical characterization. Chem Mater 13(11):4355–4361

Chapter 2
Surface Science Studies of Metal Oxide Gas Sensing Materials

Junguang Tao and Matthias Batzill

Abstract In this chapter we present recent advances in the study of metal oxide surfaces and put them in relation to gas sensing properties. A reoccurring scheme is the dependence of chemical surface properties on the crystallographic orientation of the surface. This dependence will become more important in gas sensing applications as nanomaterials with controlled crystal shapes are being designed. In particular we focus on differences of the surface properties of the two polar surfaces of ZnO and the two most abundant bulk terminations of rutile TiO_2, i.e. the (110) and (011) crystallographic orientations. On the example of these metal oxides, we describe the use of vacuum based surface science techniques, especially scanning tunneling microscopy and photoemission spectroscopy, to obtain structural, chemical, and electronic information.

2.1 Relation Between Metal Oxide Gas Sensors and Surface Science

Semiconducting metal oxides can exhibit a conductivity change due to the adsorption or reactions of molecules from the gas phase with the surface. Monitoring this conductivity change enables the use of this information as a gas response signal. The change in conductivity is brought about by an upward or downward shift of the Fermi-level within the band-gap of these predominantly n-type materials. The Fermi-level shift may be induced by charge transfer from the gas sensing material to an adsorbate. For macroscopic materials this induces a

J. Tao · M. Batzill (⊠)
Department of Physics, University of South Florida, Tampa 33620, USA
e-mail: mbatzill@usf.edu

M. A. Carpenter et al. (eds.), *Metal Oxide Nanomaterials for Chemical Sensors*, Integrated Analytical Systems, DOI: 10.1007/978-1-4614-5395-6_2, © Springer Science+Business Media New York 2013

band bending at the surface while for microscopic particles (smaller than the Debye screening length) the Fermi-level within the entire particle shifts. Such adsorbate induced shifts of the Fermi-level are dominant for surface sensitive gas sensing materials such as SnO_2 and ZnO. Another mechanism that can result in the shift of the Fermi-level is a variation of the bulk dopant concentration. For example, Ti-interstitials and oxygen vacancies in the bulk of TiO_2 act as intrinsic n-type dopants. The concentration of these dopants depends on the oxidation potential of the surrounding gas phase and the surface of the TiO_2 may act as a source or sink of Ti-interstitials for the bulk.

Clearly, gas sensing with semiconducting metal oxides is initiated by molecular interaction with surfaces and surface science studies therefore have played an important role in the understanding and describing of the fundamental mechanisms [1]. Recent advances in surface science now allow a molecular scale description of metal oxide surfaces and this has provided many new fundamental insights on the properties of these materials. Traditional surface science studies make two important simplifications to the general complex morphology of gas sensing materials; (1) macroscopic single crystalline materials (usually bulk samples but sometimes also high quality epitaxial thin films may be used) are studied, and (2) the surfaces are prepared and investigated under well-controlled ultra high vacuum (UHV) conditions. The consequences of these simplifications are that in surface science experiments on macroscopic single crystal samples, no size effects and no interface effects are observed. Furthermore, because the surfaces are generally prepared in vacuum, they are exposed to a reducing environment. In return for the loss of the 'real' gas sensing environment we gain control over crystallographic orientation and surface composition. Furthermore, we have the full arsenal of surface science techniques at our disposal for a detailed surface analysis. Therefore, surface science studies can address critical fundamental questions like: What role does the surface structure play in the adsorption of certain molecules, i.e. do different crystallographic orientations of the same material exhibit different gas sensitivities? What are the sites for molecule adsorption? and consequently, can these sites be controlled to obtain better sensitivity and selectivity? What is the electronic response of the surface upon molecule adsorption or reaction with the surface? Modern surface science studies can provide rich information on the atomic scale surface properties, which is not easily accessible by other methods.

Detailed reviews on surface science studies of SnO_2 have been recently published by one of the authors of this article [2–4] and interested readers are referred to those texts. Here we focus on recent studies mainly done in the authors' laboratory on the two common metal oxide gas sensing materials ZnO and TiO_2. On ZnO we demonstrate the use of high energy-resolution photoemission studies to obtain information on the fundamental stabilization mechanisms of polar surfaces and in particular investigate the role of hydrogen adsorption on different surface orientations. We also illustrate the power of photoemission spectroscopy to gain information on gas sensing reactions on the example of ZnO reacting with H_2S, a common gas sensing application. In the second part we study two surfaces of rutile TiO_2 and show that the surface structure strongly affects the adsorption of

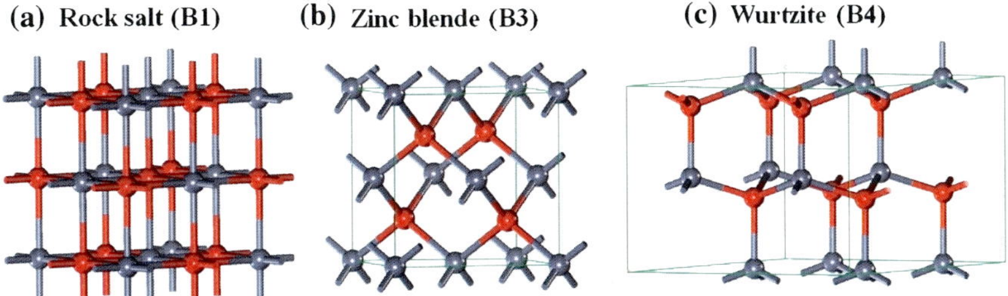

Fig. 2.1 Ball and stick representation of ZnO crystal structures. The *gray balls* are zinc atoms and the *red* ones are oxygen atoms

molecules on the example of acetic acid. In an attempt to modify surface structures and to "engineer" specific surface sites we discuss the use of grazing incidence ion irradiation to create step edges at planar surfaces. Finally, we turn to the oxidation reaction of TiO_2, which demonstrates the mechanism by which the intrinsic bulk dopant concentration is altered.

2.2 ZnO Surfaces

ZnO oxide is one of the prototypical metal oxide gas sensing materials. Its properties have obtained a lot of early attention in the surface science community and have been recently revisited [5]. Its current popularity is also related to the many different nano-structural forms and shapes ZnO self-organizes in under various growth conditions. ZnO crystallizes in three forms: cubic zinc-blende, hexagonal wurtzite and the rarely observed cubic rocksalt structure, which are schematically shown in Fig. 2.1.

Wurtzite is the thermodynamically favored form of ZnO at ambient conditions and zinc blende can be stabilized by growth on cubic substrates, while the rocksalt structure is a high-pressure meta-stable phase. The lattice constants of wurtzite ZnO are a = 3.25 Å and c = 5.2 Å. This hexagonal lattice consists of two interconnecting hexagonal-close-packed (hcp) sub-lattices of Zn^{2+} and O^{2-}, with each anion surrounded by four cations at the corners of a tetrahedron, and vice versa. The tetrahedral coordination of Zn and O is typical of sp^3 covalent bonding. The ionic character of the material gives rise to a polar repeat-unit along the c-axis. As a consequence of this polar symmetry, the (0001) and the (0001) surfaces of wurtzite ZnO exhibit different bulk terminations, with the first one terminated by Zn-atoms and the latter by O-atoms. This means that if we cleave a crystal along the (0001) plane we obtain two distinctively different surfaces. Importantly, because the repeat unit of the crystal structure perpendicular to these surfaces (along the c-axis) exhibits a dipole moment the Madelung-energy would diverge at these surfaces for an ideal bulk truncation. This is a general property of 'polar surfaces' and consequently such bulk truncated surfaces cannot be stable. Despite this inherent instability, the polar (0001) and the (0001) surfaces are among the most common crystal orientations of ZnO. This implies that there exist efficient stabilization mechanisms of these surfaces that allow a convergence of the Madelung energy (electrostatic potential). These stabilization mechanisms may also influence the gas sensing properties of ZnO and this is discussed in detail below. Other non-polar surface orientations of wurtzite are the (1120) (*a*-axis) and (1010) faces. The (1010) and (1120) surfaces are the prism faces and the (1121) surface is the pyramid face of the crystal. The non-polar ZnO(1010) surface contain equal number of Zn and O ions and has also been studied extensively by surface science methods because at this surface the coordination of the surface atoms is reduced from fourfold to threefold, thereby creating dangling bonds at the surface which makes it chemically active. Here we focus on the polar surfaces.

2.2.1 The Polar ZnO Surfaces

The two polar surfaces of ZnO are known to have different chemical and physical properties [6]. To stabilize polar surfaces, additional positive or negative charges are required to transfer to the ZnO(000 1)-O or ZnO(0001)-Zn surfaces, respectively. This can be accomplished by removal of lattice ions (surface reconstructions), formation of electronic surface states, or adsorption of impurities such as hydrogen. In the absence of any compositional and structural surface variations a metallization of the surface will occur, i.e. the Fermi-level will be pushed into the valence or conduction band. As shown in Fig. 2.2a, the electrostatic potential introduced by the internal dipole moments causes a shift of the electronic levels relative to the Fermi level in opposite directions at the two polar surfaces, see Fig. 2.2b. Once the conduction band (CB) on one side and the valence band (VB) on the opposite side of the crystal intersect the Fermi level, charges are redistributed from one side to the other. This results in a potential opposing a divergence of the surface potential due to the lattice dipole. In this simplified scheme, the potential due to the internal dipole causes a shift of the energy levels relative to the Fermi level until the VB is depleted on one side and the CB filled on the opposite side. Conceptually, this would result in a charge transfer from one side to the other and thus a potential build up that counteracts the lattice dipoles. However, such a surface metallization is energetically quite expensive and therefore polar surfaces usually stabilize by removing ionic surface charges, i.e. they form surface reconstructions with an altered surface composition compared to the bulk truncation. Such a stabilization mechanism exists for the ZnO(0001)-Zn surface. For the ZnO(000 1)-O surface, on the other hand, it is still debated if a stabilization of this surface by removal of ionic charges exist. This debate may however, mainly be of fundamental scientific interest, since under most realistic environmental conditions it appears that the O-side of the ZnO polar surfaces is terminated by hydrogen and the protons provide the necessary positive charges to converge the

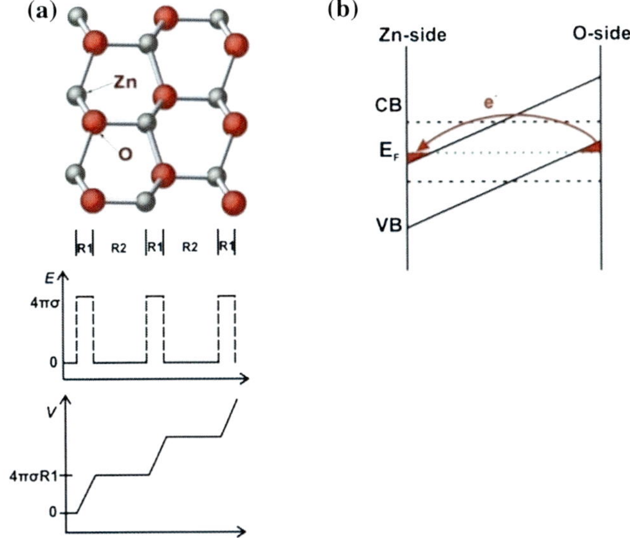

Fig. 2.2 **a** Spatial variation of the Electric Field E and of the electrostatic potential V in a sample cut along a polar direction where σ is the charge density on the planes. **b** Schematic of surface metallization. Reproduced from Ref. [7]. Copyright 2008, American Physical Society

Madelung energy. The stabilization mechanisms of the polar two surfaces are discussed next.

2.2.1.1 The ZnO(0001)-Zn Polar Surface

The surface structure of the ZnO(0001)-Zn surface prepared in UHV has been thoroughly analyzed by scanning tunneling microscopy and the typical surface morphology is shown in Fig. 2.3a. These measurements resulted in the conclusion that this polar surface is stabilized by removal of $\sim 1/4$ ML of Zn atoms from the surface in a *non-periodic* manner [8]. This is achieved by formation of high density of triangular shaped pits and ad-islands that exhibit step edges that are O-terminated as shown in Fig. 2.3b. Thus the Zn-terminated surface under UHV conditions is stabilized by a net removal of positive cations. The formation of sub-stoichiometric polar surfaces is common for the stabilization of many polar materials, however, what makes the ZnO(0001)-Zn surface special, is that it does not involve a periodic surface reconstruction and therefore this stabilization mechanism evaded detection by surface diffraction techniques for a long time. In addition to the direct visualization of the surface structure by STM, high resolution core-level photoemission spectroscopy of ionic surfaces also contains information of the coordination environment in ionic crystals. This is because the core-level binding energy measured in XPS contains a contribution of the Madelung energy at the site of the photoelectron emitting atom [9]. Surface-terrace and step-edge atoms have

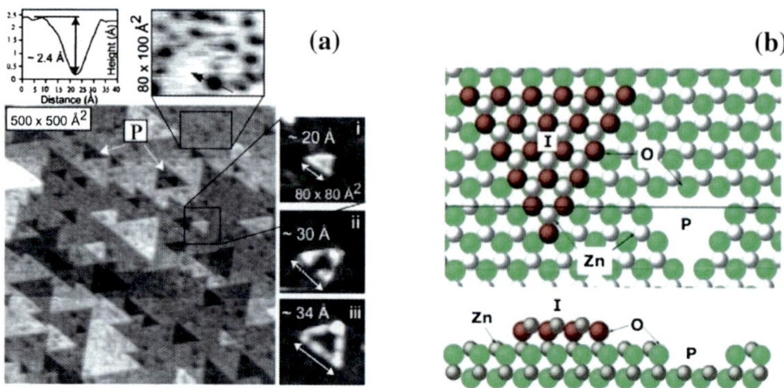

Fig. 2.3 Surface structure of ZnO(0001)-Zn. **a** STM image of surface structure, reprinted with permission from Ref. [8]. Copyright 2003, American Physical Society. **b** Ball model of *top* and cross-sectional view of the ZnO(0001)-Zn surface showing triangular pits (P) and islands (I). The step edges exhibit 3-fold coordinated oxygen atoms compared to the 4-fold coordination on the terraces. Reproduced from Ref. [7]. Copyright 2008, American Physical Society

different Madelung energy than bulk ions, due to their reduced coordination numbers. Therefore photoemitted electrons from different oxygen sites (step-edge, terrace, and bulk) on the ZnO(0001)-Zn surface exhibit slightly different binding energies and this may be used to verify the high density of O-terminated step edges on the vacuum prepared ZnO surface. Using *soft* x-ray photoemission a high surface sensitivity is obtained that enables to observe the binding energy shifts due to the high density of step edges. A de-convolution of the O-1s peak into its different components is shown in Fig. 2.4. A test to demonstrate that the different components are due to the Madelung shift at the surface is to passivate the surface with ZnS (this is discussed in detail below). Such a passivated surface exhibits a narrower O-1s peak because the (shifted) surface contributions are missing. Therefore, the detailed analysis of the O-1s peak shape in soft- x-ray photoemission studies is an elegant confirmation that this surface is stabilized by the formation of a high density of oxygen-terminated step edges in vacuum.

Although formation of high concentration of O-terminated step edges is the stabilization mechanism in vacuum, the stabilization of this polar surface in the presence of different gaseous environments may be achieved by various adsorbates. Different surface terminations have been for example predicted by ab-intio thermodynamics calculations for different hydrogen chemical potentials of the gas phase [8, 10]. Also recent STM studies of ZnO(0001)-Zn surface for exposure to water in vacuum showed that the triangular step edge structure disappears [11], indicating that even at the low water vapor pressure achievable in UHV alternative surface stabilization mechanisms by surface adsorbates become dominant. Furthermore, studies in aqueous solutions have shown that a hydroxide stabilized surface is obtained for pH values between 11 and 4 [12]. Therefore, different environments can strongly affect the surface structure and composition of metal

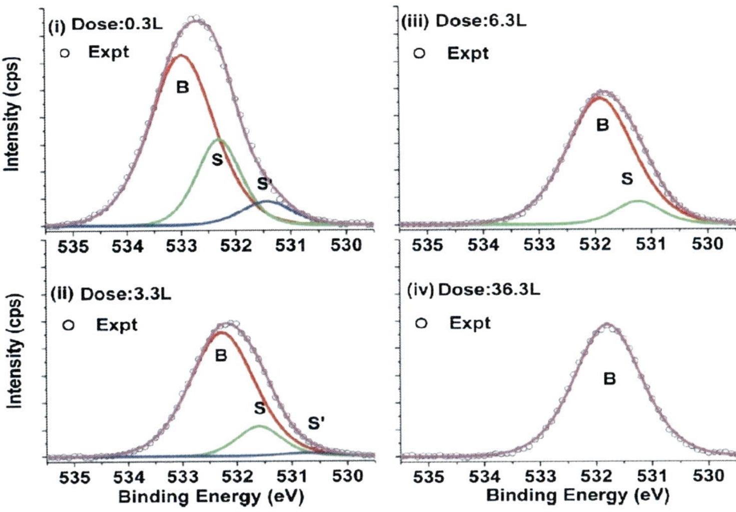

Fig. 2.4 Soft x-ray photoemission spectra of the O-1s peak on ZnO(0001)-Zn surfaces for different amount of surface passivation with ZnS. The O-1s peak can be de-convoluted into three contributions: bulk (*B*), surface terraces (4-fold coordinated) (*S*) and surface step-edge sites (3-fold coordinated oxygen) (*S'*). With increasing surface passivation the surface contribution in the O-1s signal diminishes. Reproduced from Ref. [7]. Copyright 2008, American Physical Society

oxide surfaces. Future surface science studies in other environments than UHV will be needed to advance the understanding of the role of environmental conditions on the surface properties.

2.2.1.2 The ZnO(0001)- O Polar Surface

While the stabilization of the polar ZnO(0001)-Zn surface is quite reasonably explained by removal of ¼ ML of Zn and formation of O-terminated step edges, the stabilization of the ZnO(000 1)-O polar surface is not fully understood yet. Kunat et al. observed a 1 × 3 surface reconstruction for a clean surface, which they interpreted as an ordered array of O-vacancies with 1/3 oxygen atoms missing [13]. This structure, however, does not entirely satisfy the electrostatic stabilization criteria (which requires ¼ ML of missing oxygen) and density functional theory (DFT) calculations have shown that it is not a stable surface structure [14]. Soft x-ray photoemission studies of the O-1s core level [7], shown in Fig. 2.5 have revealed the presence of OH on this surface below ~ 300 °C. The hydrogen could only be removed by annealing and holding the sample at temperatures above 300 °C. At this higher temperature, in the absence of hydroxyl groups, an upward band bending was observed, i.e. the Fermi-level was pushed into the band gap towards the valence band. This is reminiscent of the shift in the Fermi-level expected for an unreconstructed, adsorbate free surface as discussed in Fig. 2.2b.

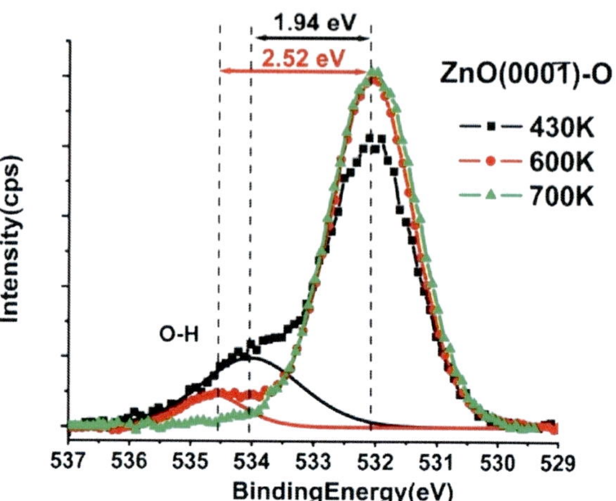

Fig. 2.5 O-1s core level photoemission spectra of the ZnO(000-1)-O surface at different sample temperatures. Only at 700 K a contribution due to surface hydroxyls has completely vanished. At lower temperatures OH groups were always detected. The binding energy of the OH group appears to shift with increasing hydrogen coverage. Reproduced from Ref. [7]. Copyright 2008, American Physical Society

However, the down-ward shift of the Fermi-level was not sufficient to actually remove electrons from the valence band. Nevertheless, especially at elevated temperatures (thermally excited) charge carriers (electrons) are depleted from the surface region by moving the Fermi-level and this can contribute to the electrostatic stabilization of the O-terminated surface. Therefore these soft x-ray photoemission studies demonstrated two important facts. Firstly, at room temperature, even under UHV conditions, the polar surface is likely (partially) hydrogen terminated and secondly, if the surface is hydrogen free (at elevated temperature) this surface is not (completely) stabilized by removal of O-ions and consequently causing a shift in the Fermi-level.

As the photoemission studies showed, hydrogen adsorption on this surface plays an important role. This obviously will affect the adsorption of molecules and thus the gas sensitivity. It is also interesting to point out that recent angle-resolved ultra violet photoemission (ARUPS) studies demonstrated the metallization of this surface due to the presence of hydrogen. ARUPS measurements of the hydrogen induced dispersed surface state at low temperatures are shown in Fig. 2.6. This metallization is due to occupation of ZnO-states in the conduction band from electrons donated from the adsorbed hydrogen [15], i.e. pushing the Fermi-level into the conduction band of ZnO and thus demonstrating a truly metallic electronic surface structure. It is important to point out though that this metallization is different to the purely electrostatic metallization due to the internal dipole moments of ZnO discussed in Fig. 2.2. In case of hydrogen adsorption the metallization is brought about by electron donation from the adsorbed hydrogen and is not in response to the stabilization of the polar surface. Consequently, similar results for surface metallization by hydrogen adsorption can also be obtained for example for the non-polar ZnO(10–10) surface [16].

In terms of gas sensing of ZnO surfaces, surface science studies have shown that the two polar surfaces behave very differently and thus a variation in the gas response is expected for molecular adsorption. Hydrogen may be inadvertently

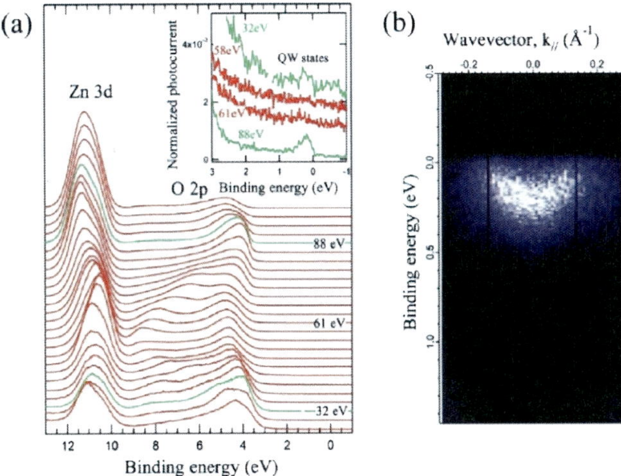

Fig. 2.6 ARUPS studies of ZnO(000-1)-O surface in the presence of hydrogen adsorbates. **a** Normal photoemission studies with the inset showing electronic states at the Fermi-level at 32 and 88 eV photon-energies, which corresponds to the Γ-point in the surface normal direction of the k-vector. **b** ARUPS photocurrent intensity measurement of the near Fermi-level emission for 90 eV photon energy along the direction. Reprinted with permission from Ref. [15]. Copyright 2010, American Physical Society

present on the O-terminated side while the Zn-terminated side is more likely to be hydrogen-free. The easy hydroxylation of the ZnO(000 1)-O causes a metallization of the surface which can affect the conductivity response of such samples. The studies of the surface properties of the polar surfaces indicate that the differences in the chemical properties of the two polar surfaces affects the chemisorption of molecules and this plays an important role in gas sensing of such systems. However, in gas sensing applications where a more drastic surface restructuring occurs, due to *reactions* of the ZnO with gas phase species, the surface orientation plays less of a role in the gas sensitivity. This is for example the case of H_2S sensing with ZnO, which is discussed in the next section.

2.2.2 Surface Reaction of ZnO with H_2S

Detection of hydrogen sulfide (H_2S) is one of the classic applications of metal oxide gas sensors. The gas sensing mechanism can be probed elegantly by surface science studies on single crystal ZnO under UHV conditions. In these studies ZnO is exposed to low background pressures of H_2S (10^{-6} Torr) within the vacuum chamber where the structural, compositional and electronic variations of the surface properties can be monitored by STM and photoemission spectroscopy. It is well known that ZnO can be easily sulfidized by H_2S [17] and therefore it is not

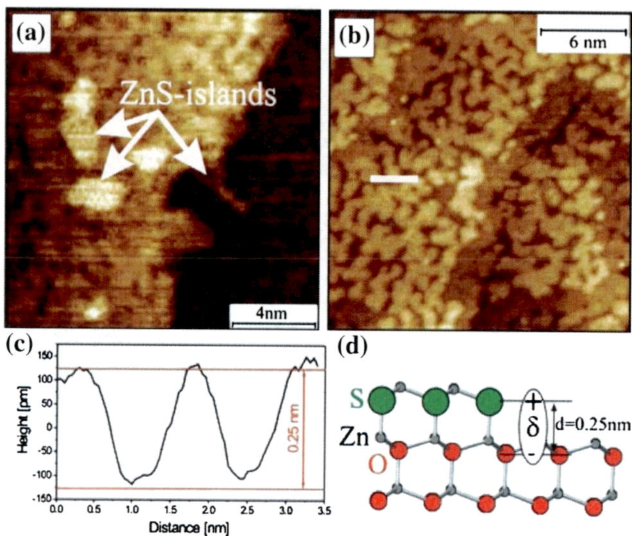

Fig. 2.7 STM images of sulfur induced surface structure for **a** 0.6 L H_2S and **b** 3 L H_2S exposure. The cross-section indicated in (**b**) is shown in (**c**). A ball-and-stick model of the surface structure is shown in (**d**) with the surface dipole moment indicated. Reprinted with permission from Ref. [18]. Copyright 2008, American Chemical Society

surprising that a ZnO sample held at 400 °C and exposed to H_2S converts to ZnS. STM studies show that two dimensional ZnS islands form at the surface of ZnO and these islands grow with exposure until the surface is covered with a dense network of meandering ZnS islands [18]. This is shown in Fig. 2.7. Although the atomic scale structure of the sulfide film cannot be unambiguously deduced from these STM studies, some islands exhibit hexagonal symmetry. Consequently, it is reasonable to assume that ZnS adopts the wurtzite structure of the ZnO substrate and an epitaxial relationship between substrate and film is established. The much larger lattice constant of wurtzite-ZnS ($a = 3.82$ Å; $c = 6.26$ Å) compared to ZnO ($a = 3.25$ Å; $c = 5.21$ Å) implies that a complete ZnS monolayer would be under considerable compressive stress. The formation of ZnS island morphology provides an efficient strain relieve mechanism for pseudomorphic ZnS layers. Thus STM allows a direct observation of the transformation of the surface of a gas-sensing material during the reaction with the environment. This real space information of the surface morphology is important for correctly interpreting spectroscopic information. Soft x-ray and ultraviolet photoemission enable monitoring the compositional and electronic-structure variation as a function of H_2S exposure and this can be correlated to the morphology changes measured by STM. The band bending, which is measured in photoemission spectroscopy by evaluating the core-level position shifts relative to the Fermi-level, is a direct measure of the gas response of the semiconducting metal oxide. Therefore the gas response

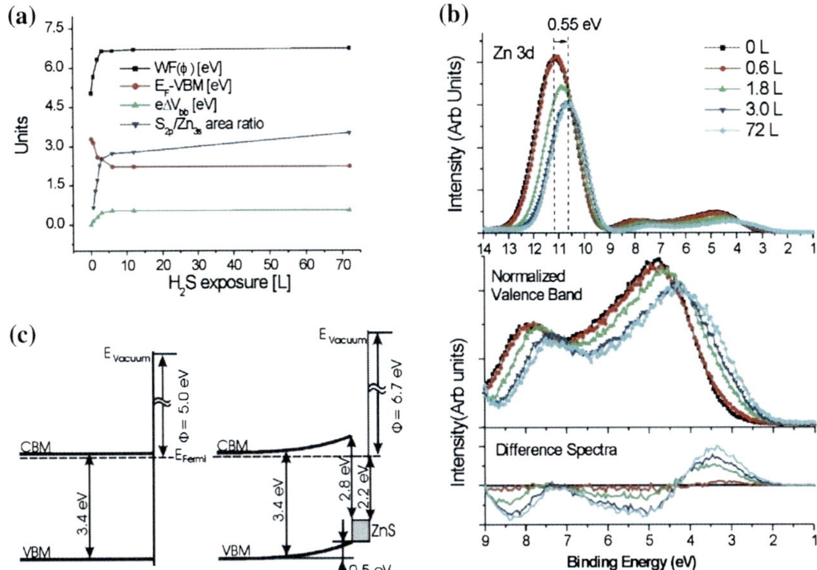

Fig. 2.8 Information obtained from photoemission spectroscopy on the surface response of ZnO to reaction with H_2S. **a** Summary of the changes in the work function, valence band maximum (relative to the Fermi level), band bending in the substrate, and the S 2p/Zn 3s peak ratios as a function of H_2S exposure. **b** Valence band and shallow Zn-3d core level spectra for different H_2S exposures. The shift in the Zn-3d core-level represents the band bending. The change in the valence band is shown in bottom panel of (**b**). The difference spectra are calculated by subtraction of the spectrum of the clean surface after shifting it to compensate for band bending effects. Therefore the positive intensity correspond to electronic states formed by ZnS formation. These electronic structure information are summarized in the band diagram displayed in (**c**). Reprinted with permission from Ref. [18]. Copyright 2008, American Chemical Society

can be measured in photoemission without directly detecting changes in the material conductivity.

Information extracted from photoemission studies of ZnO surface exposed to H_2S is summarized in Fig. 2.8a [18]. Together with STM this gives a comprehensive representation of the processes and changes of a gas sensing surface during operation. It is apparent from Fig. 2.8a that within the first 4 L (Langmuirs) of exposure most of the surface changes causing the gas sensing response have already taken place. The sulfur uptake measured by the S-2p/Zn-3s core-level ratio is rapid for the first 4 L and only very gradually increases at higher exposures. This is consistent with a 2D mono-layer growth of ZnS observed in STM. Photoemission suggests that further sulfidation upon completion of the monolayer is slow and all the gas sensing action can be attributed to the monolayer formation as evidenced by the band bending that occurs simultaneously with the initial sulfur uptake. Since photoemission is a surface sensitive technique, it detects electrostatic binding energy shifts in the Debye layer which in semiconducting oxides is much larger than the probing depths in photoemission. In the case of sulfide formation on

ZnO(0001)-Zn surface, 0.5 eV upward band bending is observed after exposure to 4 L H_2S.

In addition, valence band spectra reveal that, although ZnS is a wide band gap material, additional electronic states are formed within the band gap of ZnO. The change in the valence band is shown in Fig. 2.8b and the decrease in the separation between the valence band maximum (VBM) and the Fermi-level is also indicated in Fig. 2.8a. Finally, the work function change can be evaluated from the secondary electron cut-off in photoemission experiments. These measurements indicate that the work-function of the sample increases quite strongly upon sulfidation of the surface. This change in the work function is attributed to formation of a surface dipole as illustrated in Fig. 2.7d.

Measurement of band bending, work function, and valence band enables derivation of the surface electronic structure for a sulfidized surface. This is shown in Fig. 2.8c. Obviously the 0.5 eV upward band bending would give rise to a decrease in conductivity and thus this is the main reason for the gas response of ZnO towards H_2S. In other studies it has been shown that sulfidation can be reversed by oxidation and therefore in the absence of H_2S the pure ZnO surface can be re-established in air. The reversibility of the surface composition makes this a good sensing process.

This example of the sulfidation of ZnO by H_2S exposure illustrates how surface science can be used to measure the fundamental surface transformations responsible for the gas response by surface reactions. Because these studies are performed under UHV conditions the surfaces are only exposed to one gas. In real gas sensing applications the H_2S would be a trace amount in e.g. air. Therefore it would be interesting to investigate the role of other gases on the surface changes. Advances in 'high pressure' photoemission experimental set-ups [19], which become now more common at synchrotron facilities, will enable to do surface science experiments under more realistic pressure and gas mixture conditions in the near future and thus enable us to close the 'pressure gap' between fundamental surface science studies and 'real world' conditions. Although there have been several studies on heterogeneous catalyst surfaces [19] and even on solid-oxide electrochemical cells under operating conditions [20], there are, to the best of our knowledge, no ambient pressure studies on gas sensor surfaces been reported yet. Ambient pressure XPS studies would enable us to monitor surface properties under different chemical potentials of the gas phase and thus circumvent the inherent problem of UHV surface science studies of investigating surfaces in an unrealistically strong reducing environment.

2.3 TiO_2 Studies

Titanium dioxide is often used as a model system for transition metal oxide surfaces [21]. Consequently, the surface properties of TiO_2 are the best understood of all oxides. TiO_2 exists in three different crystal structures: rutile, anatase and brookite. Of those three, only rutile and anatase are of technological importance. Rutile is thermodynamically the most stable polymorph, however, for nano-materials anatase may become favored. Most surface studies have been performed on rutile, because of the commercial availability of single crystal samples. Experimental studies on anatase have either been performed on mineral samples or on epitaxial thin films grown on $SrTiO_3$ or $LaAlO_3$ substrates [22, 23]. In here we are focusing on the rutile phase of TiO_2.

Transition metal oxides may sense reducing and oxidizing gases due to formation and reaction with charged surface species, namely O_2^- and O^- radicals [2], or due to the reaction of lattice oxygen, i.e. bulk reduction. The former detection mechanism is dominant on materials whose conductivity may be strongly affected by charged adsorbates induced band bending such as SnO_2, ZnO or In_2O_3. The conductivity of TiO_2 on the other hand is more strongly affected by bulk reduction and oxidation. Nevertheless, the well ordered surfaces of TiO_2 make it an ideal model system for studying fundamental aspects of single molecule chemistry by e.g. STM. A review of such atomic-scale studies may be found in [24] for studies on the rutile (110) surface. The (110) surface is by far the most extensively studied surface orientation of rutile and most important properties of this surface are reviewed in [21]. Another prominent surface of rutile TiO_2 is the (011) orientation, which is the second most abundant surface in the Wulff equilibrium crystal shape construction of rutile. The (011) surface has obtained far less interest in the fundamental surface science community. The very different structural properties of the (110) and (011) surface, however, enable a nice comparison of the influence of surface structure on the surface chemical properties and thus chemical sensing behavior. Therefore in the next section we will discuss these differences.

The importance of the surface structure leads to the question on how the structure can be modified in order to achieve better selectivity and sensitivity towards target gases. Although its effectiveness for tuning gas responses is still unproven, a new ion beam method is described that enables fabrication of unique step edge structures on planar metal oxide surfaces. Finally, we end the discussion of TiO_2 by investigating the re-oxidation of $TiO_2(011)$ crystals. This is the important process of changing the intrinsic dopant concentration and thus underlies the gas response mechanism for oxidizing and reducing gases of TiO_2.

2.3.1 Comparison of the Rutile-TiO₂ (011) and (110) Surfaces

The surface structure of stoichiometric $TiO_2(110)$ is a bulk truncation with only small atomic relaxations. Figure 2.9a shows a model of the surface structure of $TiO_2(110)$. The (110) surface exhibits two-fold coordinated bridging oxygen atoms that protrude from the surface plane and are closely spaced (0.30 nm) along the [001] direction and are far apart (0.65 nm) along the [10] direction. This gives the surface a large corrugation and a strong structural anisotropy. Slight reduction of TiO_2 by vacuum annealing, causes the removal of $\sim 10\ \%$ of the bridging

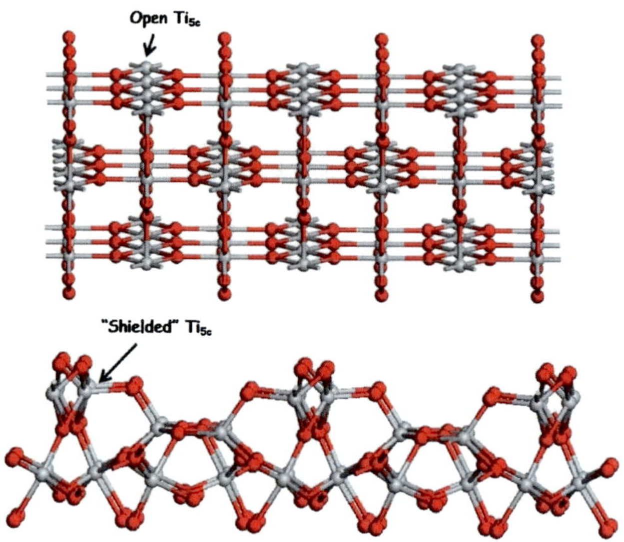

Fig. 2.9 Ball-and-stick models for TiO$_2$(110) (**a**) and (011)-2 × 1 (**b**) surfaces. *Grey* and *red* spheres are Ti and O atoms, respectively

oxygen atoms, i.e. formation of oxygen vacancies. These vacancies are very reactive surface sites and even under UHV conditions are quickly filled by adsorption of residual water from the gas phase. Water adsorbs dissociatively in these O-vacancy sites leaving two hydroxyls at the surface. This adsorption of water at defects has been studied extensively by STM [24]. Because of the very high reactivity of O-vacancies their relevance for realistic environmental conditions is doubtful. Reduction by vacuum annealing does not only induce surface O-vacancies but it also increases the number of bulk Ti-interstitials. This formation of interstitials can be reversed by annealing in an oxygen atmosphere which is discussed below.

Unlike the (110) surface, the rutile TiO$_2$(011) surface reconstructs to form a 2 × 1 periodicity. The structure of this surface reconstruction was recently resolved by surface x-ray diffraction measurements and density functional theory (DFT) calculations by two groups independently [25, 26]. The proposed structure is shown in Fig. 2.9b. This surface exhibits two different surface Ti$_{5c}$ atoms, with each of them coordinated to 2-fold coordinated oxygen (O$_{2c}$) atoms that protrude out of the surface and partially 'block' the Ti$_{5c}$ atoms from the gas phase and thus potentially hindering molecular adsorption. Although the density of Ti$_{5c}$ at the (011)-2 × 1 surface is slightly higher than on the (110) surface (0.08/Å^2 compared to 0.05/Å^2), the closest separation between Ti$_{5c}$ is similar on the (110) surface (2.959 Å) and on the (011)-2 × 1 surface (2.963 Å). However, these closest separated Ti$_{5c}$ atoms on (011)-2 × 1 are coordinated to bridging two fold coordinated O$_{2c}$ atoms while on the (110) surface the Ti$_{5c}$ sites are 'open' to the gas phase as is apparent in Fig. 2.9. Consequently, the large structural differences

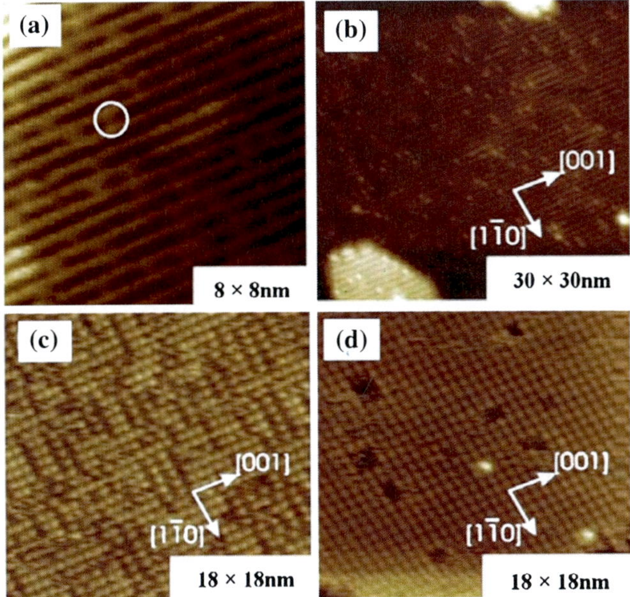

Fig. 2.10 STM images of acetic acid adsorption on $TiO_2(110)$. **a** clean, vacuum-prepared $TiO_2(110)$ (8×8 nm^2). **b** and **c** after adsorption of 0.16 and 0.43 acetate, respectively. And (**d**) close to saturation coverage of 0.5 ML forming an ordered 2×1 over layer structure. Reprinted with permission from Ref. [34]. Copyright 2011, American Chemical Society

between these two surfaces suggest different adsorption properties. One example of the largely different adsorption is discussed below on the example of acetic acid adsorption. Now that the surface structure of the (011) surface is resolved, we may expect more studies on the surface chemistry of this surface in the near future.

STM images of vacuum-annealed $TiO_2(110)$ and (011)-2×1 surfaces are shown in Figs. 2.10a and 2.11a, respectively. The bright and dark rows in Fig. 2.10a correspond to surface five-fold Ti atoms (Ti_{5c}) and two-fold bridging oxygen (O_b) sites, respectively [21]. There are several bright spots between the bright rows which are attributed to surface bridging O vacancies (O_b-vac) or -OH, where the latter is formed by water dissociation at O_b-vac sites, as is typical for this surface. In contrast to $TiO_2(110)$, the STM image of clean $TiO_2(011)$-2×1 surface exhibits rows of protrusions arranged in zigzag patterns as given in Fig. 2.11a. The bright spots in the STM image have been attributed to the geometrically protruding two-fold O (O_{2c}) atoms arranged in a zigzag pattern along the [100] direction (see Fig. 2.9b) [27].

Initial studies of water [26] and carboxylic acid [28] adsorption suggest that the (011)-2×1 surface behaves very different compared to the (110) surface. Strongly different molecular interaction with surfaces of different atomic

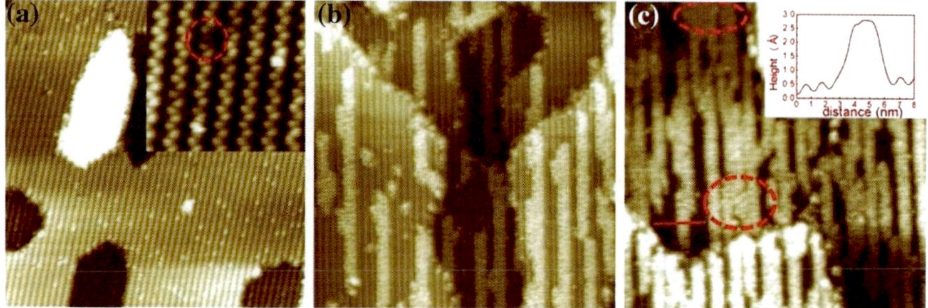

Fig. 2.11 STM images of acetic acid adsorption on $TiO_2(011)$-2 $\times$ 1. **a** clean vacuum prepared surface. **b** and **c** after exposure of 18 and 24 L acetic acid, respectively. Reprinted with permission from Ref. [34]. Copyright 2011, American Chemical Society

arrangements and coordination suggest face dependent gas responses in gas sensing materials. In the next section we investigate the role the surface structure plays for molecular adsorption on the example of acetic acid adsorption.

2.3.2 Comparative Study of Acetic Acid Adsorption on $TiO_2(110)$ vs. (011)-2 $\times$ 1

The adsorption of carboxylic acid on $TiO_2(110)$ was extensively investigated by STM, photoemission spectroscopy (PES) [29–31], and computational methods [32]. The 'open' rows of closely spaced Ti_{5c} cations are ideal sites for carboxylic acids to adsorb dissociatively by splitting-off its acidic hydrogen (which adsorbs on the surface lattice oxygen). The lattice spacing is such that the acetate adsorbs in a bidentate fashion, bridging two neighboring surface Ti_{5c} [33]. On the $TiO_2(011)$-2 $\times$ 1 surface, on the other hand, the O_{2c} are bridging the nearest Ti_{5c} atoms preventing a bridge bidentate adsorption of acetate.

Figure 2.10 shows STM images for various stages of acetic acid adsorption on $TiO_2(110)$ at room temperature. The vacuum-annealed clean surface is shown in Fig. 2.10a. The bright and dark rows correspond to surface Ti^{4+} and bridging O^{2-} sites, respectively [20]. Figure 2.10b, c and d show the $TiO_2(110)$ surface with increasing amounts of acetic acid coverage. Adsorbed acetate appears in the STM images as bright spots on top of the Ti_{5c} rows. At room temperature these individual acetate molecules are mobile and subsequent images of the same area show the acetate at different positions along the Ti_{5c} rows. The mobility and diffusion of acetate on this surface has been studied by Onishi and Iwasawa [33]. Also it is worth pointing out that at this low coverage the acetate never comes very close to each other indicating a repulsive interaction between individual acetates along the Ti_{5c} rows. With increasing acetic acid exposure the density of the adsorbed acetate increases. In Fig. 2.10c the surface is covered with ~ 0.43 $ML_{(110)}$ acetate. At this coverage different domains with ordered superstructures are observed. The formation of these structures can be understood from the repulsive interactions along the [001] direction, i.e. along the Ti_{5c} rows, which causes the acetates to maximize their separation in this direction, and an attractive force for acetates perpendicular to the Ti_{5c} rows, i.e. in the $[1\bar{1}0]$ direction. This latter explains why acetates preferentially form next-neighbor configurations in the $[1\bar{1}0]$ direction [33]. At saturation coverage, shown in Fig. 2.10d, the acetate over layer arranges in a 2×1 superstructure with 0.5 $ML_{(110)}$ coverage.

The well established adsorption of acetate on the $TiO_2(110)$ surface is contrasted to that on the less studied $TiO_2(011)$-2×1 surface [34]. STM images of the clean $TiO_2(011)$-2×1 surface show rows of protrusions arranged in zigzag patterns, as can be seen in Fig. 2.11a. Compared to the $TiO_2(110)$ surface these Ti_{5c} atoms are less accessible because they are coordinated to the protruding O_{2c} atoms. Consequently, the O_{2c} atoms may 'block' the adsorption of acetic acid or form a kinetic barrier for adsorption of acetate to the Ti_{5c} sites on this surface. Indeed, STM images of acetate adsorbed on the (011)-2×1 surface shows a very different adsorption behavior compared to the (110) surface. Figure 2.11b shows an STM image taken after a nominal acetic acid exposure of 18 L. Contrary to the homogenous acetate adsorption on the $TiO_2(110)$ surface, acetate clusters appear on the (011)-2×1 surface. Initial adsorption of acetate on the (011)-2×1 surface is limited to surface defects, such as step edges and antiphase domain boundaries in the 2×1 surface reconstruction. This indicates that acetates do not adsorb on perfect defect free (011)-2×1 surface under vacuum conditions at room temperature. Defect sites are needed to form nuclei of acetate islands that then can grow onto the defect free terraces. This indicates that the adsorbed acetates in the cluster facilitate further adsorption of acetic acid molecules to the edges of the cluster. Also, the clusters form one dimensional (1D) chains along the $[0\bar{1}1]$ direction, suggesting that diffusion of individual acetates along those rows is the preferential diffusion direction and/or the acetic acid molecules preferentially attach to this side of the clusters only. With increasing acetic acid exposure, the acetate clusters grow in length and the density of the cluster increases until about 77 % of the surface is covered. An STM image of this 'saturation' coverage is shown in Fig. 2.11c for a nominal acetic acid exposure of 24 L. Even much higher doses do not increase the surface coverage significantly anymore, probably because of the lack of sites for acetic acid to attach to the existing acetate clusters. The absolute number of this 'saturation-coverage' may however sensitively depend on the initial defect concentration on the surface. At high coverage many acetate chain-islands merge as indicated by the dashed circles in Fig. 2.11c. As shown in the inset of the Fig. 2.11c, the line profile shows that the 1D cluster chain has corrugation of ~ 2.5 Å. The quasi-1D clusters predominantly consist of three lines of bright protrusions that are assigned to adsorbed acetates. The protrusions in the center of these clusters align with the 'dark' rows of the substrate and the protrusions to the left and right are situated over the zigzag rows of the substrate. Furthermore, the center row is always imaged brighter in STM with an apparent height difference between the centre and outer rows of ~ 0.3 Å. Because of tip convolution effects the width of these clusters is overestimated in STM images. From analysis of clusters with a larger width and correlation of the protrusions with the periodicity of the substrate it was concluded that the width of the three-row clusters corresponds to the width of a 2 × 1 substrate unit cell. The acetate ions on the left and right rows are shifted in $[0\bar{1}1]$ direction by half a substrate unit cell, indicating that they are bonded to the lattice Ti_{5c} in the zigzag pattern. Besides the well ordered three-row structure, most of the clusters exhibit a high density of acetate 'vacancies'. Most of these vacancies are located in the two outer rows. For a densely packed cluster we count three protrusions per 2 × 1 substrate unit cell, i.e. 1.5 $ML_{(011)}$ equivalent or 6.0 molecules/nm^2.

In conclusion, these STM studies indicate a quite different acetate adsorption on these two TiO_2 surfaces. While on the $TiO_2(110)$ surface acetate adsorbs readily, acetic acid does not seem to adsorb dissociatively on the perfect $TiO_2(011)$-2 × 1 surface and initial adsorption is limited to defect sites at room temperature. This suggests a low initial sticking coefficient of acetic acid on the (011)-2 × 1 surface. However, once acetate clusters have nucleated at defects the experiments show that clusters grow and eventually cover large portions of the substrate. Thus the adsorption structure and kinetics is very different on these two surfaces. Further differences in the charge transfer between the TiO_2 substrate and the adsorbed acetic acid can be detected in UPS measurements. In these studies defect states induced by electron donation to the TiO_2 substrate mainly from adsorbed hydrogen plays an important role. This is introduced first before we turn to the charge transfer induced by dissociative acetic acid adsorption.

2.3.2.1 Defect and Adsorbate Induced Band-Gap States in TiO_2

Excess-charge induced Ti-3d states within the band gap of TiO_2 have been studied extensively on the (110) surface. These excess charges can originate from surface oxygen vacancies O_{vac} [35], which introduces two excess electrons, or adsorbed hydrogen [36], which introduces one excess electron. These excess charges are transferred to the Ti-3d states, which has been verified by resonant photoemission studies [37, 38]. The occupied Ti-3d states are observed in ultraviolet photo-emission spectroscopy (UPS) [35], as a band-gap state (BGS) about 0.9 eV below the Fermi level (E_F) for the (110) surface.

Figure 2.12a and b compare the differences between the defect states on (110) and (011) surface for vacuum prepared samples. In Fig. 2.12a, besides the main feature of predominantly O 2p derived valence band states of bulk $TiO_2(110)$, the Ti-3d derived band gap state (BGS) (located ∼0.9 eV below E_F) is clearly observed. On the vacuum prepared (011)-2 × 1 surface, on the other hand, the BGS is very small (see Fig. 2.12b). The BGS intensities are 1.9 and 0.17 % of the total valence band intensity for the (110) and (011) surfaces, respectively. This is consistent with STM observations of fewer point-defects on the (011)-2 × 1 surface compared to the (110) surface, which have been counted in STM images to

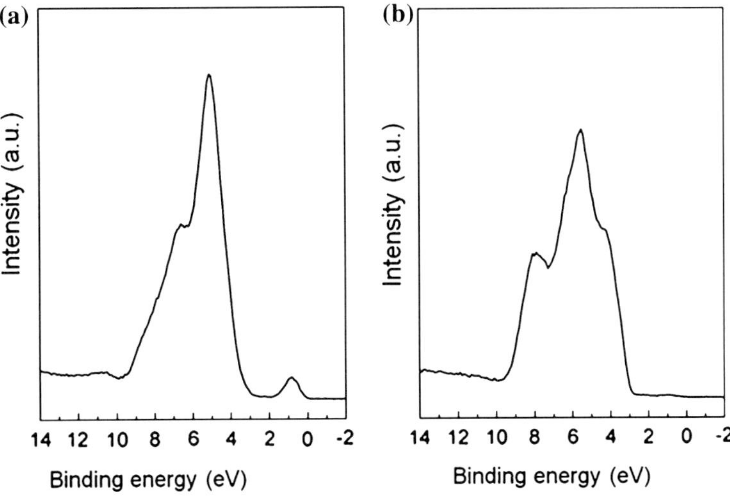

Fig. 2.12 UPS spectra of vacuum prepared rutile surfaces: **a** $TiO_2(110)$ and **b** (011)-2 × 1. Reprinted with permission from Ref [41]. Copyright 2010, American Chemical Society

typically amount to $\sim 10\,\%$ and $\sim 1\,\%$ for (110) and (011)-2 × 1 surfaces, respectively. Therefore the BGS intensity ratio in UPS agrees with the point-defect density on these two surfaces.

2.3.2.2 UPS Measurements of Acetic Acid Adsorption

To investigate the effect adsorbates may have on the defect states, acetic acid has been adsorbed on the two surfaces and the change in the electronic structure measured by UPS. The UPS spectra for (110) surface before and after acetate adsorption is shown in Fig. 2.13a. As shown by the difference spectrum in the lower panel of Fig. 2.13a, the acetate orbitals are mainly located at binding energies of ~ 10.4, ~ 8.1 and ~ 5.7 eV, which agrees for example with the reported values for bi-dentate adsorbed acetate on Cu(110) surfaces [39]. The higher binding energy peaks are dominated by the methylic carbon and carboxylic carbon orbitals while the low binding energy peak originates from carboxylic oxygen atoms, which have direct bonding with the substrate through the oxygen lone pair [39, 40]. In addition to the acetate molecules, contributions from OH_σ are expected. However, the binding energy of the OH_σ bond overlaps with the acetate peak at ~ 10.4 eV and therefore OH formation cannot be unambiguously confirmed from the UPS spectra.

Interestingly the UPS studies show that although the adsorbed hydrogen donates its electron to the substrate, the BGS on the (110) surface is decreased by $\sim 40\,\%$ upon carboxylic acid adsorption. This indicates that on the (110) surface, the adsorbed acetate is taking up the charges donated from the hydroxyls—formed

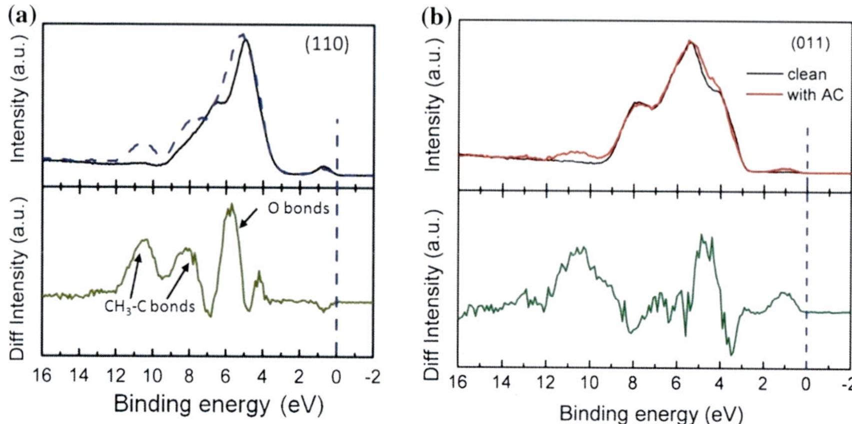

Fig. 2.13 UPS valence band spectra of **a** TiO$_2$(110) and **b** TiO$_2$(011)-2 × 1 before (*black solid curve*) and after monolayer acetate adsorption (*blue/red curves*). The difference spectra between the clean and acetate covered surfaces are shown in the lower panels. The defect states just below the Fermi-level (0 eV binding energy), behaves oppositely on the two surfaces. On the (110) surface the defect state decreases while on the (011)-2 × 1 surface the defect state increases upon acetic acid adsorption. Reprinted with permission from Ref [41]. Copyright 2010, American Chemical Society

by the dissociative adsorption of acetic acid—and even reduces the excess charges in the sample that were present prior to acetic acid adsorption. This contrasts observations on the (011)-2 × 1 surface. Adsorption of acetic acid on TiO$_2$(011)-2 × 1 at room temperature also results in dissociative adsorption. However, contrary to the (110) surface, the BGS is increasing on the (011)-2 × 1 surface upon acetate adsorption, as shown in Fig. 2.13b. This implies that the acetate is not able to take up all the excess charges donated by the hydroxyls, leaving charges in Ti-3d states. This difference may originate from the adsorption coordination of the acetate with a monodentate vs bidentate adsorption on the (011)-2 × 1 and (110) surface, respectively. Having two acetate-oxygen atoms coordinated to the (110) surface enables acetate-molecules to overcompensate for the electrons donated by hydroxyl formation and consequently resulting in the observed suppression of the BGS. In contrast, coordination of only one oxygen atom of the acetate to the substrate on the (011)-2 × 1 surface leaves excess charges (donated by –OH groups) in the TiO$_2$ substrate. Understanding the differences in charge transfer from the metal oxide to the adsorbate has important implications for gas sensing mechanisms, since the surface charge are largely responsible for the gas response in conductivity sensing materials. Therefore, as more crystallographically defined gas sensing materials are being developed; surface science studies of face-dependent charge transfer mechanisms can provide answers to fundamental gas sensing mechanisms.

In addition to the change in intensity of the BGS for the two surfaces a difference in the binding energy for the two surfaces was also observed. Comparing

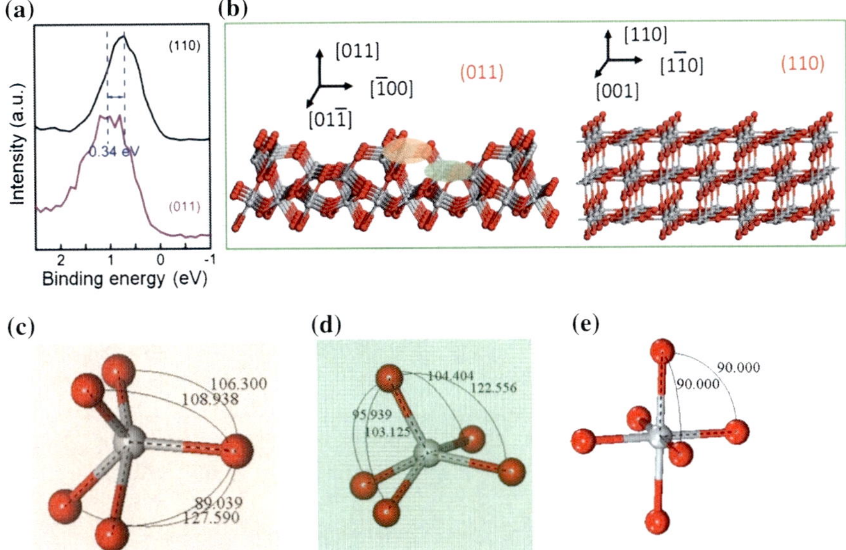

Fig. 2.14 **a** The BGS for the TiO_2(110) and (011)-2 × 1 surfaces are compared. The *black curve* is for the clean TiO_2(110) surface and the *pink curve* is measured on the (011)-2 × 1 surface after acetate adsorption, when the BGS becomes noticeable. The BGS for the (011) surface is shifted ~0.34 eV to higher binding energies compared to the (110) surface. **b** *Ball*-and-*stick* models for TiO_2(011)-2 × 1 and (110) surfaces. Grey and red spheres are Ti and O atoms, respectively. **c** and **d** show two different local Ti–O environments on the (011)-2 × 1 surface with the same color code specified in (**b**). **e** shows the octahedral coordination environment for bulk TiO_2. The values shown in (**c**), (**d**) and (**e**) are the bond angles

the BGS binding energy for the (110) and (011)-2 × 1 surface reveals that the binding energy of the BGS for the (011)-2 × 1 surface is ~ 0.34 eV higher than for the (110) surface, as emphasized in Fig. 2.14a. This is a significant energy difference for excess charges at Ti lattice sites and suggests that electrons are more effectively trapped on the (011)-2 × 1 surface than on the (110) surface.

This higher binding energy of Ti-3d states on the (011)-2 × 1 surface has been explained by the geometry of the surface reconstruction and the resulting crystal field at the surface Ti-sites [41]. Due to the surface reconstruction, the Ti–O coordination environment at (011)-2 × 1 surface is changed to two differently distorted square pyramid Ti–O coordination as schematically shown in Figs. 2.14c, d. Generally, a square pyramid coordination causes a larger crystal field splitting and stronger stabilization of the d_{xz} and d_{yz} orbitals compared to an octahedral configuration. Only Ti-atoms in the topmost surface layer exhibit this unique coordination environment and therefore the observed higher binding energy of the Ti-3d state implies that the defect-electrons are localized in the surface layer.

2.3.3 Surface Modification by Grazing Incidence Low Energy Ion Sputtering

In the previous section it has been shown that surface structures and defects play an important role in the adsorption of molecules and thus in the gas sensitivity of these materials. Therefore, it would be important to be able to modify surface structures to enable tuning of surface sensitivity to specific target gases. This requires a surface modification at the molecular scale since this is the relevant length scale for adsorption. On oxide surfaces, under-coordinated sites can be particularly reactive. Such sites are, for instance, step edges. Consequently, creating surfaces with a high density of defined step edge structures may be an approach for tuning gas sensitivities of materials. In the following we describe a method that enables such a surface structuring on the example of $TiO_2(110)$. This method employs grazing incidence low energy ion beams and exploits a self organization of the surface into nano-ripple structures that exhibit a high density of step edges.

It has been demonstrated that the sputter yield of a grazing incidence low energy ion beam varies largely at flat terraces compared to defect sites such as step edges [42, 43]. Thus, ion sputtering at grazing angles is fundamentally different from sputtering at steeper angles. At steeper angles, the ions deposit their energy in the bulk. At grazing angles, on the other hand, planar surfaces re-arrange to minimize step-edge sputtering. This results in an alignment of step edges in the direction of the azimuth direction of the ion beam. The alignment of step-edges combined with a sputter induced surface roughening gives rise to the formation of nanoscale ripples at the surface. This process is discussed next followed by a demonstration that this ion beam techniques also allows the formation of high energy step edge orientations that otherwise cannot be observed.

2.3.3.1 Nanoripple Formation by Grazing Ion Sputtering

Grazing incidence sputtering by low-energy ions can be applied to many planar single crystal substrates. With increasing ion-incident angle (measured from the surface normal) the sputter yield initially increases because the ion energy is deposited closer to the surface and thus collision cascades cause the ejection of more surface atoms. However, at a critical grazing incidence angle most of the incident ions will be quasi-elastically reflected from the surface without penetrating the surface. Furthermore, for low energy ions, the momentum transfer to the surface during the reflection is too small to induce lattice defects at defect-free terraces. At surface defects, and in particular at step edges, the incident ions can, however, interact more strongly with the substrate and thus induce sputter damage at these pre-existing surface defects. This results in a sputter yield that varies locally on the surface and this gives rise to a self-organization of the surface in nanostructures.

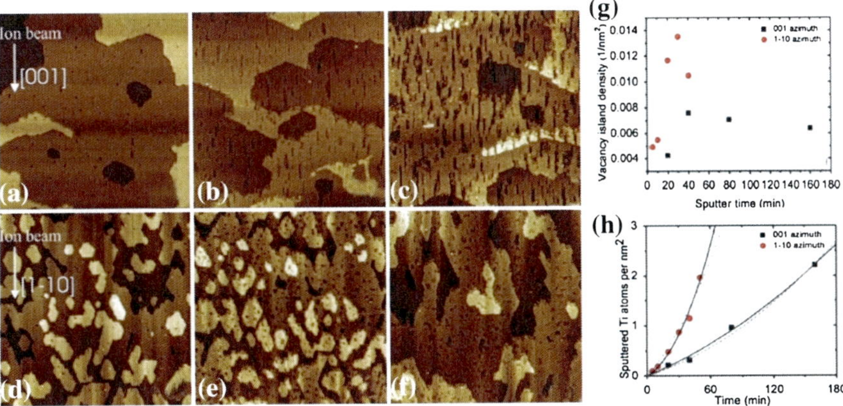

Fig. 2.15 Shown are 200 × 200 nm STM images of $TiO_2(110)$ after grazing incidence irradiation along the <001>azimuth for **a** 20 min, **b** 80 min, and **c** 160 min, and along the <10> azimuth for **d** 5 min, **e** 20 min, and **f** 30 min. **g** shows the evolution of the vacancy island density with sputter time. The totaled sputtered atoms derived from the total vacancy island area as a function of sputter time is shown in (**h**). Reprinted from Ref. [44]. Copyright 2010, American Physical Society

The initial surface evolution, i.e. for low ion fluences, is shown in Fig. 2.15 for two different ion beam azimuths at 8° grazing incidence angle. In the low fluence regime the ion beam induced formation of vacancy islands and their gradual evolution into vacancy islands with an elongation along the ion beam azimuth can be observed. Formation of vacancy islands is due to random terrace sputtering of atoms which agglomerate by surface diffusion into islands at the processing temperature of $\sim$400 °C. Once the vacancy islands are formed increased sputtering is observed along the step-edges facing the ion beam. This has two consequences; firstly the total sputter yield from the surface increases with an increasing step edge density and secondly the vacancy islands form an anisotropic shape due to the preferential sputtering at step edges exposed to the ion beam. It is apparent from Fig. 2.15 that this mechanism is also dependent on the crystallographic azimuth of the ion beam irradiation. Especially on $TiO_2(110)$ where the surface structure is very anisotropic the sputter yield of a grazing ion beam will also be anisotropic. Furthermore, variations of the step edge formation energies (see below) prefer certain step edge orientations which can favor vacancy islands with distinct shapes.

The total amount of sputtered atoms can be deduced from the surface area covered with vacancy islands that enables a measure of the sputter yield. Figure 2.15g and h plots the vacancy island density and the number of sputtered atoms versus the ion fluence for the two azimuths directions investigated. The non-linear change in sputtered atoms with ion fluence indicates that the sputter yield increases with irradiation time. This can only be explained if the changing surface morphology affects the sputter yield. This is further indication that step edge sputtering is a crucial component of the total sputter yield. For both azimuths the step edge density increases with sputter time, due to the formation of vacancy islands, and consequently increases the global step edge sputter contribution. As shown in Ref. [44] fitting a quadratic expression to the sputtered atoms vs ion fluence plot and estimating the step edge density from STM-images, the relative contributions from step edge and terrace sputtering can be extracted. This kind of analysis shows that the step edge sputtering is more than ten times that of the terrace sputtering.

With increasing ion fluence the surfaces increase in roughness as the STM images shown in Fig. 2.16 indicate. However, instead of forming a randomly rough surface, nanoripples form. These nanoripples are aligned with ion beam azimuth and thus exhibit mainly step edges that are in the direction of the ion beam. Furthermore, the ripples exhibit a fairly uniform separation of 10–20 nm and a peak-to-valley roughness of ~ 4 nm. Therefore this technique of grazing ion beams allows the formation of surfaces with step edges with defined crystallographic orientation and high density. This is a new method for modifying surfaces with under-coordinated sites which may be exploited for tuning chemical and gas sensing properties in the future. This method can be employed not only on single crystals but also on thin epitaxial films and thus this may be exploited for new gas sensing materials. In addition, this ion beam method also has the advantage of being capable of forming step edges that are meta-stable, i.e. step edges with high formation energy. Such steps are often the most chemically active step edges and thus preparation of surfaces with a high density of such steps is especially interesting for obtaining surfaces with new chemical functionalities. This is briefly discussed in the next section.

2.3.3.2 Formation of Meta-Stable Step Edge Orientations

Step edges on oxides and other covalent and ionic materials have strong orientation dependent formation energies. This causes the presences of only step edge orientations with the lowest formation energies to be present on a surface in thermodynamic equilibrium. If the step edge formation energies are known the shape of atomic-layer high adislands can be derived from a 2D analogue of the crystal shape Wulff construction where the surface energies are replaced by step edge energies [45]. For a vacuum annealed $TiO_2(110)$ surface only step edges along the [11] and [001] directions are observed but not in the [10] direction, which would be the third low index direction within the (110) surface. The observed step-edge orientations are illustrated in Fig. 2.17. This implies that the latter direction is an energetically unfavorable direction for step edge formation. Can this direction be prepared using the above described grazing ion beam method?

Directing the ion beam along the [10] azimuth indeed results in step edges aligned in this direction and STM allows the determination of the structure of this step edge [46]. Figure 2.18b and e show atomic resolved STM images. It is

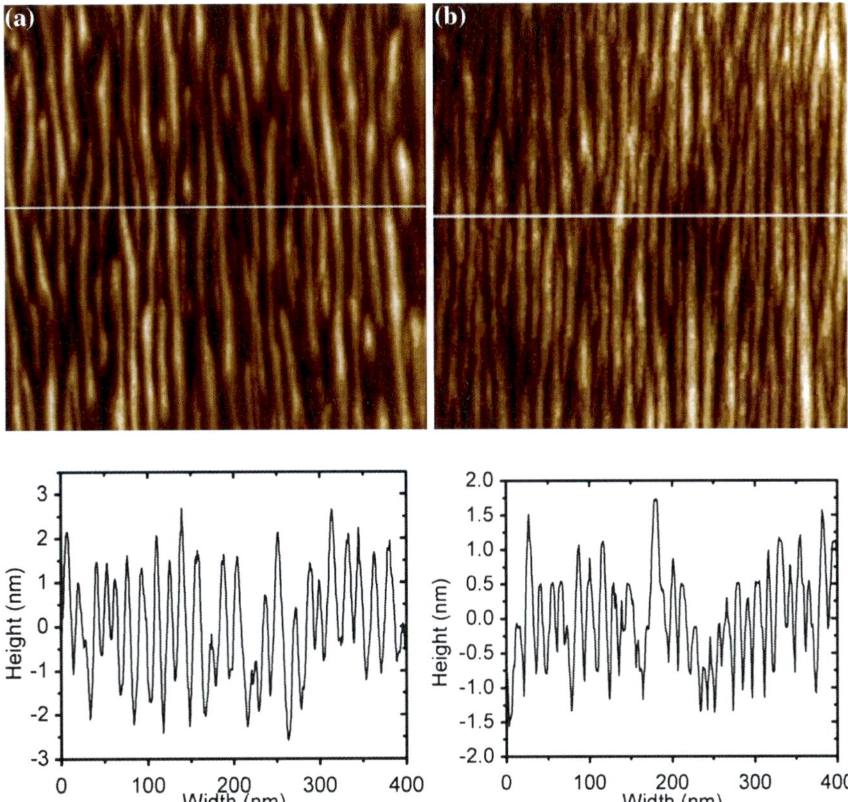

Fig. 2.16 STM images (400 × 400 nm) of the surface morphology after extended periods of grazing incidence irradiation in **a** <110> azimuth- and **b** <001> azimuth-direction. Nano-ripples are formed that are aligned parallel to the ion-beam directions. The corrugation of these ripples is ∼4 nm as can be judged from the shown cross-sections. Reprinted from Ref. [44]. Copyright 2010, American Physical Society

apparent that the periodicity of this step edge is twice that of the terrace. This suggests a step edge structure as shown in Fig. 2.18d and DFT simulations of a STM image of such a step edge results in very good agreement with the measurement. The fact that this step edge is only meta-stable is highlighted by the fact that annealing of the surface to ∼600 °C results in a reformation of energetically more favorable step edges along the [11] azimuths. This is shown in Fig. 2.18c by the formation of zigzagging steps.

In conclusion, the use of grazing ion beam step structuring allows not only the preparation of surfaces with a high step edge density but it also enables the stabilization of steps that are otherwise not present at the surface. Therefore this method may be used to structurally alter the surface and thus tune the sensing properties of metal oxide surfaces.

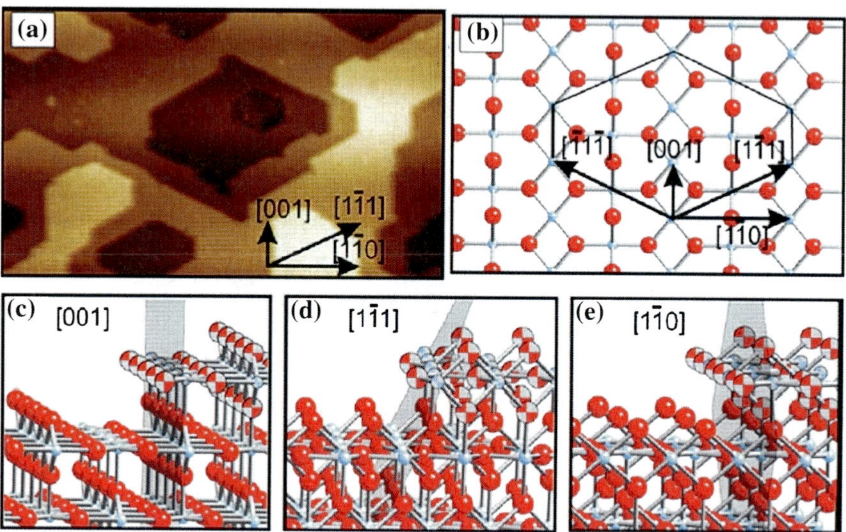

Fig. 2.17 Low energy step-edge structure on $TiO_2(110)$. **a** STM image of a typical surface structure of clean $TiO_2(110)$ surface, **b** orientations of the step edges on the surface (top view), and (**c**)–(**e**) illustrative structures of the different step edges (side views). In (**c**)–(**e**), the shaded area represents micro facets of an extended low index crystallographic plane corresponding to the step edge. Ti and O atoms are represented by small and big balls, respectively. Reprinted from Ref. [46]. Copyright 2009, American Physical Society

2.3.4 Reaction of Sub-Surface Ti Interstitials with Gas Phase Oxygen

When Titania is reduced, excess titanium occupies interstitial sites. These interstitials are fairly mobile and at elevated temperatures can readily diffuse into the bulk. The opposite is also observed. If a reduced TiO_2 crystal is annealed in an oxidizing atmosphere the Ti-interstitials are diffusing to the surface where they can react with oxygen [47, 48]. The formation and annihilation of Ti-interstitials at the surfaces of TiO_2 is directly related to the gas response of TiO_2 to oxidizing and reducing gases. Ti-interstitials act as n-type dopants in TiO_2 and thus changes in their concentration affect the conductivity of the material which can be used as the gas response signal.

The phenomenon of re-oxidation of Ti-interstitials can be directly observed by surface science techniques in vacuum. Annealing of a slightly reduced crystal in $\sim 10^{-6}$ Torr O_2 pressure results in the reaction of Ti-interstitials at the surface with oxygen. As a consequence new layers of TiO_2 are growing at the surface. This has been thoroughly studied on the more frequently investigated rutile $TiO_2(110)$ surface [49, 50]. Recently, the re-oxidation of a slightly reduced $TiO_2(011)$ surface has been studied [51]. Similar to the (110) surfaces oxidation of Ti-interstitials results in the formation of new TiO_2 layers. However, on the (011)

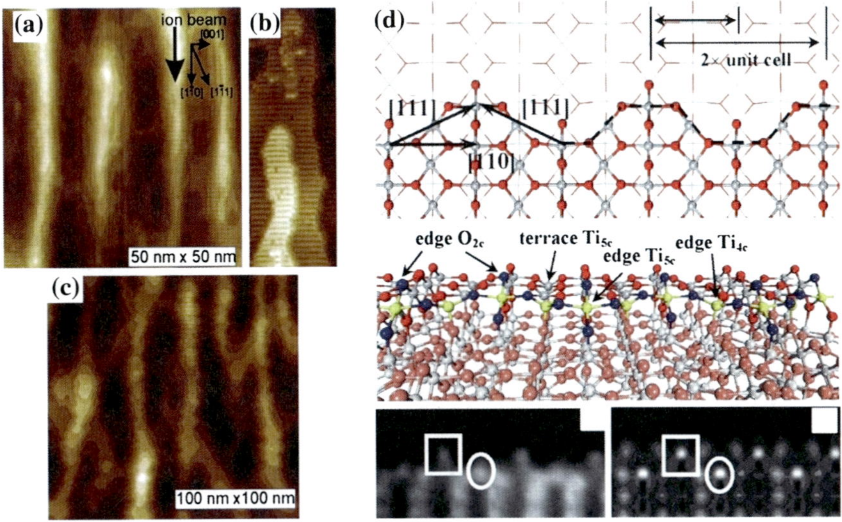

Fig. 2.18 Fabrication of meta-stable step edges by grazing ion beam irradiation along the [10] azimuth of TiO$_2$(110). (**a**) and (**b**) STM images of surface morphology after grazing ion beam irradiation with the majority of steps aligned with the ion beam azimuth direction along [10]. Annealing to 600 °C results in the restructuring in energetically more favorable [11] steps as evidenced by the zigzagging steps in (**c**). (**d**) shows an atomic scale model of the [10] steps and (**e**) and (**f**) is a comparison of the experimental STM image with a DFT-simulated STM image. Reprinted from Ref. [46]. Copyright 2009, American Physical Society

surface this new TiO$_2$ surface layer is different in its structural and electronic properties compared to the original surface. This new surface structure and its properties are discussed in the following.

2.3.4.1 Formation of New Metastable Surface Structure on TiO$_2$(011)

Vacuum prepared clean TiO$_2$(011) surface undergoes 2 × 1 reconstruction which show rows of protrusions arranged in zigzag patterns in STM images, as discussed above and shown in Figs. 2.9b, 2.19a and b. When this surface is exposed to O$_2$ at ∼300–400 °C, high resolution STM images illustrate that islands with a different structure are formed on the surface as shown in Fig. 2.19c and d. Although the periodicity of the new surface structure can also be described as a (2 × 1) surface unit cell, the structure of these islands is different from the original TiO$_2$(011)-(2 × 1) surface or a (1 × 1) bulk truncation. In particular, the new surface phase is less corrugated on the atomic scale along the long axis of the surface unit cell than the 'original' surface structure, while having a similar corrugation along the short axis of the unit cell. The corrugation measured in the STM images is shown as the line scan in Fig. 2.19e.

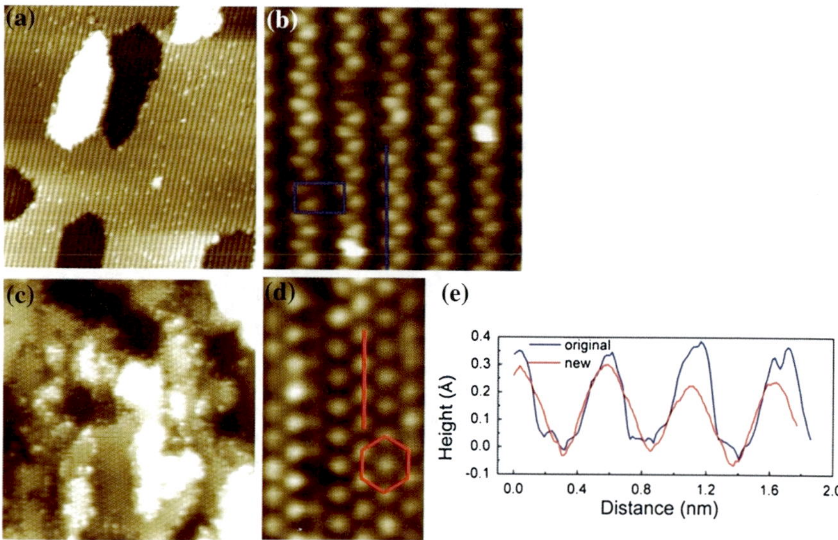

Fig. 2.19 Atomic-resolution STM images of the rutile $TiO_2(011)$ surface. The (2×1) reconstructed surface is shown in (**a**) and (**b**) and the new TiO_2 phase formed after annealing in 10^{-6} Torr O_2O_2 atmosphere is shown in (**c**) and (**d**). The image size for (**a**) and (**c**) are both $(50 \times 50 \text{ nm}^2)$. The surface unit cell for the rectangular (2×1) reconstruction is indicated in (**b**) and the quasi-hexagonal symmetry of the new TiO_2 phase is shown in (**d**). A line defect in the new TiO_2 phase can also be seen in (**d**). This defect is an anti-phase boundary due to the registry of the new phase with the $TiO_2(011)$ substrate. The line-profiles indicating the atomic corrugation along the indicated lines in (**b**) and (**d**) are shown in (**e**). Adapted from Ref. [51]

This new surface structure is thermally stable to ~ 500 °C and only annealing in vacuum above this temperature causes the reformation of the 'original' surface structure. Therefore this new structure is only metastable, however, its fairly high thermal stability suggests that it may be important in applications.

In addition to the structural variation the electronic properties of this new surface structure is also surprisingly different. In Fig. 2.20a and b, scanning tunneling spectroscopy (STS) spectra obtained for the original $TiO_2(011)$-(2×1) surface structure and the new TiO_2 phase are shown. A comparison of STS spectra of the two phases shows that the new phase exhibits filled electronic states that are in the band gap region of bulk rutile TiO_2. Also the empty states exhibit a weak variation between the two surfaces with the conduction band minimum slightly shifted towards the Fermi level for the new phase. Therefore, these spectra demonstrate that the new TiO_2 phase has a smaller band gap than bulk rutile TiO_2.

For a more accurate determination of the filled states of the new TiO_2 structure and to obtain information on the electronic states that make up the new valence band, UPS studies were performed using synchrotron radiation. Figure 2.20c and d show valence band photoemission spectra of the TiO_2 sample. After formation of the new surface structure, a new state within the bulk band gap of TiO_2 is observed with a maximum at 2.1 eV below the Fermi level. From comparison with STS

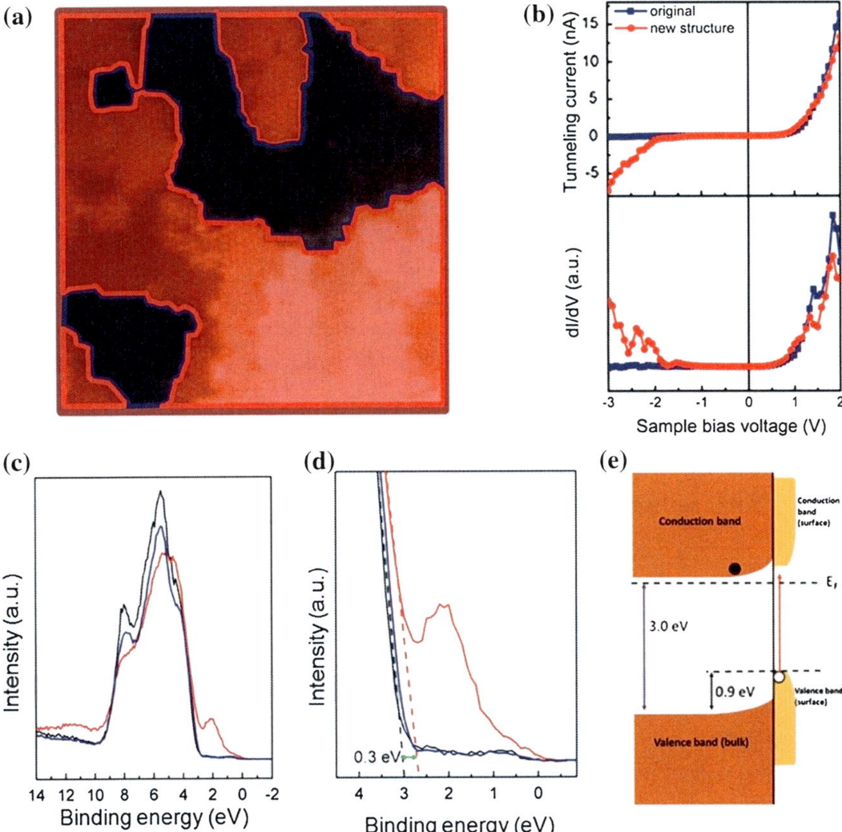

Fig. 2.20 Electronic structure determination of new structure on $TiO_2(011)$ surface. Scanning tunneling spectroscopy (STS) measurements. In (**a**), the original (2 × 1) reconstructed surface is coloured in blue, while the new TiO_2 phase is indicated in *red*. The color coded *I–V* spectra taken in the respective surface areas are shown in (**b**) and the numerically differentiated dI/dV curves are also shown in (**b**). Photoemission spectra confirm the formation of new electronic states within the band gap. UPS spectra taken with an 80 eV photon energy are shown in (**c**) and (**d**). The *black curve* corresponds to the photoemission spectrum for the original (2 × 1) reconstructed surface. The *red curve* has been acquired after low pressure oxidation and (partial) formation of a new TiO_2 phase. The *blue line* indicates a photoemission spectrum that has been taken after annealing the sample to ~600 °C in vacuum and re-formation of the (2 × 1) surface structure. The band diagram is schematically shown in (**e**). Adapted from Ref. [51]

measurements these filled states are assigned to the new surface phase of TiO_2. After annealing the sample to above 500 °C, the valence band structure of the original $TiO_2(011)$-(2 × 1) surface is regained (see Fig. 2.20c), demonstrating the reversibility of the structural and electronic surface transformation in agreement with the STM studies. Also, a ~0.3 eV upward shift of the bulk bands is observed due to band bending at the surface indicating a depletion of the surface regions of charge conduction electrons. Therefore, the formation of this new structure may aid the gas response not just by oxidation of Ti-interstitials but also by the band bending effect induced by the new surface structure.

Based on the experimental results, the electronic structure energy diagram for this new surface phase can be drawn as shown in Fig. 2.20e. The new electronic state penetrates into the bulk band gap and reduces the surface band gap by 0.9 eV. This observation of the formation of a new surface structure with unique electronic properties illustrates that oxide surfaces, although often considered as static, may exhibit a surprising variability and that their remains many unknowns about their atomic scale behavior especially if in contact with different gas phases. Therefore, oxide surfaces will remain a challenging topic for fundamental surface science studies for some time to come. The better understanding of structural, chemical, and electronic properties of these surfaces will ultimately aid the understanding and tuning of metal oxide gas sensors.

2.4 Summary and Perspective

This chapter has given examples of how surface studies of metal gas sensing materials can add to our fundamental understanding of the gas response of semiconducting metal oxides. We intended to show that surfaces can behave quite differently not just between different materials but also for different surface orientations of the same material and even the same surface orientation under different oxidation potentials. In addition, minority sites, like surface defects, can be crucial in determining the surface reactivity and may play a dominant role in the adsorption of molecules and consequently in the gas response of metal oxides. Therefore, from a surface science perspective, an increase in selectivity and sensitivity may be obtained by designing materials with single surface orientations [52] and by controlling defect constellations by e.g. doping or step edge orientations. For the latter point a method has been described that may be used for designing surfaces with defined step edge orientations and this may provide a way forward, at least for fundamental studies, to examine the role of step edges on gas responses. With the progress in real space analysis of surfaces by STM and other scanning probe techniques we may see more studies of the role of molecular scale surface features on chemical surface properties in the future.

Another important advancement in the surface science of gas sensing materials will be to investigate surfaces under more relevant environments with similar precision as is currently possible by traditional vacuum technology. To achieve this goal some of the surface science techniques that rely on vacuum need to be sacrificed but others will substitute for them and some can be modified to allow for higher pressure conditions. We are at the brink of getting out of vacuum and therefore we may expect to learn a big deal more from surface science studies about the fundamentals of gas sensing in the near future. In particular photoemission experiments can now be performed at elevated pressures [19] at a few selected beamlines at synchrotron facilities around the world. The merger of vacuum surface science with ambient pressure characterization of surfaces will ultimately enable us to determine what surface processes are central for the gas response.

Acknowledgments Support from the National Science Foundation under grant CHE-0840547 is acknowledged.

References

1. Göpel W (1985) Chemisorption and charge transfer at ionic semiconductor surfaces: implications in designing gas sensors. Prog Surf Sci 20:9
2. Batzill M, Diebold U (2005) The surface and materials science of tin oxide. Prog Surf Sci 79:47–154
3. Batzill M, Diebold U (2007) Surface studies of gas sensing metal oxides. Phys Chem Chem Phys 9:2307
4. Batzill M (2006) Surface science studies of gas sensing materials: SnO_2. Sensors 6:1345
5. Wöll C (2007) The chemistry and physics of zinc oxide surfaces. Prog Surf Sci 82:55
6. Chevtchenko SA, Moore JC, Özgür U, Gu X, Baski AA, Morkoc H, Nemeth B, Nause JE (2006) Comparative study of the (0001) and (000–1) surfaces of ZnO. Appl Phys Lett 89:182111
7. Lahiri J, Senanayake S, Batzill M (2008) Soft x-ray photoemission of clean and sulfur-covered polar ZnO surfaces: a view of the stabilization of polar oxide surfaces. Phys Rev B 78:155414
8. Dulub O, Diebold U, Kresse G (2003) Novel stabilization mechanism on polar surfaces: ZnO(0001)-Zn. Phys Rev Lett 90:016102
9. Bagus PS, Illas F, Pacchioni G, Parmigiani F (1999) Mechanisms responsible for chemical shifts of core-level binding energies and their relationship to chemical bonding. J Electron Spectrosc Relat Phenom 100:215
10. Kresse G, Dulub O, Diebold U (2003) Competing stabilization mechanism of the polar ZnO(0001)-Zn surface. Phys Rev B 68:245409
11. Önsten A, Stoltz D, Palmgren P, Yu S, Göthelid M, Karlsson UO (2010) Water adsorption on ZnO(0001): transition from triangular surface structure to a disordered hydroxyl terminated phase. J Phys Chem C 114:11157
12. Valtiner M, Torrelles X, Pareek A, Borodin S, Gies H, Grundmeier G (2010) In Situ study of the polar ZnO(0001)-Zn surface in Alkaline electrolytes. J Phys Chem C 114:15440
13. Kunat M, Gil Girol S, Becker T, Burghaus U, Wöll C (2002) Stability of the polar surfaces of ZnO: A reinvestigation using He-atom scattering. Phys Rev B 66:081402
14. Meyer B, Marx D (2003) Density-functional study of the structure and stability of ZnO surfaces. Phys Rev B 67:035403
15. Piper LFJ, Preston ARH, Fedorov A, Cho SW, DeMasi A, Smith KE (2010) Direct evidence of metallicity at ZnO(000–1)-(1 × 1) surfaces from angle resolved photoemission spectroscopy. Phys Rev B 81:233305
16. Ozawa K, Mase K (2010) Metallization of ZnO(10–10) by adsorption of hydrogen, methanol, and water: angle-resolved photoelectron spectroscopy. Phys Rev B 81:205322
17. Li L, King DL (2006) H2S removal with ZnO during fuel processing for PEM fuel cell applications. Catal Today 116:537–541
18. Lahiri J, Batzill M (2008) Surface functionalizationSurface functionalization of ZnO photocatalysts with monolayer ZnS. J Phys Chem C 112:4304
19. Knop-Gericke A, Kleimenov E, Haevecker M, Blume R, Teschner D, Zafeiratos S, Schloegl R, Bukhtiyarov VI, Kaichev VV, Prosvirin IP, Nizovskii AI, Bluhm H, Barinov A, Dudin P,

Kiskinova M (2009) X-Ray photoelectron spectroscopy for investigation of heterogeneous catalytic processes. Adv Catal 52:213

20. Zhang C, Grass ME, McDaniel AH, DeCaluwe SC, El Gabaly F, Liu Z, McCarty KF, Farrow RL, Linne MA, Hussain Z, Jackson GS, Bluhm H, Eichhorn BW (2010) Measuring fundamental properties in operating solid oxide electrochemical cells by using in situ X-ray photoelectron spectroscopy. Nature Mat. 9:944

21. Diebold U (2003) The surface science of titanium dioxide. Surf Sci Rep 48:53

22. Liang Y, Gan SP, Chambers SA, Altman EI (2001) Surface structure of anatase $TiO_2(001)$: reconstruction, atomic steps, and domains. Phys Rev B 63:235402

23. Chambers SA, Wang CM, Thevuthasan S, Droubay T, McCready DE, Lea AS, Shutthanandan V, Windisch CF (2002) Epitaxial growth and properties of MBE-grown ferromagnetic Co-doped TiO_2 anatase films on $SrTiO_3(001)$ and $LaAlO_3(001)$. Thin Solid Films 418:197

24. Dohnálek Z, Lyubinetsky I, Rousseau R (2010) Thermally-driven processes on rutile $TiO_2(110)$-1 × 1: a direct view at the atomic scale. Prog Surf Sci 85:161

25. Torrelles X, Cabailh G, Lindsay R, Bikondoa O, Roy J, Zegenhagen J, Teobaldi G, Hofer WA, Thornton G (2008) Geometric structure of $TiO_2(011)(2 × 1)$. Phys Rev Lett 101:185501

26. Gong X-Q, Khorshidi N, Stierle A, Vonk V, Ellinger C, Dosch H, Cheng H, Selloni A, He Y, Dulub O, Diebold U (2009) The 2 × 1 reconstruction of the rutile $TiO_2(011)$ surface: a combined density functional theory, X-ray diffraction, and scanning tunneling microscopy study. Surf Sci 603:138

27. He Y, Li W-K, Gong X-Q, Dulub O, Selloni A, Diebold U (2009) Nucleation and growth of 1D water clusters on Rutile TiO_2 (011)-2 × 1. Phys Chem C 113:10329–10332

28. Beck TJ, Klust A, Batzill M, Diebold U, Di Valentin C, Selloni A (2004) Surface structure of $TiO_2(011)$-(2 × 1). Phys Rev Lett 93:036104

29. Guo Q, Cocks I, Williams EM (1997) The orientation of acetate on a $TiO_2(110)$ surface. J Chem Phys 106:2924

30. Thevuthasan S, Herman GS, Kim YJ, Chambers SA, Peden CHF, Wang Z, Ynzunza RX, Tober ED, Morais J, Fadley CS (1998) The structure of formate on $TiO_2(110)$ by scanned-energy and scanned-angle photoelectron diffraction. Surf Sci 401:261

31. Gutierrez-Sosa A, Martinez-Escolano P, Raza H, Lindsay R, Wincott PL, Thornton G (2001) Orientation of carboxylates on $TiO_2(110)$. Surf Sci 471:163

32. Bates SP, Kresse G, Gillan MJ (1998) The adsorption and dissociation of ROH molecules on $TiO_2(110)$. Surf Sci 409:336

33. Onishi H, Iwasawa Y (1996) STM observation of surface reactions on a metal oxide. Surf Sci 357:773

34. Tao J, Luttrell T, Bylsma J, Batzill M (2011) Adsorption of acetic acid on rutile $TiO_2(110)$ vs (011)-2 × 1 surfaces. J Phys Chem C 115:3434

35. Henrich VE, Dresselhaus G, Zeiger HJ (1976) Observation of two-dimensional phases associated with defect states on the surface of TiO_2. Phys RevLett 36:1335

36. Henderson MA, Epling WS, Peden CHF, Perkins CL (2003) Insights into photoexcited electron scavenging processes on TiO_2 obtained from studies of the reaction of O_2 with OH groups adsorbed at electronic defects on $TiO_2(110)$. J Phys Chem B 107:534–545

37. Kurtz RL, Stockbauer R, Madey TE, Roman E, Desegovia JL (1989) Synchrotron radiation studies of H_2O adsorption on $TiO_2(110)$. Surf Sci 218:178

38. Zhang Z, Jeng S-P, Henrich VE (1991) Cation-ligand hybridization for stoichiometric and reduced TiO_2 (110) surfaces determined by resonant photoemission. Phys Rev B 43:12004

39. Bao S, Liu G, Woodruff DP (1988) Angle-resolved polarised light photoemission study of the formation and structure of acetate on Cu(110). Surf Sci 203:89

40. Karis O, Hasselstrom J, Wassdahl N, Weinelt M, Nilsson A, Nyberg M, Pettersson LGM, Stohr J, Samant MG (2000) The bonding of simple carboxylic acids on Cu(110). J Chem Phys 112:8146

41. Tao J, Batzill M (2010) Role of surface structure on the charge trapping in TiO_2 photocatalysts. J Phys Chem Lett 1:3200
42. Redinger A, Hansen H, Linke U, Rosandi Y, Urbassek HM, Michely T (2006) Superior regularity in erosion patterns by planar subsurface channeling. Phys Rev Lett 96:106103
43. Redinger A, Rosandi Y, Urbassek HM, Michely T (2008) Step-edge sputtering through grazing incidence ions investigated by scanning tunneling microscopy and molecular dynamics simulations. Phys Rev B 77:195436
44. Luttrell T, Batzill M (2010) Nanoripple formation on $TiO_2(110)$ by low-energy grazing incidence ion sputtering. Phys Rev B 82:035408
45. Gong XQ, Selloni A, Batzill M, Diebold U (2006) Steps on anatase $TiO_2(101)$. Nat Mater 5:665
46. Luttrell T, Li WK, Gong XQ, Batzill M (2009) New directions for atomic steps: step alignment by grazing incident ion beams on $TiO_2(110)$. Phys Rev Lett 102:166103
47. Li M, Hebenstreit W, Gross L, Diebold U, Henderson MA, Jennison DR, Schultz PA, Sears MP (1999) Oxygen-induced restructuring of the $TiO_2(110)$ surface: a comprehensive study. Surf Sci 437:173
48. Bowker M, Bennett RA (2009) The role of Ti^{3+} interstitials in $TiO_2(110)$ reduction and oxidation. J Phys Condens Matt 21:474224
49. Stone P, Bennett RA, Bowker M (1999) Reactive re-oxidation of reduced $TiO_2(110)$ surfaces demonstrated by high temperature STM movies. New J Phys 1:8
50. McCarty KF (2003) Growth regimes of the oxygen-deficient $TiO_2(110)$ surface exposed to oxygen. Surf Sci 543:185
51. Tao J, Luttrell T, Batzill M (2011) A two-dimensional phase of TiO_2 with a reduced bandgap. Nat Chem 3:296
52. Gurlo Nanosensors A (2010) Does crystal shape matter? Small 6:2077

Chapter 3
Design, Synthesis and Application of Metal Oxide-Based Sensing Elements: A Chemical Principles Approach

Valery Krivetskiy, Marina Rumyantseva and Alexander Gaskov

Abstract The chemical approaches to improvement of selectivity of semiconductor metal oxide gas sensors are the main subject of this chapter. Current concepts of interrelationships between metal oxide chemical composition, crystal and surface structure and its activity in the reaction with gas phase components are considered. Application of such concepts to the design of sensor materials based on nanocrystalline SnO_2 is discussed thoroughly. Experimental data concerning chemical composition, solid–gas chemical interaction activity and sensor properties is given and critically analysed. The possibility of utilization of solid–gas chemical reaction activity concepts for directed synthesis of new metal oxide semiconductor sensor materials with selective response to given gases is highlighted.

3.1 Introduction

A major shortcoming of SnO_2 as a material for gas sensors is its low selectivity, due to the presence of a wide range of adsorption sites on its surface that cannot distinguish the contribution of each type of molecule in the gas phase to the total electrical signal. One of the ways to improve its selectivity is the surface modification of a highly dispersed oxide matrix with clusters of transition metals or their oxides, which may affect the electrophysical and chemical properties of the surface. Quite a number of approaches has been suggested to date to direct the choice of modifiers for sensor materials, using the values of electronegativity [1], changes in work function [2], as well as structural properties of oxides of d-elements [3] as correlation parameters.

For sensors targeting sulfur-containing gases (in particular hydrogen sulfide) it was proposed [1] to select a modifier based on the electronegativity values of the

V. Krivetskiy · M. Rumyantseva · A. Gaskov (✉)
Lomonosov Moscow State University, Moscow, Russia
e-mail: gaskov@inorg.chem.msu.ru

M. A. Carpenter et al. (eds.), *Metal Oxide Nanomaterials for Chemical Sensors*, Integrated Analytical Systems, DOI: 10.1007/978-1-4614-5395-6_3,

corresponding cations. It was assumed that the bond formation between the sulfur ion and metal cation with low electronegativity leads to a weakening of the S–H bond and facilitates the dissociation of the molecule. Another approach is based on the change in work function during the chemisorption of molecules from the gas phase on metal surfaces [2]. The magnitude of this change depends on the nature of the metal and the gas molecules, and likely enables the selective detection of gases using planar structures of a metal/semiconductor interface, as well as with polycrystalline systems in which metal clusters are distributed on the surface of the semiconductor.

Certain predictions on sensitivity and selectivity of sensor materials can be made based on the catalytic activity of transition metals and their oxides. Possible mechanisms for the catalytic effect of additives on the sensor properties of the material were proposed [4]:

- the effect on the rate of Redox reaction, which leads to higher sensitivity and better dynamic properties at low operating temperatures;
- formation of intermediates, which can change the conductivity of SnO_2;
- the impact on the surface coverage and rate of adsorption.

Different concepts of heterogeneous catalysis allow linking the physical, chemical and electronic properties of materials and their catalytic activity in different processes [5]:

- acid–base catalysis theory, based on the number and strength of acidic and basic centers on the catalyst surface;
- the theory of "structural matching" assumes that an active center—an ensemble of atoms having some specific size and structure—is needed for the reaction on the surface to take place;
- the electronic theory of catalysis on semiconductors [6], the concept, based on the correlation of catalytic activity and electronic structure of the material, which includes consideration of the quantum-size effects with decreasing particle size to the nanometer range.

However, there is scarce number of works that use the reasoned choice of modifiers for sensor materials surfaces. In the development of SnO_2-based gas sensors the representatives of all families of chemical elements: noble metals and oxides of s-, p-, d- and f-elements, have already been tested as modifiers (Fig. 3.1). In most cases, the results are obtained by "trial and error" effort and do not imply correlations between the properties of the modifier and sensor performances to the target gas.

3.2 Reactivity of Oxide's Surface in Solid-Gas Interaction

This section discusses the approaches described in the literature to assess the acid–base and Redox surface properties of metal oxides, based on their chemical composition, oxidation state of the metal, coordination environment and crystal structure.

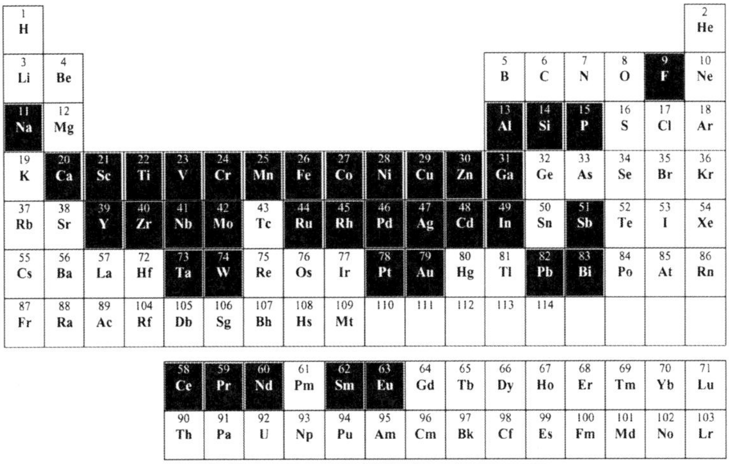

1	2	3	4	5	6	7	8	9	10	11	12	13	14	15	16	17	18
H																	He
Li	Be											B	C	N	O	F	Ne
Na	Mg											Al	Si	P	S	Cl	Ar
K	Ca	Sc	Ti	V	Cr	Mn	Fe	Co	Ni	Cu	Zn	Ga	Ge	As	Se	Br	Kr
Rb	Sr	Y	Zr	Nb	Mo	Tc	Ru	Rh	Pd	Ag	Cd	In	Sn	Sb	Te	I	Xe
Cs	Ba	La	Hf	Ta	W	Re	Os	Ir	Pt	Au	Hg	Tl	Pb	Bi	Po	At	Rn
Fr	Ra	Ac	Rf	Db	Sg	Bh	Hs	Mt	110	111	112	113	114				

Ce	Pr	Nd	Pm	Sm	Eu	Gd	Tb	Dy	Ho	Er	Tm	Yb	Lu
Th	Pa	U	Np	Pu	Am	Cm	Bk	Cf	Es	Fm	Md	No	Lr

Fig. 3.1 Elements that have been tested as modifiers in the development of SnO_2 based gas sensors

3.2.1 Adsorption–Desorption Interactions, Acid–Base Properties

The key chemical process that leads to the emergence of a sensor response in the presence of any detectable component is the adsorption (chemisorption) of gas molecules on the surface. This process is usually considered in terms of a Lewis acid–base interaction [7]. The active sites of adsorption on a metal oxide surface are coordinatively unsaturated metal cations and oxygen anions. Metal cations as electron-deficient atoms having vacant orbitals, exhibit Lewis acid properties, while oxygen anions act as bases.

The main problems in the synthesis of new materials with a well defined surface acidity are (1) a quantitative description of the changes in acidity made by modification and especially (2) predicting the reactivity of the newly created composites. The advances in IR-spectroscopic methods of probing molecules [8, 9], their thermo-programmed desorption [10] and microcalorimetric methods [11], allow a quantitative comparison of the acidity of the synthesized materials. It is important to emphasize that the acidity of the metal cation in the oxide surface is defined by a set of parameters: the type and oxidation state of the cation, the presence of defects on the surface, and the number of vacancies in the anion sublattice, which is different in the oxidized and reduced state of the surface [7]. At the same time, the acidity and reactivity of the metal oxide surface are closely associated with the "bulk" properties of the phases present [5].

In the 1970s Duffy proposed a universal approach which, allows ranking of various metal oxides on their Lewis acidity and introduced a new notion of the so called optical basicity of oxides [12]. This value characterizes the degree of "red" shift (towards longer wavelengths) of characteristic bands in UV absorption spectra corresponding to the $^1S_0 \rightarrow {}^3P_1$ transitions of probe ions (Tl^+, Pb^{2+} or Bi^{3+}), embedded in an oxide structure. This shift is caused by the expansion of the outer electron orbitals of the cations (acidic centers), surrounded by anions (basic centers). This so called nephelauxetic effect is based on the covalent interaction between the central cation and its surrounding ligands, which leads to the formation of molecular orbitals with a predominant contribution of orbitals from the central cation. The expansion is due to the fact that the formed boundary molecular orbitals are antibonding in nature, and also due to increased screening of the positively charged nucleus of the cation. On the example of complex-forming metal cations, this effect is manifested as a change in the absorption spectra corresponding to the energy transitions of d-electrons. Orbital expansion is accompanied by the weakening of electron repulsion forces and, consequently, the partial degeneracy of the energy sublevels of the excited states surrounded by the ligands (Lewis bases), one can estimate the degree of this acid–base interaction. According to this concept, the bulk Lewis acidity of the cations forming the oxide can be estimated to be inversely proportional to the basicity of oxygen anions. It can be determined by measuring the magnitude of the nephelauxetic effect arising in its absorption spectrum with introducing a probe d-element cation into the oxide matrix. The more the red shift of spectral bands caused by the expansion of the outer orbitals of the cation, the more the basicity of oxide and the weaker the Lewis acidity of its constituent cations.

By generalizing the experimental data obtained in the course of studying the optical basicity of silicate oxide glasses of different compositions, it has been shown [13] that the optical basicity Λ is an additive quantity, and therefore, the majority of individual oxides can be classified based on measured values of optical basicity of glasses of known composition:

$$\Lambda_{th} = X_{AO_{a/2}}\Lambda(AO_{a/2}) + X_{BO_{b/2}}\Lambda(BO_{b/2}) + \ldots, \tag{3.1}$$

$\Lambda(AO_{a/2}), \Lambda(BO_{b/2})\ldots$ optical basicity of oxides $AO_{a/2}$, $BO_{b/2} \ldots$
$X_{AO_{a/2}}, X_{BO_{b/2}}\ldots$ proportion of oxygen atoms, provided by each oxide

Optical basicity of simple oxides, of the substance, is correlated with other fundamental parameters. It has been shown [14] that Λ is proportional to the electronegativity of the metal, which is forming the oxide:

$$\Lambda = 0.75/(\chi_M - 0.25) \tag{3.2}$$

χ_M Pauling electronegativity of metal cation

Another type of correlation is related to the so-called polarizability of the oxygen atom in the oxide, α_O^{2-} [13]:

$$\Lambda = 1.67\left(1 - 1/\alpha_O^{2-}\right) \tag{3.3}$$

The polarizability of the oxygen atom, reflecting the tendency of electrons involved in chemical bonding to move from of the ionic core in the presence of an external electric field, is calculated from the molar polarizability of simple oxides:

$$\alpha_m\left(A_i B_j\right) = i\alpha_A + j\alpha_B, \tag{3.4}$$

A   :cation
B   :anion
i and j :their stoichiometric coefficients in oxide

Molar polarizability of oxide can be calculated from the refractive index of the oxide according to the formula of Lorentz-Lorenz (also known as Clausius-Mozotti equation) [13, 15]:

$$\alpha_m = \frac{3V_m}{4\pi N}\frac{n^2 - 1}{n^2 + 2} \tag{3.5}$$

V_m -molar volume
N -Avogadro number
n r-efractive index

Thus, it is possible to estimate the strength of acid centers on the surface of an oxide material using the bulk properties of the compound and the fundamental parameters of its constituent atoms. This simple approach has its drawbacks. The more important one is that the calculations are based on Pauling electronegativity values of the cations with no regard of the structural features of oxides and, consequently, neglects their effect on the electronegativity. As a result this approach was suitable only for compounds with ionic metal—oxygen bonds.

In 1982 Zhang for the first time made an attempt to create a quantitative Lewis acidity scale of various cations [16]. He found the approach to the quantitative description of the ion contribution to the metal—oxygen bond through the so-called polarizing power of the cation. The polarizing power P can be calculated as:

$$P = Z/r_k^2 \tag{3.6}$$

Z formal cation charge
r_k ionic radius

At the same time Zhang has developed a scale of electronegativities of the elements [17], calculated from the thermodynamic parameters using the formula:

$$\chi_Z = \frac{0.24n^*(I_z/R)^{1/2}}{r^2} + 0.775 \tag{3.7}$$

r covalent metal radius
I_z ionization potential

R   Rydberg constant

n^* effective main quantum number

Covalent components of the metal–oxygen bond in the metal oxide compounds, according to Zhang, are determined by their electronegativity. The value of Lewis acidity Z is determined by the formula:

$$Z = P - 7.7\chi_Z + 8.0 \tag{3.8}$$

Zhang was able to rank a large number of cations. However, this scale is not valid for small cations with high nuclear charge. Portier et al. [18] proposed a new parameter related to acid strength, taking into account the ionic-covalent character of the metal–oxygen bond in oxides (ionic-covalent parameter, ICP):

$$ICP = \log P_i - A\chi_i + B \tag{3.9}$$

P_i polarizing power

χ_i electronegativity

The coefficients A and B are chosen, so that the ICP of Au^+ cation was zero [19]. This cation acts as a reference due to its extremely low acidity and instability of compounds, even with very strong bases. Using this parameter authors have been able to carry out a quantitative classification of Lewis acidity of about 500 cations in different oxidation states and coordination environments.

Further attempts were made to develop the concept of optical basicity on the basis of a quantitative understanding of the ionic-covalent character of metal–oxygen bond in oxides. These approaches are based on the linear correlation of the optical basicity of alkali and alkaline earth metal oxides with ICP. Using the method of linear regression five different linear correlations between the optical basicity and the ionic-covalent parameter of different types of cations (s-p, $d^{10}s^2$, d^0, d^1–d^9, d^{10}) were subsequently established [20]. The quantitative relationship between the Lewis acidity and the fundamental parameters of the substance identified in this way can be used to predict the reactivity of not only the individual oxides, but also composite materials based on them Fig. 3.2.

Another approach that is widely used in predicting the reactivity of oxide materials, is the principle of electronegativity equalization developed by Sanderson, and the corresponding scale of electronegativities of the elements [21, 22]. Sanderson started from the principle that the formation of the ionic part in the chemical bonding between atoms of different nature is inevitably accompanied by a decrease in its covalent component. This is due to the redistribution of electron density between the atoms that form a chemical compound. This redistribution, in turn, induces a charge which balances the difference in the electronegativity of atoms. Thus, the compound acquires certain electronegativity, which is the geometric mean of electronegativities of its constituent atoms.

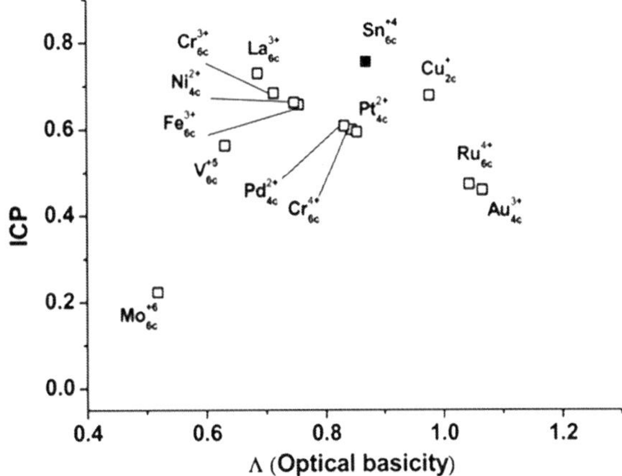

Fig. 3.2 The scale of optical basicity of selected cations based on their oxidation state, coordination number and electronegativity [20]

Sanderson's concept allows us to calculate the partial charges on the atoms forming the compound, reflecting the contribution of the ionic component of a chemical bond directly from the values of the electronegativities of the cations.

$$\sigma_M = 0.48 \times \left(\sqrt[x+y]{(S_{M^+})^x (S_{Ch})^y} - S_{M^+} \right) \times S_{M^+}^{-1/2} \tag{3.10}$$

σ_M cation partial charge
S_{M^+} cation electronegativity
S_{Ch^-} chalcogen electronegativity (for oxygen $S = 3.61$)
x, y   stoichiometric coefficients

To date, these values are known for a large number of cations in various formal oxidation states, so one can easily rank the corresponding oxides by their Lewis acidity (Fig. 3.3). The most comprehensive work in the field [23] has shown that the difference between the degree of oxidation of the cation and the calculated partial charges reflects the value of Lewis acidity. This approach was also developed further to calculate partial charge distributions in complex oxide compounds on the basis of their structural parameters [24].

It is important to note that metal oxides possess, in addition to Lewis acidity, also Brønsted acidity, defined as the ability to protonate bases (H^+-transfer). This type of acidity is the result of dissociative adsorption of water molecules on the oxide surface to form protons (H^+) and hydroxyl groups (OH^-). The driving force of these processes is the interaction between a Lewis acid (cation on the metal surface) and its conjugate base (the oxygen atom in a molecule of H_2O). A proton (Lewis acid), formed as a result of this dissociation, is captured and tightly binds the surface lattice oxygen atom of the metal oxide (Lewis base) while a hydroxyl group (base) is fixed to the metal cation (acid). It is the latter type of hydroxyl groups on the surface that are capable to exhibit acidic properties. The strength of acid sites is determined by the energy of deprotonation, which is fairly easy to

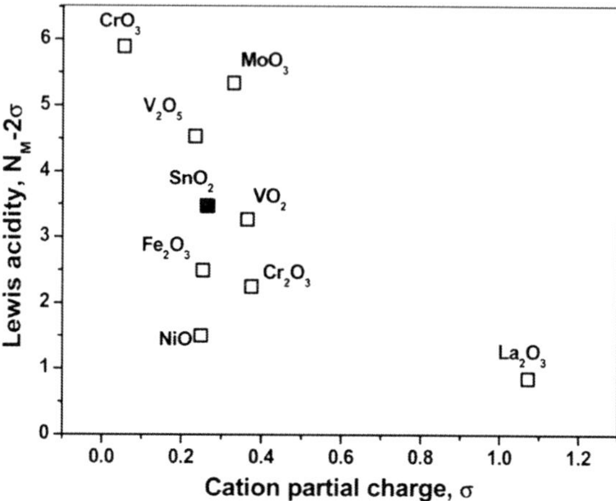

Fig. 3.3 Lewis acidity scale of simple oxides calculated on the basis of Sanderson electronegativity table [23]

quantify with the help of quantum chemical calculations [25–27]. This method is often used for catalytic materials that have shown linear correlation between the strength of Brønsted acid centers and the partial electric charges on the atoms (oxygen anions and metal cations) [24].

3.2.2 Reactivity in Oxidation Process

In most cases sensors are designed to operate in excess of oxygen in the gas phase (mainly in humid air) which means that for the majority of molecules, adsorbed on the SnO_2 surface, the next chemical step will be the interaction with different oxygen species. They are represented as chemisorbed forms of oxygen: molecular O_2^- or atomic O^-, or lattice, oxygen on the SnO_2 surface. The former are known as highly reactive electrophilic forms of oxygen, readily interacting with almost all adsorbed molecules, while the latter, lattice oxygen, considered to be less reactive nucleophilic form of oxidant. While it is widely accepted, that the concentration of chemisorbed oxygen species is the main parameter, affecting nanocrystalline SnO_2 conductance in air, little is reported about their relative importance in the process of chemical interaction of SnO_2 with gas molecules.

3.2.2.1 Oxidation by Lattice Oxygen

Oxidation of chemisorbed molecules with lattice oxygen of the oxide material was firstly described by Mars and van Krevelen [28]. For a large number of oxide systems this mechanism has been confirmed by kinetic methods as well as using direct methods of investigation of the oxides surface structure [29–31]. This process is due to lability of nucleophilic anions, which can attack the adsorbed molecules, breaking the existing chemical bonds in these molecules and forming the new ones. The adsorption of gas molecules, which occurs with a partial transfer of electron density to the metal cation on the surface, leads to the activation of these adsorbed species—the creation of additional centers with a partial positive charge, which later may become centers for nucleophilic attack by the anion O^{2-}, located in position of lattice oxygen of the oxide compound on the surface of the material. Newly formed molecules can re-enter the process of oxidation or desorb from the oxide surface. As a result, the surface oxide is transferred into the so-called "reduced" state, characterized by a large number of oxygen vacancies and metal cations in lower oxidation states as compared with the cations in the bulk material.

If the gas phase contains a significant amount of oxygen, then re-oxidation of surface occurs through the processes: adsorption of molecular oxygen, its dissociation, and incorporation into the position of oxygen vacancies, in accordance with the following equations:

$$O_{2(g)} \leftrightarrow O_{2(ads)} \tag{3.11}$$

$$O_{2(ads)} + e^- \leftrightarrow O_{2(ads)}^- \tag{3.12}$$

$$O_{2(ads)}^- + e^- \leftrightarrow 2O_{(ads)}^- \tag{3.13}$$

$$O_{(ads)}^- + e^- \leftrightarrow O_{(ads)}^{2-} \tag{3.14}$$

$$O_{(ads)}^{2-} \leftrightarrow O_{lat}^{2-} \tag{3.15}$$

where O_{lat}^{2-} is an anion located in position of lattice oxygen of the oxide compound on the surface of the material.

Thus there is a restoration of the original chemical state of the surface and the completion of the catalytic cycle of oxidation. Processes involving nucleophilic oxygen species can be described by the following general schemes:

$$\text{Oxidation} \quad M^{n+} - O^{2-} + R \rightarrow M^{(n-x)+} + RO \tag{3.16}$$

$$\text{Regeneration} \quad M^{(n-x)+} + (x/4)O_2 \rightarrow M^{n+} + (x/2)O^{2-} \tag{3.17}$$

The use of these schemes allowed the authors [32] to describe the activity of oxides in a catalytic cycle on the basis of thermodynamic parameters. Although such a fundamental approach has revealed the energy functions that characterize the reactivity of the oxides in Redox processes, the authors could not find parallels with the real catalytic activity of oxide compounds even in very simple transformations. The reason for the failure of this, at first glance a promising concept, was the large number of additional factors influencing the Redox processes on the oxide surface, which are not taken into account in the proposed scheme of oxidation. These factors are hard to account for in a quantitative way so they are mainly used as qualitative concepts.

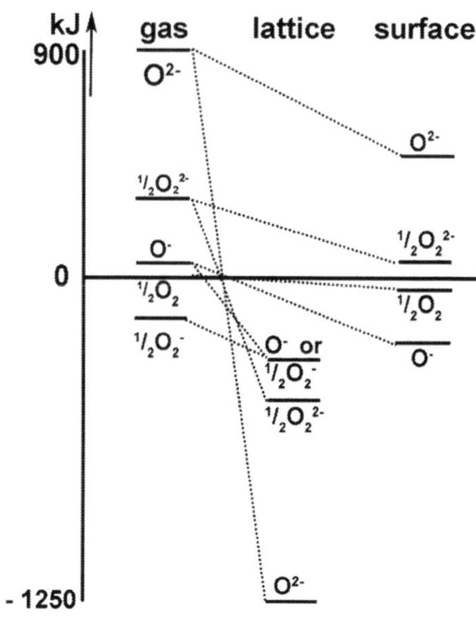

Fig. 3.4 Survey of oxygen species (From [87]. With permission.)

3.2.2.2 Oxidation via Adsorbed Oxygen

It should be taken into account that oxidation by nucleophilic lattice O^{2-} is not the only kind of oxidation reactions on the surface of oxides. The adsorption of oxygen in accordance with equations (3.11–3.14) can lead to the formation of electrophilic anions, O_2^{-} and O^{-}, on the oxide surface. The high mobility and reactivity of such species leads to oxidation processes on the surface with chemisorbed gas molecules, not subject to the "Mars-van-Krevelen" mechanism.

The nature, mechanisms of formation and chemical properties of such electrophilic oxygen anions, adsorbed on oxide surfaces, which are the subject of decade-long discussions, are still poorly known. According to the energy diagram, proposed by the authors [33], the higher the charge of the ionized forms of adsorbed oxygen, the lower its stability Fig. 3.4.

The ionized form of oxygen O_2^{-} in the gas phase is energetically more favorable compared to the electrically neutral molecular form. The latter can be stabilized on the oxide surface via electrostatic interaction. This chart suggests that the ionized O_2^{-} and O^{-} can exist in a mobile form on the oxide surface for a long time. The O^{2-} ion can be stabilized on the surface only in the position of the oxygen vacancy, since it is energetically very unstable species. The numerical values listed in the chart, are approximations, since they are dependent on the exact position of the ionized atoms of oxygen adsorbed on the surface, as well as their surroundings.

The ability of the oxide surface to bear chemisorbed oxygen species can only be described in a qualitative manner. For these purposes a division of oxides into three groups was proposed [34]:

The first group with a high tendency to adsorb oxygen is presented by oxides formed by the metal cations which are easily oxidized into a higher oxidation state, and therefore can easily donate electron density to the adsorbed oxygen species. A huge quantity of electrophilic anions (O^-, O_2^-) quickly form on the surface of these oxides (Examples include: NiO, MnO, CoO, Co_3O_4) at room temperature in an atmosphere with oxygen excess. With the rise of surface temperature the speed of this transformation increases considerably, and also adsorbed O^{2-} is formed, which is involved in the formation of additional lattice planes on the surface with utilization of cations, migrating from the bulk to the surface.

The second group of oxides (ZnO, TiO_2, V_2O_5, SnO_2) also possess the ability of donating electron density to the adsorbed oxygen molecules, arising primarily due to bulk non-stoichiometry within the oxygen lattice. The main difference between these two groups of oxides is the latter contains a relatively small number of active centers on the surface, through which the electron density transfer to the adsorbed oxygen molecules could pass. Formation of adsorbed O^{2-} ions is far less probable, whereas almost all the ionized particles of oxygen adsorbed on the surface are represented by O^- and O_2^-. The ability of the oxide surface to stabilize electrophilic adsorbed oxygen species may largely depend on the structural factors, namely the type of coordination environment of the metal cation and the quantity of surface oxygen vacancies. For this reason, the nature and structural features of the defects on the surface of such materials, which includes nanocrystalline SnO_2, thoroughly discussed in this chapter, can play a crucial role in the interaction of gas molecules and the subsequent sensing characteristics of metal oxide nanomaterials.

The third group of oxides is represented by compounds that do not have the ability to donate electron density to adsorbed oxygen molecules and therefore stable adsorbed electrophilic ionized forms of oxygen do not form on the stoichiometric surface of these oxides. Examples of such oxides are different molybdates, tungstates, vanadates, and other complex oxides with cations of transition elements in their highest oxidation state. Oxidation of gas molecules on their surface, as well as brief formation of adsorbed electrophilic oxygen species, can be described as a mechanism of the "Mars-Van-Krevelen"-type. Formed on the reduced surface of such oxide systems, electrophilic forms of oxygen very quickly become the nucleophilic form O^{2-}, occupying the oxygen vacancies in the crystallographic positions.

This classification is not valid for non-transition metal oxides, in which cations can exist in only a single oxidation state, and therefore do not possess electrondonor activity. The use of such oxides as heterogeneous oxidation catalysts requires extra effort to create defects in their crystal structure [34] and oxygen vacancies (F-centers) on the surface. They act as centers of adsorption and formation of electrophilic active oxygen species. Such oxides can be attributed to group II described above.

Currently, there are only a limited number of works devoted to a detailed study of the mechanisms of formation and reactivity of electrophilic forms of adsorbed oxygen on the oxide surface [35]. It is believed that these species are much more reactive than the nucleophilic lattice ions O^{2-} on the surface and have a much greater mobility. For this reason, oxides, included in the first group mentioned above, and materials thereof, are most commonly used as catalysts for the complete and non-selective oxidation of adsorbed molecules on the surface [36]. At the same time, a clear consensus on the selectivity of oxidation processes involving electrophilic forms of adsorbed oxygen on the surface of oxides in the scientific community is currently unavailable. In some studies the role of adsorbed electrophilic O^- ions in the selective oxidation of organics is underlined [37, 38].

3.2.3 Implications for Materials Design

The synthesis of materials with well defined activity in chemical interactions with gas molecules, should be based on a theory that quantitatively describes the above mentioned considerations in clearly defined physical concepts. Such a theory so far does not exist, due to a lack of analytical techniques to examine each elementary step of interaction of adsorbed molecules on solid surfaces. These in situ experiments, including identification of the dynamic structure transformation of solid surfaces in real conditions of chemical interactions with gases, are particularly complicated in the case of oxide systems [39]. At the same time the theoretically calculated equilibrium state of the surface and molecular complexes formed on it during reaction may indeed differ from the real ones. Moreover, measurement of the kinetics of these transformations generally is beyond the capacity of modern theoretical calculations.

These reasons are responsible for the impossibility of a sensing material's design from first principle calculations. Thus, the theoretical basis for the development of new oxide materials with improved properties in Redox reactions (higher activity and/or selectivity)—is still a huge accumulated empirical experience, combined with the qualitative concepts of reactivity given above. Therefore, one of the possible approaches for a partially directed synthesis of new oxide materials is a modification of the existing oxide systems by the introduction of additives [40]. Such modifiers may occupy different positions, both on the surface of the material and in its structure, providing a complex effect on its activity in interaction with the gas phase [41]. Depending on the concentration and type of introduced additional components, they can form solid solutions based on the main phase or create a segregation of the modifier on the surface of crystalline grains of the main phase. In some cases the stabilization of thermodynamically unstable phases can be observed. So, the modifiers cause changes in the parameters of the crystal structure itself, as well as in related properties (for example energy and length of the metal–oxygen bond), and also change the concentration of point defects, both in the cation and anion sublattices. Being distributed on the surface, these additives can also change the local structure and properties of active sites, through formation their own two-dimensional oxide structure and thus "block" the pre-existing active sites.

To check the validity of the proposed approach to selected modifiers that provide increased sensitivity and selectivity of tin dioxide towards gases of different chemical nature, we experimentally examined the modifiers with variable acid–base properties and activity in the oxidation process such as noble metals and metal oxides: Au, PdO_x, RuO_2, NiO, CuO, Fe_2O_3, La_2O_3, V_2O_5, MoO_3, Sb_2O_5.

3.3 Influence of Modifiers on SnO$_2$ Surface Acidity

Thermoprogrammed desorption of ammonia (TPD–NH_3), along with methods of microcalorimetry and IR spectroscopy is one of the most commonly used methods to assess the acidity of the surface—the strength and the concentration of both Lewis and Brønsted centers. The number of acid sites and their distribution over the activation energy of desorption of ammonia can be carried out using a model [42] with the assumption that one molecule of NH_3 desorbs from a single acid site.

For most of the materials, discussed in this chapter, two types of adsorption sites on the surface, possessing different acidity, were found—weak ($T_{des} \sim 120\ °C$) and strong ($T_{des} \sim 480\ °C$) (Fig. 3.5). Weak acid–base interaction of NH_3 with the material's surface corresponds to the Brønsted acidity centers—the hydroxyl groups on the surface which formed partly due to the adsorption and dissociation of molecular water from the air. It is reasonable, however, to judge that the number of these hydroxyl groups is governed by the heat treatment conditions of the precursor—α-stannic acid [43]. Complete removal of water is achieved at temperatures above 600 °C, which is reflected in a gradual decrease in the desorption peak in the spectra of TPD-NH_3 at 120 °C, (Acquisition of this TPD followed the sample pre-treatment for 1 h at 300 °C in dry air with subsequent cooling down to room temperature in same gas flow; saturation with 10 % NH_3 in N_2 for 0.5 h and further TPD in a N_2 flow), for materials based on pure tin dioxide produced by higher annealing temperatures in an air atmosphere. Strong adsorption sites are coordinatively unsaturated metal cations on the surface of materials, which have the capacity to interact with Lewis bases due to unfilled positions in an oxygen coordination. There are two types of such cations on the surface—Sn^{4+} with 5 oxygen atoms in the coordination sphere and Sn^{2+} with 4 oxygen atoms in the anion environment. In this case, cations Sn_{4c}^{2+} have greater bond strength with adsorbed NH_3 molecules due to the additional contribution of covalent interactions [44]. Displacement of the TPD-NH_3 maxima at higher temperatures with increasing annealing temperature is probably due to the effects of long-range order, manifested in a decrease in the degree of crystal lattice defects on the surface of the grains. For this reason, a valid comparison between Brønsted and Lewis acidity of the surface of modified materials with each other is only possible for substances produced using the same annealing temperature and with the same annealing gas conditions. Concentrations of Brønsted and Lewis acid sites for the materials under discussion are presented in Table 3.1.

Introduction of modifiers with lower optical basicity (higher Lewis acidity of cations) to the SnO_2 matrix and vice versa is accompanied, as in the case of Brønsted acid centers, with the corresponding change in their concentration on the surface (Fig. 3.6). We emphasize that such a law is valid only for materials with

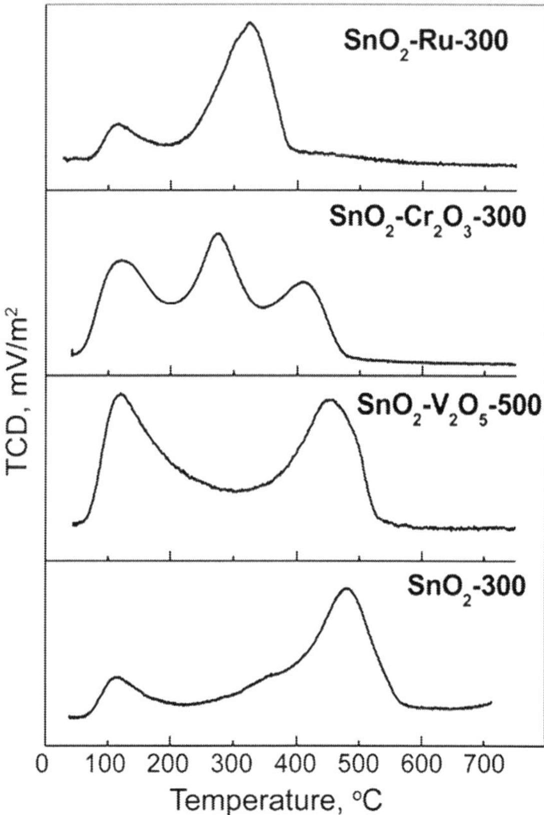

Fig. 3.5 Examples of TPD-NH$_3$ spectra of SnO$_2$-based sensor materials

the same thermal treatment, indicating the decisive contribution of crystal lattice defects on the surface in the amount of acid sites.

For a number of materials (SnO$_2$–Cr$_2$O$_3$, SnO$_2$–Pd and SnO$_2$–Ru) the desorption peak at about 300 °C instead of 480 °C was observed. One can attribute the peaks at 300 °C not to the desorption of ammonia, but to the products of NH$_3$ dissociation and oxidation (molecular nitrogen or nitrogen oxides). This is due to pronounced catalytic activity of deposited clusters of noble metals, manifested in many oxidation reactions of adsorbed molecules with oxygen. Also the possibility to create the additional chromyl oxygen anions (Cr = O) on the surface of Cr$_2$O$_3$, actively involved in the oxidation of adsorbed molecules [45, 46], may be the reason of such NH$_3$-TPD dependencies in the case of Cr-modified SnO$_2$.

Table 3.1 Parameters of SnO$_2$-based semiconductor sensor materials

Sample[a]	T_{an} °C	d_{XRD}, nm	S, m^2/g	Brønsted acidity, µmol NH$_3$/m^2	Lewis acidity, µmol NH$_3$/m^2
SnO$_2$-300	300	4	65	0.554 ± 0.05	3.1 ± 0.3
SnO$_2$-500	500	9	27	0.147 ± 0.01	1.58 ± 0.2
SnO$_2$-700	700	13	15	0.135 ± 0.01	1.03 ± 0.1
SnO$_2$-Fe$_2$O$_3$-500	500	5	40	0.385 ± 0.04	3.73 ± 0.4
SnO$_2$-V$_2$O$_5$-500	500	6	42	1.56 ± 0.15	7.05 ± 0.7
SnO$_2$-Cr$_2$O$_3$-500	500	9	40	1.64 ± 0.16	–
SnO$_2$-MoO$_3$-500	500	5	75	0.915 ± 0.09	2.22 ± 0.2
SnO$_2$-Cr$_2$O$_3$-300	500	4	114	1.11 ± 0.1	–
SnO$_2$-Pd-500	500	8	27	–	–
SnO$_2$-Pd-300	300	4	61	0.633 ± 0.06	–
SnO$_2$-Au-300	300	4	60	0.188 ± 0.01	1.96 ± 0.2
SnO$_2$-NiO-Au-350	350	5	75	0.457 ± 0.05	1.86 ± 0.2
SnO$_2$-Pt-500	500	13	15	0.673 ± 0.07	–
SnO$_2$-Ru-300	300	5	67	0.334 ± 0.03	–
SnO$_2$-Ru-700	700	20	13	–	–
SnO$_2$-La$_2$O$_3$-700	700	5	27	0.032 ± 0.03	1.53 ± 0.2
SnO$_2$-CuO-700	700	13	15	–	–
SnO$_2$-Sb-300	300	4	71	0.76 ± 0.08	2.33 ± 0.2

[a] Sample naming: Base material-Modifier-Calcination temperature

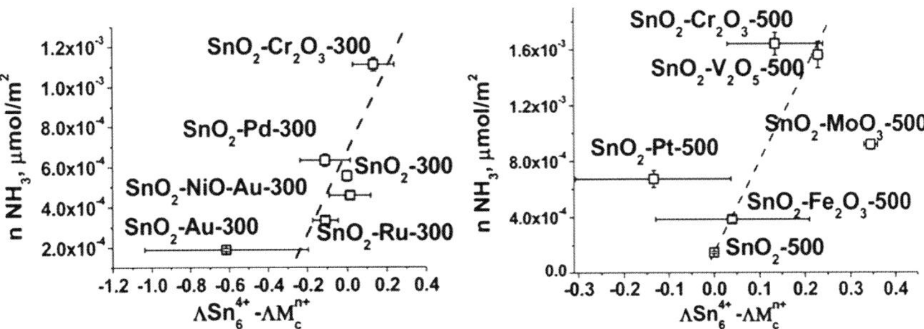

Fig. 3.6 Density of brønsted-type acid sites on the surface of differently modified nanocrystalline SnO$_2$ calcined at 300 °C (*left*) and 500 °C (*right*)

3.4 Redox Properties of Modified Materials

3.4.1 Chemisorbed Oxygen

Information regarding the Redox properties of surface modified tin dioxide can be obtained by the in situ investigation of a material's interaction with oxygen and temperature-programmed reduction within a H_2 environment.

The dependence of the SnO_2 electrical conductivity on oxygen partial pressure was investigated in detail in a wide temperature range [47, 48]. The electrical behavior of these semiconductor oxides at low temperatures is controlled by acceptor impurities—various species of chemisorbed oxygen. At 200 – 400 °C the interaction of atmospheric oxygen with an n-type semiconductor oxide surface leads to the formation of molecular (O_2^-) and atomic (O^-, O^{2-}) chemisorbed species. In general, the oxygen chemisorption can be described as [49]:

$$\frac{\beta}{2} O_{2(gas)} + \alpha \cdot e^- + S \leftrightarrow O_{\beta(ads)}^{-\alpha}, \qquad (3.18)$$

where $O_{2(gas)}$ is an oxygen molecule in the ambient atmosphere, e^- is an electron, which can reach the surface, meaning it has enough energy to overcome the electric field resulting from the negative charging of the surface. Their concentration is denoted as n_S; $n_S = [e^-]$; S is an unoccupied site or other surface defects suitable for oxygen chemisorption; $O_{\beta(ads)}^{-\alpha}$ is a chemisorbed oxygen species with $\alpha = 1$ and 2 for a singly and doubly ionized forms, $\beta = 1$ and 2 for atomic and molecular forms respectively.

Following the reasoning presented by authors [50] and considering the Weisz limitation [50], one can obtain

$$\lg n_S = const' + \lg\left(1 - \frac{n_S}{n_b}\right) - m \lg p_{O_2} \qquad (3.19)$$

where n_b is the concentration of charge carriers in the bulk, m $= 1, 0.5,$ or 0.25 for O_2^-, O^- and O^{2-} respectively.

If the sensing element is a porous film formed by small grains with grain size $d_{XRD} < 50$ nm, complete grain depletion and a flat band condition can be expected [50] and the conductivity is proportional to n_S. Thus, Eq. (3.19) can be presented as:

$$\lg G = const1 + \lg\left(1 - \frac{G}{G_0}\right) - m \lg p_{O_2} \qquad (3.20)$$

In this case G is the material conductivity at the given partial oxygen pressure, G_0 – material conductivity at $p_{O_2} \to 0$. Taking G_0 to be equal to the material conductivity under an inert atmosphere (for example in pure Ar, containing no more than 0.002 vol. % O_2), it is possible to determine experimentally the value of $\lg\left(1 - \frac{G}{G_0}\right)$ for any fixed p_{O_2}. If the experimental plots of G vs. p_{O_2} in coordinates

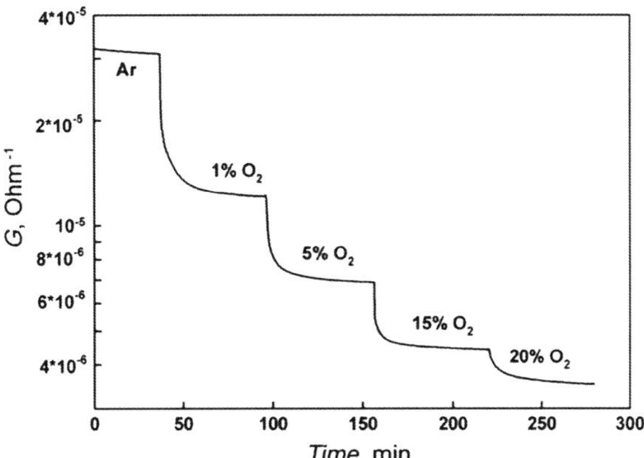

Fig. 3.7 Conductivity of SnO$_2$-500 sample at step-by-step increase of oxygen partial pressure at 400 °C

$\left(\lg G - \lg\left(1 - \frac{G}{G_0}\right)\right)$ against $(\lg p_{O_2})$ are linear, one can determine from the curve slope the predominant form of chemisorbed oxygen [51].

For pure and chemically modified SnO$_2$ the reversible decrease in electrical conductivity G is observed with increasing the partial pressure of oxygen in the gas phase (Fig. 3.7). The values of G for a fixed partial pressure of oxygen can be used for the analysis of the $\left(\lg G - \lg\left(1 - \frac{G}{G_0}\right)\right)$ against $(\lg p_{O_2})$ dependence in the temperature range 200–400 °C. In the coordinates corresponding to Eq. (3.20), the obtained dependencies of conductivity vs. oxygen partial pressure are linear (Fig. 3.8). That allows us to determine the coefficient m in the equation for the slope and to suppose the predominant form of chemisorbed oxygen. Thus, the values of $m = 1$, 0.5 and 0.25 suggest that the predominant form is $O_{2(ads)}^-$, $O_{(ads)}^-$ and $O_{(ads)}^{2-}$ respectively. The calculated values of m for unmodified SnO$_2$ are summarized in Table 3.2.

At T = 200 °C oxygen chemisorption on SnO$_2$ surface proceeds mainly with formation of the molecular $O_{2(ads)}^-$ form ($m \approx 1$). A decrease of m with the temperature increase points to an increase in the amount of atomic $O_{(ads)}^-$. With the increase of SnO$_2$ grain size this effect becomes more significant.

At T = 200 °C the introduction of modifiers does not affect the type of the predominant form of chemisorbed oxygen. Thus, the chemisorption of oxygen at this temperature occurs mainly with the formation of the molecular form, regardless of the type of modifier. For higher temperatures, the value of m decreases for the modified tin dioxide, which indicates an increasing contribution of atomic $O_{(ads)}^-$ form (Fig. 3.9).

The observed trend in increasing the influence of modifiers on the type of the predominant form of chemisorbed oxygen in the series Fe < In < La < Ru < Ni < Au < Pd < Pt can be due to a combination of several factors. The formation of a monatomic form of chemisorbed oxygen can occur in dissociative adsorption (reaction (3.11–3.15)), as well as due to the decomposition of unstable

Fig. 3.8 Conductivity of SnO_2-500 ($d_{XRD} = 14$ nm) vs. oxygen partial pressure in coordinates of Eq. (3.20)

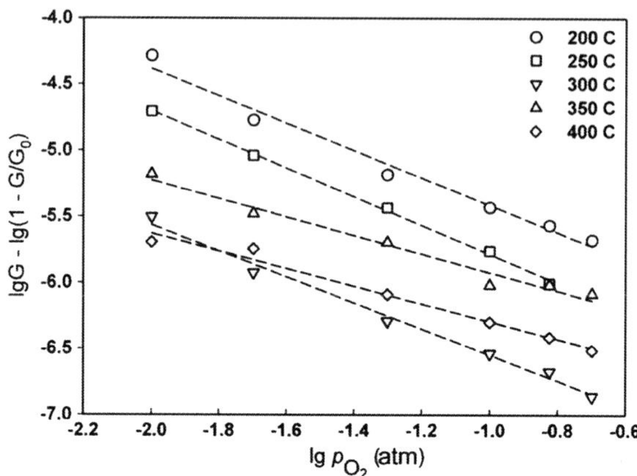

Table 3.2 Calculated values of coefficient m for unmodified SnO_2

d_{XRD}, nm	m				
	200 °C	250 °C	300 °C	350 °C	400 °C
4	0.98 ± 0.05	1.03 ± 0.02	1.02 ± 0.05	1.02 ± 0.03	0.80 ± 0.04
14	1.03 ± 0.07	1.03 ± 0.03	0.98 ± 0.05	0.88 ± 0.06	0.69 ± 0.05
28	0.94 ± 0.05	1.05 ± 0.02	1.02 ± 0.05	0.70 ± 0.06	0.67 ± 0.06
43	–	0.94 ± 0.03	0.89 ± 0.05	0.68 ± 0.09	0.65 ± 0.03

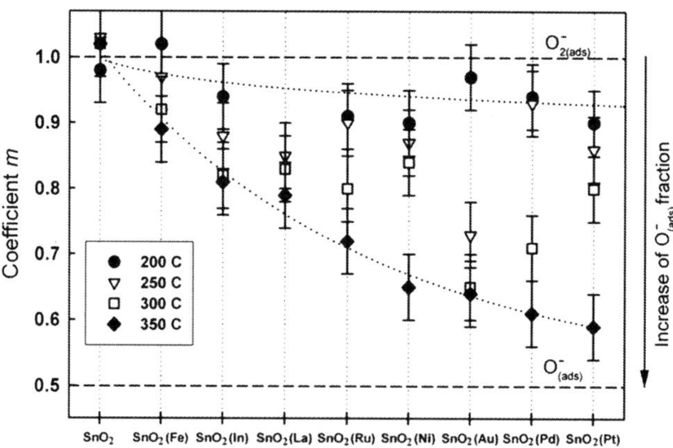

Fig. 3.9 Influence of modifier on the oxygen chemisorption on SnO_2 surface

molecular form $O^{2-}_{2(ads)}$ that can be formed by the localization of the second electron on molecular $O^{-}_{2(ads)}$ ion:

$$O^{-}_{2(ads)} + e^{-} \leftrightarrow O^{2-}_{2(ads)}. \tag{3.21}$$

Table 3.3 Distribution of oxides between the groups [5] in accordance with qo dependence on $\Delta\chi$ value

Group					
I		II		III	
Oxide	qo	Oxide	qo	oxide	qo
RuO_2	-0.74	SnO_2	-0.86	La_2O_3	-1.08
PdO	-0.89	Fe_2O_3	-0.87		
PtO_2	-1.00	In_2O_3	-0.87		
		NiO	-1.02		

Thus, the conditions of $O^-_{(ads)}$ formation may be the dissociation of oxygen (reaction (3.13)) or increase of negative charge on the adsorbed molecular oxygen species.

The authors of review [5] showed that, depending on the ratio of the charge qo of the lattice oxygen atom and the difference in electronegativities of elements $\Delta\chi$, crystalline oxides can be generally divided into three groups. The first group includes oxides of platinum metals. For this group of oxides, the minimum value q_O from -0.8 to -1.0 is typical. The second group includes oxides of p-and d-elements, and uranium oxides. The charge of oxygen varies in the range q_O from -0.6 to -1.3. The third group consists of oxides of alkali, alkaline earth and rare earth elements. This group can be characterized by the highest negative charge of oxygen q_O from -1.0 to -1.9 and the maximum $\Delta\chi$ value. In all groups, the negative charge of oxygen decreases with $\Delta\chi$ increasing.

Tin dioxide and oxides of the above modifiers can be assigned to the groups described in Table 3.3. Within each group the increase of negative charge on the oxygen anions within the modifier oxide results in the growth of oxygen chemisorbed on the surface of nanocrystalline SnO_2 as the atomic ion, $O^-_{(ads)}$. Thus, the partial charge on oxygen in the corresponding modifier oxide can be used as a criterion for assessing the ability of the cation to the transfer of electron density to oxygen.

At the same time, with similar values qo, the modifiers of group I have the greatest influence on the chemisorption of oxygen. Probably, this may be due to the fact that clusters of platinum metals and their oxides additionally catalyze the dissociation of oxygen molecules by reaction (3.13), which leads to a significant increase in the atomic form of chemisorbed oxygen. The high activity of platinum group metals in this process is shown earlier in experiments on isotopic exchange of oxygen O^{16}/O^{18} [52, 53] in oxides SiO_2, Al_2O_3, ZrO_2, TiO_2.

Oxygen adsorption on gold particles occurs without dissociation [54]. Thus, the mechanism of the effect of gold on the dominant type of particles of chemisorbed oxygen differs from that discussed above. Adsorption of oxygen on the surface of heterocontact Au/SnO_2 was studied by the authors [55] through electrical conductivity and calorimetry measurements. It was found that the heat of oxygen adsorption on the Au/SnO_2 is significantly higher than the corresponding values obtained for the adsorption of O_2 on the SnO_2 and Au separately. It was suggested that at the interface gas—Au–SnO_2 there are forms of chemisorbed oxygen with a higher negative charge compared with the species chemisorbed on the surface of the unmodified oxide. The mechanism proposed in [56], includes the exchange of electrons between the Au and SnO_2. Thus, one can assume that the formation of oxygen particles on the Au/SnO_2 occurs through an intermediate stage of formation of unstable molecular $O_{2(ads)}^{2-}$ forms.

Overall, the observed tendency to increase the fraction of atomic form of chemisorbed oxygen is in a good agreement with the increase of optical basicity or decrease of Lewis acidity of the cations of modifiers (Fig. 3.10). Thus, these approaches can be used for the qualitative prediction of the influence of modifiers on the oxidative activity of oxygen chemisorbed on SnO_2 surface.

3.4.2 Temperature Programmed Reduction by H_2

Hydrogen consumption during TPR-H_2 (preannealing at 300 °C in dry air with further cooling in the same flow, followed by TPR in 5 % mixture of H_2 in Ar) goes through two pronounced maxima, the first of which is in the low temperature range of 100–350 °C, while the second, more pronounced - in the high temperature range 380–660 °C (Fig. 3.11). High-temperature maximum corresponds to the hydrogen absorption during reduction of material to metallic tin, the low-temperature one reflects the absorption of hydrogen by chemisorbed oxygen species and the terminal atoms of the lattice oxygen on the surface of SnO_2. This process, called "reduction of the surface", results in an increase of the number of lattice oxygen vacancies in the positions of the bridging oxygen, and the proportion of cations Sn_{4c}^{2+} on the surface. The presence of the shoulder in the high-temperature peak of hydrogen consumption reflects the transition of tin dioxide SnO_2 into SnO, followed by full reduction to metallic tin. Reduction to an intermediate state is ensured by reaction with surface O^{2-} anions among the total number of oxygen atoms in the lattice. With increasing SnO_2 grain size, the fraction of surface oxygen atoms is sharply reduced, as evidenced by the decrease in the low-temperature peak for hydrogen consumption. At the same time the shoulder on high-temperature peak almost disappears and becomes shorter, indicating that the processes of reducing the tin dioxide matrix to SnO and then to metallic tin occur in parallel.

For materials modified with Ru, Pt and, in particular, Pd, the high-temperature hydrogen consumption maximum shifts toward lower temperatures (Fig. 3.12). This may be caused by the catalytic activity of noble metal clusters in the matrix of nanocrystalline tin dioxide. The most likely mechanism is a joint spillover of hydrogen and oxygen on the SnO_2 crystal lattice through the PdO_x clusters [56]. Reduction of SnO_2 at a lower temperature becomes possible due to the dissociation of hydrogen molecules on Pd or PdO_x clusters. Examples illustrating this

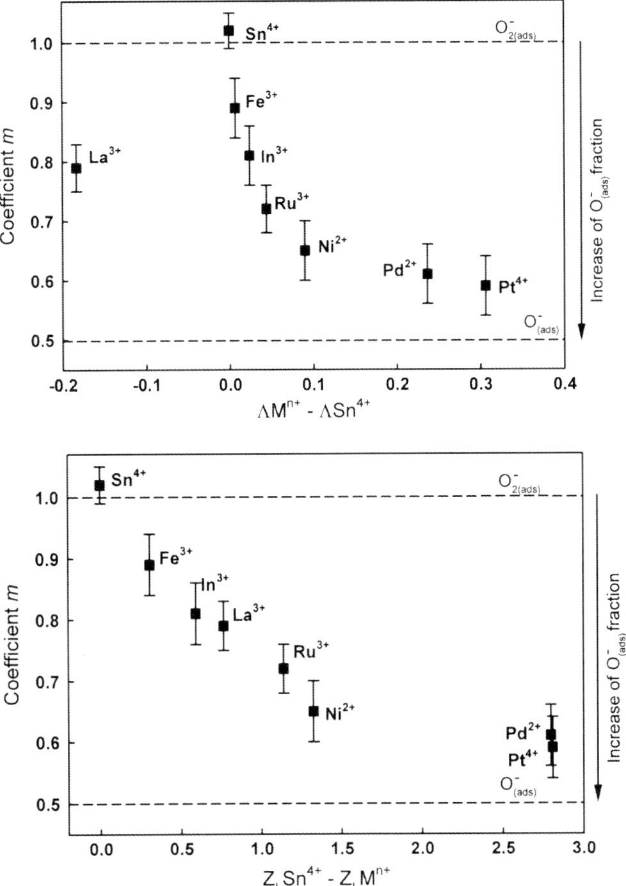

Fig. 3.10 The relationship between nanocrystalline SnO_2 surface modifier Lewis acidity, defined in terms of optical basicity (*top*) and Zhang electronegativity (*bottom*), and ratio between different forms of adsorbed oxygen

mechanism of interaction with the gas phase for systems Rh/TiO_2, Rh/Al_2O_3, Rh/CeO_2, Rh/SiO_2, Pt/Al_2O_3, Pd/Al_2O_3, Pd/Mn_2O_3, are presented in the review [57]. The onset of high temperature maxima, representing hydrogen consumption, for other materials is manifested nearly at the same temperature (~ 420 °C), which indicates the energy barrier for Mars van Krevelen oxidation of adsorbed molecules [57].

For pure SnO_2 with small crystallite size of 4 nm the amount of chemisorbed oxygen estimated from TPR-H_2, assuming the interaction

$$O^-_{2(ads)} + 2H_2 = 2H_2O + e^- \tag{3.22}$$

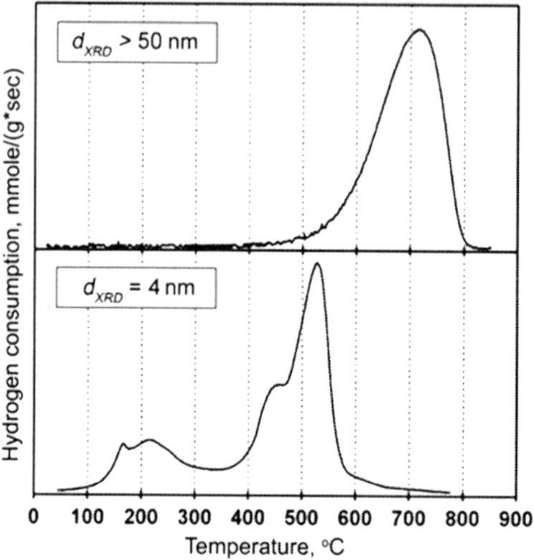

Fig. 3.11 Typical TPR spectra for unmodified SnO_2 with different crystallite size d_{XRD}

is $\sim 5 \times 10^{-6}$ mol/m^2 or 5×10^{-2} of monolayer. According to Weisz limitation [51], this quantity corresponds to the upper boundary of the coverage density of the semiconductor surface by charged particles in equilibrium conditions.

In most cases surface modification of tin dioxide leads to a decrease of hydrogen consumption at low temperatures. This may be indirect evidence of increasing the atomic anionic form of chemisorbed oxygen on the SnO_2 surface (see Fig. 3.9). In fact, if one compares the reaction (3.22) and

$$O^-_{(ads)} + H_2 = H_2O + e^- \tag{3.23}$$

for the surfaces with the same negative charge caused by oxygen acceptor species the hydrogen consumption is less in the latter case. However, comparing the amount of consumed hydrogen and the part of atomic form of chemisorbed oxygen for tin dioxide modified with various catalysts, does not show unequivocal correlations. The decrease in the total amount of chemisorbed oxygen may be additionally due to blockage of adsorption centers by the cations of modifiers. It is impossible to take into account this factor quantitatively because of the complicated distribution of modifiers between surface and bulk of SnO_2 crystallites.

Analysis of the concentration of highly reactive oxygen species on the surface of materials can be done by TPR-H_2 after pre-treatment in different atmospheres (Fig. 3.13). Annealing in nitrogen flow at 300 °C for 1 h, in contrast to the pre-treatment in synthetic air under the same conditions, is intended to remove weakly bonded forms of chemisorbed oxygen from the surface. The chemical inertness of nitrogen allows exclusion of the possible surface reduction in the chemical reaction. At the same time, desorption of mobile oxygen species from the surface is sustained due to the fact that the nitrogen molecule has a polarizability close to that of molecular oxygen. This property makes the N_2 molecule competitive in the process of adsorption at the centers occupied by chemisorbed oxygen particles [34].

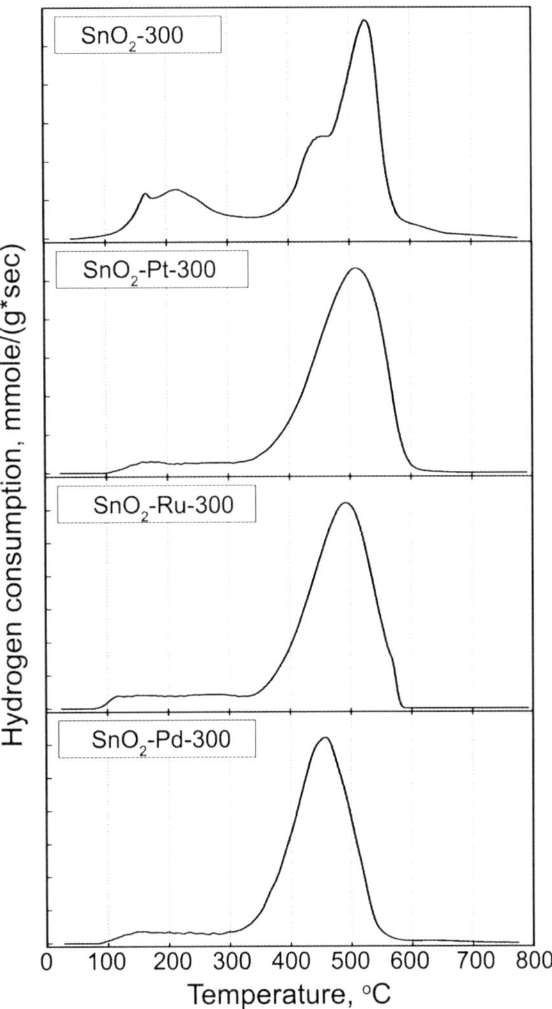

Fig. 3.12 TPR spectra of SnO$_2$-based sensor materials

The decrease of low-temperature peak of hydrogen consumption after sample pre-treatment in nitrogen indicates the desorption of mobile forms of chemisorbed oxygen. The difference of TPR-H$_2$ spectra, taken under different pre-treatment conditions, allows us to estimate the concentration of these oxygen species on the surface of materials (Fig. 3.14).

For the samples, obtained in the same conditions, the maximum concentration of mobile oxygen species is observed for materials modified with Cr$_2$O$_3$ and PdO$_x$. In both cases a likely explanation may be explained via the chemical oxidation of the modifier in an air atmosphere to the higher oxidation state. In the former case the oxidation goes through formation of chromyl group Cr = O and at least partial oxidation of metallic Pd clusters to PdO in the latter. Both materials are presumed

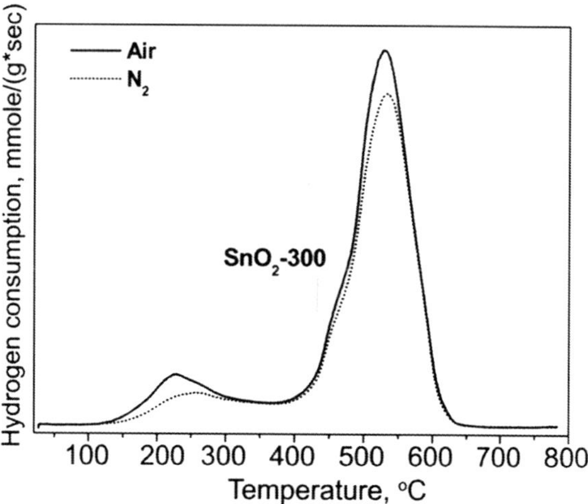

Fig. 3.13 TPR-H$_2$ spectra of SnO$_2$-300 material evaluated after sample pretreatment in different gas atmospheres

thus to be especially active in the surface oxidation processes. Overall the introduction of modifiers leads in general to a slight increase in adsorbed oxygen content, perhaps due to the introduction of additional point defects on the surface acting as additional adsorption and dissociation sites for oxygen molecule.

3.5 Classification of Gases

Properties of certain molecules: first ionization potential, electron affinity, proton affinity, which have an impact on the nature of their interaction with the surface of a semiconducting oxide, are shown in Fig. 3.15 [58]. The compounds are grouped in order of increasing proton affinity and reducing first ionization potential.

A pre-selection of the modifier to create a selective sensing material can be made on the basis of analyzing the properties of gas molecules to be detected. Target gases can be divided into groups as shown in Fig. 3.15 based on the mode of interaction with the semiconductor oxide surface. First it is necessary to distinguish oxidizing (electron accepting) and reducing (electron donating) ones. Gases should be considered as oxidizing agents when their adsorption on the surface leads to a decrease in charge carrier concentration in the surface layer of n-type semiconductor oxides due to electron capture. This group includes gas molecules with high electron affinity and high ionization potential (Group I, Fig. 3.15): oxygen, nitrogen dioxide, ozone, chlorine.

The reducing agents are gases, for which the interaction with the surface of n-type semiconductor oxide results in an increase of the charge carrier concentration in the surface layer of the latter. This effect is caused by oxidation of the reducing gas molecules by oxygen chemisorbed on the oxide's surface or lattice oxygen as it is described in the previous paragraph. This mechanism of a change in charge carrier

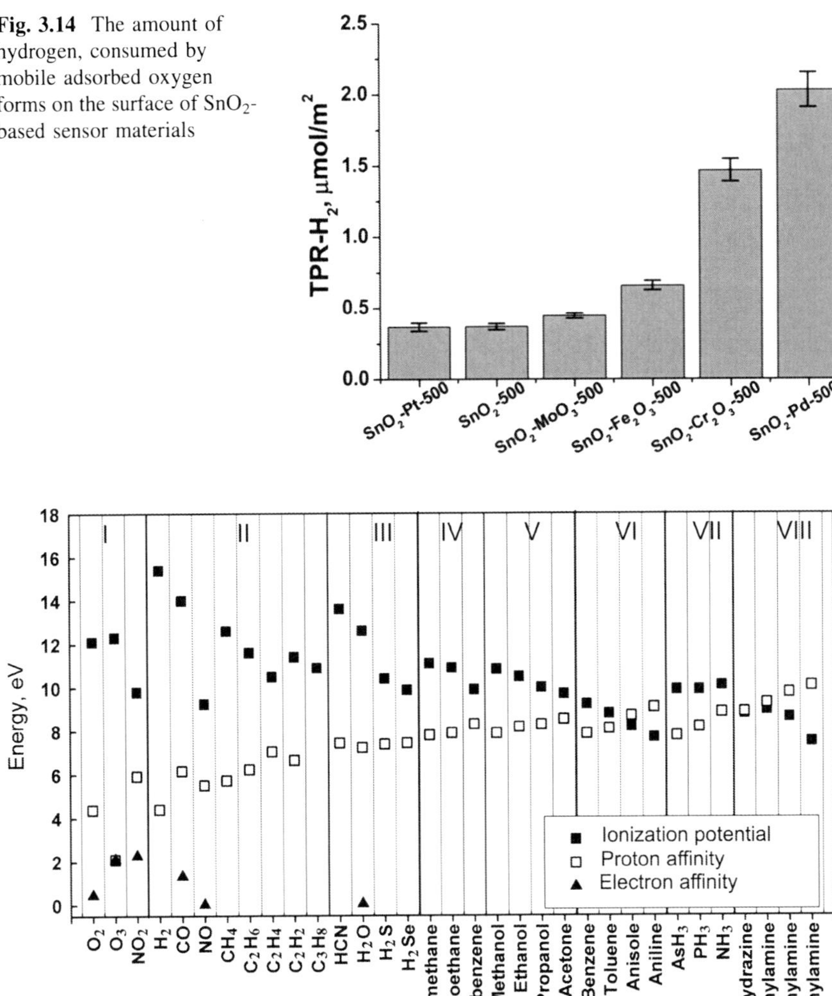

Fig. 3.14 The amount of hydrogen, consumed by mobile adsorbed oxygen forms on the surface of SnO_2-based sensor materials

Fig. 3.15 Classification of gases

concentration is typical, especially, in the oxidation of gases that do not possess pronounced electron-donor properties and have low proton affinity: CO, H_2, saturated hydrocarbons (Group II, Fig. 3.15). In contrast, the reactions of gas molecules having pronounced acidic (Group III, Fig. 3.15) or basic (Groups VII, VIII, Fig. 3.15) properties with the surface of sensitive material may include an interaction of the donor-acceptor mechanism in addition to oxidation. The reactions of complex organic molecules containing various functional groups (Groups IV–VI, Fig. 3.15), on the surface of a semiconductor oxide can proceed via different routes, depending

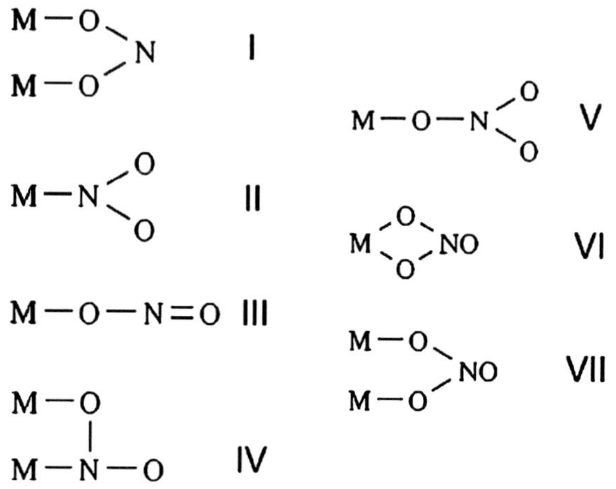

Fig. 3.16 Various NO_2 derived species, formed on oxide surface

$$M-O \diagdown$$
$$\qquad\; N \qquad I$$
$$M-O \diagup$$

$$M-N \diagup O \qquad II$$
$$\qquad\;\; \diagdown O$$

$$M-O-N=O \qquad III$$

$$M-O$$
$$\;\;\;\, |$$
$$M-N-O \qquad IV$$

$$M-O-N \diagup O \qquad V$$
$$\qquad\qquad\;\; \diagdown O$$

$$M \diagup O \diagdown NO \qquad VI$$
$$\;\;\, \diagdown O \diagup$$

$$M-O \diagdown$$
$$\qquad\; NO \qquad VII$$
$$M-O \diagup$$

on the concentration of basic and acid adsorption centers of different strengths on the surface of the semiconductor.

In Sects. 5.1–5.5 these principles will be discussed in more detail. Specific attention will be paid to material and chemical characteristics which lead to a general selection of sensor materials (and modifiers) for detection of typical representatives of the different groups of gases mentioned above—NO_2 (Group I), CO (Group II), NH_3 (Group VII), H_2S (Group III), and acetone vapor (Group V). Section 6 will highlight the sensor operating characteristics of the selected sensor materials (and modifiers) towards these specific target gases.

3.5.1 Oxidizing Gases: NO_2

Interaction of NO_2 molecules with the surface of oxides can proceed by the mechanism of molecular and dissociative adsorption. In the first case the formation of various nitrite (NO_2) and nitrate (NO_3) surface groups [59] is possible (Fig. 3.16)

Numerous literature data provide evidence that the interaction of n-type semiconductor oxides with NO_2 is accompanied by a decrease in electrical conductivity, i.e., like oxygen, nitrogen dioxide is an electron acceptor. Thus, the interaction of NO_2 with surface centers—Lewis acids with the formation of nitronium ion (Fig. 3.16b)

$$NO_2 + M^{n+} \rightarrow NO_2^+ + M^{(n-1)+} \tag{3.24}$$

can not be responsible for changing the electrical conductivity of materials.

The electron affinity of NO_2 molecule is 2.27 eV [60], which considerably exceeds the analogous value for the oxygen molecule (0.44 eV [61]). This determines the possibility of detecting low concentrations of NO_2 in the air in the presence of ~ 20 vol. % of oxygen. Authors [62] using Raman spectroscopy showed that the electrical conductivity of SnO_2 in the presence of NO_2 correlates with the concentration of surface bidentate nitrite group. Ab initio calculations [63] indicate that the most energetically favoured is the molecular adsorption of nitrogen dioxide with the coordination of one or both of the oxygen atoms of NO_2 molecules on the vacancies of bridging oxygen atoms on a partially reduced surface of tin dioxide.

Study by infrared spectroscopy and programmed desorption in the temperature range 25–400 °C [64] showed the possibility of dissociation of NO_2 during interaction with the surface of tin dioxide. The presence of NO in the thermal desorption products can serve as indirect evidence of the occurrence of following reactions:

$$NO_{2(gas)} \leftrightarrow NO_{2(ads)}, \tag{3.25}$$

$$NO_{2(ads)} + e^- \leftrightarrow NO_{2(ads)}^-, \tag{3.26}$$

$$NO_{2(ads)}^- \leftrightarrow NO_{(gas)} + O_{(ads)}^-. \tag{3.27}$$

To increase the SnO_2 sensor signal towards NO_2 it seems to be reasonable to introduce electron donor modifiers that increase the concentration of electrons with sufficient energy to overcome the barrier created by the negatively charged surface. Alternatively the use of modifiers that are able to transfer electron density to the oxygen belonging to the acceptor chemisorbed species through an increase in the degree of oxidation should also enhance the sensing signal towards NO_2.

3.5.2 Reducing Gases: CO

The mechanism of CO interaction with the surface of metal oxides depends primarily on temperature. Since the CO molecule can act as both the donor and acceptor of electrons, it can react with acidic surface centers—coordinatively unsaturated metal cations (Lewis acids) and oxygen ions, which have basic properties according to acid–base classification, given by Lewis. At low temperatures, CO adsorbs on metal cations. The structure of the molecular orbitals of CO is that the electron density transfer is possible from both the CO molecule to the free orbital of the metal (σ-donor), and vice versa—from the d-orbitals of the metal to unoccupied antibonding orbitals of CO (π-acceptor) [60]. At elevated temperatures carbon monoxide on the surface of oxide reacts with chemisorbed oxygen species or with oxygen anions in the crystal lattice. The final product of oxidation is CO_2.

The main reaction responsible for sensor signal generation, is namely the oxidation of CO by chemisorbed oxygen. Therefore, modifiers that increase the concentration and mobility of oxygen species on the surface of SnO_2, are of greatest interest for the development of highly sensitive materials for CO detection.

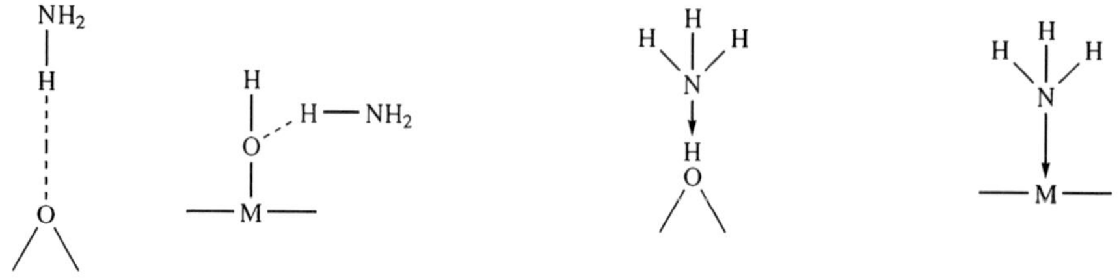

Fig. 3.17 Scheme of possible ammonia coordination on oxide surface

3.5.3 Reducing Bases: NH₃

Ammonia molecules can adsorb on the oxide surface through formation of hydrogen bonds with surface atoms of oxygen or oxygen from hydroxyl groups (Fig. 3.17a), a hydrogen bond between the nitrogen and hydrogen of the hydroxyl group (Fig. 3.17b) or a coordinative donor-acceptor bond with an unsaturated surface metal cation (Fig. 3.17c) [60]:

The formation of NH_4^+ ion is a criterion for the presence of surface Brønsted acid centers—hydroxyl groups, whereas the coordination of the NH_3 molecules indicates the presence of Lewis acid centers. Dissociative adsorption of ammonia leads to the formation of NH_2-and OH-groups on the surface:

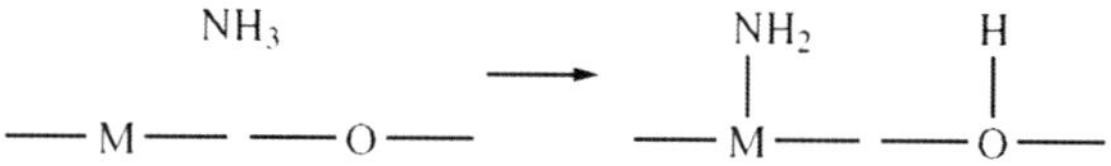

$$\ce{NH3 \atop -M- -O-} \quad\longrightarrow\quad \ce{NH2 \quad H \atop -M- -O-} \tag{3.28}$$

Based on the presented data one can suggest that the increase in NH_3 adsorption on the surface of tin dioxide and, possibly, enhancement of sensor response can be achieved by increasing the number of acidic adsorption sites.

3.5.4 Reducing Acids: H₂S

Interaction of gaseous hydrogen sulfide with the oxide surface is characterized by two factors. First, hydrogen sulfide is a strong reducing agent: the value of the ionization potential of the H_2S molecule is 4.10 eV. Secondly, hydrogen sulfide is a Brønsted acid, i.e. heterolytic cleavage of the S–H bond is quite easy, especially via the formation of new donor-acceptor bonds. Thus, H_2S can interact with various centers on the oxide surface (Fig. 3.18) [60]:

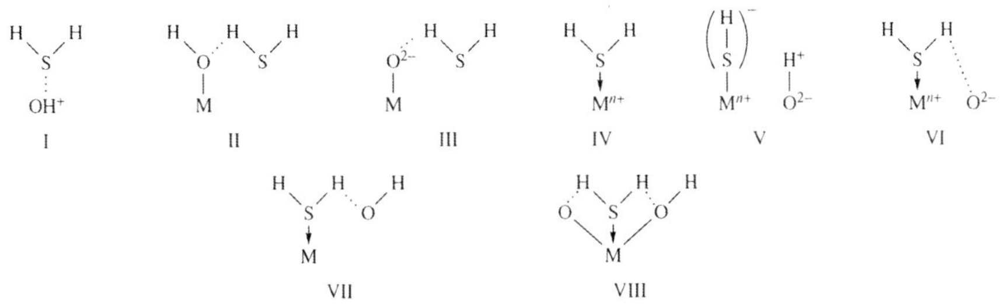

Fig. 3.18   Scheme of possible H_2S coordination on oxide surface

The possible reactions are [60]:

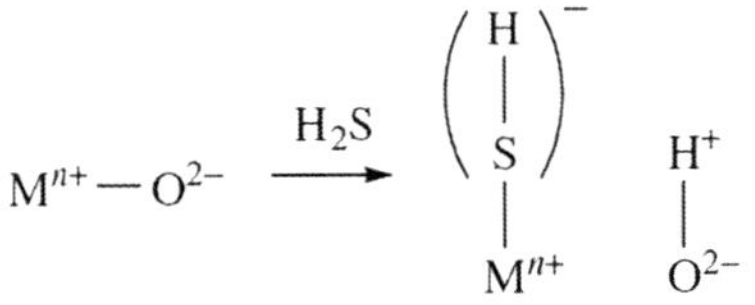

$$M^{n+} - O^{2-} \xrightarrow{\ H_2S\ } \left(\begin{array}{c} H \\ | \\ S \\ | \\ M^{n+} \end{array}\right)^{-} \quad \begin{array}{c} H^+ \\ | \\ O^{2-} \end{array} \tag{3.29}$$

$$M^{n+} - O^{2-} \xrightarrow{\ H_2S\ } \overset{HOH}{-\!\!-M-\!\!-S^{2-}} \tag{3.30}$$

$$\begin{array}{c} H \\ | \\ O \cdots H \\ | \quad\ \diagdown\ \diagup H \\ M -\!\!-\!\!- [\ ]-\!\!-O^{2-} \end{array} \longrightarrow \begin{array}{c} H \\ | \\ O \cdots H \quad\ H \\ | \qquad\ \vdots \\ M-\!\![S_x] \quad O^{2-} \end{array} \tag{3.31}$$

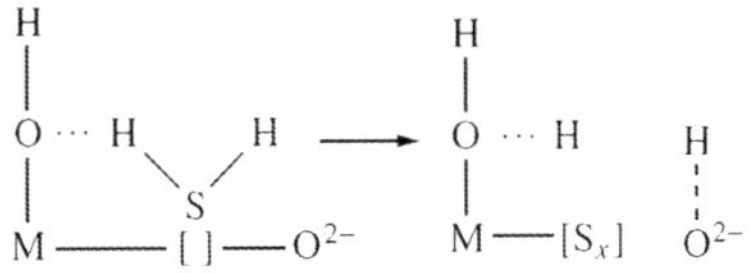

$$M^{n+}O^{2-} + H_2S \rightarrow [S_\chi] + M^{(n-2)+} + H_2O \tag{3.32}$$

As a result of interaction with gaseous hydrogen sulfide the following changes were found at the oxides surface [60]:

- formation of sulfides;
- partial blocking of Lewis acid sites;
- reduction of metal cations with variable valence;
- oxidation of sulfur and increasing acidity of the surface through formation of SO_x groups.

An increase in H_2S adsorption on the surface of tin dioxide can be achieved with the introduction of modifiers that increase the electron-donor ability of surface basic centers (oxygen anions). In this case the H_2S adsorption is accompanied by heterolytic reaction with the Lewis acid/base pair on the oxide surface. The maximum augmentation of sensor signal is possible if the interaction of a high-resistance modifier oxide with hydrogen sulfide results in the reversible formation of a highly conducting sulfide.

3.5.5 Complex Organics: Acetone

Acetone represents a complicated case for gas sensing since it can interact with metal oxide surfaces via a number of reaction routes. Each of these mechanisms for sorption affects further transformation steps, so there are number of products expected to form prior to desorption. In this sense the assumption, that the conversion of organic molecules to CO_2 and H_2O is the process, which solely governs the sensor response, seems to be an oversimplification. The polar carbonyl group is a moiety, which in the first step, could take part in the adsorption on a metal oxide surface. According to IR data it usually goes through formation of an enol species [65, 66] with further dissociation and formation of propene-2-olate strongly bonded to the surface metal cation and a proton bound to surface oxygen with the formation of a hydroxyl group. This transformation may lead to acrolein or acrylic acid. At the same time the highly electrophilic carbonyl carbon atom may be attacked by nucleophylic O^{2-} with further carbon chain cleavage and partial oxidation to acetic acid [67] or acetic aldehyde. It was shown also that the condensation of two nearby adsorbed enolates gives rise to complex molecules (oxidative coupling) [66, 67]. The interaction of adsorbed enolate with Brønsted acid sites may give a dehydration product—propene—and water as a result of the slight Brønsted basicity of the enol group. The ratio of these reaction rates at the solid surface is governed mainly by surface cation acidity and anion basicity and may be shifted to either of the processes via surface modification [37].

Also the interaction of adsorbed organic molecules with highly reactive electrophilic adsorbed oxygen species should be taken into account. It is generally referred to as an unselective interaction, giving rise to a number of products of partial oxidation or complete burning to CO_2 and H_2O. At this point it is hard to propose any restricted number of surface modification agents for SnO_2 to achieve sensor response enhancement without additional information regarding the mechanism of conversion and the sensor response to the vapor of this compound.

3.6 Sensor Properties Studies

3.6.1 CO Sensing

The benchmark study of sensor sensitivity to CO, a common reducing gas, is of special interest since it allowed for the formation of general trends with respect to mechanisms of sensor sensitivity towards reducing gases [68].

The synthesized materials showed a different temperature dependence for sensor sensitivity to CO (40 ppm in air RH = 30 % Fig. 3.19). The material modified by catalytic clusters of Pd, had the largest sensor response at a low operating temperatures (200–250 °C). For most other materials, the sensor response, in contrast, increases with increasing operating temperature. Only in the case of SnO_2-Ru-700, this dependence passes through a maximum at 300 °C. At the high measurement temperature, 350 °C, the maximum sensor response was observed for the materials modified with catalytic clusters of gold. These differences indicate the various chemical mechanisms that underlie sensor sensitivity to CO.

The process of CO oxidation by different adsorbed oxygen species on the surface (O_2^-, O^- and O^{2-}) has a decisive contribution to the formation of an electrical signal in relation to CO in the case of materials based on nanocrystalline SnO_2 :

$$\beta \cdot CO_{(gas)} + O_{\beta(ads)}^{-\alpha} \rightarrow \beta \cdot CO_{2(gas)} + \alpha \cdot e^- \qquad (3.33)$$

where $CO_{(gas)}$—the CO molecule in the gas phase, $O_{\beta(ads)}^{-\alpha}$—an atomic or molecular form of oxygen, e^-—an electron that is injected into the conduction band of the semiconductor, $CO_{2(gas)}$—the molecular reaction product desorbed from the surface.

In this context the low-temperature CO sensitivity of the Pd modified material is reasonable to connect to reaction (3.33) involving the above-discussed mobile forms of oxygen. Oxidation takes place according to the Eley-Riedel "collision" mechanism, which involves the oxidation of CO molecules in the gas phase, by atomic oxygen species, located on the PdO surface. This reaction occurs during rapprochement of two species, which in this case when a CO molecule passes near a solid surface with a partially oxidated Pd cluster [69]. We propose that the compensation of chemisorbed ionized forms of O^- expended on the oxidation of CO runs due to the phenomenon of reverse spillover—the transfer of already existing adsorbed O^- ions on the surface of SnO_2 to the clusters of Pd. This process leads to a change in the concentration of adsorbed oxygen on the surface of the semiconductor and, consequently, the change in electrical conductivity of the material that causes the sensor response to CO. With temperature increase, the process of re-oxidation of PdO_x accelerates due to the dissociation of molecular oxygen, adsorbed directly onto noble metal clusters. This phenomenon leads to a complete localization of the CO oxidation process on catalytic clusters and a

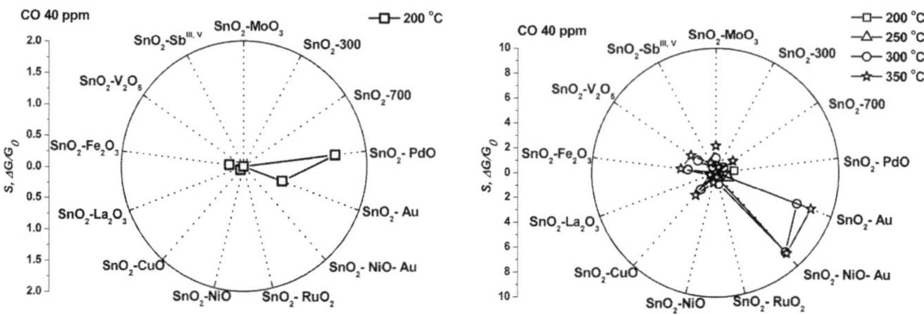

Fig. 3.19 Pattern of SnO_2-based materials sensitivity to CO (40 ppm in 30 %RH air) at low (200 °C) (*left*) and high (200–350 °C) working temperatures (*right*)

cessation of reverse spillover transport, which is accompanied by the disappearance of the sensor response to this gas.

A similar combination of factors underlies the sensor response to CO in the case of the material, modified by RuO_x clusters [70]. Displacement of the sensor response maxima in this case to the higher temperatures in comparison with the SnO_2-Pd-500 may be due to additional unusually high diffusion barriers in the case of the RuO_2 surface [71].

According to monotonous growth of the sensor response with increasing operating temperature in the case of other materials, the oxidation of adsorbed CO molecules occurs on the surface of the matrix SnO_2. The sharp rise in sensor response at a 300 °C working temperature in this context corresponds to the transition of molecular chemisorbed forms of O_2^- into the reactive atomic O^- discussed above. Thus we can assume that the participation of lattice oxygen in CO oxidation is secondary, because the growth of sensor response is observed at significantly lower operating temperatures of sensors than the temperature used to determine the activation barrier for Mars—van Krevelen type oxidation mechanism. The role of catalytic gold clusters in the case of materials showing the highest sensor response to CO, apparently, is activation of the oxygen species chemisorbed in the close vicinity to triple metal-oxide-gas interface [72, 73]. Despite the fact that this special form of oxygen is not dominant on the surface, it has extremely high reactivity. More accurate experimental data on this form of adsorbed oxygen on the surface of such materials to date are not available.

3.6.2 NH3 Sensing

The main process that determines the sensor response of semiconductor metal oxide sensors to ammonia is usually postulated to be the oxidation of surface adsorbed NH_3 [74]:

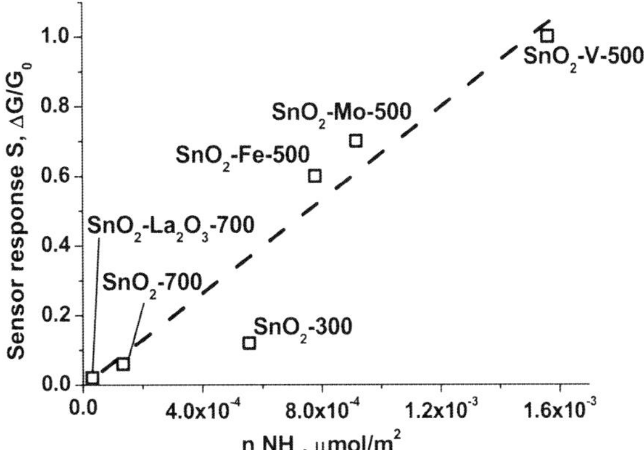

Fig. 3.20 Correlation between SnO$_2$-based materials surface brønsted acidity and sensor response towards NH$_3$ (20 ppm in humid air RH = 30 %, T = 200 °C)

$$2NH_{3(gas)} + \frac{3}{\beta}O^{-\alpha}_{\beta(ads)} \rightarrow N_{2(gas)} + 3H_2O + 3e^- \tag{3.34}$$

In accordance with the interconnection between surface acidity, molecular adsorption and sensor response the increase in the concentration of Brønsted acid sites on the surface in the case of materials modified with such acidic oxides as MoO$_3$ and V$_2$O$_5$ is accompanied by an increase in sensor response with respect to NH$_3$ gas (Fig. 3.20).

It must be noted, however, that this advantage in the magnitude of sensor signal is minuscule which respect to otherwise modified materials, which to some extent can be attributed to the competition of water vapor and NH$_3$ for surface adsorption on the same acidic sites. This phenomenon decreases ammonia adsorption thus limiting further steps for its oxidation on the surface, which leads to a reduction in the sensor response. At the same time in accordance to the data on the CO detection, materials, modified by gold, also express pronounced sensor response to ammonia since it exhibits a reducing nature as well as basic one (Fig. 3.21). The fall of the sensor response with increasing operating temperature, which is observed in the case of all tested materials, is likely due to a reduction in NH$_3$ adsorption [75]. It is important to note that the highest response towards NH$_3$ is for the case of SnO$_2$-Ru-700. The nature of this phenomenon could be explained with additional considerations on the chemical form of the modifier (Ru) on the sensor material surface. As it is introduced in the form of Ru(III) acetylacetonate with further annealing, an EPR study of this material (SnO$_2$-Ru-300) reveals a distinct signal, that corresponds to EPR-active Ru^{3+}. At the same time the interaction with reducing gas (NH$_3$) leads to an order of magnitude decrease of this signal indicating that Redox catalytic activity of this modifier is due to transition to oxidation states, other than Ru^{+3}. That might be Ru^0 and Ru^{+4} [76]. In this sense the presence of metallic form of Ru on the sensor material surface may activate the additional path of adsorbed ammonia conversion.

Fig. 3.21 Pattern of SnO$_2$-based materials sensitivity to NH$_3$ (20 ppm in humid air, RH = 30 %)

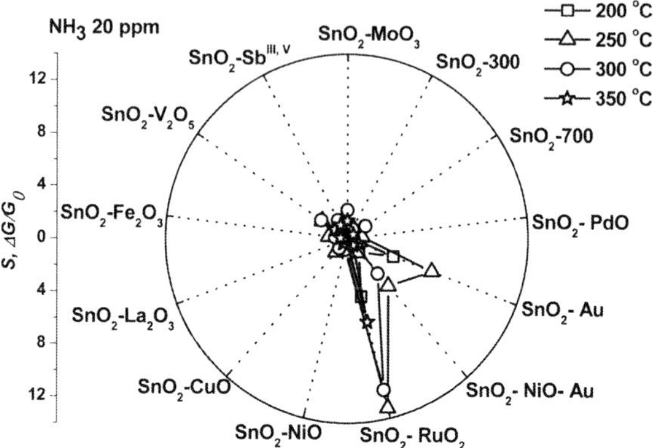

The metal surface of Ru is active in the catalytic dissociation of adsorbed molecular ammonia [77, 78] with formation of atomic nitrogen and hydrogen, desorbed in the molecular form:

$$NH_{3(gas)} + Ru(0001) \underset{k_{des}}{\overset{k_{ads}}{\longleftrightarrow}} NH_{3(ads)} + * \underset{k_{det\,r}}{\overset{k_{diff}}{\longleftrightarrow}} NH_3 - * \overset{k_{diss}}{\longrightarrow} N - * + 3H_{(ads)} \overset{k_{Ndiff}}{\longrightarrow} N_{(ads)} + 3H_{(ads)}$$

$$(3.35)$$

with NH$_{3(gas)}$—a molecule of ammonia in the gas phase, Ru (0001)—Ru metal surface, NH$_{3(ads)}$—ammonia molecule adsorbed on a metal surface *—active site of dissociation of molecular ammonia, NH$_3$−*—ammonia molecule adsorbed at the center of the dissociation on the surface of metallic Ru, N−*—nitrogen atom adsorbed at the center of the dissociation on the surface of metallic Ru, N$_{(ads)}$—the nitrogen atom, which left the center of the dissociation, H$_{(ads)}$−adsorbed hydrogen atom.

Thus, despite the lack of experimental evidence it is reasonable to propose the formation of small amounts (clusters) of metallic ruthenium on the surface of Ru-modified SnO$_2$ during the interaction with reducing agents as a reason for especially pronounced response of this material towards ammonia. In addition, significant contributions to the chemical reactions on the surface are possible through the interaction of adsorbed species and atomic nitrogen via the formation of the nitrosyl species, NO, which is a chemical radical [79]:

$$N_{(ads)} + O^-_{(ads)} \rightarrow NO_{(ads)} + e^-$$

$$(3.36)$$

With its increased reactivity these newly formed nitrosyl species, NO, on the surface may play a role for additional centers of adsorption and chemical reaction of NH$_3$. These chemical processes may contribute significantly to the sensor response of SnO$_2$-Ru-700 with respect to ammonia, but this issue is generally unexplored as of yet.

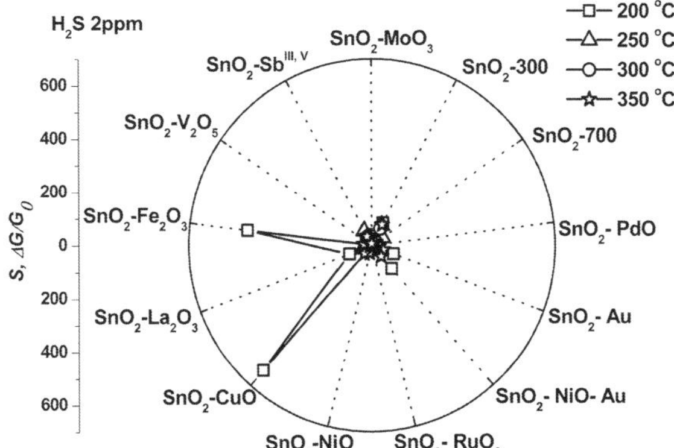

Fig. 3.22 Pattern of SnO$_2$-based sensor materials sensor response to H$_2$S (2 ppm in humid air, RH = 30 %)

3.6.3 H$_2$S Sensing

Materials modified with catalytic Au clusters, showing an increased sensor sensitivity to reducing gases, also have an elevated sensor response in the case of H$_2$S (Fig. 3.22), which is associated with the occurrence of the Redox process:

$$\beta \cdot H_2S_{(gas)} + 3O^{-\alpha}_{\beta(ads)} \rightarrow \beta \cdot SO_{2(gas)} + \beta \cdot H_2O_{(gas)} + 3\alpha \cdot e^- \tag{3.37}$$

The largest sensor response, however, was achieved for the case of the copper and iron oxide modified material, which in this case is likely caused by the chemical transformation of the material's surface.

In the case of nano SnO$_2$–CuO a significant change in resistance in the presence of H$_2$S should be attributed to the formation of copper (I) sulfide [80], which is a narrow-gap semiconductor:

$$6CuO + 4H_2S_{(gas)} = 3Cu_2S + SO_{2(gas)} + 4H_2O_{(gas)} \tag{3.38}$$

As a result of this reaction the energy barrier at the grain boundaries p-CuO/n-SnO$_2$ is removed and electrical conduction increases. A similar mechanism of sensor signal formation can be attributed to tin oxide modified with Fe$_2$O$_3$.

With an increase of the sensor working temperature, the sensor response of the material drops sharply, indicating that in these circumstances, a decisive contribution to the sensor response is made only by the process of oxidation of hydrogen sulfide to sulfur oxide by active species of oxygen adsorbed on the surface, preventing the formation of sulfides.

3.6.4 NO$_2$ Sensing

Despite the fact that the detailed mechanism of sensor response with respect to NO$_2$ is still a subject of debate, it was shown [63] that the change of the

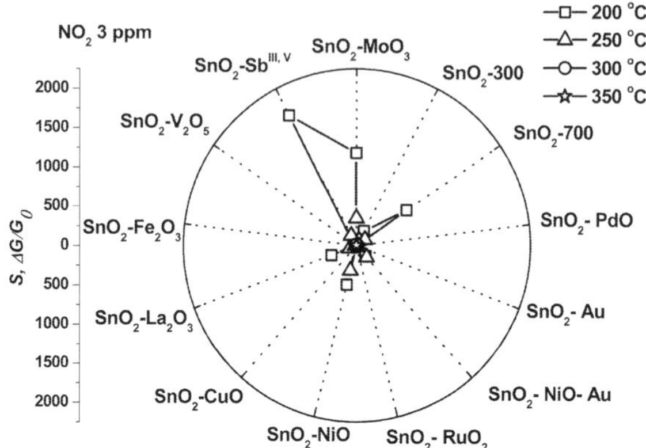

Fig. 3.23 Pattern of SnO$_2$-based sensor materials sensor response to NO$_2$ (3 ppm in humid air RH = 30 %)

conductivity of nanocrystalline SnO$_2$ in the presence of NO$_2$ molecules corresponds to the change in the concentration of adsorbed bidentate nitrite species:

$$NO_{2(gas)} + e^- \leftrightarrow NO^-_{2(ads)} \tag{3.39}$$

The increase in operating temperature sensor causes a decrease in sensor response due to desorption of NO$_2$ (Fig. 3.23).

In our study the increase of sensor signal in the case of several materials with the comparison to pure nanocrystalline tin dioxide was observed:

1. The most pronounced signal to nitrogen dioxide is obtained at 200 °C by means of SnO$_2$ doped with Sb. In this case the signal may be due to an electronic effect. Antimony introduction into tin dioxide is a well known route for increasing the density of electrons in the conduction band of semiconductor [81]. After the adsorption on the surface, the NO$_2$ molecule performs as an electronic trap, profoundly depleting the electron density in the conductance band and leading to significant decrease of nanocrystalline material conductance. So, the increase of concentration of electrons, which have enough energy to overcome the electric field resulting from the negative charging of the surface, favors reaction (3.39) that leads to the growth of sensor signal. The increase in working temperature greatly decreases the sensor response perhaps by a shift in the surface adsorption-desorption equilibrium of NO$_2$. This process also goes same way in the case of pure nanocrystalline SnO$_2$ that appears as the increase of sensor signal with SnO$_2$ crystallite size growth. Response growth with the transition from SnO$_2$-300 to the SnO$_2$-700, as described in the literature [82], has not yet found its comprehensive explanation, but it is proposed that this effect is caused by an increase in the charge carrier concentration in the conduction band of a semiconductor, which occurs with an increase in the crystallite size of SnO$_2$. This may be responsible for the shift of adsorption equilibrium (3.41). It cannot be excluded that the difference in charge carrier concentration in the presence of NO$_2$ molecules on the surface and in the pure air becomes greater for materials with large grain size and higher initial

concentration of free electrons in the surface layer in the atmosphere of pure air. It must be assumed that the same size effect is also responsible for increased sensitivity to NO_2, exhibited by the material SnO_2-La_2O_3-700;

2. in the case of SnO_2-MoO_3 the high sensitivity of this material with respect to NO_2 can be linked to the ability of Mo atoms to penetrate the surface layers of the SnO_2 grains and occupy lattice Sn positions in a variable oxidation state Mo^{+5} or Mo^{+6} [83–85]. Re-oxidation of Mo^{+5} to Mo^{+6} upon NO_2 adsorption, probably leads to the formation of additional acceptor energy levels in the band gap, which are accompanied by a sharp drop in the material's surface conductivity. For materials modified with NiO a similar effect of "receptor" sensitivity may underlie their high sensitivity to NO_2.

3.6.5 Acetone Sensing

The similarity of the sensor sensitivity diagrams for sensor materials studied in relation to acetone (Fig. 3.24) and to CO indicates a possible similarity of mechanisms for the sensor response.

However, the low-temperature sensor response mechanism, observed in the case of CO for the materials modified by clusters of Pd, is not found during the detection of acetone. This may be related to the blocking of the reverse spillover mechanism caused by the decrease in concentration or mobility of chemisorbed oxygen on the surface of the oxide matrix. This restriction is lifted out at higher sensor operating temperature that allows SnO_2-Ru-700 material to show a maximum sensitivity to acetone vapor at 300 °C, similar to the process of detecting CO.

The maximum sensor signal with respect to acetone is exhibited by Au-modified materials, but only at a 350 °C working temperature. This indicates the complexity of the conversion process of acetone on the surface of materials flowing through a set of parallel mechanisms, rather than the simple oxidation to CO_2 and water.

The concentration dependence of the sensor response toward acetone (Fig. 3.25) has a linear form in double logarithmic coordinates at low concentrations.

The position of the tipping point of this linear dependence is proportional to the rate constants of interaction between the reducing gas and ionized adsorbed oxygen forms. In the simplest case it can be stated that the O^- oxygen formed on the surface of materials governs the conductivity of sensor layer [86]. The concentration of O^- adsorbed on the surface is governed by the interrelationship of oxygen dissociative adsorption on the surface of semiconductor oxide, O^- recombination and desorption and the process of interaction between O^- and adsorbed gas molecules.

$$O_{2(gas)} + 2e^- \underset{k_{-1}}{\overset{k_1}{\longleftrightarrow}} 2O^-_{(ads)} \tag{3.40}$$

$$qO^-_{(ads)} + pA_{(ads)} \overset{k_2}{\longrightarrow} products_{(ads)} + qe^- \tag{3.41}$$

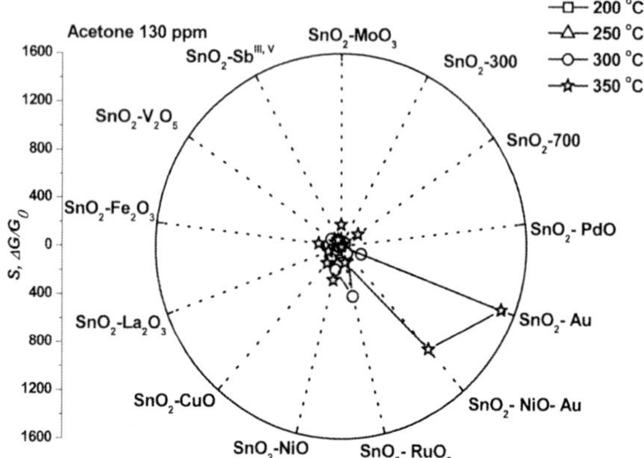

Fig. 3.24 Pattern of SnO$_2$-based materials sensor response to acetone (130 ppm in humid air, RH = 30 %)

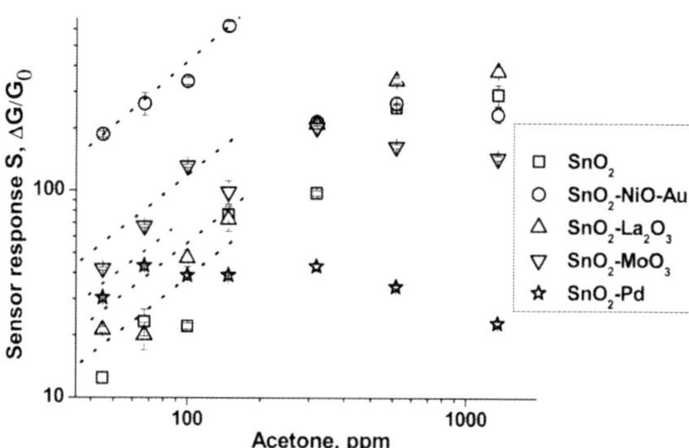

Fig. 3.25 *Calibration curves of SnO$_2$-based sensor materials sensor response to acetone (T = 400 °C, dry air, RH = 4 %)*

$$\frac{d[O_{ads}^-]}{dt} = k_1 P_{O_2}[e]^2 - k_{-1}[O_{ads}^-]^2 - k_2 P_A{}^p[O_{ads}^-]^q \tag{3.42}$$

where $O_{2(gas)}$ oxygen molecule in the gas phase, $O_{(ads)}^-$—ionized atomic form of adsorbed oxygen on the SnO$_2$ surface, $A_{(ads)}$- reducing gas molecule, adsorbed on the surface, P_{O2}- oxygen partial pressure in the gas phase, P_A- reducing gas partial pressure in the gas phase.

Since the sensor response is calculated at kinetic equilibrium state

$$\frac{d[O_{ads}^-]}{dt} = 0 \tag{3.43}$$

and the equation governing the [O$^-$] concentration could be written as

$$\frac{k_1}{k_{-1}} P_{O_2}[e]^2 = [O^-]^2 \left(1 + \frac{k_2}{k_{-1}} P_A{}^p[O^-]^{q-2}\right) \tag{3.44}$$

For the detailed analysis of this equation and the interrelationships between sensor material conductance, sensor response and gases partial pressures the reader is kindly asked to follow the [86] reference. For simplicity sake we must say that the sensor layer conductivity and sensor response is the interplay of semiconductor oxide physics and surface reactions chemistry. The relationship between the resistance of "ideal" SnO_2 with clean surface without any adsorbed oxygen or reducing gases (R_0) and actual resistance in air atmosphere with reducing gases present (R) is expressed as follows [86]:

$$\frac{R}{R_0} = \exp\left(\frac{m^2}{2}\right) \tag{3.45}$$

where

$$m = \frac{w}{L_D} \tag{3.46}$$

where w is the electron depletion layer thickness, L_D- Debye length—physical values, governing electric conductance of sensor layer. Further:

$$[e] = N_d \exp\left(-\frac{m}{2}\right) \tag{3.47}$$

$$[O^-] = N_d w \tag{3.48}$$

where N_d—concentration of donors, giving rise of electrons in conduction band of semiconductor oxide. In the case of SnO_2 them are oxygen vacancies in oxide lattice, serving as electron donors. The application of such notation gives one a possibility to introduce new form of Eq. 3.44:

$$\frac{\left(\frac{k_1}{k_{-1}}P_{O_2}\right)^{1/2}}{L_D} \cdot \exp\left(-\frac{m^2}{2}\right) = m\left(1 - \frac{k_2}{k_{-1}}P_A{}^p(N_d L_D m)^{q-2}\right)^{1/2} \tag{3.49}$$

Thus using new function, which is dependent only on reducing gas partial pressure at given reaction constants:

$$y = \frac{k_2}{k_{-1}}\left(N_d L_D\right)^{q-2}{}^{1/p} P_A \tag{3.50}$$

it is possible to numerically solve the relationship between R/R_0 (sensor response) and reducing gas concentration. The examples for different x (different k_1 to k_{-1} ratio) and different p and q (gas "reducing power") are shown on Fig. 3.26.

It must be noted that Yamazoe and Shimanoe [86] used R_0 as a resistance of semiconductor in a "flat band mode"—i.e. when nothing is adsorbed on the surface whereas we use R_0 for sensor response calculation as a resistance of semiconductor with surface covered with adsorbed oxygen to the highest extent possible in the measuring conditions. That's why the Fig. 3.25 looks like Fig. 3.26 rotated upside down.

According to this simplified model the increase of reducing gas partial pressure leads to a linear increase of sensor response (in double logarithmic coordinates)

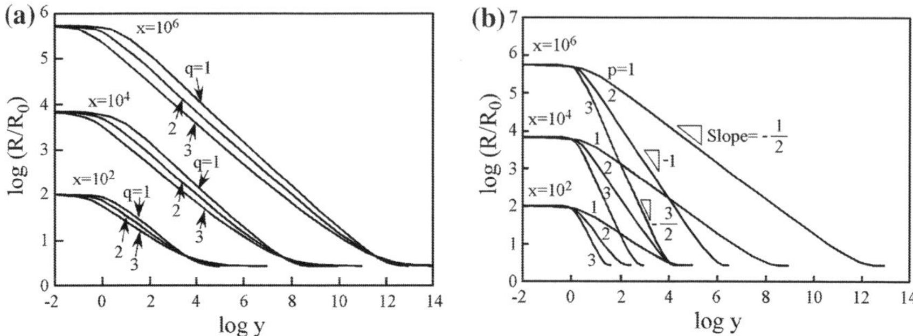

Fig. 3.26 Reduced resistance, R/R_0, as correlated with reduced reactivity of reducing gas, $y = \frac{k_2}{k_{-1}} (N_d L_D)^{\frac{q-2}{p}} P_A$ on logarithmic scales for complex surface reactions where q (a) and p (b) are varied. Variable $x = \frac{\frac{k_1}{k_{-1}} P_{O_2}}{L_D}$ reflects kinetic constant ratio given oxygen partial pressure is constant. ([86] With permission.)

only in restricted range of reducing gas partial pressure (i.e. reducing gas concentration). The k_2/k_{-1} ratio and p coefficient, given the oxygen partial pressure and k_1/k_{-1} ratio are fixed, governs the value of reducing gas concentration above which the further increase in concentration leads to no further increase in conductivity. Accordingly the sensor response also stops to rise. The height of linear slope above the X-axis (sensor response value) is governed by k_1/k_{-1} ratio—the higher the ratio, the bigger the response. The concentration regions of linear sensor response growth have the same slope in the case of all materials, with SnO_2-NiO-Au, demonstrating a significantly higher sensor response, which could be associated with a higher rate constant for dissociation of molecular O_2 on the surface of the material. In this sense the Au clusters role on the surface of SnO_2 in the increase of sensor response towards reducing gases is in the increase of k_1 constant—acceleration of molecular oxygen diffusion on the surface of oxide.

At the same time SnO_2-Pd-500 material manifests pronounced deviations in the form of calibration curve and the length of the concentration range with linear growth of sensor response. This notion is in line with considerations of differences in sensor response mechanisms for this material and other tested samples, based on oxidation processes carried on top of the Pd clusters coupled with adsorbed O^- transfer via reverse spillover mechanism.

In line with these arguments gas chromatographic analysis has shown a connection between a sharp rise in the sensor response with the degree of conversion of acetone (Fig. 3.27) as the working temperature rises. The analysis was made through the use of a special setup which allows the detection of products formed via acetone interaction with the sensor materials under the same conditions as the sensor experiments are made. This type of experiment can be generally characterized as an 'operando' experiment as discussed in the Gurlo chapter.

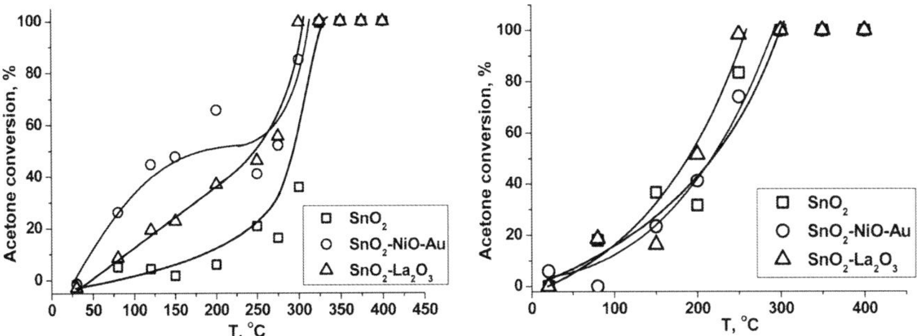

Fig. 3.27 Acetone conversion with temperature rise—1,800 (*left*) and 200 (*right*) ppm in dry air

Analysis of the chromatographic peaks (Fig. 3.28) suggests the formation of a large number of products of incomplete conversion of acetone at low operating temperatures and a decisive contribution of their complete oxidation to CO_2 and H_2O towards the sensor response magnitude. This in total can be described by the equation:

$$(CH_3)_2CO_{gas} + \frac{8}{\beta}O^{-\alpha}_{\beta(ads)} \rightarrow 3CO_{2gas} + 3H_2O_{(gas)} + \frac{8\alpha}{\beta}e^- \qquad (3.51)$$

At the same time GC/MS analysis showed a number of complex compounds, formed during interaction even at maximum operating temperatures that have bigger molar masses than that of acetone.

Two of such condensation products has been identified as 2,4-dimethylfuran and 3-pentene-2-on :

$$(3.52)$$

$$(3.53)$$

The accumulation of such substances on the sensor material's surface leads to the blocking of the molecular O_2 adsorption and dissociation centers, which is

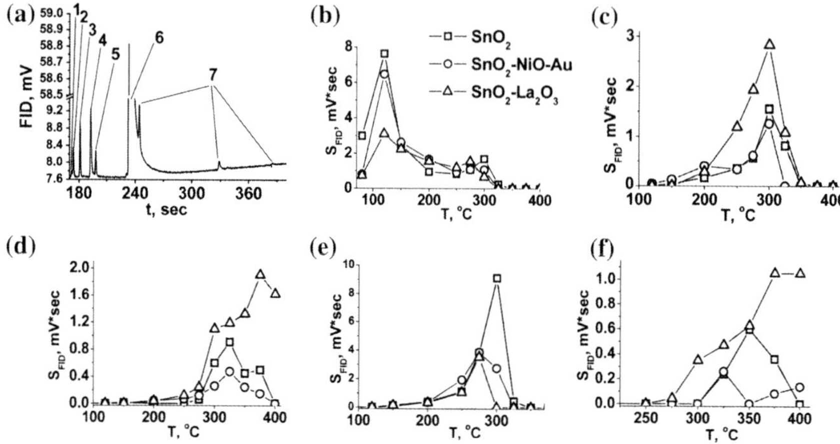

Fig. 3.28 **a** Chromatogram of products mixture obtained after acetone (1,800 ppm) interaction with sensor material (SnO$_2$-500) at T = 300 °C. **b–f** Comparative graphs of absolute content in products mixture exhausted from reactor vs. temperature of acetone (1800 ppm) interaction with sensor material (SnO$_2$-500)

Table 3.4 Selection of modifiers for SnO$_2$-based gas sensors with enhanced selectivity

Target gas	Interaction with the surface of n-type semiconductor oxide, resulting in conductance change	Suitable modifier
Reducing gases		
without pronounced acid/base properties: CO, H$_2$, CH$_4$	Oxidation by chemisorbed oxygen	Noble metals and their oxides, Au
Lewis bases: NH$_3$, amines	Oxidation by chemisorbed oxygen; Adsorption on surface acid sites	Metal oxides more acidic than SnO$_2$: V$_2$O$_5$, MoO$_3$, WO$_3$
Acids: H$_2$S	Oxidation by chemisorbed oxygen; Heterolytic reaction with Lewis acid/base pair on oxide surface	Metal oxides more basic than SnO$_2$: Fe$_2$O$_3$, In$_2$O$_3$, La$_2$O$_3$
Complex organic molecules with different functional groups:	Oxidation by chemisorbed oxygen;	Noble metals and their oxides, Au
Oxidizing gases		
O$_2$, NO$_2$, O$_3$	Chemisorption with localization of electrons from oxide conduction band	Noble metals and their oxides, Au; Electron donor additives: Sb(V); Modifiers, which can increase their oxidation state during interaction with oxidizing gas Mo(V), Ni(II)

accompanied by decrease in the conversion of acetone via the total oxidation reaction (3.40), leading to a decrease in sensor response.

Comparative analysis has shown that introduction of catalytic clusters of Au on the SnO_2 surface, leads to an increase in acetone conversion, a reduction in the formation of heavy products of the acetone conversion on the surface of the SnO_2-based sensor materials and an increase in the value of the sensor signal perhaps due to enhanced acetone combustion to CO_2 and H_2O.

3.7 Concluding Remarks

On the basis of the sensor response patterns towards CO, acetone, NH_3, H_2S and NO_2 one can conclude that chemical modification of SnO_2 with different catalysts provides an effective route for tailoring sensor materials with enhanced selectivity towards the detection various gas molecules. To predict the most suitable modifier to maximize the SnO_2 sensor response to any particular gas it is necessary to analyze the chemical nature of the interaction between semiconductor matrix, modification agent and the target gas molecule. These conclusions are summarized in Table 3.4:

However, it is currently impossible to establish a parameter, which would clearly link the fundamental properties of modifiers with their catalytic activity and their influence on SnO_2 sensor properties. The final optimization of the sensor material should be based on the requirements determined by practical tasks: sensitivity, response and recovery time, energy consumption, working temperature, precision of measurements, long term stability. In each case it is necessary to take into account the following factors:

- Tin oxide nanostructure;
- Concentration of modifier and its distribution between surface and bulk of tin oxide crystallites;
- Humidity;
- Target gas concentration range;
- Possible poisoning of sensor material.

Accounting for these factors still require detailed experimental studies to resolve specific practical problems.

References

1. Takahata K (1988) Tin oxide sensors—development and applications, highly sensitive SnO_2 gas sensors for volatile sulfides. In: Seiyama T (ed) Chemical sensor technology. Elsevier, Amsterdam, pp 39–55
2. Souteyrand E (1997) Transduction electrique pour la detection de gaz. Les capteurs chimiques, CMC2, Lyon, pp 15–35

3. Gouma PI (2003) Nanostructured polymorphic oxides for advanced chemosensors. Rev Adv Mater Sci 5:147–154
4. Norris JOW (1987) The role of precious metal catalysts in solid state gas sensorsgas sensors. In: Mosely PT, Tofield BC (eds) Alam higher. Bristol and Philadelphia, pp 124–138
5. Idriss H, Barteau MA (2000) Active sites on oxides: from single crystalsingle crystals to catalysts. Adv Catal 45:261–331
6. Vol'kenstein FF (1963) The electronic theory of catalysis on semiconductors. Pergamon, Oxford
7. Calatayud M, Markovits A, Menetrey M, Mguig B, Minot C (2003) Adsorption on perfect and reduced surfaces of metal oxides. Catal Today 85:125–143
8. Busca G (1999) The surface acidity of solid oxides and its characterization by IR spectroscopic methods. An attempt at systematization. Phys Chem 1:723–736
9. Barthomeuf D (1984) Conjugate acid-base pairs in zeolites. J Phys Chem 88:42–45
10. Niwa M, Habuta Y, Okumura K, Katada N (2003) Solid acidity of metal oxide monolayer and its role in catalytic reactions. Catal Today 87:213–218
11. Auroux A, Gervasini A (1990) Microcalorimetric study of the acidity and basicity of metal oxide surfaces. J Phys Chem 94:6371–6379
12. Duffy A, Ingram MD (1971) Establishment of an optical scale for Lewis basicity in inorganic oxyacids, molten salts, and glasses. J Am Chem Soc 93:6448–6454
13. Duffy A (1993) A review of optical basicity and its applications to oxidic systems. Geochim Cosmochim Acta 57:3961–3970
14. Duffy JA (1986) Chemical bonding in the oxides of elements: a new appraisal. J Solid State Chem 62:145–157
15. Tessman JR, Kahn AH, Shockley W (1953) Electronic polarizabilities of ions in crystals. Phys Rev 92:890–895
16. Zhang Y (1982) Electronegativities of elements in valence states and their applications. 1 Electronegativities of elements in valence states. Inorg Chem 21:3886–3889
17. Zhang Y (1982) Electronegativities of elements in valence states and their applications. 2 A scale fro strength of Lewis acids. Inorg Chem 21:3889–3893
18. Portier J, Campet G, Etourneau J, Tanguy B (1994) A simple model for the estimation of electronegativities of cations in different electronic states and coordinations. J Alloys Compd 209:285–289
19. Portier J, Campet G, Etorneau J, Shastry MCR, Tanguy B (1994) A simple approach to materials design: role played by an ionic-covalent parameter based on polarizing power and electronegativity. J Alloys Compd 209:59–64
20. Moriceau P, Lebouteller A, Bodres E, Courtine P (1999) A new concept related to selectivity in mild oxidation catalysis of hydrocarbons: the optical basicity of catalyst oxygen. Phys Chem 1:5735–5744
21. Sanderson RT (1951) An interpretation of bond lengths and a classification of bonds. Science 114:670–672
22. Sanderson RT (1975) The interrelationship of bond dissociation energies and contributing bond energies. J Am Chem Soc 97:1367–1372
23. Jeong NCh, Lee JS, Tae EL, Lee YoJ, Yoon KB (2008) Acidity scale for metal oxides metal oxides and sanderson's electronegativities of Lanthanide Elements. Angew Chem Int Ed 120:10282–10286
24. Henry M (1994) Partial charges distributions in crystalline materials through electronegativity equalization. Mater Sci Forum (152–153), pp 355–358
25. Nortier P, Borosy AP, Allavena M (1997) ab initio Hartree-Fock study of bronsted acidity at the surface of oxides. J Phys Chem B 101:1347–1354
26. Berholic J, Horsley JA, Murrel LL, Sherman LG, Soled S (1987) Bronsted acid sites in transition metal transition metal oxide catalysts: modeling of structure, acid strengths and support effects. J Phys Chem 91:1526–1530
27. Shiga A, Katada N, Niva M (2006) A theoretical study on bronsted acidity of WO_3 clusters supported on metal oxide supports by "paired interacting orbitals" (PIO) analysis. Catal Today 111:333–337

28. Mars P, van Krevelen DW (1954) Oxidations carried out by means of vanadium oxide catalysts. Chem Eng Sci (Spec Suppl) 3:41–59
29. Ali AM, Emanuelsson EAC, Patterson DA (2010) Photocatalysis with nanostructured zinc oxide thin films: the relationship between morphology and photocatalytic activity under oxygen limited and oxygen rich conditions and evidence for a Mars—Van Krevelen mechanism. Appl Catal B(97):168–181
30. Finocchio E, Busca G, Lorenzelli V, Willey RJ (1994) FTIR Studies on the selective oxidation and combustion of light hydrocarbons at metal oxide surfaces. J Chem Soc Faraday Trans 90:3347–3356
31. Over H, Kim YD, Seitsonen AP, Wendt S, Lundgren E, Schmid M, Varga P, Morgante A, Ertl G (2000) Atomic-scale structure and catalytic reactivity of the $RuO_2(110)$ surface. Science 287:1474–1476
32. Han W-P, AI M (1982) The φ-classification of metal oxides for heterogeneous oxidation catalysts. J Catal 78:281–288
33. Bielański A, Haber J (1979) Oxygen in catalysis on transition metal oxides. Cat Rev—Sci Eng 19:1–41
34. Catlow CRA, Jackson RA, Thoma JM (1990) Computational studies of solid oxidation catalystst. J Phys Chem 94:7889–7893
35. Reddy BM (2006) Redox properties of metal oxides. In: Fierro JLG (ed) Metal oxides: chemistry and applications. CRC Press, Taylor & Francis Group, Boca Raton, pp 215–246
36. Busca G, Finocchio E, Ramis G, Ricchiardi G (1996) On the role of acidity in catalytic oxidation. Catal Today 32:133–143
37. Panov GI, Dubkov KA, Starokon EV (2006) Active oxygen in selective oxidation catalysis. Catal Today 117:148–155
38. Liu H-F, Liu R-S, Liew KY, Johnson RE, Lunsford JH (1984) Partial oxidation of methane by nitrous oxide over molybdenum on silica. J Am Chem Soc 106:4117–4121
39. Schlogl R (2001) Theory in heterogenous catalysis. Cattech 5:146–170
40. Grzybowska-Swierkosz B (2002) Effect of additives on the physicochemical and catalytic properties of oxide catalysts in selective oxidation reactions. Top Catal 21:35–46
41. Rumyantseva MN, Gaskov AM (2008) Chemical modification of nanocrystalline metal oxides: effect of the real structure and surface chemistry on the sensor properties. Russ Chem Bull 57:1106–1125
42. Yushchenko VV (1997) Calculation of the acidity spectra of catalysts from temperature-programmed ammonia desorption data. Russian J Phys Chem 71:547–550
43. Burmistrov VA (1986) Hydrated oxides of IV and V groups (in Russian). Nauka, Moscow, p 160
44. Abee MW, Cox DF (2002) NH_3 chemisorption on stoichiometric and oxygen-deficient SnO_2 (110) surfaces. Surf Sci 520:65–77
45. York SC, Abee MW, Cox DF (1999) α-Cr2O3 (#): surface characterizationcharacterization and oxygen adsorption. Surf Sci 437:386–396
46. Abee MW, Cox DF (2001) BF3 adsorptionadsorption on r-Cr2O3 (#): probing the lewis basicity of surface oxygen anions. J Phys Chem B 105:8375–8380
47. Maier J, Göpel W (1988) Investigations of the bulk defect chemistry of polycrystalline Tin(IV) oxide. Thin Solid Films 72:293–302
48. De Wit JHW, Van Unen G, Lahey M (1977) Electron concentration and mobility in In_2O_3. J Phys Chem Solids 38:819–824
49. Barsan N, Weimar U (2001) Conduction model of metal oxide gas sensors. J Electroceram 7:143–167
50. Morrison SR (1977) The chemical physics of surfaces. Plenum, New York
51. Rumyantseva MN, Makeeva EA, Badalyan SM, Zhukova AA, Gaskov AM (2009) Nanocrystalline SnO_2 and In_2O_3 as materials for gas sensors: the relationship between microstructure and oxygen chemisorption. Thin Solid Films 518:1283–1288
52. Descorme C, Duprez D (2000) Oxygen surface mobility and isotopic exchange on oxides: role of the nature and the structure of metal particles. Appl Catal A 202:231–241

53. Martin D, Duprez D (1996) Mobility of surface species on oxides. 1 Isotopic exchange of $^{18}O_2$ with ^{16}O of SiO_2, Al_2O_3, ZrO_2, MgO, CeO_2, and CeO_2-Al_2O_3. Activation by noble metals. Correlation with oxide basicity. J Phys Chem 100:9429–9438

54. Haruta M (2004) Nanoparticulate goldgold catalysts for low-temperature CO oxidation. J New Mat Electrochem Systems 7:163–172

55. Montmeat P, Marchand J-C, Lalauze R, Viricelle J-P, Tournier G, Pijolat C (2003) Physico-chemical contribution of goldgold metallic particles to the action of oxygen on tin dioxide sensors. Sens Act B 95:83–89

56. Conner WC, Falconer JL (1995) Spillover in heterogeneous catalysis. Chem Rev 95:759–788

57. Aksel S, Eder D (2010) Catalytic effect of metal oxides on the oxidation resistance in carbon nanotube-inoragnic hybrids. J Mater Chem 20:9149–9154

58. http://webbook.nist.gov

59. Davydov AA (2003) Molecular spectroscopy of oxide catalyst surfaces. Wiley, Chichester, p 641

60. Zhou Z, Gao H, Liu R, Du B (2001) Study of structure and property for the electron transfer system. Theochem 545:179–186

61. Šulka M, Pitoňák M, Neogrády P, Urban M (2008) Electron affinity of the O2 molecule: CCSD(T) calculations using the optimized virtual orbitals space approach. Int J Quantum Chem 108:2159–2171

62. Sergent N, Epifani M, Comini E, Faglia G, Pagnier T (2007) Interactions of nanocrystalline tin oxide powder with NO_2: a Raman spectroscopic study. Sens Act B 126:1–5

63. Prades JD, Cirera A, Morante JR, Pruned JM, Ordejón P (2007) Ab initio study of NOx compounds adsorption on SnO_2 surface. Sens Act B 126:62–67

64. Leblanc E, Perier-Camby L, Thomas G, Giber G, Primet RM, Gelin P (2000) NOx adsorption onto dehydroxylated or hydroxylated tin oxide surface. Application to SnO2-based gas sensorsgas sensors. Sens Act B 62:67–72

65. Golodets GI, Borovik VV, Vorotyntsev VM (1986) Mechanism and kinetics of selective catalytic oxidation of acetone. Theor Exper Chem 22:235–237

66. Harrison PG, Thornton EW (1976) Tin oxide surfaces. Part 7—an infrared study of the chemisorption and oxidation of organic Lewis base molecules on tin (IV) oxide. J Chem Soc Faraday Trans I(72):2484–2491

67. Pierce KG, Barteau MA (1995) Ketone coupling on reduced TiO_2 (001) surfaces: evidence of pinacol formation. J Org Chem 60:2405–2410

68. Barŝan N, Koziej D, Weimar U (2007) Metal oxide-based gas sensor research: How to? Sens Act B 121:18–35

69. Hirvi JT, Kinnunen T-JJ, Suvanto M, Pakkanen TA, Nørskov JK (2010) CO oxidation on PdO surfaces. J Chem Phys 133:084704

70. Šljivančanin Ž, Hammer B (2010) CO oxidation on fully oxygen covered Ru (0001): role of step edges. Phys Rev B 81:121413 R

71. Seitsonen AP, Over H (2009) Intimate interplay of theory and experiments in model catalysiscatalysis. Surf Sci 603:1717–1723

72. Wang S, Wang Y, Jiang J, Liu R, Li M, Wang Y, Su Y, Zhu B, Zhang S, Huang W, Wu S (2009) A DRIFTS study of low-temperature CO oxidation over AuAu/SnO_2 catalyst prepared by co-precipitation method. Catal Commun 10:640–644

73. Hakkinen H, Abbet S, Sanchez A, Heiz U, Landman U (2003) Structural, electronic and impurity-doping effects in nanoscale chemistry: supported goldgold nanoclusters. Angew Chem Int Ed 42:1297–1300

74. Rout CS, Hegde M, Rao CNR (2007) Ammonia sensors based on metal oxide nanostructures. Nanotechnology 18:205504

75. Mortensen H, Diekhoner L, Baurichter A, Jensen E, Luntz AC (2000) Dynamics of ammonia decomposition on Ru (0001). J Chem Phys 113:6882–6887

76. Marikutsa AV, Rumyantseva MN, Gaskov AM, Konstantinova EA, Grishina DA, Deygen DM (2011) CO and NH_3 sensor properties and paramagnetic centers of nanocrystalline SnO_2 modified by Pd and Ru. Thin Solid Films. Accepted for publication

77. Ganley JC, Thomas FS, Seebauer EG, Masel RI (2004) A priori catalytic activity correlations: the difficult case of hydrogen production from ammonia. Catal Lett 96:117–122
78. Hansgen DA, Vlachos G, Chen JG (2010) Using first principles to predict bimetallic catalysts fort he ammonia decomposition reaction. Nat Chem 2:484–489
79. Boccuzzi F, Guglielminotti E (1994) IR study of TiO_2-based gas-sensor materials: effect of ruthenium on the oxidation of NH3, $(CH_3)_3N$ and NO. Sens Act B 21:27–31
80. Pagnier T, Boulova M, Galerie A, Gaskov A, Lucazeau G (2000) Reactivity of SnO_2–CuO nanocrystalline materials with H_2S: a coupled electrical and Raman spectroscopic study. Sens Act B 71:134–139
81. Jeon H, Jeon M, Kang M, Lee S, Lee Y, Hong Y, Choi B (2005) Synthesis and characterization of antimony doped tin oxide (ATO) with nanometer-sized particles and their conductivities. Mater Lett 59:1801–1810
82. Rumyantseva MN, Gaskov AM, Rosman N, Pagnier T, Morante JR (2005) Raman surface vibration modes in nanocrystalline SnO_2 prepared by wet chemical methods: correlations with the gas sensors performances. Chem Mater 17:893–901
83. Kaur J, Vankar VD, Bhatnagar MC (2008) Effect of MoO_3 addition on the NO_2 sensing properties of SnO_2 thin films. Sens Act B 133:650–655
84. Ivanovskaya M, Lutynskaya E, Bogdanov P (1998) The influence of molybdenum on the properties of SnO_2 ceramic Sensors. Sens Act B 48:387–391
85. Chiorino A, Ghiotti G, Prinetto F, Carotta MC, Gallana M, Martinelli G (1999) Characterization of materials for gas sensors: surface chemistry of SnO_2 and MoO_x–SnO_2 nano-sized powders and electrical responses of the related thick films. Sens Act B 59:203–209
86. Yamazoe N, Shimanoe K (2008) Theory of power laws for semiconductor gas sensors. Sens Act B 128:566–573
87. Gellings PJ, Bouwmeester HJM (2000) Solid state aspects of oxidation catalysis. Catal Today 58:1–53

Chapter 4
Combinatorial Approaches for Synthesis of Metal Oxides: Processing and Sensing Application

Clemens J. Belle and Ulrich Simon

Abstract This chapter gives an overview about the application of metal oxides in chemiresistors. A generalized model of working principle and the influence of particle size, microstructure, volume and surface doping are discussed. The quality factors of sensor performance and the necessity of high-throughput experimentation and combinatorial techniques for the development of new sensor materials are explained. In this context high-throughput impedance spectroscopy is presented as a rapid characterization method of a large number of samples. The complete workflow is introduced involving material synthesis and analysis, layer preparation by a laboratory robot, impedometric characterization and automated data evaluation. As examples two series of surface and volume doping demonstrate the systematic identification of new sensor materials.

4.1 Introduction

4.1.1 Sensor Concepts and Fields of Application

Gas sensor elements are employed in a widespread application spectrum, e.g. intelligent process management, environmental protection and medical diagnostic, in order to detect and to monitor a multitude of pure and mixed gases. The reliable detection, particularly of explosive, toxic or hazardous components is essential due to the continuously increasing requirements in safety and environmental regulations. Thus, the worldwide growing sensor technology market expands in both, the private

C. J. Belle · U. Simon (✉)
Institute of Inorganic Chemistry and JARA—Fundamentals of Information Technology,
RWTH Aachen University, Landoltweg 1, 52074 Aachen, Germany
e-mail: usimon@ac.rwth-aachen.de

M. A. Carpenter et al. (eds.), *Metal Oxide Nanomaterials for Chemical Sensors*,
Integrated Analytical Systems, DOI: 10.1007/978-1-4614-5395-6_4,
© Springer Science+Business Media New York 2013

and the industrial sectors and concerns nearly all areas of daily life, such as the domestic (e.g. automobile, fire alarm or humidity control), the food (e.g. fermentation or maturation), the medical (e.g. diagnostics or patient monitoring), the security (e.g. explosive or hazardous substances) and the industrial (e.g. technical processes or exhaust gases).

Already for a long period the conventional analytical methods such as mass spectroscopy, gas chromatography or optical techniques (e.g. IR and UV/Vis) are capable of precisely and selectively measuring traces of a single compound in a complex gas mixture. However, the main drawbacks of these extractive methods are the stationary operation due to their physical size of instrumentation and their total cost linked with manpower and logistics. Therefore, modern application fields require portable, user-optimized and economic devices that are adapted to mass production markets. These practical requirements are fulfilled by solid state gas sensors developed so far, because they allow in situ and immediate measurements at varying locations. Nevertheless, with respect to the above mentioned application fields in the private as well as in the industrial sector the demand for new, highly sensitive and selective gas sensor materials is steadily growing.

All existing sensor concepts are based on the transformation of chemical or rather physical interactions of the sensor material with the target analyte in the surrounding gas phase into an analytically manageable signal. Examples of different detecting principles are calorimetric sensors (pellistors) [1], electrochemical cells (e.g. λ-probe) [2], field-effect transistors [3], acoustic wave devices [4] and quartz microbalances [5 , 6]. The electrical output signal as the final result of solid–gas interactions consists of a change in capacitance, potential or conductance of active sensing elements. The electrochemical sensor devices are commonly classified according to the respective type of measuring principle as amperometric, potentiometric or conductometric gas sensors [7–10].

4.1.2 Chemiresistors

Conductometric or resistive gas sensors, so-called chemiresistors, are typically based on metal oxides. This type of solid-state gas sensor is the most extensively applied type because of the thermal and chemical stability of metal oxides. Generally, the gas sensitivity of metal oxides is connected with the change of electrical resistance (i.e. conductivity) upon surface reactions with the reactive analyte gas (es). This effect was initially reported by Brattain and Bardeen [11] in 1953 for the semiconducting metal Ge and by Heiland [12, 13] shortly afterwards for the semiconducting ceramic ZnO. During the following decade Seiyama [14, 15] and Taguchi [16] did a lot of pioneering work that finally yielded the first commercial product a fire alarm sensor. The metal oxides which are today typically used as gas sensors are in decreasing order of their frequency of occurence: SnO_2 (35 %), ZnO (10 %), TiO_2 (7 %), WO_3 (7 %), In_2O_3 (5 %), Nb_2O_5 (3 %), Ga_2O_3 (3 %), Fe_2O_3 (3 %), various mixed metal oxides as $BaTiO_3$ and Ba_2WO_5 as well as mixtures of metal oxides as SnO_2-In_2O_3

and ZnO-SnO$_2$ (a total of 13 %) [17]. This order of occurence is basically the same regarding their percentages of publications during the last three decades [18].

The example from this group of initially investigated binary oxides is In$_2$O$_3$. The n-type semiconductor possesses a high electrical conductivity due to a large number of oxygen defects in the lattice [19]. Because of it's history it is, besides SnO$_2$, the most studied resistive gas sensor material and is reported to be sensitive towards small concentrations of oxidizing (NO [20], NO$_2$, O$_3$ [21], SO$_2$ and Cl$_2$) as well as reducing gases (H$_2$ [22] and CO [22–24]).

While the first types of gas sensors were composed of binary metal oxides more recent examples of chemiresistors are mainly based on ternary oxides, such as rare-earth perovskites, LnBO$_3$ (Ln = lanthanide series; B = Cr and Fe), which are also applied as materials in electrodes of fuel cells [25], (photo) catalysts [26, 27] and magneto-optics [28]. In 1976 perovskites have been reported as quantitative gas sensing materials towards C$_2$H$_5$OH (ethanol) at 150–400 °C due to their thermal and chemical inertness [29]. In particular, several orthoferrites of which LaFeO$_3$ and SmFeO$_3$ are the best studied materials among this class of substances [30, 31] have been investigated concerning their sensitivity towards CH$_3$OH (methanol) [32] and NO$_2$ as well [33]. Most recently LaFeO$_3$ has been the subject of first-principle studies of the molecular oxygen adsorption on Fe ion sites [34] and LnFe$_{0.9}$Mg$_{0.1}$O$_3$ (Ln = Nd, Sm, Gd and Dy) has been tested for C$_2$H$_5$OH sensitivity [35]. Unlike the group of the orthoferrites, the group of the orthochromites, with LaCrO$_3$ [36, 37] as the most common composition, has widely been left unattended [32]. Ilmenite oxides, as for example CoTiO$_3$, have been investigated in respect of humidity detection in combination with Ta$_2$O$_3$ [38], but they are also of interest as catalytic [39, 40] or high-κ dielectric [41] materials. The p-type semiconductor CoTiO$_3$ was first published as an ethanol detecting material in 1999 [42].

In the last decades many attempts have been made to overcome the limitations of the first unmodified oxides and to discover new sensor materials with improved or even tailored properties. Two common pathways to strongly influence the sensor performance of a given base material are surface and volume doping by various elements or oxides. But for detailed investigations of the influence of a specific parameter or dopant the discrimination of all other side effects is essential. Thus, novel methods, such as high throughput experimentation (HTE) and combinatorial techniques, are applied which allow for keeping most parameters constant while selectively varying just one. The application of these methods accelerate both, the sample preparation and characterization by ensuring identical sample processing as well as the systematic evaluation of reliable statistical data. In order to rapidly classify and compare the obtained gas sensor performances basically four essential quality factors are used which are specified in the following section.

4.1.3 Quality Factors of Sensor Performance

A perfect gas sensor acts precisely and immediately to the prevailing ambient chemical or atmospheric conditions. The four crucial characteristics which determine the sensing quality of a certain material are the sensitivity, the selectivity, the response time and the stability.

The sensitivity S is defined as the slope of a calibration curve and thus, is a function of the specific analyte concentration. It is described as the ratio of the change in the output signal y (i.e. the change in electrical resistance in the case of chemiresistors) to the concentration x of the analyte gas (Eq. 4.1). Hence, the sensitivity describes the sensibility of a sensor material to detect a certain analyte concentration from an ambient atmosphere [43]. However, the sensor response is more commonly used to express different calculated ratios of the analyte resistance to the reference resistance (cf. Sect. 5.1.2).

$$S = \frac{dy}{dx} \tag{4.1}$$

The selectivity Q reflects the ability of a sensor to differentiate between the specific gas x to be detected and the other components of the gaseous environment x', and expresses the cross-sensitivity (Eq. 4.2).

$$Q(\%) = 100 \cdot \frac{dy/dx'}{dy/dx} \tag{4.2}$$

The response behavior expresses numerically the attended response or recovery time $\Delta t_{90\ \%}$ of the output signal to reach 90 % of its saturation value after turning on or off an analyte gas [43]. The response behavior is determined by the chemical reaction kinetics between material surface and analyte molecules, and therefore, by diffusion processes and surface reactions.

The stability describes the endurance of a sensor material to maintain its output signal over a long period of time and to target systematically varying concentrations of analyte gas. This reproducibility can be affected by (thermal) aging of the sensor layer as well as by poisoning of the sensor's surface comparable to the properties of catalysts. The stability concerns both, the structural and the electrophysical properties of the sensor material during long-term operation which is an often neglected demand in research but essential for commercial products.

4.1.4 High Throughput and Combinatorial Approaches

Facing the rapid increasing economic demands and challenging technical standards the research denoted to the development of new functional sensor materials has attracted great attention to the materials research community [44]. Thereby, the fundamental challenge consists in identifying appropriate sensor materials among the almost infinite pool of possible material compositions with respect to the possible combinations of all chemical elements that form ternary or even multinary metal oxides, independently of further modifications. Furthermore, the significant but unpredictable impact of various parameters deriving e.g. from electrode layout, preparation history or after treatment (e.g. electrode parameters, precursors, chemical composition, synthesis conditions, additives, doping, modification, annealing conditions etc.) on the sensor properties requires novel working methods that accelerate the classical one-by-one approach in finding new improved materials.

Since 1970 a "multiple sample concept (MSC)" or rather the vision of a high-throughput experimentation (HTE) technique is successfully applied in contrast to the conventional "one sample at a time" method. The MSC enables the simultaneous processing of "expensive and time-consuming" steps in an integrated material development workflow [45]. A high degree of automation and robotics, ranging from material synthesis and sensor fabrication, up to analytical methods and property characterizations, provides the comparability of samples by varying single parameters while keeping all others constant. Thus, the HTE procedure implements sample plates which facilitate the parallel processing and easy handling of up to several dozen samples which are distinguished by a composition spread (gradient library of one parameter) [46]. However, the large quantity of samples produced by the HTE procedure on a short time scale requires at the same time adequate rapid analytic and characterization methods. As such a rapid testing method, in particular the impedometric characterization, which is also applied in the material development including superconducting [47], dielectric [48], magneto-resistive [49], photochemical active [50] and catalytically active materials [51, 52], has proven to be an extremely valuable tool in the evolution of novel gas sensing materials [53].

Besides HTE, the usage of combinatorial approaches via the predecessor MSC advances the conventional "empirical trial and error process" by intelligent combination of different materials or components of the synthesis workflow [54]. For the combinatorial purpose as a first step, an expanded diverse parameter space is constituted by the discovery strategy (primary screening) to find novel or alternative materials. Then, the restriction to a realistic well-defined space is applied by the material development (secondary screening) in order to optimize hits of the previous generation or parameters of established materials. A statistical and intelligent design of experiments (DoE) should minimize the experimental operating expense and maximize the knowledge gained on the basis of multidimensional parameter optimization by descriptors, genetic algorithms (GA), artificial neural networks (ANN) and last but not least by experience [55, 56]. Initial experiments utilizing the combinatorial approach have been published in the 1960s in the context of material science. In the 1990s these strategies have been established by the pharmaceutical and biotechnological industries [57].

Combining the HTE and the combinatorial approach accelerates the improvement of present sensor materials as well as the development of novel complex sensor materials by a factor of about 10–100 and leads to extensive material libraries [55]. Such material libraries differ from each other in the systematic variations of a single or a few specific parameters (e.g. synthesis condition, analyte gas, calcination temperature etc.) or in the base material (e.g. different volume additives) and the surface dopant, respectively.

The large amount of individual information recorded throughout all working and measuring steps is collected in a database which is used for statistical interrogation of quality factors as well as for data mining. The nontrivial data interpretation via statistical analyses leads to a knowledge discovery (KD) that enables the mapping of composition-structure–property relationships, a process which was first published in 1999 for the catalytic activity towards the oxidative dehydrogenation of ethane and its dependence on the chemical composition of doped Mo-V-Nb-oxides [58].

In summary, the combination of the HTE and the combinatorial approaches in materials and sensors research, as recently reviewed [55, 59], has a huge impact on the research of enhanced sensor materials. Specifically, it helps to understand correlations between chemical composition, microstructure and sensor performance. But until now findings related to this correlation are not very common for chemiresistors. For the future it is expected that the KD will help to draw conclusions about the sensing mechanism in a qualitative and quantitative way, which is actually of much academic interest and studied by novel in situ or operando spectroscopic methods [9, 10, 60, 61].

4.2 Metal Oxide Gas Sensors

This subchapter addresses the basic working principles of chemiresistors exemplified by SnO_2 as sensor material and CO as reducing analyte gas. The overall parameters which influence the sensor performance refer to the structural properties of sensor layers and the chemical composition of sensor materials. Within the following two subchapters the particle size and the microstructure of the sensor material as well as the variation of the material composition by volume and/or surface doping are discussed in more detail [62].

4.2.1 Working Principle

The common working principle model for resistive gas sensors was developed concerning unmodified binary metal oxides, e.g. ZrO_2, TiO_2, WO_3, Fe_2O_3, In_2O_3, ZnO and SnO_2 [17, 63]. The latter two are well-known as polycrystalline thin-film materials in Figaro [14] and Taguchi sensors [16] (Figaro Inc.) and represent the first commercial application of resistive gas sensors in the 1960s. Since that time the wide-bandgap semiconductor SnO_2 is the best studied material in this field beside In_2O_3. [64–66] All these gas sensing materials of the so-called "first generation" have been n-type semiconductors in which charge transport results from

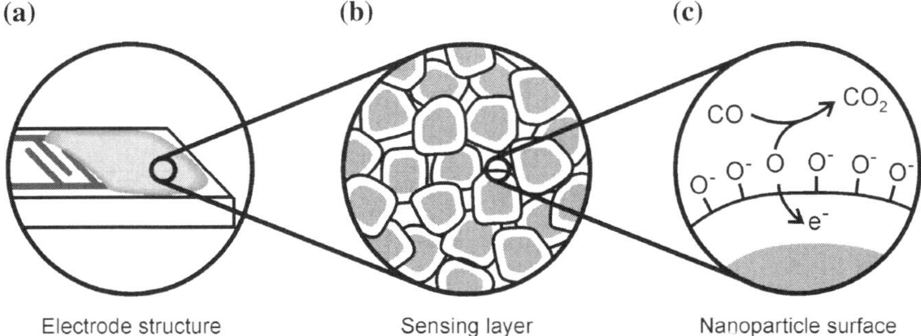

Fig. 4.1 Working principle of a semiconductor gas sensor: **a** output function (electrode structure). **b** transducer function (sensing layer) and **c** receptor function (nanoparticle surface) [adapted from 67]

mobile electrons. In contrast to that, the charge transport takes place via holes in p-type semiconductors as for example CoO and perovskite structures as $LnBO_3$ (Ln = lanthanide series, B = Cr and Fe) [17].

By zooming in on the active sensing element of a metal oxide gas sensor (Fig. 4.1), which is responsible for the output change in resistance (i.e. the sensor response) via an electrode structure (Fig. 4.1a), the metal-oxidic sensing layer upon it reveals a microstructure of nanoscaled particles (Fig. 4.1b). The gas-penetrable particle assembly is associated with a transducer function and accordingly the sensing surface of the metal oxide nanoparticles with a receptor function (Fig. 4.1c) [67, 68]. The specific interaction of the surface with target gas molecules causes a change in the electrical resistance of each grain. The transduction of this information on the molecular level results in a macroscopically accessible signal by the electrodes in terms of a change in the electrical conductivity of the entire layer.

A closer look at the sensing mechanism on the surface of nanoparticles reveals the following processes: first, ambient oxygen is physisorbed by van der Waals forces and then chemisorbed. The nature of the predominantly ionosorbed species has been detected by IR, TPD and EPR, and changes from molecular O_2^- (<420 K) to atomic O^- (420–670 K) and O^- parallel with O^{2-} (>670 K) before finally, the direct incorporation of the adsorbed species into the lattice takes place (>870 K) depending on the temperature [62, 63].

During the associated exothermal oxidation the adsorbed species become chemically bound as electron acceptors with a relative energetic position below that of the Fermi level E_F of the solid (Fig. 4.2, bottom). The required electrons are extracted from the conducting band E_C of the donor-solid which are intrinsic oxygen vacancies and trapped from the acceptor adsorbate at the surface leaving an electron-depleted region behind. [69–72] Thus, the depth of the positively-charged space charge layer Λ_{air} is simultaneously enlarged due to the negative surface potential. The maximum surface coverage is determined by the Weisz limitation to be about 10^{-3}–10^{-2} cm^{-1} ions denoting the equilibrium between the

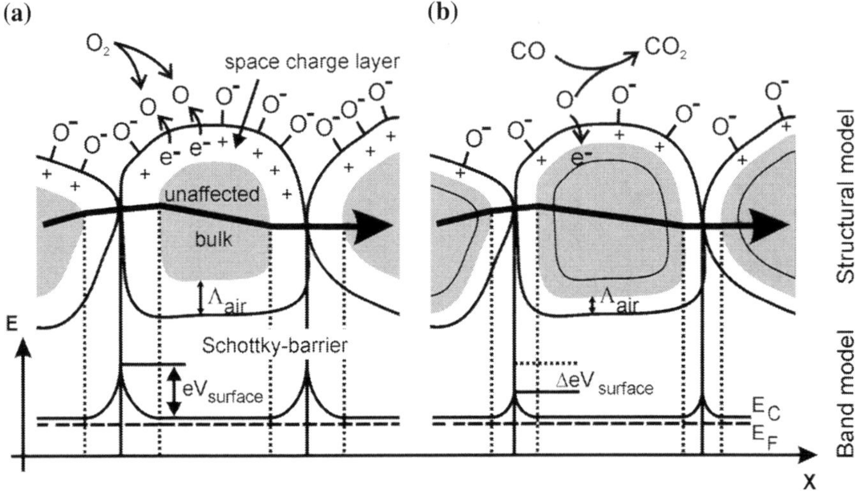

Fig. 4.2 Structural model (*top*) and band model (*bottom*) of the intergranular percolation path regarding a polycrystalline nanoparticle layer: **a** initial state upon exposure of air and **b** final state upon exposure of reducing CO gas on potential barrier eV_{sur} and space-charge layer Λ_{air} [adapted from 81, 82]

Fermi level and the site energy [73, 74]. Transferred to the energy-band model this negative surface potential causes a band bending of both, the valence band E_V (not shown) and the conducting band E_C, with a potential height of the surface potential barrier eV_{sur} of about 0.5–1.0 eV at the grain boundaries. Both, the depth and the height of this layer are affected by the amount and the nature of ionosorbed oxygen at the surface and their associated charges. Additionally, the depth of the layer corresponds to the Debye-length L_D which is characteristic for every semiconductor material and dependent on the temperature at a particular donor concentration (Eq. 4.3).

$$L_D = \sqrt{\frac{\varepsilon_0 \varepsilon k_B T}{e^2 n_d}} \tag{4.3}$$

The permittivity of free space is given by ε_0, the dielectric constant by ε, the Boltzmann constant by k_B, the operating temperature by T, the electron charge by e and the carrier concentration by n_d which corresponds to the donor concentration assuming full ionisation.

However, it is worth noting, that this idealised model conception [75] assumes the complete absence of humidity and the sufficient presence of oxygen (e.g. in synthetic air). But in every real system [76] the presence and the influence of water deriving from hydroxyl groups has to be taken into account of respective surface processes.

Besides the surface chemistry, the microstructure of sensing layer bridging two electrodes (Fig. 4.2a, top) plays a key role in the sensing mechanism. In the polycrystalline sensing layer the electrical conductivity occurs along percolation paths by means of grain-to-grain contacts of sintered nanoparticles. As a consequence of the increased potential barriers at the grain boundaries, so-called Schottky-barriers, the free electron flow within the layer is hindered. The height of the Schottky-barriers corresponds directly to the thermodynamic equilibrium of the oxygen adsorption and desorption reactions. This phenomenon results in an increased resistance. Overall, according to the total percolation paths along adjacent grains, a specific base resistance occurs which depends on the surface potential of the ambient atmosphere. The Maxwell–Boltzmann distribution describes the number of electrons which succeed in overcoming the potential barrier [77] and accordingly, the conductance G of the sensing layer is expressed by Eq. 4.4 [78].

$$G \approx exp\left(\frac{-eV_{sur}}{k_B T}\right) \tag{4.4}$$

Depending on the prevalent atmosphere the types of pre-adsorbed species (e.g. oxygen, hydroxyl groups, carbonates etc.) have a tremendous impact on the successive surface reactions and therefore on the sensing mechanism.

In the case of reducing gases (e.g. H_2, H_2S, CO, CH_4 and C_2H_5OH) first adsorption and then consecutive reactions via intermediates occur which result in a total oxidation to CO_2 and H_2O for hydrocarbons. For instance, in the presented case of CO (Fig. 4.2b, top) mono-dentate and/or bi-dentate bound carbonate groups are formed by reaction with the ionosorbed oxygen species and finally desorb as CO_2 (but for better clarity only O^- species are displayed). [79] In the last reaction step the previously localized electrons are released back to the bulk of the metal oxide by desorption of the reaction products. Thus, even traces of an analyte gas significantly increase the electron density and decrease the potential barrier at the intergrain contacts in comparison to the equilibrium state without the analyte. The material conductance of the entire layer and its change are reflected on the partial pressure of CO $[p_{CO}]$ in terms of a power-law dependence (Eq. 4.5) wherein, the exponent n depends on the morphology of the sensor layer and the properties of the sensor material [80].

$$G \approx [p_{CO}]^n \tag{4.5}$$

In contrast, for oxidizing gases (e.g. NO, NO_2, SO_2 and O_3) direct chemisorption leads to the opposite reaction, because then additional oxygen atoms are incorporated into the lattice by acceptance of electrons from oxygen vacancies of the solid. The additional electrons cause a decrease in conductivity. In the case of p-type semiconductors these two effects are vice versa. On closer inspection one has to consider the specific gas to be detected and the chemical reactions of the formed intermediates.

Summarized, the sensing mechanism is microscopically restricted to the particle surface. For polycrystalline materials the sensing effect macroscopically results in a measurable signal that is a change in resistance. This change strongly depends on the grain size of interconnected particles. Furthermore, since the working temperature influences the reaction rate of all surface processes as well as the interpenetration of the gas molecules into the porous sensing layer, every analyte gas exhibits a specific temperature of maximum sensitivity which depends on its surface coverage and the conversion rate of ionosorbed oxygen. At this favorable temperature the relative cross sensitivity towards other gases is suppressed to its lowest level [83, 84].

Concluding, the sensitivity of a given base material depends on several critical parameters, such as microstructure, chemical composition, after treatment and working temperature. The microstructure includes the particle size, the thickness and the porosity of the sensor layer. The chemical composition involves surface and volume doping of the base material with various other components. These parameters and their complex interplay on the sensing property are described in the following subchapters.

4.2.2 Particle Size

Considering that the sensing mechanism is directly linked to the surface reactions at the interface between the near-surface metal oxide and the ambient analyte molecules the sensitivity behavior is expected to increase with increasing surface area and consequently, decreasing particle size. Therefore, the particle size is one critical parameter to influence the sensing properties of a given base material. Nanoparticulate layers are inherently tailored for sensing applications, because of their tremendous surface-to-bulk ratio compared to their microcrystalline counterparts. Additionally, taking into account that the electrical properties are directly related to the intergranular percolation paths consisting of depleted zones near the surface and unaffected bulks in the centre of particles (Fig. 4.2) the response behavior should depend on the surface-to-bulk-ratio. Hence, new approaches follow the development of a hollow or hierarchical particle design to eliminate the ineffective core and to improve the sensing characteristics. These approaches have most recently been reviewed [85].

However, as the particle size is reduced, the radius will approach the dimension of the space-charge layers, eventually leading to their overlap. The progressive shrinking of the ineffective bulk also involves the Schottky-barriers at the grain boundaries. Accordingly, by further lowering the intergrain distance the energetic difference becomes smaller than the thermal energy and the heterogeneity of both regions vanishes. In this case the conductance G becomes proportional to the difference between the Fermi level E_F and the lower edge of the conduction band E_C (Eq. 4.6) [71, 81]

$$G \approx exp\left(\frac{-(E_C - E_F)}{k_B T}\right) \tag{4.6}$$

The special case, when the energy bands are nearly depleted of mobile charge carriers, is called "flat-band condition" ($eV_{sur} \leq k_B T$). At this point the bulk conductivity of the interconnected grains becomes negligible and the conductivity of the space-charge layers dominates the sensory properties [86, 87].

Based on the theoretical considerations related to the fundamental scale impact on the sensing performance, systematic investigations have been performed by Yamazoe and co-workers in the early 1990s in order to prove the theoretical model of the transducer function up to the range of flat-band conditions [66, 88–90] For that purpose various sintered SnO_2 elements have been produced which differ in the sensor material SnO_2 (foreign-oxide-stabilized, pure and impurity-doped SnO_2) and in the particle size (4–32 nm). Thereby, for nanoparticles bigger than about 20 nm an approximate independent linear correlation between the average particle diameter D and the sensor response (expressed as R_{gas}/R_{air}) towards H_2 (800 ppm) and CO (800 ppm) was determined at 573 K. For smaller particles the sensitivity starts to rise and then, a steep increase for crystallite sizes below about 10 nm was observed. This value exactly coincides with the range in which the diameter becomes equal and smaller than twice the depleted zone.

The authors explain the three different regions of grain size influence by a sensing layer model of partially sintered crystallites (Fig. 4.3). This model is based on the fact that the particles form smaller aggregates and that these aggregates are interconnected by grain boundary contacts (Fig. 4.3a, top). Thus, generally in the case of larger grains ($D \gg 2\,\Lambda$), the sensitivity or rather the conductivity is nearly independent of the grain size and mostly dominated by the bulk conductivity. In the case of middle-size grains ($D > 2\,\Lambda$) the sensitivity is affected by both, the grain boundaries and the grain size, because the depleted regions form conduction channels within every aggregate (constriction effect) (Fig. 4.3b, middle). At last, in the case of small grains ($D < 2\,\Lambda$) the whole crystallite is depleted, the conduction channels vanish and the conductivity becomes grain size controlled (Fig. 4.3, bottom) [66, 91].

Besides the above described grain model, another model defines the inter-granular necks between the polycrystalline grains as a further important parameter which controls the sensoric properties (Fig. 4.3b) [66, 92]. According to this model the width of necks, which depending on the sample history starts forming between partly sintered particles at higher annealing temperatures (>700–800 °C), [66] is responsible for the height of the potential barrier. The length of necks determines the width of the potential barrier and thus, restricts the total conductivity along the percolation paths [93].

In conclusion, an adequate application of either the grain or the neck model strongly depends on the presence of necks, that is insignificant for materials of separated grown grains (e.g. thin films) or materials which have not been thermally treated [92].

Based on the experimental data of Yamazoe and co-workers [66, 89] the effective carrier concentration was calculated as a function of the surface state density for spherical SnO_2 crystallites between 5 and 80 nm [90]. The quantitative model described the transducer function based on the assumption that the

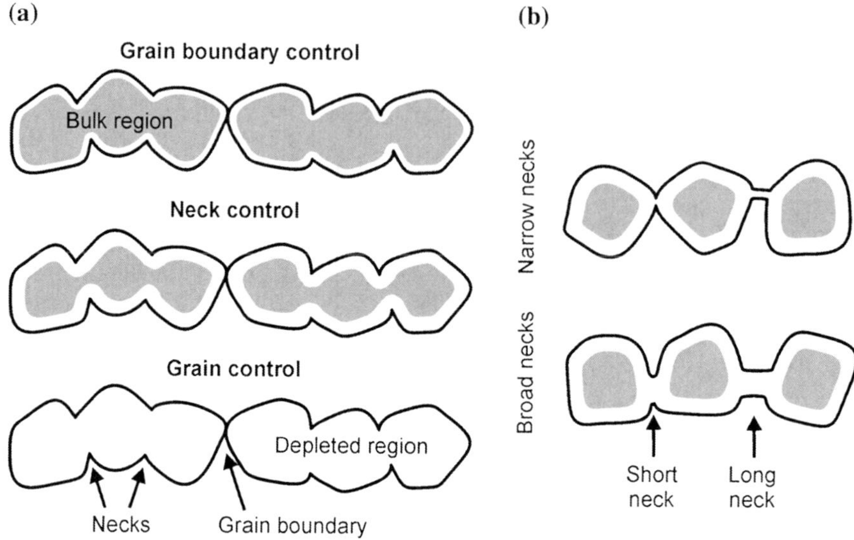

Fig. 4.3 **a** Sensing layer model (adapted from [90] and neck control model [adapted from 92] of partly sintered crystallites

variations in the surface state density were caused by trapping electrons due to surface-gas interactions. The theoretical considerations (i.e. numerical computation) led to response curves which exhibited a steep decrease of carrier concentration in the range in which the surface state density reaches the condition of fully depleted grains (i.e. all electrons are trapped at the surface). These theoretical findings explain the linear increase in conductivity with decreasing trapped charge carriers. According to the simulations the measured sensor responses (i.e. R_{gas}/R_{air}) of SnO_2 towards H_2 (800 ppm) and CO (800 ppm) at 573 K are linear functions of $1/D$ where D is the average grain size (Fig. 4.4). Thus, the experimentally found correlation in which the sensitivity is proportional to the surface to volume ratio was confirmed and the sensitivity can be calculated to $6/D$ for spherical-like nanocrystallites of SnO_2 [90].

However, experimental investigations published so far, generally confirm—even though the sample history and analytic methods differ in each case and thus, do not permit a quantitative comparison—that the material properties are complementarily affected by the number of both, structural and electronic defects. These kinds of defects are stabilized with decreasing particle size as known for spherical particles, although the form also plays a role [94]. The active sensing surface exhibits a multitude of various defects and crystallographic faces which are energetically unequal, so that the reaction kinetics of the adsorption of gas molecules, the consecutive reactions of intermediates and the desorption of reaction products all differ. [95, 96] Hence, a decrease of the particle diameter linked with an increasing surface-to-bulk ratio leads to a significant increase in surface reactivity due to a growing amount of defects on surfaces with greater curvature [92].

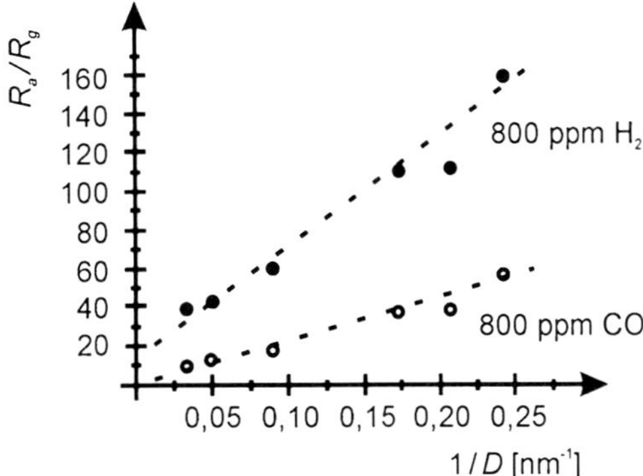

Fig. 4.4 Influence of particle diameter D on sensor response (R_{gas}/R_{air}) of SnO_2 towards **a** H_2 (800 ppm) and **b** CO (800 ppm) at 573 K [adapted from 90]

This improved surface reactivity upon decreasing particle size accompanied with a decrease of the working temperature was published for catalytically active nanoparticles [97–99]. Additionally, the accessibility of the gas detecting surface represents a key parameter in controlling the sensitivity properties of the sensor material. Thus, the microstructure including the porosity and the thickness of the sensing layer which determines the diffusion processes has a profound impact on the sensor performance.

4.2.3 Microstructure

Beside the fact that the size of particles determines the surface area and thus, the ratio between the depleted zones and the unaffected bulk, it also determines the internal microscopic surface as well [71, 100]. Additionally, the thickness and the porosity of the sensing layer affect the total surface area accessible for interpenetrating gas molecules.

The layer thickness is among the dimensional factors as the already mentioned grain and neck size and thus, affects the sensor performance as well. Due to different design approaches, sensor layers are classified as thick film (d > 100 nm) or thin film (d < 100 nm) types. In contrast to ceramic thick films, where synthesis conditions and thermal treatment are the main tuning parameters, the particle size of thin films is found to depend directly on the deposited film thickness [92]. Thus, for thin films a change in layer thickness leads to a change in the magnitude and temperature of the maximum sensor response correlated with the shift in particle size and gas-permeability. For thick films the gas accessibility depends on the nature of target analyte molecules and dominates the sensing effect, while other dependencies (e.g. the sensitivity towards reducing gases) are contradictorily reported. Due to the accessibility of all percolation paths contributing to the sensor signal thin films are found to be more sensitive and faster in response and recovery behavior than thick films [101].

In contrast to the layer thickness, the layer porosity displays a critical morphological factor which directly determines the accessible surface area. With decreasing particle size, in principle, the layers become more compact and the porosity decreases. Additionally, with increasing layer thickness the interpenetration of gas molecules becomes more hindered so that a concentration gradient occurs within the layer. This gradient strongly depends on the diffusion rates of the target analyte and its reaction products as well as on the interactions of analyte gas and reaction products with the metal oxide surface. As a result a stationary equilibrium is formed that determines the response and recovery time of the sensor [102]. In the case of strong interactions, the analyte gas reacts within short diffusion distances while in the case of relatively weak interactions, a longer diffusion distance can be covered [99].

Summarized, the sensor performance is critically influenced by the grain size, which controls the degree of analyte-surface interactions (e.g. adsorption/desorption, catalysis, reduction/reoxidation, charge transfer and chemical reactions) as well as by the layer thickness and the porosity, which affect the diffusion processes (e.g. intergrain oxygen diffusion, bulk oxygen diffusion, gas diffusion inside the metal oxide matrix and surface diffusion of adsorbed species). [92] Hence, an ideal sensing layer consists of nanoparticles embedded in a thin film featuring macropores, so that all percolation paths contribute to the overall conductivity [103].

However, one has to keep in mind that the sensing effect always results from a complex interplay of various structural parameters. These parameters can be classified by four basic factors: the *geometrical* factor including the sensor and the electrode set-up, the *dimensional* factor including grain size, neck size and film thickness, the *crystallographic* factor including crystal shape, film texture, surface geometry and surface stoichiometry and lastly the *morphological* factor including porosity, agglomeration, grain networks and intergrain contacts (Fig. 4.5). The fundamental influence of every single structural feature on the overall sensor performance complicates a direct comparison of publications and a rash transfer of results in this field even for material samples of the identical chemical composition [92].

A new approach towards the enhancement of diffusion rate and the improvement of sensing materials utilizes one-dimensional non-spherical nanostructures such as wires, rods, belts, fibers and (carbon nano) tubes. To accomplish such anisotropic morphologies, bottom-up strategies are applied in order to gain control over the nano and the microstructure at the same time. Achievements in this novel field of research are discussed in the corresponding literature [104–107], and have recently been reviewed [18, 108], including those related to carbon nanotubes [109–112].

Beside the fundamental influence of structural parameters on the performance of an ideal sensor, the working temperature can be varied to decrease cross-sensitivity characteristics, as previously mentioned. Additionally, the chemical composition by addition of various dopants to the sensor material itself can lead to an enhanced selectivity or reduced cross-sensitivity, as discussed in the following.

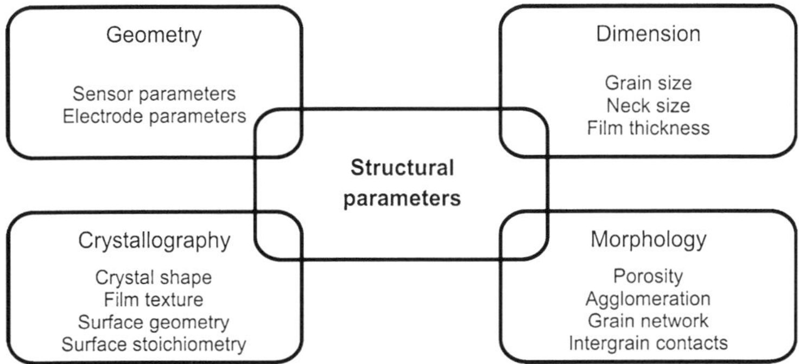

Fig. 4.5 Four basic structural parameters affecting sensing properties of gas sensor materials

4.2.4 Doping

Though, commercial chemiresistors have already been optimized concerning preparation and structural parameters, most simple metal oxide based sensor materials, suffer from limitations with respect to cross-sensitivity, thermal aging and long-term instability. In order to overcome these inherent limitations the chemical composition of the sensing material is modified by doping.

On the one hand active surface dopants enhance the sensor performance by enabling selective interactions with the target analyte and thus, increase sensitivity and selectivity while at the same time decrease reaction time and working temperature, which has a pronounced effect on the long-term stability. On the other hand volume dopants are incorporated into the lattice and thus, are present in surface sites and influence the material and sensor properties of the base material itself. A control mode to shift between the bulk modification of homogeneous or heterogeneous base materials and the surface modification by discrete particles is realized by basically two different methods for the introduction of dopants. By volume doping the metals or metal oxides are incorporated as additives during the material synthesis whereas by surface doping the catalysts are deposited subsequent to the thick film preparation.

4.2.4.1 Volume Doping

Volume-dopants are introduced into the gas sensing matrix of the base material during the preparation process. Thereby complex multi-component systems are formed with modified chemical, physical and electronic properties in the bulk and the surface states. For this purpose catalytically active (e.g. transition metals) or inert additives (e.g. metal oxides) are applied in quantities of about 1–10 wt %. However, their impact on the sensing properties does not always correlate with their catalytic activity as a result of different mechanisms. Although minor quantities of catalytic additives are sufficient for forming completely modified

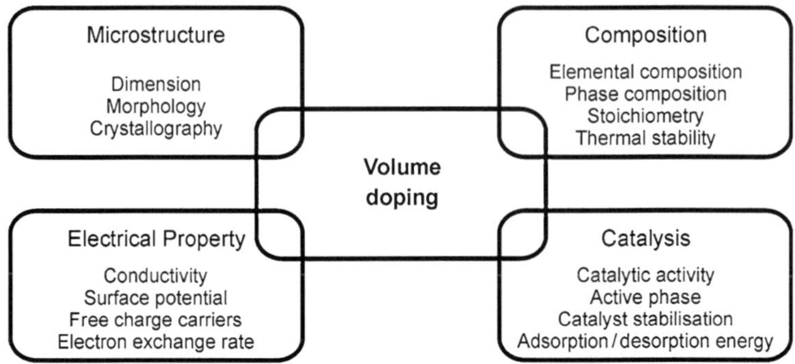

Fig. 4.6 Four basic factors affected by volume doping of base materials [adapted from 111]

materials with respect to their sensing properties, a rising amount of additives can also result in the formation of mixed layers or even segregated phases. Depending on the mutual solubility of the additive (guest) and base material (host) the formation of discrete particles, solid solutions or even foreign phases occurs at the surface [113].

The incorporation of volume dopants can have a significant influence on the sensing performance via a multitude of interacting parameters. The parameters can be arranged under four basic aspects: the *microstructure* including dimension (grain size), morphology and crystallography, the *composition* including elemental composition, surface potential, free charge carriers and electron exchange rates, the *catalysis* including catalytic activity, active phase, catalyst stabilization and adsorption/desorption energy and lastly the *electrical properties* including conductivity, surface potential, free charge carriers and electron exchange rate (Fig. 4.6). The dopants are capable of improving the sensor properties by formation and stabilization of smaller grains, by increasing the nanostructure porosity and by enhancement of long-term stability. Nevertheless, the preparation of an ideal sensor material still remains challenging, since progress of a positive-contributing parameter might simultaneously degrade another or promote a negative-contributing characteristic [111].

4.2.4.2 Surface Doping

Surface dopants are deposited on top of the pre-fabricated gas sensing materials in a working step (impregnation, spraying, etc.) subsequent to the base material preparation. This type of doping leads to a modification of the active sensing surface of the base material. Typical doping materials are noble metals (e.g. Rh, Pd, Ag, Pt, Au, etc.) or oxides of transition metals (e.g. Fe, Co, Cu, Zn, etc.) or main group elements (e.g. B, Se, etc.). Upon annealing the noble metals form distributed metallic or oxidic clusters on the sensing surface which may act catalytically as promoters or inhibitors of surface reactions and thus, shift the sensitivity maximum to lower operating

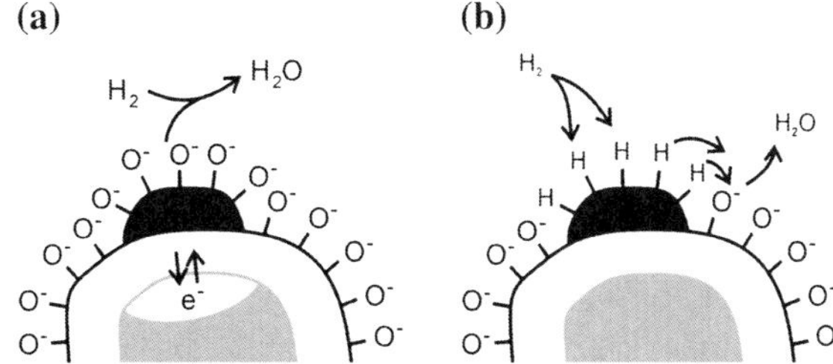

Fig. 4.7 Mechanism of **a** electronic sensitization (redox state changing) and **b** chemical sensitization (analyte activation) by surface doping of metal oxides [adapted from 66]

temperatures [111]. Generally, two mechanisms can be distinguished, electronic and chemical sensitization: [114–116].

In the electronic sensitization (Fig. 4.7a), the surface dopant acts as a stronger electron acceptor and consequently an enlarged electron depleted space-charge area is formed near the interface. Upon reaction with reducing analytes (e.g. H_2) the redox state or rather chemical potential, is changed, the trapped electrons are released back into the bulk and the conductivity significantly increases. Examples for this mechanism are AgO/Ag, PdO/Pd and CuO/Cu [66]. In the chemical sensitization (Fig. 4.7b), the mechanism refers solely to the catalytic surface reactions, whereby the deposited dopant acts as a catalyst and remains unaltered. The clusters therefore act as adsorption and activation sites for the target gas molecules (e.g. H_2) and subsequential pass the activated gas fragments down to the ionosorbed oxygen. Thus, the average oxygen coverage and therefore eV_{sur} is reduced and the conductance or the sensitivity is increased (according to Eq. 4.4). This mechanism is widely described as the "spill-over" effect [66].

Moreover, the optimization of catalytic parameters can be achieved by varying the clusters dimension, the inter cluster distance and additionally, the catalytic material itself. The sensitization is not only restricted to superficial clusters of catalytically active dopants as proposed by the mentioned models but also to atomically distributed Pt and Pd on surface and sub-surface sites as demonstrated for the enhanced sensitivity of noble metal doped SnO_2 towards CO [117, 118]. Alternatively, the diffusion of the dopant into the metal oxide and the alloy formation of reduced metal oxide in the near surface region is possible, as observed for Pd or Rh doped SnO_2 [119, 120]. Furthermore, the cooperative effect between several dopants affects the surface doping as reported from studies of heterogeneous synthesis [121, 122].

However, sometimes sufficient selectivity can only be achieved by the in-front-arrangement of additional physical auxiliaries. Specifically, these can be realized in terms of porous or rather catalytic filter layers as well as oriented or functionalized membranes as recently surveyed for zeolites [123]. Thereby, a group of diverse and (non-)reactive microporous zeolites are the most popular examples exhibiting physical (i.e. precisely adjustable pore sizes) as well as chemical filtering (frameworks for catalysts) [124]. Beside zeolites, other materials can be used which act as passive membranes or catalytic sieves for blocking parasitical interferant or material-poisoning gas molecules. Several examples in this context are multilayer sensors containing Pd doped Al_2O_3 [125], zeolite filters containing hydrophobic TiO_2 [126], mesoporous filters of Pd/Pt doped SiO_2 [127] or metallic films of Ru electrically isolated by SiO_2 [128].

In summary, the significant impact of synthesis involving the various doping possibilities combined with various measurement conditions vastly extends the experimental parameter space. Thus, characterization of new gas-sensing materials requires novel strategies which allow the investigation of correlations between composition, structure and sensing properties. Due to the tremendous influence of every single parameter on sensor performance, the systematic investigation of a single parameter requires a workflow which keeps a defined number of parameters/ conditions of the sensor preparation constant while a small and distinct number of parameters are varied. Therefore, high-throughput experimentation and statistical investigation are combined with rapid characterization via impedance spectroscopy and infrared thermography to discriminate between electrical and thermal effects [129–132]. The following chapter will highlight the first steps towards the use of such a high-throughput approach for the development of novel gas sensing materials.

4.3 High Throughput Screening

For a rapid identification of new sensor materials with improved sensor performance a high-throughput impedance spectroscopy (HT-IS) workflow including material synthesis, layer preparation, property screening and data evolution has been established. The approach focuses on semiconducting nanoscaled binary or ternary metal oxides which are derived from a polyol method and modified via volume or surface doping. The preparation via the polyl-mediated synthesis has several advantages compared to solid-state reaction techniques which typically suffer from by-product formation and high calcination temperatures. The polyol-mediated process is derived from the sol–gel process and gains access to ultrapure nanoscaled products at relative low processing temperatures. Furthermore, it allows the easy control of chemical composition on a molecular level and suitable access to materials of different size and shape. The term "polyol" refers to high-boiling polyalcohols, such as diethylene glycol which dissolve the inorganic precursor salts by a chelating effect and which serves as stabilizing ligands to avoid particle agglomeration during the synthesis. This procedure yields nanoscaled metal and metal alloy particles [133–137] as well as metal oxide nanoparticles [138] with very high specific surface areas. As mentioned before, the gas-sensitive properties of the obtained materials can be modified by volume doping during the synthesis or by surface doping prior to calcination.

The automation of the material synthesis and layer preparation procedure utilizing robotics (i.e. pipetting robot system) combined with the impedometric screening and the automatized data evaluation opens the access to detailed

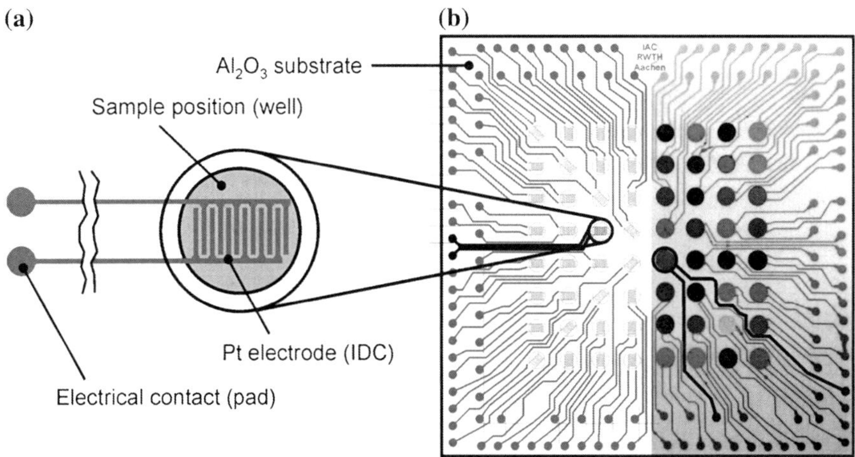

Fig. 4.8 Top view of a multi-electrode substrate (*MES*): **a** layout of a single interdigital capacitor in detail (position 34) and **b** layout without (*left part*) and photography with deposited thick films (*right part*). The positions 34 and 55 as well as their conducting leads and contact pads are highlighted in black [adapted from 141]

libraries of functionalized materials and their corresponding gas sensor properties. Thus, the following subchapter describes the characteristics of such a HT-IS material and sensor screening set-up [127–129, 139, 140].

4.3.1 Multi-Electrode Array

For impedimetric screening multi-electrode substrates (MES) have been developed (Fig. 4.8b). The substrate (101.6 mm side length and 0.7 mm thickness) consists of Al_2O_3 which is chemically inert, exhibits high thermal and mechanical stability up to 600 °C and is a good electrical insulator. The top-side of the Al_2O_3 substrate is screen-printed with platinum leads to form an 8×8 array of interdigital capacitors (IDCs) (Fig. 4.8a). These 64 positions (wells) are labeled with numbers from 1 to 8 for horizontal rows (from left) and vertical columns (top down), respectively (e.g. 34 marks the third position from left and the fourth from above which is enlarged in the detail). Conducting leads connect each IDC-structure with two contact pads near the edges of the plate for addressing the electrical properties of a respective sample position. After screen-printing the Al_2O_3 substrates have typically been calcinated at 1350 °C for 48 h prior to their use [127, 128].

A particular IDC structure composed of two interdigitating finger electrodes and their corresponding structural parameters are depicted in Fig. 4.9. The type and design of electrode was selected with a focus on the applied impedometric characterization method, which measures the resistive and capacitive properties of the samples. Since the number of IDCs within the array is limited by the size of the

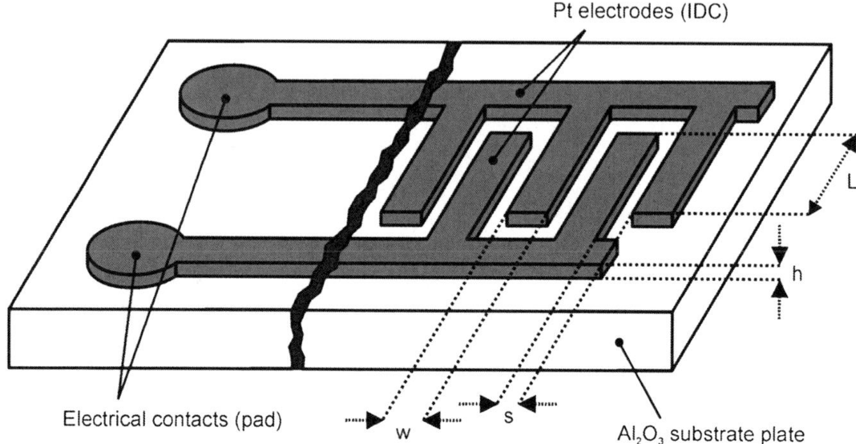

Fig. 4.9 Layout of electrode structure (*IDC*) and corresponding structural parameters [127]

substrate, the electrode parameters have been optimized by previous calculations related to the measurability of high sample resistivities and small capacitance changes. The structural parameters required for this calculation include the number of fingers (N), the finger length (L), the finger height (h), the space between two fingers (s) and the finger width (w). Additionally, the calculations consider that the preparation of the substrate, the accuracy of the screen printing process (i.e. grain size of platinum paste) and the volume contraction during the sintering process which may cause small irregularities of the electrode structures [127].

Generally, the optimal structural parameters (e.g. width of IDC) are determined by the permittivity of the substrate material and the thickness of the sensor layer. Furthermore, the relative sample resistance rises with increasing electrode distance, whereas the electrical field between the electrodes decays within the sensor layer, if the layer thickness is of the same magnitude as the electrode distance [142]. Depending on synthesis and sintering conditions the thickness of sample layers ranges between 50 and 200 μm. Simulations of vaccum capacitance for different possible set-ups of the two electrodes revealed that the effect on the capacitance is negligible, if the substrate is much thicker than the width and gap of the electrodes.

Consequently, as a result of these considerations the chosen parameters are: 12 electrode fingers with a width of 135 μm and an electrode distance of 140 μm. The whole spatial arrangement of a single electrode structure amounts to 3.4 mm in length and 2.3 mm in width [127, 137].

The compact set-up of the entire electrode array causes different lengths and distances of the conductor pathways between the IDCs in the centre and the pads at the edges. Therefore it has to be considered that different capacitive amounts of these pathways and other parasitic effects add to the fundamental capacitance of each IDC. In order to compensate the capacitance distribution of different positions, corresponding circuit equivalents are used to accurately describe the

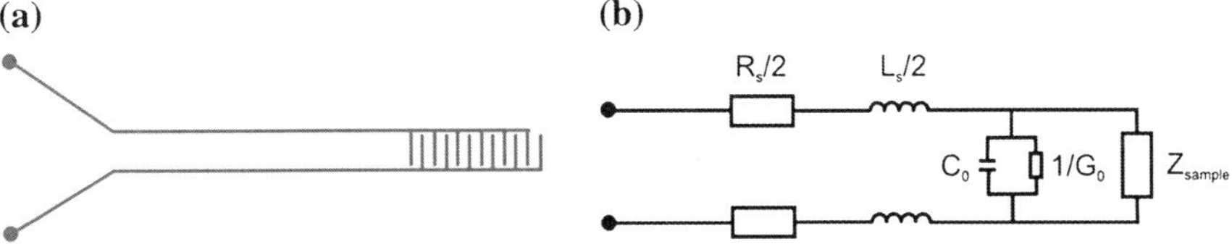

Fig. 4.10 Scheme of interdigital capacitor (*IDC*): **a** single structure with nearly parallel supply leads and **b** corresponding equivalent circuit [adapted from 127]

individual electrical properties of IDCs [127, 143]. Figure 4.10a depicts the case of an IDC structure with almost parallel supply leads as for example found at position 34 of the MES and Fig. 4.10b displays the corresponding circuit equivalent. On the contrary, the parasitic effects are negligible in the case of more nearly orthogonal pathways as for example existent at position 55 of the MES. Thus, the last mentioned position exhibits the smallest capacitance deriving only from the IDC itself and it is used to determine the parasitic effects of the other positions. The displayed circuit equivalent consists of one contact resistance R_S and one pathway inductance L_S in series for each pathway, and an RC-element in parallel to the sample impedance Z_{Sample}. The RC-element considers the stray capacitance C_O of the IDC and the substrate resistance between both pathways as $1/G_O$ in parallel to the sample impedance. The resistances of conductor paths (<10 mΩ) and the conductivity of the substrate material itself (>20 mΩ) are neglected during the evaluation steps. The resistive contribution to the parasitic effects can be eliminated by an offset adjustment (shortcut of electrodes) and the capacitive part by a compensation measurement (blank electrodes), so that consequentially the sample impedance can be calculated. In the present case the inductive parasitic effects L_S of the measured sample impedance are eliminated like the capacitive ones during the step of data fitting. Therefore the presented MES is applicable for resistive as well as for capacitive measurements.

4.3.2 High Throughput Impedance Spectroscopy Set-up

The electrical sample characterization is performed by HT-IS over a broad frequency range. The set-up of this station basically consists of the gas supply (Fig. 4.11a) (i.e. mass flow controllers and water reservoir) the measuring station (Fig. 4.11b) (i.e. measuring head with MES and furnace) and the instrumentation (Fig. 4.11c) (i.e. two relay matrices, impedance analyzer, source meter and personal computer). The relative humidity of the applied test gas atmospheres can be adjusted from 0 to 90 % by bubbling through a temperature controlled water reservoir. During alternating conditioning times and measurements a variable gas flow (standard 100 sccm) is applied by several mass flow controllers (models 1179/2179) and a multi-gas controller (647 B, all MKS Instruments).

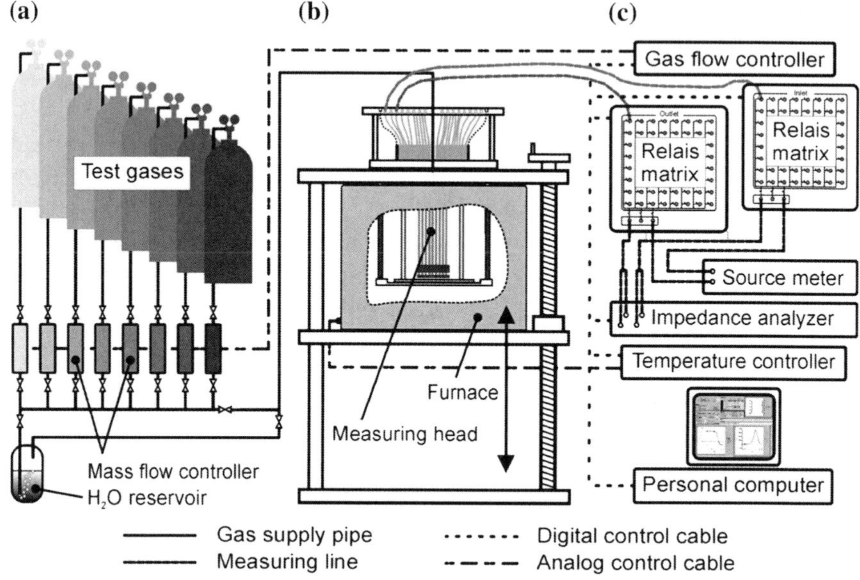

Fig. 4.11 Configuration of HT-IS set-up: **a** gas supply. **b** measuring station and **c** instrumental equipment. For clarity only two measuring lines are exemplified due to two electrodes of a single IDC (i.e. position 11)

Two relay matrices (KRE-2450-TFCU multiplexers, MTS-Systemtechnik) connect the coaxial plugs of the measuring head by means of multi-channel switching via high frequency capable coaxial wires to the analytic instruments. For reasons of clarity only two of the overall 128 wires (i.e. two wires due to two electrodes of every IDC) are displayed in Fig. 4.11.

Two-probe conductivity measurements can be carried out either in AC-mode using an impedance analyzer (Agilent 4192 A, Agilent Technologies; 5–$1.3 \cdot 10^6$ Hz and 0.1–1.299 MΩ) or in DC-mode using a source meter (Keithley 2400, Keithley Instruments Inc.; 0.2–200 V, $1 \cdot 10^{-6}$–1 A and $2 \cdot 10^{-4}$–200 MΩ). A self-developed software (Visual Basic 6, Microsoft Corporation) and a simple script language allows for combined monitoring, controlling, measuring and data recording via digital and analog interfaces (RS232 interface by GPIB protocol and IEEE interface by ASCII protocol) in all devices.

The coated MES is placed on a sample rack which is integrated into the measuring head (Fig. 4.12). The gas stream inlet is distributed via a binary tree construction [144], 64 hollow Al_2O_3-tubes and a hole-plate with individual headspaces. Thus, an equal amount of gas per position is guaranteed by a similar flow resistance for each individual channel [137, 138]. The sample rack is thermally decoupled and enables electrical contact of the substrate via Al_2O_3-covered platinum wires. The spring-loaded contact tips are equipped with coaxial plugs and cables and thus, ensure a steady electrical contact by compensation of the thermal expansion of materials at elevated temperature. All materials used for the

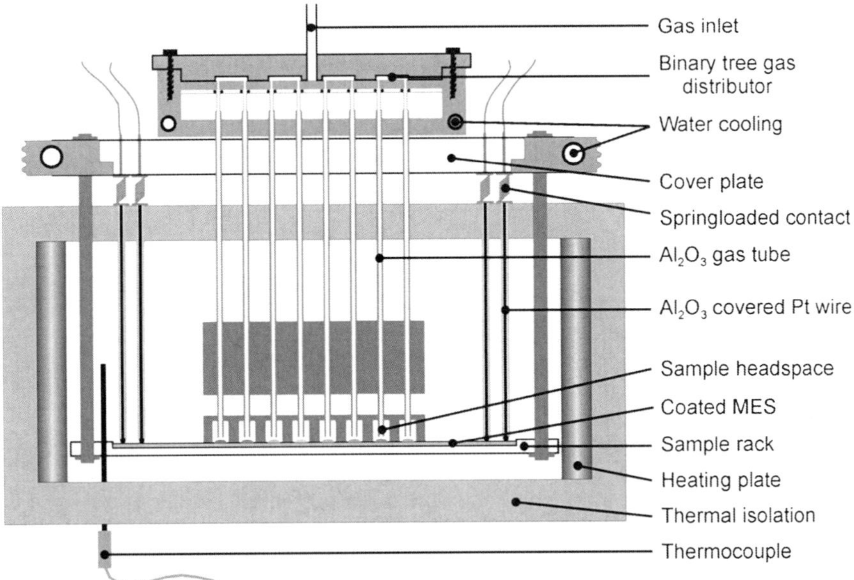

Fig. 4.12 Configuration of HT-IS measuring head and furnace with embedded *MES* [adapted from 137, 138]

measuring head are high-temperature resistant steels (dark gray) or pore-free Macor© ceramics (white) which exhibit low thermal expansion coefficients, long-term thermal stability and mechanical stress tolerance. Additionally, these materials possess a persistent chemical inertness and a low absorbance/desorbance capacity for gas. Both, the gas distribution and the cover plate of the measuring head are cooled by water and air due to the high temperatures during measurement. In order to heat the sample plate in a range from room temperature up to 600 °C a furnace can be moved over the measuring head. The furnace is equipped with four heating elements which are controlled by a thermo-controller (Eurotherm 2416) combined with a thermocouple (Ni–Cr/Ni, Thermocoax®). Steady gas and temperature conditions for all samples of one MES as well as for all measurements of different MES are crucial to obtain reproducible results and to build up comparable material libraries.

This set-up allows fully automated gas mixing, temperature controlling, sample addressing and data recording. Thereby, all of the time-consuming steps, such as heating/cooling or conditioning are simultaneously proceeded for all sample positions, which enables sequential electrical measurements.

In order to organize and to manage the large quantity of data which is recorded during the workflow (i.e. material synthesis and layer preparation as well as measuring conditions and characterization data) a flexible database has been developed in advance. That database enables the creation of material libraries which contain all of the characteristic sensing and material properties for a given sample from a MES. For data mining of the gas-sensing properties, the values of resistance and capacitance are deduced from the frequency dependence of the real and imaginary part of the fitted impedance spectrum and these fits are also included in the database. This procedure enables statistical analyses of the quality factors and the critical parameters as well as systematic data mining.

Altogether, the described procedure allows the investigation of correlations between the sensor properties, the microstructures and the chemical compositions and to trace them back to single variable parameters. [130] The application of this procedure in order to detect specified materials with desired properties from a multitude of samples is demonstrated in the following section by examples of the volume and surface doping influence on binary or ternary base materials.

4.4 Experimental Workflow

The polyol method [145–148] allows for both, the synthesis of various base metal oxides (e.g. $LnCrO_3$ and $LnFeO_3$, $Ln = La$, Pr, Nd, Sm, Eu, Gd, Tb, Dy, Ho, Er, Tm, Yb and Lu [149]) and the incorporation of various volume dopants into different base materials (e.g. $CoTiO_3$ with 2 at. % Li, Na, K, Sb, La, Sm, Gd, Ho and Pb [150]) during the synthesis. This enables the systematic study of the Ln-variation effect on the gas sensing behavior, predominantly of materials that had not been studied so far. Finally, the study of the nearly complete Ln-series will enable the comparison of structure–property relations and the development of basic design rules for specific sensors.

The wet impregnation technique facilitates the precipitation of hundreds of different surface dopant solutions on the surface of a base material (e.g. La doped $CoTiO_3$ further denoted as $CoTiO_3$:La [151]) subsequently to the preparation.

The following section describes the experimental procedures for material synthesis, layer preparation, material analysis and data mining. In this HTE approach the preparation steps from synthesis to doping are proceeded automatically by a pipetting robot system in the case of surface doping. Thereby the above mentioned examples of volume and surface doped materials are introduced to exemplify the single workflow steps. A more detailed description of the experimental elaborations has been published elsewhere [136, 152, 153].

4.4.1 Sample Preparation

Typically, for the synthesis of the base or the volume doped materials soluble metal salts (e.g. acetates, nitrates, carbonates etc.) are applied in the respective stoichiometric ratio of the desired product. The precursors are dispersed in diethylene glycol (DEG) as a chelating solvent under vigorous mechanical stirring. The reaction suspension is heated up to T_1 (80–170 °C) to dissolve the reactants

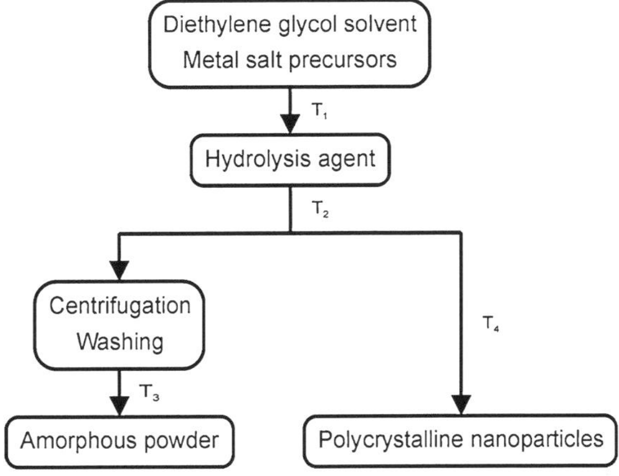

Fig. 4.13 Workflow of polyol-mediated synthesis to yield base and volume doped metal oxides

and, if necessary, a sufficient surplus of hydrolysis agent (e.g. H_2O, KOH, NH_3, H_2O_2, CH_3COOH, HNO_3 etc.) is added. The emerging suspension is heated up to T_2 (160–190 °C) and maintained at this temperature for several hours (5 h). The obtained stable suspensions exhibit a standard concentration of 1 wt % solid to DEG solution. For analytical purposes a small part of the suspension is processed by centrifugation, washing (i.e. water and acetone) and drying at T_3 (60 °C) in order to obtain the amorphous powder product ("as-synthesized"). The nano-crystalline sensor material is obtained by calcination of the suspensions at a maximum T_4 (400–900 °C for 2–12 h). The reaction parameters of every synthesis require modification and optimization regarding particle size, purity and product yield. The general synthesis workflow for the preparation of (volume doped) metal oxides is schematically represented in Fig. 4.13.

Characterization of the obtained sample materials is routinely performed by X-ray diffraction (XRD) and by scanning electron microscopy (SEM). Additionally, energy dispersive X-ray spectroscopy (EDX), transmission electron microscopy (TEM) and other characterization methods are applied as referred in the following subsections.

4.4.1.1 XRD-Characterization

The initial characterization concerning the phase and the structure of synthesized metal oxides is conducted by powder XRD on thin films in transmission mode. Generally, the "as synthesized" materials are amorphous right after synthesis and calcination yields in nanocrystalline materials. Figure 4.14 exemplifies two series of XRD patterns of $LnCrO_3$ (Fig. 4.14a) and $LnFeO_3$ (Fig. 4.14b) (Ln = La, Pr, (Nd,) Sm, Eu, Gd, Tb, Dy, Ho, Er, Tm, Yb and Lu) which illustrate that the annealed materials are nearly single-phase compounds [traces of Ln_2O_3 impurities with Ln = Tb, Ho, Tm and Yb are marked by asterisks around $\Theta = 29°$ in (a)] [147]. The observed patterns are in good agreement with the lattice parameters of orthorhombic phases reported by the Inorganic Crystal Structure Database (ICSD).

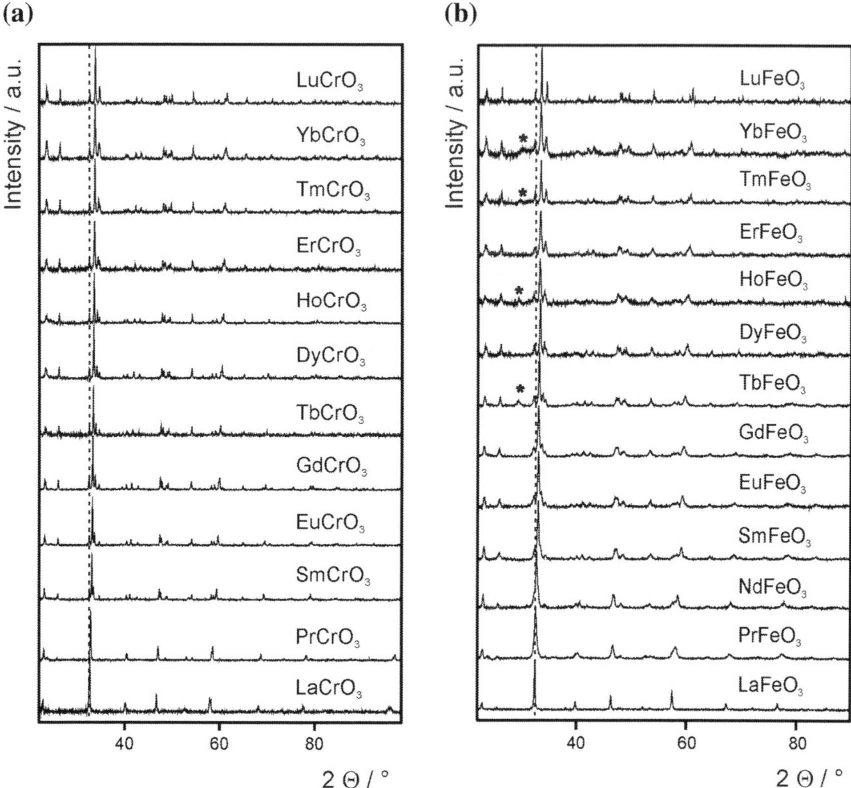

Fig. 4.14 Two series of XRD patterns for analysis of phase and structure: **a** LnCrO$_3$ and **b** LnFeO$_3$ (Ln = La, Pr, (Nd,) Sm, Eu, Gd, Tb, Dy, Ho, Er, Tm, Yb and Lu) materials [adapted from 147]

The variation of the cubic symmetry (such as SrTiO$_3$) increases with the atomic number from LaMO$_3$ to LuMO$_3$ in respect of the decreasing Ln^{3+} ion radius [154]. The linear dependence of crystal lattice parameter and the composition of mixed oxide is known as Vegard's rule (marked by dotted lines). The broadened peaks compared to the bulk material indicate the formation of nanometer sized particles after the calcination step, while the phenomenon of peak broadening has been previously reported for crystallites smaller than about 5 • 10^{-7} m [155]. The measured broadening of single or multiple peaks can be utilized to determine the crystallite size applying the Scherrer equation, although the crystallite size does not have to be congruent to the grain size. [156, 157].

4.4.1.2 SEM-Characterization

The surface morphology of materials, including the particle size and the microstructure, is visualized by a field emission scanning electron microscope (FE-SEM).

(a)

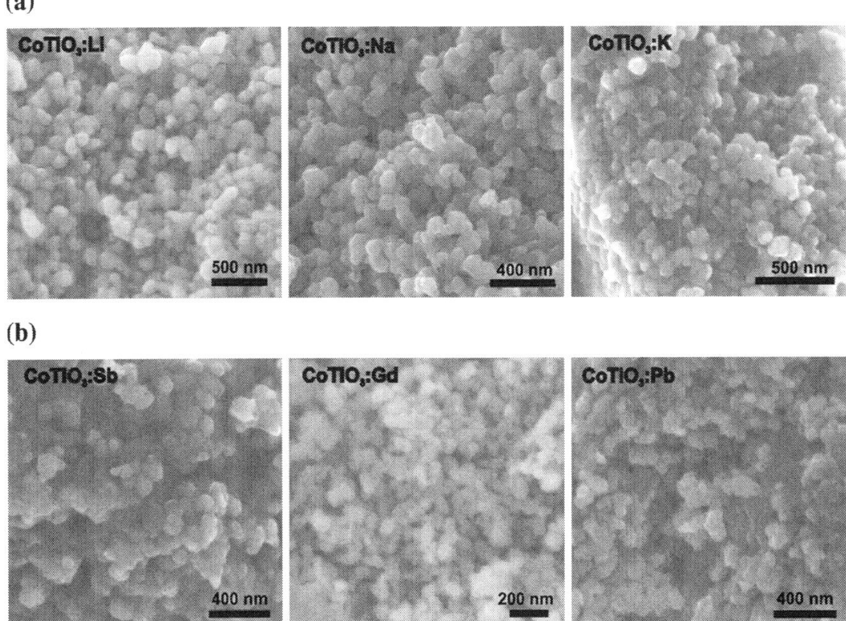

(b)

Fig. 4.15 SEM micrographs for analysis of particle size and morphology: volume doped $CoTiO_3$ with 2 at.% of **a** Li, Na and K and **b** Sb, Gd and Pb supposed in A-position and B-position of ABO_3 structure; respectively [adapted from 148]

SEM micrographs of annealed volume doped $CoTiO_3$ samples (2 at.% of Li, Na, K, Sb, Gd and Pb) are exemplified in Fig. 4.15 whereas the doping is supposed to be in A-position (Fig. 4.15a) and B-position of the ABO_3 structure (Fig. 4.15b), respectively. As a result of the thermal treatment all micrographs typically reflect agglomerations (secondary particles) of interconnected spherical particles that form open porous networks. Such a surface morphology guarantees abundant interaction opportunities between the metal oxide surface and the surrounding gas atmosphere. The irregular packed agglomerates sometimes complicate the size estimation of the primary particles (50–100 nm in diameter) which are in good agreement with the crystallite sizes determined from XRD analysis [148].

Furthermore, the elemental composition and its corresponding distribution can be observed by energy dispersive X-ray spectroscopy (EDX).

4.4.1.3 TEM-Characterization

High resolution TEM (HRTEM) analysis involving electron diffraction images can be carried out as additional method to examine the crystallinity (i.e. lattice plane distance) of the annealed nanoparticulate materials. The respective sample is taken from suspension and directly pipetted onto holey carbon-coated copper grids. The electron

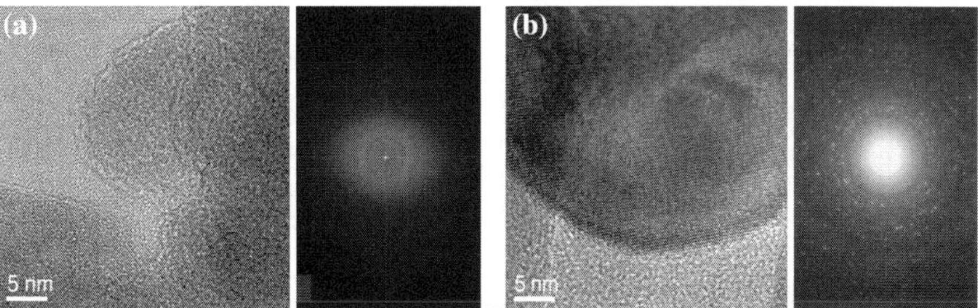

Fig. 4.16 HRTEM images and FFTs for analysis of degree of crystallinity: **a** amorphous (400 °C/1 h) and **b** crystalline (700 °C/12 h) CoTiO₃:La sample [partly adapted from 149, 156]

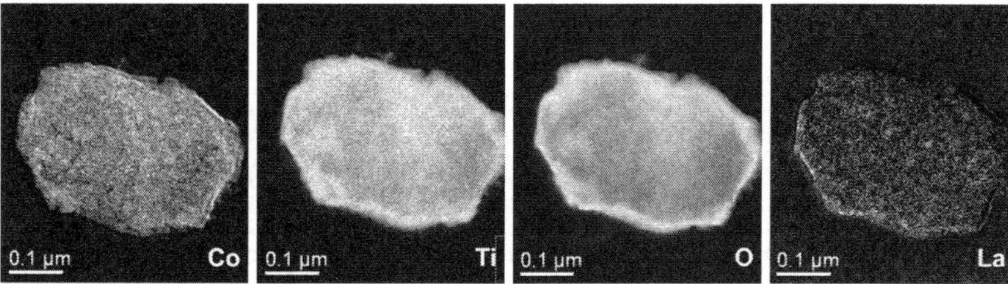

Fig. 4.17 EfTEM images for monitoring the respective elemental distribution of Co, Ti, O and La in CoTiO₃:La [adapted from 156]

diffraction pattern was subject to a fast Fourier transformation (FFT) in order to analyze the degree of crystallinity. Figure 4.16a depicts the TEM image (left) of the amorphous "as-synthesized" CoTiO₃:La sample after drying (400 °C/1 h) and Fig. 4.16b of the crystalline sample after calcination (700 °C/12 h). The electron diffraction pattern (right) illustrates in the former case the presence of an amorphous structure and in the latter case of a polycrystalline structure by concentric diffraction circles [149, 158].

The elemental distribution can be further investigated by energy-filtered TEM (Ef-TEM), in analogy to the previously mentioned EDX analysis. Thereby white domains represents the presence and black domains the absence of the respective element (i.e. Co, Ti, O and La). The EfTEM pictures of the CoTiO₃:La sample (Fig. 4.17) show a homogeneous elemental distribution of La over the sample material. Electron probe micro analysis (EPMA) confirmed the quantity of La to be 2.1 at.% (without figure) [156].

After characterization the different materials are deposited as thick films on a MES for subsequent electrical characterization. For this deposition procedure a particular thick film reactor is used as described in the following.

4.4.2 Thick Film Preparation

The individual workflow steps and operations of thick film deposition are visually summarized in Fig. 4.18. First, the powder of a nanocrystalline sample (Fig. 4.18a) is dispersed by mixing it in an agate mortar or by grinding it in a ball

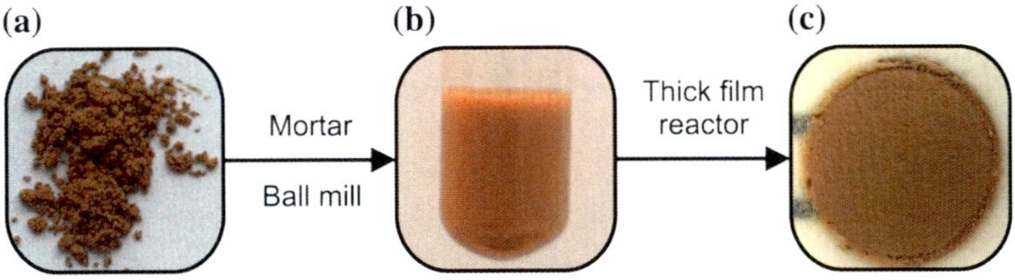

Fig. 4.18 Workflow of thick film deposition: **a** powder material. **b** material suspension and **c** dried sensing layer on top of IDC [adapted from 156]

mill with suitable amounts of water and polyethylene imine (PEI). Then, the resulting stable aqueous suspension (Fig. 4.18b) (concentration of 12.5 g L^{-1}) is deposited directly on a single IDC position of the MES by using an Eppendorf pipette (50 µL suspension) and a thick film reactor attached to the MES. Finally, the MES-reactor assembly is placed in a drying cabinet (60 °C) overnight in order to evaporate the solvent and to deposit a homogeneous sensing layer (50–100 µm) on the electrode structure (Fig. 4.18c).

The thick film reactor basically consists of a Teflon mask (plate thickness 17 mm and hole diameter 4 mm), fixed on a metallic substrate plate in such a way, that wells (volume 214 µL) are formed at each position of the MES (Fig. 4.19a and b). Every single reactor vessel is sealed by a viton O-ring. Four screws mount the metallic base and the top plate stationary by means of four alignment pins and a rubber mat for uniform pressure distribution. During the coating process each material suspension is statistically applied three times on the MES to compensate for potential gradients of furnace temperature or gas concentration as well as possible electrical contact failures.

The described thick film reactor has been designed for use with a fully automated pipetting robot (regard also Fig. 4.23). After the deposition and drying procedure the base metal oxide layers are ready for further modifications by systematic surface doping via a wet impregnation method. Therefore the aqueous metal salt solutions are prepared and the impregnation is performed by a laboratory robotic system.

The distribution of doping elements and their concentrations on a MES for the purpose of surface doping is illustrated in Fig. 4.20. All of the MES positions have been identically coated with volume doped $CoTiO_3$:La as the base material. For the wet impregnation procedure various metal salt solutions (e.g. Au as $HAuCl_4 \bullet 3H_2O$, Ce as $Ce(NH_4)_2(NO_3)_6$, Ir as $Ir(C_5H_7O_2)_3$, Pd as $Pd(NO_3)_2 \bullet 2\,H_2O$, Pt as $Pt(NH_3)_4(NO_3)_2$, Rh as $Rh(NO_3)_3 \bullet 2\,H_2O$ and Ru as $Ru(NO)(O_2C_2H_3)_3$) of three different concentrations (e.g. 0.2, 0.4 and 0.6 at.%) have been used to gain diversity within the material library. These surface dopants have been chosen in the first approach because of their known enhancement of the sensitivity and the selectivity of metal oxide semiconductors. [159, 160] In the present case a triple statistical distribution of the identical material composition (e.g. 0.2 at.% Au marked in light gray and 0.2 at.% Ce in dark gray) was performed on the MES for reproducibility purposes of the respective experiment [156].

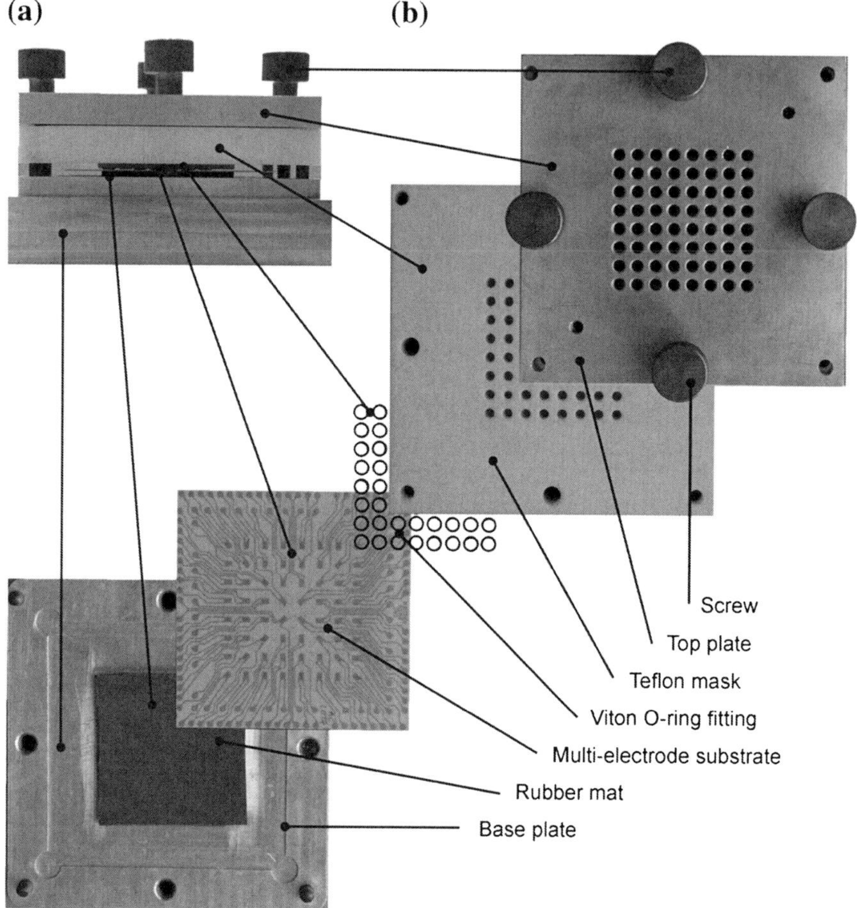

Fig. 4.19 Set-up of thick film reactor: **a** ready for use reactor (*side view*) and **b** single parts (exploded *top view*) [adapted from 138]

Subsequent thermal treatment (700 °C for 12 h) removes organic residues derived from the preparation and prevents the ready-made layers (about 0.625 mg) from significant sintering processes or changes in the microstructure during the subsequent electrical measurements at elevated temperature.

4.4.3 Laboratory Robot System

For the automated material synthesis, thick film deposition and surface doping a commercially available laboratory robot system (Lissy, Zinsser Analytic GmbH) was adapted (Fig. 4.21). The system consists of a magnetic stirrer equipped with a multiple synthesis reactor and a temperature control unit, a shaker for suspensions,

	1	2	3	4	5	6	7	8
1	Ru 0.6 at%	Pd 0.4 at%	Ce 0.2 at%	Ru 0.2 at%	Ce 0.6 at%	Pt 0.2 at%	Pd 0.2 at%	Ru 0.4 at%
2	Rh 0.2 at%	Pt 0.2 at%	Rh 0.4 at%	Pt 0.4 at%	un-doped	Pt 0.6 at%	Rh 0.6 at%	Ru 0.2 at%
3	Rh 0.6 at%	Pt 0.6 at%	Ir 0.2 at%	Ir 0.4 at%	Ir 0.6 at%	Ce 0.2 at%	Pt 0.4 at%	Ir 0.2 at%
4	Pd 0.2 at%	Ce 0.4 at%	Ce 0.6 at%	Au 0.2 at%	Au 0.4 at%	Au 0.6 at%	Ce 0.4 at%	Ir 0.4 at%
5	Pd 0.4 at%	Rh 0.4 at%	Ru 0.2 at%	Ru 0.4 at%	Ru 0.6 at%	Rh 0.2 at%	Rh 0.4 at%	Ir 0.6 at%
6	Au 0.2 at%	un-doped	Rh 0.6 at%	Pd 0.2 at%	Pd 0.4 at%	Pt 0.2 at%	Pt 0.6 at%	Au 0.4 at%
7	Au 0.6 at%	Ce 0.2 at%	Pt 0.6 at%	Pt 0.4 at%	un-doped	Ce 0.4 at%	Ir 0.2 at%	Ce 0.6 at%
8	Au 0.4 at%	Ir 0.6 at%	Ru 0.6 at%	Rh 0.2 at%	Au 0.6 at%	Ru 0.4 at%	Ir 0.4 at%	Au 0.2 at%

Fig. 4.20 Diagram of the triple statistical distribution of various surface doping elements (i.e. Ru, Rh, Pd, Ce, Ir, Pt and Au) of different concentrations (i.e. 0.2, 0.4 and 0.6 at.%) on a MES coated with volume doped $CoTiO_3$:La as base material. Two examples are emphasized in light and dark gray [adapted from 149]

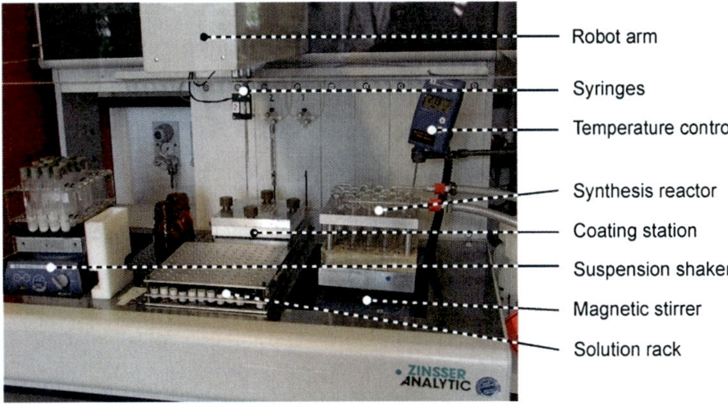

Fig. 4.21 Image of modified commercial laboratory pipetting robot system (Lissy)

a rack for volume and surface doping solutions and a coating station (i.e. thick film reactor) [137].

The synthesis station consists of a magnetic stirrer with a heating plate, a heating block, a thermal isolation block, 21 reactor vessels equipped with magnetic stir bars and water-cooled reflux condensers (Fig. 4.22). This set-up enables 21 parallel syntheses under inert gas atmosphere [137].

The set-up of the surface doping station is depicted in Fig. 4.23 in a lateral cut (cf. Fig. 4.19) [137].

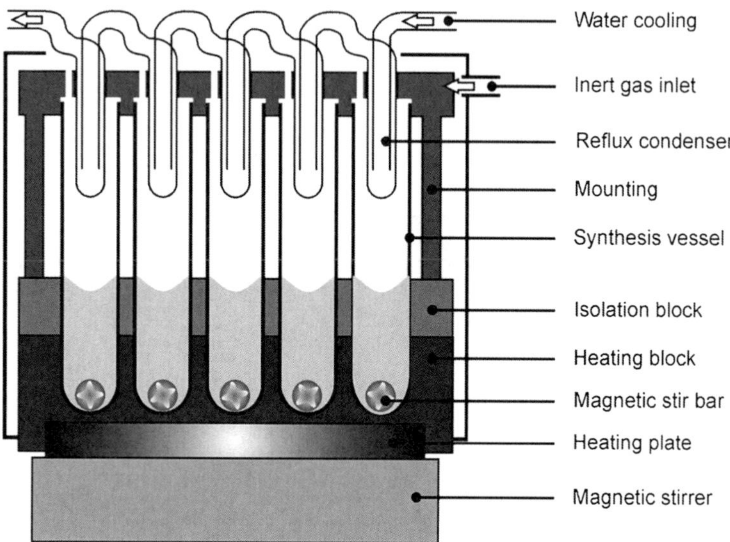

Fig. 4.22 Configuration (lateral cut) of robot synthesis station (max. 21 parallel synthesis) on top of a magnetic stirrer [adapted from 137]

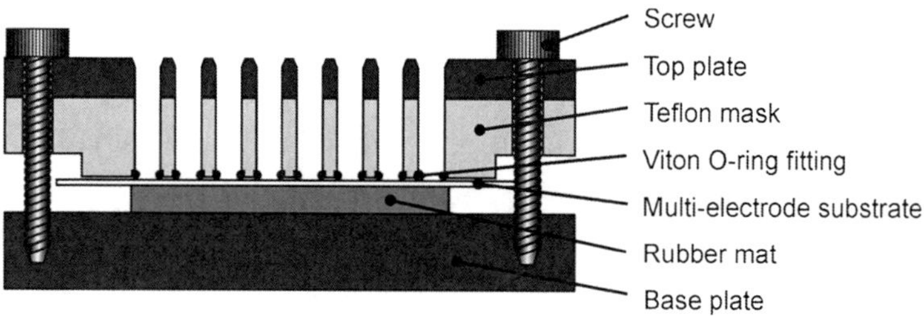

Fig. 4.23 Configuration (lateral cut) of robot coating station (i.e. thick film reactor) fitting a MES [adapted from 137]

Complex impedance spectroscopy is employed for the electrical characterization of the sensing layers deposited on the MES. The measured data is fitted automatically and visualized in terms of Trellis plots and bar diagrams in order to easily identify the samples with the best relative sensitivities and selectivities, respectively.

4.5 Impedance Spectroscopic Screening

The experiments are typically carried out within a preassigned temperature range in 25 °C steps to find out the best working temperature of the sensor material. Starting with the highest temperature which is well below the calcination temperature avoids material sintering and aging during the subsequent experiments.

Prior to the electrical measurement the respective test atmosphere is applied for up to 30 min to reach adsorption equilibrium except for the first measurement at the highest temperature and at every subsequent lower temperature. In these two cases this period is extended up to 240 and 90 min, respectively. At each temperature the sample characterization utilizes alternating sequences of test and reference gassing to investigate the sensitivity and the reversibility of material response in alternation.

In the examples described here the impedometric measurements of surface doped In_2O_3 (400–250 °C) as well as volume doped $LnCrO_3$ and $LnFeO_3$ (Ln = La, Pr, (Nd,) Sm, Eu, Gd, Tb, Dy, Ho, Er, Tm, Yb and Lu) (500–200 °C) were carried out in a frequency range from 10 to 10^7 Hz with an amplitude of 0.1 V_{rms} (15 measuring points per frequency decade). Synthetic air was humidified (RH = 45 % at 25 °C) and used as the reference gas for representative ambient conditions. As certified test gases 25 ppm H_2, 50 ppm CO, 5 ppm NO, 5 ppm NO_2, 40 ppm C_2H_5OH (not applied in the case of In_2O_3) and 25 ppm C_3H_6 (propene) each in synthetic air (consistent volume flow 100 sccm) were applied. However, for NO tests N_2 was chosen as carrier gas to avoid NO_2 formation. The electrical screening of one MES with 64 samples positions requires about 25 min at a particular atmosphere and temperature.

4.5.1 Data Handling

In order to reduce and to evaluate the huge quantity of measured data, a fitting software has been developed for automated curve fitting. The fitted data are visualized in terms of Trellis plots (color dot arrays) and fingerprints (bar diagrams) for easy identification of the materials with the best gas-sensing performances.

4.5.1.1 Data Fitting

The recorded impedance spectra are described by means of impedance functions of circuit equivalents adopted to their measuring data and visualized as so-called Argand diagrams, where the real part Z' of the impedance is plotted versus the imaginary part $-Z''$. A representative plot of the impedance spectra concerning the p-conducting $PrFeO_3$ under propylene exposure at 300 °C is displayed in Fig. 4.24a. Thereby the automated data fitting of the semicircle was performed by adapting R and C of a parallel RC element. Hence, the circuit equivalent (Fig. 4.24b) describes the impedance spectra as an impedance function via a resistor R and a capacitor C in parallel. Thus, the electrical properties of the sample material are reduced to one value for the resistance R and one for the capacitance C, respectively, resulting in a tremendous reduction of the data set. In this typical procedure C is associated with the geometric capacitance of the IDC and R with

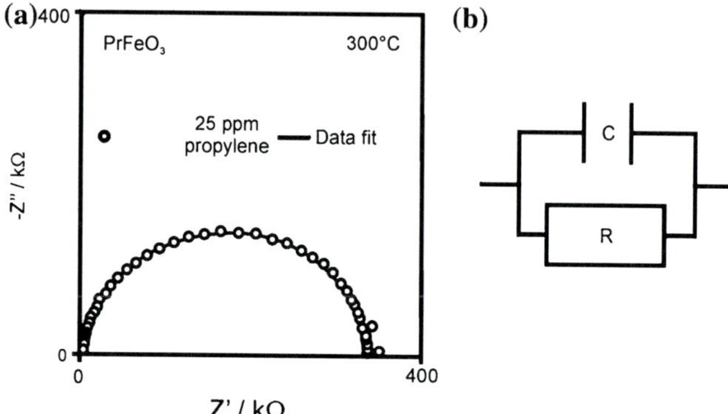

Fig. 4.24 a Argand plot (real part Z' of the impedance versus imaginary part $-Z''$) using the example of PrFeO₃ under propylene gassing at 300 °C and **b** corresponding circuit equivalent (parallel RC element) of one resistor R and one capacitor C in parallel [adapted from 138]

the resistance value under the respective testing conditions. Thus, a change in R related to the reference is therefore taken as the determination of the material sensitivities towards a particular gas exposure. The ohmic resistance of metal oxides and the related diameter of semicircle drops with an increasing temperature as expected for semiconductors. Generally in this work, the fitting with a single RC circuit was found to be sufficient. Nevertheless, other elements, such as inductances or constant phase elements (CPE), and their combinations can be implemented for the fitting as well.

4.5.1.2 Data Evaluation and Visualization

Since the obtained values of R are related to the specific material sensitivity under the respective measuring conditions the material response to a specific test gas is given by the sensor response whereas R_{TG} denotes the resistance fit under test gas conditions and R_{RG} the one under reference gas conditions.

$$\text{Sensor response} = +\frac{R_{RG} - R_{TG}}{R_{RG}} \quad |R_{RG} > R_{TG} \quad \xrightarrow{\text{Decrease of } R}$$

$$\text{Sensor response} = -\frac{R_{TG} - R_{RG}}{R_{TG}} \quad |R_{RG} < R_{TG} \quad \xrightarrow{\text{Increase of } R}$$

In the case of a decreasing resistance under test gassing the values of sensor response range from 0 to 1 and on the contrary in the case of an increasing resistance the values reside between 0 and -1. The application of the sensor response regarding the initial reference atmosphere compared to the more common formulae (i.e. sensor response $= R_{RG}/R_{TG}$ for $R_{RG} > R_{TG}$ and sensor response $= R_{TG}/R_{RG}$ for

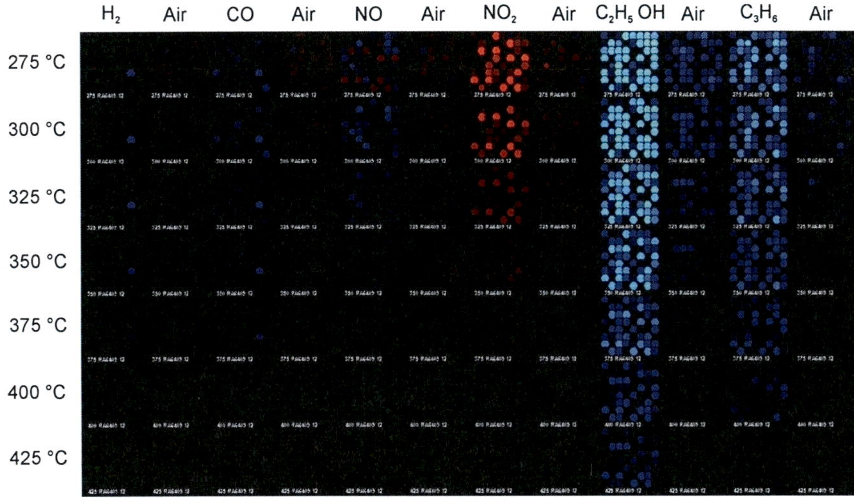

Fig. 4.25 Trellis plot of sensor responses for a substrate coated with various volume doped LnCrO₃ and LnFeO₃ materials

$R_{RG} < R_{TG}$, respectively), allows for the immediate discrimination between oxidizing and reducing test gases or n-type and p-type semiconductors, respectively. Furthermore, these changes in resistance are translated into a multicolored code (Eqs. 4.7 and 4.8) depending on the direction and extent of the change. This color change is visualized in terms of a Trellis plot, whereby the dot diameter scales down with the measuring failure resulting either from the scattering of data or the exceeding of measurement limits.

A Trellis plot for seven different temperatures (275–425 °C) and the sequence of six test gases (H_2, CO, NO, NO_2, C_2H_5OH and C_3H_6) is shown in Fig. 4.25 belonging to a substrate coated with various volume doped LnCrO₃ and LnFeO₃ materials. Thereby each 8 × 8 square matrix illustrates the sensor response of the 64 substrate positions at a distinct temperature and test gas atmosphere and thus, reflects the changes towards the initial state (reference measurement) for the respective gases. According to the definition of sensor response the perovskite materials (LnMO₃) show p-type semiconducting behavior. Most of the tested samples exhibit a pronounced sensitivity towards the applied hydrocarbons, while the sensitivity towards C_2H_5OH is much higher than towards C_3H_6, and a good reversibility on the timescale of the experiment. Further the samples exhibit less cross sensitivities towards the other test gases and especially NO_2 which vanish at higher temperatures.

The array of fingerprints in Fig. 4.26 represents the entire substrate plate of the LnCrO₃ and LnFeO₃ materials at a distinct temperature, here 350 °C. The measured sensor responses of the gas sequence (i.e. alternating reference gas and test gas, cf. Fig. 4.25) are plotted as bar diagrams which form a characteristic pattern for each material composition. In the present case some positions of the MES

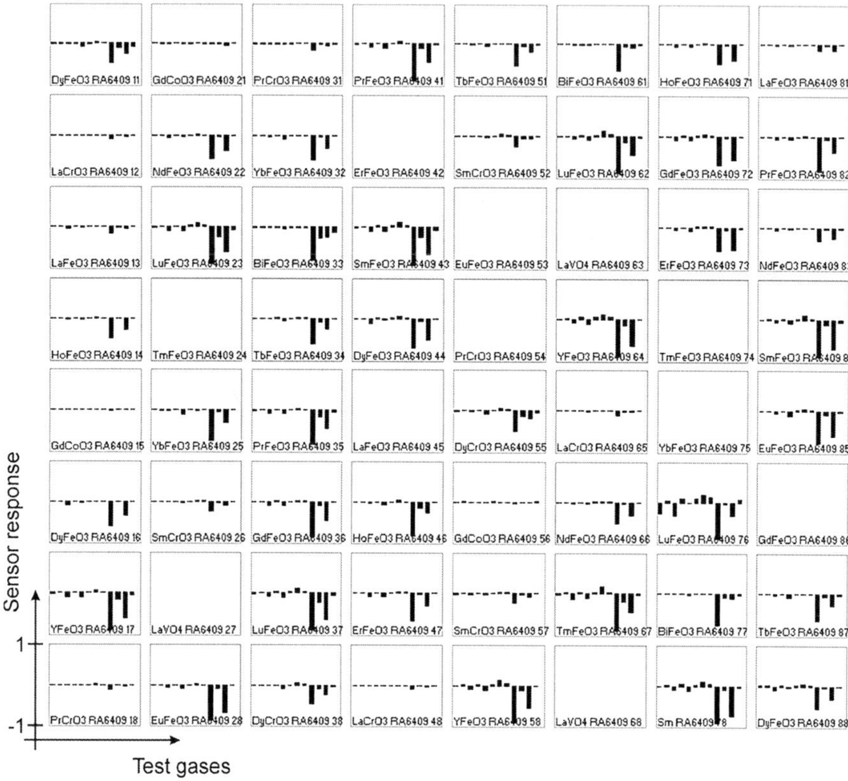

Fig. 4.26 Fingerprints of volume doped $LnCrO_3$ and $LnFeO_3$ materials at 350 °C including some positions (e.g. 24, 27, 42, 45 etc.) with only capacitive behavior due to defects in electrical contacting. The alternating test gas sequence is shown in Fig. 4.25

(e.g. 24, 27, 42, 45 etc.) exhibited solely capacitive behavior, because of insufficient electrical contacts or defect conductor paths, so that no resistivities or rather sensor responses could be determined. But in most instances a good reproducibility of the respective experiment was observed regarding the triple statistical distribution of equally coated positions (e.g. 33, 61 and 77, cf. the two reference gas steps after the C_3H_6 application at 275 and 300 °C in Fig. 4.25. Hence, slight differences in the behavior of each layer matched with variations in the microstructure caused by the layer preparation. In comparison to the rare-earth orthoferrites, the orthochromites show lower sensor responses, whereby $LaFeO_3$ and $LaCrO_3$ represent the lowest sensor responses of their group towards C_2H_5OH. The best sensing performances for all temperatures were found for $DyCrO_3$ and for $SmFeO_3$, $TmFeO_3$ and $LuFeO_3$, respectively.

In the following two subsections selected examples will illustrate the material screening by introducing results for surface and volume doping representing the reproducibility of preparation and measurements that finally lead to the discovery of new gas sensing properties and structure–property relations.

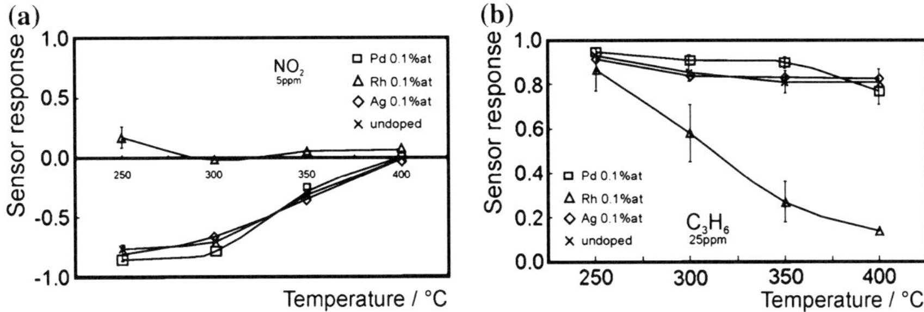

Fig. 4.27 Mean values of sensor response towards **a** 5 ppm NO_2 and **b** 25 ppm C_3H_6 and corresponding error bars each regarding 16 equally prepared In_2O_3 materials (i.e. 0.1 at.% Ag, Rh, Pd doped and undoped). [adapted from 151]

4.5.1.3 Example for Surface Doping

Based on previous works [161] a material library for the n-type semiconductor In_2O_3 was created. This library was generated from experiments on three MES plates which achieved material diversity by various surface doping elements, while the base material originated from a single synthesis. The three sample plates focused on the reproducibility of identically doped sample materials, on the diversity of doping elements and on the sensor property dependence on various dopant concentrations, respectively [151].

The homogeneity and reproducibility of sensing properties were investigated by comparing the sensor responses of 16 identically prepared sample positions of three doped (i.e. 0.1 at.% Pd, Rh and Au, respectively) and the undoped base material In_2O_3 as shown in Fig. 4.27. Generally, the minor standard deviation corresponds to a consistent sensor performance of the respective material towards H_2, NO, NO_2 (Fig. 4.27a) and C_3H_6 (Fig. 4.27b). But for CO significant variations were observed which could be traced back to a concentration gradient of CO on the sample plate caused by its catalytic conversion with oxygen resulting from the catalytic activity of some sample materials. The undoped In_2O_3 shows a higher sensor response towards H_2 and C_3H_6 by means of a resistance decrease and towards NO and NO_2 by means of a resistance increase at lower temperatures. The most pronounced effect on the sensor response results from doping with Rh whereby the cross sensitivity towards NO_2 is almost eliminated and the character of NO is changed from oxidizing to reducing at lower temperatures. Furthermore, the Rh doped samples exhibit the highest observed H_2 relative sensitivity at 250 °C which decreases with rising temperature.

A broad screening of the influence of surface doping was achieved by a huge quantity of doping elements (i.e. 0.1 at.%) on the second sample plate. All tested materials exhibit a high sensitivity towards the reducing gases H_2 and C_3H_6, whereas only for the Rh, Pd, Pt, Ag and Au doped materials an increased sensitivity towards CO was found. The typical sensor response towards the oxidizing gases NO and NO_2 was an increase in resistance which was irreversible on the time scale of the reference gassing (15 min) at 250 °C, but vanishes at higher

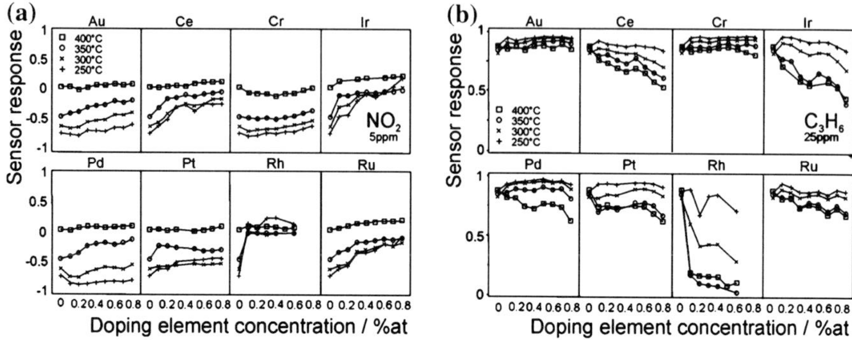

Fig. 4.28 Dependence of sensor response towards **a** 5 ppm NO_2 and **b** 25 ppm C_3H_6 for surface doped In_2O_3 (i.e. Au, Ce, Ir, Pd, Pt, Rh, Ru and Cr) at 250 °C depending on the doping concentration [adapted from 151]

temperatures leading to a full recovery. In contrast to that for the Ir and Rh doped samples of which Ir shows an increased resistance in the reversibility measurements a decrease in resistance towards NO was observed. Ru and Ce doping decreased the NO and NO_2 sensitivity of the base material.

The sensor properties of doping elements which were found to influence the sensing properties of In_2O_3 (i.e. Au, Ce, Ir, Pd, Pt, Rh, Ru and Cr) were investigated by the variation of doping concentration (i.e. 0–0.8 at.% in 0.2 steps) on the third sample plate. Due to their similar behavior Rh, Ir, Ru and Ce can be grouped together and sorted by a similar but descending effect. The oxidizing character of NO and NO_2 (Fig. 4.28a) is converted towards reducing and suppressed by an increasing element concentration, whereas the reducing character of H_2 and C_3H_6 (Fig. 4.28b) is suppressed at the same time. Pt doping reduces the sensor response except for C_3H_6 at 250 °C, whereas Cr, Au and Pd typically enhance the sensor response towards C_3H_6.

In order to progress the stages from the HT material screening to a ready-made sensor device for technically relevant application conditions the sensor preparation and surface doping were transferred to a single sensor substrate. The IDC layout of the substrate is quite analogous to the one described for the MES but features an integrated heater on the backside of the substrate (substrate 25 × 4 × 0.75 mm). The function and the layout of the single sensor and the corresponding working station have been published in previous works [137]. Under these close to reality testing conditions the sensor characteristics correlated well with those determined from the MES based measurements, i.e. the NO_2-tolerance in NO sensing at temperatures between 200 and 300 °C.

Other base materials, as for example ZnO [162] and $CoTiO_3$:La [149], were investigated and optimized as well towards their gas sensing properties by surface doping with various elements, combinations and concentrations.

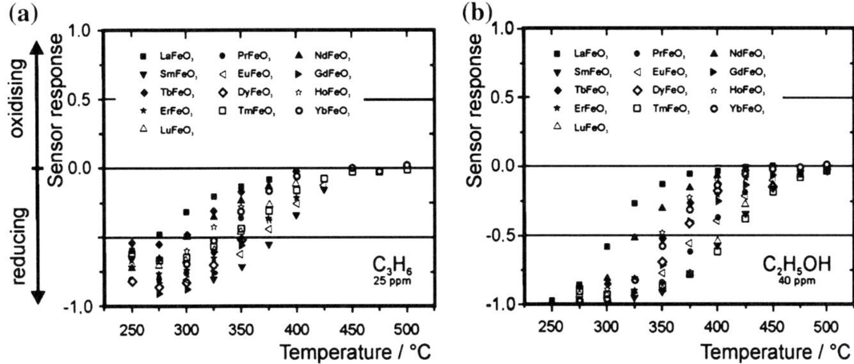

Fig. 4.29 Temperature dependence of sensor response towards **a** C_3H_6 and **b** C_2H_5OH regarding LnFeO₃ series (Ln = La, Pr, Nd, Sm, Eu, Gd, Tb, Dy, Ho, Er, Tm, Yb and Lu). [adapted from 161]

4.5.1.4 Example for Volume Doping

For detailed investigations of previous measurements, [147] the sensor responses and response times of two series of various volume doped $LnFeO_3$ and $LnCrO_3$ materials towards two different hydrocarbons were tested [163].

Regarding the $LnFeO_3$ series, the temperature dependence of the sensor response towards C_3H_6 and C_2H_5OH is illustrated in Fig. 4.29a and b, respectively. The operating temperature (200–500 °C) affects the sensing properties of all tested materials, of which $LaFeO_3$ exhibits the largest lanthanoid radius and the smallest sensitivity towards the applied hydrocarbons. The experiments concerning $YbFeO_3$ could only be evaluated above 300 °C, because the high material resistance exceeded the limits of the measurement device (R > 20 MΩ). $SmFeO_3$, $TmFeO_3$ and $LuFeO_3$ were found to be the most sensitive materials for temperatures up to 450 °C.

Compared to the $LnFeO_3$ series the temperature of maximum sensor response is shifted to lower values for the $LnCrO_3$ series. However, the sensitivity of both groups towards C_2H_5OH is higher than that towards C_3H_6 and exhibits a stronger effect in the case of the orthochromites. In this group $LaCrO_3$ and $DyCrO_3$ showed the lowest and highest sensitivity, respectively. The stronger effect of the C_2H_5OH test gas on the material sensitivities in both cases is presumably caused by the OH-functional group [164].

Besides characterization of the sensor response and selectivity, the high-throughput approach enables direct comparison of a third quality factor of gas sensors via the MES. Hence, concerning the $LnFeO_3$ series the response and recovery times upon 60 ppm C_2H_5OH were examined at 350 °C. For example the response time of $TmFeO_3$ upon 40 ppm C_2H_5OH was about 20 s and the recovery time about 44 s at 325 °C, whereas both parameters remained unaffected during several cycles [161].

Following up these investigations with the most sensitive materials for example, $SmFeO_3$ was volume doped by Co, K, Li, La, Pb, Mn and others in concentrations from 2 to 20 at. %. Thereby 10 at.% Co doped $SmFeO_3$ turned out to possess an enhanced activity towards CO up to 375 °C in contrast to the undoped

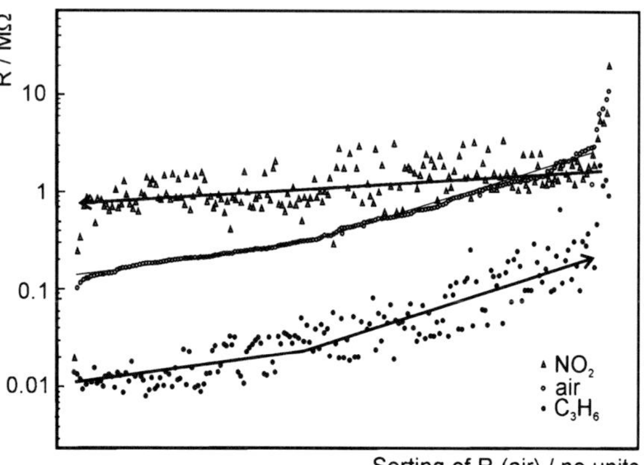

Fig. 4.30 Ascending assorted resistances of all surface doped In_2O_3 materials in the reference synthetic air versus their corresponding resistances in 5 ppm NO_2 and 25 ppm C_3H_6 at 250 °C [adapted from 151]

correspondent which showed only low sensitivity between 250 and 325 °C. The temperature dependent sensor performance of this Co doped $SmFeO_3$ sample was examined at the most sensitive temperature and at different CO concentrations from 15 to 50 ppm. The Co volume doped $SmFeO_3$ revealed a reversible response behavior featuring response and recovery times of 24 s or less, whereas the response time at higher CO concentrations was faster than at lower CO concentrations. The increase in CO sensitivity as a result of Co doping is confirmed by literature examples as well [138, 165, 166].

4.5.1.5 Structure–Property Relation

A thorough investigation of the screening data of the In_2O_3 material library via visual data mining was conducted in order to discover general correlations or systematic trends between the sensitivity and the material properties, besides the revelation of new significant sensing properties of individual materials [151].

Concerning the In_2O_3 library a plot of the ascending order of all material resistances in the reference synthetic air (45 % RH) versus their corresponding resistances in NO_2 and C_3H_6, respectively (Fig. 4.30), revealed a correlation between the sensor response and the electrical properties in reference gas at 250 °C. Generally, the resistance is typically increased by a higher concentration of the respective doping element (not displayed). Since the resistance of undoped In_2O_3 belongs to the small linear slope of lower resistances, a significant influence of the surface doping elements on the base resistance can be excluded. But for higher resistance values the rising slope contains Zr and Rh doped materials beside Ce, Cr, Pt, Ru, Au, Ir, Co and Ni. Two similar slopes are found for the sensitivities towards reducing C_3H_6 showing an offset factor of about 10 to lower resistances, whereas for oxidizing NO_2 only one smaller slope of the resistance is found showing a slightly lower offset factor to higher resistances. This circumstance lead to an approach of the respective curves for reference gas and NO_2, so that higher sensitivities are observed for smaller resistances in the reference gas. Thus, the

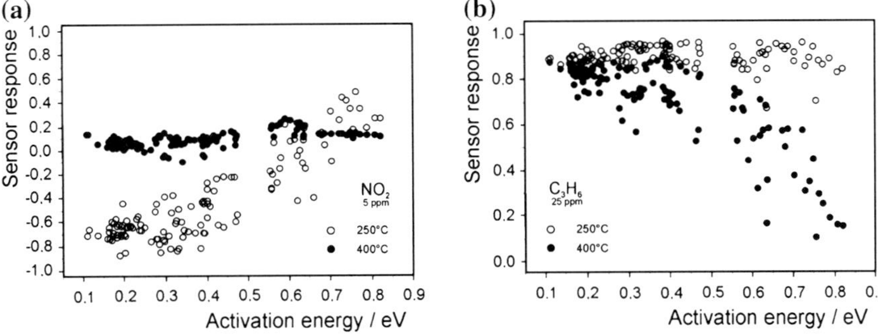

Fig. 4.31 Sensor response versus activation energy of conductivity in air at 250 and 400 °C. [adopted from 151]

sensitivity towards NO_2 and similar towards NO can be predicted on the basis of the material resistance in the reference gas air.

Additionally, a correlation between the sensitivity towards testing gases and the temperature dependence of the electrical conductivity in reference gas (i.e. activation energy) was discovered. The values of activation energy which does not necessarily represent the thermal activation of a specific electrical process were calculated by means of the Arrhenius equation on the basis of the first resistance in the reference gas at three selected temperatures (i.e. 300, 350 and 400 °C). Since most doping elements gave values comparable to the undoped In_2O_3, the doping by Ce, Au, Pt, Pd, Ru, Ir and Rh resulted in increasing activation energies. Plotting the sensor responses versus the activation energy of the conductivity in the reference gas for different temperatures exhibits increasing sensor responses towards NO and NO_2 (Fig. 4.31a) with increasing activation energies at 250 °C, whereby the sensor responses turn from oxidizing to reducing character by 0.5 and 0.7 eV, respectively. This correlation vanishes by rising temperature until no trend is observed at 400 °C due to decreasing sensor responses compared to other test gases at the same time. An opposite effect of decreasing sensor responses with rising values of the activation energy is found for H_2 and C_3H_6 (Fig. 4.31b).

These correlations indicate that the sensitivity of surface doped In_2O_3 sensors mainly depends on the electrical properties under reference gas exposure instead of the one under testing gas atmosphere. This finding is more pronounced for oxidizing gases at low temperature than for reducing gases at elevated temperatures.

Concerning the shown example of volume doped orthoferrites and orthochromites the screening of a multitude of chemically and structurally closely related compounds resulted in the finding of this structure–property relation [161]. In the case of rare-earth orthoferrites the published correlation between the sensing activity towards methanol and the lanthanoide ion radius (i.e. Gd > Eu > Sm > Nd > Pr > La) for the n-type correspondents [32] could not be confirmed towards the test gases (i.e. H_2, CO, NO, NO_2 and C_3H_6) for the 13 characterized p-type orthoferrites. But the previously published sensing activity of different p-type $LnFeO_3$ towards NO_2 (i.e. Sm > Dy > Gd $\approx$ Nd > La)

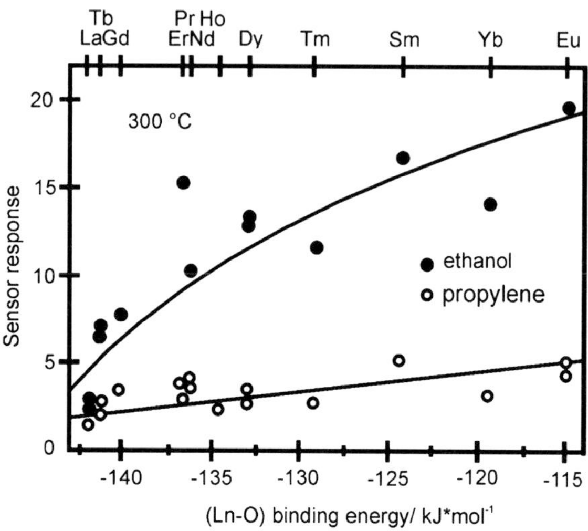

Fig. 4.32 Structure-property correlation between binding energy ΔH(Ln–O) of Ln cations (Ln = La, Pr, Nd, Sm, Eu, Gd, Tb, Dy, Ho, Er, Tm and Yb) and adsorbed oxygen versus sensor response of LnFeO$_3$ series towards C$_3$H$_6$ and C$_2$H$_5$OH at 300 °C. The curve does not represent a fit to the data but suggested subsidiary lines. [adapted from 161]

depending on the Ln species and its surface distribution [33] corroborates with our results obtained via the HT-IS. In literature the high Sm surface concentration and low Sm–O dissociation energy is reported to cause the high sensitivity of SmFeO$_3$. In our investigations an increasing sensitivity towards C$_3$H$_6$ and C$_2$H$_5$OH at 300 °C which is associated with a decreasing bond strength to chemisorbed oxygen at the near surface sites for the rare-earth series [147] was determined (Fig. 4.32), an effect that was already predicted towards CH$_3$OH [167].

Generally, from our investigations the significant dependence of the binding energy $\Delta H_{(Ln-O)}$ between lanthanoids and oxygen or rather the lanthanoid ion versus the sensor response under C$_3$H$_6$ and C$_2$H$_5$OH exposure, respectively, can be calculated by Eq. 4.9. Thereby, CN denotes the coordination number of the metal ions, H_f the molar formation energy of the oxide M$_m$O$_n$ including the indices m and n, H_s the sublimation energy of the metal and D_0 the dissociation energy of O$_2$ [147].

$$\Delta H_{(M-O)} = \frac{1}{CN\, m} \left(H_f - H_s m - \frac{n}{2} D_O \right) \tag{4.9}$$

The experiments have shown that an increase in the binding strength between metal ion and oxygen can be correlated with a smaller sensor response towards the test gases (except of LuFeO$_3$). This concept is transferable to rare-earth orthochromites LnCrO$_3$ and also to volume doped SmFeO$_3$ [139, 147].

4.6 Conclusion

The described correlations could only be discovered by the quantitative analysis of data and the associated statistical examination. Therefore the HT-IS screening represents a valuable tool for determining the crucial electrical properties of various surface and volume doped materials which define their gas sensing properties. The applied methods are accompanied with the obvious acceleration in the finding of tailored materials, systematic trends and property-structure relations.

The combined MES and HT-IS approach coupled with automated sample preparation, impedometric characterization and data evaluation has successfully been established as a beneficial preparative and analytical tool for a rapid identification of gas sensitive semiconductors. In this way a great amount of different samples were prepared and screened in a relatively short period, e.g. 832 IS measurements with 91 single frequency measurements could be performed within 124 h each under various gas atmospheres and at different temperatures including preliminary conditioning and subsequent cooling down [160].

The screening of surface doped $CoTiO_3$:La samples affirmed the reproducibility of both, the preparation and the measurements with low deviations. Using the HT-IS set-up a new CO-sensitive material with NO-tolerant NO_2-sensing behavior at 325 °C was found. Additionally, the screening of volume doped $LnFeO_3$ samples revealed a correlation between the sensitivity and the (Ln–O) binding energy.

Concluding, the compilation of huge material libraries by the HT-IS application offers the opportunity to discover correlations between the sensitivity and the binding energy Therefore this approach is a valuable tool for the systematic detection of new gas sensing materials or more generally, electrically functional materials in the future.

Acknowledgments We thank Dr. Melanie Homberger and Jutta Kiesgen for their great technical and graphical support for preparing this book chapter.

References

1. Debéda H, Rebière D, Pistré J, Ménil J (1995) Thick film pellistor array with a neutral network post-treatment. Sens Actuat B 27:297–300
2. Brailsford AD, Yussouff M, Logothetis EM (1992) Technical digest of the 4th international meeting on chemical sensors. In: Yamazoe N (ed), Japan association of chemical sensors, Tokyo
3. Ostrik B, Fleischer M, Meixner H, Kohl D (2000) Investigation of the reaction mechanism in work function type sensors at room temperature by studies of the cross-sensitivity to oxygen and water: the carbonate-carbon dioxide system. Sens Actuat B 68:197–202
4. Lucklum R, Hauptmann P (2000) The quartz crystal microbalance: mass sensitivity, viscoelasticity and acoustic amplification. Sens Actuat B 70:30–36
5. Chang SM, Kim YH, Kim JM, Chang YK, Kim JD (1995) Development of environmental monitoring sensor using quartz crystal micro-balance. Mol Cryst Liq Cryst 267:405–410
6. Schramm U, Meinhold D, Winter S, Heil C, Müller-Albrecht J, Wachter L, Hoff H, Roesky CEO, Rechenbach T, Boeker P, Lammers PS, Weber E, Bargon J (2000) A QMB-based temperature-modulated ammonia sensor for humid air. Sens Actuat B 67(3):219–226
7. Stetter JR, Li J (2008) Amperometric gas sensors—a review. Chem Rev 108:352–366
8. Pasierb P, Rekas R (2009) Solid-state potentiometric gas sensors—current status and future trends. J Solid State Electrochem 13(1):3–25
9. Bârsan N, Koziej D, Weimar U (2007) Metal oxide-based gas sensor research: how to? Sens Actuat B 121(1):18–35

10. Moos R, Sahner K, Fleischer M, Guth U, Bârsan N, Weimar U (2009) Solid state gas sensor research in Germany—a status report. Sensors 9:4323–4365
11. Brattain WH, Bardeen J (1953) Surface properties of germanium. Bell Syst Technol J 32:1–41
12. Heiland G (1954) Zum Einfluß von adsorbiertem Sauerstoff auf die elektrische Leitfähigkeit von Zinkoxydkristallen. Z Phys 138(3–4):459–464
13. Heiland G (1957) Zum Einfluß von Wasserstoff auf die elektrische Leitfähigkeit an der Oberfläche von Zinkoxydkristallen. Z Phys 148(1):15–27
14. Seiyama T, Kato A, Fujiishi K, Nagatami M (1962) A new detector for gaseous components using semiconductive thin films. Anal Chem 34(11):1502–1503
15. Seiyama T, Kagawa S (1966) Detector for gaseous components with semiconductive thin films. Anal Chem 38:1069–1073
16. Taguchi N (1962) Jpn. Pat. 45-38200 1962, Jpn Pat 47-38840, US Patent 3644795 1962
17. Eranna G, Joshi BC, Runthala DP, Gupta RP (2004) Oxide materials for development of integrated sensors—a comprehensive review. Crit Sol State Mater Sci 29:111–188
18. Choi KJ, Jang HW (2010) One-dimensional oxide nanostructures as gas-sensing materials: review and issue. Sensors 10:4083–4099
19. Lang O, Pettenkofer C, Sánchez-Royo JF, Segura A, Klein A, Jaegermann W (1999) Thin film growth and band lineup of In_2O_3 on the layered semiconductor InSe. J Appl Phys 86(10):5687–5691
20. Manno D, Micocci G, Serra A, Di Guilo M, Tepore A (2000) Structural and electrical properties of In_2O_3-SeO_2 mixed oxide thin films for gas sensing applications. J Appl Phys 88(11):6571–6577
21. Epifani M, Capone S, Rella R, Siciliano P, Vasanelli L (2003) In2O3 thin films obtained through a chemical complexation based sol-gel process and their application as gas sensor devices. J Sol-Gel Sci Technol 26(1–3):741–744
22. Korotcenkov G, Brinzani V, Cerneavschi A, Ivanov M, Cornet A, Morante J, Cabot A, Arbiol J (2004) In2O3 films deposited by spray pyrolysis: gas response to reducing (CO, H2) gases. Sens Actuat B 98(2–3):122–129
23. Belyesheva TV, Kazachkov EA, Gutman EE (2001) Gas sensing properties of In_2O_3 and Au-doped In_2O_3 films for detecting carbon monoxide in Air. J Anal Chem 56(7):676–678
24. Ivanovskaya M, Kotsikau D, Fagglia G, Nelli P, Irkaev S (2003) Gas-sensitive properties of thin film heterojunction structures based on Fe2O3–In2O3 nanocomposites. Sens Actuat B 93(1–3):422–430
25. Yokokawa H, Sakai N, Horita T, Yamaji K (2001) Recent developments in solid oxide fuel cell materials. Fuel cells 1(2):117–131
26. Xinshu N, Honghua L, Guogang L (2005) Preparation, characterization and photocatalytic properties of $REFeO_3$ (RE = Sm, Eu, Gd). J Mol Catal A 232(1–2):89–93
27. Peña MA, Fierro JLG (2001) Chemical structures and performance of perovskite oxides. Chem Rev 101:1981–2017
28. Keller N, Mistrik J, Visnovsky S, Schmool DS, Dumont Y, Renaudin P, Guyot M, Krishnan R (2001) Magneto-optical Faraday and Kerr effect of orthoferrite thin films at high temperatures. Eur Phys J B 21(1):67–73
29. Obayashi H, Sakurai Y, Gejo T (1976) Perovskite-type oxides as ethanol sensors. J Solid State Chem 17:299–303
30. Martinelli G, Carotta MC, Ferroni M, Sadaoka Y, Traversa E (1999) Screen-printed perovskite-type thick films as gas sensors for environmental monitoring. Sens Actuat B 55:99–110
31. Niu X, Du W, Du W (2004) Preparation, characterization and gas-sensing properties of rare earth mixed oxides. Sens Actuat B 99:399–404
32. Arakawa T, Tsuchi-ya S, Shiokawa J (1981) Catalytic activity of rare-earth orthoferrites and orthochromites. Mat Res Bull 16:97–103
33. Aono H, Traversa E, Sakamoto M, Sadaoka Y (2003) Crystallographic crystallization and NO_2 gas sensing property of $LnFeO_3$ prepared by thermal decomposition of Ln-Fe hexacyanocomlexes, $Ln[Fe(CN)_6]*nH_2O$, Ln = La, Nd, Sm, Gd, an Dy. Sens Actuat B 94:132–139

34. Liu X, Hu J, Cheng B, Qin H, Zhao M, Yang C (2009) First-principles study of O_2 adsorption on the $LaFeO_3$ (010) surface. Sens Actuat B 139:520–526
35. Liu X, Hu J, Cheng B, Qin H, Jiang M (2009) Preparation and gas sensing characteristics of p-type semiconducting LnFe0.9Mg0.1O3 (Ln = Nd, Sm, Gd and Dy) materials. Curr Appl Phys 9:613–617
36. Fino D, Russo N, Saracco G, Specchia V (2003) The role of suprafacial oxygen in some perovskites for the catalytic combustion of soot. J Catal 217:367–375
37. Huang R-F, Howng W-Y (1996) Effect of defect structure on gas sensitivity of $LaCrO_3$. J Mater Res 11(12):3077–3082
38. Toyama Prefecture (1984) Jpn Kokai Tokyo Koho, JP 59067601
39. Yamamoto A, Takada T, Nakamura T, Sato A, Endo E (2003) Sony Corp, Japan, Jpn. Kokai Tokkyo Koho, JP 2003200051
40. Brik Y, Kacimi M, Ziyad M, Bozon-Verduraz F (2001) Titania-supported cobalt and cobalt-phosphorus catalysts: characterization and performances in ethane oxidative dehydrogenation. J Catal 202(1):118–128
41. Pan TM, Lei TF, Chao TS (2001) High-κ cobalt-titanium oxide dielectrics formed by oxidation of sputtered Co/Ti or Ti/Co films. Appl Phys Lett 78(10):1439–1441
42. Chu X, Liu X, Wang G, Meng G (1999) Preparation and gas sensing properties of nano-$CoTiO_3$. Mat Res Bull 34(10/11):1789–1795
43. Kohl C-D (2005) Electronic noses. In: Waser R (ed) Nanoelectronics and information technology. Wiley-VCH, Berlin
44. Schüth F (2004) Hochdurchsatzuntersuchungen, Chemische Technik: Prozesse und Produkte, In: Neue Technologien, Dittmeýer R, Keim W, Kreysa G, Oberholz A (eds), Winnacker/Küchler, vol 2. Wiley-VCH, Weinheim, Germany
45. Hanak JJ (1970) The "multiple-sample concept" in materials research: synthesis, compositional analysis and testing of entire multicomponent systems. J Mater Sci 5:964–971
46. Hanak JJ (2004) A quantum leap in the development of new materials and devices. Appl Surf Sci 223(1–3):1–8
47. Xiang X-D, Schultz PG (1997) The combinatorial synthesis and evaluation of functional materials. Physica C 282–287:428–430
48. van Dover RB, Schneemeyer RF, Fleming RM (1998) Discovery of a useful thin-film dielectric using a composition-spread approach. Nature 392:162
49. Briceño G, Shang H, Sun X, Schultz PG, Xiang X-D (1995) A class of cobalt oxide magnetoresistance materials discovered with combinatorial synthesis. Science 270:273–275
50. Baeck SH, Jaramillo TF, Brändi C, McFarland EW (2002) Combinatorial electrochemical synthesis and characterization of tungsten-based mixed metal oxides. J Comb Chem 4: 563–568
51. Reichenbach HM, McGinn PJ (2001) Combinatorial synthesis of oxide powders. J Mater Res 16(4):967–974
52. Hagemeyer A, Stasser P, Volpe AF Jr (eds) (2004) High-throughput screening in chemical catalysis. Wiley-VCH, Weinheim
53. Aramova MA, Chang KS, Tageuchi I, Jabs H, Westerheim D, Gonzalez-Martin A, Kim J, Lewis B (2003) Combinatorial libraries of semiconductor gas sensor as inorganic electronic noses. Appl Phys Lett 83(6):1255–1257
54. Dagani R (1999) A faster route to new materials. Chem Eng News 77(10):51–60
55. Maier WF, Stöwe K, Sieg S (2007) Kombinatorische und Hochdurchsatz-Techniken in der Materialforschung. Angew Chem 119:2–60
56. Maier WF, Stöwe K, Sieg S (2007) Combinatorial and high-throughput materials science. Angew Chem Int Ed 46(32):6016–6067
57. Czarnik AW, DeWitt SH (eds) (1997) A practical guide to combinatorial chemistry. Am Chem Soc, Washington
58. Cong P, Dehestani A, Doolen R, Giaquinta DM, Guan S, Markov V, Poojary D, Self K, Turner H, Weinberg WH (1999) Combinatorial discovery of oxidative dehydrogenation catalysts within the Mo-V-Nb-O system. Proc Nat Acad Sci USA 96:11077–11080

59. Potyrailo RA, Mirsky VM (2008) Combinatorial and high-throughput development of sensing materials: the first 10 years. Chem Rev 108(2):770–813
60. Gurlo A (2006) Interplay between O_2 and SnO_2: oxygen ion sorption and spectroscopic evidence for adsorbed oxygen. Chem Phys Chem 7:2041–2052
61. Gurlo A, Riedel R (2007) In situ und Operando-Spektroskopie zur Untersuchung von Mechanismen der Gaserkennung. Angew Chem 119:3900–3923
62. Franke ME, Koplin TJ, Simon U (2006) Metal and metal oxide nanoparticles in chemiresistors: does the nanoscale matter? Small 2(1):36–50
63. Bârsan N (1994) Conduction models in gas-sensing SnO2 layers: grain-size effects and ambient atmosphere influence. Sens Actuat B 17(3):241–246
64. Bârsan N, Schweizer-Berberich M, Göpel W (1999) Fundamental and practical aspects in the design of nanoscaled SnO_2 gas sensors: a status report. Fresenius J Anal Chem 365(4):287–304
65. Baraton MI, Merhari L (2004) Advances in air quality monitoring via nanotechnology. J Nanoparticle Res 6(1):107–117
66. Korotcenkov G (2007) Practical aspects in design of one-electrode semiconductor gas sensors: status report. Sens Actuat B 121(2):664–678
67. Yamazoe N (1991) New approaches for improving semiconductor gas sensors. Sens Actuat B 5(1–4):7–19
68. Yamazoe N, Sakai G, Shimanoe K (2003) Oxide semiconductor gas sensor. Catal Surv Asia 7:63–75
69. Samson S, Fonstad CG (1973) Defect structure and electronic donor levels in stannic oxide crystals. J Appl Phys 44(10):4618–4621
70. Jarzebski ZM, Marton JM (1976) Physical properties of SnO_2 materials. J Electrochem Soc 123:299C–310C
71. Maier J, Göpel W (1988) Investigations of the bulk defect chemistry of polycrystalline tin (IV) oxide. J Solid State Chem 72(2):293–302
72. Göpel W, Schierbaum KD (1995) Chemisorption and charge transfer at ionic semiconductor surfaces: imaging in designing gas sensors. Sens Actuat B 26–27:1–12
73. Weisz PB (1953) Effects of electronic charge transfer between adsorbate and solid on chemisorption and catalysis. J Chem Phys 21:1531–1538
74. Lampe U, Fleischer M, Reitmeier N, Meixner H, McMonagle JB, Marsch A (1997) New metal oxide sensors: materials and properties. In: Göpel W, Hesse J, Zemel JN (eds) sensors, vol 2. Wiley-VCH, Weinheim, pp 29–30
75. Ogawa H, Nishikawa M, Abe A (1982) Hall measurement studies and an electrical conduction model of tin oxide ultrafine particle films. J Appl Phys 53:4448–4454
76. Bârsan N, Weimar U (2003) Understanding the fundamental principles of metal oxide based gas sensors; the example of CO sensing with SnO_2 sensors in the presence of humidity. J Phys: Condens Matter 15(20):R813–R839
77. Barton MI, Merhari L, Ferkerl H, Catagnet JF (2002) Comparsion of the gas sensing properties of tin, indium and tungsten oxides nanopowders: carbon monoxide and oxygen detection. Mater Sci Eng, C 19:315–321
78. Madou J, Morrison SR (1989) Chemical sensing with solid state devices. Academic Press, New York
79. Lenaerts S, Honore M, Huyberechts G, Roggen J, Maes G (1994) In situ infrared and electrical characterization of tin dioxide gas sensors in nitrogen/oxygen mixtures at temperatures up to 720 K. Sens Actuat B 19:478–482
80. Bârsan N, Weimar U (2001) Conduction model of metal oxide gas sensors. J Electroceram 7:143–167
81. Öberg PÅ, Togawa T, Spelman FA (2004) Sensors in medicine and health Care. In: Hesse J, Gardner J, Göpel W (eds) Sensors Application, vol 3. Wiley-VCH, Weinheim
82. Schierbaum KD, Weimar U, Göpel W, Kowalowski R (1991) Conductance, work function and catalytic activity of SnO_2-based gas sensors. Sens Actuat B 3:205–214

83. Nemov TG, Yordanov SP (1996) Ceramic sensors—technology and application. Technomic Publishing Company Inc, Lancaster, Pennsylvania, U S A 138
84. Pardo M, Sberveglieri G (2004) Electronic olfactory systems based on metal oxide semiconductor arrays. MRS Bull 29(19):703–708
85. Lee J-H (2009) Gas sensors using hierarchical and hollow oxide nanostructures: overview. Sens Actuat B 140:319–336
86. Xu C, Tamaki J, Miura N, Yamazoe N (1989) International symposium on fire ceramics, Arita, Japan
87. Göpel W, Hesse J, Zemel JN (1991) Sensors—a comprehensive survey, vol 2. Wiley-VCH, Weinheim
88. Xu C, Tamaki J, Miura N, Yamazoe N (1990) Correlation between gas sensitivity and crystallite size in porous SnO_2-based sensors. Chem Lett 19(3):441–444
89. Xu C, Tamaki J, Miura N, Yamazoe N (1990) Relationship between gas sensitivity and microstructure of porous stannic oxide. J Electrochem Soc Jpn 58(12):1143–1148
90. Xu V, Tamaki J, Miura N, Yamazoe N (1991) Grain size effects on gas sensivity of porous SnO_2-based elements. Sens Actuat B 3(2):147–155
91. Rothschild A, Komem Y (2004) The effect of grain size on the sensitivity of nanocrystalline metal-oxide gas sensors. J Appl Phys 95:6374–6380
92. Wang X, Yee SS, Carey WP (1995) Transition between neck-controlled and grain-boundary-controlled sensitivity of metal-oxide gas sensors. Sens Actuat B 24–25:454–457
93. Korotcenkov G (2008) The role of morphology and crystallographic structure of metal oxides in response of conductometric-type gas sensors. Mat Sci Eng R 61:1–39
94. Schmidt-Mende L, MacManus-Driscoll JL (2007) ZnO—nanostructures, defects, and devices. Mater Today 10(5):40–48
95. Gurlo A (2010) Nanosensors: does crystal shape matter? Small 6(11):2077–2079
96. Seyed-Razavi A, Snook IK, Barnard AS (2010) Origin of nanomorphology: does a complete theory of nanoparticle evolution exist? J Mater Chem 20:416–421
97. Abbet S, Heiz U (2004) The chemistry of nanomaterials. In: Rao CNR, Müller A, Cheetham AK (eds), vol 2. Wiley-VCH, Weinheim, 551–588
98. Wang Q, Ostafin AE (2004) Metal nanoparticles in catalysis. In: Nalwa HS (ed) Encyclopedia of nanoscience and nanotechnology vol 5. American scientific: Stevenson Ranch, CA, 475–503
99. Panchapakesan B, De Voe DL, Widmaier MR, Cavicchi R, Semancik S (2001) Nanoparticle engineering and control of tin oxide microstructures for chemical microsensor applications. Nanotechnology 12:336–349
100. Benkstein KD, Semancik S (2006) Mesoporous nanoparticles TiO2 thin films for conductometric gas sensing on micro hotplate platforms. Sens Actuat B 113:445–453
101. Hyodo T, Nishida N, Shimizu Y, Egashira M (2002) Preparation and gas sensing properties of thermally stable mesoporous SnO_2. Sens Actuat B 83:209–215
102. Becker T, Ahlers S, Bosch von Braunmühl C, Müller G, Kisewetter O (2001) Gas sensing properties of thin- and thick-film tin-dioxide materials. Sens Actuat B 77(1–2):55–61
103. Shimizu Y, Jono A, Hyodo T, Egashira M (2005) Preparation of large mesoporous SnO_2 powder for gas sensor application. Sens Actuat B 108:56–61
104. Shen G, Chen P-C, Ryu K, Zhou C (2009) Devices and chemical sensing applications of metal oxide nanowires. J Mater Chem 19:828–839
105. Raible I, Burghard M, Schlecht U, Yasuda A, Vossmeyer T (2005) V_2O_5 nanofibres: novel gas sensors with extremely high sensitivity and selectivity to amines. Sens Actuat B 106(2):730–735
106. Jiaqiang X, Yuping C, Yadong L, Jianian S (2005) Gas sensing properties of ZnO nanorods prepared by hydrothermal method. J Mater Sci 40(11):2919–2921
107. Baratto C, Comini E, Faglia G, Sberveglieri G, Zha M, Zappettini A (2005) Metal oxide nanocrystals for gas sensing. Sens Actuat B 109(1):2–6
108. Comini E, Baratto C, Faglia G, Ferroni M, Vomiero A, Sberveglieri G (2009) Quasi-one dimensional metal oxide semiconductors: preparation, characterization and application as chemical sensors. Prog Mater Sci 54:1–67

109. Cao Q, Rogers JA (2009) Ultrathin films of single-walled carbon nanotubes for electronics and sensors: a review of fundamental and applied aspects. Adv Mater 21:29–53
110. Maiti A (2004) Electrochemical and chemical sensing at the nanoscale: molecular modelling applications. Mol Sim 30(4):191–198
111. Balasubramanian K, Burghard M (2005) Chemically functionalized carbon nanotubes. Small 1(2):180–192
112. Korotcenkov G (2005) Gas response control through structural and chemical modification of metal oxide films: state of the art and approaches. Sens Actuat B 107:209–232
113. Morrison SR (1987) Selectivity in semiconductor gas sensors. Sens Actuat 12:425–440
114. Kohl D (1990) The role of noble metals in the chemistry of solid-state gas sensors. Sens Actuat B 1:158–165
115. Yamazoe N (1991) New approaches for improving semiconducting gas sensors. Sens Actuat B 5:7–19
116. Kappler J, Bârsan N, Weimar U, Diéguez A, Alay JL, Romano-Rodriguez A, Morante JR, Göpel W (1998) Correlation between XPS, Raman and TEM measurements and the gas sensitivity of Pt and Pd doped SnO_2 based gas sensors. Fresenius J Anal Chem 361(2):110–114
117. Cabot A, Arbiol J, Morante JR, Weimar U, Bârsan N, Göpel W (2000) Analysis of the noble metal catalytic additives introduced by impregnation of as obtained SnO_2 sol-gel nanocrystals for gas sensors. Sens Actuat B 70(1–3):87–100
118. Tsud N, Johanek V, Stara I, Veltruska K, Matolin V (2001) ISS, and TPD study of Pd-Sn interactions on Pd-SnOX systems. Thin Solid Films 391:204–208
119. Nehasil V, Janecek P, Korotcenkov G, Matolin V (2003) Investigation of behaviour of Rh deposited onto polycrystalline SnO_2 by means of TPD, AES and EELS. Surf Sci 532–535:415–419
120. Ruiz AM, Cornet A, Shimanoe K, Morante JR, Yamazoe N (2005) Effects of various metal additives on the gas sensing performances of TiO_2 nanocrystals obtained from hydrothermal treatments. Sens Actuat B 108(1–2):34–40
121. Mohr C, Hofmeister H, Radnik J, Claus P (2003) Identification of active sites in gold-catalysed hydrogenation of acrolein. J Am Chem Soc 125:1905–1911
122. Fong YY, Abdullah AZ, Ahmad AL, Bhatia S (2007) Zeolite membrane based selective gas sensors for monitoring and control of gas emissions. Sens Lett 5(3–4):485–499
123. Sahner K, Moos R, Matam M, Tunney JJ, Post M (2005) Hydrocarbon sensing with thick and thin film p-type conducting perovskite materials. Sens Actuat B 108:102–112
124. Sahm T, Weizhi R, Bârsan N, Mädler L, Weimar U (2007) Sensing of CH_4, CO and ethanol with in situ nanoparticle aerosol-fabricated multilayer sensors. Sens Actuat B 127(1):63–68
125. Trimboli J, Mottern M, Verweij H, Dutta PD (2006) Interaction of water with Titania: implications for high-temperature gas sensing. J Phys Chem 110(11):5647–5654
126. Cabot A, Arbiol J, Cornet A, Morante JR, Chen F, Liu M (2003) Mesoporous catalytic filters for semiconductor gas sensors. Thin Solid Films 436(1):64–69
127. Pijolat C, Viricelle JP, Tournier G, Montmeat P (2005) Application of membranes and filtering films for gas sensors improvements. Thin Solid Films 490(1):7–16
128. Simon U, Sanders D, Jockel J, Heppel C, Brinz T (2002) Design strategies for multielectrode arrays applicable for high-throughput impedance spectroscopy on novel gas sensor materials. J Comb Chem 4:511–515
129. Frantzen A, Sanders D, Jockel J, Scheidtmann J, Frenzer G, Maier WF, Brinz T, Simon U (2004) Hochdurchsatzmethode zur impedanzspektroskopischen Charakterisierung resistiver Gassensoren. Angew Chem 116(6):770–773
130. Frantzen A, Sanders D, Jockel J, Scheidtmann J, Frenzer G, Maier WF, Brinz T, Simon U (2004) High-throughput method for the impedance spectroscopic characterization of resistive gas sensors. Angew Chem Int Ed 43(6):752–754
131. Simon U, Sanders D, Jockel J, Brinz T (2005) Setup for high-throughput impedance screening of gas-sensing materials. J Comb Chem 7(5):682–687
132. Frantzen A, Sanders D, Scheidtmann J, Simon U, Maier WF (2005) A flexible database for combinatorial and high-throughput materials science. QSAR Comb Sci 24:22–28

133. Figlarz M, Fiévet F, Lagier JP (1982) Process for reducing metallic compounds using polyols, and metallic powders produced thereby, European patent 0113281

134. Toneguzzo P, Viau G, Acher O, Guillet F, Bruneton E, Fievet-Vincent F, Fiévet F (2000) CoNi and FeCoNi fine particles prepared by the polyol process : physico-chemical characterization and dynamic magnetic properties. J Mater Sci 35:3767–3784

135. Poul L, Ammar S, Jouini N, Fievet F, Villain F (2003) Synthesis of inorganic compounds (metal oxide and hydroxide) in medium. A versatile route related to the sol-gel process. J Sol-Gel Sci Technol 26:261–265

136. Jézéquel D, Guenot J, Jouini N, Fiévet F (1995) Submicrometer zinc oxide particles: elaboration in polyol medium and morphological characteristics. J Mater Res 10:77–83

137. Feldmann C, Jungk H-O (2001) Polyol-mediated preparation of nanoscale oxide particles. Angew Chem Int Ed 40(2):359–362

138. Siemons M, Weirich T, Maier J, Simon U (2004) Preparation of nanosized perovskite-type oxides via polyol method. Z Anorg Allg Chem 630(12):2083–2089

139. Koplin TJ (2006) Entwicklung und Anwendung von Hochdurchsatztechniken zur Darstellung und Untersuchung neuer nanostrukturierter Sensormaterialien, PhD thesis, RWTH Aachen University, Aachen

140. Siemons M, Koplin TJ, Simon U (2007) Advances in high throughput screening of gas sensing materials. Appl Surf Sci 254(3):669–676

141. Siemons M, Simon U (2006) High throughput screening of the sensing properties of doped SmFeO3. Sol State Phenom 128:225–236

142. Scheibe C, Obermeier E, Maunz W, Plog C (1994) Development of a high-temperature basic device for chemical sensors based on an IDC with on-chip heating. Sens Actuat B 25(1–3):403–406

143. Sanders D (2004) Entwicklung von Gassensoren auf Indiumoxid-Basis mittels Hochdurch¬satz-Impedanzspektroskopie, PhD thesis, RWTH Aachen University, Aachen

144. Bergh SH, Guan S (2000) Fluid distribution for chemical processing microsystems, US Patent 6890493

145. Feldmann C (2003) Polyol-mediated synthesis of nanoscale functional materials. Adv Funct Mater 13(2):101–107

146. Feldmann C (2004) Darstellung und Charakterisierung der nanoskaligen Vb-Metalloxide M2O5 (M = V, Nb, Ta). Z Anorg Chem 630:2473–2477

147. Fièvet F, Sugimoto T (eds) (2000) Polyol process, in fine particles: synthesis, characterization, and mechanisms of growth. Marcel Dekker, New York, pp 460–496

148. Poul L, Jouini N, Fièvet F (2000) Layered hydroxide metal acetates (metal = zinc, cobalt and nickel): elaboration via hydrolysis in polyol medium and comparative study. Chem Mater 12:3123–3132

149. Siemons M, Leifert A, Simon U (2007) Preparation and gas sensing characteristics of nanoparticulate p-type semiconducting rare-earth orthoferrites $LnFeO_3$ and orthochromites $LnCrO_3$ (Ln = La, Pr, Nd, Sm, Eu, Gd, Tb, Dy, Ho, Er, Tm, Yb, Lu). Adv Funct Mater 17:2189–2197

150. Siemons M, Simon U (2007) Gas sensing properties of volume-doped $CoTiO_3$ synthesized via polyol method. Sens Actuat B 126:595–603

151. Siemons M, Simon U (2006) Preparation and gas sensing properties of nanocrystalline La-doped $CoTiO_3$. Sens Actuat B 120(1):110–118

152. Frenzer G, Frantzen A, Sanders D, Simon U, Maier WF (2006) Wet chemical synthesis and screening of thick porous oxide films for resistive gas sensing applications. Sensors 6: 1568–1586

153. Sanders D, Simon U (2007) High-throughput gas sensing screening of surface doped In_2O_3. J Comb Chem 9:53–61

154. Lyubutin IS, Dmitrieva TV, Stepin AS (1999) Dependence of exchange interactions on chemical bond angle in a structural series: cubic perovskite-rhombic orthoferrite-rhombohedral hematite. J Exp Theor Phys 88(3):590–597

155. Krischner H, Koppelhuber-Bitschnau B (1994) Röntgenstrukturanalyse und Rietveldmethode, eine Einführung, Vieweg Verlag
156. Scherrer P (1918) Bestimmung der Größe und der inneren Struktur von Kolloidteilchen mittels Röntgenstrahlen, Nachrichten von der Gesellschaft der Wissenschaften zu Göttingen, Mathematisch-Physikalische Klasse 98–100
157. Birks LS, Friedman H (1946) Particle size determination from X-ray line broadening. J Appl Phys 17(8):687–692
158. Siemons M (2006) High throughput methods for synthesis and impedance characterization of ABO_3 gas sensing materials, PhD thesis, RWTH Aachen University, Aachen
159. Elbe D, Brinz T, Ullmann I, Krummel C, Schelling C, Heppel C, Robert Bosch GmbH (2004) DE 10319193 A1 20041118
160. Shimizu Y, Egashira M (1999) Basic aspects and challenges of semiconductor gas sensor. MRS Bull 24(6):18–24
161. Sanders D, Siemons M, Koplin TJ, Simon U (2005) Development of a high-throughput impedance spectroscopy screening system (HT-IS) for characterization of novel nanoscaled gas sensing materials, Mater. Res. Soc. Symposium Proceedings 876E, R6.1.1–R6.1.6
162. Koplin TJ, Siemons M, Océn-Valéntine C, Sanders D, Simon U (2006) Workflow for high-throughput screening of gas sensing materials. Sensors 6:298–307
163. Siemons M, Simon U (2007) High throughput screening of the propylene and ethanol sensing properties of rare-earth orthoferrites and orthochromites. Sens Actuat B 126(1):181–186
164. Heinert L (2000) Systematische Struktur-Wirkungs-Untersuchungen zwischen halbleitenden Metalloxidsensoren und Kohlenwasserstoffen, PhD thesis, Justus-Liebig-Universität Giessen
165. Song P, Qin H, Zhang L, Liu X, Huang S, Hu J, Jiang M (2005) Electrical and CO gas-sensing properties of perovskite-type La0.8Pb0.2Fe0.8Co0.2O3 semiconductive material. Phys B 368(1–4):204–208
166. Lee H-J, Song J-H, Yoon Y-S, Kim T-S, Kim K-J, Choi W-K (2001) Enhancement of CO sensitivity of indium oxide-based semiconductor gas sensor through ultra-thin cobalt adsorption. Sens Actuat B 79:200–205
167. Arakawa T, Kurachi H, Shiokawa J (1985) Physicochemical properties of rare earth perovskite oxides used as gas sensor material. J Mater Sci 20:1207–1210

Chapter 5
Selective Crystal Structure Synthesis and Sensing Dependencies

Lisheng Wang and Perena Gouma

Abstract Chemo-resistive sensors utilizing meal oxides form a very important type of sensors for gas detection. They are based on the interaction between gas molecules and surface ionosorbed oxygen species accompanied by electron transfer, which eventually leads to the change of material resistance. This process is controlled by a few external parameters (working temperature) and internal parameters (microstructure, chemical composition and crystal structure). While most parameters have been paid sufficient attention to, the influence of crystal structures is still largely unexplored. On the other hand, metal oxides exist in more than one crystalline form. The structural and property difference between different structures is expected to affect the sensing behavior of the material. Taking TiO_2 and WO_3 as examples, this chapter reviews how to selectively synthesize desired crystal structures and how they are related to the performance as a gas sensor. TiO_2 exists in two major polymorphs, with rutile being the thermodynamically stable phase and anatase being the metastable one. Compared to rutile, anatase is more open-structured and more chemically active and has lower surface energy. The hydrothermal method has been proved to be very effective in anatase synthesis as long as particle size is well controlled (normally under 20 nm) and dopants could stabilize this phase. Studies have found that anatase shows higher sensitivity as a gas sensor which is believed to be attributed to its higher chemical activity. WO_3 undergoes a series of phase transition when it is cooled down and γ-WO_3 is usually the room-temperature (RT) stable phase. The low-temperature stable phase, ε-WO_3, is the least symmetric among all the phases and is the only one with a ferroelectric feature. By a rapid

L. Wang
Department of Electrical and Computer Engineering,
University of British Columbia, Vancouver, BC, V6T 1Z4, Canada

P. Gouma (✉)
Department of Materials Science and Engineering,
SUNY at Stony Brook, Stony Brook, NY 11794-2275, USA
e-mail: pgouma@notes.cc.sunysb.edu

M. A. Carpenter et al. (eds.), *Metal Oxide Nanomaterials for Chemical Sensors*, Integrated Analytical Systems, DOI: 10.1007/978-1-4614-5395-6_5, © Springer Science+Business Media New York 2013

solidification method called flame spray pyrolysis, ε-WO_3 is able to be synthesized in high purity at RT. Doping with silicon and chromium could effectively stabilize this phase up to 500 °C by forming boundary domains or surface layers. The dopant-stabilized ε-WO_3 shows high sensitivity and unique selectivity to polar gas molecules, esp. acetone, which may be due to the strong interaction between the ε-WO_3 surface dipole and polar molecules.

5.1 Introduction

Chemo-resistive gas sensors based on semiconducting metal oxides (SMOs) are one of the most used and investigated type of gas sensors due to their low cost, flexibility, simplicity of their use and large number of detectable gases etc. The first reported work in this area dated back to 1957, in which Bielanski et al. [1] showed the electrical conductivity change of a few SMOs, including n-type ZnO, and p-type NiO etc., when they were exposed to ethanol. In 1962, Seiyama et al. [2] used ZnO thin films to detect several kinds of gases and volatile organic compounds (VOCs), including toluene, benzene and CO_2 etc. The decisive step was taken when Taguchi brought SMO sensors into industry (Taguchi-type sensors). To date, there are many companies offering this type of sensors, such as Figaro, FIS, MICS, UST, CityTech, Applied-Sensors, NewCosmos, etc. Their applications span from "simple" explosive or toxic gases alarms to air intake control in cars to components in complex chemical sensor systems. On the scientific research side, there have been numerous publications during the half century discussing new detection systems, improved sensing properties, influence of various parameters and so on.

Despite the diversity mentioned above, the underlying working mechanism of an SMO sensor remains the same and is well known. From the point view of chemistry, in the atmosphere, a SMO material tends to chemically adsorb surrounding oxygen molecules onto its surface, forming ionosorbed oxygen species (O^{2-}, O^-, and O_2^-), accompanied by electron trapping, which could be expressed by the following equations.

$$\begin{cases} O_2(\text{gas}) \Leftrightarrow O_2(\text{adsorbed}) \\ O_2(\text{adsorbed}) + e \Leftrightarrow O_{2\text{ads}}^- \\ O_{2\text{ads}}^- + e \Leftrightarrow 2O_{\text{ads}}^- \\ O_{\text{ads}}^- + e \Leftrightarrow 2O_{\text{ads}}^{2-} \end{cases} \tag{5.1}$$

When the sensor is further surrounded by a certain kind of oxidizing (NO_x and O_3, etc.) or reducing (H_2, CO, NH_3, most VOCs, etc.) gas, it will adsorb gas molecules onto its surface through weak van der Waals forces (called physisorption), covalent bonding (called chemisorption) and even electrostatic attraction. Those adsorbed molecules tend to extract electrons from the sensor surface or provide electrons to the sensor surface through the reaction with ionosorbed oxygen species. The following two reactions represent a simplified description of such process.

$$\begin{cases} H_2 + O_{ads}^- \rightarrow H_2O + e \\ NO_2 + e \rightarrow NO + O_{ads}^- \end{cases} \tag{5.2}$$

Such process will inevitably change the density of the charge carriers near the surface and leads to the change of surface resistance and finally the resistance of the entire material. For example, an n-type SMO whose major charge carriers are electrons will show a decrease of resistance if exposed to reducing gases and an elevation of resistance if exposed to oxidizing gases.

To serve as feasible gas sensors in practice, one metal oxide must meet a few requirements: high chemical activity, a stable structure and modest conductivity. Besides external conditions, including working temperature and humidity, the most important internal conditions which are able to influence the behavior of an SMO sensor include:

1. Chemical composition. Korotcenkov et al. [3] have reviewed the choice of sensing material through the properties of chemical species. In general, only transition-metal oxides with d^0 configurations of cations (TiO_2, WO_3, perovoskites, etc.) and post-transition-metal oxides with d^{10} configurations of cations (ZnO, SnO_2, In_2O_3 etc.) find their real gas sensor application [4]. They have modest band gaps (3–4 eV) and they are active in redox reactions. Among them, SnO_2-based sensors are the most extensively studied and the only type which has been commercialized. Pre-transition-metal oxides (e.g. MgO, Al_2O_3, etc.) usually are good insulators and inert to redox gas molecules. Transition-metal oxides with d^n ($0 < n < 10$) configuration of cations (e.g. Cr_2O_3, VO_2, NiO, RuO_2 etc.), although sensitive to the change of outside ambient, are not structurally stable (susceptible to oxidation or reduction) and usually metallic. However, high catalytic activity of transition metal elements makes them perfect doping elements to modify the surface activity of the base materials, either enhancing their sensitivity or improving their selectivity [5].

2. Microstructure. Microstructure determines the surface area of a material. As the grain size of a crystal decreases, its surface/boundary region becomes more and more dominant compared to bulk region. It has been determined that when the grain size is smaller than two times of Debye Length, exposure to surrounding gases will lead to change of charge carrier density within the entire grain. [6] Not taking other factors into account, smaller grain sizes will always lead to higher sensitivity. Since Debye Length of most materials is less than 100 nm, this effect is also called "nano effect". There have been extensive studies on utilizing different types of SMO nanomaterials, e.g. nanoparticles [7] and one-dimensional nanostructures [4, 8, 9] etc. to enhance sensor sensitivity. Another effective means to achieve high surface area is to create void areas in the material, i.e. porous structures [10], which also have found their great potential in highly sensitive gas detection.

Table 5.1 Structural information of TiO_2 polymorphs

Structure	Name	Existing conditions	Space group	Lattice parameters
Tetragonal	Rutile	Stable <1843 °C	P4$_2$/mnm	$a = 4.584$ Å, $c = 2.953$ Å, $Z = 2$
Tetragonal	Anatase	Metastable	I4$_1$/amd	$a = 3.793$ Å, $c = 9.51$ Å, $Z = 4$
Orthorhombic	Brookite	Metastable rare	Pbca	$a = 5.456$ Å, $b = 9.182$ Å, $c = 5.143$ Å, $Z = 8$

3. Crystal structure. Even if the chemical composition and microstructure has been determined, an SMO may still exhibit different thermodynamically stable or metastable crystal structures depending on its external environment (temperature and pressure) and crystallization process, called polymorphism. For example, it is well known that TiO_2 occurs naturally in several phases with the most common being rutile (stable) and anatase (metastable). WO_3 experiences a series of phase transition from γ-WO_3 to ε-WO_3 as it cools down [11] and there also exists a metastable h-WO_3 [12]. Two common forms of ZnO are: hexagonal wurtzite (stable) and cubic zinc blende (metastable) [13]. The differences between polymorphs of an SMO are mainly expressed by their physical properties and sometimes chemical properties as well. In particular, the arrangement of atoms near the material surface is expected to affect its interaction with surrounding gas molecules. However, to date there have been only a few studies focusing on the influence of crystal structure on material sensing behaviors [14]. Possible reasons may involve the difficulty to selectively synthesize and stabilize desired metastable crystal structure and lack of evidence showing the gas-surface interactions.

This chapter will take WO_3 and TiO_2 as examples to study how metastable crystal structures of SMOs could be selectively synthesized and how they affect the sensing performance of a material.

5.2 TiO$_2$ Gas Sensors: From Rutile to Anatase

5.2.1 Phase Evolution of TiO$_2$

Titanium dioxide exists in nature in three major forms whose structural information is listed in Table 5.1. There also exist a series of synthetic metastable phases and high-pressure phases. Among those structures, only rutile and anatase have been extensively studied and widely brought into application.

In spite of the structural diversity, the basic unit of all TiO_2 polymorphs remains the same, which is a slightly distorted octahedron consisting of a central titanium atom surrounded by six corner oxygen atoms. The average Ti–O bond length is

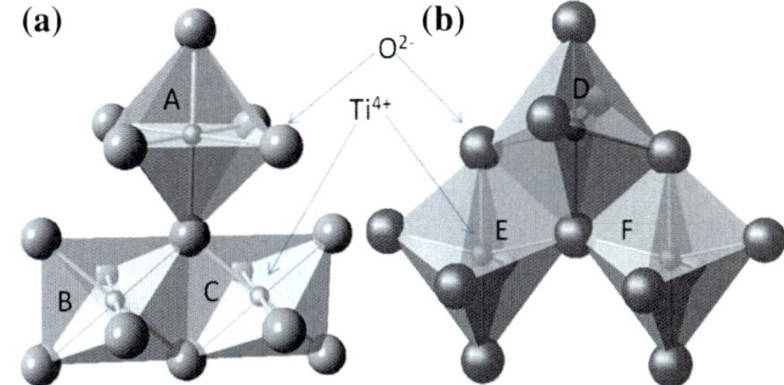

Fig. 5.1 Three adjacent [TiO$_6$] octahedron units in **a** rutile and **b** anatase. Among them, A–B, A–C and E–F are corner sharing while B–C, D–E and D–F are edge sharing

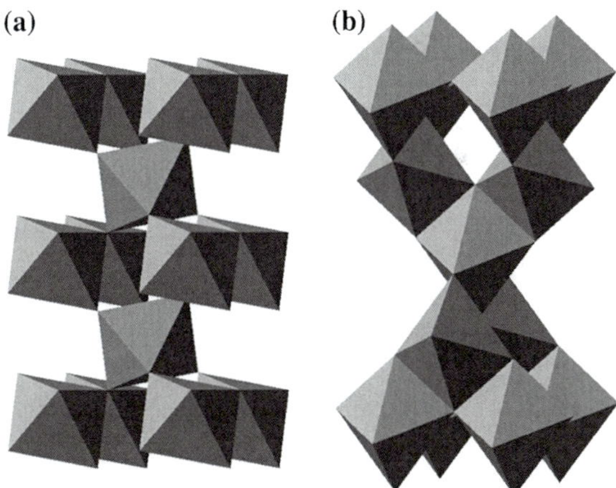

Fig. 5.2 Crystal structure comparison of **a** rutile and **b** anatase

about 1.96 Å. However, the bonds between titanium and two opposite oxygen atoms (forming apices) are slightly longer than the bonds between titanium and other four oxygen atoms (forming base). In TiO$_2$, adjacent [TiO$_6$] units connect each other through two ways: corner oxygen sharing and edge oxygen sharing, illustrated in Fig. 5.1.

It is the way of sharing oxygen atoms between adjacent [TiO$_6$] octahedron units that differs each structure from others. In rutile, corner sharing occurs along the <110> directions in which the apices of one unit are the base points of its neighbors (Fig. 5.2a). Such arrangement is stacked along the [001] direction, and edge sharing occurs during the stacking. Overall, two out of twelve edges in one octahedron are shared with others, all of which are base edges. In anatase, neighboring [TiO$_6$] units corner-share base oxygen atoms on (001) planes. Adjacent (001) octahedron planes connect each other by edge-sharing (Fig. 5.2b). The number of shared edges in one octahedron is four, all of which are side edges.

Further analysis indicates that rutile sand anatase have the same degree of symmetry, with point group 4/mmm in Hermann–Mauguin notation or D$_{4h}$ in Schönflies notation. However, anatase ($V_m = 20.1$ cm^3/mol) has a more open structure than rutile ($V_m = 18.7$ cm^3/mol) which leads to a smaller hardness (5.5–6.0 for anatase versus 6.0–6.5 for rutile). For the same reason, rutile is chemically and thermodynamically more stable while anatase is the more chemically active phase with lower surface energy. Therefore, although rutile is more commonly used in most areas, anatase is considerably preferable in the applications such as phtocatalysts [15], dye-sensitized solar cells [16], etc.

5.2.2 Selective Synthesis of Anatase

As we have already addressed, anatase exists in nature, indicating its fairly high thermodynamic stability. Therefore it is not surprising to detect anatase in synthetic TiO_2 products. In particular, a few strategies have been employed to selectively synthesize high-content or even pure anatase in various microstructures.

Among different techniques, solution-based methods, esp. hydrothermal methods are the most commonly used and most extensively studied ones. In a general term, in this type of methods, titanium precursors, e.g. titanium (IV) isopropoxide [17, 18] or TiF_4[19] etc., react with certain acids or bases in aqueous/organic solutions which forms amorphous TiO_2 via sol–gel hydrolysis precipitation and eventually transforms to anatase via hydrothermal treatment in controlled conditions [20]. The synthesized products are usually ultrafine particles [18, 20, 21] (Fig. 5.3a) or thin films on certain substrates [19, 22].

Other microstructures of anatase could also be synthesized via hydrothermal method with some modifications. Wang et al. [23] synthesized anatase TiO_2 three-dimensional hierarchical nanostructures by combining hydrothermal synthesis and following annealing process at 550 °C (Fig. 5.3b). The hierarchical nanostructures were formed by the self-organization of several tens of radially distributed thin petals with a thickness of several nanometers. Nian et al. [24] used titanate nanotubes as precursors and successfully synthesized single-crystalline anatase nanorods (Fig. 5.3c). Peng et al. [25] grew mesoporous anatase nanoparticles by using cetyltrimethylammonium bromide (CTAB) as surfactant-directing agent and pore-forming agent (Fig. 5.3d).

Other reports on selective fabrication of anatase by solution-based methods include, but not limited to: anodic oxidative hydrolysis of $TiCl_3$ within a hexagonal close-packed nanochannel alumina to grow well-aligned anatase nanowire arrays [26] and ultrasound irradiation of titanium precursor solutions for anatase nanoparticles [27], etc.

Although rutile is the stable phase of TiO_2 studies have demonstrated that solution-based methods generally favor the anatase phase. These observations are believed to be attributed to two main effects: surface energy and precursor chemistry. At very small particle dimensions, the surface energy is an important part of the total energy and it has been found that the surface energy of anatase is the lowest in all types of TiO_2 [28]. That explains the observation of anatase–rutile phase transition when nanoparticles are larger than a certain size [28]. Similar situation also occurs in t-ZrO_2 and γ-Al_2O_3 [29], etc. In addition, studies have also found that precursors and other experimental conditions also influence the

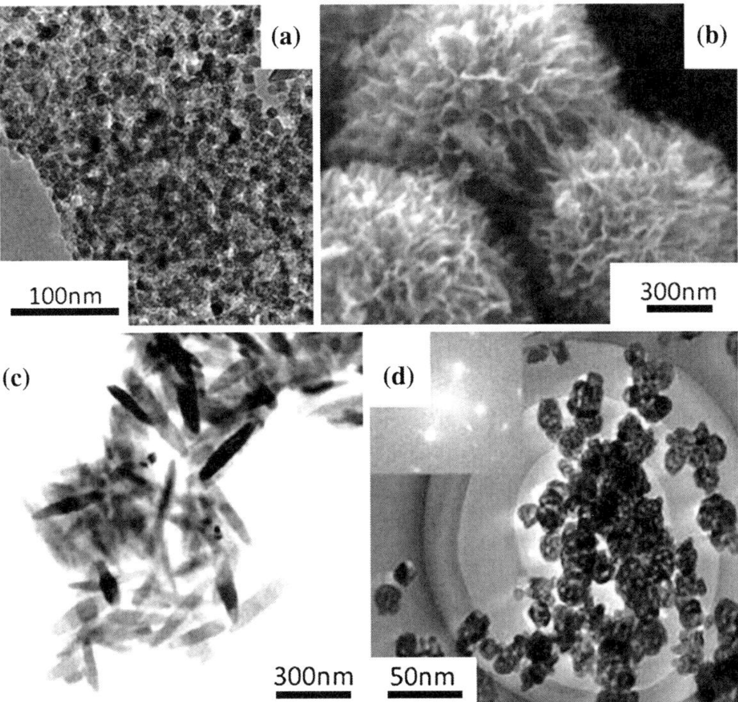

Fig. 5.3 Different microstructured anatase: **a** nanoparticles (reproduced from Reyes-Coronado et al. [20] by permission from an IOP Publishing Ltd); **b** three-dimensional hierarchical nanostructures (reproduced from Wang et al. [23] by permission from American Chemical Society); **c** nanorods (reproduced from Nian and Teng [24] by permission from American Chemical Society); and **d** mesoporous particles (reproduced from Peng et al. [25] by permission from American Chemical Society)

structure of the final product significantly as the nucleation and growth of the different polymorphs of TiO_2 are determined by the precursor chemistry, which depends on the reactants used. For example, stronger acids often lead to formation of brookite [30]. The underlying mechanism, however, has not been fully comprehended.

In addition to solution-based methods, dry methods were also developed to synthesize anatase. For example, anatase single-crystal nanotubes were fabricated by annealing highly ordered titania arrays in air at 450 °C [31] and nanostructured anatase were obtained by chemical vapor deposition using titanium precursors as source materials [32].

Anatase is only a type of metastable phase which means it will transform to rutile irreversibly at elevated temperatures, therefore its thermal stability is always a concern [33], esp. when TiO_2 is used for high-temperature applications, e.g. exhaust monitoring. Reported phase transition temperatures range from 400 to 1000 °C [34], greatly depending on experimental conditions and particle sizes.

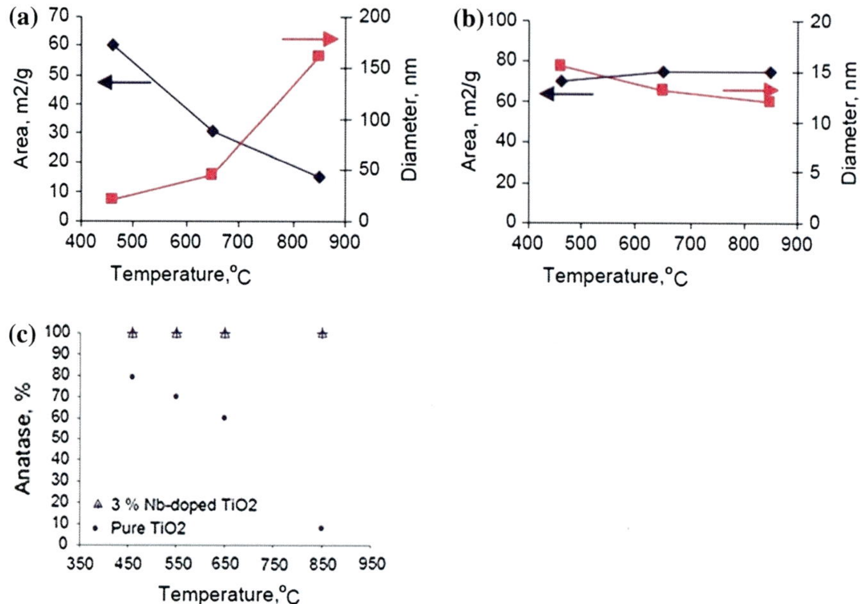

Fig. 5.4 Structural comparison of pure and Nb-doped anatase during annealing: crystal size change in **a** undoped TiO_2 and **b** 3 at % Nb-doped TiO_2; **c** anatase–rutile transition in pure and doped TiO_2 (reproduced from [36] by permission from an IOP Publishing Ltd)

The major strategy to stabilize as-made anatase is to dope foreign atoms or compounds into the matrix. Various dopants, e.g. Nb [35, 36], Ce [37], SiO_2 [37] and Al [38] etc. In particular, SiO_2-stabilized anatase was able to preserve the metastable structure up to 1200 °C, the highest temperature for anatase ever reported.

As an example, Fig. 5.4 shows the effect of Nb doping on the stability of antase reported by Anukunprasert et al. [36] Upon annealing from 450 to 850 °C, pure TiO_2 experienced huge crystal growth and significant phase transition from anatase to rutile, whereas Nb-doped TiO_2 clearly hindered such phase transition and inhibited the grain growth

5.2.3 Sensing Behaviors of Anatase

Undoped TiO_2 is typically an n-type semiconductor with the indirect band gap of 3.03 eV for rutile and 3.18 for anatase. Titanium cations in TiO_2 have d^0 configurations which makes TiO_2 suitable for gas sensing applications.

Development of TiO_2-based chemo-resistive gas sensors could date back to 1975, in which TiO_2 was used to detect oxygen in automobile exhausts [39]. Other detected gases include: H_2 [40, 41], CO [40, 41], NO_x [42] and VOCs [23, 43] etc.

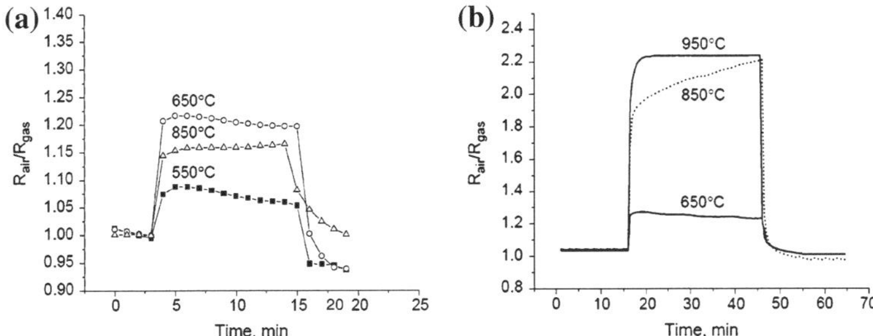

Fig. 5.5 Comparison of sensing response to 1000 ppm CO between pure and 3 at % Nb-doped TiO$_2$ (reproduced from [36] by permission from an IOP Publishing Ltd)

Although most sensors were based on TiO$_2$ nanoparticles [44] or thin films [43], mesoporous-structured [23, 40] and one-dimensional (1D) nanostructured TiO$_2$ [45, 46] were also reported in gas detection applications. Regarding crystal structure, sensing studies on pure rutile, pure anatase and rutile/anatase mixtures have all been reported, some of which compared the effect of phase content on sensing behaviors.

Anukunprasert et al. [36] compared the high-temperature sensing behavior of undoped and 3 at % Nb-doped TiO$_2$ upon exposure to CO (Fig. 5.5). It was found that the sensitivity of CO was significantly increased with an increase of temperature in Nb-doped TiO$_2$, which kept its anatase during annealing. In contrast, undoped TiO$_2$, which underewent large-scale anatase–rutile transition upon annealing, didn't show clear sensitivity enhancement at higher temperatures.

To investigate such phenomenon, three factors have to be considered: Nb dopants, smaller particle size of Nb-doped TiO$_2$ and higher content of anatase in Nb-doped TiO$_2$. However, Teleki et al. has demonstrated that Nb dopants higher than 2 % would lead to the decrease of sensitivity (Fig. 5.6a) [44]. Therefore Nb dopants could not be the reason for the observed sensitivity enhancement. In addition, we have already mentioned that anatase preferably exists in smaller sizes, which means particle size effect and anatase effect is actually one effect, which leads to enhanced sensitivity to CO. Similar results were also reported by Teleki et al. [44] when TiO$_2$ was used to detect ethanol (Fig. 5.6b).

We believe the higher sensitivity of anatase could be attributed to its higher activity induced by its more open structure. In fact, the catalytic effect difference of rutile and anatase in photo catalysis has been extensively studied and most reports have shown a higher chemical activity in anatase. Since Ti^{4+} cations also act as catalysts in the reaction of gas molecules with surface-adsorbed oxygen species, it is very likely that anatase also possesses higher activity during the gas sensing process. Therefore, a higher sensitivity in anatase is expected.

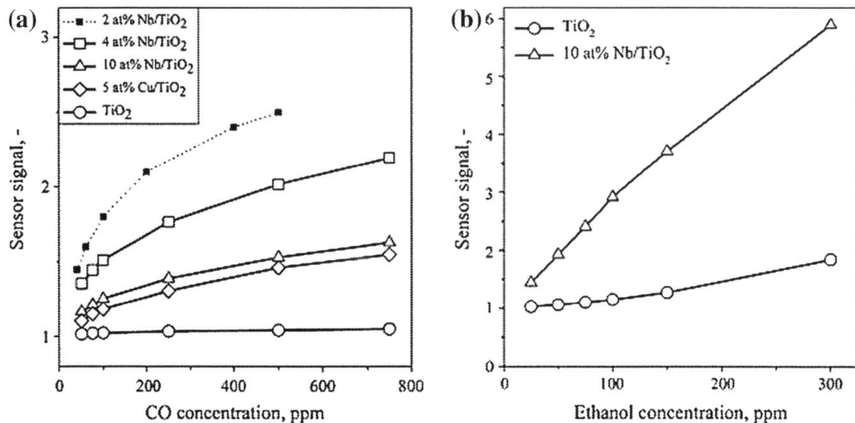

Fig. 5.6 Sensor signals as a function of **a** CO and **b** ethanol concentrations for undoped and doped TiO_2 (reproduced from [44] by permission from Elsevier Ltd.)

Fig. 5.7 Basic structure
of WO_3: Two $[WO_6]$
octahedron units sharing
corner oxygen

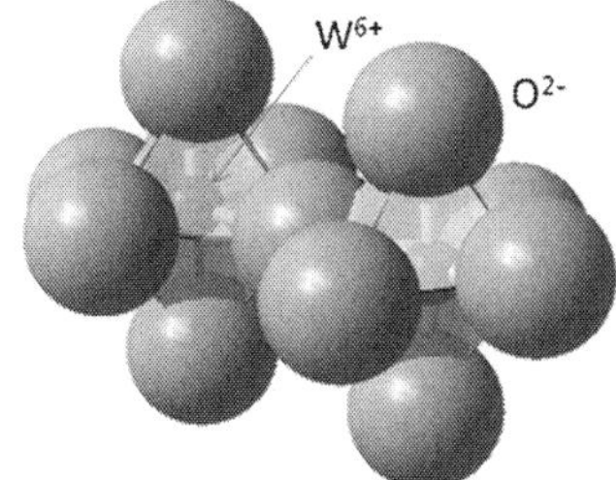

5.3 WO_3 Gas Sensors: From the γ Phase to the ε Phase

5.3.1 Phase Evolution of WO_3

All tungsten trioxides are based on the $[WO_6]$ octahedron units which is similar to $[TiO_6]$ units in TiO_2. The average W–O bond length is about 1.90 Å which however, is alterable according to different structures [47]. Different from TiO_2, the only way of connection between neighboring $[WO_6]$ units is by corner sharing, shown in Fig. 5.7.

Since these units could connect along different directions, WO_3 has a type of amorphous phase (a-WO_3) in which $[WO_6]$ units do not construct a regular pattern, as well as several crystalline phases. Stable crystalline WO_3 phases include: triclinic, monoclinic, orthorhombic and tetragonal ones. Their parameters are shown in Table 5.2.

Generally speaking, all stable-phased WO_3 have a similar ReO_3 structure (also a perovskite-like structure). An ideal ReO_3 structure can be treated as the repeat and extension of an octahedron unit along a, b, c axes vertical to each other. Every

Table 5.2 Structural information of stable WO_3 phases [11]

Structure	Symbol	Existing temperature (°C)	Space group	Lattice parameters
Monoclinic	ε	<-40	Pc	$a = 5.278$ Å, $b = 5.156$ Å, $c = 7.664$ Å, $\beta = 91.762°$
Triclinic	δ	$-40 \sim 17$	$P\bar{1}$	$a = 7.310$ Å, $b = 7.524$ Å, $c = 7.685$ Å $\alpha = 88.850°$, $\beta = 90.913°$, $\gamma = 90.935°$
Monoclinic	γ	17–320	$P2_1/n$	$a = 7.301$ Å, $b = 7.538$ Å, $c = 7.689$ Å, $\beta = 90.893°$
Orthorhombic	β	320–720	Pmnb	$a = 7.341$ Å, $b = 7.570$ Å, $c = 7.754$ Å
Tetragonal	α	720–900	P4/nmm	$a = 5.250$ Å, $c = 3.915$ Å

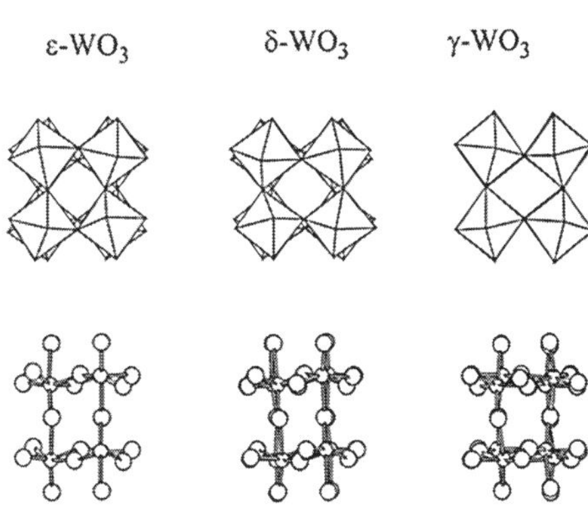

Fig. 5.8 Primary features of the structures of γ, δ, ε-WO_3 showing the tilting of the WO_6 octahedra (*top*) and the W–O bonds (*bottom*) (reproduced from [50] by permission from an IOP Publishing Ltd)

stable WO_3 phase could be recognized as a distortion of the above ReO_3 structure (Fig. 5.8).

By carefully examining the direction and magnitude of the tungsten shifts in each $[WO_6]$ octahedron of the three phases listed in Fig. 5.8, one can find that every tungsten atom always has a slight shift from its central position along every direction. Whereas in the γ and δ phases the magnitude of the shifts in every direction (x, y and z) is all roughly the same, in the ε-phase the shifts in the negative z direction are larger than those in the positive z direction. Because of the inequality of shifts in the z direction, a net spontaneous polarization develops, which leads to ferroelectricity in the ε phase.

It is easy to conclude that the symmetry of WO_3 is lowered from tetragonal α phase to triclinic phase as temperature goes down. Accordingly, its change of physical properties can be observed. Although monoclinic ε phase seems to be more symmetric than triclinic δ phase, considering the ferroelectric feature, its symmetry is the lowest. In fact, the most significant property changes occur when WO_3 undergoes a δ-ε transition at around -40 °C. Besides the appearance of ferroelectricity [48, 49], a remarkable volume contraction occurs upon transformation into the ε phase. In addition, the resistivity increases 20–30 times, according to Salje et al.'s research. The band gap increases from 2.6 eV to an unknown value >2.85 eV, resulting in a color change from pale green to bluish white [48].

5.3.2 Selective Synthesis of ε-WO$_3$

Different from anatase in TiO_2, ε-WO$_3$ is a stable phase of WO_3, but it is only stable below $-40\ ^{\circ}C$ and does not exist in nature. The first report on synthetic ε-WO$_3$ at RT was published in 1990 [51]. In this work, 100 nm-sized WO_3 microcrystals were prepared by burning a tungsten wire in a gas mixture of Ar and O_2. The ε phase of WO_3 was found to be mixed with γ-WO$_3$ in the final product at RT according to the Raman spectra. In a follow-up research, the same group suggested that larger microcrystals tend to take the γ phase and smaller microcrystals tend to take the ε phase [52]. Since then, ε-WO$_3$ at RT was occasionally reported in several other works, mixed with γ-WO$_3$ [53–55]. However, they did not study the high temperature stability of ε-WO$_3$ nor did they give any explanations why ε-WO$_3$ exists.

Recently, Wang et al. [56] have achieved synthesizing RT-stabilized ε-WO$_3$ for the first time by using a rapid solidification technique, i.e. flame spray pyrolysis (FSP). It is a very effective method to synthesize nano-sized oxide particles in a large amount and an excellent quality. [57] In simple terms, this method includes rapid decomposition of tungsten-containing precursors into tungsten captions and further oxidation into WO_3 assisted by high-temperature flame. The doping of chromium atoms was used to stabilize the ε phase at RT. Inspired by Wang et al.'s work, Righettoni et al. [58]. successfully produced silicon-stabilized ε-WO$_3$.

It was discovered (Fig. 5.9) [56, 58] that as-synthesized undoped products were a mixture of γ/ε-phased WO_3 particles with a fairly high percentage of ε-WO$_3$ (around 70 %). Both chromium and silicon dopants would substantially increase the ε phase percentage, 80 % in 10 at.% Cr-doped WO_3 and 100 % in 10 at.% Si-doped WO_3. Heat treatment would lead to the growth of particles and a ε-γ phase transition. In undoped WO_3 there was only 30 % ε phase after 500 $^{\circ}C$ heat treatment. However, doping with silicon or chromium would significantly inhibit such transition up to 500 $^{\circ}C$, i.e. there was still 80 % ε phase in Cr, Si-doped WO_{-3} after 500 $^{\circ}C$ annealing.

Further structural analyses suggested that both chromium and silicon dopants did not form any stable crystalline compounds in either as-synthesized or heat-treated WO_3 particles. Instead, the majority of chromium atoms existed in the form of $Cr = O$ terminal bonds (Fig. 5.10a) implying that they favored attachment on the particle surfaces to form chromate species. In Si-doped particles, amorphous SiO_2 domains formed near the boundary areas of WO_3 nanocrystals (Fig. 5.10b).

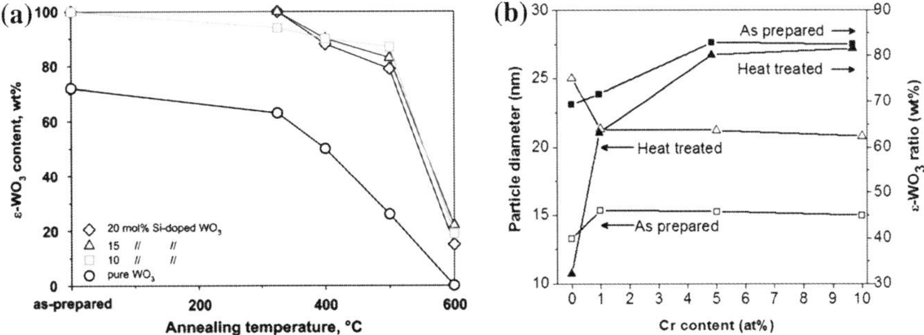

Fig. 5.9 The influence of doping and heat treatment on particle size and ε phase content in FSP-made **a** Si-doped WO_3 (reproduced from [58] by permission from American Chemical Society); **b** Cr-doped WO_3 (reproduced from [56] by permission from American Chemical Society)

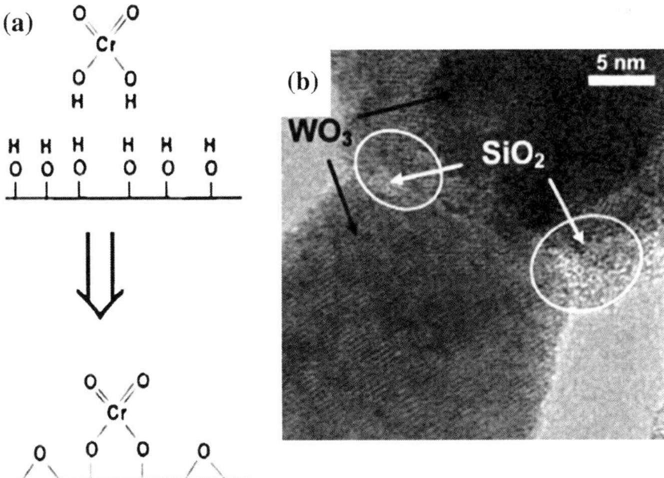

Fig. 5.10 Existence of chromium and silicon dopants in WO_3: **a** Illustration of the formation of surface chromates (reproduced from [64] by permission from American Chemical Society); **b** TEM image showing the amorphous SiO_2 domains (reproduced from [58] by permission from American Chemical Society)

It is interesting to observe the appearance of the ε phase in WO_3 considering it is only stable below −40 °C. As is known, ε-WO_3 has an acentric structure which is the least symmetric in the WO_3 family. Therefore the only proposed reason for its existence above RT is that tungsten atoms are repelled from their thermodynamically-dictated positions under certain circumstances. In previous reports, strong milling treatment [53] and high pressure [59] introduced external forces to force atoms to leave their stable positions. Gas evaporation in Ref. [51], hot-wire chemical vapor deposition in Ref. [60], pulsed laser deposition in Ref. [61] and

FSP [56] were all rapid solidification techniques in which tungsten atoms may not have enough time to settle down in their stable positions. In contrast, ε-WO$_3$ doesn't appear in WO$_3$ nanoparticles produced by some other techniques such sol–gel coating [62] or acid precipitation [63], etc.

Such hypothesis also explains why smaller microcrystals tended to take the ε phase as observed by Arai et al. [52] since smaller crystals tend to contain higher density of defects and deformation during the crystal growth. Doping foreign atoms is another means to increase the ε phase content because they introduce defects into the lattice and or pin the particle surfaces reducing the driving force for a phase transition, similar to anatase stabilization by doping as in Sect. 5.2.2.

As-synthesized ε-WO$_3$ is only a metastable structure above RT, which will irreversibly transform to the more thermodynamically stable γ phase due to the relief of internal stress. Doping with foreign atoms, however, may form a protective layer or pins, such as the surface chromates [56] or amorphous SiO$_2$ domains [58], preventing, or at least deferring the phase transition. This is in accordance with reports on other systems such as Vemury et al. who observed Si-doping of TiO$_2$ preventing its transformation from anatase to the more dense rutile phase [65].

5.3.3 Sensing Behaviors of ε-WO$_3$

Undoped WO$_3$ is typically an n-type semiconductor with the indirect band gap ranging from 2.6 eV (highly crystalline) to 3.3 eV (amorphous). Similar to TiO$_2$, the d^0 configurations and strong catalytic effect of tungsten cations enables WO$_3$ to be an excellent material for gas sensing purposes. Most WO$_3$ sensors are based on γ-WO$_3$ since it is the stable phase of WO$_3$ from RT to 320 °C, or sometimes a mixture of β/γ or γ/δ phases due to the similarity of those phases and therefore difficulty of complete phase transition.

The first work with respect to the WO$_3$ gas sensors was reported by Shaver [66] in 1967, in which a Pt-activated γ-WO$_3$ thin film was developed to detect airborne H$_2$ with enhanced sensitivity. In 1991, γ-WO$_3$ was found to have an excellent sensitivity to NO$_x$ [67]. Up to date γ-WO$_3$ is still being extensively studied for NO$_x$ detection. Most of reported WO$_3$ sensors used small-grained thin/thick films as materials [68–70]. Recently, WO$_3$ 1D nanostructures have also been successfully synthesized and applied to gas sensing [71, 72]. To modify the sensing properties, other elements or compounds were usually added into the system. Existing doping elements include Cu [73], noble metals [74–76] and Cr [77] etc. Besides NO$_x$, WO$_3$ sensors for other gases such as O$_3$ [78–81], H$_2$S [72, 82–85] NH$_3$ [86, 87] and VOCs [88, 89], etc. However, the sensitivity to these gases is generally less than to NO$_x$ and doping is usually needed to enhance selectivity [88, 89].

The recent achievement of selective synthesis and stabilization of ε-WO$_3$ provides the basis of using this material for gas sensing applications although there have been only limited reports discussing the sensing behavior of this phase. Both Wang et al. [56] and Righettoni et al. [58] reported the sensitive detection of

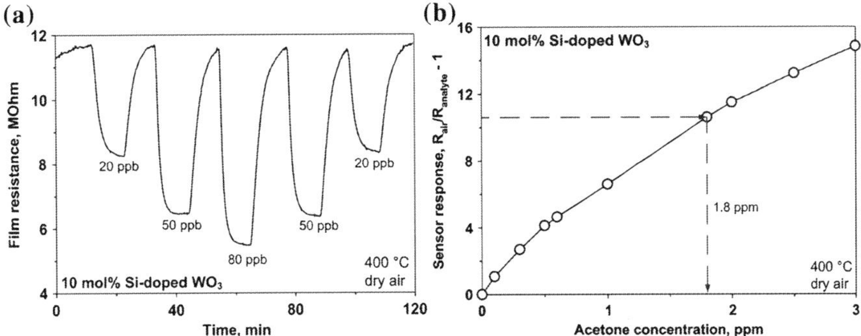

Fig. 5.11 10 at % Si-doped WO_3 as sensing material for acetone detection: **a** resistance change; **b** Relationship between sensing response and acetone concentration domains (reproduced from [58] by permission from American Chemical Society)

ε-WO_3 to acetone. 10 at % Cr-doped WO_3 (80 % ε-WO_3) was able to detect 1 ppm acetone with the sensing response up to 2 (sensing response is defined as R_0/R_g-1). 10 at % Si-doped WO_3 (87 % ε-WO_3) had even a better performance, whose sensing response was as high as 6 upon exposure to 1 ppm acetone (Fig. 5.11b) and its detection limit went as low as 20 ppb (Fig. 5.11a). Such sensitivity enhancement is believed to be due to smaller particle size and higher ε-WO_3 content. Both sensors are remarkably more sensitive than screen-printed γ/δ-WO_3 thick films whose response was only 3.5 towards 50 ppm acetone [90].

Resistance change open acetone exposure (Fig. 5.11a) indicates that the ε-WO_3 sensor behaves as a typical n-type semiconductor, similar to γ-WO_3. Like other metal oxides, the sensing mechanism of ε-WO_3 sensors towards acetone involves gas molecule adsorption, reaction with ionosorbed oxygen species and electron extraction processes. Detailed reaction formulas were described in Ref. [90].

Wang et al. [56] also studied the cross sensitivity of ε-phased WO_3, whose result is shown in Fig. 5.12a. The ε-phased WO_3 showed a fairly selective detection to acetone compared to other gases. Ethanol and methanol also belong to VOCs. However, the responses to 1 ppm of these gases were even lower than that of 0.2 ppm of acetone. More interestingly, although WO3 is generally considered to be very sensitive to NOx gases, the ε-phased WO3 sensor appeared quite inert to both NO and NO2.

Such results indicate that Cr, Si-doped ε-WO_3 have quite different sensing behaviors from commonly used γ-WO_3. To investigate such difference, we have to consider two factors: the first would be the effect of dopants. In fact, Cr-doped γ/δ-WO_3 has already been synthesized in the forms of nanopowders [91] and mesoporous structures [77] which showed enhanced sensitivity to either NH_3 [91] or NO_2[77] whereas Cr-doped ε-WO_3 were almost inert to NH_3 and NO_2 according to Wang et al. [56] 's research. Therefore, dopants were unlikely to be the reason. The other factor has to be the effect of ε-WO_3. If taking a closer look, one may find that the sensitivity of ε-WO_3 increases as the dipole moment of sensed gas

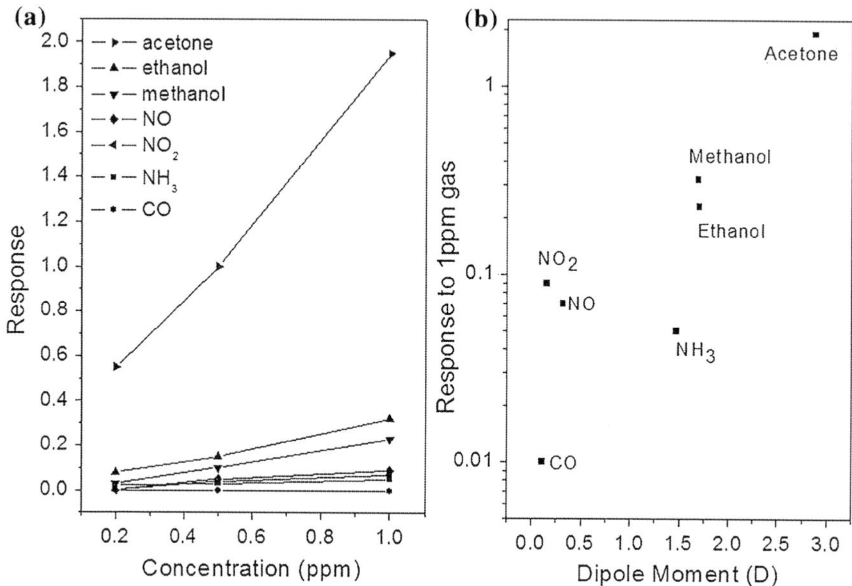

Fig. 5.12 a Sensitivity comparison of 10 at % Cr-doped WO_3 (here, sensitivity is defined as R_0/R_g-1) (reproduced from [56] by permission from American Chemical Society). **b** Relationship between sensor response and dipole moment of detected gases

molecules increases with the only exception of NH_3 (Fig. 5.12b). In other words, ε-WO_3 favors polar molecules. As we already know, ε-WO_3 is a type of ferroelectric material meaning that it has a spontaneous electric dipole moment. As a consequence, we believe that the interaction between the ε-WO_3 surface dipole and polar molecules (e.g. acetone) will contain electrostatic attraction. Such interaction could be much stronger than that between the ε-WO_3 surface and non-polar gases, which only contains van der Waals forces and covalent bonding, leading to the observed selectivity. Recently studies on ferroelectric surfaces have partially verified the above theory [92].

5.4 Summary and Outlooks

Chemo-resistive sensors utilizing meal oxides form a very important type of sensors for gas detection. Although the basic underlying mechanism could be simplified as the reaction between redox gas molecules and material surface species accompanied by electron transfer, the actual process, however, could be very complicated, which may be strongly influenced by external conditions e.g. working temperature and humidity, etc. and internal parameters, e.g. microstructure, chemical composition and crystal structure, etc. While most parameters have been thoroughly studied, the influence of crystal structures is still largely unexplored.

On the other hand, many metal oxides exist in more than one crystalline form. Some oxides will have metastable polymorphs besides their thermodynamically stable phases, e.g. h-WO_3, β, h-MoO_3, γ-Al_2O_3 and anatase-type TiO_2 etc. Some oxides will undergo a series of phase transition when they are cooled down from melting point to 0 K, e.g. ZrO_2, WO_3 and VO_2 etc. However, if one phase appears out of its own phase zone under certain conditions, it is also in a metastable state.

Chemo-resistive SMO gas sensors usually work at high temperatures to obtain sufficient surface activity, and successful fabrication of metastable-structured metal oxides includes two aspects, selective synthesis and stabilization. Hydrothermal growth is a very effective method of synthesizing materials with controlled phases, both stable and metastable, as long as particle sizes are well controlled or precursors are carefully selected. Examples include selective synthesis of different phases of TiO_2 by controlling particle sizes and reactants [20] and synthesis of h-WO_3 by using similar structured precursor [12], synthesis of m, t-ZrO_2 by controlling particle size [93], etc. Rapid solidification methods, such as FSP, pulsed laser deposition are also able to produce metastable phases. They are more suitable for synthesizing less symmetric metastable phases such as ε-WO_3 or preserving high-temperature phases such as t-ZrO_2 [94]. To stabilize synthesized metastable metal oxides, foreign elements or compounds are usually doped into the matrix material, forming surface/boundary barriers or lattice pins to prevent or defer the phase transition, with the most famous example being yttria-stabilized zirconia, as well as dopant-stabilized anatase and ε-WO_3 described in the chapter.

Through the examples of anatase and ε-WO_3, we have demonstrated that sensing behavior of a metal oxide could be strongly influenced by its crystal structure. A metastable phase is sometimes preferable in gas sensing applications as they may provide unique physical and chemical properties. The open structure of anatase makes it more chemically active than rutile, therefore more sensitive upon exposure to gases. The ferroelectric feature of ε-WO_3 increases its interaction with polar gas molecules which leads to a more selective detection to those gases.

However, crystal structure influence on sensor behaviors is largely unexplored because the surface chemistry during sensing is rather complicated and each case may have its own unique characteristic. More studies need to be conducted in the topic before it is fully comprehended.

References

1. Bielanski A, Deren J, Haber J (1957) Electric conductivity and catalytic activity of semiconducting oxide catalysts. Nature 179(4561):668–679. doi:10.1038/179668a0
2. Seiyama T, Kato A, Fujiishi K, Nagatani M (1962) A new detector for gaseous components using semiconductive thin films. Anal Chem 34(11):1502–1503. doi:10.1021/ac60191a001
3. Korotcenkov G (2007) Metal oxides for solid-state gas sensorsgas sensors: what determines our choice? Mater Sci Eng B 139(1):1–23. doi: 10.1016/j.mseb.2007.01.044
4. Comini E (2006) Metal oxide nano-crystals for gas sensinggas sensing. Anal Chim Acta 568(1–2):28–40. doi:10.1016/j.aca.2005.10.069

5. Krivetskiy VV, Ponzoni A, Comini E, Badalyan SM, Rumyantseva MN, Gaskov AM (2010) Materials based on modified SnO_2 for selective gas sensorsgas sensors. Inorg Mater 46(10):1100–1105. doi:10.1134/S0020168510100134

6. Schierbaum KD, Weimar U, Gopel W, Kowalkowski R (1991) Conductance, work function and catalytic activity of SnO_2-based gas sensorsgas sensors. Sens Actuators B Chem 3(3):205–214. doi: 10.1016/0925-4005(91)80007-7

7. Franke ME, Koplin TJ, Simon U (2006) Metal and metal oxide nanoparticles in chemiresistors: does the nanoscale matter? Small 2(1):36–50. doi:10.1002/smll.200500261

8. Huang XJ, Choi YK (2007) Chemical sensors based on nanostructured materialsmaterials. Sens Actuators B Chem 122(2):659–671. doi: 10.1016/j.snb.2006.06.022

9. Comini E, Baratto C, Faglia G, Ferroni M, Vomiero A, Sberveglieri G (2009) Quasi-one dimensional metal oxide semiconductors: preparation, characterizationcharacterization and application as chemical sensorschemical sensors. Prog Mater Sci 54(1):1–67. doi:10.1016/j.pmatsci.2008.06.003

10. Tiemann M (2007) Porous metal oxidesmetal oxides as gas sensorsgas sensors. Chem Eur J 13(30):8376–8388. doi:10.1002/chem.200700927

11. Woodward PM, Sleight AW, Vogt T (1997) Ferroelectric tungsten trioxide. J Solid State Chem 131(1):9–17. doi:10.1006/jssc.1997.7268

12. Gerand B, Nowogrocki G, Guenot J, Figlarz M (1979) Structural study of a new hexagonal form of tungsten trioxide. J Solid State Chem 29(3):429–434. doi:10.1016/0022-4596(79)90199-3

13. Ashrafi A, Jagadish C (2007) Review of zincblende ZnO: stability of metastable ZnO phases. J Appl Phys 102(7):071101. doi:10.1063/1.2787957

14. Gouma PI, Prasad AK, Iyer KK (2006) Selective nanoprobes for 'signalling gases'. Nanotechnology 17(4):S48–S53. doi:10.1088/0957-4484/17/4/008

15. Wold A (1993) Photocatalytic properties of TiO_2. Chem Mater 5(3):280–283. doi:10.1021/cm00027a008

16. Park NG, van de Lagemaat J, Frank AJ (2000) Comparison of dye-sensitized rutile- and anatase-based TiO_2 solar cells. J Phys Chem B 104(38):8989–8994. doi:10.1021/jp9943651

17. Zaban A, Aruna ST, Tirosh S, Gregg BA, Mastai Y (2000) The effect of the preparation condition of TiO_2 colloids on their surface structuresurface structures. J Phys Chem B 104(17):4130–4133. doi:10.1021/jp993198m

18. Wang CC, Ying JY (1999) Sol-gel synthesissynthesis and hydrothermal processing of anatase and rutile titania nanocrystals. Chem Mater 11(11):3113–3120. doi:10.1021/cm990180f

19. Shimizu K, Imai H, Hirashima H, Tsukuma K (1999) Low-temperature synthesissynthesis of anatase thin films on glass and organic substrates by direct deposition from aqueous solutions. Thin Solid Films 351(1–2):220–224. doi:10.1016/S0040-6090(99)00084-X

20. Reyes-Coronado D, Rodriguez-Gattorno G, Espinosa-Pesqueira ME, Cab C, de Coss R, Oskam G (2008) Phase-pure TiO_2 nanoparticles: anatase, brookite and rutile. Nanotechnology 19(14):145605. doi:10.1088/0957-4484/19/14/145605

21. Tomita K, Petrykin V, Kobayashi M, Shiro M, Yoshimura M, Kakihana M (2006) A water-soluble titanium complex for the selective synthesissynthesis of nanocrystalline brookite, rutile, and anatase by a hydrothermal method. Angew Chem Int Ed 45(15):2378–2381. doi:10.1002/anie.200503565

22. Bosc F, Ayral A, Albouy PA, Guizard C (2003) A simple route for low-temperature synthesissynthesis of mesoporousmesoporous and nanocrystalline anatase thin films. Chem Mater 15(12):2463–2468. doi:10.1021/Cm031025a

23. Wang CX, Yin LW, Zhang LY, Qi YX, Lun N, Liu NN (2010) Large scale synthesissynthesis and gas-sensing properties of anatase TiO_2 three-dimensional hierarchical nanostructuresnanostructures. Langmuir 26(15):12841–12848. doi:10.1021/La100910u

24. Nian JN, Teng HS (2006) Hydrothermal synthesissynthesis of single-crystalline anatase TiO_2 nanorods with nanotubes as the precursor. J Phys Chem B 110(9):4193–4198. doi:10.1021/Jp0567321

25. Peng TY, Zhao D, Dai K, Shi W, Hirao K (2005) Synthesis of titanium dioxide nanoparticles with mesoporousmesoporous anatase wall and high photocatalytic activity. J Phys Chem B 109(11):4947–4952. doi:10.1021/Jp044771r
26. Zhang XY, Yao BD, Zhao LX, Liang CH, Zhang LD, Mao YQ (2001) Electrochemical fabrication of single-crystalline anatase TiO_2 nanowire arrays. J Electrochem Soc 148(7):G398–G400. doi:10.1149/1.1378293
27. Huang WP, Tang XH, Wang YQ, Koltypin Y, Gedanken A (2000) Selective synthesissynthesis of anatase and rutile via ultrasound irradiation. Chem Commun 15:1415–1416. doi:10.1039/B003349I
28. Zhang HZ, Banfield JF (1998) Thermodynamic analysis of phase stability of nanocrystalline titania. J Mater Chem 8(9):2073–2076. doi:10.1039/A802619J
29. Navrotsky A (2003) Energetics of nanoparticle oxides: interplay between surface energy and polymorphism. Geochem Trans 4:34–37. doi:10.1039/b308711e
30. Cassaignon S, Koelsch M, Jolivet JP (2007) Selective synthesissynthesis of brookite, anatase and rutile nanoparticles: thermolysis of $TiCl_4$ in aqueous nitric acid. J Mater Sci 42(16):6689–6695. doi:10.1007/s10853-007-1496-y
31. Zhao JL, Wang XH, Sun TY, Li LT (2005) In situ templated synthesissynthesis of anatase single-crystal nanotube arrays. Nanotechnology 16(10):2450–2454. doi:10.1088/0957-4484/16/10/077
32. Goossens A, Maloney EL, Schoonman J (1998) Gas-phase synthesissynthesis of nanostructured anatase TiO_2. Chem Vap Deposition 4(3):109–114. doi:10.1002/(SICI)1521-3862(199805)04:03<109:AID-CVDE109>3.0.CO;2-U
33. Gouma PI, Dutta PK, Mills MJ (1999) Structural stability of titania thin films. Nanostruct Mater 11(8):1231–1237. doi:10.1016/S0965-9773(99)00413-4
34. Gouma PI, Mills MJ (2001) Anatase-to-rutile transformation in titania powders. J Am Ceram Soc 84(3):619–622. doi:10.1111/j.1151-2916.2001.tb00709.x
35. Arbiol J, Cerda J, Dezanneau G, Cirera A, Peiro F, Cornet A, Morante JR (2002) Effects of Nb dopingdoping on the TiO_2 anatase-to-rutile phase transition. J Appl Phys 92(2):853–861. doi:10.1063/1.1487915
36. Anukunprasert T, Saiwan C, Traversa E (2005) The development of gas sensor for carbon monoxidecarbon monoxide monitoring using nanostructure of Nb-TiO_2. Sci Technol Adv Mater 6(3–4):359–363. doi:10.1016/j.stam.2005.02.020
37. de Farias RF, Airoldi C (2005) A study about the stabilization of anatase phase at high temperatures on sol-gel cerium and copper doped titania and titania-silica powders. J Non-Cryst Solids 351(1):84–88. doi:10.1016/j.jnoncrysol.2004.09.015
38. Perera S, Gillan EG (2005) High-temperature stabilized anatase TiO_2 from an aluminum-doped $TiCl_3$ precursor. Chem Commun 48:5988–5990. doi:10.1039/B512148e
39. Tien TY, Stadler HL, Gibbons EF, Zacmanidis PJ (1975) TiO_2 as an air-to-fuel ratio sensor for automobile exhausts. Am Ceram Soc Bull 54(3):280–285
40. Devi GS, Hyodo T, Shimizu Y, Egashira M (2002) Synthesis of mesoporousmesoporous TiO_2-based powders and their gas-sensing properties. Sens Actuators B Chem 87(1):122–129. doi: 10.1016/S0925-4005(02)00228-9
41. Birkefeld LD, Azad AM, Akbar SA (1992) Carbon-monoxide and hydrogen detection by anatase modification of titanium-dioxide. J Am Ceram Soc 75(11):2964–2968. doi:10.1111/j.1151-2916.1992.tb04372.x
42. Ruiz AM, Sakai G, Cornet A, Shimanoe K, Morante JR, Yamazoe N (2003) Cr-doped TiO_2 gas sensor for exhaust NO_2 monitoring. Sens Actuators B Chem 93(1–3):509–518. doi: 10.1016/S0925-4005(03)000183-7
43. Garzella C, Comini E, Tempesti E, Frigeri C, Sberveglieri G (2000) TiO_2 thin films by a novel sol-gel processing for gas sensor applications. Sens Actuators B Chem 68(1–3):189–196. doi: 10.1016/S0925-4005(00)00428-7
44. Teleki A, Bjelobrk N, Pratsinis SE (2008) Flame-made Nb- and Cu-doped TiO_2 sensors for CO and ethanolethanol. Sens Actuators B Chem 130(1):449–457. doi: 10.1016/j.snb.2007.09.008

45. Lu HF, Li F, Liu G, Chen ZG, Wang DW, Fang HT, Lu GQ, Jiang ZH, Cheng HM (2008) Amorphous TiO_2 nanotube arrays for low-temperature oxygen sensors. Nanotechnology 19(40). doi: 10.1088/0957-4484/19/40/405504

46. Francioso L, Taurino AM, Forleo A, Siciliano P (2008) TiO_2 nanowires array fabrication and gas sensinggas sensing properties. Sens Actuators B Chem 130(1):70–76. doi: 10.1016/j.snb.2007.07.074

47. Chatten R, Chadwick AV, Rougier A, Lindan PJD (2005) The oxygen vacancyoxygen vacancy in crystal phases of WO_3. J Phys Chem B 109(8):3146–3156. doi:10.1021/Jp045655r

48. Salje E, Viswanathan K (1975) Physical-properties and phase-transitions in WO_3. Acta Crystallogr Sect A Found Crystallogr A31(May1):356–359. doi: 10.1107/S0567739475000745

49. Matthias BT, Wood EA (1951) Low temperature polymorphic transformation in WO_3. Phys Rev 84(6):1255. doi:10.1103/PhysRev.84.1255

50. Salje EKH, Rehmann S, Pobell F, Morris D, Knight KS, Herrmannsdorfer T, Dove MT (1997) Crystal structure and paramagnetic behaviour of -WO_{3-x}. J Phys Condens Matter 9(31):6563–6577. doi: 10.1088/0953-8984/9/31/010

51. Arai M, Hayashi S, Yamamoto K, Kim SS (1990) Raman studies of phase transitions in gas-evaporated WO_3 microcrystals. Solid State Commun 75(7):613–616. doi:10.1016/0038-1098(90)90429-F

52. Hayashi S, Sugano H, Arai H, Yamamoto K (1992) Phase transitions in gas evaporated WO_3 microcrystals: a Raman study. J Phys Soc Jpn 61(3):916–923. doi:10.1016/0038-1098(90)90429-F

53. Cazzanelli E, Vinegoni C, Mariotto G, Kuzmin A, Purans J (1999) Low-temperature polymorphism in tungsten trioxide powders and its dependence on mechanical treatments. J Solid State Chem 143(1):24–32. doi:10.1006/jssc.1998.8061

54. Souza AG, Mendes J, Freire VN, Ayala AP, Sasaki JM, Freire PTC, Melo FEA, Juliao JF, Gomes UU (2001) Phase transition in WO_3 in microcrystals obtained by sintering process. J Raman Spectrosc 32(8):695–699. doi:10.1002/jrs.727

55. Khatko V, Guirado F, Hubalek J, Llobet E, Correig X (2005) X-ray investigations of nanopowder WO_3 thick films. Phys Status Solidi A-Appl Mat 202(10):1973–1979. doi:10.1002/pssa.200520071

56. Wang L, Teleki A, Pratsinis SE, Gouma PI (2008) Ferroelectric WO_3 nanoparticles for acetone selective detection. Chem Mater 20(15):4794–4796. doi:10.1021/Cm800761e

57. Madler L, Stark WJ, Pratsinis SE (2002) Flame-made ceria nanoparticles. J Mater Res 17(6):1356–1362. doi:10.1557/JMR.2002.0202

58. Righettoni M, Tricoli A, Pratsinis SE (2010) Thermally stable, silica-doped ε-WO_3 for sensing of acetone in the human breath. Chem Mater 22(10):3152–3157. doi:10.1021/Cm1001576

59. Boulova M, Rosman N, Bouvier P, Lucazeau G (2002) High-pressure Raman study of microcrystalline WO3 tungsten oxide. J Phys Condens Matter 14(23):5849–5863. doi: 10.1088/0953-8984/14/23/314

60. Mahan AH, Parilla PA, Jones KM, Dillon AC (2005) Hot-wire chemical vapor deposition of crystalline tungsten oxide nanoparticles at high density. Chem Phys Lett 413(1–3):88–94. doi:10.1016/j.cplett.2005.07.037

61. Baserga A, Russo V, Di Fonzo F, Bailini A, Cattaneo D, Casari CS, Bassi AL, Bottani CE (2007) Nanostructured tungsten oxide with controlled properties: synthesis and Raman characterizationcharacterization. Thin Solid Films 515(16):6465–6469. doi:10.1016/j.tsf.2006.11.067

62. Srivastava AK, Agnihotry SA, Deepa M (2006) Sol-gel derived tungsten oxide films with pseudocubic triclinic nanorods and nanoparticles. Thin Solid Films 515(4):1419–1423. doi:10.1016/j.tsf.2006.03.055

63. Supothina S, Seeharaj P, Yoriya S, Sriyudthsak M (2007) Synthesis of tungsten oxide nanoparticles by acid precipitation method. Ceram Int 33(6):931–936. doi:10.1016/j.ceramint.2006.02.007

64. Weckhuysen BM, Wachs IE, Schoonheydt RA (1996) Surface chemistry and spectroscopyspectroscopy of chromium in inorganic oxides. Chem Rev 96(8):3327–3349. doi:10.1021/cr940044o
65. Vemury S, Pratsinis SE (1995) Dopants in flame synthesissynthesis of titania. J Am Ceram Soc 78(11):2984–2992. doi:10.1111/j.1151-2916.1995.tb09074.x
66. Shaver PJ (1967) Activated tungsten oxide gas detectorsdetectors. Appl Phys Lett 11(8):255–257. doi:10.1063/1.1755123
67. Akiyama M, Tamaki J, Miura N, Yamazoe N (1991) Tungsten oxide-based semiconductor sensor highly sensitive to NO and NO_2. Chem Lett 9:1611–1614
68. Gouma PI, Kalyanasundaram K (2008) A selective nanosensing probe for nitric oxide. Appl Phys Lett 93(24):244102. doi:10.1063/1.3050524
69. Xie GZ, Yu JS, Chen X, Jiang YD (2007) Gas sensing characteristics of WO_3 vacuum deposited thin films. Sens Actuators B Chem 123(2):909–914. doi: 10.1016/j.snb.2006.10.059
70. Ashraf S, Blackman CS, Palgrave RG, Naisbitt SC, Parkin IP (2007) Aerosol assisted chemical vapour deposition of WO_3 thin films from tungsten hexacarbonyl and their gas sensinggas sensing properties. J Mater Chem 17(35):3708–3713. doi:10.1039/B705166B
71. Piperno S, Passacantando M, Santucci S, Lozzi L, La Rosa S (2007) WO_3 nanofibers for gas sensinggas sensing applications. J Appl Phys 101(12):124504. doi:10.1063/1.2748627
72. Ponzoni A, Comini E, Sberveglieri G, Zhou J, Deng SZ, Xu NS, Ding Y, Wang ZL (2006) Ultrasensitive and highly selective gas sensorsgas sensors using three-dimensional tungsten oxide nanowire networks. Appl Phys Lett 88(20):203101. doi:10.1063/1.2203932
73. Rossinyol E, Prim A, Pellicer E, Rodriguez J, Peiro F, Cornet A, Morante JR, Tian BZ, Bo T, Zhao DY (2007) Mesostructured pure and copper-catalyzed tungsten oxide for NO_2 detection. Sens Actuators B Chem 126(1):18–23. doi: 10.1016/j.snb.2006.10.017
74. Penza M, Martucci C, Cassano G (1998) NO_x gas sensinggas sensing characteristics of WO_3 thin films activated by noble metals (Pd, Pt, AuAu) layers. Sens Actuators B Chem 50(1):52–59. doi: 10.1016/S0925-4005(98)00156-7
75. Ivanov P, Llobet E, Blanco F, Vergara A, Vilanova X, Gracia I, Cane C, Correig X (2006) On the effects of the materialsmaterials and the noble metal additives to NO_2 detection. Sens Actuators B Chem 118(1–2):311–317. doi: 10.1016/j.snb.2006.04.036
76. Su PG, Wu RJ, Nieh FP (2003) Detection of nitrogen dioxide using mixed tungsten oxide-based thick film semiconductor sensor. Talanta 59(4):667–672. doi:10.1016/S0039-9140(02)00582-9
77. Rossinyol E, Prim A, Pellicer E, Arbiol J, Hernandez-Ramirez F, Peiro F, Cornet A, Morante JR, Solovyov LA, Tian BZ, Bo T, Zhao DY (2007) Synthesis and characterizationcharacterization of chromium-doped mesoporousmesoporous tungsten oxide for gas-sensing applications. Adv Funct Mater 17(11):1801–1806. doi:10.1002/adfm.200600722
78. Gillet M, Aguir K, Bendahan M, Mennini P (2005) Grain sizeGrain size effect in sputtered tungsten trioxide thin films on the sensitivity to ozone. Thin Solid Films 484(1–2):358–363. doi:10.1016/j.tsf.2005.02.035
79. Labidi A, Gillet E, Delamare R, Maaref M, Aguir K (2006) Ethanol and ozone sensing characteristics of WO_3 based sensors activated by AuAu and Pd. Sens Actuators B Chem 120(1):338–345. doi: 10.1016/j.snb.2006.02.015
80. Korotcenkov G, Blinov I, Ivanov M, Stetter JR (2007) Ozone sensors on the base of SnO_2 films deposited by spray pyrolysis. Sens Actuators B Chem 120(2):679–686. doi: 10.1016/j.snb.2006.03.029
81. Vallejos S, Khatko V, Aguir K, Ngo KA, Calderer J, Gracia I, Cane C, Llobet E, Correig X (2007) Ozone monitoring by micro-machined sensors with WO_3 sensing films. Sens Actuators B Chem 126(2):573–578. doi: 10.1016/j.snb.2007.04.012
82. Barrett EPS, Georgiades GC, Sermon PA (1990) The mechanism of operation of WO_3-based H_2S sensors. Sens Actuators B Chem 1(1–6):116–120. doi: 10.1016/0925-4005(90)80184-2

83. Ionescu R, Hoel A, Granqvist CG, Llobet E, Heszler P (2005) Low-level detection of ethanolethanol and H_2S with temperature-modulated WO_3 nanoparticle gas sensors. Sens Actuators B Chem 104(1):132–139. doi: 10.1016/j.snb.2004.05.015

84. Rout CS, Hegde M, Rao CNR (2008) H_2S sensors based on tungsten oxide nanostructuresnanostructures. Sens Actuators B Chem 128(2):488–493. doi: 10.1016/j.snb.2007.07.013

85. Geng LN, Huang XL, Zhao YQ, Li P, Wang SR, Zhang SM, Wu SH (2006) H_2S sensitivity study of polypyrrole/WO_3 materialsmaterials. Solid-State Electron 50(5):723–726. doi:10.1016/j.sse.2006.04.024

86. Llobet E, Molas G, Molinas P, Calderer J, Vilanova X, Brezmes J, Sueiras JE, Correig X (2000) FabricationFabrication of highly selective tungsten oxide ammonia sensors. J Electrochem Soc 147(2):776–779. doi:10.1149/1.1393270

87. Xu CN, Miura N, Ishida Y, Matsuda K, Yamazoe N (2000) Selective detection of NH_3 over NO in combustion exhausts by using Au and MoO_3 doubly promoted WO_3 element. Sens Actuators B Chem 65(1–3):163–165. doi: 10.1016/S0925-4005(99)00413-X

88. Kanda K, Maekawa T (2005) Development of a WO_3 thick-film-based sensor for the detection of VOC. Sens Actuators B Chem 108(1–2):97–101. doi: 10.1016/j.snb.2005.01.038

89. Hubalek J, Malysz K, Prasek J, Vilanova X, Ivanov P, Llobet E, Brezmes J, Correig X, Sverak Z (2004) Pt-loaded Al_2O_3 catalytic filters for screen-printed WO_3 sensors highly selective to benzene. Sens Actuators B Chem 101(3):277–283. doi: 10.1016/j.snb.2004.01.015

90. Khadayate RS, Sali V, Patil PP (2007) Acetone vapor sensing properties of screen printed WO_3 thick films. Talanta 72(3):1077–1081. doi:10.1016/j.talanta.2006.12.043

91. Jimenez I, Centeno MA, Scotti R, Morazzoni F, Arbiol J, Cornet A, Morante JR (2004) NH_3 interaction with chromium-doped WO_3 nanocrystalline powders for gas sensinggas sensing applications. J Mater Chem 14(15):2412–2420. doi:10.1039/B400872C

92. Garrity K, Kolpak AM, Ismail-Beigi S, Altman EI (2010) Chemistry of ferroelectric surfaces. Adv Mater 22(26–27):2969–2973. doi:10.1002/adma.200903723

93. Tani E, Yoshimura M, Somiya S (1983) Formation of ultrafine tetragonal ZrO_2 powder under hydrothermal conditions. J Am Ceram Soc 66(1):11–14. doi:10.1111/j.1151-2916.1983.tb09958.x

94. Jossen R, Heine MC, Pratsinis SE, Akhtar MK (2006) Thermal stability of flame-made zirconia-based mixed oxides. Chem Vap Deposition 12(10):614–619. doi:10.1002/cvde.200506380

Chapter 6
Synthesis of Metal Oxide Nanomaterials for Chemical Sensors by Molecular Beam Epitaxy

Manjula I. Nandasiri, Satyanarayana V. N. T. Kuchibhatla and Suntharampillai Thevuthasan

Abstract In order to develop next generation chemical sensors using nano-scale materials, we need to understand the sensing mechanisms at atomic level. This requires synthesizing chemical sensing materials with controlled structure, chemical composition and surface morphology. Although the commonly used wet chemical synthesis methods provide quality materials for large-scale production of materials, alternative thin film deposition techniques such as sputtering, chemical vapor deposition (CVD), and molecular beam epitaxy (MBE) can also be useful to achieve atomic-scale control over the structure and composition over a large fabrication area for potential device fabrication as well as to gain an understanding of the chemical sensing properties of nano-scale materials. Especially, MBE has been used to synthesize metal oxide thin films with ultra-pure, well-ordered surfaces, which can be used to understand the effect of surface morphology, structure, and composition on the gas sensing properties. In this chapter, we provide a detailed discussion of thin film growth using MBE along with some in situ characterization capabilities such as reflection high energy electron diffraction (RHEED) and low energy electron diffraction (LEED). In addition, this chapter focuses on the discussion of the growth, characterization and gas sensing properties of metal oxide thin films such as doped CeO_2 and SnO_2. The chapter also emphasizes the significance of various in situ and ex situ characterization techniques to understand the material properties there by developing methodologies to synthesize better materials with tunable characteristics for sensing applications.

M. I. Nandasiri · S. Thevuthasan (✉)
EMSL, Pacific Northwest National Laboratory, Richland, WA 99354, USA
e-mail: theva@pnl.gov

M. I. Nandasiri
Department of Physics, Western Michigan University, Kalamazoo, MI 49008, USA

S. V. N. T. Kuchibhatla
Battelle Science and Technology India, Pune, MH 411057, India

M. A. Carpenter et al. (eds.), *Metal Oxide Nanomaterials for Chemical Sensors,*
Integrated Analytical Systems, DOI: 10.1007/978-1-4614-5395-6_6,
© Springer Science+Business Media New York 2013

6.1 Introduction

Since the industrial revolution, materials for detection and monitoring of toxic matter, chemical wastes, and air pollutants have become important components of environmental research. This led to the development of chemical sensors for various target analytes including hydrogen, oxygen, NO_x, and CO_x, which have a potential impact on the environment and thus these sensors are being studied extensively. Chemical sensors operate on a simple principle that the certain property of the active sensing element (i.e. conductance, color, density etc.) undergoes a detectable change when exposed to the target substance to be sensed. Among all the types of chemical sensors, solid state gas sensors have attracted a great deal of attention due to their advantages such as high sensitivity, greater selectivity, portability, high stability and low cost [1, 2]. Especially, semiconducting metal oxides such as SnO_2, TiO_2, and WO_3 have been widely used as the active sensing platforms in solid state gas sensors [3]. For the enhanced properties of solid state gas sensors, finding new sensing materials or improvement in the performance of existing materials is desired. Thus, nanostructured materials such as nanotubes [4–6], nanowires [7–11], nanorods [12–15], nanobelts [16, 17], and nano-scale thin films [18–23] have been synthesized and studied for chemical sensing applications.

Nanoscale materials have unique characteristics such as high surface area-to-volume ratio and enhanced reactivity which are desired for ideal sensing materials. Thus, they are well suited for sensor development and hence, there has been considerable focus on the development of nanomaterials with tunable functionalities for sensing applications. With the existing technology, nanoscale sensors can be successfully fabricated using advanced nanomaterials. However, synthesis and characterization of these nanomaterials is a challenging task, partly because of the sensitivity of these materials to the environment. In addition, the capabilities with unprecedented spatial, energy and temporal resolutions are needed to characterize these materials.

In order to successfully utilize various materials in gas/chemical sensor applications, it is quite essential to understand the physico-chemical changes that occur on the sensing surface. In order to achieve this there is a growing need for establishing ultra-pure surfaces with controlled orientation and chemical identities. As opposed to most of the solution based methods which are used to synthesize materials for bulk-scale manufacturing, thin film techniques, especially molecular beam epitaxy (MBE) offers an ideal platform for gaining such an understanding as these thin films are deposited with atomic-scale control over structure and composition. Thus, various chemical sensing materials have been synthesized by MBE, which are listed in Table 6.1. In these studies, both the single-crystal and polycrystalline metal oxide thin films have been synthesized and characterized to investigate their chemical sensing properties. This chapter summarizes the significance of MBE based growth of thin films for sensing applications with the primary focus aimed at gaining a fundamental understanding of the influence of various experimental parameters on the sensing material characteristics.

Table 6.1 Chemical sensing materials synthesized by molecular beam epitaxy for detection of various gases including H_2, O_2, NO_x, CO_2, SO_2, and CO

Material	Detectable gases	Type of crystallinity	Comments	References
SnO_2	H_2, CO, CH_4	Single crystalline	$SnO_2(110)$ thin films were grown on $TiO_2(110)$ substrates to study the properties of SnO_2/TiO_2 interface	[131]
		Single crystalline	Pure and cobalt-doped $SnO_2(101)$ thin films were grown on Al_2O_3	[134]
		Single crystalline	Different growth regimes were identified at different growth conditions	[132, 133]
		Single crystalline	Use of Sn-doped $TiO_2(110)$ single crystal substrate favored the epitaxial film growth	[135, 138]
		Single and Poly crystalline	Sensing properties of mono and poly crystalline SnO_2 films were compared in the detection of different gases	[124]
In_2O_3	O_3, NO_x, Cl_2, CO, CO_2, NH_3	Polycrystalline	In_2O_3 thin films showed a high sensitivity to NO_2, CO and CO_2 at room temperature	[139]
		Single crystalline	High quality epitaxial In_2O_3 thin films were grown on YSZ and Al_2O_3 single crystal substrates	[140–142]
ZnO	CH_4, O_3, NO_x, H_2, O_2	Single crystalline	High quality ZnO single crystal thin films were grown on $Al_2O_3(0001)$ substrates	[143–145]
		Single crystalline	Nanotubes were fabricated and growth mechanism of ZnO nanotubes was studied	[146, 147]
		Single crystalline	ZnO nanorods showed high sensitivity to O_3 at room temperature	[148, 149]
WO_3	O_3, NO_x, H_2S, SO_2, NH_3	Single and poly crystalline	The microstructure of WO_3 thin films depends on deposition conditions	[150–152]
MoO_3	NH_3, NO_2	Single and poly crystalline	The structure of the MoO_3 films was strongly depend on the growth temperature	[40]
TiO_2	O_2, CO, SO_2	Single crystalline	High quality epitaxial TiO_2 films with both rutile and anatase polymorphs were deposited	[153–156]
CeO_2	O_2	Single crystalline	Influence of dopant (samaria) concentration and film thickness on the O_2 gas sensing properties was investigated	[19, 23, 99]

6.1.1 Molecular Beam Epitaxy

MBE is a widely used deposition technique for the growth of high purity metal oxide thin films [24]. In particular, MBE has been used to understand the properties of oxide films, which can be used for carrying out fundamental studies of chemical sensing materials [19, 23, 25–30], high temperature superconductors [31–34], surface geochemical reactions [35–39], and catalysis [40–46]. It is imperative to briefly discuss the fundamentals of MBE, along with the two most important techniques that are essential for understanding the characteristics of MBE grown films, (1) reflection high-energy electron diffraction (RHEED), and (2) low energy electron diffraction (LEED). By no means, is the following discussion comprehensive and the authors suggest the reader to utilize the references there-into expand their knowledge on these three techniques.

One of the essential aspects of efficiently utilizing various properties of a materials system is to first understand the fundamental characteristics of the material. This often requires materials in pristine form with controlled chemical and structural features at the atomic level. MBE offers such a control and has been shown to be quite useful in the case of III-V semiconductor research [47]. "Epitaxy" has its origin from Greek where *epi* means "above" and *taxis* means "in ordered manner" leading to the meaning "arranged above in ordered manner". High quality epitaxial films can be grown on a bulk, single crystal substrate of the same material (e.g. GaAs on GaAs substrate) or on a different material substrate (e.g. CeO_2 film on Al_2O_3 substrate). The latter is known as "hetero" epitaxial growth in which the growth and orientation of the film are dictated by the substrate properties. It is often true that having in-plane crystal symmetry, minimal in-plane lattice mismatch, and large negative free energy change are essential for having a layer-by-layer epitaxial growth. Most of the oxide films grown using MBE fall under this category. Chambers [47] has written an extensive review on the epitaxial growth of oxide thin films where growth and characterization of MgO, NiO, CoO, TiO_2, Fe_xO_Y, CeO_2 and other oxides have been discussed in detail.

During a typical MBE growth process, the source—i.e. the high purity metal— is evaporated with the help of effusion cells, also known as Knudsen cells (k-cells) or with the help of an electron beam assisted heating in an oxidizing environment. In the case of doped or complex oxide films, more than one source could be used with precise control of the flux with the help of pre-calibrated quartz crystal oscillators (QCOs). In order to effectively grow oxide films without leading to intermediate metallic oxide phases, it has been suggested to use atomic oxygen by utilizing either electron cyclotron resonance (ECR) oxygen plasma generator or NO_2. While the latter dissociates near the substrate (usually at an elevated temperature) into $NO + O$; the former produces a mix of O ions, radicals and some un-dissociated oxygen. The use of activated O from either an ECR oxygen plasma or NO_2 over the molecular O_2 leads to well controlled oxidation of metals and the overall growth process. As most of the films discussed in this chapter are grown with the help of ECR oxygen plasma generator, we describe the process as oxygen plasma-assisted MBE.

In this chapter, the growth and characterization of Sm doped CeO_2 (SDC) nano-scale thin films on Al_2O_3(0001) and Cu_2O nanoclusters on strontium titanate (STO) substrates carried out as a part of various user projects in the Environmental Molecular Sciences Laboratory (EMSL) are discussed in detail. All of the high quality thin films were synthesized using oxygen plasma-assisted MBE and investigated by various in situ and ex situ characterization techniques. The growth, characterization and gas sensing properties of MBE grown SnO_2 thin films will also be reviewed. Before discussing the details, the principles of RHEED and LEED are discussed as the two most commonly used techniques for structural analysis of surfaces, while RHEED is used in situ to monitor the growth mode of the films in MBE due to the convenient optics, LEED is used for studying any possible reconstructions on the surface of the as-grown films immediately after the growth.

6.1.2 Electron Diffraction

For many decades, electron diffraction has been an indispensable technique for understanding the crystal structure of surfaces, over-layers and thin films and is widely used as a tool for monitoring the quality of epitaxial films as they are grown by molecular beam epitaxy. Both LEED and RHEED are conceptually similar to the x-ray diffraction (XRD) used for the bulk crystal structure determination. The fundamental difference between the electron diffraction(ED) techniques and the XRD is in the much lower penetration depth in the former case. While the electron penetration depth is confined to the order of few monolayers (1–3 nm), x-rays can penetrate through the bulk of a crystal (up to $\sim 10^5$ nm). LEED plays a crucial role in understanding the surface structure, in particular the over-layer structure of adsorbates covered surfaces, surface relaxation and reconstruction. Due to the excellent surface sensitivity and the ease with which the instrument can be used in situ to monitor the film growth, RHEED is an important tool for thin film research.

Typical instrument configurations for LEED and RHEED are shown in Fig. 6.1. The main difference between LEED and RHEED is the beam energy and the scattered electron path. While LEED uses perpendicular incidence and backscattered electrons, RHEED uses glancing incidence ($\leq 3°$) and forward scattered electrons. Due to the backscattered geometry, LEED instrumentation is generally difficult to incorporate for monitoring the in situ growth. Hence, RHEED, with no physical interference with the deposition process, became a more popular in situ technique during MBE growth of thin films.

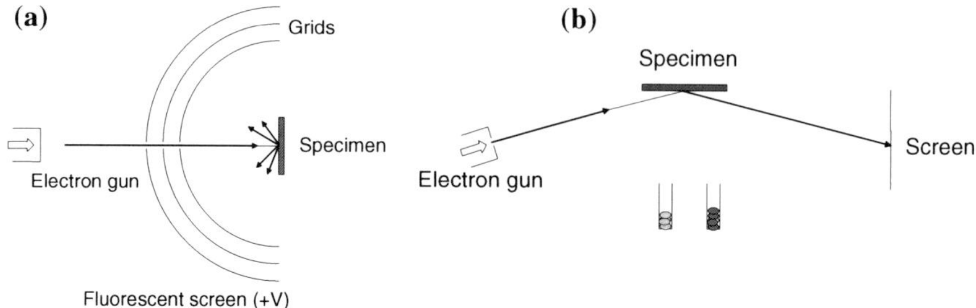

Fig. 6.1 Typical instrument configuration for **a** LEED, **b** RHEED

6.1.2.1 LEED

The position and intensity of diffraction peaks can be used to understand the structural characteristics of the surface or over-layer under investigation and LEED instrumentation can be used to obtain both these components as a function of electron energy. The basic LEED set-up consists of an electron source, usually producing an electron beam 1 mm in diameter and aligned perpendicular to the sample surface, a series of hemispherical grids oriented with their origin at the sample surface and a phosphor screen held at high positive potential as shown in Fig. 6.1a. The hemispherical grids are placed at different potentials to provide field-free path to the diffracted electrons and to allow only the elastically scattered electrons contribute to the diffraction pattern. The phosphorous coated screen is kept at a high potential (several kilovolts) to show visible fluorescence when accelerated electrons strike the screen. These fluorescent spots from various diffracted beams will result in a periodic diffraction pattern, which can be captured with the help of a CCD camera. LEED is often used in two qualitatively different modes, one to determine or verify the symmetry of the surface and the second to determine the surface structure in some detail. The diffraction spots seen in a LEED pattern [48] (see Fig. 6.2) provide information about the symmetry of the surface structure of the sample (they do not directly provide information about the actual atomic positions). LEED is frequently used in this qualitative mode to verify sample surface cleanliness, surface orientation, surface relaxation, and the presence of ordered adsorbate layers. Photometric measurements on the phosphor screen or a Faraday cup (adjustable to follow changes in the diffracted beam position with changing primary electron energy) can be used to determine the intensity of the diffraction spots as a function of energy. In order to obtain information about the actual atomic positions, the intensity versus primary electron energy curves (i.e. I–V curves) can be collected and diligently analyzed using models of the surface structure along with the calculations involving multiple electron scattering. Despite the complicated nature of multiple electron scattering at low energies, the use of I–V curves and theoretical modeling has continued to be useful for detailed analyses of surface structures and is a major tool for examining surface reconstructions. Most real surfaces are not "ideal" because atoms at surfaces relax from their

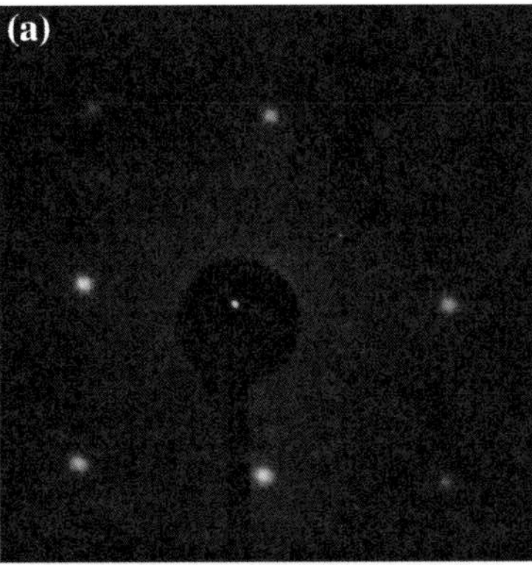
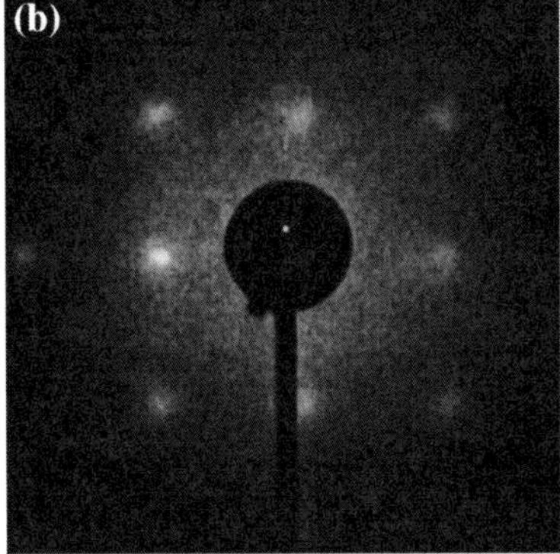

Fig. 6.2 Typical LEED patterns of **a** STO (001) substrate (~ 98 eV incident beam), **b** CeO_2 (001) film deposited at 650 °C (~ 165 eV incident beam) [48]. (Reprinted from [48] with permission from American vacuum society)

regular bulk positions in order to reduce the surface energy. This process is commonly known as surface reconstruction. Understanding the surface reconstruction using LEED can provide information both on the symmetry of the final structure and the interactions between the surface and over-layer.

6.1.2.2 RHEED

In addition to surface crystal structure, RHEED can be effectively used to understand the growth mode and the deposition rate of the films in situ. RHEED involves reflection of electrons incident at grazing angles onto the film/substrate surface. When the films grow as layer-by-layer in a smooth fashion, a RHEED pattern appears as lines or streaks. On the other hand, if the growth is associated with development of three-dimensional structures, typical diffraction spots appear indicating a rough surface. Polycrystalline films are generally associated with rings in the diffraction pattern. Therefore RHEED is useful to evaluate the surface morphology of a growing film in situ (as shown in Fig. 3) [47]. The semicircular zones of scattered beams in Fig. 6.3a are a result of coherent diffraction over length scales comparable to the electron coherence length ($\sim 1,000$Å for a well

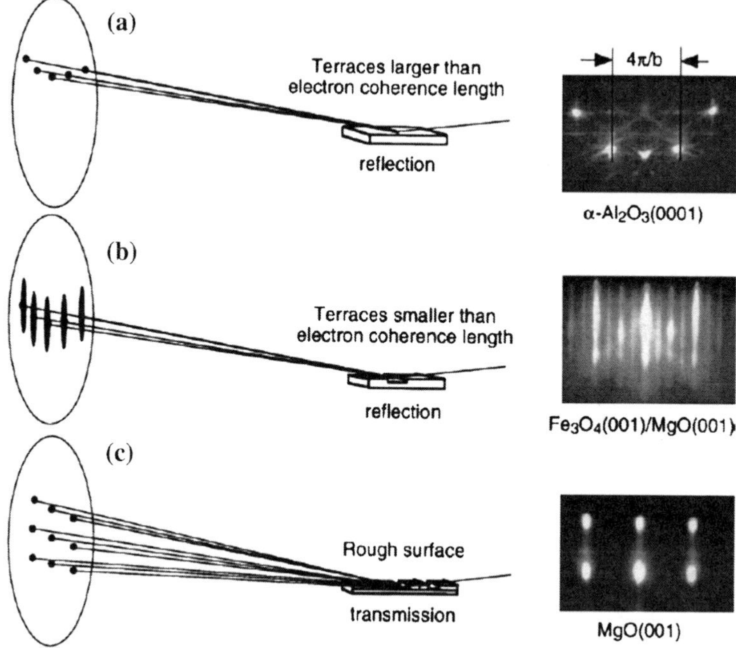

Fig. 6.3 Representative RHEED pictures along with sample possible surface characteristics. Evidently, the RHEED pattern is dependent on the terrace widths on (substrate/film) and the aspect ratio of the film [47]. (Reprinted from [47] with permission from Elsevier)

collimated beam). These scattered beams consist of spots for a crystalline surface. These spots, as presented in Fig. 6.3, are the diffracted beams within the zeroth-order Laue zone of Al$_2$O$_3$(0001). The spacing between two spots or two streaks is inversely proportional to the in-plane lattice parameter perpendicular to the beam.

Under specific conditions, where the terrace widths are smaller than the electron coherence length, the spots in a RHEED pattern become streaks. This leads to a spread in the electron beam in the incident direction causing streaks to appear in the diffraction pattern. Further, the dimensions of clusters growing on a substrate can primarily result in the transmission of electrons (predominantly elastic scattering) rather than reflection, which leads to a diffraction pattern with spots as shown in Fig. 6.3c.

6.1.2.3 Understanding Crystallographic Orientations Using RHEED

As indicated by the forgoing discussion, RHEED is a very powerful technique for in situ characterization during MBE growth. One of the major factors that control a material's sensing characteristics is the surface termination, as often the reactivity on the surface determines the response and sensitivity. Hence, while growing nanoclusters using MBE, it becomes necessary to understand their crystallographic orientation. Kuchibhatla et al. have shown that in situ RHEED analysis in combination with atomic force microscopy (AFM) can be effectively used not only to determine the growth mode but also to understand the orientation relationship between the clusters and the substrate [49]. The following section is an example of such studies based on Cu_2O nanocluster growth on STO substrate.

The RHEED pattern, along the $\langle 100 \rangle$ direction from the clean STO(100) surface before film deposition shows streaks indicating a well ordered surface of the single crystal substrate. Following the Cu_2O deposition, spots displace the streaks in the RHEED pattern, and persist until the end of deposition as shown in Fig. 6.4a [49]. AFM image of the film grown for 30 min is shown in Fig. 6.5 along with the corresponding RHEED pattern [inset (a)] and a high magnification image [inset (b)] [49]. Spots in this RHEED pattern are due to transmission diffraction indicating the three-dimensional (3-D) cluster/island growth. Clustered growth occurs due to the high lattice mismatch between the substrate and the film. Thirty minutes of Cu_2O growth results in faceted, isolated island structures with occasional agglomeration, which are identified as truncated octahedra. In the case of materials with cubic lattice structure, octahedral shape is a quite common morphology with $\{111\}$, $\{100\}$ surface habit planes. AFM images [Fig. 6.5, inset (b)] also indicate that most of the truncated octahedra and other agglomerated structures have a preferred in-plane orientation at a rotation of $45°$ with respect to the STO $\langle 100 \rangle$ direction. The Cu_2O clusters can be identified with $\{100\}$ basal planes and $\{111\}$ facets and the edges with $\langle 110 \rangle$ orientation. Based on the AFM images and the RHEED patterns obtained, the pattern emerges with $45°$ in-plane rotation compared to a RHEED pattern expected for Cu_2O $\langle 110 \rangle$ //STO $\langle 110 \rangle$. Hence, the epitaxial (in-plane) orientation relationship between Cu_2O nanoclusters and STO(100) is Cu_2O(100)//STO(100) and Cu_2O $\langle 100 \rangle$ //STO $\langle 110 \rangle$ as represented in Fig. 6.4 [49]. It is hypothesized that, while the lattice mismatch between the STO(100) and Cu_2O(100) leads to the formation of clusters, the inherent epitaxial nature of the individual clusters leads to some strain which is further relaxed through in-plane rotation. In summary, such a detailed characterization of surface layers is critical towards an accurate understanding of material properties which may govern sensing mechanisms.

6.2 Doped Cerium Oxide Thin Films for Oxygen Sensing

Oxygen sensors have attracted a great deal of attention in the recent years, due to their wide range of applications in the automobile industry, chemical industry, medicine and life science [50–55]. Especially, they have been used in internal combustion engines to control the air–fuel ratio, where oxygen sensors are used to identify the oxygen concentration in the exhaust gas [53, 56–58]. This process minimizes the generation of pollutants like carbon monoxide and increases the thermal efficiency of the engine [59–61]. Oxygen sensors have also been used in

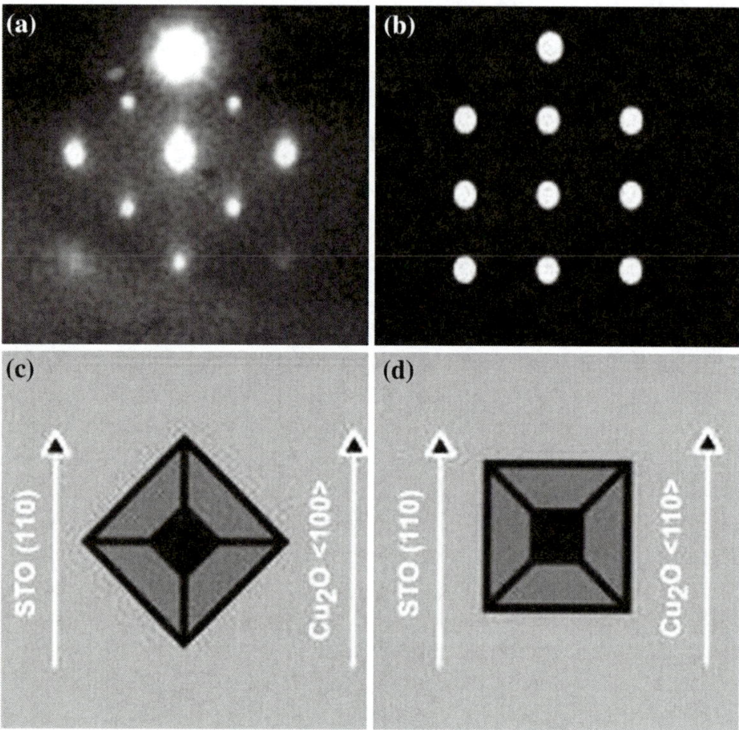

Fig. 6.4 **a** Experimental RHEED pattern corresponding to epitaxial, cuprous oxide clusters on STO(100) with Cu_2O ⟨100⟩ //STO ⟨110⟩, **b** empirical RHEED pattern that might be observed for Cu_2O ⟨110⟩ //STO ⟨110⟩ (or Cu_2O ⟨100⟩ //STO ⟨100⟩, and **c** and **d** schematics showing the structural arrangement of the cuprous oxide clusters corresponding to **a** and **b**, respectively [49]. (Reprinted from [49] with permission from American institute of physics)

Fig. 6.5 AFM images showing the morphology of the as deposited Cu_2O on STO(100) at 2 Å/min for 30 min along with a RHEED pattern [inset (**a**)] and high magnification image [inset (**b**)]. The AFM images were acquired using a DI Nanoscope IIIa multimode scanning probe microscope in tapping mode. RHEED measurements were carried out with 15 kV e-beam at an incidence angle of ∼3–5° [49]. (Reprinted from [49] with permission from American institute of physics)

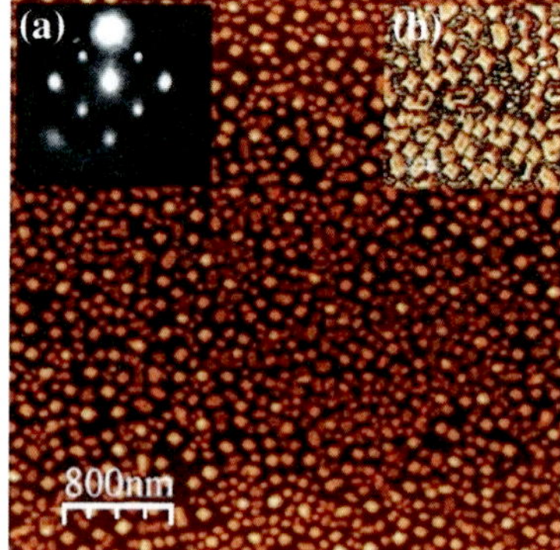

metallurgical heat furnaces to monitor and control the gas purity [56, 62]. Most of these applications need oxygen gas sensors with a large dynamic range, fast response and portability, which can be operated at elevated temperatures. Among all the types of oxygen gas sensors based on different principles, resistive-type oxygen gas sensors have become more popular due to their excellent properties [53, 63]. Metal oxide semiconductors (MOS) have been used as the sensing material in resistive-type oxygen sensors, which operate on the principle of change in their overall conductance with the oxygen gas under ambient conditions [19]. Thus, nano-scale metal oxide thin films with high conductivity can be used as the active sensing material, where the sensitivity can be enhanced by doping.

Recently, pure and doped CeO_2 have been investigated for oxygen gas sensing, because of their lower activation energy for oxygen ion conduction, which facilitates ionic conductivity at intermediate temperatures (500–800 °C) [19, 23, 64–68]. Furthermore, CeO_2 based materials have the ability to store or release oxygen, in response to the ambient oxygen concentration. In addition, these materials exhibit high corrosion resistance to reactive gases and durability against the harsh environment [64, 69]. Due to these unique properties, they have also been used in catalysis and solid oxide fuel cells [70–74]. The ionic conductivity and redox properties of CeO_2 depend on the defect concentration and oxygen ion migration. CeO_2 doped with divalent or trivalent cations has high ionic conductivity over extended temperature and oxygen partial pressure ranges (500–900 °C and 1–10^{-15} atm), due to the creation of oxygen vacancies [73, 75–80]. The ideal choice of dopant depends on the ability of the dopant to minimize the internal strain in the lattice [80–82]. Thus, defect cluster formation and vacancy-dopant interactions of doped CeO_2 materials have been investigated by various theoretical methods, in order to identify the ideal dopants for CeO_2 [83–86].

In general, the ionic conductivity (σ) can be expressed as an exponential function of the activation energy for oxygen vacancy diffusion (E_a),

$$\sigma = \sigma_0 \cdot \exp(-E_a/K_B T)/T$$

Where,

T   -temperature

K_B -Boltzmann constant

σ_0 -temperature independent pre-factor

Materials like rare-earth doped CeO_2 with lower E_a facilitate higher ionic conductivity at lower temperatures. For doped CeO_2, E_a can be expressed as a sum of the association binding energy of vacancy-dopant cluster (E_{ass}) and the migration barrier (E_m) [83]. E_m is associated with vacancy-dopant interactions, which can be divided into repulsive elastic and attractive electronic parts. The highest ionic conductivity can be obtained at minimum E_{ass}, wherein the repulsive elastic and attractive electronic interactions are balanced with each other, which occurs for dopants with atomic numbers close to 61 [83]. Furthermore, Andersson et al. have noticed an existence of the crossover for the migration barrier close to atomic number 61. These results suggest a minimum in E_a between atomic numbers 61 (Pm) and 62 (Sm), which should result in films with the highest conductivity. However, Sm was considered as the ideal dopant for CeO_2, since Pm is a radioactive element.

Recently, enormous attempts were made to optimize the ionic conductivity of SDC by using different growth methods and varying dopant concentration [74, 79, 87–92]. The optimum $SmO_{1.5}$ concentration values for the highest conductivity of SDC show some discrepancy due to the difference in synthesis techniques and texture of the samples. However, most of the researchers have observed 20 mol % $SmO_{1.5}$ as the optimum concentration [87–89, 91, 92]. For highly-oriented SDC thin films grown by MBE, it was found to be 12–15 mol % $SmO_{1.5}$ [19, 74]. The increase in the conductivity with the $SmO_{1.5}$ concentration up to the optimum limit is due to the introduction of oxygen ion vacancies, while defect association and ordering at higher dopant concentration leads to a decrease in conductivity [88].

Here, we discuss a study [19, 23] that deals with the effect of dopant concentration and film thickness on the conductivity and structure–property relationship of SDC nano-scale thin films as an oxygen sensing platform. First, the optimum dopant concentration was established for the conductivity of SDC epitaxial thin films grown by oxygen plasma-assisted MBE. Then, the influence of film thickness on the properties of SDC epitaxial thin film with optimum concentration was studied using different in situ and ex situ characterization capabilities. For all the SDC nano-scale thin films with different dopant concentrations and film thicknesses, conductance was measured as a function of temperature and oxygen partial pressure to optimize the ionic conductivity of SDC oxygen sensing platform.

Growth of SDC thin films on Al_2O_3 (0001) substrates with different film thicknesses and various dopant concentrations of Sm and in situ characterization were carried out in a dual-chamber ultrahigh vacuum (UHV) system described in detail elsewhere [47]. High purity Ce rods were used as the source material in an e^--beam evaporator. Sm was evaporated from an effusion cell. Growth rate of the films was monitored using a pre-calibrated quartz crystal oscillator (QCO). Al_2O_3 (0001) substrates were ultrasonically cleaned in acetone prior to insertion into the dual-chamber UHV system. Once in the MBE chamber, the substrates were cleaned by annealing for several minutes at 650 °C in the oxygen plasma operating at 200 W under a chamber pressure of $\sim 2 \times 10^{-5}$ Torr of oxygen. The growth of the film was monitored using in situ RHEED with a 15 kV electron beam. The substrate temperature, Sm and Ce deposition rates and oxygen partial pressure were systematically varied and the resulting films were characterized to establish optimum growth conditions for high quality, epitaxial SDC thin films with controlled chemistry [93]. After establishing the optimum growth conditions, 110 nm SDC thin films were grown with different Sm concentrations (0–14 atom %) [19]. For the second stage of the experiment, SDC film thickness was varied from 50 to 300 nm at a step size of 50 nm, while keeping the Sm concentration constant at the optimum value [23]. Following the growth, SDC thin films with various dopant concentrations and film thicknesses were characterized using XRD, x-ray

Fig. 6.6 Schematic of the experimental set up used for the conductance measurements of SDC thin films [23]. (Reprinted from [23], © 2011 IEEE)

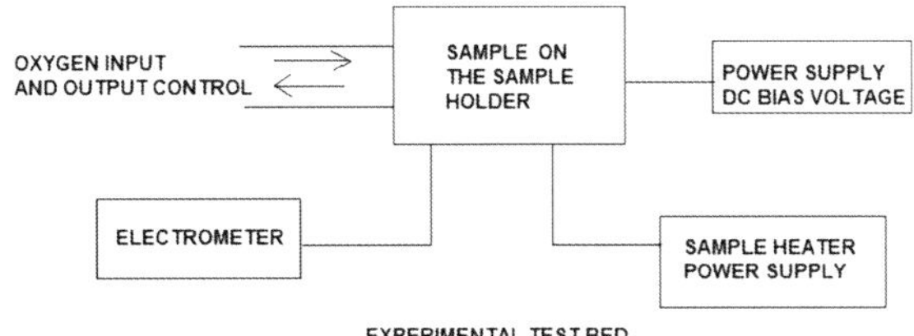

photoelectron spectroscopy (XPS), and Rutherford backscattering spectrometry (RBS) [19, 23].

The electrical resistance of SDC thin films was measured by two probe conductivity measurement technique. The current through the SDC thin films was measured at a fixed bias voltage as a function of temperature and oxygen partial pressure in an experimental test bed [19, 23]. A schematic of the experimental test bed used to measure the conductance of SDC thin films is shown in Fig. 6.6 [23]. For each SDC thin film, the current was measured at different oxygen partial pressures (1 mTorr–100 Torr), by varying the temperature up to 973 K. A lower supply voltage is desirable to reduce the power consumption for most of the real world applications. This consideration becomes more stringent when a panel of such sensors is to be used for a particular application. Thus, a reasonably low but sufficient bias voltage (2 V) was used to measure the conductance of SDC thin films.

The RHEED pattern from the $Al_2O_3(0001)$ substrate following the oxygen plasma treatment is shown in Fig. 6.7a, which indicates a high quality substrate surface which is essential for epitaxial growth of thin films. Following the deposition of SDC thin film, the RHEED streaks associated with the substrate changed to streaks associated with the SDC film as shown in Fig. 6.7b, indicating an epitaxial layered growth [47]. High resolution XRD pattern for the as-grown SDC film with 6 atom % Sm concentration (Fig. 6.8a) clearly shows (111), (222), and (333) reflections from SDC film, indicating the SDC(111) preferred orientation of the film. In addition, it shows the reflections coming from Al_2O_3 (0001) substrate. Glancing incidence (as the incidence angle is fixed @ $\sim 5°$ the single crystal substrate and any single crystal component in the film will not be under Bragg's condition and hence will not result in any peaks in the pattern—only the polycrystalline grains will show peaks in such configuration) XRD pattern confirmed the absence of any polycrystalline or secondary phases in the thin film within the detection limit as shown in Fig. 6.8b. However, SDC thin films with higher Sm concentrations have exhibited polycrystalline phases in contrast to the films with lower dopant concentrations [74].

XPS data have been used to determine the concentration and oxidation state of each element in the SDC films [19]. The high resolution core-level Ce 3d and Sm 3d spectra collected from the SDC film surface are shown in Fig. 6.9. The Ce 3d spectrum exhibited the peaks associated with Ce^{4+} oxidation state [94, 95], indicating a completely oxidized SDC film. Furthermore, Sm 3d spectrum confirmed

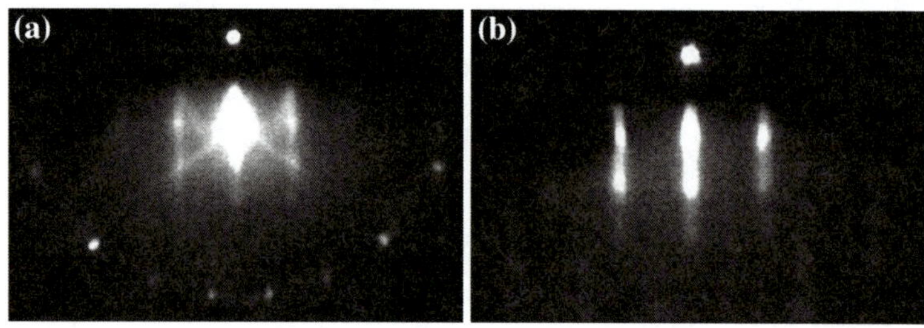

Fig. 6.7 The in situ RHEED patterns collected from **a** $Al_2O_3(0001)$ substrate surface after the oxygen plasma treatment, **b** SDC thin film grown on $Al_2O_3(0001)$

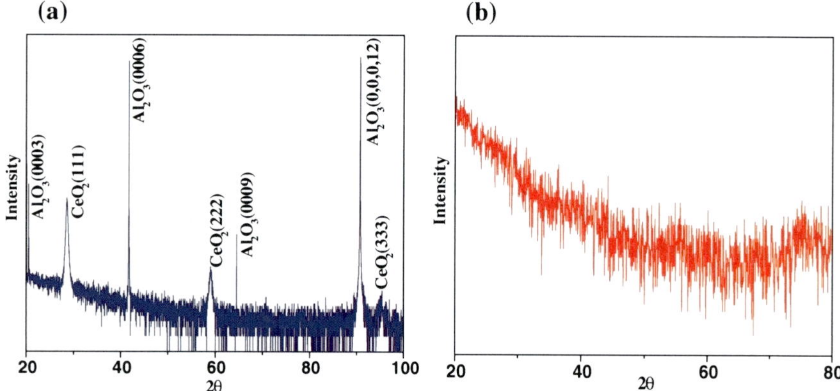

Fig. 6.8 Representative **a** HRXRD pattern, **b** GIXRD pattern from 6 atom % Sm SDC 100 nm thin film on $Al_2O_3(0001)$

that, Sm is in the 3+ oxidation state for as-grown SDC thin films. The XPS depth profile data were collected by sputtering the SDC films with 2 kV Ar^+ rastered over a 2×2 mm area of the sample. The XPS depth-profile for SDC film clearly showed the uniform distribution of Sm in cerium oxide lattice throughout the bulk of the film (Fig. 6.10a). The sharp interface in the XPS depth profile indicated that, there was no inter-diffusion of metal atoms at the film/substrate interface.

RBS measurements in random and channeling directions were carried out on various SDC thin films to further understand the film quality and interface characteristics [96]. The experimental and simulated RBS spectra along the random geometry is shown in Fig. 6.10b with arrows indicating the channel numbers for elements Ce, Sm, Al and O. Since the Sm and Ce peaks are overlapped, it is difficult to determine the exact Sm concentration in the film based on the RBS data. The RBS spectrum was simulated using SIMNRA [97, 98] to find the film thickness. Based on the simulations, thicknesses of the SDC films were determined with a standard deviation of 10 %, which are consistent with the XPS depth-profile and x-ray reflectivity data. The sharp Ce peak and Al edge in the RBS spectra

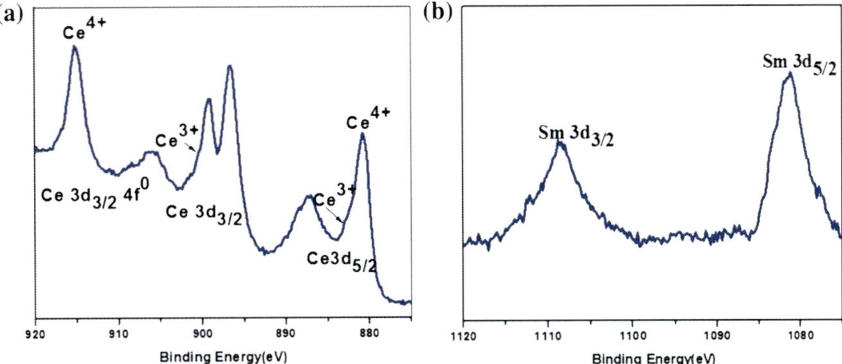

Fig. 6.9 High-resolution XPS from SDC (6 atom %) thin films **a** Ce 3d and, **b** Sm 3d regions. The oxidation states are 4+ and 3+ for Ce and Sm, respectively [19]. (Reprinted from [19] with permission from Elsevier)

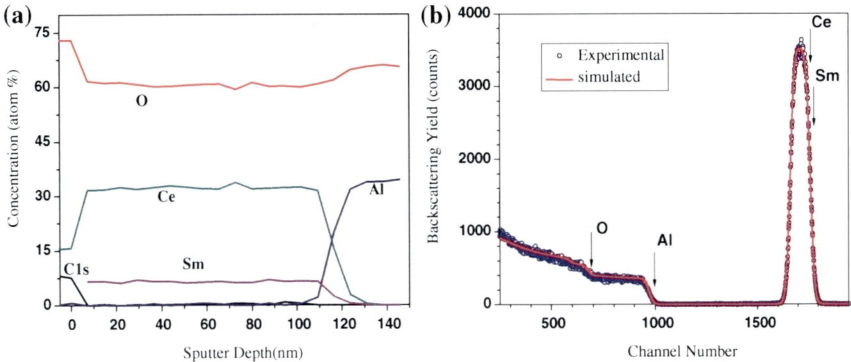

Fig. 6.10 **a** XPS depth profile, and **b** RBS spectrum in random direction along with simulated spectra for SDC (6 atom %) thin film. The incident energy of the He^+ beam was 2.04 MeV and the scattering angle was 150° [19]. (Reprinted from [19] with permission from Elsevier)

confirmed that there is no inter-diffusion of metal atoms across the film/substrate interface as indicated by XPS depth profile. The RBS spectrum in the channeling direction confirmed the high quality of the films with a low minimum yield. In addition to the as-grown films, XRD, XPS, and RBS measurements were carried out on some of the SDC films following the conductance measurements to identify any changes in the film properties and the results were identical to that of as-grown films [19]. It suggested that, the film properties did not change significantly during the conductance measurements, which is essential for long term use of SDC thin films as an oxygen sensing platforms at high temperatures.

The design of SDC thin films for sensor applications requires the optimization of various chemical and physical parameters. These include the Sm doping level as well as the SDC film thickness and conductance studies as a function of operation temperature and oxygen concentration and as such these are critical fundamental studies. The conductance of various SDC thin films in this study was obtained by measuring the current at a fixed bias voltage (2 V) using two-probe method under different oxygen pressures (1mTorr–100 Torr) with increasing temperature up to 973 K. Figure 6.11 shows the resistance of the SDC films with different Sm concentration (0–14 atom %) as a function of temperature at an oxygen pressure of 10 Torr [19]. The resistance of the SDC thin films decreased with the increase in temperature regardless of the dopant concentration. The activation energy for the oxygen ion conduction was determined from the slope of these Arrhenius plots in Fig. 6.11 using the Arrhenius equation, which was described earlier in this section. The SDC films with 3 and 6 atom % Sm concentrations showed the lowest activation energy of ~ 0.6 eV and the SDC film with 14 atom % Sm concentration showed the highest activation energy of ~ 0.9 eV. It is worth mentioning that, these activation energies are independent of oxygen pressure. Figure 6.12 shows the effect of oxygen pressure on the conductance of the 6 atom % SDC thin film, which can be described as the input and output relationship of the sensing plat-form, where the input quantity is the oxygen pressure and output is measured in terms of current across the SDC thin film [99]. Figure 6.13 shows the current measured across the SDC film as a function of Sm concentration at an oxygen pressure of 100 Torr, where the current increased with the Sm concentration up to 6 atom % at all temperatures from 773 to 973 K [19]. The current measured across the SDC film started to decrease with the increase of Sm concentration beyond 6 atom %. The increase in conductance with increasing Sm dopant concentration up to 6 atom % is attributed to an increase of oxygen ion vacancies in CeO_2 fluorite lattice due to the doping, which facilitate oxygen ion diffusion through a hopping process at elevated temperatures. In order to maintain the electrical neutrality and structural stability, every two Sm^{3+} ions result in one oxygen vacancy in the CeO_2 fluorite lattice, which enhanced the oxygen ion conductance. Higher dopant concentrations may lead to structural defects and polycrystalline features [74], which disturbs the oxygen vacancy ordering and increases the activation energy of the films. Thus, the higher activation energy in SDC films above 6 atom % Sm concentration can be attributed to the loss of epitaxial quality of the films. It is therefore concluded that the optimum Sm concentration of SDC is about 6 atom % for oxygen sensor applications [19].

After establishing the optimum concentration for SDC thin films, the influence of film thickness on the conductance of SDC thin film with 6 atom % Sm concentration was investigated [23, 99]. It is also important to observe the temperature dependency of the conductivity of SDC films with different thicknesses, since the overall conductivity of the SDC thin film used for oxygen sensing is dependent on its operating temperature [100]. Figure 6.14 shows the change in overall conductance of SDC thin films with the increasing temperature at thicknesses ranging from 50 up to 300 nm at an oxygen pressure of 10 Torr [99]. An identical trend of increase in the overall conductance with an increase in the operating temperature from 673 to 973 K was observed for each SDC film thickness. When the SDC film thickness is increased from 50 up to 200 nm, a significant gain in overall

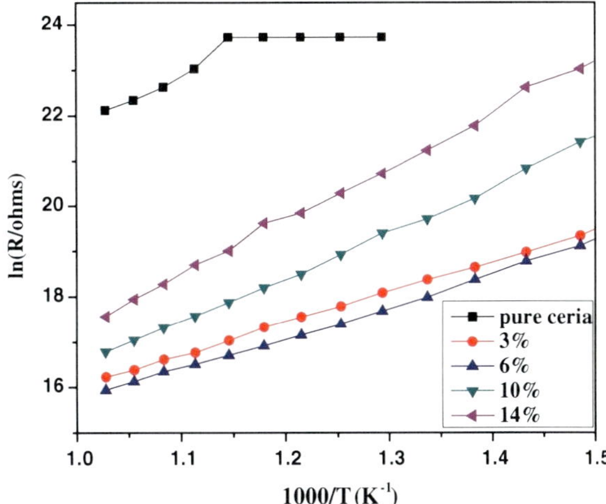

Fig. 6.11 The resistance (R) of the SDC film as a function of temperature (T) with different Sm concentration (0–14 atom %) at oxygen pressure of 10 Torr [19]. (Reprinted from [19] with permission from Elsevier)

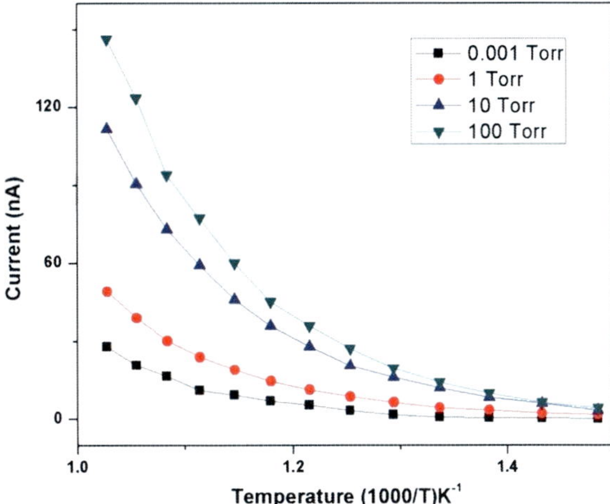

Fig. 6.12 The effect of oxygen pressure on the current measured across the 50 nm SDC (6 atom %) thin film [99].([99], reproduced with permission from Cambridge University Press)

conductance of the material system is achieved. However, there is saturation in the overall conductance increment with the increasing SDC film thickness beyond 200 nm. As a representative example, Fig. 6.15 clearly shows the saturation in the overall conductance increment of SDC thin films beyond the critical thickness (200 nm) at 873 K for oxygen pressures of 10 and 100 Torr at a fixed bias voltage [23]. This characteristic thickness dependent behavior of SDC films is also observed experimentally for oxygen pressures of 0.001, 1, and 100 Torr in the temperature range of 673–973 K.

To understand this thickness dependent behavior of the conductance of SDC thin films, Sanghavi et al. have proposed a hypothesis with a possible mechanism, which will be summarized here [23]. According to this hypothesis, the number of oxygen vacancies in the SDC film increases with an increase in the thickness,

Fig. 6.13 The current measured across the SDC films as a function of Sm concentration at different temperatures, indicating that 6 atom % is the optimum Sm concentration for the conductance of the film [19]. (Reprinted from [19] with permission from Elsevier)

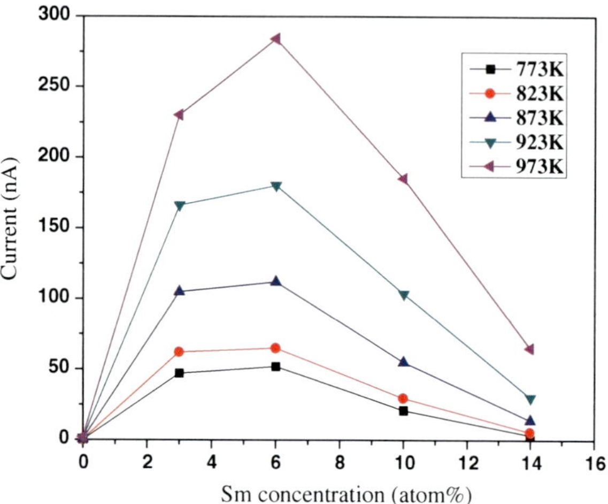

Fig. 6.14 The current measured across the SDC (6 atom %) films as a function of temperature at each film thickness at oxygen pressure of 10 Torr [99]. ([99] reproduced with permission from Cambridge University Press)

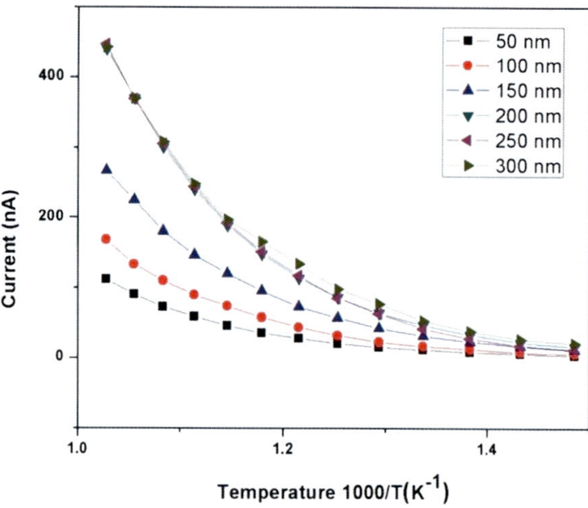

which are also well aligned due to the highly oriented nature of the SDC films. In the presence of an applied electric field, the oxygen ions hop through these aligned oxygen vacancies, completing the electrical circuit to produce a measurable electrical signal. Thus, with the increase in SDC film thickness, the number of conduction channels for the flow of ions in the SDC thin film increases, which enhances the overall conductivity. Above a particular SDC film thickness, the saturation in the increment of the overall conductance with the increase in film thickness was explained with respect to the supplied bias voltage. For a constant voltage applied across the two terminals on the surface of the sensing film, there is a corresponding voltage drop along the thickness axis perpendicular to the SDC film surface. Beyond a particular film depth, the electromotive force is not suffi-cient to drive the oxygen ions from one terminal to another under the influence of

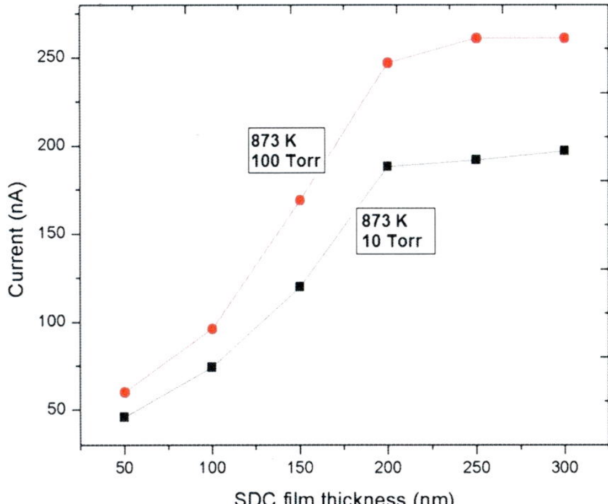

Fig. 6.15 The change in overall conductance as a function of the film thickness measured at 873 K for oxygen pressures of 10 and 100 Torr [23]. (Reprinted from [23], © 2011 IEEE)

applied bias voltage. Thus, there is saturation in the current measured across the SDC thin film beyond critical thickness at a fixed bias voltage. Furthermore, Sanghavi et al. have suggested that, the saturation limit for the conductance increment can be increased by increasing the bias voltage and they have experimentally observed this behavior, where the saturation limit for the conductance increment was increased at a bias voltage of 5 V [23].

For developing a sensor platform for real world applications, there is a need to test the hysteresis error performance of the potential sensing platform, which operates in an ever-changing environment. Thus, Sanghavi et al. have analyzed the hysteresis error performance of 6 atom % SDC films of thickness ranging from 50 up to 300 nm at two representative temperatures of 798 and 898 K [23]. In this hysteresis test, the SDC thin films were subjected to a continuous cyclic process of increasing and decreasing oxygen pressures at a fixed operating temperature and hysteresis error is measured as the difference in the measured overall conductance for consecutive cycles of pressure variation. The hysteresis error performance of 300 nm SDC thin film at operating temperature of 798 K is shown in Fig. 6.16 in terms of the current through the material system with cyclic oxygen pressure variations from 0.001 to 100 Torr [23]. All the SDC films with different thicknesses have showed similar behavior with respect to hysteresis error observed, suggesting the hysteresis error performance is independent of the sensing film thickness. This experimental analysis showed a maximum hysteresis error of 5–7 % for any SDC film under consideration.

In this study, Sanghavi et al. have also tested the dynamic response of the SDC thin film in terms of its overall conductance by subjecting it to rapidly changing oxygen pressure environments. The dynamic response observed for the 300 nm SDC thin film is shown in Fig. 6.17 in terms of current through the material system to random oxygen pressure variations from 0.001 to 100 Torr, at a fixed operating temperature of 798 K [99]. In the Fig. 6.17, the vertical axis is shared by the input oxygen

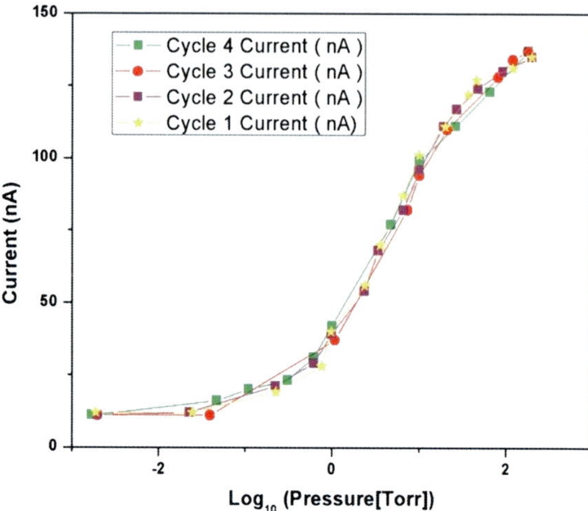

Fig. 6.16 Hysteresis error performance of 300 nm SDC film at 798 K for 4 cycles of oxygen pressure variation from 0.001 to 100 Torr, indicating a tolerable hysteresis for the SDC film [23]. (Reprinted from [23], © 2011 IEEE)

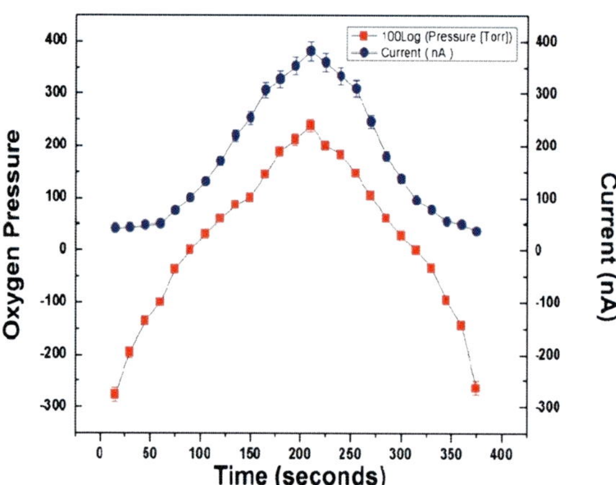

Fig. 6.17 Dynamic response of 300 nm SDC film for rapid oxygen pressure variation at a time interval of 15 s [99].([99] reproduced with permission from Cambridge University Press)

pressure in logarithmic scale and corresponding output current. The horizontal axis is the time axis with units in seconds. In this experiment, the SDC film was exposed to randomly changed values of oxygen pressures at every 15 s. The data showed a good proportionality between oxygen pressure variation and the overall conductance change of the SDC film. It was also observed that the dynamic time response of SDC thin films is independent of the sensing film thickness. They have further checked the sensing performance of SDC thin films for its reproducibility by running experimental cycles with the same set of parameters, which showed consistency in the obtained results with a standard deviation of 1–4 % [23].

In this section, we discussed the oxygen sensing properties of SDC nano-scale thin films grown by MBE. As we mentioned in the introduction, various kinds of thin films have been synthesized by MBE to understand their gas sensing properties. While MBE grown SDC thin films with controlled surface morphology, crystalline orientation, film thickness, and composition have provided a better platform to understand their oxygen sensing properties with respect to the film properties and operating conditions, it is worthwhile to evaluate the other sensing materials synthesized by MBE. In the next section, we will summarize the surface, structural, and gas sensing properties of SnO_2 thin films grown by MBE.

6.3 SnO_2 Thin Films Grown by Molecular Beam Epitaxy for Gas Sensing Applications

SnO_2 has been widely used in semiconducting metal oxide gas sensors as an active sensing material for detection of reducing gases such as CO, H_2, and CH_4 [3]. SnO_2 is a wide band gap n-type semiconductor with the rutile structure [101], which shows enhanced electrical, optical and electrochemical properties due to its dual valency and variability of surface composition [102–104]. As a gas sensing material, SnO_2 exhibits high sensitivity and better stability towards reducing gases at low temperatures [3]. SnO_2 based gas sensors operate on a principle that, their conductance changes when the adsorbed oxygen ions on the surface react with the gas molecules [105]. As a pure n-type semiconductor, exposure of SnO_2 to air leads to adsorption of oxygen atoms on the surface. Then, the transfer of an electron from SnO_2 to the adsorbed oxygen results in various oxygen anion species as follows, which cover the SnO_2 particles on the surface [106, 107].

$$O_2(gas) \Longleftrightarrow O_2(ad) \Longleftrightarrow O_2^-(ad) \Longleftrightarrow O^-(ad) \Longleftrightarrow O^{2-}(ad) \Longleftrightarrow O^{2-}(lattice)$$

Due to the depletion of electrons, a positively charged layer is also formed below the particle surface leading to space charge layers near the SnO_2 surface [108–111]. This process leads to high electrical resistance of SnO_2 in air [112]. However, when SnO_2 is exposed to reducing gases at elevated temperatures, oxygen adsorbates react with the adsorbed reducing gas molecules and release the electrons to the conduction band increasing the electronic conductance of SnO_2 [109, 110]. The overall reaction takes place on the SnO_2 surface between oxygen adsorbates and adsorbed reducing (R) gas molecules can be given as follows.

$$O_2(gas) \longrightarrow O_2(ad)$$
$$R(gas) \longrightarrow R(ad)$$
$$R(ad) + O_2(ad) \longrightarrow RO_2 + 2e^-$$

Thus, the adsorption of reducing gas species controls the conductance of SnO_2, which is strongly influenced by the surface morphology, composition and structures of SnO_2 [113–116]. Furthermore, the sensitivity of the SnO_2 gas sensors can be defined as the ratio of its conductance in a test gas with reducing gas species to that in air. Up to date, several models have been proposed for SnO_2 gas sensors to understand their sensing mechanism [117–121], which investigated the influence of the test gas concentration and operating temperature on the sensitivity of SnO_2 gas sensors.

In general, commercially available SnO_2 gas sensors are either bulk ceramic or thick film based devices [105]. In addition, nano-scale thin film SnO_2 gas sensors have also been studied [122, 123]. In most cases, SnO_2 thin films have been used in polycrystalline form for gas sensing applications, which suffer from the lack of long-term stability [124]. On the other hand, single-crystal (epitaxial) SnO_2 thin films show better homogeneity, compactness, and stability in contrast to poly-crystalline films [125]. Single-crystal films are also useful for understanding the gas sensing mechanism of SnO_2 by eliminating the grain boundary effect due to more uniform adsorption and desorption kinetics [126]. Furthermore, nanomate-rials and epitaxial thin films with dimensions smaller than space charge layers may exhibit a strong change in the conductance with exposure to reducing gases [104]. Recently, Kroneld et al. have reported the growth of SnO_2 single-crystal films, which exhibited greater potential for continuous gas detection [124].

Due to the limited availability of commercial single-crystal SnO_2, epitaxial SnO_2 thin films have been fabricated by using various thin film deposition tech-niques including sputtering [127], chemical vapor deposition [125], pulsed laser deposition [128], atomic layer deposition [129, 130], and molecular beam epi-taxy[124, 131, 132] for gas sensing applications. However, studies on high quality epitaxial SnO_2 thin films are limited by their poorly controlled impurity levels [132, 133]. As we discussed in the previous sections, MBE is a useful technique to grow high quality single crystal metal oxide thin films with controlled impurity levels and surface properties. In this section, we will discuss the gas sensing properties of MBE grown SnO_2 single-crystal films.

Kroneld et al. have conducted a comparative study of the gas sensing properties of epitaxial and polycrystalline SnO_2 thin films deposited on r-plane sapphire substrates by MBE technique [124]. The optimum growth conditions for high quality SnO_2 were established in terms of the growth temperature, growth rate and source materials, which determined the crystalline nature of the films. In this study, SnO_2 thin films were deposited on r-plane sapphire substrates by varying the film thickness and growth temperature as 30–100 nm and 260–550 °C, respec-tively. The growth rate was also varied between 0.02 and 0.1 nm/s. Following the growth, SnO_2 thin films were characterized by RHEED, XRD, and AFM. The electrical properties of SnO_2 thin films were investigated by using a Hall effect measurement system and the conductivity was measured by the four-point Van der Pauw method. H_2, CO, O_2, and NO_2 were used as the test gases with various concentrations for gas sensing studies under different operating temperatures. XRD patterns for the films grown at 230 and 550 °C are shown in Fig. 6.18. At 230 °C, the XRD pattern showed broad peaks with low intensities, indicating a polycrystalline film. However, at high temperature, only $SnO_2(101)$ peak existed in the XRD pattern in addition to a substrate peak, suggesting the growth of highly-oriented $SnO_2(101)$ thin film. All the films were n-type semiconductors as revealed by the Hall effect measurements. Both epitaxial and polycrystalline SnO_2 thin films were investigated for gas sensing properties and polycrystalline films

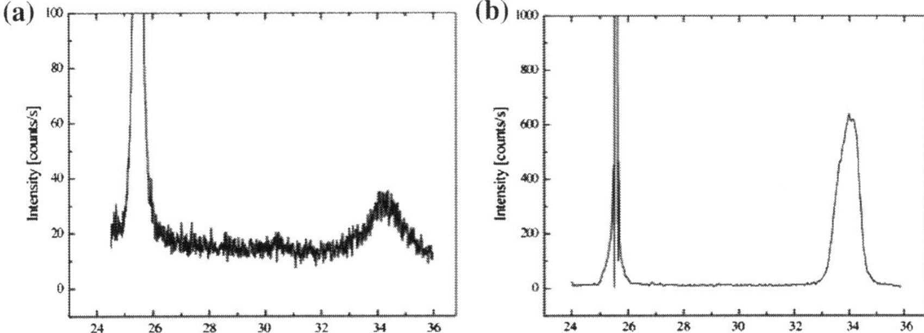

Fig. 6.18 XRD $2\theta - \omega$ scans of the SnO_2 thin films grown at **a** 230 °C and, **b** 550 °C [124]. (Reprinted from [124] with permission from Elsevier)

showed better sensitivity towards the test gases. For hydrogen detection, there was no significant difference in the reaction time for epitaxial and polycrystalline films. However, epitaxial films showed relatively high sensitivity towards ethanol at concentrations of 100, 200, and 500 ppm at 400 °C as shown in Fig. 6.19b. Better response of the films towards ethanol was observed with the increasing gas concentration while the sensitivity saturated. Furthermore, slower reaction time was observed for polycrystalline film and the recovery was not nearly complete in the reported interval (Fig. 6.19a). The epitaxial SnO_2 thin films deposited by magnetron sputtering also showed high sensitivity towards ethanol with good stability and fast response [127]. Thus, it has been concluded that, epitaxial SnO_2 thin films exhibit greater potential for continuous ethanol detection.

White et al. have also deposited epitaxial SnO_2 thin films on r-plane sapphire by MBE and extensively investigated the surface morphology, epitaxial relationship, and growth behavior of the films [132]. $SnO_2(101)$ epitaxial films exhibited Volmer-Weber growth indicated by the spotty RHEED pattern (Fig. 6.20a). However, initial 3-D island growth transformed to a 2-D layer growth during the deposition as observed by RHEED patterns in Fig. 6.20. Furthermore, the initial rough surface became smoother with the increasing thickness of the films. $SnO_2(101)$ and (202) reflections in the HRXRD pattern confirmed the epitaxial growth of phase-pure $SnO_2(101)$. They have also used the HRXRD data to obtain information about the crystalline quality of SnO_2 thin films by comparing the full width at a half maximum (FWHM) value of $SnO_2(101)$ peak. The FWHM value of the $SnO_2(101)$ ω-scan decreased with the increasing substrate temperature. The surface roughness of the films measured by AFM was also decreased with the increasing substrate temperature as shown in Fig. 6.21. Thus, it was concluded that the surface morphology and crystalline quality of $SnO_2(101)$ thin films strongly depend on the growth temperature. TEM images showed dislocations within the $SnO_2(101)$ films, which are originated at the film-substrate interface. The growth behavior of SnO_2 showed two distinct growth regimes, namely O_2-rich regime and Sn-rich regime. Secondary ion mass spectrometry (SIMS) data showed low impurity concentration in SnO_2 films in contrast to SnO_2 films deposited by other

Fig. 6.19 The relative conductivity change of the sample sensing film upon the exposure of 100, 200 and 500 ppm C_2H_5OH in air with 40 % humidity. **a** Sample with smaller grains, and **b** a monocrystalline sample. The working temperature of the samples was 400 °C. Clearly the monocrystalline sample shows shorter reaction times, and in this case, higher sensitivity towards the test gas [124]. (Reprinted from [124] with permission from Elsevier)

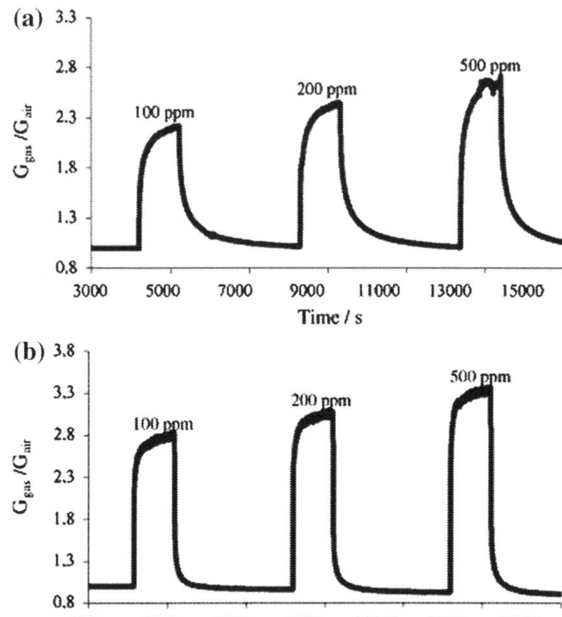

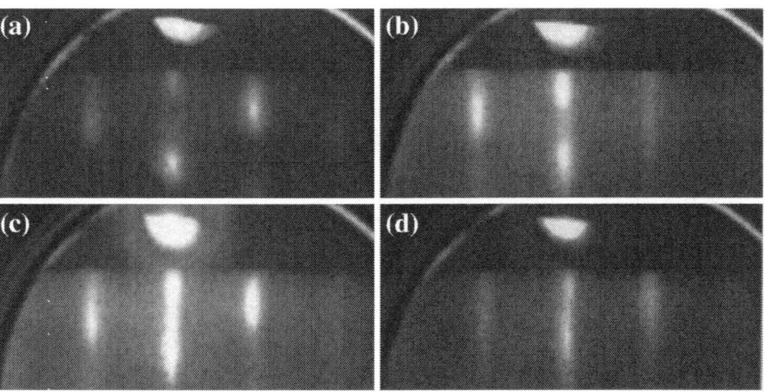

Fig. 6.20 Evolution of RHEED observed along the <101> azimuth during growth of SnO_2 on r-sapphire. Images **a–d** were recorded at 10 s, and 3.25, 4.75, and 27.25 min, respectively, after beginning of growth. A transition from a 3D initial surface to a 2D surface was observed [132]. (Reprinted from [132] with permission from American vacuum society)

growth techniques. It emphasizes the importance of MBE to deposit high quality SnO_2 epitaxial thin films.

MBE can also be used to grow doped SnO_2 thin films with a controlled dopant level by co-evaporation of metal sources. The MBE growth of pure and cobalt doped SnO_2 thin films on Al_2O_3 has been reported [134]. These results can be reviewed in the Chapter authored by Batzill in this book.

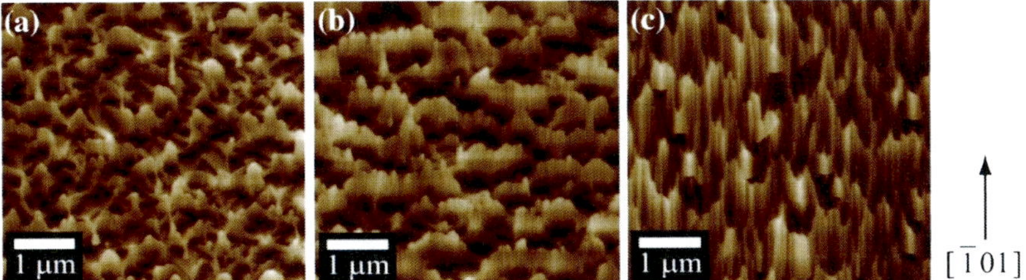

Fig. 6.21 5×5 μm^2 AFM images of the SnO$_2$ surface for three samples grown in an oxygen plasma pressure of 1.2×10^{-5} Torr at substrate temperatures of **a** 700 °C, **b** 775 °C, and **c** 850 °C. The surface roughness values for each sample (a-c) were 15, 21, and 9 nm, respectively. The height scale for each sample (**a–c**) was 0–100, 0–130, and 0–70 nm [132]. (Reprinted from [132] with permission from American vacuum society)

TiO$_2$ has also been used as a substrate for the MBE growth of SnO$_2$ thin films other than Al$_2$O$_3$ [131, 133, 135]. Tsai et al. have reported the heteroepitaxial growth of SnO$_2$ films on TiO$_2$(110) with a detailed study of the film properties, which strongly depended on the MBE growth parameters [133]. In this study, high purity liquid Sn was evaporated from a Knudsen cell in the presence of oxygen plasma to deposit SnO$_2$. The substrate temperature and both tin and oxygen beam equivalent pressure values were varied to investigate the growth behavior of SnO$_2$ thin films. The crystalline quality of the films improved with the increased substrate temperature as indicated by FWHM values of the rocking curve with respect to SnO$_2$(110) reflection. However, the optimum growth temperature was set to 625 °C, since the growth of SnO$_2$ was not observed at 700 °C. The growth of SnO$_2$ was also not observed in the oxygen-insufficient environment. The initial spotty RHEED pattern of the SnO$_2$ thin films indicated a rough surface with the formation of 3-D islands. However, the transition of the spotty RHEED pattern to a streaky RHEED pattern after 17 min of growth suggested a smooth surface with 2-D layers, which confirmed the coalescence of the SnO$_2$ islands. Furthermore, the smoothest surface of the SnO$_2$ thin film was observed in the oxygen-rich regime as confirmed by RHEED and AFM data. HRXRD patterns revealed the growth of epitaxial SnO$_2$(110) thin films as shown in Fig. 6.22. The x-ray rocking curve with respect to SnO$_2$(110) reflection was used to evaluate the crystalline quality of the films as shown in the inset of Fig. 6.22. The broadness of the rocking curve was influenced by the lattice mismatch between the substrate and the film and a narrower rocking curve was observed in the oxygen-rich regime. A growth diagram for the MBE growth of epitaxial SnO$_2$ thin films on TiO$_2$(110) substrates was created based on these results, which demonstrates the influence of growth temperature and oxygen partial pressure on the crystalline quality and surface morphology of the films.

The interface properties of the SnO$_2$ thin films grown on TiO$_2$(110) substrates by MBE have also been investigated by using XPS and SIMS, which showed the inter-diffusion of metal atoms at the film/substrate interface [131]. XPS revealed a changing surface composition of SnO$_2$ thin films with increasing growth temperature as shown in Fig. 6.23. For the films grown at 500 and 750 °C, Sn, O and

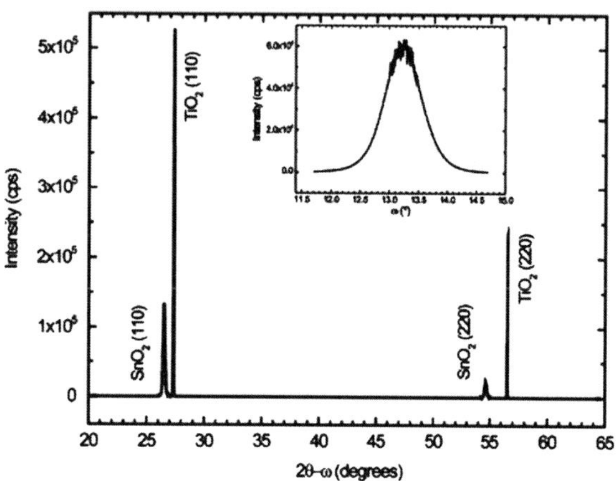

Fig. 6.22 X-ray diffraction $2\theta - \omega$ scan of a SnO$_2$ film with a thickness of 625 nm deposited on a TiO$_2$(1 1 0) substrate at 625 °C. The inset shows the ω-rocking curve of SnO$_2$(1 1 0) [133]. (Reprinted from [133] with permission from Elsevier)

low amounts of Ti were detected. The film grown at 775 °C had a mixed composition, with a Sn:Ti ratio of 55:45 and XPS also indicated chemical interaction between Sn and Ti in this film. At higher growth temperatures (800 and 900 °C), no Sn was detected on the substrate surface within the XPS sampling depth as shown in Fig. 6.23. According to SIMS depth profiles of Sn and Ti (Fig. 6.24), film/substrate interface for the films grown at temperatures below 600 °C was quite abrupt. However, an extended weak tail in the Sn depth profile indicated penetration of some Sn into the substrate. This diffusion tail could also result from mixing caused by the incident ion beam during the SIMS analysis. More pronounced diffusion tails in the Sn and Ti depth profiles for the films grown at 750 and 763 °C confirmed the inter-diffusion of metal atoms at the interface between the film and substrate. In addition, SnO$_2$ thin films grown at temperatures between 500 and 763 °C showed a distinct Sn-rich layer on the TiO$_2$ substrate. SIMS depth profiles of Sn and Ti for the film grown at 775 °C showed a Sn/Ti mixed surface layer, which was comparable with the XPS data described above. At growth temperatures above 775 °C, Sn diffused rapidly into the substrate, which resulted in Sn-doped TiO$_2$ rather than a distinct SnO$_2$ epilayer. All the films were highly (110)-oriented as confirmed by the XRD patterns. However, the lattice parameter of the SnO$_2$(110) films decreased with increasing growth temperature, which indicates the formation of Ti$_x$Sn$_{1-x}$O$_2$ solid solutions at growth temperatures of 750 and 763 °C. Thus, all the XRD, XPS and SIMS data revealed the interfacial diffusion of metal atoms in SnO$_2$ thin films grown on TiO$_2$(110) in a narrow growth temperature regime between 750 and 775 °C. It is reported that, the stability and selectivity of SnO$_2$ gas sensors can be improved by addition of Ti [136, 137]. Furthermore, Ti-doped SnO$_2$ can be used as a substrate for the epitaxial growth of SnO$_2$ to improve the crystalline quality by reducing the lattice mismatch between the substrate and film [104, 138]. Thus, the formation of Ti$_x$Sn$_{1-x}$O$_2$ by inter-diffusion of SnO$_2$ and TiO$_2$ will be important for those applications.

It is evident from the above studies that the growth conditions strongly affect the surface properties and crystalline orientation of epitaxial SnO$_2$ thin films. Especially,

Fig. 6.23 X-ray photoelectron spectra of SnO_2 thin films grown at the temperatures indicated showing Sn 3d and Ti 2p core lines. The *dashed lines* represent the binding energy expected for the Sn $3d_{5/2}$ (*higher binding energy*) and the Ti $2p_{3/2}$ (*lower binding energy*) core lines of SnO_2 and TiO_2, respectively [131]. (Reprinted from [131] with permission from American chemical society)

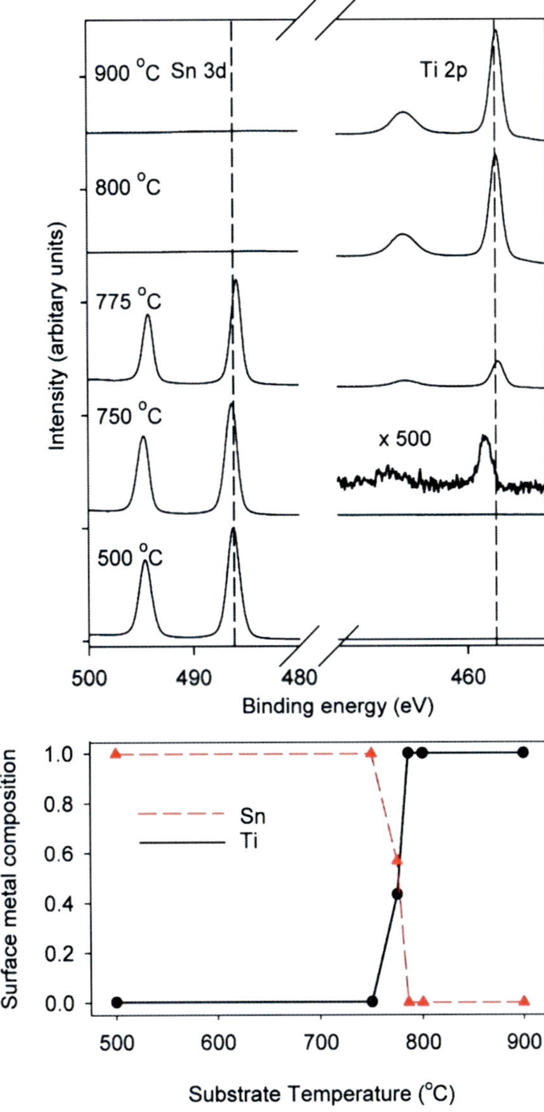

high quality epitaxial SnO_2 thin films were obtained in a very narrow temperature regime, which could be a result of large lattice mismatch between the film and substrate. Hishita et al. have demonstrated the use of Sn-doped TiO_2 as the substrates to overcome this issue by minimizing the lattice mismatch between SnO_2 film and the substrate [138]. They have prepared $Sn_{0.075}TiO_{0.925}O_2(110)$ substrates at 1,100 °C by using a vacuum deposition-and-annealing method to deposit SnO_2 thin films by MBE. The calculated lattice parameters of the $Sn_{0.075}TiO_{0.925}O_2$ (110) substrate were a = = 0.4603 and c = 0.2976 nm, compared to that of pure TiO_2 substrate (a = 0.4584 and c = 0.2953), which reduces the lattice mismatch between the film

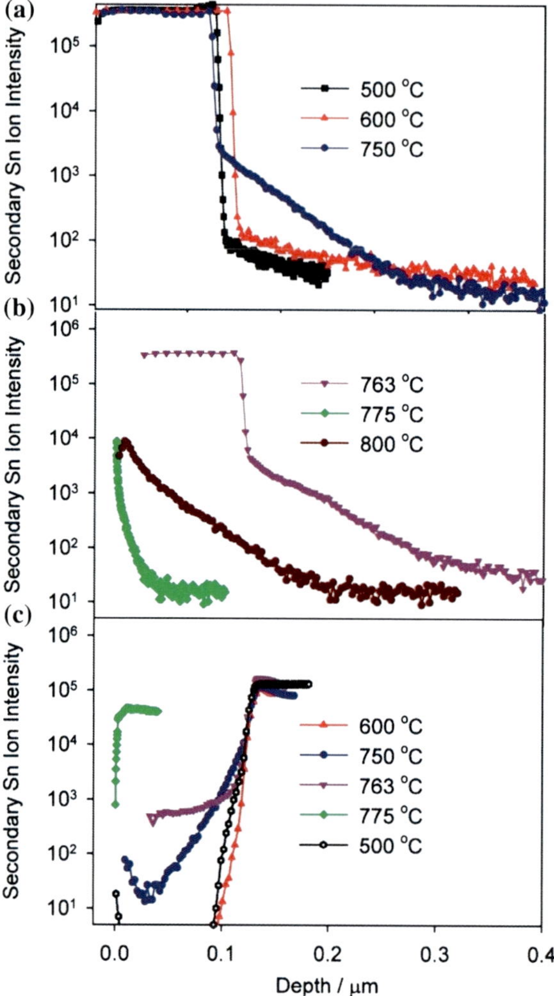

Fig. 6.24 SIMS Sn ion depth profiles of SnO_2 thin films grown at **a** 500, 600, and 750 °C, **b** 763, 775, and 800 °C, and **c** SIMS Ti ion depth profiles showing inter-diffusion of Ti and Sn. Distinct Sn-rich layers can be discerned in samples grown at 763 °C and below [131]. (Reprinted from [131] with permission from American chemical society)

and substrate. Furthermore, the effect of the growth temperature on the surface morphology of the SnO_2 films grown on Sn-doped TiO_2 were compared with those grown on pure $TiO_2(110)$ substrates. It was concluded that, the temperature range for the lateral growth of SnO_2 films can be extended to a lower temperature with Sn-doped TiO_2 substrates.

6.4 Summary

Future trends in the sensor technology include the miniaturization of the sensor devices and making the sensors harsh environment compatible. The use of nano-scale thin films as the sensing materials is the key to develop small scale chemical sensors. To develop chemical sensing nano-scale materials and understand their structure–property relationship, we need a deposition technique with an atomic-scale control over the chemical and structural properties. MBE is a widely used technique for the deposition of metal oxide thin films with ultra pure surfaces to understand their fundamental properties. Due to the slow growth rate and UHV deposition conditions, MBE has various advantages over the other deposition techniques; (1) control of the growth conditions results in high quality and more accurate stoichiometry, (2) less surface contamination or oxidation, and (3) minimal inter-diffusion at the interfaces of the grown thin films. In addition, MBE growth chambers can be easily incorporated with limited in situ characterization capabilities such as XPS and RHEED like sputter and pulsed laser deposition growth capabilities. Thus, various chemical sensing metal oxides including SnO_2, CeO_2, TiO_2, ZnO, In_2O_3, and WO_3 have been synthesized by MBE. In this book chapter, we have discussed the growth, characterization, and gas sensing properties of metal oxide thin films grown by MBE including Sm doped CeO_2 and SnO_2. In addition, we have briefly discussed the MBE deposition technique along with in situ electron diffraction techniques such as RHEED and LEED. Furthermore, we have explained the use of in situ reflection high energy electron diffraction along with atomic force microscopy to understand the growth modes and orientation relationship between the clusters and the substrate using Cu_2O nanoclusters grown on STO substrates. In general, we have shown that MBE is a useful technique to understand the fundamental properties of chemical sensing thin films.

Acknowledgments   A portion of this research was performed using EMSL, a national scientific user facility sponsored by the Department of Energy's Office of Biological and Environmental Research and located at Pacific Northwest National Laboratory (PNNL). PNNL is operated for the U.S. DOE by Battelle Memorial Institute under contract number DE-AC05-76RL01830.

References

1. Capone S, Forleo A, Francioso L, Rella R, Siciliano P, Spadavecchia J, Presicce DS, Taurino AM (2004) Solid state gas sensors: state of the art and future activities. ChemInform 35(29). doi:10.1002/chin.200429283
2. Moseley PT (1997) Solid state gas sensors. Meas Sci Technol 8(3):223
3. Korotcenkov G (2007) Metal oxides for solid-state gas sensors: what determines our choice? Mater Sci Eng, B 139(1):1–23
4. Baughman RH, Zakhidov AA, de Heer WA (2002) Carbon nanotubes–the route toward applications. Science 297(5582):787–792
5. Iijima S (1991) Helical microtubules of graphitic carbon. Nature 354(6348):56–58
6. Kong J, Franklin NR, Zhou C, Chapline MG, Peng S, Cho K, Dai H (2000) Nanotube molecular wires as chemical sensors. Science 287(5453):622–625
7. Comini E et al (2007) Single crystal ZnO nanowires as optical and conductometric chemical sensor. J Phys D Appl Phys 40(23):7255
8. Hwang IS, Kim SJ, Choi JK, Choi J, Ji H, Kim GT, Cao G, Lee JH (2010) Synthesis and gas sensing characteristics of highly crystalline ZnO–SnO$_2$ core–shell nanowires. Sens Actuators B: Chem 148(2):595–600

9. Kolmakov A, Zhang Y, Cheng G, Moskovits M (2003) Detection of CO and O_2 using Tin Oxide Nanowire sensors. Adv Mater 15(12):997–1000
10. Kumar M et al (2009) Tunable synthesis of indium oxide octahedra, nanowires and tubular nanoarrow structures under oxidizing and reducing ambients. Nanotechnology 20(23):235608
11. Li C, Zhang D, Liu X, Han S, Tang T, Han J, Zhou C (2003) In2O3 nanowires as chemical sensors. Appl Phys Lett 82(10):1613–1615
12. Chu D, Zeng YP, Jiang D, Masuda Y (2009) In2O3-SnO$_2$ nano-toasts and nanorods: precipitation preparation, formation mechanism, and gas sensitive properties. Sens Actuators B: Chem 137(2), 630–636. 32
13. Forleo A, Francioso L, Capone S, Casino F, Siciliano P, Tan OK, Hui H (2011) Fabrication at wafer level of miniaturized gas sensors based on SnO$_2$ nanorods deposited by PECVD and gas sensing characteristics. Sens Actuators B: Chem 154(2):283–287
14. Zhang Y, Yu K, Jiang D, Zhu Z, Geng H, Luo L (2005) Zinc oxide nanorod and nanowire for humidity sensor. Appl Surf Sci 242(1–2):212–217
15. Zhi-Peng S et al (2006) Rapid synthesis of ZnO nano-rods by one-step, room-temperature, solid-state reaction and their gas-sensing properties. Nanotechnology 17(9):2266
16. Comini E, Faglia G, Sberveglieri G, Pan Z, Wang ZL (2002) Stable and highly sensitive gas sensors based on semiconducting oxide nanobelts. Appl Phys Lett 81(10):1869–1871
17. Pan ZW, Dai ZR, Wang ZL (2001) Nanobelts of semiconducting oxides. Science 291:1947–1949
18. Gao L, Li Q, Song Z, Wang J (2000) Preparation of nano-scale titania thick film and its oxygen sensitivity. Sens Actuators B: Chem 71(3):179–183
19. Gupta S, Kuchibhatla SVNT, Engelhard MH, Shutthanandan V, Nachimuthu P, Jiang W, Saraf LV, Thevuthasan S, Prasad S (2009) Influence of samaria doping on the resistance of ceria thin films and its implications to the planar oxygen sensing devices. Sens Actuators B: Chem 139(2):380–386
20. Hu Y, Tan OK, Pan JS, Huang H, Cao W (2005) The effects of annealing temperature on the sensing properties of low temperature nano-sized SrTiO3 oxygen gas sensor. Sens Actuators B: Chem 108(1–2):244–249
21. Ichimura M, Baoleer A, Sueyoshi T (2010) Properties of gas sensors based on photochemically deposited nano- crystalline SnO$_2$ films. Phys Status Solidi (c) 7(3–4), 1168–1171
22. Ogita M, Higo K, Nakanishi Y, Hatanaka Y (2001) Ga2O3 thin film for oxygen sensor at high temperature. Appl Surf Sci 175–176:721–725
23. Sanghavi R, Nandasiri M, Kuchibhatla S, Weilin J, Varga T, Nachimuthu P, Engelhard MH, Shutthanandan V, Thevuthasan S, Kayani A, Prasad S (2011) Thickness dependency of thin-film samaria-doped ceria for oxygen sensing. Sens J, IEEE 11(1), 217–224
24. Chambers SA (2010) Epitaxial growth and properties of doped transition metal and complex oxide films. Adv Mater 22(2):219–248
25. DiMeo JF, Cavicchi RE, Semancik S, Suehle JS, Tea NH, Small J, Armstrong JT, Kelliher JT (1998) In situ conductivity characterization of oxide thin film growth phenomena on microhotplates. J Vac Sci Technol A: Vac, Surf, Films 16(1):131–138
26. LeGore LJ, Lad RJ, Moulzolf SC, Vetelino JF, Frederick BG, Kenik EA (2002) Defects and morphology of tungsten trioxide thin films. Thin Solid Films 406(1–2):79–86
27. Moulzolf SC, Ding S-a, Lad RJ (2001) Stoichiometry and microstructure effects on tungsten oxide chemiresistive films. Sens Actuators B: Chem 77(1–2):375–382
28. Moulzolf SC, Frankel DJ, Lad RJ (2002) In situ four-point conductivity and hall effect apparatus for vacuum and controlled atmosphere measurements of thin film materials. Rev Sci Instrum 73(6):2325–2330
29. Poirier GE, Cavicchi RE, Semancik S (1993) Ultrathin heteroepitaxial SnO$_2$ films for use in gas sensors. AVS, Chicago, pp 1392–1395

30. Vetrone J, Chung YW, Cavicchi R, Semancik S (1993) Role of initial conductance and gas pressure on the conductance response of single-crystal SnO_2 thin films to H_2, O_2, and CO. J Appl Phys 73(12):8371–8376
31. Goldman AM (2006) Oxide heterostructures grown by molecular beam epitaxy: spin injection in superconductors and magnetic coupling phenomena. Appl Surf Sci 252(11):3928–3932
32. Oh S, Di Luccio T, Eckstein JN (2005) T linearity of in-plane resistivity in $Bi_2Sr_2CaCu_2O_{8+\delta}$ thin films. Phys Rev B 71(5):052504
33. Parendo KA, Sarwa B, Tan KH, Goldman AM (2006) Hot-electron effects in the two-dimensional superconductor-insulator transition. Phys Rev B 74(13):134517
34. Parendo KA, Tan KHSB, Goldman AM (2006) Electrostatic and parallel-magnetic-field tuned two-dimensional superconductor-insulator transitions. Phys Rev B 73(17):174527
35. Chambers SA, Liang Y (1999) Growth of β-MnO_2 films on TiO_2(110) by oxygen plasma assisted molecular beam epitaxy. Surf Sci 420(2–3):123–133
36. Gao Y, Chambers SA (1997) Heteroepitaxial growth of α-Fe_2O_3, γ-Fe_2O_3 and Fe_3O_4 thin films by oxygen-plasma-assisted molecular beam epitaxy. J Cryst Growth 174(1–4):446–454
37. Guo LW, Peng DL, Makino H, Inaba K, Ko HJ, Sumiyama K, Yao T (2000) Structural and magnetic properties of Mn3O4 films grown on MgO(001) substrates by plasma-assisted MBE. J Magn Magn Mater 213(3):321–325
38. Lind DM, Berry SD, Chern G, Mathias H, Testardi LR (1992) Growth and structural characterization of Fe3O4 and NiO thin films and superlattices grown by oxygen-plasma-assisted molecular-beam epitaxy. Phys Rev B 45(4):1838–1850
39. Peacor SD, Hibma T (1994) Reflection high-energy electron diffraction study of the growth of NiO and CoO thin films by molecular beam epitaxy. Surf Sci 301(1–3):11–18
40. Altman EI, Droubay T, Chambers SA (2002) Growth of MoO3 films by oxygen plasma assisted molecular beam epitaxy. Thin Solid Films 414(2):205–215
41. Chen PJ, Goodman DW (1994) Epitaxial growth of ultrathin Al2O3 films on Ta(110). Surf Sci 312(3):L767–L773
42. Freund HJ (1995) Metal oxide surfaces: electronic structure and molecular adsorption. Phys Status Solidi (b) 192(2), 407–440
43. Ohsawa T, Lyubinetsky I, Du Y, Henderson MA, Shutthanandan V, Chambers SA (2009) Crystallographic dependence of visible-light photoactivity in epitaxial TiO_2-xNx anatase and rutile. Physical Review B 79(8):085401
44. Peden CHF, Herman GS (1999) Z. Ismagilov, I.; Kay, B. D.; Henderson, M. A.; Kim, Y.-J.; Chambers, S. A., Model catalyst studies with single crystals and epitaxial thin oxide films. Catal Today 51(3–4):513–519
45. René F (2000) Growth of thin, crystalline oxide, nitride and oxynitride films on metal and metal alloy surfaces. Surf Sci Rep 38(6–8):195–294
46. Street SC, Xu C, Goodman DW (1997) The physical and chemical properties of ultrathin oxide films. Annu Rev Phys Chem 48(1):43–68
47. Chambers SA (2000) Epitaxial growth and properties of thin film oxides. Surf Sci Rep 39(5–6):105–180
48. Kim YJ, Gao Y, Herman GS, Thevuthasan S, Jiang W, McCready DE, Chambers SA (1999) Growth and structure of epitaxial CeO_2 by oxygen-plasma-assisted molecular 35 beam epitaxy. J Vac Sci Technol A: Vac, Surf, Films 17(3):926–935
49. Kuchibhatla SVNT, Hu SY, Yu ZQ, Shutthanandan V, Li YL, Nachimuthu P, Jiang W, Thevuthasan S, Henager CH, Sundaram SK (2009) Morphology, orientation relationship, and stability analysis of Cu_2O nanoclusters on $SrTiO_3$(100). Appl Phys Lett 95(5), 053111–053111-3
50. Francioso L, Presicce DS, Taurino AM, Rella R, Siciliano P, Ficarella A (2003) Automotive application of sol-gel TiO_2 thin film-based sensor for lambda measurement. Sens Actuators B: Chem 95(1–3):66–72

51. Izu N, Shin W, Matsubara I, Murayama N (2004) Development of resistive oxygen sensors based on cerium oxide thick film. J Electroceram 13(1):703–706
52. Papkovsky DB (1995) New oxygen sensors and their application to biosensing. Sens Actuators B: Chem 29(1–3):213–218
53. Ramamoorthy R, Dutta PK, Akbar SA (2003) Oxygen sensors: materials, methods, designs and applications. J Mater Sci 38(21):4271–4282
54. Smiddy M, Fitzgerald M, Kerry JP, Papkovsky DB, O' Sullivan CK, Guilbault GG (2002) Use of oxygen sensors to non-destructively measure the oxygen content in modified atmosphere and vacuum packed beef: impact of oxygen content on lipid oxidation. Meat Sci 61(3):285–290
55. Tsukada K, Sakai S, Hase K, Minamitani H (2003) Development of catheter-type optical oxygen sensor and applications to bioinstrumentation. Biosens Bioelectron 18(12):1439–1445
56. Benammar M (1994) Techniques for measurement of oxygen and air-to-fuel ratio using zirconia sensors: a review. Meas Sci Technol 5(7):757
57. Lari A, Khodadadi A, Mortazavi Y (2009) Semiconducting metal oxides as electrode material for YSZ-based oxygen sensors. Sens Actuators B: Chem 139(2):361–368
58. Ogita M, Yuasa S, Kobayashi K, Yamada Y, Nakanishi Y, Hatanaka Y (2003) Presumption and improvement for gallium oxide thin film of high temperature oxygen sensors. Appl Surf Sci 212–213:397–401
59. Balducci G, Islam MS, Kaspar J, Fornasiero P, Graziani M (2000) Bulk reduction and oxygen migration in the ceria-based oxides. Chem Mater 12(3):677–681
60. Mamontov E, Egami T, Brezny R, Koranne M, Tyagi S (2000) Lattice defects and oxygen storage capacity of nanocrystalline ceria and ceria-zirconia. J Phys Chem B 104(47):11110–11116
61. Maskell WC (1987) Inorganic solid state chemically sensitive devices: electrochemical oxygen gas sensors. J Phys E: Sci Instrum 20(10):1156
62. Dietz H (1982) Gas-diffusion-controlled solid-electrolyte oxygen sensors. Solid State Ionics 6(2):175–183
63. Gerblinger J, Lohwasser W, Lampe U, Meixner H (1995) High temperature oxygen sensor based on sputtered cerium oxide. Sens Actuators B: Chem 26(1–3):93–96
64. Beie HJ, Gnörich A (1991) Oxygen gas sensors based on CeO_2 thick and thin films. Sens Actuators B: Chem 4(3–4):393–399
65. Izu N, Itoh T, Shin W, Matsubara I, Murayama N (2007) The effect of hafnia doping on the resistance of ceria for use in resistive oxygen sensors. Sens Actuators B: Chem 123(1):407–412
66. Izu N, Oh-hori N, Itou M, Shin W, Matsubara I, Murayama N (2005) Resistive oxygen gas sensors based on Ce1-xZrxO2 nano powder prepared using new precipitation method. Sens Actuators B: Chem 108(1–2):238–243
67. Izu N, Shin W, Matsubara I, Murayama N (2006) Evaluation of response characteristics of resistive oxygen sensors based on porous cerium oxide thick film using pressure modulation method. Sens Actuators B: Chem 113(1):207–213
68. Jasinski P, Suzuki T, Anderson HU (2003) Nanocrystalline undoped ceria oxygen sensor. Sens Actuators B: Chem 95(1–3):73–77
69. Várhegyi EB, Perczel IV, Gerblinger J, Fleischer M, Meixner H, Giber J (1994) Auger and SIMS study of segregation and corrosion behaviour of some semiconducting oxide gas-sensor materials. Sens Actuators B: Chem 19(1–3):569–572
70. Agrafiotis C, Tsetsekou A, Stournaras CJ, Julbe A, Dalmazio L, Guizard C (2000) Deposition of nanophase doped-ceria systems on ceramic honeycombs for automotive catalytic applications. Solid State Ionics 136–137(1301–1306):37
71. Bera D, Kuchibhatla SVNT, Azad S, Saraf L, Wang CM, Shutthanandan V, Nachimuthu P, McCready DE, Engelhard MH, Marina OA, Baer DR, Seal S, Thevuthasan S (2008) Growth and characterization of highly oriented gadolinia-doped ceria (111) thin films on zirconia (111)/sapphire (0001) substrates. Thin Solid Films 516(18):6088–6094

72. Fergus JW (2006) Electrolytes for solid oxide fuel cells. J Power Sources 162(1):30–40
73. Kharton VV, Figueiredo FM, Navarro L, Naumovich EN, Kovalevsky AV, Yaremchenko AA, Viskup AP, Carneiro A, Marques FMB, Frade JR (2001) Ceria-based materials for solid oxide fuel cells. J Mater Sci 36(5):1105–1117
74. Yu ZQ, Kuchibhatla SVNT, Saraf LV, Marina OA, Wang CM, Engelhard MH, Shutthanandan V, Nachimuthu P, Thevuthasan S (2008) Conductivity of oriented samaria-doped ceria thin films grown by oxygen-plasma-assisted molecular beam epitaxy. Electrochem Solid-State Lett 11(5):B76–B78
75. Bellino MG, Lamas DG, Walsöe de Reca NE (2006) Enhanced ionic conductivity in nanostructured, heavily doped ceria ceramics. Adv Funct Mater 16(1):107–113
76. Esposito V, Traversa E (2008) Design of electroceramics for solid oxides fuel cell applications: playing with ceria. J Am Ceram Soc 91(4):1037–1051
77. Omar S, Wachsman ED, Nino JC (2008) Higher conductivity Sm3+ and Nd3+ co-doped ceria-based electrolyte materials. Solid State Ionics 178(37–38):1890–1897
78. Sanna S, Esposito V, Pergolesi D, Orsini A, Tebano A, Licoccia S, Balestrino G, Traversa E (2009) Fabrication and electrochemical properties of epitaxial samarium-doped ceria films on SrTiO3-buffered MgO substrates. Adv Funct Mater 19(11):1713–1719
79. Zha S, Xia C, Meng G (2003) Effect of Gd(Sm) doping on properties of ceria electrolyte for solid oxide fuel cells. J Power Sources 115(1):44–48
80. Kilner JA (2008) Ionic conductors: feel the strain. Nat Mater 7(11):838–839
81. Kilner JA, Brook RJ (1982) A study of oxygen ion conductivity in doped non-stoichiometric oxides. Solid State Ionics 6(3):237–252
82. Kim DJ (1989) Lattice parameters, ionic conductivities, and solubility limits in fluorite-structure MO2 oxide [M = Hf4 + , Zr4 + , Ce4 + , Th4 + , U4 +] solid solutions. J Am Ceram Soc 72(8):1415–1421
83. Andersson DA, Simak SI, Skorodumova NV, Abrikosov IA, Johansson B (2006) Optimization of ionic conductivity in doped ceria. Proc Nat Acad Sci USA 103(10):3518–3521
84. Gerhardt-Anderson R, Nowick AS (1981) Ionic conductivity of CeO2 with trivalent dopants of different ionic radii. Solid State Ionics 5:547–550
85. Hayashi H, Sagawa R, Inaba H, Kawamura K (2000) Molecular dynamics calculations on ceria-based solid electrolytes with different radius dopants. Solid State Ionics 131(3–4):281–290
86. Minervini L, Zacate MO, Grimes RW (1999) Defect cluster formation in M2O3-doped CeO2. Solid State Ionics 116(3–4):339–349
87. Eguchi K (1997) Ceramic materials containing rare earth oxides for solid oxide fuel cell. J Alloy Compd 250(1–2):486–491
88. Fu Y-P, Wen S-B, Lu C-H (2008) Preparation and characterization of samaria-doped ceria electrolyte materials for solid oxide fuel cells. J Am Ceram Soc 91(1):127–131
89. Jung G-B, Huang T-J, Chang C-L (2002) Effect of temperature and dopant concentration on the conductivity of samaria-doped ceria electrolyte. J Solid State Electrochem 6(4):225–230
90. Mansilla C, Holgado JP, Espinós JP, González-Elipe AR, Yubero F (2007) Microstructure and transport properties of ceria and samaria doped ceria thin films prepared by EBE-IBAD. Surf Coat Technol 202(4–7):1256–1261
91. Yahiro H, Eguchi Y, Eguchi K, Arai H (1988) Oxygen ion conductivity of the ceria-samarium oxide system with fluorite structure. J Appl Electrochem 18(4):527–531
92. Zhan Z, Wen T-L, Tu H, Lu Z-Y (2001) AC impedance investigation of samarium-doped ceria. J Electrochem Soc 148(5):A427–A432
93. Yu ZQ, Kuchibhatla SVNT, Engelhard MH, Shutthanandan V, Wang CM, Nachimuthu P, Marina OA, Saraf LV, Thevuthasan S, Seal S (2008) Growth and structure of epitaxial $Ce_{0.8}Sm_{0.2}O_{1.9}$ by oxygen-plasma-assisted molecular beam epitaxy. J Cryst Growth 310(10):2450–2456

94. Henderson MA, Perkins CL, Engelhard MH, Thevuthasan S, Peden CHF (2003) Redox properties of water on the oxidized and reduced surfaces of CeO2(1 1 1). Surf Sci 526(1–2):1–18

95. Kim YJ, Thevuthasan S, Shutthananadan V, Perkins CL, McCready DE, Herman GS, Gao Y, Tran TT, Chambers SA, Peden CHF (2002) Growth and structure of epitaxial Ce1-xZrxO2 thin films on yttria-stabilized zirconia (111). J Electron Spectrosc Relat Phenom 126(1–3):177–190

96. Thevuthasan S, Peden CHF, Engelhard MH, Baer DR, Herman GS, Jiang W, Liang Y, Weber WJ (1999) The ion beam materials analysis laboratory at the environmental molecular sciences laboratory. Nucl Instrum Methods Phys Res, Sect A 420(1–2):81–89

97. Mayer M (1997) SIMNRA user's guide. Tech Rep IPP 9/113

98. Mayer M (1999) SIMNRA, a simulation program for the analysis of NRA, RBS and ERDA. In: Proceedings of the 15th international conference on the application of accelerators in research and industry

99. Sanghavi RP, Nandasiri M, Kuchibhatla S, Nachimuthu P, Engelhard MH, Shutthanandan V, Jiang W, Thevuthasan S, Kayani A, Prasad S (2009) Performance evaluation of an oxygen sensor as a function of the samaria doped ceria film thickness. MRS Online Proc Libr 1209:P03–07

100. Moos R, Menesklou W, Schreiner H-J, Härdtl KH (2000) Materials for temperature independent resistive oxygen sensors for combustion exhaust gas control. Sens Actuators B: Chem 67(1–2):178–183

101. Dolbec R, El Khakani MA, Serventi AM, Trudeau M, Saint-Jacques RG (2002) Microstructure and physical properties of nanostructured tin oxide thin films grown by means of pulsed laser deposition. Thin Solid Films 419(1–2):230–236

102. Kim S, Oliver M (2010) Structural, electrical, and optical properties of reactively sputtered SnO2 thin films. Met Mater Int 16(3), 441–446

103. Kim TW, Lee DU, Lee JH, Choo DC, Jung M, Yoon YS (2001) Structural, electrical, and optical properties of SnO2 nanocrystalline thin films grown on p-InSb (111) substrates. J Appl Phys 90(1):175–180

104. Batzill M, Diebold U (2005) The surface and materials science of tin oxide. Prog Surf Sci 79(2–4):47–154

105. Saukko S, Lassi U, Lantto V, Kroneld M, Novikov S, Kuivalainen P, Rantala TT, Mizsei J (2005) Experimental studies of O2 − SnO2 surface interaction using powder, thick films and monocrystalline thin films. Thin Solid Films 490(1):48–53

106. Batzill M, Diebold U (2007) Surface studies of gas sensing metal oxides. Phys Chem Chem Phys 9(19):2307–2318

107. Madou MJ, Morrison SR (1989) Chemical sensing with solid state devices. Academic, San Diego

108. Schierbaum KD, Wiemhöfer HD, Göpel W (1988) Defect structure and sensing mechanism of SnO_2 gas sensors: comparative electrical and spectroscopic studies. Solid State Ionics 28–30(Part 2):1631–1636

109. Seal S, Shukla S (2002) Nanocrystalline SnO gas sensors in view of surface reactions and modifications. JOM 54(9):35–38

110. Yamazoe N, Fuchigami J, Kishikawa M, Seiyama T (1979) Interactions of tin oxide surface with O2, H2O and H2. Surf Sci 86:335–344

111. Yamazoe N, Sakai G, Shimanoe K (2003) Oxide semiconductor gas sensors. Catal Surv Asia 7(1):63–75

112. Chang S-C (1980) Oxygen chemisorption on tin oxide: correlation between electrical conductivity and EPR measurements. J Vac Sci Technol 17(1):366–369

113. Barsan N, Weimar U (2001) Conduction model of metal oxide gas sensors. J Electroceram 7(3):143–167

114. Korotcenkov G (2005) Gas response control through structural and chemical modification of metal oxide films: state of the art and approaches. Sens Actuators B: Chem 107(1):209–232

115. Göpel W (1994) New materials and transducers for chemical sensors. Sens Actuators B: Chem 18(1–3):1–21
116. Morrison SR (1981) Semiconductor gas sensors. Sens Actuators 2:329–341
117. Gardner JW (1990) A non-linear diffusion-reaction model of electrical conduction in semiconductor gas sensors. Sens Actuators B: Chem 1(1–6):166–170
118. Geistlinger H (1993) Electron theory of thin-film gas sensors. Sens Actuators B: Chem 17(1):47–60
119. McAleer JF, Moseley PT, Norris JOW, Williams DE (1987) Tin dioxide gas sensors. Part 1. Aspects of the surface chemistry revealed by electrical conductance variations. J Chem Soc, Faraday Trans 1: Phys Chem Condens Phases 83(4):1323–1346
120. Morrison SR (1987) Mechanism of semiconductor gas sensor operation. Sens Actuators 11(3):283–287
121. Srivastava RK, Lal P, Dwivedi R, Srivastava SK (1994) Sensing mechanism in tin oxide-based thick-film gas sensors. Sens Actuators B: Chem 21(3):213–218
122. Sberveglieri G (1995) Recent developments in semiconducting thin-film gas sensors. Sens Actuators B: Chem 23(2–3):103–109
123. Korotcenkov G, Cho BK, Tolstoy V (2010) SnO2-based thin film gas sensors with functionalized surface. Adv Mater Res 93–94:145–148
124. Kroneld M, Novikov S, Saukko S, Kuivalainen P, Kostamo P, Lantto V (2006) Gas sensing properties of SnO2 thin films grown by MBE. Sens Actuators B: Chem 118(1–2):110–114
125. Feng X, Ma J, Yang F, Ji F, Luan C (2008) Preparation and characterization of single crystalline SnO2 films deposited on α-Al2O3 (0001) by MOCVD. Mater Lett 62(12–13):1809–1811
126. Semancik S, Cavicchi RE (1991) The growth of thin, epitaxial SnO2 films for gas sensing applications. Thin Solid Films 206(1–2):81–87
127. Lee DS, Rue GH, Huh JS, Choi SD, Lee DD (2001) Sensing characteristics of epitaxially-grown tin oxide gas sensor on sapphire substrate. Sens Actuators B: Chem 77(1–2):90–94
128. Ohgaki T, Matsuoka R, Watanabe K, Matsumoto K, Adachi Y, Sakaguchi I, Hishita S, Ohashi N, Haneda H (2010) Synthesizing SnO$_2$ thin films and characterizing sensing performances. Sens Actuators B: Chem 150(1), 99–104
129. Kim DH, Kim W-S, Lee SB, Hong S-H (2010) Gas sensing properties in epitaxial SnO$_2$ films grown on TiO$_2$ single crystals with various orientations. Sens Actuators B: Chem 147(2):653–659
130. Rosental A, Tarre A, Gerst A, Sundqvist J, Hårsta A, Aidla A, Aarik J, Sammelselg V, Uustare T (2003) Gas sensing properties of epitaxial SnO$_2$ thin films prepared by atomic layer deposition. Sens Actuators B: Chem 93(1–3):552–555
131. Palgrave RG, Bourlange A, Payne DJ, Foord JS, Egdell RG (2009) Interfacial diffusion during growth of SnO2(110) on TiO2(110) by oxygen plasma assisted molecular beam epitaxy. Cryst Growth Des 9(4):1793–1797
132. White ME, Tsai MY, Wu F, Speck JS (2008) Plasma-assisted molecular beam epitaxy and characterization of SnO2(101) on r-plane sapphire. J Vac Sci Technol A: Vac, Surf, Films 26(5):1300–1307
133. Tsai MY, White ME, Speck JS (2008) Plasma-assisted molecular beam epitaxy of SnO$_2$ on TiO$_2$. J Cryst Growth 310(18):4256–4261
134. Batzill M, Burst JM, Diebold U (2005) Pure and cobalt-doped SnO$_2$(101) films grown by molecular beam epitaxy on Al$_2$O$_3$. Thin Solid Films 484(1–2):132–139
135. Hishita S, Janecek P, Haneda H (2010) Epitaxial growth of tin oxide film on TiO2(1 1 0) using molecular beam epitaxy. J Cryst Growth 312(20):3046–3049
136. Chen JS, Li HL, Huang JL (2002) Structural and CO sensing characteristics of Ti-added SnO$_2$ thin films. Appl Surf Sci 187(3–4):305–312
137. Zakrzewska K, Radecka M (2007) TiO2-SnO2 system for gas sensing-Photodegradation of organic contaminants. Thin Solid Films 515(23):8332–8338
138. Hishita S, Janecek P, Haneda H (2009) Epitaxial growth of SnO$_2$ film on Sn-doped TiO$_2$(110). Vacuum 84(5):597–601

139. Winter R, Scharnagl K, Fuchs A, Doll T, Eisele I (2000) Molecular beam evaporation-grown indium oxide and indium aluminium films for low-temperature gas sensors. Sens Actuators B: Chem 66(1–3):85–87
140. Bourlange A, Payne DJ, Palgrave RG, Foord JS, Egdell RG, Jacobs RMJ, Schertel A, Hutchison JL, Dobson PJ (2009) Investigation of the growth of In_2O_3 on Y-stabilized ZrO2(100) by oxygen plasma assisted molecular beam epitaxy. Thin Solid Films 517(15), 4286–4294
141. Mei ZX, Wang Y, Du XL, Zeng ZQ, Ying MJ, Zheng H, Jia JF, Xue QK, Zhang Z (2006) Growth of In_2O_3 single-crystalline film on sapphire(0 0 0 1) substrate by molecular beam epitaxy. J Cryst Growth 289(2):686–689
142. Taga N, Maekawa M, Shigesato Y, Yasui I, Haynes TE (1998) Deposition of hetero-epitaxial In_2O_3 thin films by molecular beam epitaxy. Jpn J Appl Phys 37(12A), 6524–6529
143. Chen Y, Bagnall DM, Zhu Z, Sekiuchi T, Park KT, Hiraga K, Yao T, Koyama S, Shen MY, Goto T (1997) Growth of ZnO single crystal thin films on c-plane (0 0 0 1) sapphire by plasma enhanced molecular beam epitaxy. J Cryst Growth 181(1–2):165–169
144. Fons P, Iwata K, Niki S, Yamada A, Matsubara K (1999) Growth of high-quality epitaxial ZnO films on α-Al_2O_3. J Cryst Growth 201–202:627–632
145. Heo YW, Ip K, Pearton SJ, Norton DP, Budai JD (2006) Growth of ZnO thin films on c-plane Al2O3 by molecular beam epitaxy using ozone as an oxygen source. Appl Surf Sci 252(20):7442–7448
146. Jian-Feng Y, You-Ming L, Hong-Wei L, Yi-Chun L, Bing-Hui L, Xi-Wu F, Jun-Ming Z (2005) Growth and properties of ZnO nanotubes grown on Si(1 1 1) substrate by plasma-assisted molecular beam epitaxy. J Cryst Growth 280(1–2):206–211
147. Liang HW, Lu YM, Shen DZ, Li BH, Zhang ZZ, Shan CX, Zhang JY, Fan XW, Du GT (2006) Growth of vertically aligned single crystal ZnO nanotubes by plasma-molecular beam epitaxy. Solid State Commun 137(4):182–186
148. Kang BS, Heo YW, Tien LC, Norton DP, Ren F, Gila BP, Pearton SJ (2005) Hydrogen and ozone gas sensing using multiple ZnO nanorods. Appl Phys A Mater Sci Process 80(5):1029–1032
149. Tien LC, Norton DP, Pearton SJ, Wang HT, Ren F (2007) Nucleation control for ZnO nanorods grown by catalyst-driven molecular beam epitaxy. Appl Surf Sci 253(10): 4620–4625
150. Greenwood OD, Moulzolf SC, Blau PJ, Lad RJ (1999) The influence of microstructure on tribological properties of WO3 thin films. Wear 232(1):84–90
151. Lad RJ (2002) Heteroepitaxy of tungsten oxide films on sapphire and silicon for chemiresistive sensor applications. Proc IEEE, Sens 1(393–397):44
152. LeGore LJ, Greenwood OD, Paulus JW, Frankel DJ, Lad RJ (1997) Controlled growth of WO3 films. AVS, Philadelphia, pp 1223–1227
153. Gao W, Klie R, Altman EI (2005) Growth of anatase films on vicinal and flat LaAlO3 (110) substrates by oxygen plasma assisted molecular beam epitaxy. Thin Solid Films 485(1–2):115–125
154. Gao Y, Chambers SA (1996) MBE growth and characterization of epitaxial TiO2 and Nb-doped TiO2 films. Mater Lett 26(4–5):217–221
155. Shao R, Wang C, McCready DE, Droubay TC, Chambers SA (2007) Growth and structure of MBE grown TiO2 anatase films with rutile nano-crystallites. Surf Sci 601(6):1582–1589
156. Weng X, Fisher P, Skowronski M, Salvador PA, Maksimov O (2008) Structural characterization of TiO_2 films grown on $LaAlO_3$ and $SrTiO_3$ substrates using reactive molecular beam epitaxy. J Cryst Growth 310(3):545–550

Chapter 7
Atomic Layer Deposition for Metal Oxide Nanomaterials

Xiaohua Du

Abstract Solid state gas sensors based on semiconducting metal oxides have been widely investigated and utilized in environmental monitoring, chemical process controls and personal safety. In recent years, one dimensional nanostructures, such as nanowires, nanorods, nanotubes and nanobelts, have attracted much attention due to their great potential application in gas sensing, and for overcoming fundamental limitations due to their ultra high surface-to-volume ratio. A variety of methods have been developed to fabricate these nanostructures. The nanostructure based gas sensors demonstrated excellent response and recovery characteristics. However, the developed methods are not convenient for mass production and improvements on sensitivity, selectivity and long term stability are still needed. Atomic layer deposition (ALD) is a film deposition technique based on the sequential use of self-terminating surface reactions. Due to the unique nature of the reaction process, ALD becomes an ideal deposition technique to form atomic thin films and nanolaminate structures. ALD is finding ever more applications for emerging nanodevices. The potential to control thickness at the sub-nm level, and the ability to deposit thin films over highly corrugated substrates with high aspect ratio topography makes ALD of great interest in fabrication of one dimensional nanomaterial. Utilizing fabrication through nanotechnology, ALD has found new opportunities in gas sensors based on metal oxide semiconductors. In this chapter, the general characteristics of atomic layer deposition, the sensing performance enhancements by quasi-1 dimensional nanostructures and nanomaterials, the method to fabricate such nanostructures and the recent exploration of ALD in gas sensing studies are reviewed.

X. Du (✉)
HGST, a Western Digital Company, 5601 Great Oaks Pkwy,
San Jose, CA 95119 USA
e-mail: xiaohua.du@hgst.com

M. A. Carpenter et al. (eds.), *Metal Oxide Nanomaterials for Chemical Sensors*,
Integrated Analytical Systems, DOI: 10.1007/978-1-4614-5395-6_7,
© Springer Science+Business Media New York 2013

7.1 Functional Nanomaterials for Chemical Sensing Technologies

Among all the solid state gas sensor systems, compact metal oxide solid state (SS) chemical sensors are the most promising and have been widely used in the detection of toxic pollutants (CO, H_2S, NO_x, SO_2 etc.) and inflammable gases (H_2, CH_4, hydrocarbons etc.). However, disadvantages such as lack of reproducibility, poor selectivity for specific gases and insufficient sensitivity are frequently observed [1–3]. Improving the performance of these gas sensors is difficult, and requires a better understanding of the sensing mechanism, an active research field in of itself [4].

The most commonly accepted operation principle for nanostructure thin or thick film n-type semiconductor gas sensors involves the changing of carrier density in a space-charge region and is based on the variation in the potential barrier height at the grain boundary, which is induced by the surrounding gases [5–11]. Two main sensing mechanisms are proposed for the semiconductor metal oxide gas sensors based on the ionosorption model and oxygen-vacancy model, as shown in Fig. 7.1. A space-charge region is formed under the surface of either the ionosorbed oxygen species in the ionosorption model or the ionized oxygen vacancies in the oxygen-vacancy model. The depth of the space-charge region is called the Debye length and is typically only several nanometers [12, 13].

It is known that the sensitivity of semiconducting nanomaterials is determined by two basic factors, namely, the ratio of the surface area (S) to volume (V) and the ratio of the Debye length (DL) to the effective radius of conducting channels (R). All other conditions being the same, the larger the S/V and DL/R ratios the higher the material sensitivity. This expectation is supported by gas sensor sensitivity studies on nanoparticles [14–16], as shown in Fig. 7.2. Numerous studies of thin nanostructure SnO_2 films showed that the sensitivity of samples with respect to detected gases dramatically increased as the grain size decreased to several nanometers [17–20]. Thus, the importance of the Debye length has been established for the sensitivity of sintered nanoparticles and nanocrystalline-grained thin films. Nowadays, the technological challenge moves to the fabrication of materials with nanocrystals. However, efforts based on solution methods (such as Sol–gel technique) can be challenged due to the difficulty in narrowly controlling the finite grain sizes comparable to Debye length. Additionally, further enhancement of both sensitivity and response in gas detection is cumbered with the limited surface-to-volume ratio (SVR) of nanocrystalline or polycrystalline based film.

To overcome the limitations from low SVR films, nanomaterials and nanostructures are developed for the purpose of high-performance chemical sensing. Quasi-one-dimensional nanomaterials have attracted much attention and have been deemed to be one of the best candidates for realizing ultrasensitive solid state sensors, due to its large SVR and Debye length comparable dimensions. In recent years, superior sensor characteristics, such as high sensitivity, high response speed and quick recovery, have been observed for nanowires [21–23], nanobelts [24],

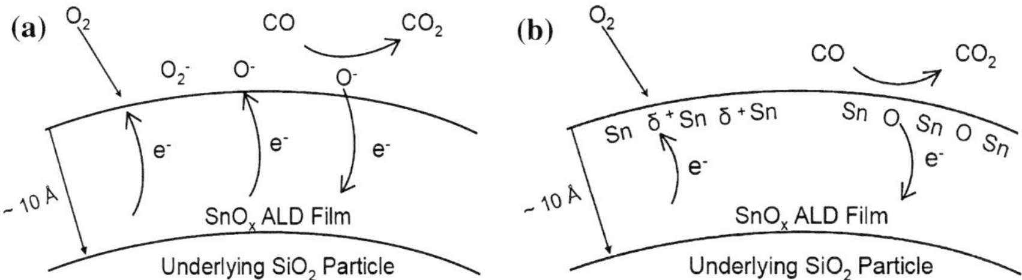

Fig. 7.1 Schematics of sensing mechanism of n-type metal oxide materials: **a** ionosorption model and **b** oxygen-vacancy model, illustrated by interactions between O_2 and CO gases on surface of SnO_2 films/particles. Reprinted from Ref. 68, with permission from Elsevier, Copyright (2008)

Fig. 7.2 The effect of SnO_2 particle/crystallite size D on the sensor response [expressed as (Rgas/Rair)] upon exposure to **a** 800 ppm H2 or **b** CO in air at an operating temperature of 573 K. Reprinted from Ref. 14, with permission from Elsevier, Copyright (1991)

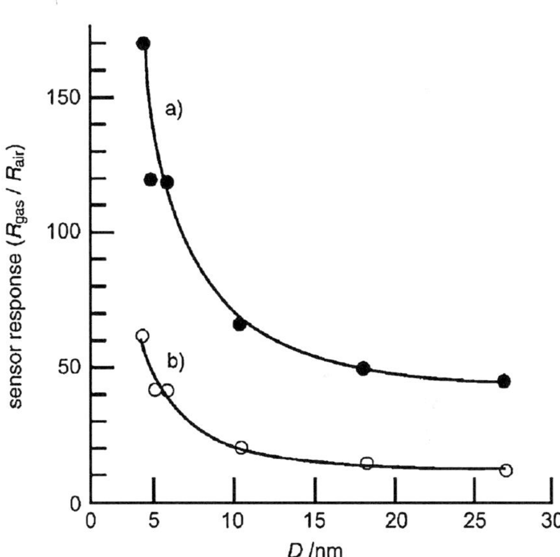

carbon nanotubes (CNTs) [25, 26] and core/shell structures built on CNTs [27]. Pioneered by Lieber et al. nanowire sensors made from materials such as metal oxides [28–30], silicon [31–33], metals [34–40] and conducting polymers [41, 42] have been intensively investigated and used as chemical and biosensors. Since then the number of reports on gas sensors based on other 1D metal oxide nanostructures have been growing exponentially every year.

Nanomaterial-based sensors are normally fabricated from the "top-down" or "bottom-up" approaches. However, the fabrication techniques used are not convenient for mass production. First of all, it is a high cost process to fabricate an individual nanowire/nanobelt; it is also difficult to manipulate an individual nanowire/nanobelt to form FET structures, which can be used to manipulate the sensing properties. Second, improvements are still to be made regarding the reproducibility of the chemical composition of the nanomaterial surface. This is a critical parameter that controls the reproducibility of the gas sensor performances. Besides the above limitations, the reproducibility and repeatability are not guaranteed when using nanomaterial arrays/films to fabricate sophisticated sensor structures and often require extensive calibration procedures.

Thin film deposition techniques provide another suitable way to confine the grain size. Different methods have been employed for the production of thin films of semiconducting oxides with the aim to obtain nanostructure, stable, cheap, low power consuming, and reliable gas sensors. To take advantage of the high sensitivity, metal oxide films with exact nanometer film thicknesses would be desirable for experimental investigations of metal oxide gas sensors. Techniques that can be performed with precise film thicknesses will be able to overcome the variability in sensitivity produced by the finite Debye length.

Techniques and methods with reduced fabrication complexity and production cost, are still highly desirable and demanded in both fundamental sensing studies and mass production of gas detectors. The techniques emerging shall have the capability to straightforwardly produce 1D metal oxide nanostructures with well controlled diameter, length, composition, and morphology, and be compatible to a desirable surface structure made by modern techniques. Atomic layer deposition is such a deposition technique. Ultrathin and conformal metal oxide films with thicknesses on the order of the Debye length can now be precisely deposited using the enabling method of atomic layer deposition (ALD) techniques.

7.2 Exploration of ALD in Application of Solid State Gas Sensors

7.2.1 Atomic Layer Deposition, an Enabling Tool for Nanotechnology and Nanofabrication

Atomic layer deposition (ALD) is a recently developed thin film deposition technique. ALD occurs through alternating reactions between gaseous precursor molecules and the surface active sites on substrate [43–49]. In general, a typical ALD cycle is composed of four essential steps, which is illustrated in Fig. 7.3. (1) The first precursor is introduced into the reactor and allowed to chemisorb on the substrate until the surface is saturated, leaving a new type of species on the surface. (2) The excess precursor and any by-product are then purged away from the deposition system. (3) The second precursor is then delivered to the reactor and allowed to react with the adsorbed surface species that are left from the 1st step, regenerating the same surface species as on the initiation surface. (4) Excess precursor and reaction by-products are again removed from the system. The result is a surface saturated with up to a monolayer of the desired compound. Sequential repetition of the ALD cycle results in layer-by-layer growth with precise thickness control. As an example

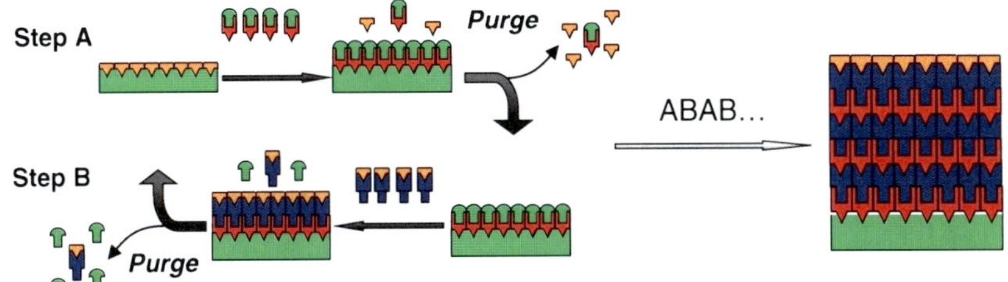

Fig. 7.3 Cartoon illustration of a typical ALD process. ALD deposition process is composed of 4 essential steps and conducted on a substrate functionalized with reactive surface species. **a** The first precursor is introduced into the reactor and allowed to chemisorb on the substrate until the surface is saturated, leave a new type of species on the surface. **b** The excess precursor and any by-product are then purged away from the deposition system. **c** The second precursor is then delivered to the reactor and allowed to react with the adsorbed surface species that are left from the 1st step, regenerating the same surface species as on the initiation surface. **d** Excess precursor and reaction by-products are again removed from the system. The result is a surface saturated with up to a monolayer of the desired compound. Sequential repetition of the ALD cycle results in layer-by-layer growth with precise thickness control

of the ALD process, consider the following binary reaction sequence for SnO_2 in which the surface species are designated by asterisks:

$$(A) \; Sn - OH^* + SnCl_4(g) \rightarrow Sn - O - SnCl_3^* + HCl(g)$$
$$(B) \; Sn - Cl^* + H_2O(g) \rightarrow Sn - OH^* + HCl(g)$$

The substrate surface is initially covered with hydroxyl (OH) groups. In reaction (A), tin tetrachloride ($SnCl_4$) reacts with the hydroxyl groups to deposit a monolayer of tin-chlorine groups producing hydrogen chloride (HCl) as a by-product. Because the chlorine-terminated surface is inert to $SnCl_4$, further exposure to $SnCl_4$ yields no additional growth beyond one monolayer. In reaction (B), this new surface is exposed to water, regenerating the initial hydroxyl-terminated surface and again releasing hydrogen chloride The net effect of one (A)–(B) cycle is to deposit one monolayer of SnO_2 on the surface. The overall reaction for the ALD process is:

$$SnCl_4(g) + 2 \, H_2O \, (g) \rightarrow SnO_2(s) + 4HCl \, (g)$$

Due to the unique nature of this process, ALD possesses many intrinsic advantages over traditional deposition processes such as chemical vapor deposition (CVD) and physical vapor deposition (PVD). With an optimized ALD process, excellent conformality, good areal uniformity, precise thickness controllability, excellent film-adhesion and repeatability have been observed. In addition, the alternating reaction strategy eliminates the "line of site" or "constant exposure" requirements that limit conventional methods such as CVD and PVD, resulting pure, dense, smooth and highly conformal films. These attributes make ALD an ideal method for applying precise and conformal coatings over nanomaterials with sophisticated geometry, offering perfect step-coverage on high aspect ratios features [50, 51].

Besides the opportunity for atomic level thickness control and perfect step-coverage, the step by step growth mode also makes atomic level composition control feasible. Future advances in ALD may enable the nanoscale manipulation and engineering of desired film properties at the interface and in the thin film through deposition of nanolaminates or mixed component alloys. Several early reports on fabrication of nanolaminates through ALD were reviewed in Ref. [52].

Atomic layer deposition was originated in 1977 for fabrication of electroluminescent flat panel displays. In mid 1990s, its unique features make ALD attractive for integrated circuit and data storage applications when the transistor dimension was shrunken to sub micron scale. ALD is now being exploited for fabrication of high-k dielectric and metal films, seed films, diffusion barriers, etch-stop layers and a variety of gap layers for semiconductor, magnetic head, non-volatile memory, organic light emissive diode (OLED) and solar applications.

ALD is also emerging as a key enabling technology for applications outside of the semiconductor industry. Other applications include microelectromechanical systems fabrication, organic light emitting diodes displays, catalysts and electronic and optoelectronic materials. Lower dimensional nanomaterials, such as nanowires, nanobelts, nanotubes and nanodots, are emerging as basic building blocks for the realization of nanotechnology [53–55]. The use of ALD for the fabrication of nanolaminates and nanomaterials, or use these nanomaterial as templates, would be an important contribution to nanotechnology, as well as the gas sensing field for fundamental study and fabrication of high performance gas sensing devices.

7.2.2 Template Assisted Fabrication of Nanostructures using ALD

Two trends exist in the exploration of ALD in gas sensing field. One is the fabrication of metal oxide nanotubes by templates using ALD. In this approach, positive templates and negative templates are used as scaffolds. When the positive type is used, the conformal metal oxide film is deposited onto the template's outer surface; subsequently, they are removed by chemical etching and/or pyrolysis. The remaining shell forms the metal oxide nanotubes. The templates are typically made from nanorods, nanowires, block copolymer nanorods, poly (vinyl alcohol) fibers (cellulose). The template's composition determines the methodology of its removal, which includes chemical etching, thermal decomposition, or dissolution. Porous membranes with monodisperse cylindrical pores, such as aluminum oxides (AAOs) and track etched polymers [mostly, polycarbonate (PC)], are used as negative templates. Oxide materials are deposited on the surfaces of their inside pores. Once metal oxide is deposited onto the porous template, the template and unnecessary parts of the nanostructure are removed by mechanical polishing or chemical treatment to obtain single strands of oxide nanotubes.

Combining the positive template strategy with ALD, the formation of oxide nanotubes with controllable wall thickness has recently been reported by many researchers. In an early study [56], an easily accessible polycarbonate (PC) filter was

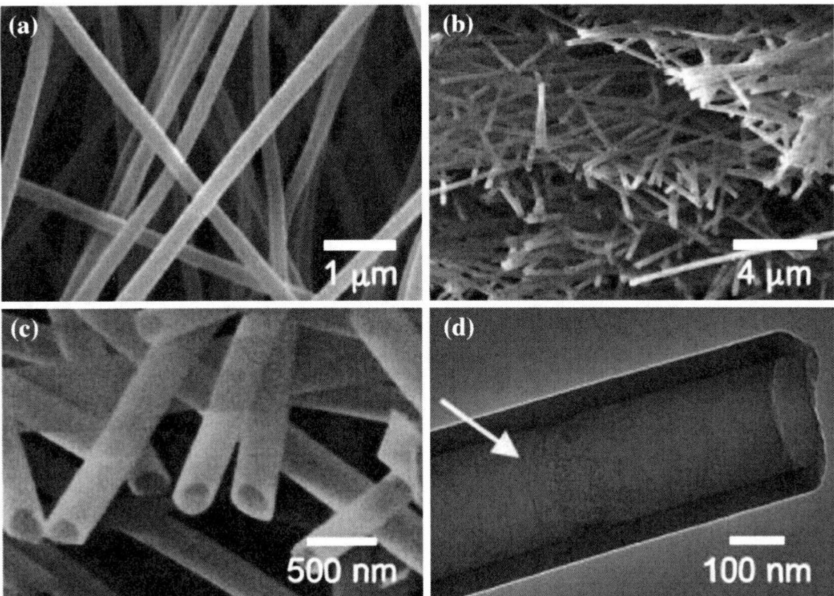

Fig. 7.4 Representative SEM images of positive templates and alumina microtubes fabricated by ALD. **a** Electrospun PVA fibers, **b, c** Al$_2$O$_3$ microtube replicas the fibers by wrapping on the outer surface through ALD, **d** TEM image of a single Al$_2$O$_3$ microtube, the arrow identifies a peripheral circle formed near an apparent neck in the fiber template. Reprinted from Ref. 58, with permission from Elsevier, Copyright (2007)

used as a template for making nanotubes by ALD. As shown in Fig. 7.4, Al$_2$O$_3$ nanotubes were prepared by ALD with trimethylaluminum and water as source materials after chemical etching, thermal decomposition, or dissolution of the organic cores [57–59]. TiO$_2$ and ZrO$_2$ nanotubes with precisely controlled thickness were synthesized by ALD on the inner wall of PC filter nanoholes, followed by subsequent etching by chloroform. George and co-workers reported the ALD process to conformally and completely coat inside nanopores with a high aspect ratio [60]. Independently, Sander et al. conducted ALD of TiO$_2$ on the highly ordered AAO as a template and obtained TiO$_2$ nanotubes [61]. Figure 7.5a illustrates the process adopted to fabricate the titania nanotubes and Fig 7.5b shows TEM images of individual TiO$_2$ nanotubes in various orientations. Gu et al. fabricated highly ordered zirconia and hafnia nanotubes by atomic layer deposition within the anodic alumina oxide (AAO) template. The diameters of the AAO pores are in the range of 200–300 nm with a thickness of 60 um. The results indicated that the free-standing nanotubes were uniformly grown through the entire template thickness. The ALD process conformally replicated the AAO template dimensions, and the wall thickness of the resultant nanotube was precisely controlled by the number of ALD cycles [62].

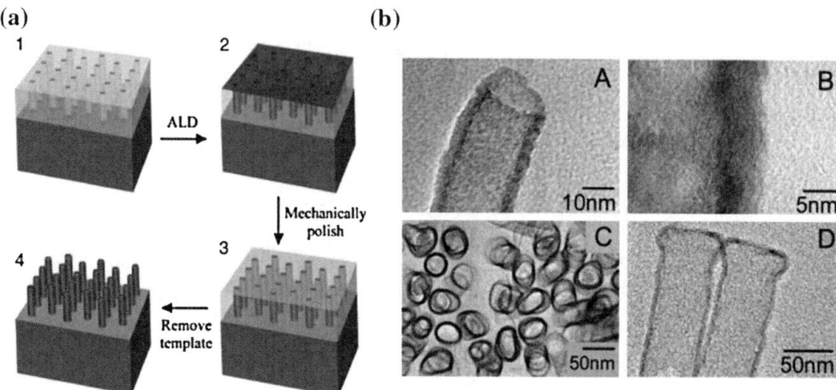

Fig. 7.5 Process schematic for fabrication TiO$_2$ nanotubes and TEM images of such. **a** Schematic of process to create TiO$_2$ nanotube arrays on substrates through etchable negative templates. *1* Nanoporous-alumina template created by anodization of an Al film, *2* TiO$_2$ was filled in the inside of the nanopores by ALD, *3* Top layer of TiO$_2$ was removed by mechanical polish, *4* Alumina template was chemically etched away to reveal TiO$_2$ nanotube arrays. **b** TEM images of TiO$_2$ nanotubes. *A* Open end of a single tube, *B* High magnification image of tube edge, *C* Top view of several nanotubes oriented in cross-section, *D* Closed ends of two nanotubes. Reprinted from Ref. 61, with permission from Elsevier, Copyright (2004)

The nanotube/nanorod structures grown using ALD are currently being investigated by researchers for gas sensing purposes. Kim et al. fabricated a Ru nanowire array device for sensor applications, using a similar process as in Ref. 61. Figure 7.6 shows a tilt FE-SEM image and a schematic of the nanowire array device on Ti/Si [63]. The I–V test on the nano-device confirms good electrical contact between the Ru nanowire and the Ti bottom electrode. A high sensitivity is expected for the device with an enhanced SVR. Elam et al. coated AAO membranes with Pd metal films using ALD to form a prototype gas sensor [64]. The Pd/AAO samples exhibited rapid and reversible conductivity changes upon hydrogen exposure and showed great promise as hydrogen gas sensors with fast response and high sensitivity. Willinger and co-workers applied a homogeneous coating of vanadium oxide on carbon nanotubes using an ALD process. The high surface area hybrid materials have an unprecedented quality since the ALD technique permits the coating of the inner and outer surface of the CNTs with a highly conformal film of controllable thickness. The ALD-coated tubes were tested as the active component in gas sensing devices. They show electric responses that are directly related to the peculiar structure, i.e., the p-n heterojunction formed between the support and the film [65].

Kim et al. reported a process combining electrospinning and ALD to fabricate nanotubes used as sensing materials [66]. Electrospun polyacrylonitrile (PAN) nanofibers of 100–200 nm diameters are used as a template after stabilization at 250 °C. A uniform and conformal SnO$_2$ coating on the nanofiber template is achieved by ALD and the wall thickness is precisely controlled by adjusting the number of ALD cycles. Calcination at 700 °C transforms the amorphous

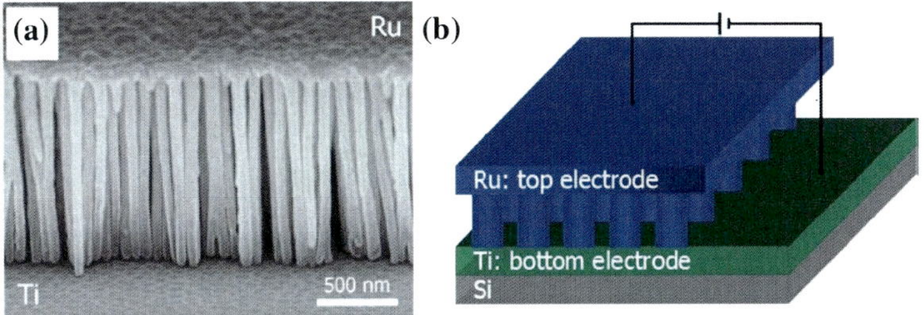

Fig. 7.6 A tilt view FE-SEM image of Ru nanowire array device on Ti/Si substrate (**6a**), and schematic of the geometry of the Ru nanowire sensor (**6b**). Reprinted from Ref. 63, with permission from Elsevier, Copyright (2008)

nanofibers into SnO_2 nanotubes composed of several nanometer-sized crystallites. The SnO_2 nanotube network sensor responds to ethanol, H_2, CO, NH_3 and NO_2 gases, and has proven to exhibit excellent sensitivity to ethanol owing to its hollow, nanostructure character. The results demonstrate that the combination of electrospinning and ALD is a very effective and promising technique to fabricate long and uniform metal oxide nanotubes for gas sensing applications.

Despite the shortcomings in the maximum attainable nanotube length, the ALD-powered directed method has much to offer in terms of simplicity of the processes involved and uniformity of the resulting nanotubes. Therefore, it is considered to be highly favorable for mass production environments where high yield is desired. Another advantage granted by ALD is the applicability of this process to a wide range of materials; Finally, ALD also enables us to make nanotubes with a multiwall structure [67].

7.2.3 Ultra-Thin Gas Sensing Films Deposited by ALD

The other approach is straight forward. Pinhole free dense metal oxide films with thicknesses comparable to the Debye length can be deposited by ALD and utilized for sensing applications. For compact thin films, the interaction with surrounding gases takes place only at the geometric surface. The best sensitivity can be obtained when the depletion layer extends through the whole film; when the depletion layer is thinner than the whole film, the film with higher resistivity shows higher response.

Du and George conducted a thorough fundamental study on the CO gas sensing processes on ultrathin SnO_2 film surface by FTIR [68]. Ultrathin tin oxide films were deposited on SiO_2 nanoparticles using ALD techniques. These SnO_2 films were then exposed to O_2 and CO gas pressure at 300 °C to measure and understand their ability to serve as CO gas sensors. The response of the ultrathin film was consistent with both the ionosorption and oxygen-vacancy models for chemi-resistant semiconductor gas sensors. As shown in Fig. 7.7a, Du's study demonstrated that O_2 presence

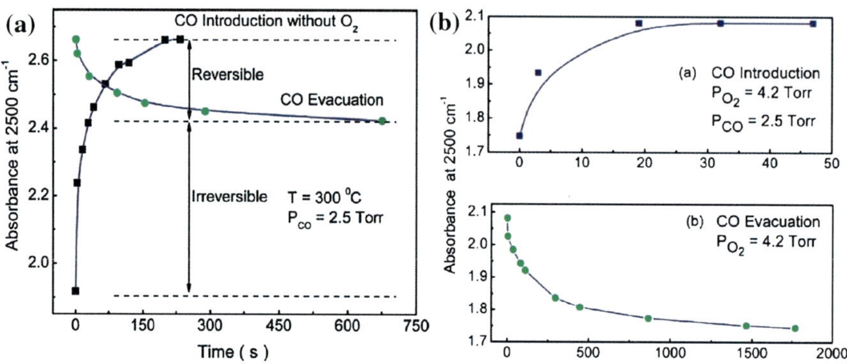

Fig. 7.7 According to Drude-Zener theory, the absorbance at a given wavelength is directly related to the electrical conductivity of materials. **a** the background infrared absorbance versus time for CO introduction without O_2 and then was evacuated from the reactor. This behavior argues that CO introduction produces both reversible and irreversible effects on the SnO_x ALD film. These vacancies produced when exposed to CO remain after CO evacuation and will not be filled until O2 is reexposed to the SnOx ALD film. **b** investigation of the transient for CO introduction and evacuation in the presence of O_2. The different time scales observed in **b** reflect the different kinetics for either the CO reaction with surface oxygen to produce CO_2 or O_2 refilling oxygen vacancies or forming ionosorption species on the surface of the SnO_x ALD film. Reprinted from Ref. 68, with permission from Elsevier, Copyright (2008)

is not necessary for the SnO_2 films to detect CO. The study indicates that CO can produce oxygen vacancies on the SnO_2 surface that can be ionized and release electrons which increase the SnO_2 film conductivity as suggested by the oxygen-vacancy model. By introduction and evacuation of CO with presence of O_2, the surface reaction of the ultrathin SnO_2 film to both CO and O_2 were investigated. It shows that the surface response to CO is one magnitude faster than that to O_2. The results were summarized in Fig. 7.7b. Du and George also conducted a systematic study regarding the thickness dependence on CO gas sensing [69]. Ultrathin tin oxide films with a wide range of thicknesses were deposited on flat hotplate templates by ALD. As seen in Fig. 7.8, the CO gas sensitivity increased for thicknesses between 15.9 and 26.2 Å and then decreased for thicknesses between 26.2 and 58.7 Å. The results were interpreted in terms of the Debye length and resistance for the SnO_2 ALD films. The gas sensor sensitivity was temperature dependent and displayed its highest sensitivity at temperatures between 250 and 325 °C. The response times of the SnO_2 ALD gas sensors were also faster at operating temperatures above 260 °C.

Besides the work done by Du and George, a comparative study was performed by Utriainen et al. [70]. The sensitive layer, 20–50 nm thick SnO_2, was deposited by ALD and PVD on top of a micro hotplate structure. The results were compared to a conventional thick film SnO_2-based sensor. The study found that the gas sensitivity characteristics between PVD and the thick film sensor are quite similar, but ALD SnO_2 possessed differences and interesting possibilities especially in terms of selectivity, baseline stability and recovery speed. Niskanen produced the

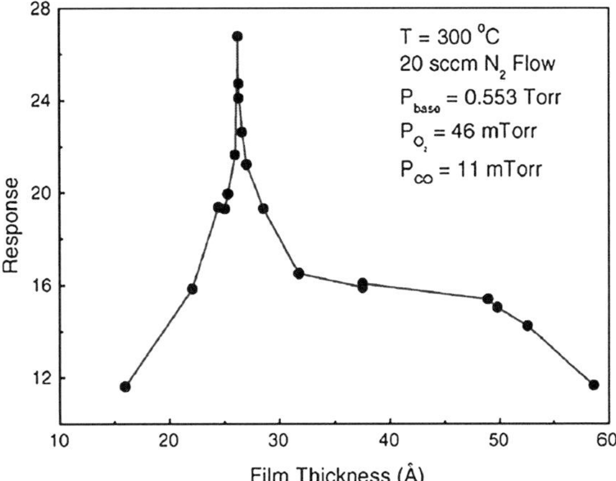

Fig. 7.8 Sensor responses to a CO pressure of 11 mTorr at 300 °C for various ALD film thicknesses grown on separate hotplate templates. Reprinted from Ref. 69, with permission from Elsevier, Copyright (2008)

gas-sensitive tin dioxide film in a micro-hotplate gas sensor [71] by ALD. The performance of the device was demonstrated using ethanol, acetone and acrylonitrile as model analytes. The fast response times and low drift rates of the output signal indicates a structurally stable tin dioxide film again reflecting the capabilities of ALD in gas sensor applications.

A recent report also confirms Du's results. Natarajan reported the influence of film thickness on the electrical and gas sensing properties of tin oxide thin films grown by the ALD technique [72]. It is found that the sheet conductance of the films increased with the thickness and was not significantly influenced by the film thickness when over 50 nm. The authors explained this effect by oxygen depletion at the film surface, and the proposed explanation was verified by subjecting the films to different lengths of post-annealing in an oxygen depleted atmosphere. It was observed that the film thickness had a significant influence on the gas sensing property of the films. The films had maximum sensitivity to ethanol when the thickness was about 40 nm. The authors proposed a model to explain this observation based on the increase in resistance due to multiple grain boundaries.

Kim studied the influence of the structure of SnO_2 films on gas sensing properties [73]. Epitaxial SnO_2 films were deposited on TiO_2 single crystals with various orientations by plasma enhanced atomic layer deposition (PEALD). All the SnO_2 films were 90 nm thick after 1,000 ALD cycles and epitaxially grown on TiO_2 substrates. Differently oriented epitaxial SnO_2 films showed different H_2 gas response and different temperature dependence of gas response. Lee et al. utilized a similar PEALD process to deposit SnO_2 thin films on Si (100) substrate [74]. It is found that the dominant oxygen species for post-annealed films were O_2^-, O^{2-} and O^- for 100, 200 and 400 cycles, respectively; and the film for 200 cycles has a good CO sensing property at the highest concentration of O^{2-} species.

Besides SnO_2 thin films, other gas sensing functional materials deposited using ALD are also being studied. Ra presented the gas sensing characteristics of the individual ZnO nanowires with single-crystalline and multiple grain boundaries (GBs)

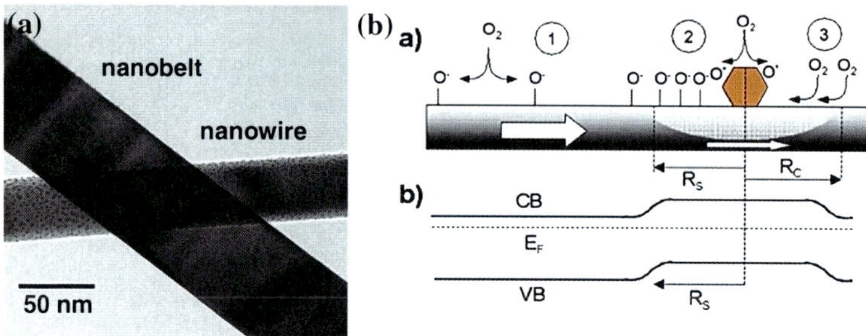

Fig. 7.9 Illustration of loading nanoslusters on the surface of sensing materials. **a** TEM images of SnO_2 nanostructures produced using vapor synthesis method, covered with vapor-deposited Pd clusters. **b** schematic depiction of the accelerated dissociation of molecular oxygen on Pd nanoparticles; as well as the band diagram of the pristine SnO_2 nanostructure and in the vicinity/ beneath a Pd nanoparticles. Reprinted from Ref. 78, with permission from Elsevier, Copyright (2005)

fabricated using bottom-up and top-down approaches [75, 76], respectively. ZnO nanowire arrays are fabricated by nanoscale spacer lithography, a top-down paradigm consisting of photolithography, ZnO atomic layer deposition, and low-damage dry etching. ZnO nanowire devices based on this technique showed good electrical transport and gas sensing properties to various concentrations of H_2 and CO.

Aronniemi investigated an iron oxide thin film grown with ALD for gas sensor applications [77]. X-ray photoelectron spectra recorded at elevated temperature imply that the surface iron is mainly in the Fe^{3+} state and that oxygen has two chemical states: one corresponding to the lattice oxygen and the other to adsorbed oxygen species. Electric conductivity has an activation energy of 0.3–0.5 eV and almost Ohmic current–voltage dependency. When exposed to O_2 and CO, a typical n-type response is observed.

7.2.4 Potential Application of ALD on Loading Catalytic Nanoparticles/Nanoclusters

A conventional method aimed at the enhancement of sensitivity and selectivity of semiconducting metal oxide nanomaterials consists of loading the nanostructures with catalytic metal nanoparticles and nanoclusters, e.g., palladium, platinum, gold and silver. Figure 7.9a shows transmission electron micrograph of tin oxide nanowires and nanobelts covered with vapor-deposited palladium clusters. Pd nanoparticles on the surface of tin dioxide nanostructures led to the increase in the sensitivity with respect to oxygen and hydrogen [78]. This is explained by the increase in the concentration of O^- anions due to the acceleration of dissociation of oxygen molecules on palladium particles, as shown in Fig. 7.9b. In turn, the size of dopant nanoparticles also affects the sensitivity of nanomaterials with respect to

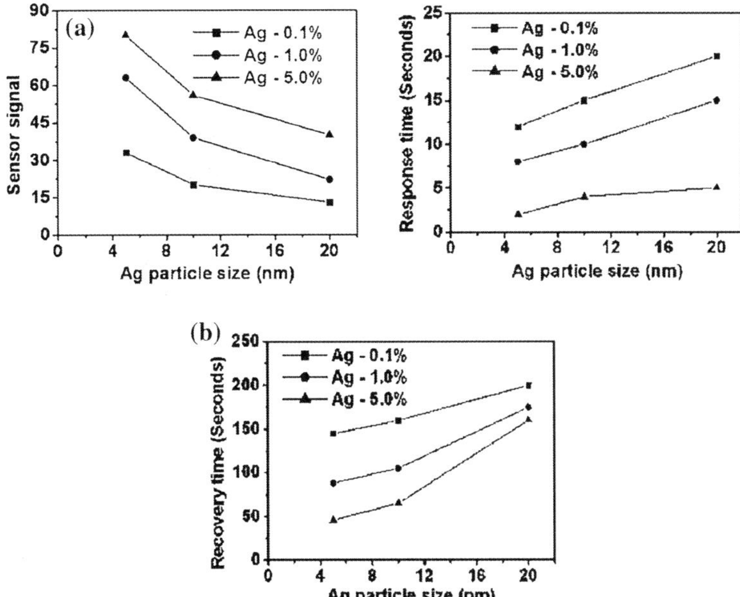

Fig. 7.10 Influence of Ag particle size on ethanol sensing of $SnO_{1.8}$:Ag nanoparticles films. **a** Variation of sensor signal with size of Ag particles at different Ag concentrations at 400 °C for 1,000 ppm ethanol, **b** Variation of response time at 400 °C for 1,000 ppm ethanol with size of Ag particles in SnO1.8 :Ag sensors, **c** variation of recovery time at 400 °C in synthetic air with size of Ag particles in SnO1.8 :Ag, sensors. Reprinted from Ref. 79, with permission from Elsevier, Copyright (2006)

gases [79]. Joshi and coworkers developed ppb-level gas sensors composed of mono-dispersed $SnO_{1.8}$:Ag nanoparticle films by varying the Ag particle size as well as its concentration through a well-defined aerosol route. As shown in Fig. 7.10, the sensitivity of this system with respect to ethanol vapor increased with a decrease in the size of silver particles deposited on tin dioxide particles; and shorter response and recovery times are also observed upon decreasing the size of Ag nanoparticles. The timescales for response and recovery observed in Fig. 7.10b and 10c also confirm Du and George's findings.

It is well known that islands are formed during the initial growth stage of ALD for some material systems. Island formation is usually observed for metal ALD on oxide substrates. For example, copper nanoparticles were deposited inside the pores of anodized aluminum oxide using atomic layer deposition [80]. Figure 7.11a–7.11c show cross-sectional SEM images of samples with copper particles deposited with 100, 200 and 400 cycles, respectively. It is clear that the particles are densely and evenly distributed along the pore walls. Roy et al. reported the nucleation of transition metals during the initial ALD process, and found that the nanoparticle size increased and coverage density also changed with the deposition numbers [81].

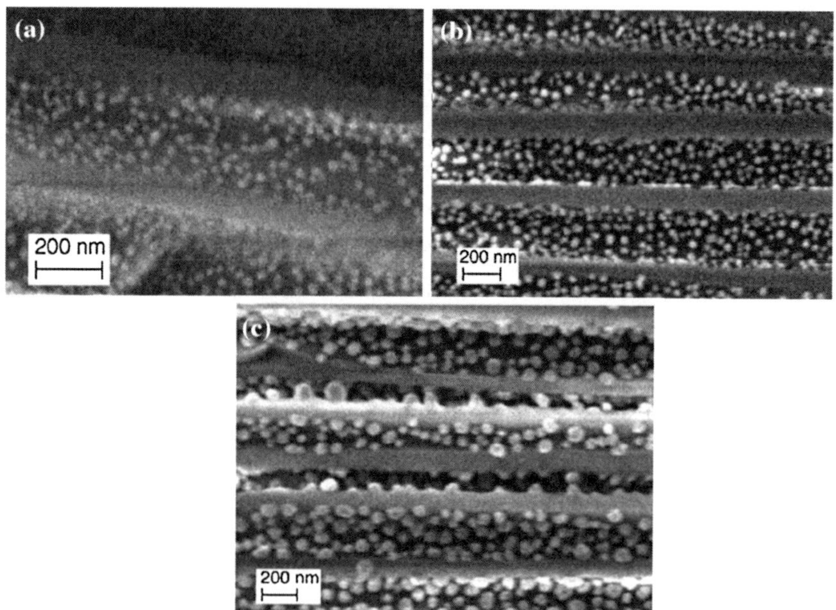

Fig. 7.11 Cross-sectional SEM image of nanoporous alumina substrates with copper nanoparticles deposited for **a** 100 cycles **b** 200 cycles and **c** 400 cycles. Reprinted from Ref. 80, with permission from Elsevier, Copyright (2003)

This unique feature is also being explored for use in catalytic reaction applications. Stair et al. fabricated a useful platform for heterogeneous catalysis [82]. Coating the AAO walls with catalytically active materials turns the nanolith into a novel catalytic system. Pellin et al. showed that atomic layer deposition on anodic aluminum oxide is a facile, flexible route to the synthesis of catalytic membranes with precise control of pore wall composition and diameters [83]. The ultra-uniform inorganic catalytic membranes exhibit a surprising specificity and conversion for ODH of cyclohexane to cyclohexene. The oxidative dehydrogenation of cyclohexane was shown to depend strongly on pore diameter and to be more specific than similarly active alumina powder catalysts. Vajda and Curtiss used ALD to coat porous AAO membranes with alumina before Pt-cluster deposition [84]. The ALD process ensures a uniform surface chemistry for the attachment of the clusters. As illustrated in Fig. 7.12, the study showed that size preselected Pt clusters stabilized on high-surface-area supports are 40–100 times more active for the oxidative dehydrogenation of propane than previously studied platinum and vanadia catalysts, while at the same time maintaining high selectivity towards formation of propylene over by-products. These results seem to form the basis for development of a new class of catalysts by providing a route to bond-specific chemistry, ranging from energy-efficient to environmentally friendly synthesis strategies and furthermore, may be very useful for the sensitive and selective detection of hydrocarbons. However, there is no report to date on the utilization of

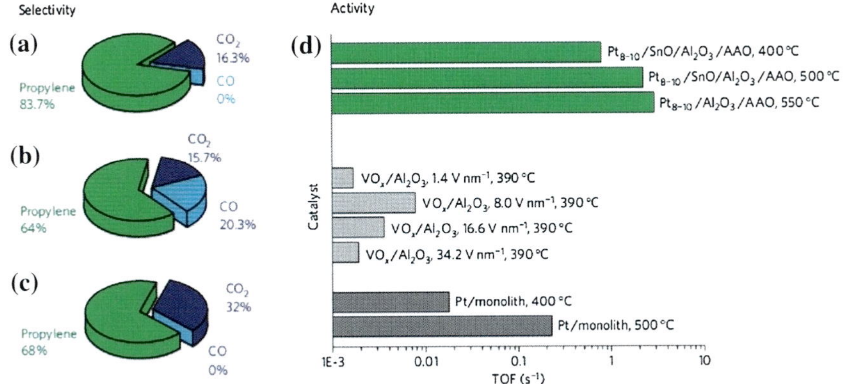

Fig. 7.12 Catalyst activity and selectivity. Selectivity of the Pt$_{8-10}$-based catalysts at various temperatures and support compositions: SnO/Al$_2$O$_3$ at 400 °C (**a**), SnO/Al$_2$O$_3$ at 500 °C (**b**) and Al$_2$O$_3$ at 550 °C (**c**). TOFs of propylene produced on the Pt$_{8-10}$ catalysts (*green*) and reference ODH catalysts (*grey*) expressed as number of propylene molecules formed per metal atom (**d**). Pt monolith and vanadia data from refs 29 and 22, respectively. Reprinted from Ref. 84, with permission from Elsevier, Copyright (2009)

ALD to load the catalytic metal nanoparticles and nanoclusters onto quasi-1 dimensional gas sensing nanostructures. There is a great opportunity to investigate the utilization of ALD to load the catalyst and study the gas sensing performance of such nanostructures.

7.3 Concluding Remarks and Future Directions

In this chapter, we summarized the atomic layer deposition and its unique features, the sensing mechanism of solid sate metal oxide gas sensors, the exploration of using quasi-1 dimensional nanostructures in improvement of gas sensing performance, and the potential of using ALD as a tool to fabricate the nanomaterials.

ALD is a surface-controlled deposition process and is renowned for its self-limiting nature to deposit films in a layer-by-layer fashion. The extremely high level of thickness control, the atomic level composition manipulation, together with the excellent conformality are the most important characteristics among the many benefits of ALD. These unique features make ALD attractive for integrated circuit and data storage, organic light emissive diode (OLED), solar as well as sensing applications. ALD is now considered as an enabling tool for nanoscience and nanotechnology.

Compact metal oxide solid state chemical sensors are the most promising and have been widely used in the detection of toxic pollutants and inflammable gases. With a large surface-to-volume ratio and Debye length comparable dimensions, quasi-1 dimensional nanomaterials have been regarded as one of the best candidates for realizing ultrasensitive solid state sensors. The growth of one-dimensional (1D) oxide nanostructures has been developed in the "bottom-up" approach which is based on the nanometer scale building blocks such as nanotubes, nanowires and nanobelts/nanoribbons; and a more expensive "top down" approach based on patterning through lithography. However, the fabrication techniques used are not convenient for mass production and the reproducibility and repeatability still need improvement when using nanomaterial arrays/films to fabricate sophisticated sensor structures.

In this respect, thin film deposition techniques provide another suitable way to confine the grain size. Different methods have been employed for the production of thin films of semiconducting oxides with the aim to obtain nanostructured, stable, cheap, low power consuming, and reliable gas sensors. ALD is being explored in the fabrication of metal oxide nanotubes/nanorods using positive templates and negative templates as scaffolds. ALD is also adopted to deposit pinhole free dense metal oxide films with thickness comparable to the Debye length for sensing mechanism study and sensitivity improvement.

As a conventional method, catalytic metal nanoparticles and nanoclusters, e.g., palladium, platinum, gold and silver are loaded onto semiconducting metal oxide nanomaterials to enhance the sensitivity and selectivity. It is well known that metal islands are formed on oxide substrates during the initial ALD growth stage. This unique feature is being explored in the catalytic applications. However, there is no report on the utilization of ALD to load the catalytic metal nanoparticles/nanoclusters onto the quasi-1 dimensional gas sensing nanostructures. There is a great opportunity to investigate the utilization of ALD to load the catalyst and great room exists for enhancement of gas sensing performance of such nanostructures.

References:

1. Sayago I, Gutierrez J, Ares L, Robla JI, Horrillo MC, Getino J, Agapito JA (1995) Long—term reliability of gas sensors for detection of nitrogen oxides. Sens Actuators B 26(1–3):56–58
2. Meixner H, Lampe U (1996) Metal oxide sensors. Sens Actuators B 33(1):198–202
3. Martinelli G, Carotta MC (1995) Thick-film gas sensors. Sens Actuators B 23:157–161
4. Barsan N, Koziej D, Weimar U (2007) Metal oxide-based gas sensor research: how to? Sens Actuators B 121:18–35
5. Barsan N, Weimar U (2003) Understanding the fundamental principles of metal oxide based gas sensors; the example of CO sensing with SnO_2 sensors in the presence of humidity. J Phys Condens Matter 15:R813–R839
6. Watsont J, Ihokura K, Colest GSV (1993) The tin dioxide gas sensor. Measur Sci Technol 4:711–719
7. Safonova O, Bezverkhy I, Fabrichnyi P, Rumyantseva M, Gaskov A (2002) Mechanism of sensing CO in nitrogen by nanocrystalline SnO_2 and $SnO_2(Pd)$ studied by Mössbauer spectroscopy and conductance measurements. J Mater Chem 12:1174–1178
8. Sahm T, Gurlo A, Bârsan N, Weimar U (2006) Basics of oxygen and SnO_2 interaction; work function change and conductivity measurements. Sens Actuators B 118(1–2):78–83
9. Chwieroth B, Patton BR, Wang Y (2001) Conduction and gas-surface reaction modeling in metal oxide gas sensors. J Electroceram 6(1):27–41

10. Barsan N, Weimar U (2001) Conduction model of metal oxide gas sensors. J Electroceram 7(3):143–167
11. Gulati S, Mehan N, Goyal DP, Mansingh A (2002) Electrical equivalent model for SnO_2 bulk sensors. Sens Actuators B 87:309–320
12. Mizsei J (1995) How can sensitive and selective semiconductor gas sensors be made? Sens Actuators B 23(2–3):173–176
13. Ogawa H, Nishikawa M, Abe A (1982) Hall measurement studies and an electrical conduction model of tin oxide ultrafine particle films. J Appl Phys 53:4448–4455
14. Xu C, Tamaki J, Miura N, Yamazoe N (1991) Grain size effects on gas sensitivity of porous SnO_2-based elements. Sens Actuators B 3:147–155
15. Yamazoe N (1991) New approaches for improving semiconductor gas sensors. Sens Actuators B 5:7–19
16. Franke ME, Koplin TJ, Simon U (2006) Metal and metal oxide semiconductor in chemiresistors: does the nanoscale matter? Small 2(1):36–50
17. Jin ZH, Zhou HJ, Jin ZL, Savinell RF, Liu CC (1998) Application of nano-crystalline porous tin oxide thin film for CO sensing. Sens Actuators B 52:188–194
18. Yoo DJ, Tamaki J, Park SJ, Miura N, Yamazoe N (1995) Effects of thickness and calcination temperature on tin dioxide sol-derived thin film sensor. J Electrochem Soc 142:L105–L107
19. Bruno L, Pijolat C, Lalauze R (1994) Tin dioxide thin film gas sensor prepared by chemical vapor deposition—influence of grain size and thickness on the electrical properties. Sens Actuators B 18:195–199
20. Kim KH, Park CG (1991) Electrical properties and gas sensing behavior of SnO_2 films prepared by chemical vapor deposition. J Electrochem Soc 138:2408–2412
21. Kolmakov A, Moskovits M (2004) Chemical sensing and catalysis by one-dimensional metal-oxide nanostructures. Ann Rev Mater Res 34:151–180
22. Kolmakov A, Zhang YX, Cheng GS, Moskovits M (2003) Detection of CO and O_2 using tin oxide nanowire sensors. Adv Mater 15:997–1000
23. Wan Q, Li QH, Chen YJ, Wang TH, He XL, Li JP, Lin CL (2004) Fabrication and ethanol sensing characteristics of ZnO nanowire gas sensors. Appl Phys Lett 84:3654–3656
24. Comini E, Faglia G, Sberveglieri G, Pan ZW, Wang ZL (2002) Stable and highly sensitive gas sensors based on semiconducting oxide nanobelts. Appl Phys Lett 81:1869–1871
25. Collins PG, Bradley K, Ishigami M, Zettl A (2000) Extreme oxygen sensitivity of electronic properties of carbon nanotubes. Science 287:1801–1804
26. Kong J, Franklin NR, Zhou CW, Chapline MJ, Peng S, Cho KJ, Dai HJ (2000) Nanotube molecular wires as chemical sensors. Science 287:622–625
27. Chen YJ, Zhu CL, Wang TH (2006) The enhanced ethanol sensing properties of multiwalled carbon nanotubes/SnO_2 Core/shell nanostructures. Nanotechnology 17:3012–3017
28. Kohl D (2001) Function and applications of gas sensors. J Phys D Appl Phys 34(19):R125–R149
29. Huang H, Lee YC, Tan OK, Zhou W, Peng N, Zhang Q (2009) High sensitivity SnO_2 single-nanorod sensors for the detection of H_2 gas at low temperature. Nanotechnology 20:115501 (5pp)
30. Francioso L, Taurino AM, Forleo A, Siciliano P (2008) TiO2 nanowires array fabrication and gas sensing properties. Sens Actuators B 130(1):70–76
31. Cui Y, Wei Q, Park H, Lieber CM (2001) Nanowire nanosensors for highly-sensitive, selective and integrated detection of biological and chemical species. Science 293:1289–1292
32. McAlpine MC, Ahmad H, Wang D, Heath JR (2007) Highly order nanowire array on plastic substrates for ultrasensitive flexible chemical sensors. Nat Mater 6:379–384
33. Eliol OH, Morisette D, Akin D, Denton JP, Bashir R (2003) Integrated nanoscale silicon sensors using top-down fabrication. Appl Phys Lett 83(11):4613–4615
34. Murray BJ, Walter EC, Penner RM (2004) Amine vapor sensing with silver mesowires. Nano Letter 4(4):665–670
35. Adeghian RB, Kahrizi M (2007) A novel miniature gas ionization sensor based on freestanding gold nanowires. Sens Actuators A 137(2):248–255

36. Liu Z, Searson PC (2006) Single nanoporous gold nanowire sensors. J Phys Chem B 110(9):4318–4322
37. Im Y, Lee C, Vasquez RP, Bangar MA, Myung NV, Menke EJ, Penner RM, Yun M (2006) Investigation of a single Pd nanowire for use as a hydrogen sensor. Small 2(3):356–358
38. Walter EC, Favier F, Penner RM (2002) Palladium mesowire arrays for fast hydrogen sensors and hydrogen-actuated switches. Anal Chem 74:1546–1553
39. Atashbar MZ, Singamaneni S (2005) Room temperature gas sensor based on metallic nanowires. Sens Actuators B 111–112(11):13–21
40. Favier F, Walter EC, Zach MP, Benter T, Penner RM (2001) Hydrogen sensors and switches from electrodeposited palladium mesowire arrays. Science 293(5538):2227–2231
41. Dan YP, Cao YY, Mallouk TE, Johnson AT, Evoy S (2007) Dielectrophoretically assembled polymer nanowires for gas sensing. Sens Actuators B 125(1):55–59
42. Huang J, Virji S, Weiller BH, Kaner RB (2003) Polyaniline nanofibers: facile synthesis and chemical sensors. J Am Chem Soc 125(2):314–315
43. George SM, Ott AW, Klaus JW (1996) Surface chemistry for atomic layer growth. J Phys Chem 100(31):13121–13131
44. Goodman CHL, Pessa MV (1986) Atomic layer epitaxy. J Appl Phys 60(3):R65–R81
45. Suntola T (1992) Atomic layer epitaxy. Thin Solid Films 216(1):84–89
46. Suntola T (1994) Atomic Layer Epitaxy. In Hurle DTJ (ed) Handbook of Crystal Growth, Part B: Growth Mechanisms and Dynamics, Vol. 3. Elsevier, Amsterdam (Chapter 14)
47. Leskela M, Ritala M (2002) Atomic layer deposition (ALD): from precursors to thin film structures. Thin Solid Films 409:138–146
48. Puurunen RL (2005) Surface chemistry of atomic layer deposition: a case study for the trimethylaluminum/water process. J Appl Phys 97(1–52):121301
49. Ritala M, Leskela M (2001) Atomic layer deposition. In: Handbook of thin film materials, Vol 1, Elsevier, San Diego (Chapter 2)
50. Elam JW, Routkevitch D, Mardilovich PP, George SM (2003) Conformal coating on ultrahigh-aspect-ratio nanopores of anodic alumina by atomic layer deposition. Chem Mater 15(18):3507–3517
51. Kucheyev SO, Biener J, Wang YM, Baumann TF, Wu KJ, Buuren TV, Hamza AV, Satcher JH (2005) Atomic layer deposition of ZnO on ultralow-density nanoporous silica aerogel monoliths. Appl Phys Lett, Vol 86(8):083108(1–3)
52. Ritala M, Leskela M (1999) Atomic layer epitaxy-a valuable tool for nanotechnology? Nanotechnology 10(1):19–24
53. Lu W, Lieber CM (2006) Semiconductor Nanowires. J Phys D Appl Phys 39(21):R387–R406
54. Xia Y, Yang P, Sun Y, Wu Y, Mayers B, Gates B, Yin Y, Kim F, Yan H (2003) One-dimensional nanostructures: synthesis, characterization, and applications. Adv Mater 15(5):353–389
55. Wolf ED (2004) Nanophysics and nanotechnology: an introduction to modern concepts in nanosciences. Wiley-VCH, Weinheim
56. Shin H, Jeong DK, Lee J, Sung MM, Kim J (2004) Formation of TiO2 and ZrO2 nanotubes using atomic layer deposition with ultraprecise control of the wall thickness. Adv Mater 16(14):1197–1200
57. Hwang J, Min B, Lee JS, Keem K, Cho K, Sung M-Y, Lee M-S, Kim S (2004) Al2O3 nanotubes fabricated by wet etching of ZnO/Al2O3 core/shell nanofibers. Adv Mater 16(5):422–425
58. Peng Q, Sun XY, Spagnola JC, Hyde GK, Spontak RJ, Parsons GN (2007) Atomic layer deposition on electrospun polymer fibers as a direct route to Al_2O_3 microtubes with precise wall thickness control. Nano Lett 7(3):719–722
59. Ras RHA, Kemell M, Wit JD, Ritala M, Brinke GT, Leskela M, Ikkala O (2007) Hollow inorganic nanospheres and nanotubes with tunable wall thicknesses by atomic layer deposition on self-assembled polymeric templates. Adv Mater 19:102–106
60. Elam JW, Routkevitch D, Mardilovich PP, George SM (2003) Conformal coating on ultrahigh-aspect-ratio nanopores of anodic alumina by atomic layer deposition. Chem Mater 15(18):3507–3517

61. Sander MS, Côté MJ, Gu W, Kile BM, Tripp CP (2004) Template-assisted fabrication of dense, aligned arrays of Titania nanotubes with well-controlled dimensions on substrates. Adv Mater 16(22):2052–2057
62. Gu DF, Baumgart H, Namkoong G, Abdel-Fattah TM (2009) Atomic layer deposition of ZrO_2 and HfO_2 nanotubes by template replication. Electrochem Solid-State Lett 12(4):K25–K28
63. Kim WH, Park SJ, Son JY, Kim HJ (2008) Ru nanostructure fabrication using an anodic aluminum oxide nanotemplate and highly conformal Ru atomic layer deposition. Nanotechnology, 19:045302 (8pp)
64. Elam JW, Xiong G, Han CY, Wang HH, Birrell JP, Welp U, Hryn JN, Pellin MJ, Baumann TF, Poco JF, Satcher JH Jr (2006) Atomic layer deposition for the conformal coating of nanoporous materials. J Nanomaterials (1), p 1–5 (Article ID 64501)
65. Willinger MG, Neri G, Rauwel E, Bonavita A, Micali G, Pinna N (2008) Vanadium oxide sensing layer grown on carbon nanotubes by a new atomic layer deposition process. Nano Lett 8(12):4201–4204
66. Kim WS, Lee BS, Kim DH, Kim HC, Yu WR, Hong SH (2010) SnO_2 nanotubes fabricated using electrospinning and atomic layer deposition and their gas sensing performance. Nanotechnology, 21:245605(1–7)
67. Bae CD, Yoon YJ, Yoo HJ, Han D, Cho JH, Lee BH, Sung MM, Lee MG, Kim JY, Shin HJ (2009) Controlled fabrication of multiwall anatase TiO2 nanotubular architectures. Chem Mater 21(13):2574–2576
68. Du X, Du Y, George SM (2008) CO gas sensing by ultrathin tin oxide films grown by atomic layer deposition using transmission FTIR spectroscopy. J Phys Chem A 112:9211–9219
69. Du X, George SM (2008) Thickness dependence of sensor response for CO gas sensing by tin oxide films grown using atomic layer deposition. Sens Actuators B 135:152–160
70. Utriainen M, Varpula A, Niskanen AJ, Sinkkonen J, Novikov S, Airaksinen VM, Einehag M, Johansson D, Nyholm S (2008) Novel hand-held chemical detector with micro gas sensors. Nordic innovation centre (NICe) project number: 06071. Online: http://www.nordicin novation.net/_img/06071_treatgarden_final_technical_report_web.pdf
71. Niskanena AJ, Varpula A, Utriainen M, Natarajan G, Cameron DC, Novikov S, Airaksinen VM, Sinkkonen J, Franssil S (2010) Atomic layer deposition of tin dioxide sensing film in microhotplate gas sensors. Sens Actuators B Chem 148(1):227–232
72. Natarajan G, Cameron DC (2009) Influence of oxygen depletion layer on the properties of tin oxide gas-sensing films fabricated by atomic layer deposition. Appl Phys A Mater Sci Process 95(3):621–627
73. Kim DH, Kim WS, Lee SB, Hong SH (2010) Gas sensing properties in epitaxial SnO_2 films grown on TiO_2 single crystals with various orientations. Sens Actuators B 147(2):653–659
74. Lee W, Hong K, Park Y, Kim NH, Choi Y, Park J (2005) Surface and sensing properties of PE-ALD SnO_2 thin film. Electron Lett 41(8):475–477
75. Ra YW, Choi KS, Kim JH, Hahn YB, Im HY (2008) Fabrication of ZnO nanowires using nanoscale spacer lithography for gas sensors. Small 4(8):1105–1109
76. Ra H-W, Khan R, Kim JT, Kang BR, Im YH (2009) The effect of grain boundaries inside the individual ZnO nanowires in gas sensing. Nanotechnology 21(8):085502(1–5)
77. Aronniemi M, Saino J, Lahtinen J (2008) Characterization and gas-sensing behavior of an iron oxide thin film prepared by atomic layer deposition. Thin Solid Films 516(18):6110–6115
78. Kolmakov A, Klenov DO, Lilach Y, Moskovits M (2005) Enhanced gas sensing by individual SnO_2 nanowires and nanobelts functionalized with Pd catalyst particles. Nano Letter 5(4):667–673
79. Joshi RK, Kruis FE (2006) Influence of Ag particle size on ethanol sensing of $SnO_{1.8}$:Ag nanoparticle films: a method to develop parts per billion level gas sensors. Appl Phys Lett 89:153116(1–3)
80. Johansson A, Törndahl T, Ottosson LM, Boman M, Carlsson JO (2003) Copper nanoparticles deposited inside the pores of anodized aluminum oxide using atomic layer deposition. J Mater Sci Eng C23(6–8):823–826

81. Lim BS, Rahtu A, Gordon RG (2003) Atomic layer deposition of transition metals. Nat Mater 2:749–754
82. Feng H, Elam JW, Libera JA, Pellin MJ, Stair PC (2009) Catalytic Nanoliths. Chem Eng Sci 64:560–567
83. Pellin MJ, Stair PC, Xiong G, Elam JW, Birrell J, Curtiss L, George SM, Han CY, Iton L, Kung H, Kung M, Wang HH (2005) Mesoporous catalytic membranes: synthetic control of pore size and wall composition. Catal Lett 102(3–4):127–130
84. Vajda S, Pellin MJ, Greeley JP, Marshall CL, Curtiss LA, Ballentine GA, Elam JW, Catillon-Mucherie S, Redfern PC, Mehmood F, Zapo P (2009) Subnanometre platinum clusters as highly active and selective catalysts for the oxidative dehydrogenation of propane. Nat Mater 8:213–219

Chapter 8
Microwave Synthesis of Metal Oxide Nanoparticles

Natalie P. Herring, Asit B. Panda, Khaled AbouZeid,
Serial H. Almahoudi, Chelsea R. Olson, A. Patel and M. S. El-Shall

Abstract This chapter summarizes microwave irradiation methods for the preparation of metal oxide nanoparticles and their catalytic and sensing properties and applications. Microwave irradiation provides rapid decomposition of metal precursors and can be extended for synthesis of a wide range of metal oxide nanoparticles with various compositions, sizes and shapes. This chapter introduces the microwave method and describes the nucleation and growth process for the formation nanocrystals. We offer a broad overview of metal oxide nanostructures synthesized by microwave irradiation including: ZnO, TiO_2, CeO_2, other rare earth metal oxides, transitional metal oxides and metal ferrite nanostructures. Finally, we describe the application of metal oxides in the photocatalytic degradation of organic dyes and gas sensing devices.

8.1 Introduction

The synthesis and characterization of nanocrystals with controlled size and shape have attracted rapidly growing interest both for fundamental scientific interest and many practical and technological applications [1–6]. The shape control and assembly of nanostructures into organized patterns provide valuable routes to the design of functional materials and to a variety of applications [1–6]. Several methods based on physical and chemical approaches have been developed for the synthesis of controlled size and shape nanostructures. Examples of these approaches include

N. P. Herring · A. B. Panda · K. AbouZeid · S. H. Almahoudi · C. R. Olson · A. Patel ·
M. S. El-Shall (✉)
Department of Chemistry, Virginia Commonwealth University,
Richmond, 23284-2006, USA
e-mail: mselshal@vcu.edu

M. A. Carpenter et al. (eds.), *Metal Oxide Nanomaterials for Chemical Sensors*,
Integrated Analytical Systems, DOI: 10.1007/978-1-4614-5395-6_8,
© Springer Science+Business Media New York 2013

solvothermal methods, template-assisted, kinetic growth control, sonochemical reactions, and thermolysis of single-source precursor in ligating solvents [1–19].

Microwave Irradiation (MWI) methods provide simple and fast routes to the synthesis of nanomaterials since no high temperature or high pressure is needed. Furthermore, MWI is particularly useful for a controlled large-scale synthesis that minimizes the thermal gradient effects [20–41]. The heating of a substance by microwave irradiation depends on the ability of the material (solvent or reagent) to absorb microwave radiation and convert it into heat. This is based on two principal mechanisms: dipole rotation and ionic conduction, that is, by reversal of solvent dipoles and the resulting replacement of charged ions of a solute [42, 43]. Polar reactants with a high microwave extinction coefficient can be excited by direct absorption of microwaves. Due to the difference in the solvent and reactant dielectric constants, selective dielectric heating can provide significant enhancement in reaction rates. By using metal precursors that have large microwave absorption cross-sections relative to the solvent, very high effective reaction temperatures can be achieved. The rapid transfer of energy directly to the reactants (faster than they are able to relax), causes an instantaneous internal temperature rise. Thus, the activation energy is essentially decreased as compared with conductive heating and the reaction rate increases accordingly. As a consequence, reactions might be performed at lower temperatures and hotspots or other temperature inhomogeneities can be prevented. Furthermore, reaction parameters such as temperature, time, and pressure can be controlled easily. This also allows the rapid decomposition of the precursors thus creating highly supersaturated solutions where nucleation and growth can take place to produce the desired nanocrystalline products. These conditions lead to the formation of very small nanocrystals since the higher the supersaturation the smaller the critical size required for nucleation. Thus, in the synthesis of nanocrystals using MWI, the reaction can be controlled very conveniently, and the resulting nanocrystals show very good monodispersity and crystallinity.

The growth of the newly formed nanocrystals can be effectively inhibited by the adsorption of ligating organic surfactants that bind strongly to the nanocrystals, thus stabilizing and passivating the surface. Furthermore, selective adsorption of the ligating organics can significantly slow down the growth of the nanocrystal in all but the favorable crystallographic plane thus resulting in a one dimensional (1D) structure. Since in MWI it is possible to quench the reaction very early on (~ 10 s), this provides the opportunity of controlling the nanostructures from small spherical nuclei to short rods to extended assemblies of nanowires by varying the MWI reaction time and the relative concentrations of different organic surfactants with variable binding strengths to the initial precursors and to the nanocrystals.

MWI methods have been demonstrated for the synthesis of a variety of high quality, nearly monodisperse semiconductor, metal and metal oxide nanoparticles as well as one-dimensional nanostructures [20–41]. In this chapter, we describe several examples of the application of MWI for the synthesis of a variety of metal oxide nanostructures of controlled size and shape. We also provide an overview of the use of metal oxide for the development of nanoparticle's formation of efficient photocatalysts and gas sensors for a variety of environmental and analytical applications. In order to provide a basic understanding of nanoparticle's formation, we first present a brief overview of the classical nucleation theory (CNT) since nucleation and growth greatly influence the control of the size and shape of nanoparticles which consequently determine the unique properties that may characterize nanoparticles [44–47].

8.2 Brief Overview of Nucleation and Growth from the Vapor Phase

Nucleation of liquid droplets from the vapor or nanocrystals from the solution phase can occur homogeneously or heterogeneously. Homogeneous nucleation occurs in the absence of any foreign particles or surfaces when the vapor molecules themselves cluster to nuclei within the supersaturated vapor or solution. According to the Classical Nucleation Theory (CNT) for vapor phase nucleation, embryonic clusters of the new phase can be described as spherical liquid droplets with the bulk liquid density inside and the vapor density outside [44, 45]. The free energy of theses clusters relative to the vapor is the sum of two terms: a positive contribution from the surface free energy and a negative contribution from the bulk free energy difference between the supersaturated vapor and the liquid. The surface free energy results from the reversible work in forming the interface between the liquid droplet and the vapor. For a cluster containing n atoms or molecules, the interface energy is given by

$$\sigma A(n) = 4\pi\sigma(3v/4\pi)^{2/3} n^{2/3} \tag{8.1}$$

Where σ is the interfacial tension or surface energy per unit area, $A(n)$ is the surface area of the clusters, and v is the volume per molecule in the bulk liquid. Since n molecules are transferred from the vapor to the cluster, the bulk contribution to the free energy is n $(\mu_l - \mu_v)$ where μ_l. And μ_v are the chemical potentials per molecule in the bulk liquid and vapor, respectively. Assuming *ideal* vapor, it can be shown that

$$(\mu_l - \mu_v) = -n\, k_B T \ln S \tag{8.2}$$

Where k_b is the Boltzmann constant, T is the temperature, and S is vapor supersaturation ration defined as $S = P/P_e$, where P is the pressure of the vapor and P_e is the equilibrium or "saturation" vapor pressure at the temperature of the vapor (T).

The sum of the contributions in Eqs. (8.1) and (8.2) is the reversible work (free energy) $W(n)$, done in forming a cluster containing n atoms or molecules. This free energy is given by

$$W(n^*) = -n\, k_B T \ln S\; + 4\pi\sigma(3v/4\pi)^{2/3} n^{2/3} \tag{8.3}$$

Because of the positive contribution of the surface free energy, there is a free energy barrier which impedes nucleation. The smallest cluster n* which can grow with a decrease in free energy is determined to be:

$$n* = 32\pi\sigma^3 v^2/3(k_B T \ln S)^3 \tag{8.4}$$

Substituting n* into Eq. 8.3 yields the barrier height W(n*), given by

$$W(n*) = 16\pi\sigma^3 v^2/3(k_B T \ln S)^2 \tag{8.5}$$

It is clear from Eq. (8.5) that increasing the supersaturation (S) reduces the barrier height and the critical cluster size (n*) and hence, fluctuations can allow some clusters to grow large enough to overcome the barrier and grow into stable droplets.

8.3 Nucleation and Growth from Supersaturated Solutions

The formation of nanocrystals in supersaturated solutions follows the basic principles of crystallization: a nucleation event precedes the growth of nanocrystals and eventually bulk crystals. The most widely accepted mechanism is known as *La Mer-mechanism* [46]. According to this mechanism, the reaction can be divided into three phases: first, the concentration of reactant increases gradually and eventually exceeds solubility. Second, the concentration of reactants reaches the critical limit of supersaturation and rapid nucleation occurs. This nucleation burst results in a sudden decrease of reactant concentration. Finally, nuclei grow slowly as the reaction solution depletes in reactants. Since growth is usually thermodynamically favored over nucleation, nanoparticles can be grown monodispersely when the second phase can be limited to a short period of time by suitable choice of reactant concentration and temperature. The overall nanocrystals' formation can be described in terms of three stages: (1) nucleation, (2) growth and (3) competitive growth which is commonly referred to as Ostwald ripening (47–49). For the formation of monodisperse nanocrystals, single rather than multiple nucleation events are necessary in order to prevent additional nucleation during the subsequent growth process [13, 15, 46]. Therefore, the key point to achieve good control over the size and size distribution of nanoparticles is to decouple the nucleation and growth processes. In the nucleation stage, the number and size of the small nuclei formed are controlled by the degree of the solution supersaturation. As shown by the CNT above, the larger the supersaturation, the smaller the critical size of the nucleus and the smaller the nanocrystals that can grow. At the second stage, nanocrystals exhibit a monotonic growth due to the addition of atoms from the solution into the nuclei which results in decreasing the degree of supersaturation with time and increasing the total volume of the nanocrystals. Finally, when the nanocrystals are large enough and the degree of supersaturation is negligible, since all atoms are already incorporated in the nanocrystals, Ostwald ripening starts to operate where competitive growth or diffusion-limited aggregation takes place [47]. This process results in an increase in the mean size of the nanocrystals due to the mass transfer from smaller to larger particles. The net result is that larger particles grow and smaller particles shrink in size. Another result is that the number of particles in a system is drastically reduced, as smaller particles vanish completely in order to donate their atoms for the continued growth of the larger particles [47]. Therefore, not only does the surface-to-volume ratio of the larger particles decrease, making them more stable, but the less stable (smaller) particles become fewer, and the total surface energy of the system decreases.

The growth of the newly formed nanocrystals can be effectively inhibited by the adsorption of ligating organic surfactants that bind strongly to the nanocrystals, thus stabilizing and passivating the surface. The ability to cap the nanocrystal's surface provides a way not only to control the surface states, but also to prevent rapid agglomeration of the particles due to Ostwald ripening. MWI in organic solutions allows the production of high degrees of supersaturation due to the rapid heating of the nanocrystals' precursors [20, 21, 24, 29, 33, 38–40]. The size of the nanocrystals is tuned by varying the concentration of the precursors and the MWI times, while the shape is controlled by varying the concentration and composition of the ligating solvents which stabilize the nanocrystals by passivating the surfaces [20, 21, 33–35, 40]. Following MWI for the desired time in an organic phase, the synthesized nanocrystals can be separated by size-selective precipitation through the gradual addition of a hydrophilic solvent such as ethanol to the toluene or hexane dispersion containing nanoparticles with various particles sizes. The large nanoparticles tend to precipitate first because of their strong van der Waals attraction.

8.4 Experimental Methods

For all the syntheses described here, a conventional microwave oven (2.45 GHz) operating at 600–1,000 W was used. In most cases, the reaction mixture was microwaved in 30 s cycles (on for 10 s, off and stirring for 20 s) for reaction times that varied from 10 s to several minutes. A schematic showing the simple microwave components used for MWI synthesis is shown in Fig. 8.1.

For the synthesis of ZnO nanocrystals, anhydrous zinc acetate (Aldrich, 99.99 %) was dissolved in a mixture of oleic acid (OAc) (Aldrich, technical grade, 90 %) and oleylamine (OAm) (Aldrich). The reaction mixture was heated to dissolve the zinc acetate and remove any residual water, in a hot oil bath with stirring to 120 °C, and this temperature was maintained for 1 h. Following the microwave reaction, the precipitate was separated from the liquid phase by centrifugation and dried at 60 °C overnight.

For the synthesis of CeO_2 nanocrystals, cerium acetate in a mixture of oleic acid and oleylamine is first heated to about 110 °C in order to remove the hydrated

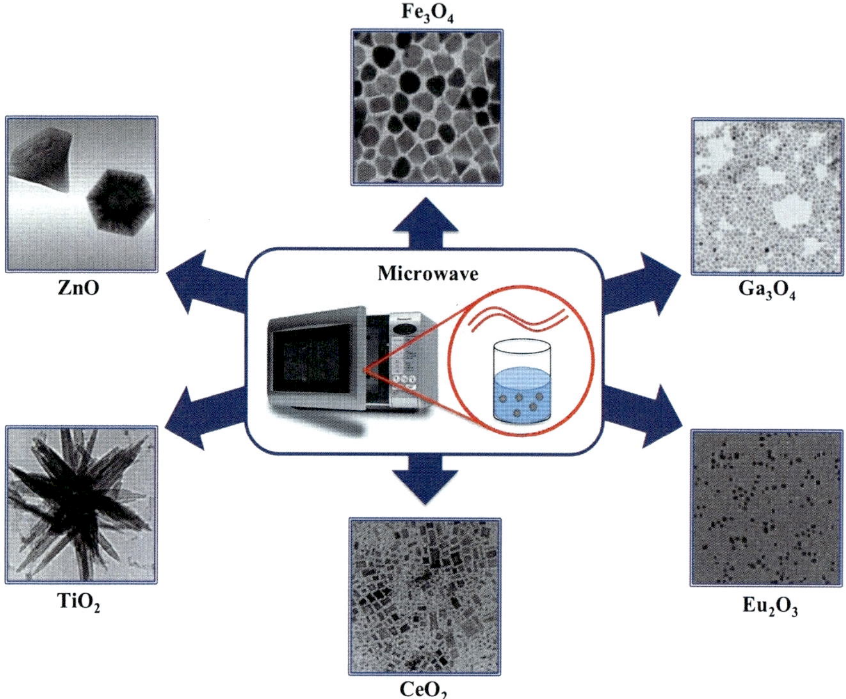

Fig. 8.1 Schematic showing the MWI experimental setup

water and obtain a clear solution. The mixture is then transferred quickly to a conventional microwave for the desired amount of time.

For the synthesis of colloidal TiO_2 nanocrystals, titanium butoxide + titanium tetrachloride were used as precursors and a mixture of OAc/OAm as ligating solvents [48, 49]. For the synthesis of rare earth oxides, metal acetate or metal acetylacetonate was used as a precursor and a mixture of OAc/OAm as ligating solvents [35]. For the synthesis of Au, Ag, Pt, Ni, and Co, the precursors used were $HAuCl_4$, silver acetate, platinum acetylacetonate, nickel acetylacetonate, and $Co_2(CO)_8$, respectively. After microwaving for the desired time, the synthesized nanocrystals were washed with ethanol, centrifuged, and re-dispersed in hydrophobic solvents such as toluene or dichloromethane.

For the synthesis of metal ferrite nanocrystals, high-purity iron(III) acetate $[Fe(acac)_3]$, cobalt(III) acetate $[Co(acac)_2]$, manganese(II) acetate $[Mn(acac)_2]$, zinc(III) acetate $[Zn(acac)_2]$, nickel(II) acetate $[Ni(acac)_2]$, oleic acid and oleylamine have been used (Aldrich and Fluka) as received without further purification. In a typical reaction for the synthesis of Fe_3O_4 nanoparticles, a mixture of oleic acid (3.6 ml, 0.011 mol) and oleylamine (3.75 ml, 0.011 mol) was heated at 110–120 °C in an oil bath and 0.19 g of $Fe(acac)_3$ was added to the solvent under vigorous stirring for 10 min, which resulted in a clear brown solution. The clear solution was placed in a conventional microwave with the power set to 60 % of 650 W and operated in 3 min cycles (on for 2 min, off and stirring for 1 min) for total microwaving times that varied from 6 to 20 min. It should be mentioned that continuous microwaving, instead of the 3 min cycles, resulted in particles with broad size distributions. After microwaving for 3–4 min the reaction mixture turned black. The progress of the reaction was monitored by taking aliquots at different microwaving times. After microwaving for a desired time, the resultant nanocrystals were precipitated from the reaction mixture using ethanol, and the precipitate was collected after centrifugation. The solid precipitate was re-dispersed in typical hydrophobic solvents such as toluene, dichloromethane, chloroform or hexane. Precipitation and dispersion were repeated for 2–3 times to remove the free ligating solvents.

For the synthesis of $CoFe_2O_4$, $MnFe_2O_4$, $NiFe_2O_4$ or $ZnFe_2O_4$ nanoparticles, the metal (II) acetylacetonate [$Co(acac)_2$, $Mn(acac)_2$, $Ni(acac)_2$ or $Zn(acac)_2$] with $Fe(acac)_3$ in a molar ratio of 1:2 was used. The other synthetic procedures were similar to that of the Fe_3O_4 nanoparticles. A major difference was observed in the synthesis of the $MnFe_2O_4$ nanoparticles where the use of 3:1 oleic acid to oleylamine ratio produced large cube-like structures with particle size ~ 120 nm.

8.5 Selected Oxide Nanostructures

8.5.1 ZnO Nanostructures

ZnO nanostructures have been extensively studied due to their unique properties which make them promising materials for potential uses in various applications including sensors, catalysis and optoelectronic devices [50–64]. Many methods to synthesize ZnO nanostructures have been reported [50–79]. These methods include chemical vapor deposition (CVD) [53, 58], chemical bath deposition, pulse laser deposition [67, 68, 72] and other high temperature methods [65, 67, 68, 72, 76], as well as, thermal [69, 71, 73, 76, 79] and microwave [66] assisted synthesis methods. Furthermore, each of these methods utilizes different starting materials and proceeds via different reaction pathways. For these reasons, each method produces particles with unique morphologies and physical and electronic properties. One challenge is the development of a rational synthesis strategy for the production of nanoparticles. We have developed a simple microwave method for the synthesis of ZnO nanoparticles with controlled morphology. Through this technique, hexagonal pyramids, nanotriangles and nanorods are synthesized by varying the capping materials.

It was observed that the resulting morphology of the ZnO nanostructures is strongly dependent on the molar ratios of oleic acid relative to oleylamine. The resulting morphologies are summarized in Table 8.1. Particles produced in the absence of oleylamine resulted in spherical and rod shaped nanoparticles, increasing the amount of oleylamine leads to pyramidal shaped particles.

Table 8.1 Summary of ZnO particle morphologies resulting from using different molar ratios of oleic acid (OAc) relative to oleylamine (OAm)

OAc: OAm	Morphology
3:1	Triangular pyramid
1:1	Triangular pyramid
1:3	Hexagonal pyramid
OAm Only	Triangular pyramid
OAc Only	Spherical and rod

Preparing the particles using equimolar proportions produces triangular pyramid morphologies. Figure 8.2 displays typical TEM images of the nanostructures produced using only oleic acid (a and b) or only oleylamine (c and d) at different MWI times. It is obvious that rod morphologies are predominant in pure oleic acid while the pyramid morphologies are dominant in pure oleylamine. The reaction is much faster in the presence of oleic acid as compared to pure oleylamine. Spherical ZnO nanoparticles are observed following the MWI for 30 s in pure oleic acid. These particles grow into long rods at longer reaction times as shown in the TEM image of Fig. 8.2b obtained following MWI of the zinc acetate solution in pure oleic acid for 60 s. In pure oleylamine, MWI for 10 min produces pyramid-shape particles as shown in the TEM images displayed in Fig. 8.2c and d. At equivalent proportions of oleic acid to oleylamine, uniform triangular pyramid particles are formed as shown in Fig. 8.3a. This morphology is still predominant with increasing the amount of oleic acid as shown in Fig. 8.3b for a mixture containing a 3:1 molar ratio of oleic acid to oleylamine. However, increasing the amount of oleylamine produces hexagonal pyramid shaped particles as shown in Fig. 8.3c and d for a mixture containing a 1:3 molar ration of oleic acid to oleylamine. Therefore, the oleic acid and oleylamine act as capping agents with different selectivity to different crystal faces of the ZnO nanostructures.

The XRD patterns of the different nanoparticle shapes are shown in Fig. 8.4. The observed diffraction peaks reveal that all the prepared morphologies have the hexagonal ZnO crystal structures (JCPDS Reference # 36–1451). It is interesting to note that even though all three morphologies have the same crystal structure, different crystal planes have different relative intensities. Differences in peak intensities, relative to the 101 peak, signify preferential growth in specific directions, which results in the various morphologies observed. Significant enhancement in the intensity of the 100 and 110 faces is observed for the hexagonal pyramid shaped particles. This is probably due to excess oleylamine, which must cap other faces more efficiently than the 100 and 110 faces. Also, for triangular prymid shaped particles, the effect of oleic acid is clearly observed as a decrease of the intensity of the 100 face, which indicated preferable capping of this face.

Figure 8.5 displays the UV-Vis absorption spectra of the spherical, pyramidal and rod shape nanoparticles dispersed in toluene. Absorption of ZnO nanoparticles can provide valuable information regarding particle size, shape and sample homogeneity (73). There exist three main components for a ZnO absorption spectrum: absorbance edge, main peak and tailing absorption. Tailing is most

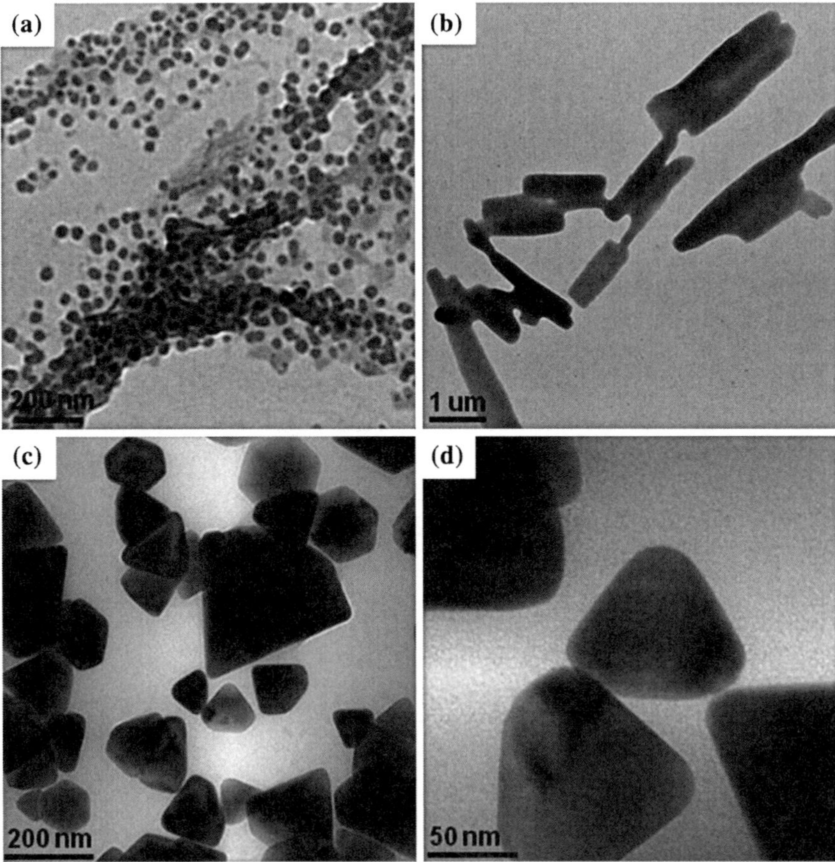

Fig. 8.2 TEM images of ZnO nanostructures prepared by MWI of 0.1 M zinc acetate solution in pure oleic acid for 30 s (**a**) and 60 s (**b**). The *triangular pyramid shape* particles shown in (**c**) and (**d**) were prepared by MWI of 0.1 M zinc acetate solution in pure oleylamine for 10 min

likely attributed to scattered light from the colloidal suspension. The absorption edge is due to electronic transitions between the valence and conduction bands. The sharpness of the edge is related to the homogeneity of the sample, with the edge broadening as the particle size distribution increases (73). However, it is important to note that the largest particles are the main contributors to the absorbance edge while all particles contribute to the maximum absorbance peak. As shown in Fig. 8.5, all absorption peaks are significantly blue-shifted from the absorption of bulk ZnO, which absorbs at approximately 385 nm (3.2 eV). Since both the triangular and hexagonal pyramidal shape particles are similar in size, they display similar onset absorption around 370 nm with tailing absorption at longer wavelengths. Rods prepared in the presence of oleic acid only show a significantly more blue-shifted absorption peak (around 350 nm) than the pyramidal structures. It is worthy to note that the rods are larger than the pyramidal

Fig. 8.3 TEM images of ZnO nanostructures prepared by MWI of 0.1 M zinc acetate solution in different molar ratios of oleic acid (OAc) to oleylamine (OAm). The ratios of OAc : OAm are: (**a**) 1:1 for 10 min, (**b**) 3:1 for 10 min and (**c**, **d**) 1:3 for 15 min

structures particles, yet their absorptions appear blue-shifted. Since smaller particles absorb at shorter wavelengths, this absorption peak may be attributed to many small particles adsorbed at the edges of the ZnO rods.

8.5.2 TiO_2 Colloidal Nanostructures

TiO_2 nanocrystals have attracted much interest due to their unique electronic, optical, catalytic, chemical, and photochemical properties which contribute to a wide variety of applications in solar cells, pigments, gas sensors, and photocatalysis [80–85]. The synthesis of colloidal TiO_2 with controlled size, shape and crystal structure is essential for many of theses applications. Bulk TiO_2 has three

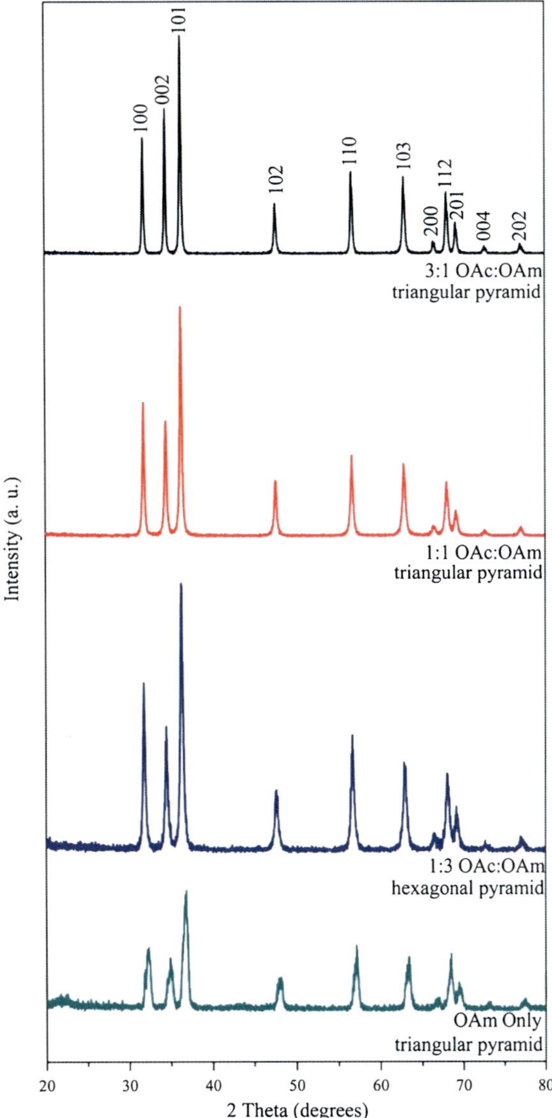

Fig. 8.4 XRD patterns of ZnO nanostructures prepared by MWI of 0.1 M zinc acetate solution in different molar ratios of oleic acid (OAc) to oleylamine (OAm) as indicated

crystalline phases: rutile is the high temperature phase, and both anatase and brookite are metastable phases and transform to rutile at high temperatures, around 700–1,000 °C [80–85].

The TiO_2 nanocrystals are prepared using titanium butoxide + $TiCl_4$ as precursors and OAc-OAm as ligating solvents [86]. Figure 8.6 displays TEM images of the TiO_2 nanocrystals prepared under different reaction conditions. When the mole ratio of the precursors to the OAm to the OAc is approximately 1:4:2 under short MWI times (5–10 min), small spherical or elongated particles are formed

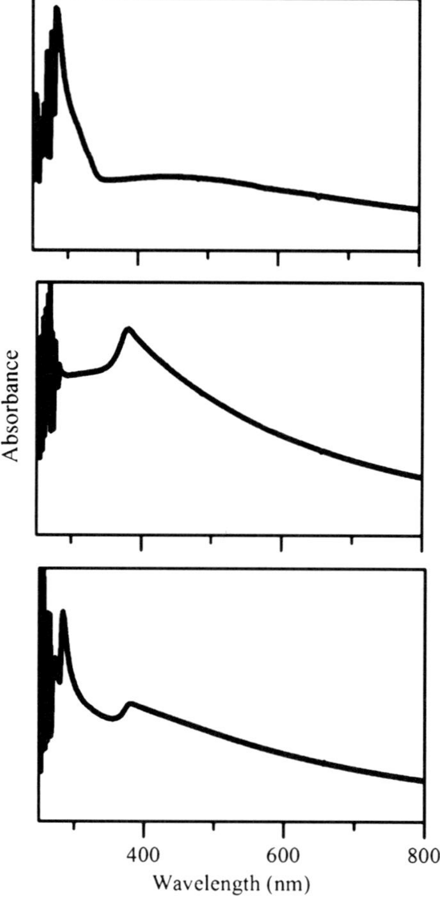

Fig. 8.5 UV-Vis absorption spectra of ZnO nanostructures prepared by MWI of 0.1 M zinc acetate solution in (*top*) pure oleic acid (30 s), (*middle*) 1:1 oleic acid to oleylamine (10 min), and (*bottom*) pure oleylamine (10 min)

(Fig. 8.6a). Rods are obtained by using only OAm with a higher concentration relative to the precursor (6:1), as shown in Fig. 8.6b. Prolonging the MWI reaction time results in growth of the nanoparticles to form cube-like (Fig. 8.6c) and star-like aggregated rods (Fig. 8.6d). For example, using relative concentrations of the precursor to the OAm to the OAc of approximately 1:6:9 under long MWI times (10–15 min) results in orientationally aggregated rods around a central particle producing a star-like shape (Fig. 8.6d). However, if the concentration of the OAc is increased (precursor: OAm: OAc = 1:6:18), most of the nanocrystals display cube-like and bipyramid shapes (Fig. 8.6c). The selective crystal growth with different shapes is due to the strong ligand binding and passivation of the oleic acid—oleylamine to the TiO_2 nanocrystal surfaces. The growth of cube-like and bipyramid shapes under high concentrations of oleic acid is consistent with its strong binding to certain surfaces of the anatase crystals [80, 83].

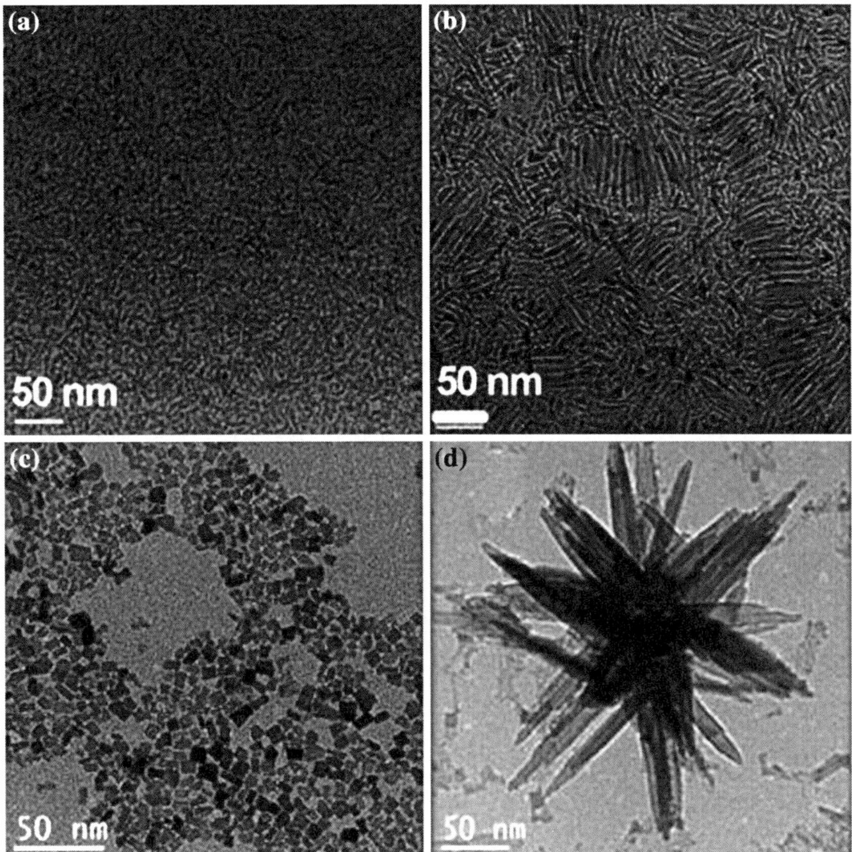

Fig. 8.6 TEM images of TiO_2 nanocrystals with different shapes: **a** spherical, **b** rods, **c** cubes and **d** star-like shape

The XRD pattern of the prepared TiO_2 nanocrystals, displayed in Fig. 8.7a, is consistent with the standard pattern of anatase (JCPDS 21–1272). No peaks are found at 27° or 31°, indicating the absence of rutile and brookite phases. However, the results indicate that depending on the precursor used, different phases of TiO_2 are obtained. For example, when titanium tetrachloride was used, a mixture of anatase and rutile was obtained.

The UV-Vis absorption and PL spectra are shown in Fig. 8.7b. The UV spectrum is typical for anatase [87, 88]. The PL spectrum exhibits two peaks due to band-to-band transitions centered around 384 nm, and surface trap states at 425 nm. No peaks are observed between 600 and 800 nm, which are associated with transitions from the conduction band edge to holes trapped at the interstitial Ti^{3+} site [81].

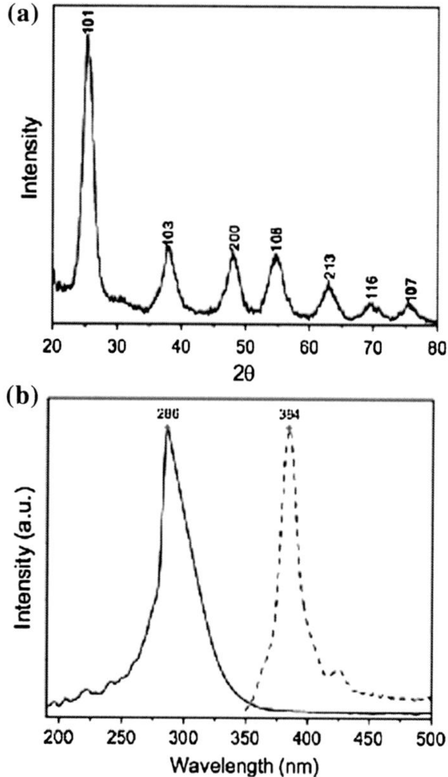

Fig. 8.7 **a** XRD pattern and **b** UV-Vis absorption (*solid line*) and PL (*dotted line*) spectra of TiO$_2$ nanocrystals

8.5.3 CeO$_2$ Nanostructures

CeO$_2$ nanostructures have wide applications in many fields such as catalysis, fuel cells, surface modifications, lubrications, gas sensors, radiation protectors, and anode materials for lithium ion battery systems [78, 89, 90]. The properties of CeO$_2$ nanocrystals are significantly different, and in most cases, superior to the properties of bulk ceria composed of micron-sized particles. For example, significant enhancement is found in the rate of CO oxidation on gold nanocatalysts supported on CeO$_2$ nanocrystals as compared to the same catalyst supported on micron-sized CeO$_2$ particles [27]. Also, the electronic conductivity of CeO$_2$ nanocrystals is much higher than in large particles [78]. It is expected that CeO$_2$ nanocrystals with different shapes would exhibit different quantum confinement effects, and therefore display different optical and PL properties.

For the MWI synthesis of CeO$_2$ nanocrystals, cerium acetate in a mixture of oleic acid and oleylamine is first heated to about 110 °C in order to remove the hydrated water and obtain a clear solution. The mixture is then transferred quickly to a conventional microwave for the desired amount of time. Different shapes of the CeO$_2$ nanocrystals can be obtained at different MWI times by using the mole ratio of 1:14:9 corresponding to Ce(ac)$_2$:OAm:OAc, respectively. For example, at MWI times of (1–4 min), (5–10 min) and (15–20 min), spherical particles, short rods and wires, respectively are formed as shown from the TEM images of Fig. 8.8a, b, and c, respectively. The 1D orientational growth leading to the formation of rods and wires is induced by the presence of oleic acid which exhibits strong binding and selective adsorption to the surface atoms of the CeO$_2$ nanocrystals. Increasing the concentration of the precursor and decreasing the concentration of OAc produces cube-like nanocrystals as shown in Fig. 8.8d. This is consistent with the nonselective adsorption of oleylamine on the CeO$_2$ nanocrystals, and thus the growth occurs in 3D to produce the cube-like nanocrystals.

The evolution of the XRD patterns of the CeO$_2$ nanoparticles prepared after 10, 20 and 30 min MWI times, respectively is shown in Fig. 8.9a. The patterns could be readily indexed to the face-centered cubic phase (space group: *Fm3 m*, JCPDS #34–0394). There is considerable broadening in the XRD peaks since the CeO$_2$ nanoparticles are very small in dimension as shown from the TEM image in Fig. 8.9a. As the MWI times increase, the diffraction peaks become sharper consistent with the shape transformation from very small spherical particles to short rods and long wires or to cube-like nanocrystals.

Figure 8.9b displays the UV-Vis and PL spectra of the CeO$_2$ cube-like nanocrystals. The UV-absorption shows a peak around 300 nm corresponding to a charge transfer transition from O to Ce in the CeO$_2$ nanocrystals. Extrapolation of the absorption edge of this band (335 nm) indicates a band gap of 3.7 eV, which confirms the strong quantum confinement in the prepared CeO$_2$ cube-like nanocrystals. This value is consistent with the reported band gap energies of ceria nanoparticles (3.03 to 3.68 eV) prepared using sonochemical synthesis and larger than the values (3.38 to 3.44 eV) reported for the particles prepared using reverse micelles [91, 92]. The PL spectrum shows a major band around 430 nm, and two low energy bands which could be attributed to surface trap states. These bands could be related to surface modifications of the nanocrystals by the oleylamine passivation.

8.5.4 Other Rare Earth Oxide Nanostructures

Rare earth oxides with one dimensional structure represent a particularly interesting class of materials due to their unique electronic, optical, magnetic and catalytic properties arising from the confinement of the 4f electrons [35, 93–96]. These properties are critical for many interesting applications involving, for example, optical displays, optical communication, UV shielding, medical diagnostics, and efficient catalysis for the oxidation of heavy oils, jet fuels and coal gasification [35, 94, 96].

Recently, we developed a rapid, simple and versatile MWI methodology for the synthesis of organically passivated uniform, single crystalline rare earth oxide (M$_2$O$_3$, M = Pr, Nd, Sm, Eu, Gd, Tb, Dy) nanorods ($\sim$1.2 $\times$ 4–5 nm), and cube-like or

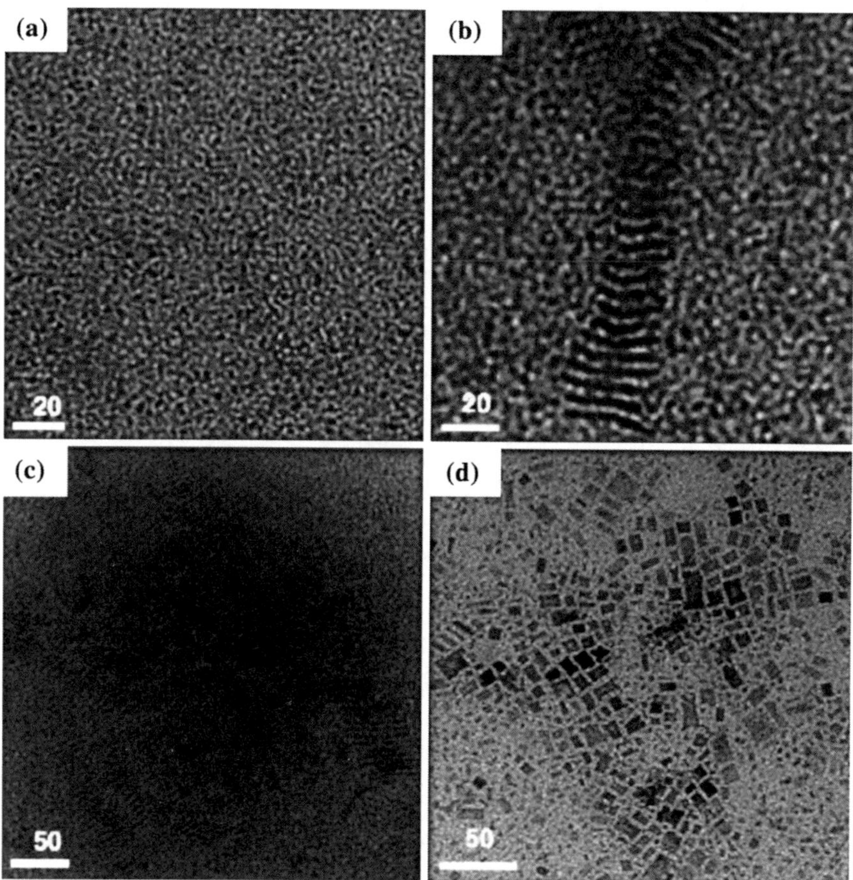

Fig. 8.8 TEM images of CeO$_2$ nanocrystals with different shapes: **a** spheres, **b** rods, **c** wires and **d** cubes

square nanoplates (6 × 6 nm) [35]. Figure 8.10a, b and c display representative examples of the TEM images of the as-synthesized Eu$_2$O$_3$, with spherical, rod and cube-like shapes, respectively. Figure 8.10d and e display TEM images of Gd$_2$O$_3$ (wires), and Pr$_2$O$_3$ (rods), respectively. The relative amount of rods and cube-like nanocrystals can be controlled by the MWI time, and the relative concentrations of the metal precursor, the oleic acid and the oleylamine. The competitive adsorption of oleic acid and oleylamine can effectively inhibit the growth of the nanocrystal in all but the favorable crystallographic plane where the growth is significantly enhanced thus resulting in a 1D structure [34, 35]. The crystal plane with a higher surface energy is expected to have a faster growth rate. By varying the relative concentrations of oleic acid and oleylamine in the MWI syntheses, the shape of the resulting nancrystals can be controlled. When the mole ratio of the metal precursor: OAc: OAm is 1:17:17, spherical particles are formed as shown in Fig. 8.10a for the Eu$_2$O$_3$ nanocrystals. This

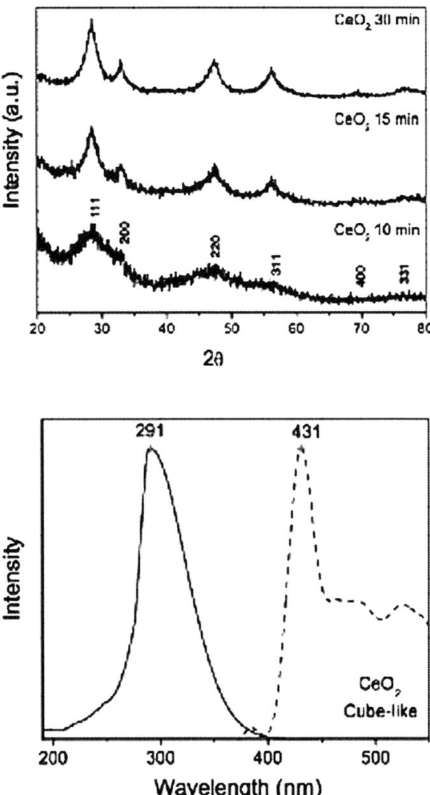

Fig. 8.9 a XRD patterns of CeO$_2$ nanocrystals prepared at different MWI times. **b** UV-Vis absorption (*solid line*) and PL (*dotted line*) spectra of CeO$_2$ nanocrystals

is consistent with the stronger binding capability of oleic acid relative to oleylamine, and therefore it binds strongly with the atoms of all the planes and the thermodynamic limit (3D) is approached in the presence of excess oleic acid. The nanorods are formed when the mole ratio of OAm: OAc is about 3:2 as shown in Fig. 8.10b. However, when oleylamine is present in excess, cube-like or square nanoplates are formed as shown for Eu$_2$O$_3$ in Fig. 8.10c. Repeated sets of experiments under the same MWI times, show that the nanorods and the cube-like (or square nanoplates) are predominantly formed when the mole ratios of the metal acetate: OAm: OAc are 1:22:14 and 1:27:9, respectively. These results demonstrate that by controlling the composition of the capping ligands good control over the shape of the resulting nanocrystals can be achieved during the growth process.

The XRD patterns of the rare earth nanorods are consistent with the cubic Ia3 space group and exhibit intensity enhancement of the (400) diffraction compared to the standard patterns for the bulk rare earth oxides [35]. This enhancement of the diffraction intensity from the (400) plane is unique to the nanorods since it is not observed in spherical or plate-like nanocrystals [94]. The high-resolution TEM images of the individual rods and wires show well-resolved lattice planes perpendicular to the long axis with an interplanar distance corresponding to the

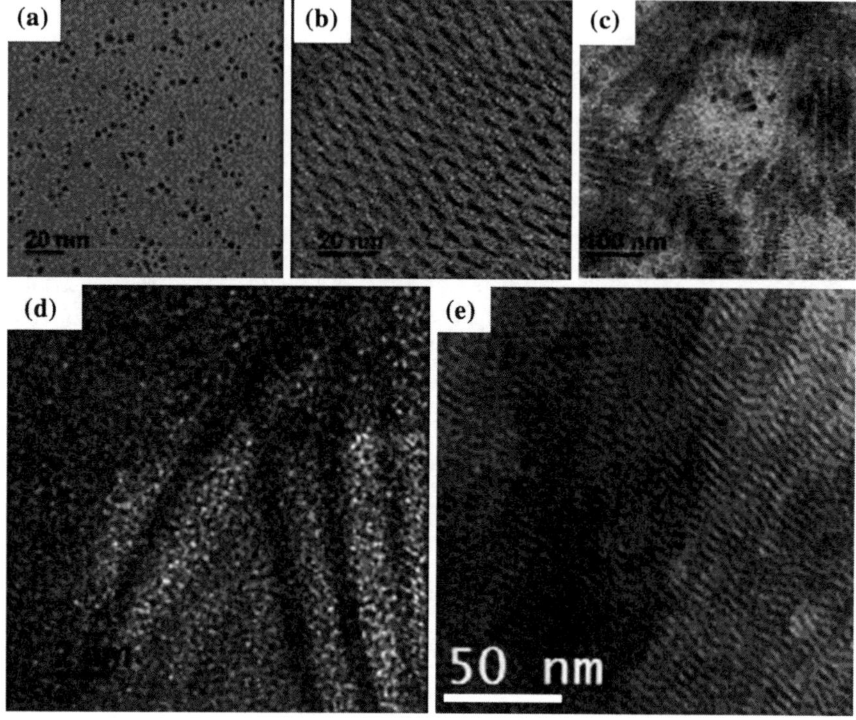

Fig. 8.10 TEM images of Eu_2O_3 **a** spheres, **b** rods and **c** plates, and **d** Gd_2O_3 wires and **e** Pr_2O_3 rods

d spacing of the (400) plane of the cubic Ia3 space group [35]. The assembly of the rods into ordered 2D supercrystals (Fig. 8.10b) has been confirmed by small angle XRD analysis [35].

The PL properties of the rare earth nanorods synthesized by MWI have been investigated [35]. For example, the PL spectrum of the Eu_2O_3 nanorods shows strong split peaks at 612 and 620 nm which can be assigned to the $^5D_0 \rightarrow {}^7F_{1,2,3,4}$ transitions of the Eu^{3+} ions [35, 97]. For spherical Eu_2O_3 nanoparticles, these peaks appear as a single broad band centered around 615 nm [97]. The difference is probably due to different surface sites occupied by the Eu^{3+} ions in the nanorods.

8.5.5 Transition Metal Oxide Nanostructures

In this section, we demonstrate the application of the MWI method for the synthesis of nearly monodisperse transition metal oxide nanocrystals. Figure 8.11 displays representative TEM images of the synthesized ZnO, Ga_3O_4, Mn_3O_4 and In_3O_4 nanocrystals. The compositions of the oxide nanocrystals have been confirmed by the XRD data. These nanocrystals are produced by the MW decomposition of the corresponding metal acetylyacetonate in oleylamine –DMF solutions (mole ratio of acetylacetonate: OAm is 1:15). Under these conditions, Ga_3O_4 and Mn_3O_4 nanocrystals form spherical particles while ZnO and In_3O_4 produce nanoprisms and nanoplates, respectively. We are currently exploring the use of mixed ligand systems such as OAm–OAc and ODA as well as different metal precursors in order to control the composition of the oxide nanocrystals. For example, in addition to Mn_3O_4, MnO nanocrystals can be synthesized by adding a few drops of formic acid to the oleylamine liganding solvent which tends to increase the reducing environment necessary for the synthesis of MnO.

8.5.6 Shape-Controlled Metal Ferrite Nanostructures

The nanocrystalline ferrite spinel-structured MFe_2O_4 materials (MFe_2O_4, M = Mn, Fe, Co, Ni, Zn, Ni) have been the subject of extensive research over the past decade due to their unique magnetic properties which make them scientifically and technologically important [97–111]. In this class of iron oxide materials, the magnetic properties can be tuned over a wide range depending on the nature of the dopant ions, M^{2+}. This tunability is critical for a variety of important applications involving ferrofluids, data storage, color imaging, catalysis, biosensors, and biomedical applications such as drug delivery, hyperthermia of cancer cells, magnetic labeling, etc. Most of these applications require particles of uniform size and shape distribution [100–102, 105, 107, 110]. In addition, depending on the particular application; the magnetic nanoparticles must possess specific properties. For example, data storage applications require particles with stable, switchable magnetic states not affected by temperature fluctuations. In biomedical applications, the nanoparticles must exhibit super paramagnetic behavior at room temperature. For high performance electromagnetic and spintronic device applications, the tunability of both the magnetic properties and the electrical resistivity are important [99, 103, 108, 111].

Several techniques have been developed to synthesize the ferrite nanocrystals including for example, sol–gel processes, hydrothermal methods, co-precipitation from solutions or reverse micelles, sonochemical reactions, solid state reactions, and mechanical ball milling [112–115]. However, many of these methods produce nanoparticles that are severely aggregated with nonuniform shape and size, which create serious disadvantages for successful applications [116]. Moreover, in some methods, the synthesized particles are poorly crystalline and calcination at high temperatures is needed to induce the highly crystalline structures [117]. This could lead to other problems since high temperature annealing often induces coalescence and coarsening of the nanoparticles, thus broadening the size distribution. Furthermore, magnetic hardening occurs after heat treatments at relatively high temperatures [118]. For these reasons, the synthesis of colloidal magnetic oxide nanocrystals by non-aqueous methods has been developed to overcome the

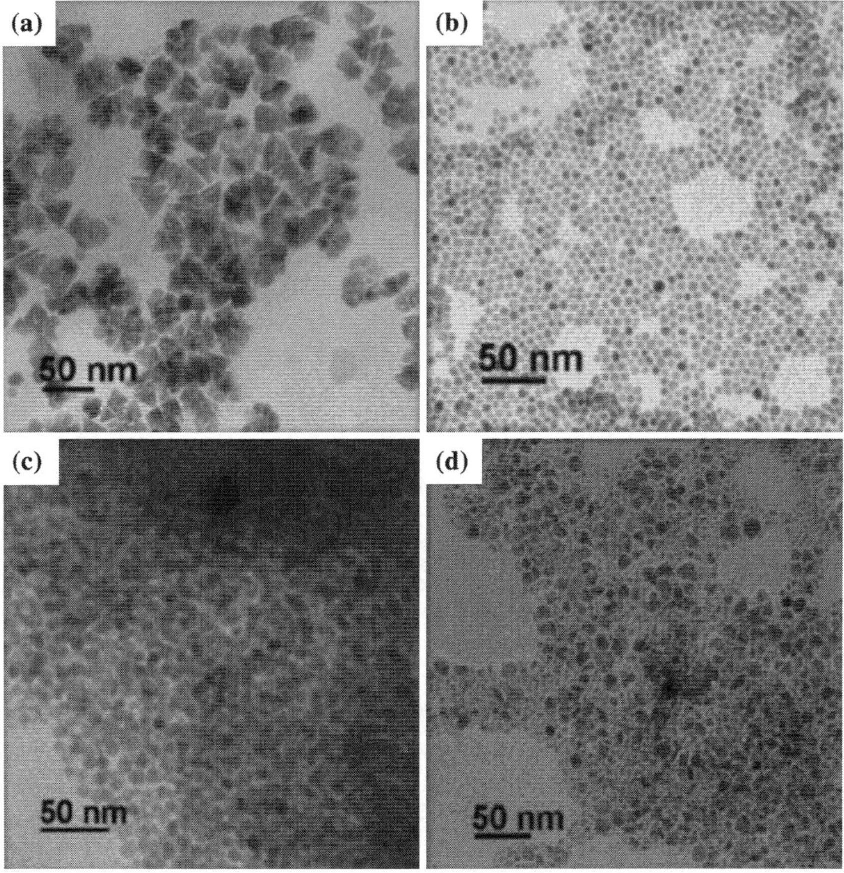

Fig. 8.11 TEM images of ZnO prisms, Ga_3O_4 spheres, Mn_3O_3 spheres and In_3O_4 cubes

disadvantages associated with the above mentioned methods. For example, Rockenburger et al. produced transition metal oxides nanocrystals by pyrolysis of organometallic precursors in long chain amines [119]. Jana et al. synthesized transition metal oxides nanocrystals by decomposition of metal carboxilates in oleic acid and octadecene [120]. Hyeon et al. reported the synthesis of γ-Fe_2O_3, Fe_3O_4, $CoFe_2O_4$, and $MnFe_2O_4$ using metal carbonyls as precursors and trimethylamine oxide as an oxidant in non-aqueous solutions [117, 121, 122]. Sun et al. demonstrated the synthesis of nearly mono-disperse Fe_3O_4, $CoFe_2O_4$, and $MnFe_2O_4$ nanocrystals using metal acetylacetonates as precursors in the presence of 1,2-hexadecanediol, oleylamine, and oleic acid [123]. All of these successful methods involve high temperature (~ 300 °C) decomposition of the metal precursor and a rigorous inert environment. The requirements of high temperature and inert atmosphere limit the large-scale production of the magnetic nanoparticles.

Clearly, there is a need to develop a general and simple synthetic method operative at near room temperature in air and applicable to a wide range of ferrite systems.

Recently, we have developed a general procedure for the synthesis of a variety of high quality, crystalline metal ferrite nanostructures (MFe_2O_4, M = Mn, Fe, Co, Ni and Zn) with narrow size distributions in a mixture of oleic acid and oleylamine using the MWI approach [124]. A brief summary of these results are described below.

The shape and size of the nanocrystals depend on the experimental conditions that influence the degree of supersaturation, the nucleation rate and the growth kinetics of the nanocrystal phase. In the MWI method, these conditions include the decomposition rate of the ferrite precursors, the effective reaction temperature during the MW process, the air oxidation process, the reaction time, and the rates of adsorption of the organic ligands onto the nanocrystal surfaces. The relative binding energies of the ligating solvents (oleylamine and oleic acid) to the precursor molecules and to the newly formed nanocrystals also affect the nucleation rate [21]. In general, ligands which bind weakly to the precursors increase the nucleation rate while ligands which bind weakly to the newly formed nanocrystals enhance the growth rate.

The competitive adsorption of oleic acid and oleylamine can effectively inhibit the growth of the nanocrystal in all but the favorable crystallographic plane where the growth is significantly enhanced. The oleylamine binds on the surface of the ferrite nanocrystals more weakly than oleic acid. A weakly binding ligand allows for further growth of the nanocrystals since the ligand can reversibly coordinate to the newly grown nanocrystals. This also allows for the nanocrystal growth via the Ostwald ripening process where large crystals grow even larger on the expense of smaller nanocrystals [47].

It was observed that microwaving the reaction mixture for 6–8 min, 10–14 min, and 16–20 min resulted in nanocrystals with diameters of 4–5 nm, 8–10 nm, and $\sim$20 nm, respectively. We have also observed that when the molar ratio of oleic acid to oleic oleylamine to $Fe(acac)_3$ is less than 3:3:1 spherical particles are produced. However, by increasing the molar ratio of the ligating solvents (the ratio of oleic acid to oleylamine to $Fe(acac)_3$ is 4:4:1 or more), large 30 nm quasi triangular and cube-like particles are produced.

Both oleic acid and oleylamine are essential for the formation of the Fe_3O_4 nanocrystals. Microwaving of $Fe(acac)_3$ in only oleic acid did not result in any particles' formation and in using only oleylamine amine resulted in very large particles. Here the amine acts as an activating agent for the decomposition of $Fe(acac)_3$ to Fe_3O_4 and not as stabilizing agent since the amine can stabilize Fe_2O_3 but not Fe_3O_4 [125, 126]. Increasing the ratio of oleic acid to oleylamine from 1:1 resulted in larger particles with different shapes. For example, using oleic acid to oleylamine ratios of 1.5:1, 2:1 and 3:1, resulted in spherical, triangular and quasi cube (polyhedron) shapes, respectively.

For Fe_3O_4, the size of the nanocrystals can be varied from 4 to 20 nm by changing the microwaving time or by changing the ratio of ligating solvents. Figure 8.12a and b and display TEM images of Fe_3O_4 nanocrystals produced by

MWI reaction times of 6–8 min, and 15–18 min; respectively using the same reaction conditions with the 1:1 ratio of oleic acid to oleyamine under the 3 min reaction cycles described in the experimental section. The resulting spherical nanoparticles with 4 and 20 nm diameters are nearly monodisperse and exhibit no sign of aggregation. In contrast to the stepwise MWI cycles, continuous microwaving irradiation results in particles with broad size distribution and some aggregated particles. The change in the particles' size and the broad size distribution is most probably due to the Oswald ripening process which becomes more significant at local higher temperatures and concentrations [47–49]. This also indicates that the crystal growth process following the nucleation of the Fe_3O_4 nanocrystals is not particularly fast since continuous MWI results in a broad size distribution. Figure 8.12c and d display TEM images of the Fe_3O_4 nanoparticles prepared using oleic acid to oleylamine ratios of 2:1 and 3:1, respectively. It is clear that by increasing the concentration of the oleic acid, triangular and quasi cube (polyhedron) shapes are produced.

Figure 8.13 represents the XRD patterns of the as synthesized Fe_3O_4 nanoparticles with particle sizes of (a) 4 nm, (b) 8 nm and (c) 20 nm corresponding to the TEM images displayed in Fig. 8.12a, b and c, respectively. The XRD pattern shown in Fig. 8.13d belongs to the quasi triangular particles with the TEM shown in Fig. 8.12d. It is clear that the particles are highly crystalline and all the characteristic peaks for the spinal Fe_3O_4 phase are present and well resolved. The gradual decrease in the peak broadenings from a to d is consistent with increasing the average particle size as confirmed by the TEM images shown in Fig. 8.12. The average particle diameter estimated from the Scherrer's formula is consistent with that determined by the analysis of the TEM images [127].

We have readily extended the developed synthesis by microwave irradiation to the synthesis of several metal ferrites MFe_2O_4 where M = Co, Mn, Ni, Zn. For the synthesis of these MFe_2O_4, we have used $M(acac)_2$ in 1:2 ratio with $Fe(acac)_3$ instead of using only $Fe(acac)_3$; other reaction conditions were maintained the same as for the Fe_3O_4 synthesis.

For all the ferrites we have synthesized nanoparticles with 4 different sizes simply by increasing the microwave irradiation times under otherwise identical experimental conditions. Examples of TEM images of $NiFe_2O_4$, $ZnFe_2O_4$ and $MnFe_2O_4$ nanocrystals produced are shown in Fig. 8.14. The TEM images show that the nanoparticles exhibit narrow size and shape distribution similar to the case of Fe_3O_4. However, the same conditions used to prepare the ∼50 nm triangular like particles of Fe_3O_4 (oleic acid to oleylamine ratio of 2:1) when applied to the $MnFe_2O_4$ synthesis resulted in the formation of ∼120 nm cube like structures as shown in Fig. 8.14. This indicates that the growth kinetics of the nanocubes is much faster for $MnFe_2O_4$ than for Fe_3O_4.

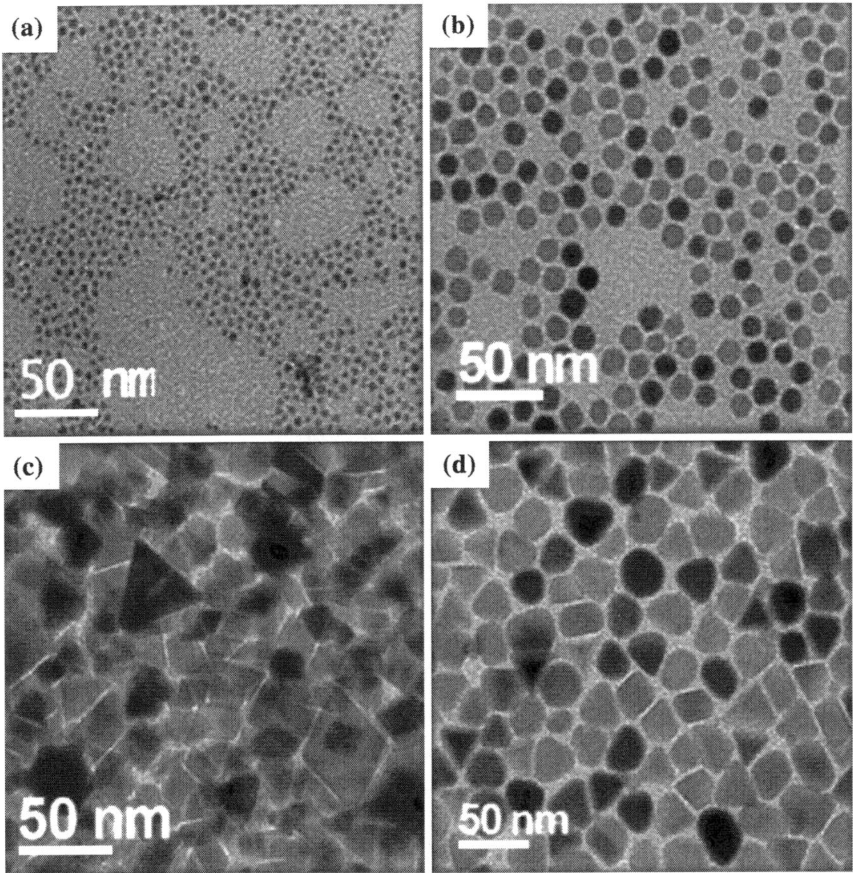

Fig. 8.12 TEM images of Fe$_3$O$_4$: nanostructures of different sizes and shapes. **a** 4 nm and **b** 20 nm prepared using the same ratio of oleic acid to oleyl amine (1:1) but different microwave times (6–8 min and 15–18 min, respectively). **c** *Triangular* and *prism* shapes, and **d** Quasi *cubes* and *triangular* shapes prepared using oleic acid to oleylamine ratios of 2:1 and 3:1, respectively

8.6 Metal Oxide Nanoparticles for Photocatalysis

Metal oxide nanoparticles are of great interest for the degradation of organic dyes [128–149]. Industrial processing and manufacturing of textiles and paper as well as printing and photography has resulted in a significant amount of brightly colored wastewater that is polluted with non-biodegradable organic dyes [128, 130, 138]. This pollution is hazardous to the environment by damaging stream ecosystems as well as consumers of contaminated fish. Therefore, a way to reduce the pollution of organic dyes from contaminated water is desirable.

The most widely used dyes are aromatic organic compounds. Many of these dyes are known to degrade into highly carcinogenic aromatic amine compounds; therefore, they must be removed from industrial waste. Unfortunately, their high

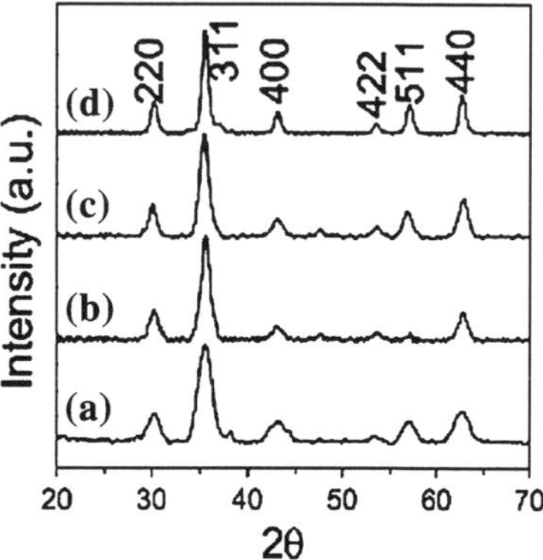

Fig. 8.13 XRD patterns of the synthesized Fe_3O_4 nanoparticles **a** 4 nm, **b** 8 nm, **c** 20 nm and **d** quasi *triangular* particles

solubility makes them difficult to remove and enhances their contamination potential [138]. Current removal techniques include chemical coagulation, air location and adsorption methods [138]. These methods simply remove the hazardous dyes from the water and do not reduce the toxicity or the compounds. Degrading the organic dyes into nontoxic species is a more appealing solution offered by photodegradation on semiconductor surfaces.

Metal oxide nanoparticles, such as TiO_2 and ZnO, are photocatalysts shown to degrade organic dyes to harmless, colorless end products in the presence of sunlight/UV-light. TiO_2 and ZnO are well suited as photocatalyts because of their high photosensitivity, stability, large band gap and relatively low cost. Compared to TiO_2, ZnO is advantageous because it absorbs over a larger region of the solar spectrum [130]. Furthermore, both TiO_2 and ZnO are not environmental hazards and may be used for in situ degradation of organic dye pollutants.

Experimentally, observing degradation of organic dyes is quite simple. Organic dyes have high molar absorbtivities and form brightly colored solutions. A traditional experiment involves introducing a known amount of catalyst particles to a solution having a fixed dye concentration. The solution is then stirred under dark conditions to allow for all of the dye molecules to adsorb to the semiconductor particles' surfaces. After approximately 30 min, the solution reaches equilibrium conditions. Then, the solution is irradiated using either a UV or solar light source. At specific time points, aliquots are removed and analyzed by UV–Visible spectroscopy, which are used to determine the rate of degradation. It is important to note that solutions are stirred magnetically during the entire photodegradation process. Figure 8.14 shows the degradation of malachite green in the presence of ZnO nanoparticles. Malachite green shows a strong absorption peak at 627 nm. As the solution is exposed to UV irradiation, the intensity of this peak decreases as seen in Fig. 8.15a. The decoloration of the solution, which is easily observable by UV–Visible spectroscopy, indicates degradation of malachite green into benign products. As irradiation time increases, the degradation of dye also increases. Degradation kinetics are easily followed by plotting absorbance intensity of the dye peak is as a function of time (Fig. 8.15b).

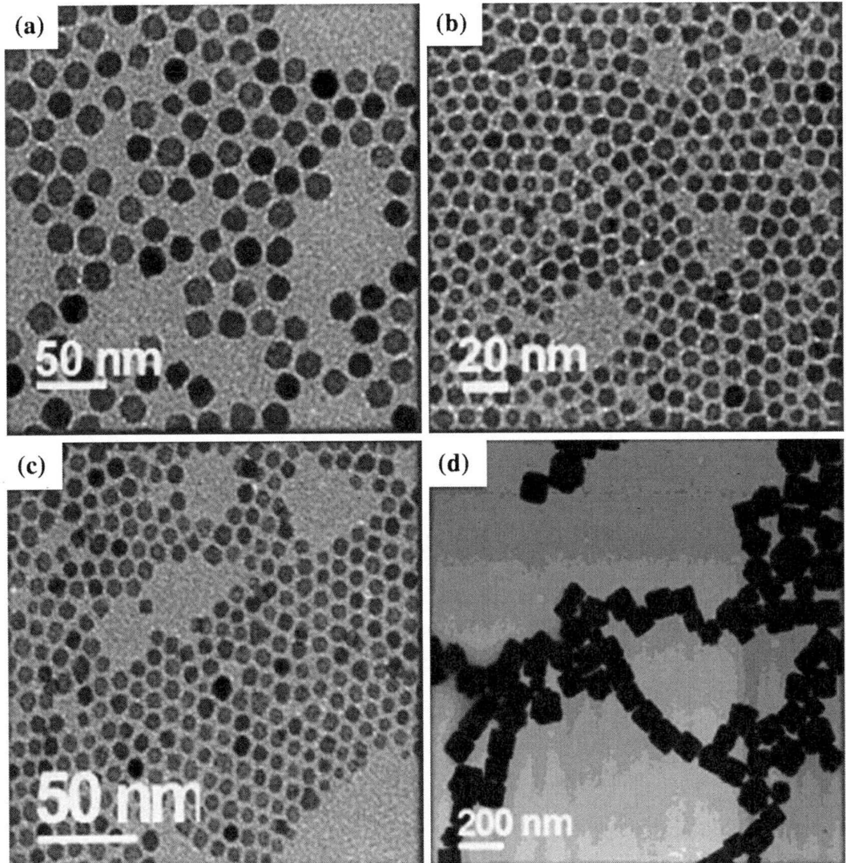

Fig. 8.14 TEM images of metal ferrite nanostructures of different sizes and shapes. a 20 nm $NiFe_2O_4$ and b 8 nm $CoFe_2O_4$, c 8 nm $MnFe_2O_4$, and d 120 nm $MnFe_2O_4$ cubes. The spherical particles and the cubes were prepared using 1:1 and 2:1 mol ratios of oleic acid to oleylamine, respectively

The mechanism by which photodegradation occurs is a result of semiconductor nanoparticles acting as sensitizers for light-reduced redox processes [130, 134]. In solution, organic dyes adsorb to the surface of metal oxide nanoparticles. Photodegradation on semiconductor surfaces occurs when the photocatalyst absorbs UV light and generates an electron/hole pair by exciting an electron from the valence band into the conduction band. The excited electron and corresponding holes then promote chemical reactions that degrade the dye. This is known as an advanced oxidation process (AOP) that generates hydroxyl radicals and destroys pollutants by oxidizing dye molecules. The reaction mechanism for ZnO photocatalysis occurs as follows. Note that a similar reaction mechanism also occurs for TiO_2 particles. First, a photon generates an electron hole pair

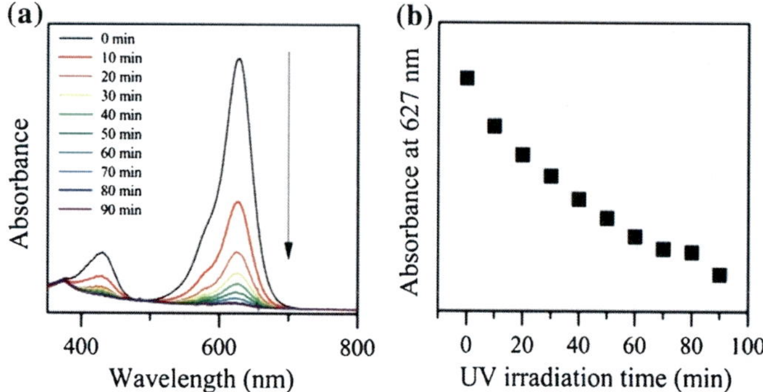

Fig. 8.15 **a** Absorbance spectra of malachite green oxidized in the presence of ZnO nanoparticles with respect to UV irradiation time. **b** Absorbance intensity of malachite *green* at 627 nm plotted as a function of UV irradiation time. ZnO acts as a photocatalyst and the strong absorbance peak of malachite *green* decreases as the organic dye is degraded into benign products

$$ZnO + h\nu \rightarrow ZnO\left(e_{CB}^{-} + h_{VB}^{+}\right). \tag{8.6}$$

If the photogenerated species does not recombine or dissipate as heat, it will react with nearby oxygen or water molecules to form oxygen and hydroxyl radicals [128, 131, 145], respectively

$$e_{CB}^{-} + O_2 \rightarrow \cdot O_2^{-} \tag{8.7}$$

$$h_{VB}^{+} + H_2O \rightarrow \cdot OH + H^{+} \tag{8.8}$$

The produced oxygen and hydroxyl radicals are strong oxidizers and will promote oxidation of the organic dyes on the surface, hence the term advanced oxidation process is given to the degradation process. Degradation occurs when the hydroxyl radicals react with dye molecules

$$\cdot OH + dye \rightarrow H_2O + dye\ intermediate \tag{8.9}$$

$$\cdot OH + dye\ intermediate \rightarrow H_2O + CO_2 + mineralization\ products. \tag{8.10}$$

The hydroxyl radical is the critical species for the degradation process; this is shown in the following discussion on reaction kinetics. Final degradation of dye intermediates is crucial for environmental safety as the incomplete oxidation leads to intermediates that are often more toxic than the parent dye molecule.

The degradation process is fairly complex. Understanding the reaction kinetics and the many parameters involved is critical to optimize the system and apply this degradation phenomenon for reducing polluted water sources.

The degradation process is strongly influenced by the initial concentration of the dye. Equilibrium of adsorption of dye molecules onto the catalyst surface must be achieved; then, the degradation rate of the dye species is related to the formation of hydroxyl radical. The formed hydroxyl radicals react rapidly with the aromatic ring compounds; therefore, formation of ·OH has been deemed the rate-determining step [144]. The overall rate constant (k) is proportional to the probability of generating hydroxyl radical on the catalyst surface ($P_{\cdot OH}$) and the radical reacting with the dye (P_{dye}) by

$$k = k_0 P_{\cdot OH} P_{dye} \qquad (8.11)$$

where k_0 is the reaction rate constant. Both the probability of generating radicals and the reaction probability are implicitly dependent on the initial concentration of the dye; k_0 is independent on the initial dye concentration. $P_{\cdot OH}$ is related to the initial concentration of the dye because the adsorbed OH- positions are replaced by dye molecules during the oxidation process; fewer OH- active sites will reduce the generation of · OH radicals and slow the degradation process [144].

The initial concentration of the dye also affects the reaction rate by influencing the path length of photons entering the solution [131, 144, 145]. Degradation occurs where the UV irradiation is most intense, which is on surface nearer to the irradiation source. As the concentration increases, the solution becomes darker and the path length decreases [144]. If the illumination time and amount of catalyst remain constant, the degradation rate is significantly decreased. Therefore, more catalyst is needed to degrade the dye.

Since the amount of catalyst relative to the concentration of dye affects the degradation rate [131, 144, 145], the affect of catalyst concentration should also be discussed. As expected, increasing the amount of catalyst surface available will increase the rate of photodegradation up to a certain limit. Once this limit is reached, increasing the amount of catalyst will lead to increased light scattering from turbidity [144] as well as a possible decrease in surface area from agglomeration [144]. Hence, increased amounts of catalysts may not increase the degradation rate as a result of aggregation and a reduced irradiation field.

It has been previously explained that the irradiation field plays a critical role in the degradation process. Since degradation does not occur in dark conditions, the catalyst surface cannot promote oxidation alone. As expected, increasing the irradiating light intensity increases the rate of degradation [144]. Thus, the intensity of irradiation affects the degree of light absorption by the catalyst surface. At low intensities, degradation is linearly dependent on the irradiation intensity [128, 144]. However, square-root dependence is observed at high intensities [128]. This is attributed to a couple of different reasons. At higher intensities, it is believed that there is increased importance on electron/hole pair recombination [128]. The square-root dependence is also attributed to a competing reaction between hydroxyl radicals that form hydrogen peroxide

$$2 \cdot OH \rightarrow H_2O_2 \qquad (8.12)$$

which dominates at higher intensities [128]. These finds are particularly important for industrial applications determining whether artificial irradiation sources or light concentrating systems would be profitable.

Table 8.2 Comparison of band gap energies of various semiconductor particles

Catalyst	Band gap energy (eV)
ZnO	3.17
TiO_2	3.10
SnO_2	3.54
ZrO_2	3.87
CdS	2.26
WO_3	2.76
α-Fe_2O_3	2.20

At this point, we have only discussed the use of ZnO and TiO_2 semiconductor nanoparticles as photocatalyts. Even though these are the most active, other semiconductor particles such as SnO_2, ZrO_2, Fe_2O_3, WO_3, CdS and ZnS have the required properties to act as photocatalysts [140, 144]. Sakthivel et. al tested a variety of semiconductor particles for the photoassisted degradation of acid brown 14 using sunlight [144]. Their results showed the order of photocatalytic activity as ZnO > TiO_2 > Fe_2O_3 > ZrO_2 > CdS > WO_3 > SnO_2. It is important to note that the particles used varied in size from 1.25 to 12.50 μm and had a broad range of surface areas. ZnO demonstrated increased photocatalytic activity, which is attributed to the increased absorption range of ZnO compared to TiO_2. The remainder of the results may be explained by the band gap energy of each catalyst, which is shown in Table 8.2. Particles with small band gaps, less than 3 eV, have a higher incidence of electron/hole pair recombination. At these energy gaps, the excited electron rapidly falls into holes and does not interact to form oxidative radial species. For particles with larger band gaps, light irradiation is not sufficient enough to excite an electron from the valence band into the conduction band and therefore no radical species are formed. These results demonstrate that the band gap between 3.1–3.17 eV may be ideal for photodegradation processes.

The surface of the metal oxide surface is critical to its performance as a photocatalyt. It is well known that metal-oxide surfaces contain a large number of point defects, the most predominant of which are oxygen vacancies [150]. Defects have a strong affect on the electronic band structure and therefore the catalytic and other properties. Oxygen vacancies are responsible for electronic structure surface band bending. High concentrations of O-vacancies leave extra electrons at the surface, which change the surface carrier concentration. This change has a strong influence on the electronic conductivity. Surface defects also influence the chemisorption of gases, resulting in a change of the surface conductivity, a requirement for ZnO gas sensors. Therefore, the amount of surface defects, namely O-vacancies, is directly related to the use of ZnO in electronic, catalytic and sensor applications. For this reason, the preparation method utilized to synthesize metal oxide nanoparticles directly affects the application.

Further tuning of photocatalytic activity is possible by doping the metal oxide particles. Adding a metal dopant has been shown to increase the photocatalytic activity of semiconductor nanoparticles by shifting the light absorption region [134]

and inhibiting electron/hole pair recombination [133, 134]. In metal–semiconductor composite materials, the metal acts as an electron sink and electrons accumulate on the metal [134]. This accumulation of electrons on the metal causes the Fermi level to shift towards the conduction band of the semiconductor [134]. By influencing the interfacial charge transfer process, interactions between the metal and semiconductor nanoparticles maximize the efficiency of photocatalytic reactions [131, 144, 145]. However, it should be noted that some cases have shown metal doping to negatively effect the photocatalytic process; the metal and semiconductor combination plays a crucial role in determining the effectiveness [133, 134].

Overall, metal oxide particles are extremely promising for degrading organic dyes that enter the environment through various industrial-processing techniques. Metal oxide particles are able to completely degrade organic dyes into non-hazardous products when irradiated with sunlight; this is very appealing because additional processing is not required as with current dye removal techniques. Tuning the system for photodegradation requires careful selection of metal oxide, dopant metal and control of experimental parameters. Even though further research is still needed to implement metal oxides as photocatalysts in industrial applications, the outlook shows progress towards wastewater clean up.

8.7 Applications in Sensors

Metal oxides, particularly ZnO, SnO_2, and TiO_2, demonstrate gas-sensing properties and have been extensively studied for their use in sensing devices. Preparing metal oxide nanoparticles for gas sensing applications requires knowledge regarding how the particle size, grain size, morphology and surface defects affect the gas sensing properties. The effect of each of these parameters has been well studied; however, the preparation method also affects semiconductor gas sensing ability. Particles synthesized using microwave irradiation have been shown to exhibit higher response and selectivity compared to particles synthesized through conventional heating methods [151]. Here, we will discuss the gas sensing mechanism and the affect of the previously mentioned parameters on gas sensing metal oxides' performance.

Most metal oxide gas sensors operate on the basis that gas adsorption onto a semiconductor surface results in a conduction change. When semiconductors are exposed to ambient atmospheres, oxygen molecules adsorb to the surface and form O_2^- species by capturing electrons from the conduction band [152]. This state containing a high concentration of adsorbed oxygen species is considered a high resistance state [152]. Then, when the semiconductor is exposed to a reductive gas, the gas molecules react with the surface oxygen species. The concentration of surface oxygen species decreases and the concentration of electrons in the conduction band increases, which results in an increase in conductivity [152]. This change in conductivity gives the foundation for semiconductor gas sensors.

The relationship between a semiconductor's conductivity (σ) and the oxygen partial pressure (P_{O2}) is expressed as [153]

$$\sigma = C \, \exp(-E_A/k_g T) P_{o2}^{1/m} \tag{8.13}$$

where C is a constant, E_A is the activation energy for bulk conduction and m is a constant that depends on the defects involved in the conduction mechanism. As the oxygen partial pressure increases the conductivity decreases. This relationship allows for the sensing of other gases by measuring change in conductivity.

The sensitivity of a particular sensor is often used to characterize the sensing ability. The sensitivity (Sg) is given by

$$S_g = R_a/R_G \tag{8.14}$$

where R_a and R_g are the electrical resistance of the sensor in air and ethanol, respectively. Occasionally, different authors will use different symbols to define sensitivity, but the overall relationship remains constant. Most experiments use this definition of sensitivity to design their experiments to measure the change in conductivity. Furthermore, the sensitivity can be represented as a function of the partial pressure of the target gas (P_g) by

$$S_g = AP_g^{\beta} \tag{8.15}$$

where A is a prefactor and β is an exponent determined by the charge of surface species and stoichiometry of the elementary reactions of the surface. The exponent β is a rational fraction usually having a value of 1 or 1/2 depending on the reaction [152].

Preparing gas sensors from semiconductor nanoparticles typically requires the particles to be deposited as thin films. The thin films will have leads at either end that are used to measure the electrical resistance of the sensor. However, most studies report results in correlation with the original particle morphology. The following discussion gives a similar account of gas sensing properties based on characterization of prepared particles.

Since the sensor response occurs on the particles' surface and the surface activity varies with size, especially at the nanoscale, the sensor response depends on the particle size. The response relation to particle size is based on the conductivity (σ) relation to the potential energy barrier (eV_S) at the interface between two neighboring particles by [154]

$$\sigma = \sigma_0 \exp\left(\frac{-eV_s}{kT}\right) \tag{8.16}$$

where σ_0 is a factor that includes the bulk intergranular conductance. The energy barrier is related to the particle size and it is therefore expected that varying particle size with change the sensor response. As expected, smaller particles show enhanced sensor abilities [155]. Zhang and Liu studied the effect of doped SnO_2 particle size on sensitivity [154]. The energy barrier decreased with increasing particle size, which means that sensitivity also increases with particle size. When designing particles for gas sensing applications it is important to control the particle size.

In ceramic films and films comprised of semiconductor particles, the electrical modification occurs on the grain boundary or porous surface [152]. Therefore, the grain size of the semiconductor material plays a critical role in controlling gas-sensing properties. Several studies [153, 156] have examined how the grain size affects the sensitivity. Therefore, controlling the grain size may enhance the sensitivity. Xu et al. utilized different preparation methods to prepare ZnO particles with varying grain sizes [157]. Particles with smaller grain sizes showed increased gas sensitivity; grain sizes greater than 40 nm showed significantly decreased sensor response. However, the results also show that using different methods produces different grain sizes as well as varying the lattice distortion. Lattice distortion will also affect the sensor response, as will be discussed later. However, it is apparent the reducing the grain size of the semiconductor crystal will enhance gas sensitivity.

A particles' morphology also affects sensor response. Different crystal faces will have different reactivity with respect to surface reactions, and varying the morphology of semiconductor particles will allow for different crystal faces to be exposed. Gas sensing properties of spherical [151, 158], rod [152, 159–162], flower [163] and plate [164] shaped semiconductors have been reported. Krishnakumar et al. reported [163] that spherical shaped semiconductor particles have a lower response to CO gas compared to flower and star-like shaped particles. This difference in sensing from different morphologies is attributed to varying crystallite size and each morphology processing characteristic defects such as oxygen vacancies or excess zinc [163].

Since the sensor response occurs on the particles' surface, the composition of the surface, which contains oxygen point defects, will affect the sensor response. For example, higher responses are observed for particles prepared at lower temperatures, which are believed to have a larger number of active sites on the surface [151]. Epifani et al. report the study of oxygen vacancies on SnO_2 surfaces and the NO_2 sensing response [165]. It is determined that surface oxygen vacancies make the surface strongly open to to oxygen ionosorption. Thus, increased concentration of oxygen vacancies enhances gas sensor response. This is an important finding since different preparation methods will result in various surface compositions that should be tailored to maximize gas sensitivity.

In order for semiconductor gas sensors to be used industrially, particle preparation needs to be cost effective and able to be produced on large scales. Many different preparation methods have been examined but the largest yield comes from wet chemical synthesis methods. This has led to several studies regarding how the preparation method affects the resulting particles' sensor ability. One approach to efficiently produce large quantities of semiconductor nanoparticles is through the use of microwave heating as opposed to conventional heating methods. Particles formed in low temperature regions may possess a higher number of oxygen point defects [160], which will affect sensitivity. The sensor response of semiconductor nanoparticles produced via microwave heating techniques has been the focus of several studies [151, 158, 159, 162–164]. Cho et al. studied the sensing behavior of SnO_2 for CO gas and determined that particles prepared by microwave heating exhibited a more rapid response. Furthermore, microwave calcinations also enhanced sensor response [159]. SnO_2 particles treated using convention calcination methods showed a 90 % response time to 30 ppm of CO of 76 s, which compares to 10–12 s for microwave calcinations [159]. Particles prepared by these methods have the same morphology, but those treated with microwave displayed lightly rougher surface, smaller sizes and significantly larger pore volumes. Srivastava reported similar results [158] and showed that SnO_2 particles prepared by microwave routes showed increased sensitivity at lower temperatures. Few comparisons for gas sensing properties among particles prepared using both methods exist, but the use of microwave heating to synthesize successful gas sensor materials have been reported.

Many gas sensors fabricated out of semiconductor materials require high operating temperatures. This is impractical for many applications, such as sensing in potential explosive environments, and researchers are striving to fabricate semiconductor gas sensors that operate at ambient temperatures. It is well known that UV irradiation generates charge carriers at the semiconductor surface by creating electron/hole pairs, which subsequently react with electron donors or acceptors. These electron hole pairs will neutralize adsorbed oxygen species that will react with the active gas species to be detected [161]. This led Peng et al. to study the effect of UV irradiation on gas sensing ability of ZnO nanorods [161]. As expected, UV irradiation enchaced the gas sensitivity. The ZnO nanowires exhibited faster recovery times that suggest UV irradiation produces a clean effect that does not poison the sensor. This result is encouraging for developing low temperature gas sensing devices.

Overall, metal oxides, such as ZnO and SnO_2, exhibit unique gas sensing abilities. The sensor response depends on various parameters, which include size, morphology and also preparation method. Therefore, the sensitivity is controllable by tuning the synthesis method. These types of sensors are promising for future applications as they demonstrate selectivity to active gases and high sensitivities.

8.8 Conclusion

It is clear from the above examples that by manipulating the nucleation and growth kinetics of the nanocrystals through the choice of the appropriate precursors, ligand systems, solvents and the reaction times, one can control the size, shape and composition of the resulting metal oxide nanocrystals.

In conclusion, the microwave irradiation method is simple, versatile and rapid. It allows the synthesis of a wide variety of metal oxide and doped metal oxide nanostructures with controlled size and shape. The important advantage of microwave dielectric heating over convective heating is that the reactants can be added at room temperature (or slightly higher temperatures) without the need for high-temperature injection. The method allows the passivation of the metal oxide nanostructures by ligating organic solvents of variable structures and surface properties.

Acknowledgments We thank the National Science Foundation (CHE-0911146) for the support of this work. We also thank NSF (OISE-0938520) for the support of the *"US-Egypt Advanced Studies Institute on Nanomaterials and Nanocatalysis for Energy, Petrochemicals and Environmental Applications"* which facilitated the preparation of this chapter.

References

1. Chow GM, Gonsalves KE (1996) Nanotechnology: molecularly designed materials. American Chemical Society, Washington
2. Edelstein AS, Cammarata RC (1996) Nanomaterials: synthesis, properties and applications. Institute of Physics Publishing, Bristol and Philadelphia
3. Liz-Marzan LM, Kamat PV (2003) Nanoscale materials. Kluwer Academic Publishers, Dordrecht
4. Nagarajan R, Hatton TA (2008) Nanoparticles: synthesis, stabilization, passivation and functionalization. American Chemical Society, Washington
5. Ozin GA, Arsenault AC (2005) Nano chemistry: a chemical approach to nanomaterials. RSC Publishing, Cambridge
6. Schmid G (2004) Nanoparticles: from Theory to Application. Wiley, Weinheim
7. Alivisatos AP (1996) Perspectives on the physical chemistry of semiconductor nanocrystals. J Phys Chem 100:13226–13239
8. Burda C, Chen XB, Narayanan R, El-Sayed MA (2005) Chemistry and properties of nanocrystals of different shapes. Chem Rev 105:1025–1102
9. Dahl JA, Maddux BLS, Hutchison JE (2007) Toward greener nanosynthesis. Chem Rev 107:2228–2269
10. El-Sayed MA (2004) Small is different: shape-, size-, and composition-dependent properties of some colloidal semiconductor nanocrystals. Acc Chem Res 37:326–333
11. Li Y, Malik MA, O'Brien P (2005) Synthesis of single-crystalline CoP nanowires by a one-pot metal-organic route. J Am Chem Soc 127:16020–16021
12. Park J, An KJ, Hwang YS, Park JG, Noh HJ, Kim JY, Park JH, Hwang NM, Hyeon T (2004) Ultra-large-scale syntheses of monodisperse nanocrystals. Nat Mater 3:891–895
13. Peng X, Wickham J, Alivisatos AP (1998) Kinetics of II-VI and III-V colloidal semiconductor nanocrystal growth: "focusing" of size distributions. J Am Chem Soc 120:5343–5344
14. Peng XG, Manna L, Yang WD, Wickham J, Scher E, Kadavanich A, Alivisatos AP (2000) Shape control of CdSe nanocrystals. Nature 404:59–61
15. Peng ZA, Peng X (2002) Nearly monodisperse and shape-controlled CdSe nanocrystals via alternative routes: nucleation and growth. J Am Chem Soc 124:3343–3353
16. Regulacio MD, Han M-Y (2010) Composition-tunable alloyed semiconductor nanocrystals. Acc Chem Res 43:621–630
17. Talapin DV, Lee J-S, Kovalenko MV, Shevchenko EV (2010) Prospects of colloidal nanocrystals for electronic and optoelectronic applications. Chem Rev 110:389–458
18. Wang X, Zhuang J, Peng Q, Li Y (2005) A general strategy for nanocrystal synthesis. Nature 437:121–124
19. Xia Y, Xiong Y, Lim B, Skrabalak SE (2009) Shape-controlled synthesis of metal nanocrystals: simple chemistry meets complex physics? Angew Chem Int Ed 48:60–103

20. Abdelsayed V, Aljarash A, El-Shall MS, Al Othman ZA, Alghamdi AH (2009) Microwave synthesis of bimetallic nanoalloys and co oxidation on ceria-supported nanoalloys. Chem Mater 21:2825–2834
21. Abdelsayed V, Panda AB, Glaspell GP, El-Shall MS (2008) In nanoparticles: synthesis, stabilization, passivation, and functionalization. In: Nagarajan R, Hatton TA (Eds) American Chemical Society, Washington
22. Boxall DL, Lukehart CM (2001) Rapid synthesis of Pt or Pd/Carbon nanocomposites using microwave irradiation. Chem Mater 13:806–810
23. Chen WX, Zhao J, Lee JY, Liu ZL (2005) Microwave heated polyol synthesis of carbon nanotubes supported Pt nanoparticles for methanol electrooxidation. Mater Chem Phys 91:124–129
24. El-Shall MS, Abdelsayed V, Khder A, Hassan HMA, El-Kaderi HM, Reich TE (2009) Metallic and bimetallic nanocatalysts incorporated into highly porous coordination polymer MIL-101. J Mater Chem 19:7625–7631
25. Gallis KW, Landry CC (2001) Rapid calcination of nanostructured silicate composites by microwave irradiation. Adv Mater 13:23
26. Gerbec JA, Magana D, Washington A, Strouse GF (2005) Microwave-enhanced reaction rates for nanoparticle synthesis. J Am Chem Soc 127:15791–15800
27. Glaspell G, Fuoco L, El-Shall MS (2005) Microwave synthesis of supported Au and Pd nanoparticle catalysts for CO oxidation. J Phys Chem B 109:17350–17355
28. Harpeness R, Gedanken A (2004) Microwave synthesis of core, áíShell Gold/Palladium Bimetallic Nanoparticles. Langmuir 20:3431–3434
29. Hassan HMA, Abdelsayed V, Khder A, AbouZeid KM, Terner J, El-Shall MS, Al-Resayes SI, El-Azhary AA (2009) Microwave synthesis of graphene sheets supporting metal nanocrystals in aqueous and organic media. J Mater Chem 19:3832–3837
30. He J, Zhao XN, Zhu JJ, Wang J (2002) Preparation of CdS nanowires by the decomposition of the complex in the presence of microwave irradiation. J Cryst Growth 240:389–394
31. Kundu S, Peng LH, Liang H (2008) A new route to obtain high-yield multiple-shaped gold nanoparticles in aqueous solution using microwave irradiation. Inorg Chem 47:6344–6352
32. Liang J, Deng Z, Jiang X, Li F, Li Y (2002) Photoluminescence of Tetragonal ZrO_2 nanoparticles synthesized by microwave irradiation. Inorg Chem 41:3602–3604
33. Mohamed MB, AbouZeid KM, Abdelsayed V, Aljarash AA, El-Shall MS (2010) Growth mechanism of anisotropic gold nanocrystals via microwave synthesis: formation of Dioleamide by Gold Nanocatalysis. ACS Nano 4:2766–2772
34. Panda AB, Glaspell G, El-Shall MS (2006) Microwave synthesis of highly aligned ultra narrow semiconductor rods and wires. J Am Chem Soc 128:2790–2791
35. Panda AB, Glaspell G, El-Shall MS (2007) Microwave synthesis and optical properties of uniform nanorods and nanoplates of rare earth oxides. J Phys Chem C 111:1861–1864
36. Pastoriza-Santos I, Liz-Marzan LM (2002) Formation of PVP-protected metal nanoparticles in DMF. Langmuir 18:2888–2894
37. Patra CR, Alexandra G, Patra S, Jacob DS, Gedanken A, Landau A, Gofer Y (2005) Microwave approach for the synthesis of rhabdophane-type lanthanide orthophosphate (Ln = La, Ce, Nd, Sm, Eu, Gd and Tb) nanorods under solvothermal conditions. New J Chem 29:733–739
38. Siamaki AR, Khder AERS, Abdelsayed V, El-Shall MS, Gupton BF (2011) Microwave-assisted synthesis of palladium nanoparticles supported on graphene: a highly active and recyclable catalyst for carbon-carbon cross-coupling reactions. J Catal 279:1–11
39. Wang HQ, Thomas T (2009) Monodisperse upconverting nanocrystals by microwave-assisted synthesis. ACS Nano 3:3804–3808
40. Zedan AF, Sappal S, Moussa S, El-Shall MS (2010) Ligand-controlled microwave synthesis of cubic and hexagonal CdSe nanocrystals supported on Graphene. Photoluminescence quenching by Graphene. J Phys Chem C 114:19920–19927
41. Zhu J, Palchik O, Chen S, Gedanken A (2000) Microwave assisted preparation of CdSe, PbSe, and Cu2-xSe nanoparticles. J Phys Chem B 104:7344–7347

42. Kappe CO (2004) Controlled microwave heating in modern organic synthesis. Angew Chem Int Ed 43:6250–6284
43. Kubrakova I, Toropchenova E (2008) Microwave heating for enhancing efficiency of analytical operations. Inorg Mater 44:1509–1519
44. Abraham FF (1974) Homogeneous nucleation theory. Academic, New York
45. Kashchiev D (2000) Nucleation, basic theory with applications. Butterworth Heinemann, Burlington
46. LaMer VK, Dinegar RH (1950) Theory, production and mechanism of formation of Monodispersed hydrosols. J Am Chem Soc 72:4847–4854
47. Ratke L, Voorhees PW (2002) Growth and coarsening—Ostwald ripening in materials processing. Springer, New York
48. Talapin DV, Rogach AL, Haase M, Weller H (2001) Evolution of an ensemble of nanoparticles in a colloidal solution: theoretical study. J Phys Chem B 105:12278–12285
49. Talapin DV, Rogach AL, Shevchenko EV, Kornowski A, Haase M, Weller H (2002) Dynamic distribution of growth rates within the ensembles of colloidal II-VI and III-V semiconductor nanocrystals as a factor governing their photoluminescence efficiency. J Am Chem Soc 124:5782–5790
50. Chang JF, Kuo HH, Leu IC, Hon MH (2002) The effects of thickness and operation temperature on ZnO: Al thin film CO gas sensor. Sens Actuators B Chem 84:258–264
51. Comini E, Faglia G, Sberveglieri G, Pan ZW, Wang ZL (2002) Stable and highly sensitive gas sensors based on semiconducting oxide nanobelts. Appl Phys Lett 81:1869–1871
52. Gao PX, Ding Y, Mai WJ, Hughes WL, Lao CS, Wang ZL (2005) Conversion of zinc oxide nanobelts into superlattice-structured nanohelices. Science 309:1700–1704
53. Gargas DJ, Moore MC, Ni A, Chang SW, Zhang ZY, Chuang SL, Yang PD (2010) Whispering gallery mode lasing from zinc oxide hexagonal Nanodisks. ACS Nano 4:3270–3276
54. Han JB, Fan FR, Xu C, Lin SS, WeiM, Duan X, Wang ZL (2010) ZnO nanotube-based dye-sensitized solar cell and its application in self-powered devices. Nanotechnology 21
55. He JH, Hsu JH, Wang CW, Lin HN, Chen LJ, Wang ZL (2006) Pattern and feature designed growth of ZnO nanowire arrays for vertical devices. J Phys Chem B 110:50–53
56. Huang MH, Mao S, Feick H, Yan HQ, Wu YY, Kind H, Weber E, Russo R, Yang PD (2001) Room-temperature ultraviolet nanowire nanolasers. Science 292:1897–1899
57. Kong YC, Yu DP, Zhang B, Fang W, Feng SQ (2001) Ultraviolet-emitting ZnO nanowires synthesized by a physical vapor deposition approach. Appl Phys Lett 78:407–409
58. Lupan O, Emelchenko GA, Ursaki VV, Chai G, Redkin AN, Gruzintsev AN, Tiginyanu IM, Chow L, Ono LK, Roldan Cuenya B, Heinrich H, Yakimov EE (2010) Synthesis and characterization of ZnO nanowires for nanosensor applications. Mater Res Bull 45:1026–1032
59. Minne SC, Manalis SR, Quate CF (1995) Parallel atomic force microscopy using cantilevers with integrated piezoresistive sensors and integrated piezoelectric actuators. Appl Phys Lett 67:3918–3920
60. Wang XD, Summers CJ, Wang ZL (2004) Large-scale hexagonal-patterned growth of aligned ZnO nanorods for nano-optoelectronics and nanosensor arrays. Nano Lett 4:423–426
61. Wang ZL (2004) Nanostructures of zinc oxide. Mater Today 7:26–33
62. Wang ZL, Song JH (2006) Piezoelectric nanogenerators based on zinc oxide nanowire arrays. Science 312:242–246
63. Zheng MJ, Zhang LD, Li GH, Shen WZ (2002) Fabrication and optical properties of large-scale uniform zinc oxide nanowire arrays by one-step electrochemical deposition technique. Chem Phys Lett 363:123–128
64. Zhu GA, Yang RS, Wang SH, Wang ZL (2010) Flexible high-output nanogenerator based on lateral ZnO nanowire array. Nano Lett 10:3151–3155

65. Bacaksiz E, Yilmaz S, Parlak M, Varilci A, Altunbas M (2009) Effects of annealing temperature on the structural and optical properties of ZnO hexagonal pyramids. J Alloy Compd 478:367–370
66. Bilecka I, Elser P, Niederberger M (2009) Kinetic and thermodynamic aspects in the microwave-assisted synthesis of ZnO nanoparticles in benzyl alcohol. ACS Nano 3:467–477
67. El-Shall MS, Graiver D, Pernisz U, Baraton MI (1995) Synthesis and characterization of nanoscale zinc oxide particles: I. laser vaporization/condensation technique. Nanostruct Mater 6:297–300
68. El-Shall MS, Li S (1998) Synthesis and characterization of metal and semiconductor nanopartilces. In: Duncan MA (ed) Advances in metal and semiconductor clusters, vol 4. JAI Press Inc, pp 115–177
69. Huang AS, Caro J (2010) Controlled growth of zinc oxide crystals with tunable shape. J Cryst Growth 312:947–952
70. Lao CS, Gao PM, Sen Yang R, Zhang Y, Dai Y, Wang ZL (2006) Formation of double-side teethed nanocombs of ZnO and self-catalysis of Zn-terminated polar surface. Chem Phys Lett 417:358–362
71. Munoz-Hernandez G, Escobedo-Morales A, Pal U (2009) Thermolytic growth of ZnO nanocrystals: morphology control and optical properties. Cryst Growth Des 9:297–300
72. Valerini D, Caricato AP, Lomascolo M, Romano F, Taurino A, Tunno T, Martino M (2008) Zinc oxide nanostructures grown by pulsed laser deposition. Appl Phys Mater Sci Process 93:729–733
73. Wahab R, Ansari SG, Seo HK, Kim YS, Suh EK, Shin HS (2009) Low temperature synthesis and characterization of rosette-like nanostructures of ZnO using solution process. Solid State Sci 11:439–443
74. Wang XD, Song JH, Wang ZL (2006) Single-crystal nanocastles of ZnO. Chem Phys Let 424:86–90
75. Zhang JW, Zhu PL, Li JH, Chen JM, Wu ZS, Zhang ZJ (2009) Fabrication of octahedral-shaped polyol-based zinc alkoxide particles and their conversion to octahedral polycrystalline ZnO or single-crystal ZnO nanoparticles. Cryst Growth Des 9:2329–2334
76. Zhang R, Kerr LL (2007) A simple method for systematically controlling ZnO crystal size and growth orientation. J Solid State Chem 180:988–994
77. Zhang ZH, Lu MH, Xu HR, Chin WS (2007) Shape-controlled synthesis of zinc oxide: a simple method for the preparation of metal oxide nanocrystals in non-aqueous medium. Chem-a Eur J 13:632–638
78. Zhou F, Zhao XM, Xu H, Yuan CG (2007) CeO_2 spherical crystallites: synthesis, formation mechanism, size control, and electrochemical property study. J Phys Chem C 111:1651–1657
79. Zhou X, Xie ZX, Jiang ZY, Kuang Q, Zhang SH, Xu T, Huang RB, Zheng LS (2005) Formation of ZnO hexagonal micro-pyramids: a successful control of the exposed polar surfaces with the assistance of an ionic liquid. Chem Commun 44:5572–5574
80. Cozzoli PD, Kornowski A, Weller H (2003) Low-temperature synthesis of soluble and processable organic-capped anatase TiO2 nanorods. J Am Chem Soc 125:14539–14548
81. Ding KL, Miao ZJ, Liu ZM, Zhang ZF, Han BX, An GM, Miao SD, Xie Y (2007) Facile synthesis of high quality TiO_2 nanocrystals in ionic liquid via a microwave-assisted process. J Am Chem Soc 129:6362
82. Huang, W. P.; Tang, X. H.; Wang, Y. Q.; Koltypin, Y.; Gedanken, A. (2000) Selective synthesis of anatase and rutile via ultrasound irradiation. Chem Commun 1415-1416
83. Jun YW, Casula MF, Sim JH, Kim SY, Cheon J, Alivisatos AP (2003) Surfactant-assisted elimination of a high energy facet as a means of controlling the shapes of TiO_2 nanocrystals. J Am Chem Soc 125:15981–15985
84. Wang C-C, Ying JY (1999) Sol-gel synthesis and hydrothermal processing of anatase and rutile Titania nanocrystals. Chem Mater 11:3113–3120
85. Wilson GJ, Will GD, Frost RL, Montgomery SA (2002) Efficient microwave hydrothermal preparation of nanocrystalline anatase TiO_2 colloids. J Mater Chem 12:1787–1791

86. Trentler TJ, Denler TE, Bertone JF, Agrawal A, Colvin VL (1999) Synthesis of TiO_2 nanocrystals by nonhydrolytic solution-based reactions. J Am Chem Soc 121:1613–1614
87. Mao YB, Wong SS (2006) Size- and shape-dependent transformation of nanosized titanate into analogous anatase Titania nanostructures. J Am Chem Soc 128:8217–8226
88. Zhu HY, Lan Y, Gao XP, Ringer SP, Zheng ZF, Song DY, Zhao JC (2005) Phase transition between nanostructures of titanate and titanium dioxides via simple wet-chemical reactions. J Am Chem Soc 127:6730–6736
89. Service RF (1996) Small clusters hit the big time. Science 271:920–922
90. Sun YG, Mayers B, Xia YN (2003) Transformation of silver nanospheres into nanobelts and triangular nanoplates through a thermal process. Nano Lett 3:675–679
91. Srivatsan S et al (1960) Reverse micellar synthesis of cerium oxide nanoparticles. Nanotechnology 2005:16
92. Yin LX, Wang YQ, Pang GS, Koltypin Y, Gedanken A (2002) Sonochemical synthesis of cerium oxide nanoparticles—Effect of additives and quantum size effect. J Colloid Interface Sci 246:78–84
93. Kompe K, Borchert H, Storz J, Lobo A, Adam S, Moller T, Haase M (2003) Green-emitting $CePO_4$: Th/$LaPO_4$ core-shell nanoparticles with 70 % photoluminescence quantum yield. Angew Chem Int Ed 42:5513–5516
94. Si R, Zhang YW, You LP, Yan C (2005) Rare-earth oxide nanopolyhedra, nanoplates, and nanodisks. Angew Chem Int Ed 44:3256–3260
95. Stouwdam JW, van Veggel F (2002) Near-infrared emission of redispersible Er3+, Nd3+, and Ho3+ doped LaF3 nanoparticles. Nano Lett 2:733–737
96. Yada M, Kitamura H, Ichinose A, Machida M, Kijima T (1999) Mesoporous magnetic materials based on rare earth oxides. Angew Chem Int Ed 38:3506–3510
97. Bazzi R, Flores-Gonzalez MA, Louis C, Lebbou K, Dujardin C, Brenier A, Zhang W, Tillement O, Bernstein E, Perriat P (2003) Synthesis and luminescent properties of sub-5 nm lanthanide oxides nanoparticles. J Lumin 102:445–450
98. Ammar S, Helfen A, Jouini N, Fievet F, Rosenman I, Villain F, Molinie P, Danot M (2001) Magnetic properties of ultrafine cobalt ferrite particles synthesized by hydrolysis in a polyol medium. J Mater Chem 11:186–192
99. Coey JMD, Berkowitz AE, Balcells L, Putris FF, Parker FT (1998) Magnetoresistance of magnetite. Appl Phys Lett 72:734–736
100. Dumestre F, Chaudret B, Amiens C, Renaud P, Fejes P (2004) Superlattices of iron nanocubes synthesized from Fe[N(SiMe3)(2)](2). Science 303:821–823
101. Hafeli U, Schutt W, Teller J, Zborowski M (1997) Scientific and clinical applications of magnetic carriers. Plenum Press, New York
102. Hergt R, Andra W, d'Ambly CG, Hilger I, Kaiser WA, Richter U, Schmidt HG (1998) Physical limits of hyperthermia using magnetite fine particles. IEEE Trans Magn 34:3745–3754
103. Kiyomura T, Maruo Y, Gomi M (2000) Electrical properties of MgO insulating layers in spin-dependent tunneling junctions using Fe[sub 3]O[sub 4]. J Appl Phys 88:4768–4771
104. Ngo AT, Pileni MP (2001) Assemblies of ferrite nanocrystals: Partial orientation of the easy magnetic axes. J Phys Chem B 105:53–58
105. Patolsky F, Weizmann Y, Katz E, Willner I (2003) Magnetically amplified DNA assays (MADA): sensing of viral DNA and single-base mismatches by using nucleic acid modified magnetic particles. Angew Chem Int Ed 42:2372–2376
106. Pham-Huu C, Keller N, Estournes C, Ehret G, Ledoux MJ (2002) Synthesis of $CoFe_2O_4$ nanowire in carbon nanotubes. A new use of the confinement effect. Chem Commun 1882–1883
107. Redl FX, Cho KS, Murray CB, O'Brien S (2003) Three-dimensional binary superlattices of magnetic nanocrystals and semiconductor quantum dots. Nature 423:968–971
108. Sorenson TA, Morton SA, Waddill GD, Switzer JA (2002) Epitaxial electrodeposition of Fe3O4 thin films on the low-index planes of gold. J Am Chem Soc 124:7604–7609

109. Sousa MH, Tonrinho FA, Depeyrot J, da Silva GJ, Lara M (2001) New electric double-layered magnetic fluids based on copper, nickel, and zinc ferrite nanostructures. J Phys Chem B 105:1168–1175

110. Tarascon JM, Armand M (2001) Issues and challenges facing rechargeable lithium batteries. Nature 414:359–367

111. Versluijs JJ, Bari MA, Coey JMD (2001) Magnetoresistance of half-metallic oxide nanocontacts. Phys Rev Lett 87:026601

112. Fatemi DJ, Harris VG, Browning VM, Kirkland JP (1998) Processing and cation redistribution of MnZn ferrites via high-energy ball milling. J Appl Phys 83:6867–6869

113. Lopez Perez JA, Lopez Quintela MA, Mira J, Rivas J, Charles SW (1997) Advances in the preparation of magnetic nanoparticles by the microemulsion method. J Phys Chem B 101:8045–8047

114. Shafi KVPM, Gedanken A, Prozorov R, Balogh J (1998) Sonochemical preparation and size-dependent properties of nanostructured $CoFe_2O_4$ particles. Chem Mater 10:3445–3450

115. Sugimoto T, Shimotsuma Y, Itoh H (1998) Synthesis of uniform cobalt ferrite particles from a highly condensed suspension of beta-FeOOH and beta-Co(OH)(2) particles. Powder Technol 96:85–89

116. Prozorov T, Prozorov R, Koltypin Y, Felner I, Gedanken A (1998) Sonochemistry under an applied magnetic field: Äâ determining the shape of a magnetic particle. J Phys Chem B 102:10165–10168

117. Hyeon T, Chung Y, Park J, Lee SS, Kim Y-W, Park BH (2002) Synthesis of highly crystalline and monodisperse cobalt ferrite nanocrystals. J Phys Chem B 106:6831–6833

118. Kang S, Harrell JW, Nikles DE (2002) Reduction of the fcc to L10 ordering temperature for self-assembled FePt nanoparticles containing Ag. Nano Lett 2:1033–1036

119. Rockenberger JR, Scher EC, Alivisatos AP (1999) A new nonhydrolytic single-precursor approach to surfactant-capped nanocrystals of transition metal oxides. J Am Chem Soc 121:11595–11596

120. Jana NR, Chen Y, Peng X (2004) Size- and shape-controlled magnetic (Cr, Mn, Fe, Co., Ni) oxide nanocrystals via a simple and general approach. Chem Mater 16:3931–3935

121. Hyeon T, Lee SS, Park J, Chung Y, Na HB (2001) Synthesis of highly crystalline and monodisperse maghemite nanocrystallites without a size-selection process. J Am Chem Soc 123:12798–12801

122. Kang E, Park J, Hwang Y, Kang M, Park J-G, Hyeon T (2004) Direct synthesis of highly crystalline and monodisperse manganese ferrite nanocrystals. J Phys Chem B 108:13932–13935

123. Sun S, Zeng H, Robinson DB, Raoux S, Rice PM, Wang SX, Li G (2003) Monodisperse MFe_2O_4 (M = Fe, Co., Mn) nanoparticles. J Am Chem Soc 126:273–279

124. Panda AB, GlaspellG, El-Shall MS (in preparation)

125. Boal AK, Das K, Gray M, Rotello VM (2002) Monolayer exchange chemistry of Œ ≥ -Fe_2O_3 nanoparticles. Chem Mater 14:2628–2636

126. Rajamathi M, Ghosh M, Seshadri R (2002) Hydrolysis and amine-capping in a glycol solvent as a route to soluble maghemite gamma-Fe_2O_3 nanoparticles. Chem Commun 1152–1153

127. Klug HP, Alexander LE (1962) X-Ray diffraction procedures for polycrystalline and amorphous materials. Wiley, New York

128. Bahnemann D, Bockelmann D, Goslich R (1991) Mechanistic studies of water detoxification in illuminated TiO_2 suspensions. Sol Energy Mater 24:564–583

129. Bohle DS, Spina CJ (2009) Cationic and anionic surface binding sites on nanocrystalline zinc oxide: surface influence on photoluminescence and photocatalysis. J Am Chem Soc 131:4397–4404

130. Chakrabarti S, Dutta BK (2004) Photocatalytic degradation of model textile dyes in wastewater using ZnO as semiconductor catalyst. J Hazard Mater 112:269–278

131. Gaya UI, Abdullah A (2008) Heterogeneous photocatalytic degradation of organic contaminants over titanium dioxide: a review of fundamentals, progress and problems. J Photochem Photobiol C-Photochem Rev 9:1–12
132. Hariharan C (2006) Photocatalytic degradation of organic contaminants in water by ZnO nanoparticles: revisited. Appl Catal a-Gen 304:55–61
133. Hoffmann MR, Martin ST, Choi WY, Bahnemann DW (1995) Environmental applications of semiconductor photocatalysis. Chem Rev 95:69–96
134. Kamat PV, Meisel D (2002) Nanoparticles in advanced oxidation processes. Curr Opin Colloid Interface Sci 7:282–287
135. Katsumata H, Kawabe S, Kaneco S, Suzuki T, Ohta K (2004) Degradation of bisphenol A in water by the photo-Fenton reaction. J Photochem Photobiol a Chem 162:297–305
136. Lakshmi S, Renganathan R, Fujita S (1995) Study on TiO_2-mediated photocatalytic degradation of methylene-blue. J Photochem Photobiol a Chem 88:163–167
137. Lu F, Cai WP, Zhang YG (2008) ZnO hierarchical micro/nanoarchitectures: solvothermal synthesis and structurally enhanced photocatalytic performance. Adv Funct Mater 18:1047–1056
138. Malik PK, Sanyal SK (2004) Kinetics of decolourisation of azo dyes in wastewater by UV/H_2O_2 process. Sep Purif Technol 36:167–175
139. Matthews RW (1984) Hydroxylation reactions induced by near-ultraviolet photolysis of aqueous titanium-dioxide suspensions. J Chem Soc Faraday Trans I 80:457–471
140. Neppolian B, Choi HC, Sakthivel S, Arabindoo B, Murugesan V (2002) Solar/UV-induced photocatalytic degradation of three commercial textile dyes. J Hazard Mater 89:303–317
141. Ollis DF (1991) Solar-assisted photocatalysis for water-purification—issues, data, questions. In: Pelizzetti E, Schiavello M (eds) Photochemical conversion and storage of solar energy. Kluwer Academic Publishing, Dordrecht, pp 593–622
142. Pirkanniemi K, Sillanpaa M (2002) Heterogeneous water phase catalysis as an environmental application: a review. Chemosphere 48:1047–1060
143. Ray AK, Beenackers A (1998) Novel photocatalytic reactor for water purification. AIChE J 44:477–483
144. Sakthivel S, Neppolian B, Shankar MV, Arabindoo B, Palanichamy M, Murugesan V (2003) Solar photocatalytic degradation of azo dye: comparison of photocatalytic efficiency of ZnO and TiO_2. Solar Energy Mater Solar Cells 77:65–82
145. Sobana N, Swaminathan M (2007) The effect of operational parameters on the photocatalytic degradation of acid red 18 by ZnO. Sep Purif Technol 56:101–107
146. Vogel R, Hoyer P, Weller H (1994) Quantum-sized PbS, CdS, Ag2S, SB2S3, and Bi2S3 particles as sensitizers for various nanoporous wide-bandgap semiconductors. J Phys Chem 98:3183–3188
147. Wang JC, Liu P, Fu XZ, Li ZH, Han W, Wang XX (2009) Relationship between oxygen defects and the photocatalytic property of ZnO nanocrystals in nafion membranes. Langmuir 25:1218–1223
148. Yuan JQ, Choo ESG, Tang XS, Sheng Y, Ding J, Xue JM (2010) Synthesis of ZnO-Pt nanoflowers and their photocatalytic applications. Nanotechnology 21:185606
149. Zeng HB, Cai WP, Liu PS, Xu XX, Zhou HJ, Klingshirn C, Kalt H (2008) ZnO-based hollow nanoparticles by selective etching: elimination and reconstruction of metal-semiconductor interface, improvement of blue emission and photocatalysis. Acs Nano 2:1661–1670
150. Henrich VE, Cox PA (1994) The surface science of metal oxides. Cambridge University Press, Cambridge
151. Chu XF, Chen TY, Zhang WB, Zheng BQ, Shui HF (2009) Investigation on formaldehyde gas sensor with ZnO thick film prepared through microwave heating method. Sens Actuators B-Chem 142:49–54
152. Wan Q, Li QH, Chen YJ, Wang TH, He XL, Li JP, Lin CL (2004) Fabrication and ethanol sensing characteristics of ZnO nanowire gas sensors. Appl Phys Lett 84:3654–3656

153. Sberveglieri G (1995) Recent developments in semiconducting thin-film gas sensors. Sens Actuators B-Chem 23:103–109
154. Zhang G, Liu ML (2000) Effect of particle size and dopant on properties of SnO_2-based gas sensors. Sens Actuators B-Chem 69:144–152
155. Franke ME, Koplin TJ, Simon U (2006) Metal and metal oxide nanoparticles in chemiresistors: Does the nanoscale matter? Small 2:36–50
156. Pan QY, Xu JQ, Dong XW, Zhang JP (2000) Gas-sensitive properties of nanometer-sized SnO_2. Sens Actuators B-Chem 66:237–239
157. Xu JQ, Pan QY, Shun YA, Tian ZZ (2000) Grain size control and gas sensing properties of ZnO gas sensor. Sens Actuators B-Chem 66:277–279
158. Srivastava A, Lakshmikumar ST, Srivastava AK, Rashmi Jain K (2007) Gas sensing properties of nanocrystalline SnO_2 prepared in solvent media using a microwave assisted technique. Sens Actuators B Chem 126:583–587
159. Cho PS, Kim KW, Lee JH (2007) Improvement of dynamic gas sensing behavior of SnO_2 acicular particles by microwave calcination. Sens Actuators B-Chem 123:1034–1039
160. Dai ZR, Pan ZW, Wang ZL (2003) Novel nanostructures of functional oxides synthesized by thermal evaporation. Adv Funct Mater 13:9–24
161. Peng L, Zhao QD, Wang DJ, Zhai JL, Wang P, Pang S, Xie TF (2009) Ultraviolet-assisted gas sensing: a potential formaldehyde detection approach at room temperature based on zinc oxide nanorods. Sens Actuators B Chem 136:80–85
162. Qurashi A, Tabet N, Faiz M, Yamzaki T (2009) Ultra-fast microwave synthesis of ZnO nanowires and their dynamic response toward hydrogen gas. Nanoscale Res Lett 4:948–954
163. Krishnakumar T, Jayaprakash R, Pinna N, Donato N, Bonavita A, Micali G, Neri G (2009) CO gas sensing of ZnO nanostructures synthesized by an assisted microwave wet chemical route. Sens Actuators B Chem 143:198–204
164. Jing ZH, Zhan JH (2008) Fabrication and gas-sensing properties of porous ZnO nanoplates. Adv Mater 20:4547–4551
165. Epifani M, Prades JD, Comini E, Pellicer E, Avella M, Siciliano P, Faglia G, Cirera A, Scotti R, Morazzoni F, Morante JR (2008) The role of surface oxygen vacancies in the NO_2 sensing properties of SnO_2 nanocrystals. J Phys Chem C 112:19540–19546

Part II
Novel Morphologies and Signal Transduction Principles in Metal Oxide-Based Sensors

Chapter 9
Metal Oxide Nanowires: Fundamentals and Sensor Applications

Zhiyong Fan and Jia G. Lu

Abstract Detection of chemicals species such as industrial toxic and inflammable gasses, chemical warfare agents, disease related chemicals, is of paramount importance to public safety and health. The driving force is to develop highly sensitive, selective, and stable sensors with rapid detection and recovery time. Metal oxide thin films have long been used as chemical sensors due to the rich oxygen vacancies on the surface that are electrically and chemically active. The chemical adsorption induces redox reactions and consequently alters the electrical conductance. However, there have been a number of limitations: relatively high operation temperature, indirect and inefficient refreshing method, and lack of chemical selectivity. In the surge of quasi-one-dimensional (Q1D) metal oxide nanowire research, it is demonstrated that the unique shape anisotropy significantly enhances the sensor performances due to the large surface-to-volume ratio and size comparable to the Debye screening length. This not only enhances the sensitivity at room temperature, but provides efficient modulation of the surface detection and desorption processes. More importantly, it opens a pathway for developing wireless low-power sensor network to transmit information from a remote site. This chapter provides a review of the state-of-the-art research covering the synthesis and fundamental properties of Q1D metal oxide systems, and focusing on their applications as chemical sensors.

Z. Fan
Department of Electronic and Computer Engineering,
Hong Kong University of Science and Technology,
Clear Water Bay, Kowloon, Hong Kong SAR, China

J. G. Lu (✉)
Department of Physics and Department of Electrophysics, University of Southern California,
Los Angeles, CA 90089-0484, USA
e-mail: jialu@usc.edu

M. A. Carpenter et al. (eds.), *Metal Oxide Nanomaterials for Chemical Sensors*,
Integrated Analytical Systems, DOI: 10.1007/978-1-4614-5395-6_9,
© Springer Science+Business Media New York 2013

9.1 Introduction

Efficient detection of chemicals species such as industrial toxic and inflammable gasses, chemical warfare agents, disease related chemicals, is of paramount importance to public safety and health. The driving force is to develop highly sensitive, selective, and stable sensors with rapid detection and recovery time. Metal oxide thin films have long been used as chemical sensors due to the significant density of oxygen vacancies on the surface that are electrically and chemically active. The chemical adsorption/dissociation of oxygen and analyte gases induced by these vacancies result in redox reactions, which consequently alter the electron/hole concentration in the film, and thus modulate the corresponding electrical conductance. However, there have been a number of practical limitations: relatively high operation temperature, inefficient refreshing method, and lack of chemical selectivity. In the surge of quasi-one-dimensional (Q1D) metal oxide nanowire research, it is demonstrated that the unique shape anisotropy significantly enhances the sensor performances due to the large surface-to-volume ratio and size comparable to the Debye screening length. This not only enhances the sensitivity at room temperature, but provides efficient modulation of the surface detection and desorption processes. More importantly, it opens a pathway for developing a wireless low-power sensor network to transmit information from a remote site.

Using zinc oxide (ZnO) nanowires as a model example, this chapter provides a review of the state-of-the-art research covering the synthesis and fundamental properties of Q1D metal oxide systems, and focusing on their applications as chemical sensors.

9.2 Synthesis of Metal Oxide Nanowires

The synthesis of metal oxide nanowires can be classified into two categories, i.e. top–down and bottom–up approaches. The top–down method is a straightforward process to produce quasi-one-dimensional (Q1D) nanostructures inheriting the lithographic technologies in microelectronics industry [1, 2]. Whereas thebottom–up method denotes the directional crystal growth or atomic self-assembly based on chemical or biochemical synthesis [3, 4]. Upon extensive efforts in the past decades, the bottom–up method has been proven to be an efficient, low cost approach for nanomaterials synthesis with controllable chemical composition, crystalline structures and morphologies [5]. According to the synthesis environments, two representative methods can be concluded, i.e. the vapor phase growth and liquid phase growth. The template-based growth, which integrates the vapor/liquid phase growth process and the dimensional confinement of templates, expands the horizon of fabrication hierarchy [6, 7].

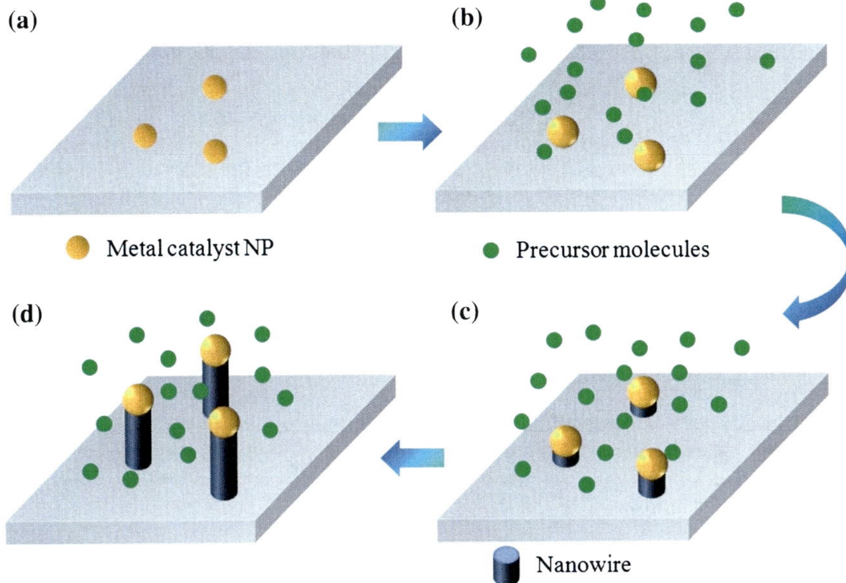

Fig. 9.1 Schematic illustrations of VLS process. **a** A substrate coated with metal catalyst nanoparticles. **b** Formation of eutectic alloy droplet at each catalyst site. **c** The supersaturation state leads to the nucleation of nanowires. **d** Growth of nanowires along one particular crystallographic orientation

9.2.1 Vapor Phase Growth

A variety of Q1D metal oxides with fruitful morphologies and high crystalline qualities have been produced by vapor phase growth processes. The nanowire growth from vapor phase is governed by vapor–liquid–solid (VLS) and vapor–solid (VS) mechanisms.

The vapor–liquid–solid mechanism is a catalyst-assisted growth phenomenon proposed by Wagner and Ellis [8] in 1960s for their work on Si whisker growth. Several decades later, Yang and co-workers provided the direct evidence of VLS growth via the real-time observation on individual Ge and GaN nanowires [9, 10]. Figure 9.1 shows a schematic of a typical VLS process. In this process, various nanoparticles or nanoclusters have been used as catalysts, such as Au [11–14], Cu [15], Co [16], and Sn [17], etc. The formation of a eutectic alloy droplet occurs at each catalyst site, followed by the nucleation and growth of a solid nanowire due to the supersaturation of the liquid droplet. Incremental growth of the nanowire taking place at the droplet interface constantly pushes the catalyst upwards. This growth method inherently provides site-specific nucleation at each catalytic site [18].

The typical experimental apparatus for vapor phase thermal CVD synthesis includes a horizontal tube furnace, gas lines with mass flow controllers, and a vacuum pump for vapor pressure control. The variations of the precursors delivery into the

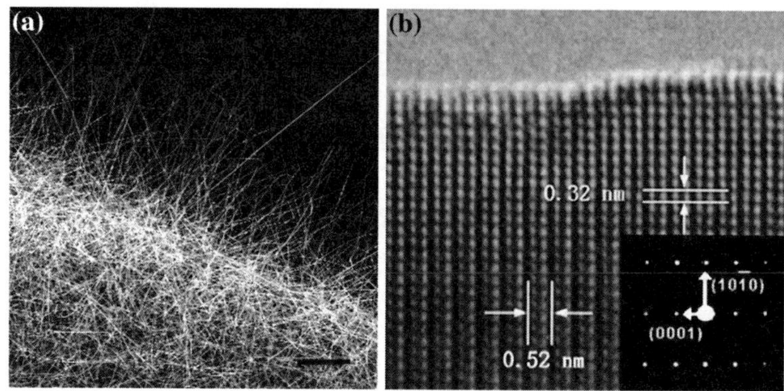

Fig. 9.2 **a** A SEM image shows the as-grown ZnO nanowires Inset: a ZnO nanowire with diameter of 50 nm terminated with an Au nanoparticle. (*scale bar* is 100 nm). **b** High resolution TEM image of a ZnO nanowire with lattice spacing 0.52 nm. The inset SAED indicates that the growth direction is [0001] along the *c* axis. Reprinted with permission from [14], Copyright 2004, American Institute of Physics

growth zone include: pulse-laser-deposition (PLD) [3, 19], molecular beam epitaxy (MBE) [20], chemical vapor deposition (CVD) [11], pulsed laser chemical vapor deposition (PLCVD) [21], metal–organic CVD (MOCVD) [22], and plasma enhanced CVD (PECVD) [23], etc. A large assortment of metal oxide nanowires have been readily synthesized using these methods, such as ZnO [24], In_2O_3 [25] SnO_2 [26] MgO [27], Ga_2O_3 [28] TiO_2 [29]. For instance, Fig. 9.2 shows the electron micrographs of ZnO nanowires synthesized by directly evaporating pure Zn powder at 700 °C in low oxygen concentration (<2 %) using Au catalyst nanoparticles [11, 30]. Figure 9.2b is a HRTEM image and the inset selected-area electron diffraction (SAED) pattern indicates single crystalline nature with [0001] growth direction. The growth direction determined in Q1D system is along the crystal plane that is the most stable and has the lowest surface free energy.

A modified vapor trapping CVD process is introduced for ZnO nanowire synthesis. It is intended to explore the growth and corresponding structure and property variations with respect to the mass transport rate (Fig. 9.3) [11]. A quartz vial is loaded into the center of a quartz tube to effectively maintain the local zinc vapor partial pressure at a relatively stable level inside the vial. Pure zinc powder is placed close to the bottom of the quartz vial (position A), serving as the source material. Chips B and C coated with catalyst (gold) nanoparticles are placed at downstream for nanowire harvesting. In a typical synthesis process, the furnace is first heated up to 600–700 °C in 10 min with a constant Ar gas flow. The oxygen gas (2 % concentration mixed with Ar) is then introduced, while the temperature is kept at the setpoint. After 30 min, the oxygen gas is switched off accompanied with furnace cool down. ZnO nanowires will be collected on the catalyst coated substrates.

Figure 9.4 shows the SEM images of ZnO nanostructures grown at different regions on chip B, where Fig. 9.4a schematically indicates the locations of b1, b2, b3, b4, b5, and b6, respectively. Due to the absence of Au catalyst in the locations

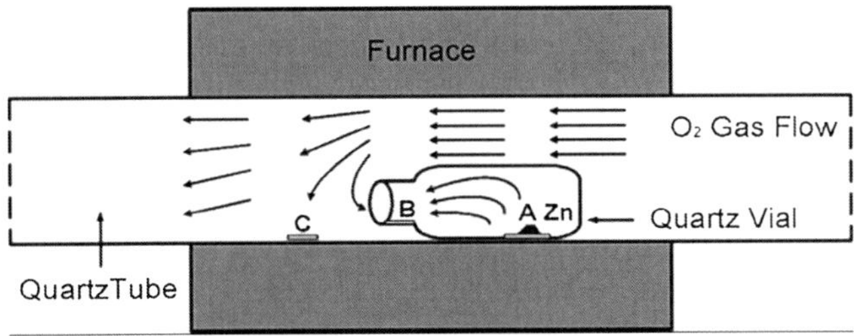

Fig. 9.3 Schematic illustration of the modified CVD system. A small quartz vial is placed inside of a *horizontal quartz* tube for zinc vapor trapping during the synthesis process. Reprinted with permission from [11], Copyright 2004, American Chemical Society

b1 and b2, the growth process follows vapor–solid (VS) mechanism. The products represent prismatic hexagonal rods on the substrate. In the regions b3–b6, a transition from nanowires to dendrite crystals can be observed. The key point of this process is that the spatial variation of Zn and O_2 vapor pressure inside the quartz vial, which provides a strategy to tune *in-situ* the morphologies of nanocrystals. More importantly, the transport properties of nanowires could be controlled by carefully changing the synthesis conditions.

Impurity doping in the synthesis process not only can alter the thermodynamic balance thus influencing the crystal formation, but can also manipulate the electron transport properties [31–34]. By adopting a laser ablation method in the CVD system, we have demonstrated successfully the incorporation of In, Ga, Pb atoms into the ZnO lattice [21]. In this doping process, pure zinc powder (99.99 %) is placed in the center of a horizontal tube furnace, while the dopant metal is loaded at the up-stream of the quartz tube. The dopant source is constantly ablated by Nd:YAG (yttrium aluminum garnet) laser pulses. The doped nanowires are synthesized via the VLS process on catalyst-coated silicon substrates. The energy dispersive X-ray (EDX) spectrum and HRTEM analyses indicate that the impurity atoms have been substitutionally doped into the wurtzite crystal lattice and the nanowires maintain the same growth direction along the c-axis ([0001] direction).

Vapor–solid (VS) process is another commonly used nanostructure growth process which does not rely on metallic catalysts. An assortment of metal oxide nanomaterials have been reported based on VS mechanism, such as CuO, FeOx, ZnO, SnO_2 nanowire/nanorods [35–41]. Figure 9.5a, and b depict the $\alpha\text{-}Fe_2O_3$ and CuO nanowires grown by direct heating of the corresponding metal substrates under appropriate flow of oxygen [40, 41]. In addition, the VS process can yield metal oxide nanomaterials with complex structures. Using ZnO material as an example, VS growth process can lead to nanowires, nanobelts, nanocombs, nanohelixes, and other more complex structures [37, 38]. It worth noting that although VS growth process can produce diversified nanostructures, the growth locations, geometry and crystallographic orientations of materials can not be controlled without the catalyst.

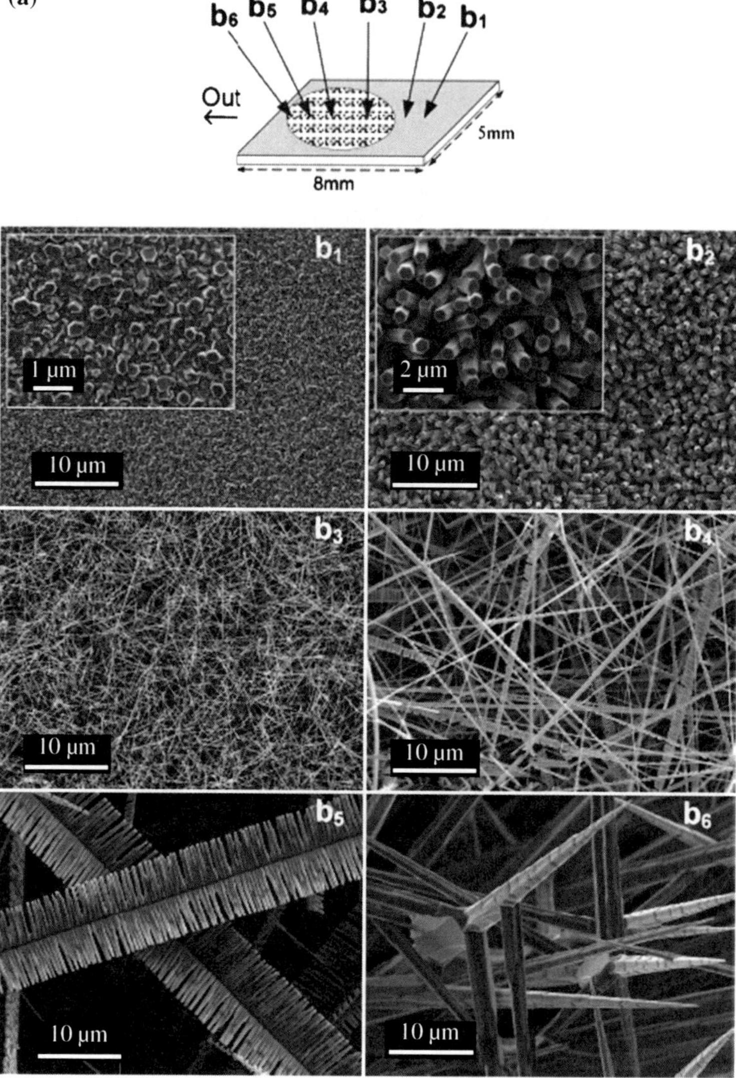

Fig. 9.4 **a** Schematic of chip B. The SEM analysis is conducted on the *six labeled areas*. The *shaded circle* indicates the region of Au catalyst deposition. SEM images representing the morphologies of ZnO nanocrystals at regions (*b*1), (*b*2), (*b*3), (*b*4), (*b*5) and (*b*6). Reprinted with permission from [11], Copyright 2004, American Chemical Society

9.2.2 Template-Based Growth

The second widely used nanowire fabrication method is template-based growth, which utilizes the electroless and electrodeposition of materials into templates. This approach has been well-developed to make monodispersed and anisotropic

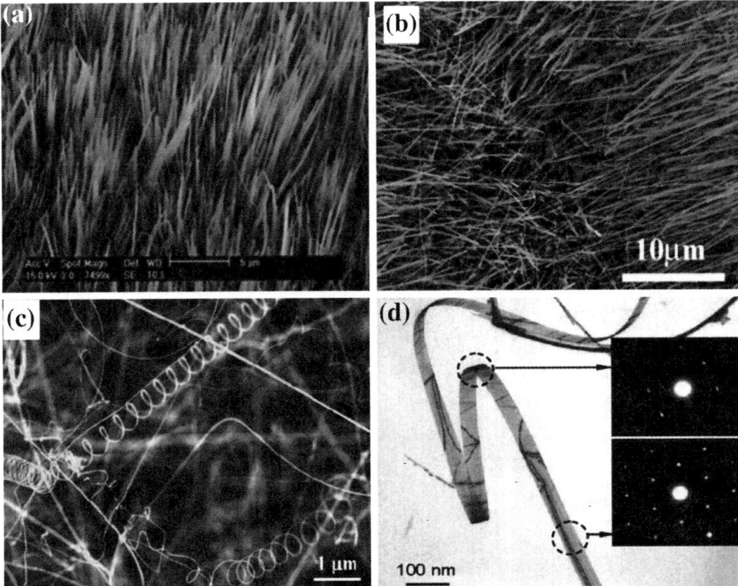

Fig. 9.5 SEM images of metal oxidesmetal oxides grown via VS process. **a** α-Fe$_2$O$_3$ nanowire arrays. Reprinted with permission from [40], Copyright 2005, American Institute of Physics. **b** CuO nanowire arrays. Reprinted with permission from [41], Copyright 2010, Institute of Physics. **c** ZnO nanohelixes. **d** TEM images of ZnO nanobelt. Reprinted with permission from [38], Copyright 2003, American Chemical Society

Q1D nanomaterial arrays [6, 42–44]. Among the various templates, robust porous anodic aluminum oxide (AAO) membranes have demonstrated to be one of the most promising candidates for nanomaterial synthesis.

In the conventional anodization condition, the pore diameter, barrier layer thickness and interpore distance can be controlled by the anodization voltage with respective proportionality constants of $\sim$0.9, 1.3 and 2.5 nm/V [45]. Figure 9.6a shows the AAO membrane with a close-packed array of regular hexagonal cells, obtained by two-step anodization under 60 V [46]. Typically, the anodization process only generates limited self-assembled area with the domain size (ordering range) in the scale of several micrometers. To implement larger domain size, Masuda et al. have developed a pretexturing method by defining the anodization sites [47, 48]. The process uses a pre-patterned mold to produce an ordered array on Al foil before anodization. However, only a few groups have attempted this method due to the time-consuming and high-cost fabrication process. A simple method has been developed to make long-range-ordered nanopore arrangement in millimeter scale. The electrochemically polished Al substrates are imprinted twice using a commercial straight-line diffraction grating under appropriated pressure, with 60° rotation between the two imprints [49, 50]. Following anodization under optimized bias voltage, a highly ordered nanopore array (Fig. 9.6b) is produced.

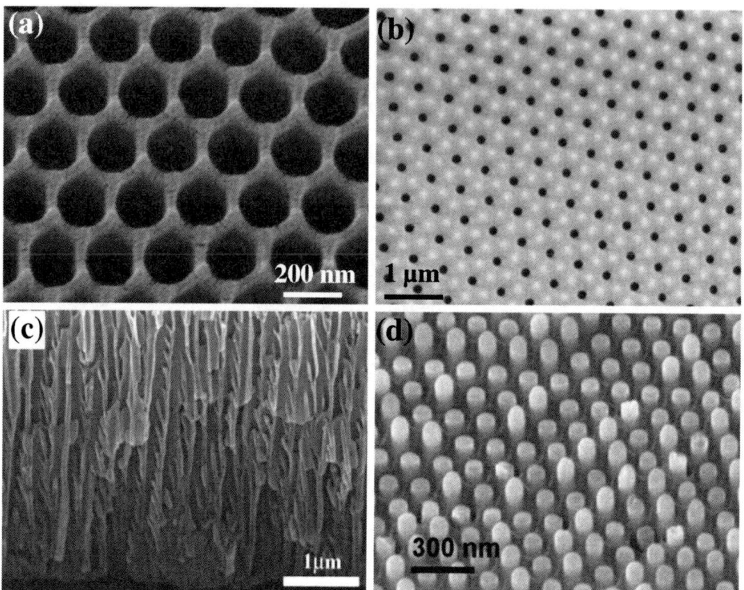

Fig. 9.6 SEM images of **a** AAO membrane obtained by 2-step-anodization under 60 V. Reprinted with permission from [46], Copyright 2008, Wiley–VCH. **b** AAO membrane obtained by pre-texturing method under 195 V. **c** Serrated anodic alumina membranes anodized at 60 V in aqueous phosphoric acid solution. Reprinted with permission from [51], Copyright 2009, American Chemical Society. Inset displays a serrated Pt nanowire obtained via electrodeposition. **d** Highly ordered Ge nanowire arrays embedded in an AAO template after partial etching of AAO. Reprinted with permission from [49], Copyright 2010, American Chemical Society

In addition, an innovative AAO template with periodic serrated inner structures was recently developed (Fig. 9.3c) using phosphoric acid electrolyte under proper condition [51, 52]. Three-dimensional structures with straight and serrated nano-channels can be subsequently fabricated by applying anodization in oxalic and phosphoric acid, alternatively. Using this unique type of template, serrated nanowires have been grown, as shown in Fig. 9.6c.

Taking advantage of the well-developed technique on AAO fabrication, many Q1D structures with tunable morphologies and properties can be routinely synthesized, such as Sb nanowires [53], Co nanowires [54], Co nanotubes [46], Pt nanowires [51], Ge nanopillars [49], SnO_2 nanowires [55], ZnO nanowires [56], CdS nanopillars [50], etc. In the synthesis techniques, deposition driven by electric field is the most commonly used approach [46, 51, 53–55, 57–59]. This method has robust capability in fabricating nanowires for elemental metals and their alloys. To obtain metal oxide nanowires, one can apply post-thermal oxidization on the as-prepared metal nanowires [55].

Even though there are some reports on single crystalline nanowires synthesized by electrodeposition, electrochemical deposition often yields polycrystals. High quality single crystal nanowires can be achieved by integrating the vapor phase or liquid phase growth into template-based method [49, 50, 56, 60–62]. For instance, electrodeposition is first adopted to fill metal catalyst segments into the bottom of the nanotemplate. Ordered nanowire arrays have been routinely formed in the nanochannels via CVD process (Fig. 9.6d) [56].

9.3 Fundamental Properties

Materials with size confinement exhibit unique properties. When the physical size approaches fundamental length scales, such as the screening length and Bohr radius, material properties will correspondingly change from their traditional bulk behaviors. In the case of Q1D structures, the contribution of surface related properties will become increasingly important due to the enhanced surface-to-volume ratio which is inversely proportional to radius.

9.3.1 Optical Properties

In order to develop metal oxide nanowires as the building blocks for nanoscale electronic devices, it is crucial to evaluate in depth their fundamental properties. The optical and electrical transport properties are closely related to the inherent native defects for the undoped nanowires and extrinsic impurities for the doped nanowires. Many photo-spectroscopy techniques such as photoluminescence (PL) [63], cathodoluminescence (CL) [64], and electroluminescence (EL) [65] have been employed to probe the optical and electronic properties of nanowires. PL measurement is one of the most sensitive tools to investigate radiative centers in the material. From the spectral peak and decay time measurements, one can determine the different emission mechanisms, the energy band gap, defect levels and their relative density.

When the size approaches the fundamental length scale, the optical properties of metal oxide Q1D materials begin to show size effects. Chang et al. [66] and Shalish et al. [67] compared the ratio of the deep level/near band edge transitions of different sized ZnO nanowire and found the ratio is directly related to the wire diameter/length ratio. The temperature dependent PL spectra of ZnO nanowires with different diameters (d) are shown in Fig. 9.7a [66]. The green luminescence band at ~ 2.5 eV, originated from bulk defects, is much stronger for wide nano-wire ($d > 100$ nm) than the thinner wires ($d < 30$ nm). Figure 9.7b shows the zoomed-in view of the band-edge transition (BET) emission at 13 K for different diameters [66]. The emission from wide nanowires shows a sharp peak at 3.362 eV arising from the donor bound excitons and a minor peak at 3.366 eV as a result of surface bound exciton emission. The emission peak at 3.366 eV becomes prevailing in the thinner wires. This blue shift is a result of the finite size effect,

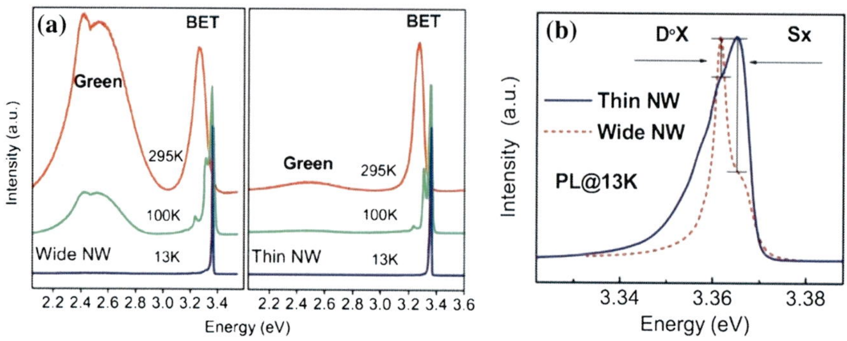

Fig. 9.7 **a** PL spectra of wide (*left*) and thin (*right*) ZnO nanowires at different temperatures show that the bulk defect states (*Green*) are more pronounced in wide nanowires. Intensity is normalized with respect to the band-edge transition peak. **b** A close-up view of band-edge transition (*BET*) of thin and wide nanowires at 13 K, indicating the *blue shift* in thinner wires due to the prevailing surface bound exciton emission. Reprinted with permission from [66]. Copyright 2007, American Institute of Physics

due to the emerging of the surface bound exciton emission that becomes dominant in smaller diameter nanowires.

9.3.2 Electrical Properties

In order to study the electrical transport properties of the nanowires, metallic electrodes are patterned onto individual nanowires. The intrinsic electrical properties are best conducted by four-probe conductivity measurement [21] and three-terminal field effect transistor (FET) response [30]. Figure 9.8a represents the logarithmic scaled conductivity of a ZnO nanowire as a function of reciprocal temperature [68]. For $T > 50$ K, the conduction shows Arrhenius behavior with thermally activated transport. The conductivity can be expressed as $\sigma \sim \exp(-E_a/k\,T)$, where E_a is the activation energy around 30 meV, attributed to the shallow donor levels below the conduction band edge. For $T < 50$ K, 3D Mott's variable range hopping model governs the transport, with conductivity expressed as $\sigma \sim \exp(-A\,T^{-1/4})$.

Figure 9.8b shows an AFM image of a ZnO nanowire FET on a SiO_2/p^{++} Si substrate [68]. Based on simple electrostatic model [69], taking an infinitely long thin wire of radius r located a distance h above a conducting plane, the capacitance per length L is obtained to be $C/L = (2\pi\,\varepsilon_r\,\varepsilon_0)/[\ln(2\,h/r)]$, where ε_r is the effective dielectric constant for the wire in between air and SiO_2 plane [21]. In general, the field effect mobility (μ) can be expressed using the following expressions:

$$\mu = \frac{\partial I_{DS}}{\partial V_g} \frac{L^2}{C V_{DS}} \tag{9.1}$$

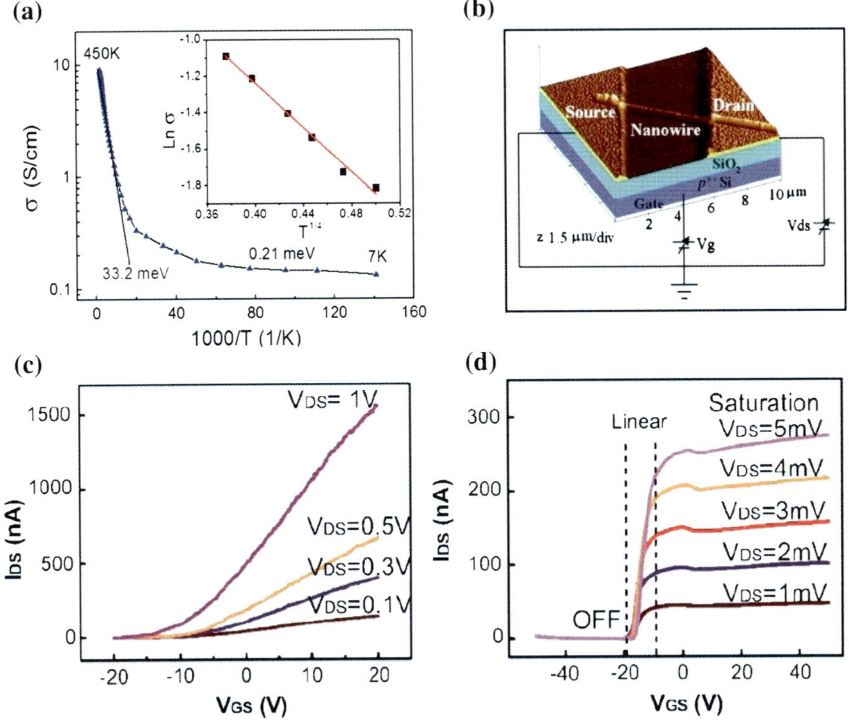

Fig. 9.8 a Conductivity *vs* reciprocal temperature. Inset: $\ln \sigma$ vs $T^{-1/4}$ plot in the low temperature range showing a linear fit to the 3D VRH model. Reprinted with permission from [68], Copyright 2008, American Institute of Physics. **b** An AFM image of a nanowire placed on a Si back gate under a SiO₂ dielectric layer with source and drain electrodes patterned to the wire ends. **c** $I_{DS} - V_{GS}$ *curves* of a ZnO nanowire FET without surface treatments showing typical *n*-type semiconducting behavior. **d** $I_{DS} - V_{GS}$ of a surface treated nanowire FET exhibits significantly enhanced ON/OFF ratio and transconductance. Reprinted with permission from [30], Copyright 2009, American Institute of Physics

From the gate voltage dependent source-drain conductance characteristics (Fig. 9.8c [30]), one estimates the field effect mobility to be around 30 cm²/Vs, and an electron concentration around $4.0 \times 10^7 \text{cm}^{-1}$ [69]. It is known that the surface defects act as scattering and trapping centers, thus affecting the charge transport. Surface passivation has been applied on metal oxide nanowires. Park et al. [70] have reported an electron mobility of 1,000 cm²/Vs after coating the nanowires with a polyimide passivation layer to reduce the electron scattering and trapping at the surface. Later, a CMOS-compatible process to coat a SiO₂/Si₃N₄ bilayer is conducted on ZnO nanowire FET. The surface treated ZnO nanowires show dramatically enhanced electron mobility up to 4,000 cm²/Vs. These results indicate that the ZnO Q1D device has promising potential in high speed electronics applications.

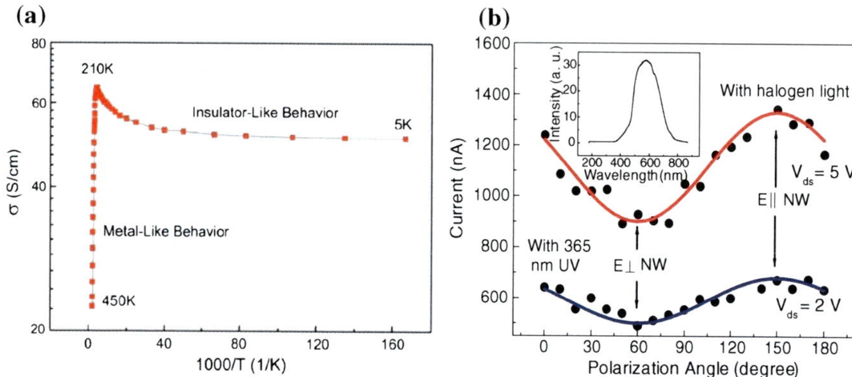

Fig. 9.9 **a** Photoconductivity of ZnO nanowire under UV irradiation reveals a metal-to-insulator transition at 210 K. Reprinted with permission from [68], Copyright 2008, American Institute of Physics. **b** Photocurrent as a function of the light polarization angle for both UV (*blue curve*) and halogen light (*red*). The inset plot shows a spectrum of halogen light source. Reprinted with permission from [93], Copyright 2004, American Institute of Physics

Under light illumination, the conductivity of semiconducting nanowires typically increases due to excess photogenerated charge carriers. Interestingly, pronounced metal–insulator transitions have been observed in undoped and doped ZnO nanowires. Figure 9.9a shows the temperature dependent conductivity of a ZnO nanowire FET upon continuous UV ($\lambda_{UV} = 365$ nm, $E_{UV} = 3.4$ eV) irradiation [68]. The conductivity is significantly increased by two orders of magnitudes at low temperature. As temperature rises, thermally activated charge carriers contribute further to an increase in conductance. Yet, as the carrier concentration approaches a critical value, an insulator-to-metal transition occurs. The strong Coulomb interactions among the dramatically enhanced charge carriers cause the conductance to decrease as the temperature rises.

In polarization measurements, it is found that the photo conductance of ZnO nanowires is sensitive to the angle of the incident illumination as shown in Fig. 9.9b. The photocurrent is maximized when the electrical field of the incident light is parallel to the nanowire long axis and minimized at 90°. The large polarization response is accounted for by the large dielectric contrast between the nanowire and surrounding air. The polarization-dependent behavior can be potentially utilized for an optical switch, photodetector and polarizer using aligned nanowire arrays.

9.4 Nanowire Sensing Applications

There are a number of unique characteristics that make metal oxide nanowires ideal candidates for chemical sensing as compared to other nanomaterials. Essentially, the chemical sensing process is a surface sensitive process which relies on molecular adsorption on the nanowire surface and electrostatic interaction between the adsorbed molecules and the semiconducting channel. It is known that the metal oxide nanowire surfaces are rich with surface states, which can serve as efficient molecular adsorption sites [71, 72]. Typically nanowires have a radius comparable to the Debye screening length, resulting in a dramatic effect on the nanowire conductance from the surface electrostatic perturbation caused by adsorbed molecules [71–73]. On the other hand, the metal oxide nanowire surface is relatively more stable, as compared to that of Si nanowires, which have been widely investigated for chemical/biological sensing. These fundamental aspects favor metal oxides as highly sensitive chemical sensors. Table 9.1 summarizes sensing applications using various types of metal oxide nanomaterials. As shown, a broad spectrum of chemical sensors has been investigated. In this section, we will introduce the conductometric chemical sensing with metal oxide nanowires. More importantly, we will describe the effects of nanowire diameter and external gate potential, on the sensor performance.

9.4.1 Conductometric Chemical Sensor

Quasi-1D geometry of nanowires makes them relatively easy to be fabricated into two terminal devices; therefore a conductometric chemical sensor device structure is a natural choice for nanowire based chemical sensors. To-date, there have been a number of different types of metal oxide nanowires configured as chemical sensors. These materials include ZnO [14, 73–75], In_2O_3 [76–78], SnO_2 [79–82], TiO_2 2, CuO [41], etc. Figure 9.10 shows the schematic of a nanowire sensor device fabricated on a Si substrate capped with a layer of thermally grown oxide. A device fabrication process can be briefly described as following. The as-synthesized nanowires are suspended first in alcohol of a controlled concentration. Then the suspension is dropped onto a degenerately doped p^{++} silicon substrate capped with 100–500 nm oxide layer. Photoresist is then spin-coated onto the substrate and photolithography is performed to define an array of 100 μm^2 pads. Finally, 10 nm thick Ni or Ti and 100 nm thick Au is evaporated in sequence onto this substrate, forming the electrical contacts to the nanowires. Individual nanowires with good contacts on both ends are located with a high magnification optical microscope or scanning electron microscope (SEM), as shown in Fig. 9.11b. Consequently, two terminal nanowire devices are obtained, with the degenerately doped Si serving as a global back gate to form field effect transistors, shown in Fig. 9.10a.

The electrical transport studies on these nanowire devices have been discussed in the previous section. In order to perform gas sensing studies, these devices are typically placed in an enclosed chamber with controlled target gas concentration, and electrical measurement can be performed at the same time to monitor the conductance change upon gas exposure. We have explored the sensing responses of a number of gases, including O_2 [14], NO_2 [73], NH_3 [74], CO [74], with ZnO nanowire devices, as shown in Fig. 9.11. Figure 9.11a shows the

Table 9.1 Metal oxide nanomaterials for sensing applications (RT denotes room temperature)

Gases	Nanomaterials	Detection limit	Response time	Operating temperature ($^{\circ}$C)	Reference
O_2	ZnO	10 ppm	20 s	RT	[14], [94]
	SnO_2	ppb level	40 ~ 200 s	127 ~ 277	[55], [95]
	TiO_2	200 ppb	2 min	RT	[96]
O_3	ZnO	75 ppb	20 s	RT	[97]
	SnO_2	75 ppb	20 s	77 ~ 177	[98]
	In_2O_3	75 ppb	20 s	77 ~ 227	[98]
NO_2	SnO_2	300 ppb	120 s	100	[99]
	In_2O_3	5 ppb	5 ~ 10 s	RT	[100]
	$Ru\text{-}SnO_2$	50 ppm	30 ~ 90 s	250	[101]
	ZnO	200 ppb	100 s	27	[73]
	Ga_2O_3	20 ppm	100 h		[102]
	WO_3	50 ppb	Several min	250	[103]
	TeO_2	10 ppm	600 s	RT	[104]
	CdO	200 ppm	200 s	-271.8 ~ 17	[105]
	WOx	3 ppm	10 min	RT	[87]
	TiO_2	500 ppb	12 ~ 60 min	150 ~ 300	[106]
NH_3	In_2O_3	0.00 %	5 ~ 20 s	RT	[76]
	ZnO	0.50 %	5 ~ 100 s	RT	[107, 108]
	TeO_2	100 ppm	600 s	RT	[104]
	WOx	100 ppm	20 ~ 400 s	20 ~ 200	[87]
	SnO_2	25 ppm	20 s ~ 80 s	60 ~ 300	[109, 110]
	V_2O_5	10 ppm			[111]
CO	In_2O_3	10 ppm	50 ~ 200 s	160 ~ 280	[112]
	SnO_2	5 ppm	100 s	20 ~ 200	[113]
	ZnO	500 ppm			[114, 115]
H_2S	WO_3	10 ppm	Several min	250	[103]

(continued)

Table 9.1 (continued)

Gases	Nanomaterials	Detection limit	Response time	Operating temperature(°C)	Reference
	TeO_2	50 ppm	600 s	RT	[104]
	ZnO_2	0.1 ppm	20 min	RT ~ 200	[116]
H_2	In_2O_3	10 ppm	5 s ~ 10 s	RT ~ 275	[112]
	Pd, Pt, Au, Ni, Ag, Ti; doped ZnO	10 ppm	600 s	RT	[117]
	ZnO	500 ppm	150 ~ 300 s	20 ~ 200	[97, 117]
	SnO_2	ppb level	20 ~ 40 s		[95]
	TiO_2	100 ppm	100 ~ 500 s	180 ~ 375	[118]
	WOx	100 ppm	50 s	RT	[119]
Ethanol	SnO_2	23 ppm	30 s	300	[79]
	ZnO	50 ppm	60 s	RT ~ 300	[120]
	V_2O_5	10 ppm	10 ~ 100 s	150 ~ 400	[121]
	In_2O_3	100 ppm	15 s	150 ~ 370	[122]
	Ga_2O_3	1500 ppm	2.5 s	As low as 100	[123]
Humidity	ZnO	20 %RH	3 s	RT	[124, 125]
	SnO_2	30 %RH	100 ~ 200 s	RT	[126]
	Al_2O_3	45 %RH	20 s	RT	[127]
	CeO_2	33 %RH	3 s	RT	[128]
Acetone	ZnO	5 ppm	30 ~ 200 s	150 ~ 350	[129]
Acetonitrile	SnO_2	2 ppm	10 min	350 ~ 500	[101]
	In_2O_3	2 ppm	10 min	350 ~ 450	[101]
Toluene	V_2O_5	1000 ppm	500 s	RT	[111]
	SnO_2	10 ppm	1 ~ 5 s	310 ~ 380	[100]
1-butylamine	V_2O_5	30 ppb	200 s	RT	[111]
1-propanol	V_2O_5	1000 ppm	300 s	RT	[111]
Formaldehyde	ZnO	30 ppm	5 ~ 10 min	200 ~ 600	[130]
Gasoline	ZnO	50 ppm		RT ~ 332	[115]

(continued)

Table 9.1 (continued)

Gases	Nanomaterials	Detection limit	Response time	Operating temperature(°C)	Reference
Dimethyl Methylphosphonat (DMMP)	SnO_2	50 ppb	6 s	500	[102]
	In_2O_3	Lower than 100 ppb		400 ~ 500	[101]
Dipropylene glycol methyl ether (DPGME)	SnO_2	Lower than 100 ppb		400 ~ 500	[101]
	In_2O_3	Lower than 100 ppb		400 ~ 500	[101]
LPG	WOx	100 ppm	38 s	RT	[119]
	ZnO	100 ppm		150 ~ 465	[115, 119]

Fig. 9.10 **a** Schematic of a metal-oxide nanowire gas sensor device structure. **b** Optical and scanning electron microscopy image (Inset) of a metal oxide nanowire in contact with two metal electrodes. Each metal square electrode has a size of 100 μm by 100 μm and the scale bar in the inset is 3 μm. Reprinted with permission from [40], Copyright 2005, American Institute of Physics

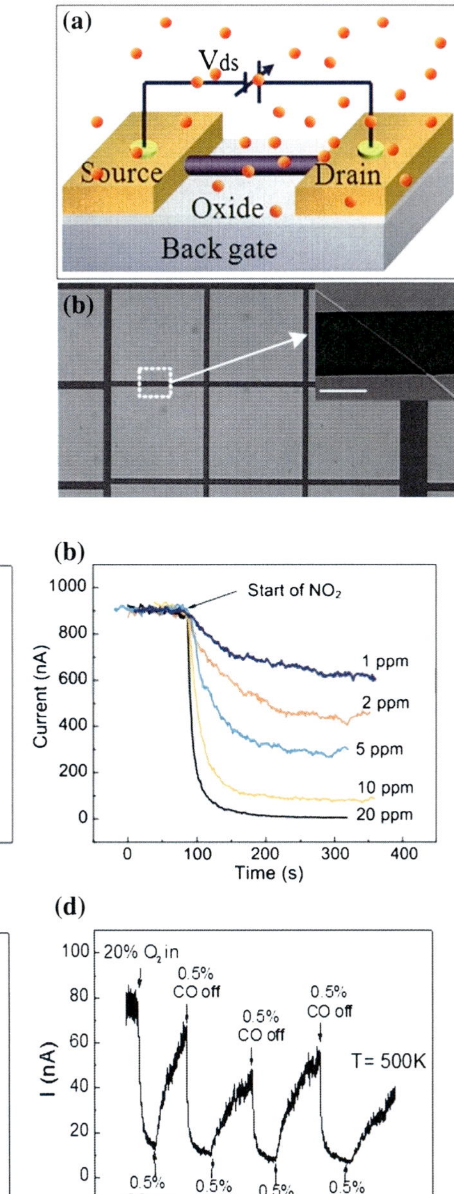

Fig. 9.11 **a** *I − V curves* of a ZnO nanowire sensor measured without presence of O_2 (0 ppm) and in 10, 25, 50 ppm O_2. Reprinted with permission from [14], Copyright 2004, American Institute of Physics, **b** time domain current response of ZnO nanowire sensor to NO_2 Reprinted with permission from [73], Copyright 2005, American Institute of Physics, **c** and **d** are time domain current response of ZnO nanowire sensor to NO_2, NH_3 and CO. Reprinted with permission from [74], Copyright 2006, Institute of Electrical and Electronics Engineers

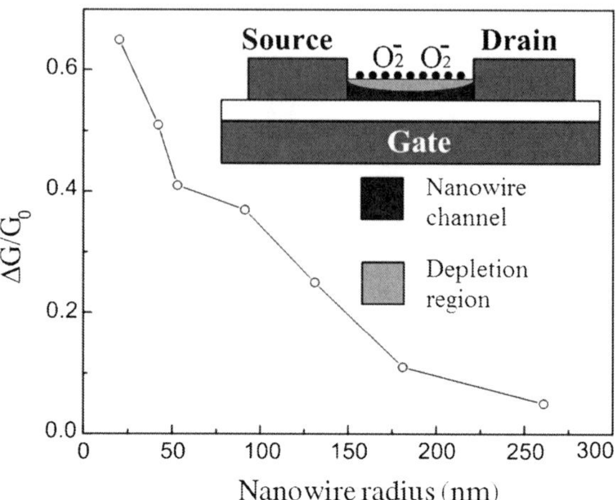

Fig. 9.12 Relative conductance change of a ZnO nanowire gas sensor exposed to 1 atmosphere air, the inset is the schematic of formation of surface depletion region by O_2 molecular adsorption. Reprinted with permission from [14], Copyright 2004, American Institute of Physics

I–V characteristics of a ZnO nanowire device measured under 0–50 ppm O_2 concentration [14]. It can be seen that exposure to O_2 decreases the conductance of the ZnO nanowire device, due to the fact that O_2 molecules adsorbed at these sites act as electron acceptors to form O_2^- at room temperature. These chemisorbed O_2^- deplete the surface electron states and consequently reduce the channel conductivity. Compared with bulk materials, such a surface effect is more significant on nanowire conductance, since the surface-to-volume ratio of the nanowire is much larger.

Besides measuring the *I–V* curves of the nanowire devices, the temporal response of nanowire conductance to gas exposure have been monitored in real-time. Figure 9.11b–d shows the conductance response to NO_2, NH_3 and CO measured at room temperature. Both NO_2 and NH_3 appear to be interact as an "oxidizing" gas as they reduce nanowire channel electron concentration and conductance. Nevertheless, ZnO nanowires have shown much higher sensitivity to NO_2 gas, as compared to O_2 and NH_3 [73, 74]. As shown in Fig. 9.11b and c, 2 ppm NO_2 causes a 50 % conductance decrease while 1 % NH_3 yields the same effect. This difference was attributed to the higher binding strength between NO_2 molecules and the nanowire surface binding site [73].

In addition to the sensing response at room temperature to O_2, NO_2, NH_3, the response of ZnO nanowires to CO gas in the presence of O_2 at 500 K is also studied [74]. Figure 9.11d demonstrates the time domain measurement of nanowire conductance when the nanowire is exposed to 0.5 % CO in 20 % O_2. It can be seen that nanowire conductance first decreases upon admittance of 20 % O_2, however, the admittance of CO increases the nanowire conductance. It is known that CO is a combustible gas, the sensing of CO is based on the reaction between CO and the pre-adsorbed oxygen species to form CO_2 [55]:

$$CO + O^- \rightarrow CO_2 + e^- \tag{9.2}$$

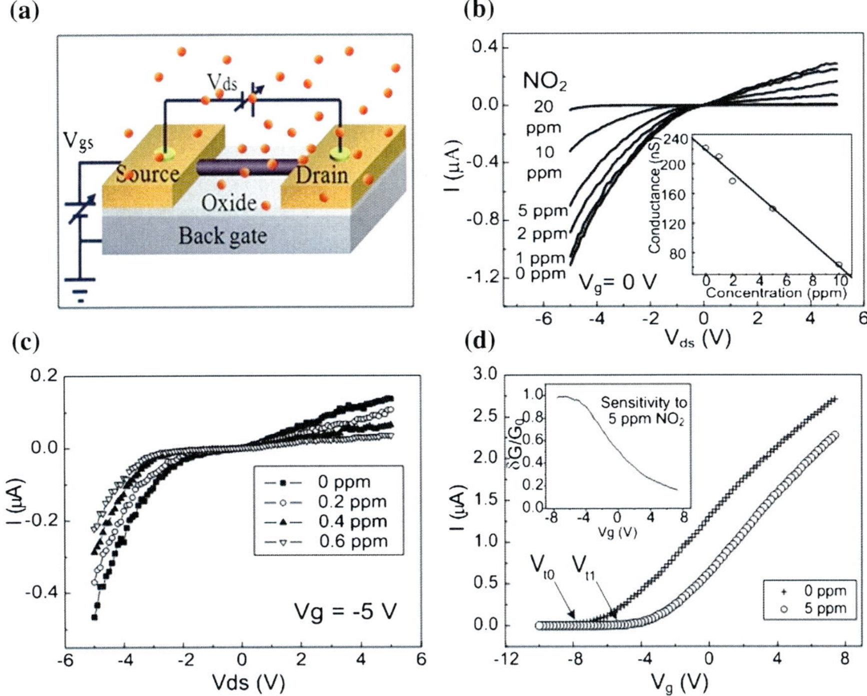

Fig. 9.13   **a** Schematic of a nanowire FET sensor, **b** and **c** are $I - V_{ds}$ *curves* of a ZnO nanowire FET sensor measured under different NO_2 exposure concentration at $V_g = 0$ and -5 V, respectively. The inset of **b** shows the nanowire conductance versus NO_2 concentration, extrapolated from linear region at $V_{ds} = 0$ V, **d** $I - V_g$ *curves* of the ZnO nanowire FET sensor obtained without NO_2 and with presence of 5 ppm NO_2. The inset shows the extrapolated relative conductance change versus gate voltage. Reprinted with permission from [73], Copyright 2005, American Institute of Physics

9.4.2 Diameter Dependent Sensitivity

Compared to metal oxide thin film, a semiconductor nanowire with a large surface-to-volume ratio, a radius comparable to the Debye length, and well defined single crystal facets, have simple and direct detection mechanisms that render them with better sensing stability [71, 73], resulting in dominant effect on nanowire con-ductance from the surface electrostatic perturbation caused by adsorbed molecules. Following this rational, it is expected to render strong diameter dependent gas sensing performance. In fact, we have performed systematic studies on selecting ZnO nanowires with diameters ranging from 50 to 500 nm [14]. In these mea-surements, the conductance changes of nanowire devices are recorded when their environment is changed from vacuum to 1 atmosphere synthetic air. As shown in Fig. 9.12, the relative conductance change $\Delta G/G_0$ increases as the nanowire radius decreases, where G_0 denotes the nanowire conductance in vacuum and ΔG is the

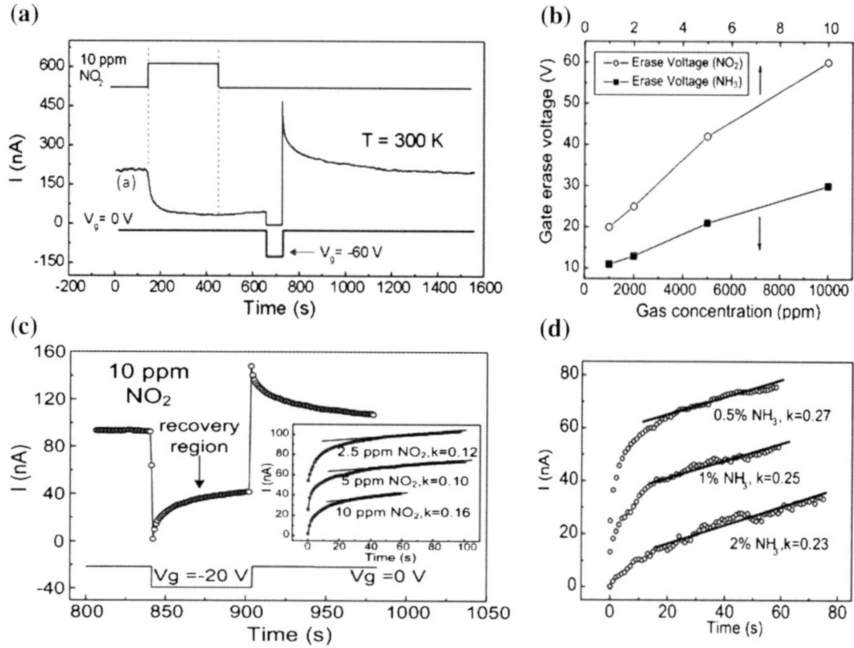

Fig. 9.14 **a** ZnO nanowire sensing response to 10 ppm NO_2 and gate voltage induced sensor refresh process. **b** Gate "refresh" voltage as a function of NO_2 and NH_3 concentration. **c** Nanowire conductance recovery under $V_g = -20$ V pulse in 10 ppm NO_2. Inset: recovery curves at different NO_2 concentrations. The solid lines represent the linear fitting in the recovery region. **d** Conductance recovery curves for NH_3 under $V_g = -20$ V pulse. Reprinted with permission from [73]. Copyright 2005, American Institute of Physics

conductance change from vacuum to atmosphere. This trend clearly supports the above rationale and demonstrates the advantage of using thinner nanowires for gas sensing. Schematically, depletion of electrons in ZnO nanowire by O_2 molecular adsorption is shown in the inset of Fig. 9.12.

9.4.3 Field Effect Sensor

Smaller diameter of nanowires leads to a higher sensitivity towards chemicals adsorbed on the surface. In addition, the conductance of nanowires is also sensitive to the electrostatic potential applied by the external gate electrode in an FET configuration shown in Fig. 9.13a. It is well known that free carrier concentration can be readily tuned by using the gate voltage in a FET which naturally results in the change of the Debye screening length, and thickness of the surface depletion region caused by molecular adsorption. One can speculate that these changes will consequently affect nanowire sensing performance. In fact, we have performed systematic studies on this aspect [73]. Figure 9.13b and c plot $I - V_{ds}$ curves of a ZnO nanowire FET obtained under different NO_2 concentrations at $V_g = 0$ and -5 V. The minimum NO_2 concentration detected with this particular device is 200 ppb. The inset of Fig. 9.13b shows a linear decrease of the nanowire conductance with increasing NO_2 concentration up to 10 ppm. It is evident that the FET device operated at lower gate voltage demonstrated higher sensitivity to NO_2. Furthermore, the dependence of gas detection sensitivity, defined as relative conductance change $S \equiv |G_{gas} - G_0|/G_0$ where G_0 and G_{gas} is the conductance of nanowire before and after NO_2 exposure, on the gate potential can be obtained from $I - V_g$ curves acquired under different target gas concentrations, shown in Fig. 9.13d. Note that exposure to NO_2 shifts the threshold voltage of the FET positively from V_{t0} (-7.9 V) in 0 ppm NO_2 to V_{t1} (-5.2 V, or $V_{t1} = 2.7$ V) in 5 ppm NO_2, indicating a decrease of electron concentration for an n-type ZnO nanowire channel [14], which is consistent with the oxidizing nature of NO_2 [73]. Analytically, the relationship between threshold voltage V_t and electron concentration n can be expressed as:

$$n = C(V_g - V_t)/e\pi R^2 L \tag{9.3}$$

where C is the nanowire capacitance with respective to the back gate, R and L are radius and length of nanowire channel, respectively. Hence:

$$\Delta n = -C\Delta V_t/e\pi R^2 L \tag{9.4}$$

Using the definition of relative sensitivity $S \equiv |G_{gas} - G_0|/G_0$, in conjunction with $G = n\mu e\pi R^2/L$,

$$S = |n_{gas} - n_0|/n_0 = |\Delta n|/n_0 = \Delta V_t/(V_g - V_{t0}) \tag{9.5}$$

The derived $S(V_g) \propto (V_g - V_{t0})^{-1}$ relationship is depicted in the inset of Fig. 9.13d and the maximum sensitivity can be achieved when the gate potential approaches V_{t0}. Qualitatively, this result can be understood as when the gate voltage is approaching the threshold, the nanowire channel is close to being fully depleted, thus the change of the surface molecular adsorption dominantly alters the channel conductance.

The above discussions reveal that the range of sensitivity measurement of a nanowire field effect sensor can be controlled by simply tuning the gate voltage. Furthermore, it has been found that a large gate field can also affect gas adsorptivity on the nanowire surface significantly [73, 83]. As shown in Fig. 9.14a, after a nanowire is first exposed to a 10 ppm NO_2 pulse, a -60 V gate voltage (much below the threshold value $V_t = -25$ V in 10 ppm NO_2 for this particular device) pulse with 60 s duration was applied. This voltage pulse leads to complete conductance depletion, followed by a recovery of conductance close to the initial level in about 4 min. Therefore, the large negative gate pulse efficiently refreshed the nanowire sensor [73].

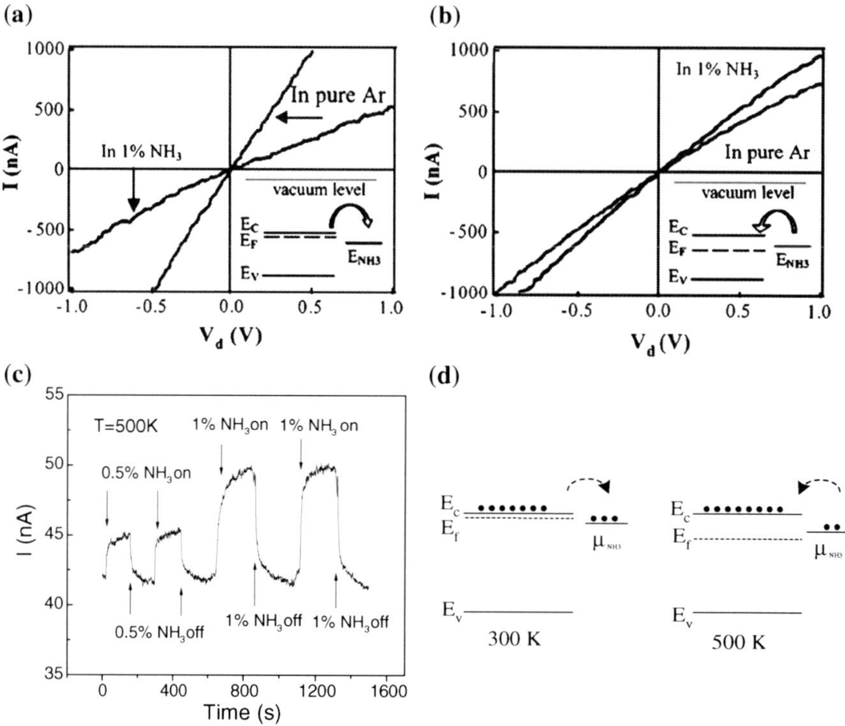

Fig. 9.15 **a** $I - V_{ds}$ curves of heavily doped In_2O_3 nanowire before and after exposure to 1 % NH3. Inset: energy band diagrams of heavily doped In_2O_3 and NH_3 molecules. **b** $I - V_{ds}$ *curves* of lightly doped In_2O_3 nanowire before and after exposure to 1 % NH_3. Inset: energy band diagrams of lightly doped In_2O_3 and NH_3 molecules. Reprinted with permission from [86], Copyright 2003, American Institute of Physics. **c** ZnO nanowire sensing response to different concentrations of NH_3 at 500 K. **d** Schematic of Fermi level shift from 300–500 K and the resulted reversal of electron transfer direction. Reprinted with permission from [74], Copyright 2006, Institute of Electrical and Electronics Engineers

The unique sensor gate refreshing behavior is rationalized with a gate-dependent surface adsorption process [73]. Although the binding energy between the adatoms and adsorption sites on the nanowire surface is usually much larger than room temperature thermal energy, a negative gate potential could deplete the electrons in the metal oxide nanowire and reduces the number of electrons available at vacancy sites, thus lowering the chemisorption rate [73]. Hole migration to the surface driven by the negative field will lead to the discharge of gas molecules. In addition, if the surface adsorbed molecules have an appreciable dipole moment [73], it is likely that the negative gate induced field can stretch and weaken the bonding. In combination, these factors lead to the electrodes orption process and the fast conductance recovery at room temperature.

From a sensor application point of view, the use of changes in gate voltage for sensor rejuvenation has certain advantages compared to the conventional heating

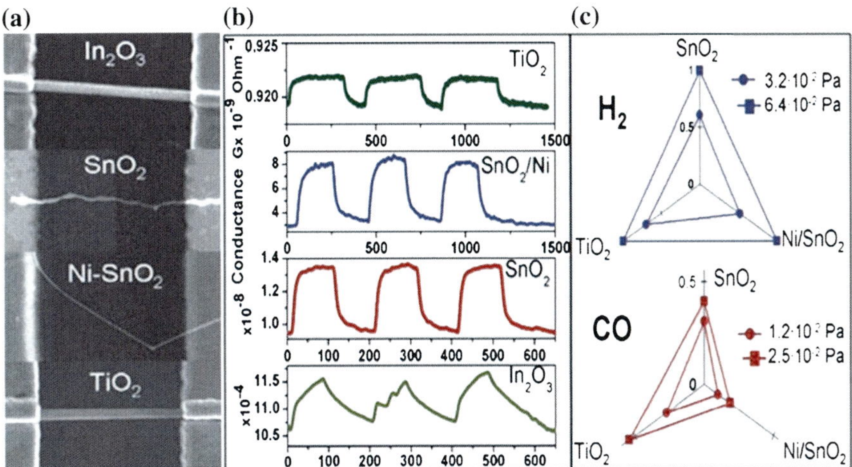

Fig. 9.16 **a** SEM images of single In_2O_3, SnO_2, Ni doped SnO_2 and TiO_2 mesowires integrated on the same chip. **b** The response of the array of the chemiresistors to hydrogen pulses with partial pressure of 6.4×10^{-2} Pa. **c** The response of a three-chemiresistor array to H_2 (*top*) and CO (*bottom*) inputs, normalized by maximum value. Reprinted with permission from [88], Copyright 2006, American Chemical Society

and UV illumination recovery methods. These methods either consume much more energy than sensor operation, or are difficult to integrate with nanowire sensors to form a compact solution. In addition, UV illumination usually generates a large number of electron–hole pairs leading to longer recovery times [84], which hinders fast sensor operation.

Besides NO_2, we have also explored refreshable sensing with NH_3, another type of toxic gas [73]. As mentioned in Sect. 9.4.1, NH_3 exposure also decreases ZnO nanowire conductance at room temperature, however, ZnO nanowire sensors are much less sensitive to NH_3 gas, due to the weaker electron capture capability of NH_3 molecules, as compared to NO_2 [73]. Interestingly, experiments have shown that the gate voltage required to refresh the adsorbed NH_3 molecules on ZnO nanowires is lower than that for NO_2. For example, as shown in Fig. 9.14b, the "refresh" or "erase" voltage for 1 % NH_3 is -30 V as compared with -60 V for 1 % NO_2. Using the same gate pulse duration (60 s), the minimum gate "refresh" voltage is determined to be concentration dependent, as plotted in Fig. 9.14b. More intriguingly, the time domain responses for NO_2 and NH_3 show a clear difference. Particularly at the negative gate pulse duration, in which nanowire conductance shows a quick transient followed by a relative steady recovery process, as expected from the dynamic adsorption–desorption process [73]. The slopes (k) of the linear recovery regions for NO_2 and NH_3 at different concentrations are shown in Fig. 9.14c inset and Fig. 9.14d, respectively. It is observed that this slope is concentration insensitive with distinguishable difference for NO_2 and NH_3, specifically, $k_{NH3} \approx 2k_{NO2}$. The fact that NH_3 is much easier to be electro-desorbed

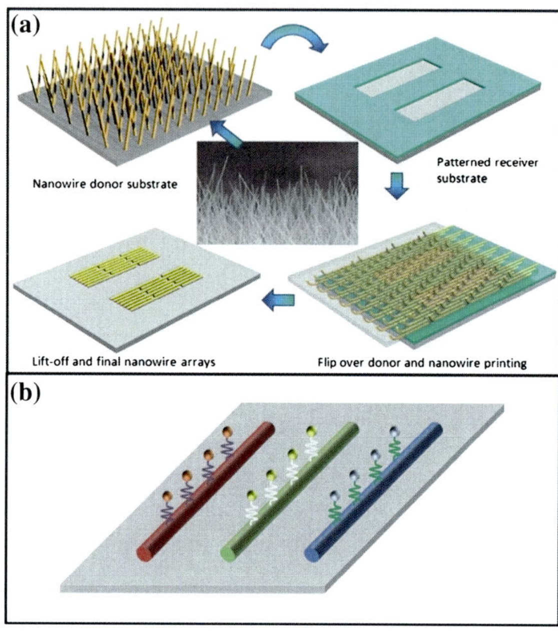

Fig. 9.17 a Schematic of nanowire contact printing process. Reprinted with permission from [91], Copyright 2008, American Chemical Society. **b** Schematic of heterogeneous nanowire integration for smart sensing application

than NO_2 renders a potential mechanism to distinguish NO_2 and NH_3 with the assistance of additional signal processing circuits and pattern recognition software.

It is noteworthy that the apparently different sensitivity of ZnO nanowire to NO_2 and NH_3 has been attributed to their different carrier activation energy on the nanowire surface under the framework of the Volkenstein model [85]. Experiments have shown that conductance activation energy in a NO_2 surface coated nanowire is much higher than that of a NH_3 surface coated nanowire. Considering the fact that activation of carriers from the adsorbed molecules will most likely result in the desorption process, such an activation energy can also be qualitatively associated to the desorption energy, which in turn indicates higher desorption energy for NO_2, as compared to NH_3.

9.4.4 Reversal Sensing Mechanism

External doping and temperature variation influence the electronic property of nanowire channels, thus the sensing response characteristics to the surface adsorbed molecules could be affected. Figure 9.15a and b represent a reversal sensing response towards 1 % NH_3 for the In_2O_3 nanowires with heavy and light oxygen vacancy doping [86]. The effective doping is achieved by tuning the oxygen partial pressures during systhesis. In the case for lightly doped nanowires, the nanowire Fermi level E_F is below the chemical potential of NH_3 (E_{NH3}), thus electrons should transfer from nanowire to the adsorbed NH_3 species, yielding a

Fig. 9.18 Prototype of a compact chemical sensor that is incorporated into a mobile phone. Courtesy of Dr. Francisco Hernandez-Ramirez

reduction of conductance (Fig. 9.15a). However, for heavily doped nanowires, the nanowire Fermi level E_F rises above E_{NH3}, thus the electron transfer direction is reversed, and the conductance increases (Fig. 9.15b).

Fermi level is temperature dependent, thus the shift in Fermi level as temperature varies can alter the sensing response [74, 87]. The ammonia chemical potential μ_{NH3} generally lies closely below the E_F level of n-channel ZnO at room temperature. Upon exposure, the adsorbed NH_3 will deplete electrons in the nanowire channel, thus inducing conduction suppression, as shown in Fig. 9.15c. Whereas at an elevated temperature of 500 K, conductance increases upon exposure to NH_3 occurs (Fig. 9.15c), indicating that E_F lowers below μ_{NH3}, giving rise to reversal of electron transfer. In detail, Fig. 9.15d depicts the metal oxide band diagram schematic upon the adsorption of NH_3 at 300 and 500 K. At 300 K, the E_F level lies above the chemical potential of ammonia gas (denoted as μ_{NH3} in Ref. [74]). To reach equilibrium upon chemisorption, the electrons are withdrawn from the ZnO, resulting in a conductance decrease. An elevated temperature yields the down-shift of both E_F and μ_{NH3}. At 500 K, the shift in E_F places it below μ_{NH3}, as depicted in Fig. 9.15d and a reversal of charge transfer occurs, i.e. the electron accumulation layer is formed on the ZnO surface, inducing an increase of conductivity.

9.4.5 Nanosensor Integration

Although single nanowires have a characteristic response to different chemical species, in order to achieve fast identification of the composition and concentration of mixed species, it is more practical to build a sensing device with arrays of nanowires made of different materials. Due to the difference in chemical compositions and surface states, nanowires made of different materials may have dissimilar responses to the target analytes. On the other hand, the response of a nanowire to specific species can be also modified via surface decoration/modification. For example, Pd decorated SnO_2 nanowire surface leads to improved H_2 sensing property [82]. The ultimate objective of fabricating a nanowire array based gas sensor is to realize the concept of an "Electronic Nose" which is composed of a large number of sensors. In fact, the development of such a device based on metal oxide nanowires has been demonstrated by Kolmakov and co-workers [88], and lately by a few other groups [89, 90]. In the first work, single SnO_2, Ni doped SnO_2, In_2O_3 and TiO_2 mesowires were integrated on the same chip to form an array of chemoresistors, as shown in Fig. 9.16a. Note that these structures were actually randomly placed on the four different regions of the same Si/SiO_2 wafer for device fabrication [88]. Figure 9.16b shows the response of these nano/mesowires to H_2 pulses after low pressure O_2 exposure. Similar responses were observed in the case of CO with an exception of In_2O_3 mesowire. At the chosen range of concentrations, CO did not cause a measurable change in conductance in this mesowire. In addition, the In_2O_3 whisker response to hydrogen was slower than that for any of the other structures. In this work, though both H_2 and CO increase the conductance of the chemiresistors, and the selectivity of the individual nanostructure is not sufficient to discriminate between them, gas distinguishability can be achieved using pattern recognition. Specifically, as shown in Fig. 9.16c, signal patterns are depicted for these three sensing elements as radial plots: each radial beam shows the signal of one single sensing element normalized to its maximum value and the radial plots of H_2 and CO gas responses over the chemiresistor array are prominently different for two gases leading to feasible pattern recognition.

The concept of smart sensing with multiple nanowires has confirmed that an "Electronic Nose" can only be realized via heterogeneous integration of nanowires. In addition, this integration has to be more or less deterministic, namely, specific types of nanowires have to be placed at the defined location, therefore, arrays of sensor elements can be easily fabricated at low cost. Recently, we have developed a scalable contact printing technique to achieve heterogeneous integration of semiconductor nanowires for sensor applications [91]. As shown in Fig. 9.17a, this method involves the directional sliding of the nanowire growth substrate with randomly aligned nanowires on top of a receiver substrate. During this process, nanowires are effectively combed (i.e., aligned) by the directional shear force, and are eventually detached from the growth substrate and transferred to the target substrate which has lithographically defined patterns. This nanowire transfer method has been proven highly generic for a number of nanowire materials, and is also compatible with various type of substrates, including Si/SiO_2, glass, plastic and paper. More importantly, all-nanowire integrated sensing circuits have been fabricated using two different types of nanowires. Therefore, this technique provides an ideal technological platform for building nanowire arrays towards selective and smart chemical sensing, as illustrated in Fig. 9.17b.

9.5 Future Outlook

We have presented in this chapter the recent progress in the studies of Q1D metal oxides for chemical sensor applications. The topics cover nanowire growth, structure and material characterizations, and their advantages as stable, sensitive, and fast chemical sensors with low-power consumption, easy integration, and potential chemical selectivity. Because of the intensive research effort in this field, it is difficult to numerate all the achievements. Thus, in this chapter, we could only highlight the essence of this research area.

Although the studies are still in the research stage, the future of nanowire based devices is promising in terms of practical sensing performance. As in the case of thin film metal oxide sensors, chemical selectivity remains a challenge for practical implementations. Possible solutions include surface coating with a chemically selective membrane, or surface modifications with specific functional groups. And a more sophisticated solution is to combine modular sensing units with different nanomaterials to form a sensor network with wireless transmission capability.

As a model for sensor applications, Morante et al. [80] have developed a portable chemical sensor by integrating SnO_2 nanowires and micromembranes with integrated heaters into one device. This device demonstrates hand held with a built-in nanowire chemical sensor. Figure 9.18 shows a photograph of the experimental setup, including the electronic integrated circuit. All of the measurements can be performed and analyzed at an external processing station through its USB connector. This small and lightweight device enables ways to perform *in-situ* and real-time monitoring and detection of hazardous gases. Such a device can be made portable and mobile, with a wireless transducer to implement a wireless sensor network [92]. Ultimately, one envisions a smart sensor system, which integrates multiple sensing elements with signal processing on a single chip. Such a system with multicomponent chemical analysis integrated with pattern recognition and signal transmission provides a promising prospect for a compact and portable chemical sensor.

In summary, the chapter has examined the fundamental properties of metal oxide nanowires and their potential application as chemical sensors. The intrinsic nature of large surface-to-volume ratio, radii comparable to the Debye length, and well defined single crystal facets make them stable sensors with tunable sensitivity, fast recovery, as well as simple integration for efficient analyte differentiation. The ultimate goal of chemical sensors is to mimic the olfactory function of mammalian systems, in detecting complex gas mixtures and odors using a network of sensor arrays and pattern recognition algorithms. The recent development of a metal oxide nanowire based "Electronic Nose" embodied into a single device integrating the sensing and signal processing functions in one chip have been described. The identifications of compounds and concentrations are obtained via pattern analysis of the measured responses carrying the "fingerprint" of a complex gas mixture. For detailed review of this particular topic, please refer to Chaps. 14, 15 in this book.

Acknowledgments The authors thank Dr. Dongdong Li at Shanghai Advanced Research Institute, Chinese Academy of Sciences and Mr. Miao Yu at Hong Kong University of Science and Technology for technical assistance on this chapter.

Reference

1. Ra Y, Choi K, Kim J, Hahn Y, Im Y (2008) Fabrication of ZnO nanowires using nanoscale spacer lithography for gas sensors. Small 4:1105–1109
2. Francioso L, Taurino AM, Forleo A, Siciliano P (2008) TiO_2 nanowires array fabrication and gas sensing properties. Sens Actuator B Chem 130:70–76
3. Duan XF, Lieber CM (2000) General synthesis of compound semiconductor nanowires. Adv Mater 12:298–302
4. Knez M et al (2003) Biotemplate Synthesis of 3-nm Nickel and Cobalt Nanowires. Nano Lett 3:1079–1082
5. Lu JG, Chang P, Fan Z (2006) Quasi-one-dimensional metal oxide materials - Synthesis, properties and applications. Mater Sci Eng R-Rep 52:49–91
6. Hurst SJ, Payne EK, Qin LD, Mirkin CA (2006) Multisegmented one-dimensional nanorods prepared by hard-template synthetic methods. Angew Chem Int Edit 45:2672–2692
7. Han S, Zhang DH, Zhou CW (2006) Synthesis and electronic properties of ZnO/CoZnO core-shell nanowires. Appl Phys Lett 88:133109
8. Wagner RS, Ellis WC (1964) Vapor-Liquid-Solid Mechanism of Crystal Growth. Appl Phys Lett 4:89–90
9. Wu YY, Yang PD (2001) Direct observation of vapor-liquid-solid nanowire growth. J Am Chem Soc 123:3165–3166
10. Stach EA et al (2003) Watching GaN nanowires grow. Nano Lett 3:867–869
11. Chang PC et al (2004) ZnO nanowires synthesized by vapor trapping CVD method. Chem Mater 16:5133–5137
12. Huang MH et al (2001) Catalytic growth of zinc oxide nanowires by vapor transport. Adv Mater 13:113–116
13. Zhao QX, Willander M, Morjan RE, Hu Q, Campbell EEB (2003) Optical recombination of ZnO nanowires grown on sapphire and Si substrate. Appl Phys Lett 83:165
14. Fan ZY, Wang DW, Chang PC, Tseng WY, Lu JG (2004) ZnO nanowire field-effect transistor and oxygen sensing property. Appl Phys Lett 85:5923–5925
15. Li SY, Lee CY, Tseng TY (2003) Copper-catalyzed ZnO nanowires on silicon (100) grown by vapor-liquid-solid process. J Cryst Growth 247:357–362
16. Lee CJ et al (2002) Field emission from well-aligned zinc oxide nanowires grown at low temperature. Appl Phys Lett 81:3648–3650
17. Gao PX, Ding Y, Wang ZL (2003) Crystallographic orientation-aligned ZnO nanorods grown by a tin catalyst. Nano Lett 3:1315–1320
18. Chik H, Liang J, Cloutier SG, Kouklin N, Xu JM (2004) Periodic array of uniform ZnO nanorods by second-order self-assembly. Appl Phys Lett 84:3376–3378
19. Li C et al (2003) Diameter-controlled growth of single-crystalline In_2O_3 nanowires and their electronic properties. Adv Mater 15:143–146
20. Wu ZH, Mei XY, Kim D, Blumin M, Ruda HE (2002) Growth of Au-catalyzed ordered GaAs nanowire arrays by molecular-beam epitaxy. Appl Phys Lett 81:5177–5179
21. Thompson RS, Li D, Witte CM, Lu JG (2009) Weak Localization and Electron-Electron Interactions in Indium-Doped ZnO Nanowires. Nano Lett 9:3991–3995
22. Park WI, Kim DH, Jung SW, Yi GC (2002) Metalorganic vapor-phase epitaxial growth of vertically well-aligned ZnO nanorods. Appl Phys Lett 80:4232–4234

23. Liu X, Wu XH, Cao H, Chang RPH (2004) Growth mechanism and properties of ZnO nanorods synthesized by plasma-enhanced chemical vapor deposition. J Appl Phys 95:3141–3147
24. Huang MH et al (2001) Room-temperature ultraviolet nanowire nanolasers. Science 292:1897–1899
25. Liang CH, Meng GW, Lei Y, Phillipp F, Zhang LD (2001) Catalytic growth of semiconducting In_2O_3 nanofibers. Adv Mater 13:1330–1333
26. Wan Q, Dattoli E, Lu W (2008) Doping-dependent electrical characteristics of SnO2 nanowires. Small 4:451–454
27. Chen YJ, Li JB, Han YS, Yang XZ, Dai JH (2002) The effect of Mg vapor source on the formation of MgO whiskers and sheets. J Cryst Growth 245:163–170
28. Chang PC, Fan ZY, Tseng WY, Rajagopal A, Lu JG (2005) beta-Ga2O3 nanowires: Synthesis, characterization, and p-channel field-effect transistor. Appl Phys Lett 87:222102
29. Wu JM, Shih HC, Wu WT, Tseng YK, Chen IC (2005) Thermal evaporation growth and the luminescence property of TiO_2 nanowires. J Cryst Growth 281:384–390
30. Chang PC et al (2006) High-performance ZnO nanowire field effect transistors. Appl Phys Lett 89:133113
31. Yuan GD et al (2008) Tunable n-type conductivity and transport properties of Ga-doped ZnO nanowire arrays. Adv Mater 20:168–173
32. Bae SY, Na CW, Kang JH, Park J (2005) Comparative structure and optical properties of Ga-, In-, and Sn-doped ZnO nanowires synthesized via thermal evaporation. J Phys Chem B 109:2526–2531
33. He H, Lao CS, Chen LJ, Davidovic D, Wang ZL (2005) Large-scale Ni-doped ZnO nanowire arrays and electrical and optical properties. J Am Chem Soc 127:16376–16377
34. Yamamoto T, Katayama-Yoshida H (1999) Solution using a codoping method to unipolarity for the fabrication of p-type ZnO. Jpn J Appl Phys 2(38):L166–L169
35. Hwang SO et al (2008) Synthesis of vertically aligned manganese-doped Fe_3O_4 nanowire arrays and their excellent room-temperature gas sensing ability. J Phys Chem C 112:13911–13916
36. Zheng CL, Wan JG, Cheng Y, Cu DH, Zhan YJ (2005) Preparation of SnO_2 nanowires synthesized by vapor-solid mode and its growth mechanism. Int J Mod Phys B 19:2811–2816
37. Pan ZW, Dai ZR, Wang ZL (2001) Nanobelts of semiconducting oxides. Science 291:1947–1949
38. Kong XY, Wang ZL (2003) Spontaneous polarization-induced nanohelixes, nanosprings, and nanorings of piezoelectric nanobelts. Nano Lett 3:1625–1631
39. Jiang XC, Herricks T, Xia YN (2002) CuO nanowires can be synthesized by heating copper substrates in air. Nano Lett 2:1333–1338
40. Fan ZY, Wen XG, Yang SH, Lu JG (2005) Controlled p- and n-type doping of Fe_2O_3 nanobelt field effect transistors. Appl Phys Lett 87:013113
41. Li D, Hu J, Wu R, Lu JG (2010) Conductometric chemical sensor based on individual CuO nanowires. Nanotechnology 21:485502
42. Possin GE (1970) A Method for Forming Very Small Diameter Wires. Rev Sci Instrum 41:772–774
43. Martin CR (1994) Nanomaterials: A Membrane-Based Synthetic Approach. Science 266:1961–1966
44. Routkevitch D, Bigioni T, Moskovits M, Xu JM (1996) Electrochemical fabrication of CdS nanowire arrays in porous anodic aluminum oxide templates. J Phys Chem 100:14037–14047
45. Nielsch K, Choi J, Schwirn K, Wehrspohn RB, Gösele U (2002) Self-ordering regimes of porous alumina: The 10 % porosity rule. Nano Lett 2:677–680
46. Li D, Thompson RS, Bergmann G, Lu JG (2008) Template-based Synthesis and Magnetic Properties of Cobalt Nanotube Arrays. Adv Mater 20:4575–4578

47. Masuda H et al (2001) Square and triangular nanohole array architectures in anodic alumina. Adv Mater 13:189–192
48. Masuda H et al (1997) Highly ordered nanochannel-array architecture in anodic alumina. Appl Phys Lett 71:2770–2772
49. Fan ZY et al (2010) Ordered Arrays of Dual-Diameter Nanopillars for Maximized Optical Absorption. Nano Lett 10:3823–3827
50. Fan ZY et al (2009) Three dimensional nanopillar array photovoltaics on low cost and flexible substrate. Nat Mater 8:648–653
51. Li D, Jiang C, Jiang J, Lu JG (2009) Self-Assembly of Periodic Serrated Nanostructures. Chem Mater 21:253–258
52. Li D, Zhao LA, Jiang CH, Lu JG (2010) Formation of Anodic Aluminum Oxide with Serrated Nanochannels. Nano Lett 10:2766–2771
53. Chang PC, Chen HY, Ye JS, Sheu FS, Lu JG (2007) Vertically aligned antimony nanowires as solid-state pH sensors. Chemphyschem 8:57–61
54. Liu ZW et al (2008) Shape anisotropy and magnetization modulation in hexagonal cobalt nanowires. Adv Funct Mater 18:1573–1578
55. Kolmakov A, Zhang YX, Cheng GS, Moskovits M (2003) Detection of CO and O_2 using tin oxide nanowire sensors. Adv Mater 15:997–1000
56. Fan ZY et al (2006) Electrical and photoconductive properties of vertical ZnO nanowires in high density arrays. Appl Phys Lett 89:213110–213112
57. Penner RM, Martin CR (1986) Controlling the Morphology of Electronically Conductive Polymers. J Electrochem Soc 133:2206–2207
58. Park S, Chung S, Mirkin CA (2004) Hybrid Organic-Inorganic, Rod-Shaped Nanoresistors and Diodes. J Am Chem Soc 126:11772–11773
59. Dan YP, Cao YY, Mallouk TE, Johnson AT, Evoy S (2007) Dielectrophoretically assembled polymer nanowires for gas sensing. Sensor Actuat B-Chem. 125:55–59
60. Cheng B, Samulski ET (2001) Fabrication and characterization of nanotubular semiconductor oxides In_2O_3 and Ga_2O_3. J Mat Chem 11:2901–2902
61. Shimizu T et al (2007) Synthesis of vertical high-density epitaxial Si(100) nanowire arrays on a Si(100) substrate using an anodic aluminum oxide template. Adv Mater 19:917–920
62. Che G, Lakshmi BB, Martin CR, Fisher ER, Ruoff RS (1998) Chemical vapor deposition based synthesis of carbon nanotubes and nanofibers using a template metho. Chem Mat 10:260–267
63. Meyer B K et al (2004) Bound exciton and donor-acceptor pair recombinations in ZnO. *Phys*. Status Solidi B-Basic Solid State Phys 241:231–260
64. Yang YH, Chen XY, Feng Y, Yang GW (2007) Physical Mechanism of Blue-Shift of UV Luminescence of a Single Pencil-Like ZnO Nanowire. Nano Lett 7:3879–3883
65. Park WI, Yi GC (2004) Electroluminescence in n-ZnO nanorod arrays vertically grown on p-GaN. Adv Mater 16:87–90
66. Chang PC, Chien CJ, Stichtenoth D, Ronning C, Lu JG (2007) Finite size effect in ZnO nanowires. Appl Phys Lett 90:113101
67. Shalish I, Temkin H, Narayanamurti V (2004) Size-dependent surface luminescence in ZnO nanowires. Phys Rev B 69:245401
68. Chang PC, Lu JG (2008) Temperature dependent conduction and UV induced metal-to-insulator transition in ZnO nanowires. Appl Phys Lett 92:212113
69. Martel R, Schmidt T, Shea HR, Hertel T, Avouris P (1998) Single- and multi-wall carbon nanotube field-effect transistors. Appl Phys Lett 73:2447–2449
70. Park WI, Kim JS, Yi GC, Bae MH, Lee HJ (2004) Fabrication and electrical characteristics of high-performance ZnO nanorod field-effect transistors. Appl Phys Lett 85:5052–5054
71. Fan ZY et al (2004) Photoluminescence and polarized photodetection of single ZnO nanowires. Appl Phys Lett 85:6128–6130
72. Zhang Y, Kolmakov A, Chretien S, Metiu H, Moskovits M (2004) Control of catalytic reactions at the surface of a metal oxide nanowire by manipulating electron density inside it. Nano Lett 4:403–407

73. Hendrich VE , Cox P A (1994) Surface Science of Metal Oxides
74. Fan ZY, Lu JG (2005) Gate-refreshable nanowire chemical sensors. Appl Phys Lett 86:123510
75. Li QH, Liang YX, Wan Q, Wang TH (2004) Oxygen sensing characteristics of individual ZnO nanowire transistors. Appl Phys Lett 85:6389–6391
76. Kalinin SV et al (2005) Electronic transport imaging in a multiwire SnO_2 chemical field-effect transistor device. J Appl Phys 98:044503
77. Wang Y et al (2008) Nanostructured sheets of Ti-O nanobelts for gas sensing and antibacterial applications. Adv Funct Mater 18:1131–1137
78. Kang BS et al (2005) Hydrogen and ozone gas sensing using multiple ZnO nanorods. Appl Phys A Mater 80:1029–1032
79. Sberveglieri G et al (2007) Synthesis and characterization of semiconducting nanowires for gas sensing. Sens Actuator B-Chem 121:208–213
80. Baratto C et al (2005) Metal oxide nanocrystals for gas sensing. Sens Actuator B Chem 109:2–6
81. Qi Q, Zhang T, Liu L, Zheng X (2009) Synthesis and toluene sensing properties of SnO_2 nanofibers. Sens Actuator B-Chem 137:471–475
82. Ponzoni A et al (2008) Metal oxide nanowire and thin-film-based gas sensors for chemical warfare simulants detection. IEEE Sens J 8:735–742
83. Yu C et al (2005) Integration of metal oxide nanobelts with microsystems for nerve agent detection. Appl Phys Lett 86:063101
84. You GF, Thong JTL (2010) Thermal oxidation of polycrystalline tungsten nanowire. J Appl Phys 108:094312
85. Liu Z et al (2007) Room temperature gas sensing of p-type TeO_2 nanowires. Appl Phys Lett 90:173119
86. Liu X, Li C, Han S, Han J, Zhou CW (2003) Synthesis and electronic transport studies of CdO nanoneedles. Appl Phys Lett 82:1950–1952
87. Kim YS et al (2005) Room-temperature semiconductor gas sensor based on nonstoichiometric tungsten oxide nanorod film. Appl Phys Lett 86:213105
88. Kim I et al (2006) Ultrasensitive chemiresistors based on electrospun TiO_2 nanofibers. Nano Lett 6:2009–2013
89. Li C et al (2003) Surface treatment and doping dependence of In_2O_3 nanowires as ammonia sensors. J Phys Chem B 107:12451–12455
90. Xue XY et al (2006) Synthesis and ethanol sensing properties of indium-doped tin oxide nanowires. Appl Phys Lett 88:201907
91. Wang XH, Zhang J, Zhu ZQ (2006) Ammonia sensing characteristics of ZnO nanowires studied by quartz crystal microbalance. Appl Surf Sci 252:2404–2411
92. Meier DC, Semancik S, Button B, Strelcov E, Kolmakov A (2007) Coupling nanowire chemiresistors with MEMS microhotplate gas sensing platforms. Appl Phys Lett 91:063118
93. Wang GX, Park JS, Park MS, Gou XL (2008) Synthesis and high gas sensitivity of tin oxide nanotubes. Sens Actuator B-Chem 131:313–317
94. Raible I, Burghard M, Schlecht U, Yasuda A, Vossmeyer T (2005) V_2O_5 nanofibres: novel gas sensors with extremely high sensitivity and selectivity to amines. Sens Actuator B Chem 106:730–735
95. Ryu K, Zhang D, Zhoua C (2008) High-performance metal oxide nanowire chemical sensors with integrated micromachined hotplates. Appl Phys Lett 92:093111
96. Hernandez-Ramirez F et al (2007) High response and stability in CO and humidity measures using a single SnO_2 nanowire. Sens Actuator B-Chem 121:3–17
97. Xu JQ, Chen YP, Li YD, Shen JN (2005) Gas sensing properties of ZnO nanorods prepared by hydrothermal method. J Mater Sci 40:2919–2921
98. Xu JQ, Chen YP, Chen DY, Shen JN (2006) Hydrothermal synthesis and gas sensing characters of ZnO nanorods. Sens Actuator B-Chem 113:526–531
99. Wang CH, Chu XF, Wu MW (2006) Detection of H_2S down to ppb levels at room temperature using sensors based on ZnO nanorods. Sens Actuator B Chem 113:320–323

100. Wang HT et al (2005) Detection of hydrogen at room temperature with catalyst-coated multiple ZnO nanorods. Appl Phys A-Mater Sci Process 81:1117–1119
101. Varghese OK, Gong DW, Paulose M, Ong KG, Grimes CA (2003) Hydrogen sensing using titania nanotubes. Sens Actuator B Chem 93:338–344
102. Rout CS, Kulkarni GU, Rao CNR (2007) Room temperature hydrogen and hydrocarbon sensors based on single nanowires of metal oxides. J Phys D-Appl Phys 40:2777–2782
103. Ying Z, Wan Q, Song ZT, Feng SL (2004) SnO$_2$ nanowhiskers and their ethanol sensing characteristics. Nanotechnology 15:1682–1684
104. Hsueh T, Hsu C, Chang S, Chen I (2007) Laterally grown ZnO nanowire ethanol gas sensors. Sens Actuator B-Chem 126:473–477
105. Liu JF, Wang X, Peng Q, Li YD (2005) Vanadium pentoxide nanobelts: Highly selective and stable ethanol sensor materials. Adv Mater 17:764–767
106. Chu XF, Wang CH, Jiang DL, Zheng CM (2004) Ethanol sensor based on indium oxide nanowires prepared by carbothermal reduction reaction. Chem Phys Lett 399:461–464
107. Yu MF, Atashbar MZ, Chen XL (2005) Mechanical and electrical characterization of beta-Ga$_2$O$_3$ nanostructures for sensing applications. IEEE Sens J 5:20–25
108. Wan Q et al (2004) Positive temperature coefficient resistance and humidity sensing properties of Cd-doped ZnO nanowires. Appl Phys Lett 84:3085–3087
109. Zhang YS et al (2005) Zinc oxide nanorod and nanowire for humidity sensor. Appl Surf Sci 242:212–217
110. Kuang Q, Lao C, Wang ZL, Xie Z, Zheng L (2007) High-sensitivity humidity sensor based on a single SnO$_2$ nanowire. J Am Chem Soc 129:6070–6071
111. Varghese OK, Grimes CA (2003) Metal oxide nanoarchitectures for environmental sensing. J Nanosci Nanotechno 3:277–293
112. Fu XQ, Wang C, Yu HC, Wang YG, Wang TH (2007) Fast humidity sensors based on CeO$_2$ nanowires. Nanotechnology 18:145503
113. Chang S et al (2008) Highly Sensitive ZnO Nanowire Acetone Vapor Sensor With Au Adsorption. IEEE Trans Nanotechnol 7:754–759
114. Han N, Tian Y, Wu X, Chen Y (2009) Improving humidity selectivity in formaldehyde gas sensing by a two-sensor array made of Ga-doped ZnO. Sens Actuator B Chem 138:228–235
115. Fan ZY, Lu JG (2006) Chemical sensing with ZnO nanowire field-effect transistor. IEEE Trans Nano 5:393
116. Rout CS, Krishna SH, Vivekchand SRC, Govindaraj A, Rao CNR (2006) Hydrogen and ethanol sensors based on ZnO nanorods, nanowires and nanotubes. Chem Phys Lett 418:586–590
117. Chen P, Shen G, Zhou C (2008) Chemical Sensors and Electronic Noses Based on 1-D Metal Oxide Nanostructures. IEEE Trans Nanotechnol 7:668–682
118. Zhang DH et al (2004) Detection of NO$_2$ down to ppb levels using individual and multiple In$_2$O$_3$ nanowire devices. Nano Lett 4:1919–1924
119. Hernandez-Ramirez F et al (2007) Portable microsensors based on individual SnO$_2$ nanowires. Nanotechnology 18:495501
120. Sysoev VV, Goschnick J, Schneider T, Strelcov E, Kolmakov A (2007) A gradient microarray electronic nose based on percolating SnO$_2$ nanowire sensing elements. Nano Lett 7:3182–3188
121. Kolmakov A, Klenov DO, Lilach Y, Stemmer S, Moskovits M (2005) Enhanced gas sensing by individual SnO$_2$ nanowires and nanobelts functionalized with Pd catalyst particles. Nano Lett 5:667–673
122. Zhang Y, Kolmakov A, Lilach Y, Moskovits M (2005) Electronic control of chemistry and catalysis at the surface of an individual tin oxide nanowire. J Phys Chem B 109:1923–1929
123. Li C et al (2003) In$_2$O$_3$ nanowires as chemical sensors. Appl Phys Lett 82:1613–1615
124. Geistlinger H (1993) Electron Theory of Thin-Film Gas Sensors. Sens Actuator B-Chem 17:47–60
125. Zhang DJ et al (2003) Doping dependent NH3 sensing of indium oxide nanowires. Appl Phys Lett 83:1845–1847

126. Sysoev VV, Button BK, Wepsiec K, Dmitriev S, Kolmakov A (2006) Toward the nanoscopic "electronic nose": Hydrogen vs carbon monoxide discrimination with an array of individual metal oxide nano- and mesowire sensors. Nano Lett 6:1584–1588
127. Chen P, Ishikawa FN, Chang H, Ryu K, Zhou C (2009) A nanoelectronic nose: a hybrid nanowire/carbon nanotube sensor array with integrated micromachined hotplates for sensitive gas discrimination. Nanotechnology 20:125503
128. Baik JM et al (2010) Tin-Oxide-Nanowire-Based Electronic Nose Using Heterogeneous Catalysis as a Functionalization Strategy. ACS Nano 4:3117–3122
129. Fan ZY et al (2008) Wafer-scale assembly of highly ordered semiconductor nanowire arrays by contact printing. Nano Lett 8:20–25
130. Yu X et al (2008) Wireless hydrogen sensor network using AlGaN/GaN high electron mobility transistor differential diode sensors. Sens Actuator B Chem 135:188–194

Chapter 10
ZnO Nanowires for Gas and Bio-Chemical Sensing

Stephen J. Pearton, David P. Norton and Fan Ren

Abstract There has been significant recent interest in the use of surface-functionalized thin film and nanowire ZnO for sensing of gases, heavy metals, UV photons and biological molecules. For the detection of gases such as hydrogen, the ZnO is typically coated with a catalyst metal such as Pd or Pt to increase the detection sensitivity at room temperature. Functionalizing the surface with oxides, polymers and nitrides is also useful in enhancing the detection sensitivity for gases and ionic solutions. The use of enzymes or adsorbed antibody layers on the ZnO surface leads to highly specific detection of a broad range of antigens of interest in the medical and homeland security fields. We give examples of recent work showing sensitive detection of glucose and lactic acid and the integration of the sensors with wireless data transmission systems to achieve robust, portable sensors.

10.1 Introduction

The explosion of interest in nanoscience coupled with growing demand for reliable, low-power chemical sensors for a wide variety of industrial applications has led to a surge in the development of nanostructured materials for gas detection [1–5]. One-dimensional (1D) nanostructures such as nanowires, nanorods, and

S. J. Pearton (✉) · D. P. Norton
Department of Materials Science and Engineering, University of Florida,
Gainesville, FL 32611, USA
e-mail: spear@mse.ufl.edu

F. Ren
Department of Chemical Engineering, University of Florida, Gainesville, FL 32611, USA

M. A. Carpenter et al. (eds.), *Metal Oxide Nanomaterials for Chemical Sensors*,
Integrated Analytical Systems, DOI: 10.1007/978-1-4614-5395-6_10,
© Springer Science+Business Media New York 2013

nanobelts are particularly suited for chemical sensing due to their large surface-to-volume ratio [6–11]. In terms of material selection, wide band-gap semiconductors are ideal for gas sensing having numerous advantageous properties including ability to operate at high temperatures (or alternatively, to have low leakage current at room temperature), radiation and environmental stability, and mechanical robustness. Literature from more than a decade of research shows improvement in consistent growth methods and potential for immediate industrial use of 1D semiconductor nanostructures. Particular semiconductor materials including III-nitrides such as GaN [12, 13] and InN [14–16], metal oxides including ZnO and SnO_2 [17, 18], and high-temperature materials such as SiC [19] have seen the greatest interest for chemical gas sensing applications, chiefly for the detection of H_2, O_2, NH_3, and ethanol.

Chemical sensors have gained in importance in the past decade for applications that include homeland security, medical and environmental monitoring and also food safety. A desirable goal is the ability to simultaneously analyze a wide variety of environmental and biological gases and liquids in the field and to be able to selectively detect a target analyte with high specificity and sensitivity. In the area of detection of medical biomarkers, many different methods, including enzyme-linked immunosorbent assay (ELISA), particle-based flow cytometric assays, electrochemical measurements based on impedance and capacitance, electrical measurement of microcantilever resonant frequency change, and conductance measurement of semiconductor nanostructures. Gas chromatography (GC), ion chromatography, high density peptide arrays, laser scanning quantitative analysis, chemiluminescence, selected ion flow tube (SIFT), nanomechanical cantilevers, bead-based suspension microarrays, magnetic biosensors and mass spectrometry (MS) have been employed. Depending on the sample condition, these methods may show variable results in terms of sensitivity for some applications and may not meet the requirements for a handheld biosensor.

For homeland security applications, reliable detection of biological agents in the field and in real time is challenging. Toxins such as ricin, botulinum toxin or enterotoxin B are environmentally stable, can be mass-produced and do not need advanced technologies for production and dispersal. A significant issue is the absence of a definite diagnostic method and the difficulty in differential diagnosis from other pathogens that would slow the response in case of a terror attack. This is a critical need that has to be met to have an effective response to terrorist attacks. Given the adverse consequences of a lack of reliable biological agent sensing on national security, there is a critical need to develop novel, more sensitive and reliable technologies for biological detection in the field. Some specific toxins of interest include Enterotoxin type B (Category B, NIAID), Botulinum toxin (Category A NIAID) and ricin (Category B NIAID).

While the techniques mentioned above show excellent performance under lab conditions, there is also a need for small, handheld sensors with wireless connectivity that have the capability for fast responses. The chemical sensor market represents the largest segment for sales of sensors, including chemical detection in gases and liquids, flue gas and fire detection, liquid quality sensors, biosensors and medical sensors. Some of the major applications in the home include indoor air quality and natural gas detection. Attention is now being paid to more demanding applications where a high degree of chemical specificity and selectivity is required. For biological and medical sensing applications, disease diagnosis by detecting specific biomarkers (functional or structural abnormal enzymes, low molecular weight proteins, or antigen) in blood, urine, saliva, or tissue samples has been established. Most of the techniques mentioned earlier such as ELISA, possess a major limitation in that only one analyte is measured at a time. Particle-based assays allow for multiple detection by using multiple beads but the whole detection process is generally longer than 2 h, which is not practical for in-office or bedside detection. Electrochemical devices have attracted attention due to their low cost and simplicity, but significant improvements in their sensitivities are still needed for use with clinical samples. Microcantilevers are capable of detecting concentrations as low as 10 pg/ml, but suffer from an undesirable resonant frequency change due to the viscosity of the medium and cantilever damping in the solution environment. Nano-material devices have provided an excellent option toward fast, label-free, sensitive, selective, and multiple detections for both pre-clinical and clinical applications. Examples of detection of biomarkers using electrical measurements with semiconductor devices include carbon nanotubes for lupus erythematosus antigen detection [20], compound semiconducting nanowires and In_2O_3 nanowires for prostate-specific antigen detection [21], and silicon nanowire arrays for detecting prostate-specific antigen [22]. In clinical settings, biomarkers for a particular disease state can be used to determine the presence of disease as well as its progress.

Semiconductor-based sensors can be fabricated using mature techniques from the Si chip industry and/or novel nanotechnology approaches. Specifically, the development of hydrogen-selective sensors has seen considerable recent attention for their potential use with proton-exchange membrane (PEM) and solid oxide fuel cells for space craft and industrial applications. Sensors in these harsh environments must selectively detect hydrogen at room temperature or below while using minimum power. Several groups have already demonstrated the application of nitride and oxide semiconductor nanostructures for exclusive H_2 sensing using carbon nanotubes (CNTs) [1, 23] and ZnO [24, 25]. Several issues must still be addressed however, including quantifying sensitivity, improving detection limits at room temperature, and reducing power consumption.

In this review we discuss the progress of ZnO-based semiconductor nanostructures for nanoelectronic devices and for chemical and gas sensing, specifically for hydrogen. Functionalizing the surface with oxides, polymers and nitrides is also useful in enhancing the detection sensitivity for gases and ionic solutions. The wide bandgaps of these materials make them ideal for solar-blind UV detection, which can be use for detecting fluorescence from biotoxins. The use of enzymes or adsorbed antibody layers on the semiconductor surface leads to highly specific detection of a broad range of antigens of interest in the medical and homeland security fields. The use of catalyst metal coatings on ZnO nanowires is found to greatly enhance the detection sensitivity for hydrogen. All sensors tested exhibited no response at room temperature upon exposure to various other gases including O_2, C_2H_5, N_2O, and CO_2. The high surface-to-volume ratio of nanowires and the ability to use simple contact fabrication schemes make them attractive for hydrogen sensing applications functionalities.

10.2 Zinc Oxide Nanostructures Based Sensors

The history of ZnO nanostructures is extensive; there is much literature dedicated to the synthesis and characterization of the various types of nanostructured ZnO morphologies grown. A review of the many forms of nano-sized ZnO including nanobelts, -cages, -combs, and -wires is provided by Wang [26]. This ease of nano-fabrication combined with the high compatibility with Si-based microelectronics afforded to semiconducting metal oxides makes ZnO a particularly interesting candidate for solid-state chemical gas sensing. ZnO is a chemically and thermally stable, inherently n-type semiconductor with a large exciton binding energy and wide bandgap. ZnO-based gas sensors have already been developed from thin films [27], nanoparticles [28], and nanowires [29]. Additionally, ZnO easily forms a heterostructure system with MgO and CdO over a limited miscibility range that is promising for blue/UV optoelectronics, transparent electronics, and sensors. Schottky diode devices using thin Pd or Pt contacts have detected hydrogen at concentrations of hundreds of ppm at temperatures as low as 100 C. It is generally accepted that H_2 is dissociated when adsorbed on Pt and Pd at room temperature. The reaction is as follows:

$$H_{2(ads)} \rightarrow 2\,H^+ + e^-$$

Dissociated hydrogen causes a change in the channel and conductance change of the nanowire by altering the zero-bias depletion depth.

10.3 Hydrogen Detection Using Zinc Oxide Nanostructures

For our work, site selective growth of ZnO nanorods was achieved by Molecular Beam Epitaxy (MBE) on Al_2O_3. Growth time was ~ 2 h at 600 °C producing single crystal nanorods with a typical length of 2–10 μm and having a diameter between 30 and 150 nm. Al/Ti/Au electrodes were deposited via e-beam evaporation with a separation distance of ~ 3 μm. Finally, Au wires were bonded to the contact pads for I–V measurements performed at temperatures between 25 and 150 °C in 10 % H_2 in a N_2 ambient. No current was measured through the discontinuous Au islands and no thin film of ZnO was observed with the nanorod growth. In some cases, nanorods were functionalized using Pd, Pt, Au, Ni, Ag or

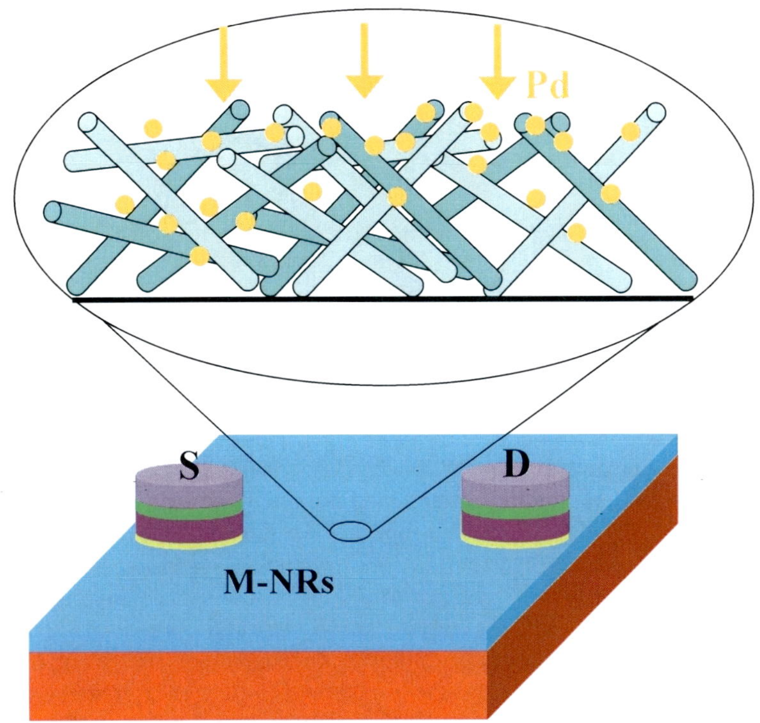

Fig. 10.1 Schematic of multiple nanowire semiconductor sensor coated with catalyst metal. The source (*S*) and drain (*D*) contacts are both Ohmic in nature

Ti discontinuous thin films (~ 100 Å thick) deposited by sputtering, shown schematically in Fig. 10.1.

Although there was no detectable change in current upon H_2 exposure at room temperature, ZnO nanorods did reflect changes in current beginning at ~ 112 °C as shown in Fig. 10.2. While change in current only approaches 16 nA at 200 °C, these changes are readily detected by conventional ammeters, however, an on-chip heater would be needed for practical application of ZnO nanorods for detection of hydrogen. Regardless, these results show the possible ability of ZnO nanorods for use as hydrogen gas sensors without deposition of an additional functional layer. In this case, the gas sensing mechanism is believed to derive from desorption of absorbed surface oxygen and grain boundaries in poly-ZnO or from changes in surface or grain boundary conduction by gas adsorption/desorption. Since the nanowires are typically single crystal, the latter effect is not likely to contribute. When detecting hydrogen with similar thin film rectifiers, observed changes in current were consistent with changes in the near surface doping of the ZnO, suggesting the introduction of a shallow donor state via hydrogen adsorption. The nanorods also showed a strong photoresponse to above bandgap UV light

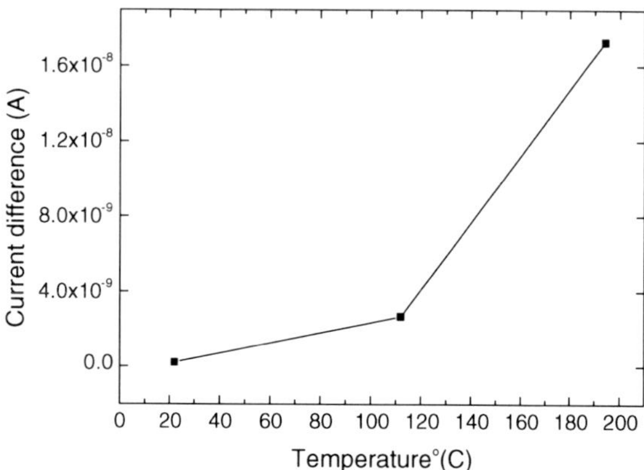

Fig. 10.2 Change in current of ZnO nanorods measured at 0.1 V in 10 % H_2 in N_2 ambient. The error bars on each point are ±10 % of the value

(366 nm). The quick photoresponse indicated that the changes in conductivity due to injection of carriers was bulk-related and not due to surface effects.

Similar to GaN and InN, previous reports have shown the addition of metallic nanoparticles on the surface increases the detection sensitivity for H_2 [30], so metal catalyst coatings were deposited to ZnO nanorods to increase the detection efficiency for hydrogen gas. Figure 10.3 shows the time dependence of relative resistance change of metal-coated multiple ZnO nanorods with various metal functionalization layers upon exposure to 500 ppm H_2. The measured bias voltage was 0.5 V. Pt-coated nanorods exhibited a relative response of up to 8 % at room temperature upon exposure to 500 ppm hydrogen concentration in N_2 ambient. Unlike the results for the nitride-based nanostructures, this was a factor of two larger than obtained with Pd-coatings. Pt-coated ZnO nanorods easily detected hydrogen down to 100 ppm (experimental limit), with relative responses of 4 % at this concentration after 10 min exposure. All other metal functional layers showed very little relative responses to hydrogen exposure.

Differences in relative response to hydrogen exposure between Pt- and Pd-coated samples were suggested to come from catalytic properties of the metals. Pd has a higher permeability than Pt, but the solubility of H_2 is larger in Pd. Additionally, bonding studies of H to Ni, Pt, and Pd surfaces have shown that the adsorption energy is lowest on Pt. As shown however, the calculated activation energy for GaN nanowires and InN nanobelts was lower for Pd-coated samples, not Pt. This is likely due to the adsorption of hydrogen at an oxide interface between the nitride surface and the metal and not from adsorption of hydrogen at the metal coating. The existence and importance of this oxide interface for hydrogen sensing on nitride-based semiconductors is well documented. Because hydrogen adsorption on the oxide side of a ZnO-based sensor would contribute to the creation of a shallow donor state, the sensing mechanism for hydrogen detection on metal coated ZnO nanostructures is probably different from that of GaN or InN.

Similar to nitride-based nanostructured sensors is the rapid, initial response of ZnO multiple nanorod sensors to hydrogen exposure. Effective nanorod resistance

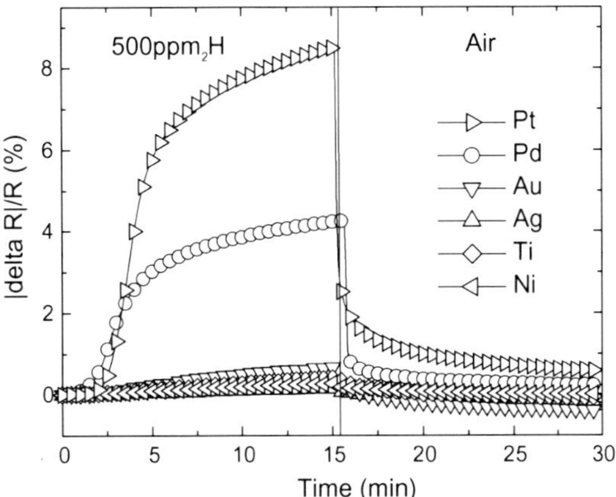

Fig. 10.3 Relative response of ZnO nanorods to 500 ppm H$_2$ using various metal functionalizations (Reprinted with permission from Ref. 89, Heo YW, Norton DP, Tien LC, Kwon Y, Kang BS, Ren F, Pearton SJ and LaRoche JR (2004) Mat Sci Eng R 47:1 Copyright Elsevier)

continues to change for >15 min exposure. This suggests chemisorption of hydrogen to the metal/ZnO interface is the rate-limiting step in conductance changes to the ZnO. Recovery of initial resistance upon removal of hydrogen from the ambient was rapid. An activation energy of 12 kJ/mole was measured for the chemisorption of hydrogen onto the metal coating surface [15].

Pt-coated ZnO nanorods are good potential candidates for detection of ppm concentrations of hydrogen at room temperature. Both initial response and recovery characteristics are rapid and likely limited by adsorption/desorption of hydrogen at the metal-ZnO interface. Because these ZnO nanorods can be placed on cheap substrates such as glass, they are well-suited for low-cost sensing applications that operate at very low power requirements. Reliability and long-term reproducibility of sensor response will still need to be considered before these sensors are considered for space applications.

10.4 UV Photodetectors

The development of UV detectors in the spectral range shorter than $\lambda \sim 400$ nm has attracted much interest recently because of potential applications in detection of biological materials and for the defense industry. In the former case, the UV photons are used to excite fluorescence at UV wavelengths from biological materials of interest and this is detected by the photodetectors. Wide bandgap detectors are very useful in bio-warfare agent detection because some pathogenic biological molecules fluoresce in the UV spectral region [31].

The most common UV detectors are based on p-i-n Si photodiodes or UV-filtered photomultiplier tubes. The use of nitride semiconductor UV detectors has advantages in terms of more precise detection windows, lower background currents due to solar fluxes and wider range of operating temperatures. Photodetectors that have no response for photons at wavelengths >290 nm are called solar-blind and are useful in applications that need to detect UV photons in the presence of sunlight, such as flame sensors, missile detectors and aircraft detection. Si detectors have very poor solar-blind performance characteristics and wide bandgap systems offer improved speed and lower dark currents.

Currently the most commonly used wide bandgap semiconductor system for UV detection is GaN/AlGaN. There is also interest in developing ZnO/ZnMgO nanowire UV detectors as a complementary technology for UV detection, with the following advantages relative to the nitrides [32–52].

1. The ZnO-based materials offer similar band-gaps to the nitrides, but can be grown at much lower-temperature on a wider range of substrates, including large area Si or cheap transparent materials such as glass. The nanowires can be transferred to any substrate for integration with other sensors and are compatible with low temperature materials such as polymers.
2. The nanowire UV detectors operate at very low power levels compared to existing nitride UV detectors.
3. The fabrication approach developed previously for ZnO nanowire gas sensors allows for a simple, low-cost, single-step approach to realizing robust UV detectors.
4. ZnO nanowire UV detectors can be readily integrated with on-chip wireless circuits to provide data transmission to a central monitoring location. Thus, it is possible to have either single detectors or arrays of detectors that operate at very low-power levels and do not need constant monitoring by humans.
5. The versatility of substrates also makes it possible to utilize 3D stacking technology developed for silicon substrates for data intensive applications. Devices stacked with overlying ZnO sensors would permit maximum sensor density and higher levels of integration with silicon or gallium arsenide electronics.

ZnO nanowires grown by site-selective MBE are single crystal and typically conducting with a carrier density in the 10^{17}–10^{18} cm^{-3} range. These nanowires can be removed by sonication from their original substrate and then transferred to arbitrary substrates where they can be contacted at both ends by Al/Pt/Au Ohmic electrodes, as shown in Fig. 10.4(top). The current–voltage and photoresponse characteristics were obtained both in the dark and with ultra-violet (254 or 366 nm) illumination. The current–voltage (I–V) characteristics are Ohmic under all conditions, with nanowire conductivity under UV exposure of 0.2 Ohm.cm. The photoresponse showed only a minor component with long decay times (tens of seconds) thought to originate from surface states. The results show the high quality of material prepared by MBE and the promise of using ZnO nanowire structures for solar-blind UV detection. Recent reports have shown the sensitivity of ZnO nanowires to the presence of oxygen in the measurement ambient and to ultra-violet (UV) illumination [51, 52]. In the latter case, above bandgap illumination was found to change the current–voltage (I–V) characteristics of ZnO nanowires grown by thermal evaporation of ball-milled powders between two Au electrodes from rectifying to Ohmic. By contrast, there was no change in the effective built-in

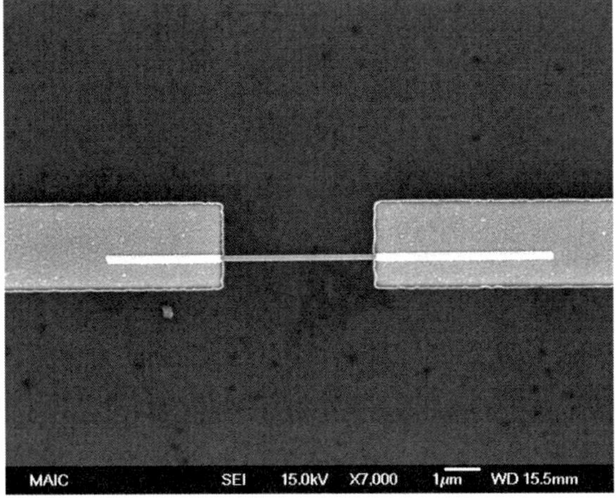

Fig. 10.4 SEM micrograph of ZnO nanowire and time dependence of photocurrent as the 366 nm light source is modulated

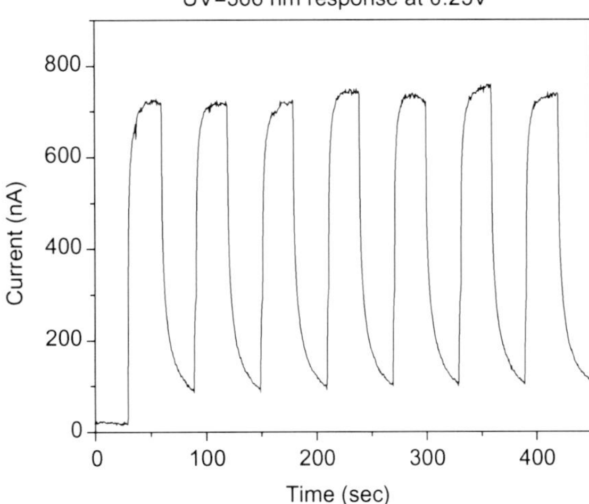

potential barrier between the ZnO nanowires and the contacts for below bandgap illumination. The slow photoresponse of these nanowires was suggested to originate in the presence of surface states which trapped electrons with release time constants from msec to hours.

By sharp contrast to these results, we have demonstrated that the photoresponse characteristics of single ZnO nanowires grown by site selective Molecular Beam Epitaxy (MBE) have relatively fast photoresponse and show electrical transport dominated by bulk conduction. Figure 10.4 (bottom) shows the change in current at fixed bias of the nanowires in the dark and under illumination from 366 nm light. The conductivity is greatly increased as a result of the illumination, as evidenced by the higher current. No effect was observed for illumination with below bandgap light. Transport measurements show that the ideality factor of Pt

Schottky diodes formed on the nanowires exhibit an ideality factor of 1.1, which suggests that there is little recombination occurring in the nanowire. It also exhibits excellent Ohmicity of the contacts to the nanowire, even at low bias. On blanket films of n-type ZnO with carrier concentration in the 10^{16} cm^{-3} range we obtained contact resistance of $3-5 \times 10^{-5}$ Ohm.cm^{-2} for these contacts. In the case of ZnO nanowires made by thermal evaporation, the I–V characteristics were rectifying in the dark and only became Ohmic during above bandgap illumination. The conductivity of the nanowire during illumination with 366 nm light was 0.2 Ohm.cm.

The photoresponse of the single ZnO nanowire at a bias of 0.25 V under pulsed illumination from a 366 nm wavelength Hg lamp shows the photoresponse is much faster than that reported for ZnO nanowires grown by thermal evaporation from ball-milled ZnO powders and likely is due to the reduced influence of the surface states seen in that material. The generally quoted mechanism for the photoconduction is creation of holes by illumination that discharge the negatively charged oxygen ions on the nanowire surface, with de-trapping of electrons and subsequent transmission to the electrodes. The recombination times in high quality ZnO measured by time-resolved photoluminescence are short, on the order of tens of ps, while the photoresponse measures the electron trapping time. There is also a direct correlation reported between the photoluminescence lifetime and the defect density in both bulk and epitaxial ZnO. In our nanowires, the electron trapping times are on the order of tens of seconds and these trapping effects are only a small fraction of the total photoresponse recovery characteristic. Note also the fairly constant peak photocurrent as the lamp is switched on, showing that any traps present have discharged in the time frame of the measurement. Once again we see an absence of the very long time constants for recovery seen in nanowires prepared by thermal evaporation.

10.5 pH Measurement

The measurement of pH is needed in many different applications, including medicine, biology, chemistry, food science, environmental science as well as oceanography. Solutions with a pH less than 7 are acidic and solutions with a pH greater than 7 are basic or alkaline. We have found that ZnO nanorod surfaces respond electrically to variations of the pH in electrolyte solutions introduced via an integrated microchannel. The ion-induced changes in surface potential are readily measured as a change in conductance of the single ZnO nanorods and suggest that these structures are very promising for a wide variety of sensor applications. An integrated microchannel was made from SYLGARD@ 184 polymer. Entry and exit holes in the ends of the channel were made with a small puncher (diameter less than 1 mm) and the film immediately applied to the nanorod sensor. The pH solution was applied using a syringe autopipette (2–20 ul). An SEM image of the completed device is shown in Fig. 10.5 (top).

Fig. 10.5 SEM of a integrated microchannel across a ZnO nanorod contacted at both ends by Ohmic contacts. The conductance of the nanorod as a function of the pH of the solution flowed across it is shown at the *bottom* of the figure

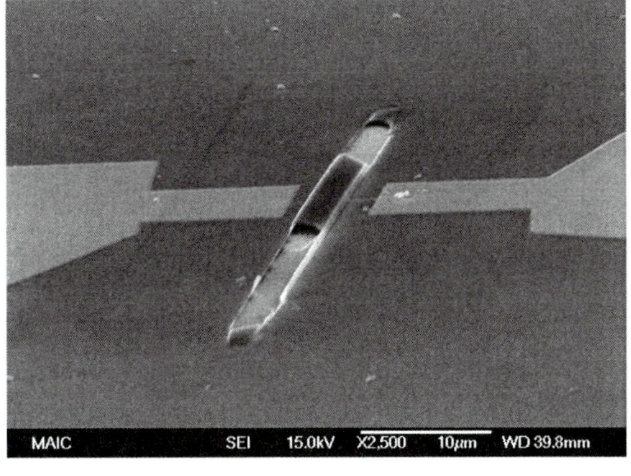

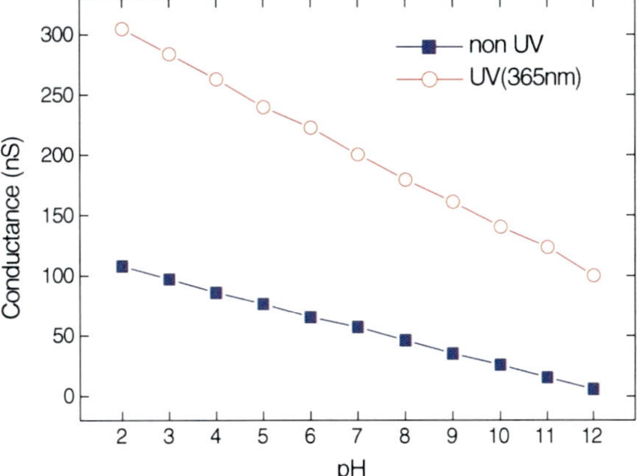

Prior to the pH measurements, we used pH 4, 7, 10 buffer solutions to calibrate the electrode and the measurements at 25 °C were carried out in the dark or under ultra-violet (UV) illumination from 365 nm light using an Agilent 4156C parameter analyzer to avoid parasitic effects. The pH of the solution was varied via titration of an aqueous solution (pH = 7) with either an aqueous HNO_3 solution or an aqueous NaOH solution using distilled water as the solvent. The electrode was a conventional Acumet standard Ag/AgCl electrode. The nanorods showed a very strong photoresponse as a function of pH and furthermore, the conductivity is greatly increased as a result of UV illumination, as evidenced by the higher current displayed in Fig. 10.5 (bottom). No effect was observed for illumination with below bandgap light. The photoconduction appears predominantly to originate in bulk conduction processes with only a minor surface trapping component. The adsorption of polar molecules on the surface of ZnO affects the surface potential and device characteristics.

The current at a bias of 0.5 V as a function of time for nanorods exposed for 60 s to a series of solutions whose pH was varied from 2 to 12 was reduced upon

exposure to these polar liquids as the pH is increased. The experiment was conducted starting at pH $= 7$ and went to pH $= 2$ or 12. The I–V measurement in air
was slightly higher than in an aqueous solution with pH $= 7(10$–20 %). The data
shows the sensor is sensitive to the concentration of the polar liquid and therefore
could be used to differentiate between liquids into which a small amount of
leakage of another substance has occurred. The conductance of the rods was higher
under UV illumination but the percentage change in conductance is similar with
and without illumination. The nanorods exhibited a linear change in conductance
between pH 2–12 of 8.5nS/pH in the dark and 20nS/pH when illuminated with UV
(365 nm) light as shown at the bottom of Fig. 10.5. The nanorods show stable
operation with a resolution of ~ 0.1 pH over the entire pH range, showing the
remarkable sensitivity to relatively small changes in concentration of the liquid.
The small size and low power requirements are the main advantages of this
approach.

10.6 Biomedical Applications

10.6.1 Glucose Detection

AlGaN/GaN HEMTs with ZnO nanowires on the gate region can be used for
measurements of pH in exhaled breath condensate (EBC) and glucose, through
integration of the pH and glucose sensor onto a single chip and with additional
integration of the sensors into a portable, wireless package for remote monitoring
applications. Figure 10.6 shows an optical microscopy image of an integrated pH
and glucose sensor chip and cross-sectional schematics of the completed pH and
glucose device. The gate dimension of the pH sensor device and glucose sensors
was $20 \times 50 \ \mu m^2$.

For the glucose detection, a highly dense array of 20–30 nm diameter and 2 μm
tall ZnO nanorods were grown on the $20 \times 50 \ \mu m^2$ gate area. The lower right
inset in Fig. 10.6 shows closer view of the ZnO nanorod arrays grown on the gate
area. The total area of the ZnO was increased significantly with the ZnO nanorods.
The ZnO nanorod matrix provides a microenvironment for immobilizing negatively charged GO_x while retaining its bioactivity, and passes charges produced
during the GO_x and glucose interaction to the AlGaN/GaN HEMT. The GOx
solution was prepared with concentration of 10 mg/mL in 10 mM phosphate buffer
saline (pH value of 7.4, Sigma Aldrich). After fabricating the device, 5 μl GO_x
(~ 100 U/mg, Sigma Aldrich) solution was precisely introduced to the surface of
the HEMT using a pico-liter plotter. The sensor chip was kept at 4 °C in the
solution for 48 h for GO_x immobilization on the ZnO nanorod arrays followed by
an extensively washing to remove the un-immobilized GO_x.

To take the advantage of quick response (less than 1 s) of the HEMT sensor, a
real-time EBC collector is needed. The amount of the EBC required to cover the

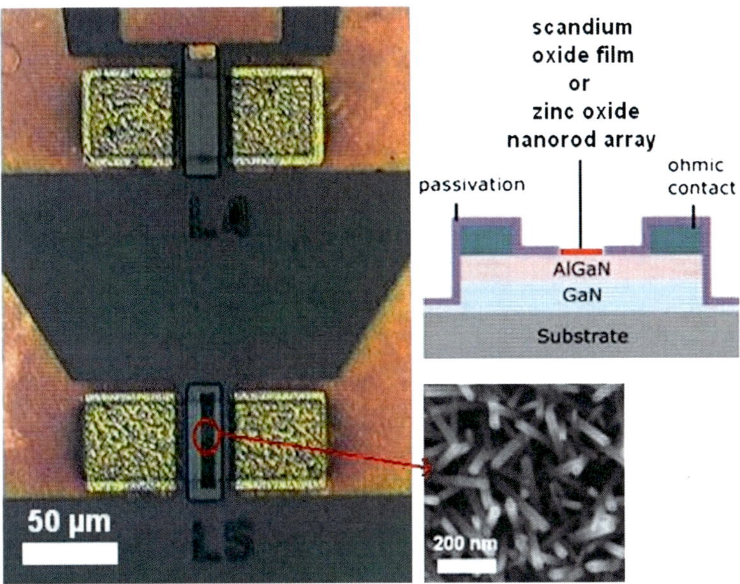

Fig. 10.6 SEM image of an integrated pH and glucose sensor. The insets show a schematic cross-section of the pH sensor and also an SEM of the ZnO nanorods grown in the gate region of the glucose sensor

HEMT sensing area is very small. Each tidal breath contains around 3 μl of the EBC. The contact angle of EBC on ZnO has been measured to be less than 45°, and it is reasonable to assume a perfect half sphere of EBC droplet formed to cover the sensing area 4×50 μm^2 gate area. The volume of a half sphere with a diameter of 50 μm is around 3×10^{-11} liter. Therefore, 100,000 of 50 μm diameter droplets of EBC can be formed from each tidal breath.

To condense the entire 3 μl of water vapor, only ~ 7 J of energy need to be removed for each tidal breath, which can be easily achieved with a thermal electric module, a Peltier device. The figure also shows a photograph and schematic of the system for collecting the EBC. The AlGaN/GaN HEMT sensor is directly mounted on the top of the Peltier unit (TB-8-0.45-1.3 HT 232, Kryotherm), which can be cooled to precise temperatures by applying known voltages and currents to the unit. During our measurements, the hotter plate of the Peltier unit was kept at 21 °C, and the colder plate was kept at 7 °C by applying a bias of 0.7 V at 0.2 A. The sensor takes less than 2 s to reach thermal equilibrium with the Peltier unit. This allows the exhaled breath to immediately condense on the gate region of the HEMT sensor.

Prior to pH measurements of the EBC, a Hewlett Packard soap film flow meter and a mass flow controller were used to calibrate the flow rate of exhaled breath. The HEMT sensors were also calibrated and exhibited a linear change in current between pH 3–10 of 37 μA/pH. Due to the difficulty to collect the EBC with different glucose concentrations, the samples for glucose concentration detection were prepare from glucose diluted in PBS or DI water.

The HEMT sensors were not sensitive to switching of N_2 gas, but responded to applications of exhaled breath pulse inputs from a human test subject, which shows the current of a ZnO coated HEMT sensor biased at 0.5 V for exposure to different flow rates of exhaled breath (0.5–3.0 l/min). The flow rates are directly proportional to the intensity exhalation. Deep breath provides a higher flow rate. A similar study was conducted with pure N_2 to eliminate the flow rate effect on sensor sensitivity. The N_2 did not cause any change of drain current, but the increase of exhaled breath flow rate decreased the drain current proportionally from 0.5 L/min to a saturation value of 1 L/min. For every tidal breath, the beginning portion of the exhalation is from the physiologic dead space, and the gases in this space do not participate in CO_2 and O_2 exchange in the lungs. Therefore, the contents in the tidal breath are diluted by the gases from this dead space. For higher flow rate exhalation, this dilution effect is less effective. Once the exhaled breath flow rate is above 1L/min, the sensor current change reaches a limit. As a result, the test subject experiences hyper ventilation and the dilution becomes insignificant. The sensor is operated at 50 Hz and 10 % duty cycle, which produces heat during operation. It only takes a few seconds for the EBC to vaporize from the sensing area and causes the spike-like response. The principal component of the EBC is water vapor, which represents nearly all of the volume (>99 %) of the fluid collected in the EBC. The measured current change of the exhale breath condensate shows that the pH values are within the range between pH 7 and 8. This range is the typical pH range of human blood.

The glucose was sensed by ZnO nanorod functionalized HEMTs with glucose oxidase enzyme localized on the nanorods, shown in Fig. 10.7. This catalyzes the reaction of glucose and oxygen to form gluconic acid and hydrogen peroxide. Figure 10.8 shows the real time glucose detection in PBS buffer solution using the drain current change in the HEMT sensor with constant bias of 250 mV. No current change can be seen with the addition of buffer solution at around 200 s, showing the specificity and stability of the device. By sharp contrast, the current change showed a rapid response of less than 5 s when target glucose was added to the surface. So far, the glucose detection using Au nano-particle, ZnO nanorod and nanocomb, or carbon nanotube material with GOx immobilization is based on electrochemical measurement. Since there is a reference electrode required in the solution, the volume of sample can not be easily minimized. The current density is measured when a fixed potential applied between nano-materials and the reference electrode. This is a first order detection and the range of detection limit of these sensors is 0.5–70 μM. Even though the AlGaN/GaN HEMT based sensor used the same GOx immobilization, the ZnO nanorods were used as the gate of the HEMT. The glucose sensing was measured through the drain current of HEMT with a change of the charges on the ZnO nano-rods and the detection signal was amplified through the HEMT. Although the response of the HEMT based sensor is similar to that of an electrochemical based sensor, a much lower detection limit of 0.5 nM was achieved for the HEMT based sensor due to this amplification effect. Since

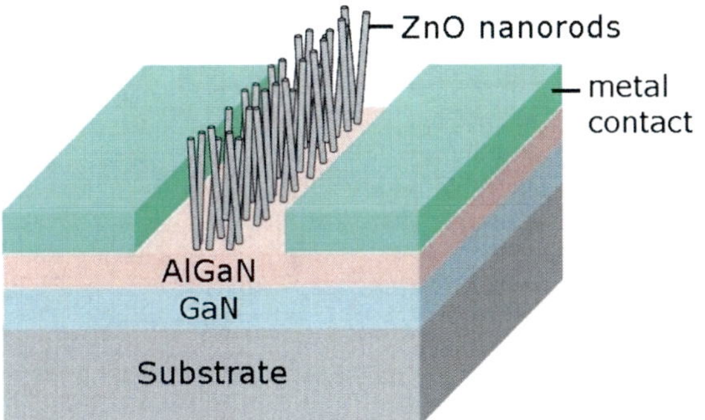

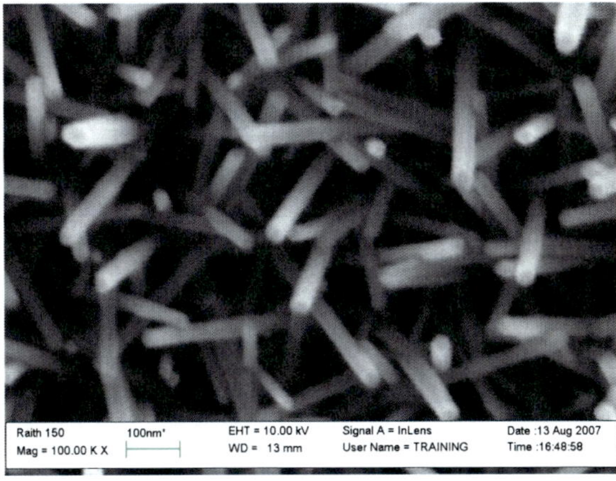

Fig. 10.7 Schematic of ZnO nanorod functionalized HEMT (*top*) and SEM of nanorods on gate area (*bottom*)

there is no reference electrode required for the HEMT based sensor, the amount of sample only depends on the area of gate dimension and can be minimized. The sensors do not respond to glucose unless the enzyme is present, as shown in Fig. 10.9.

Although measuring the glucose in the EBC is a noninvasive and convenient method for the diabetic application, the activity of the immobilized GO_x is highly dependent on the pH value of the solution. The GOx activity can be reduced to 80 % for pH = 5–6. Once the OH^- ions produce from reaction between oxygen and water diffused away the gate area, the pH value decreased. Thus around 85 min, the pH value of the glucose solution around gate area decreased low enough to allow the activity of GOx to resume and the drain current of the glucose sensor showed another sudden increase. Then, the same process happened again and drain current of the glucose current gradually decreased for a second time.

The human pH value can vary significantly depending on the health condition. Since we cannot control the pH value of the EBC samples, we needed to measure the pH value while determine the glucose concentration in the EBC. With the fast response time and low volume of the EBC required for HEMT based sensor, a handheld and real-time glucose sensing technology can be realized.

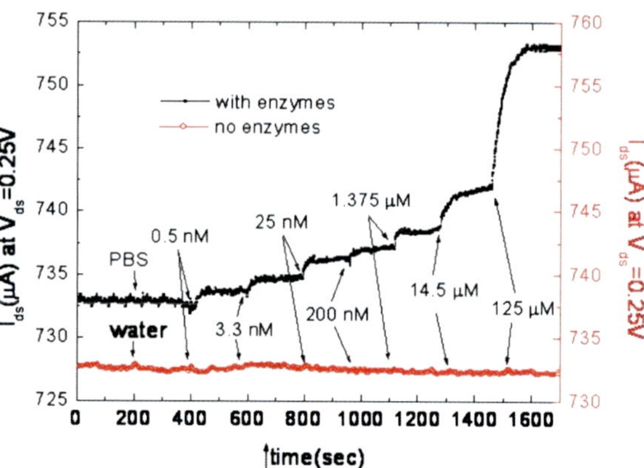

Fig. 10.8 Plot of drain current versus time with successive exposure of glucose from 500 pM to 125 μM in 10 mM phosphate buffer saline with a pH value of 7.4, both with and without the enzyme located on the nanorods

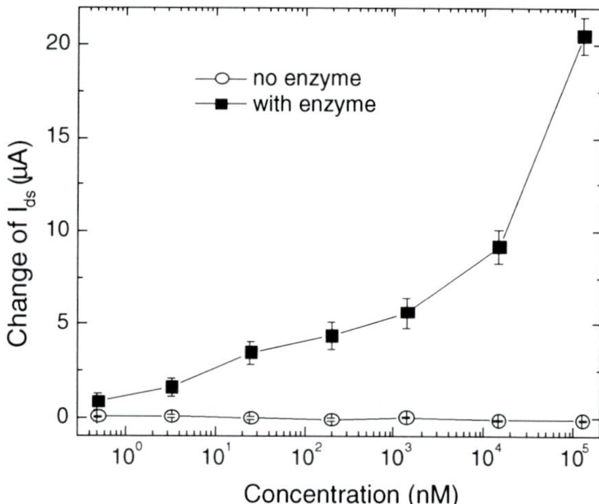

Fig. 10.9 Change in drain-source current in HEMT glucose sensors both with and without localized enzyme

10.6.2 *Lactic Acid Detection*

Lactic acid can also be detected with ZnO nanorod-gated AlGaN/GaN HEMTs. Interest in developing improved methods for detecting lactic acid has been increasing due to its importance in areas such as clinical diagnostics, sports medicine, and food analysis. An accurate measurement of the concentration of lactic acid in blood is critical to patients that are in intensive care or undergoing surgical operations as abnormal concentrations may lead to shock, metabolic disorder, respiratory insufficiency, and heart failure. Lactic acid concentration can also be used to monitor the physical condition of athletes or of patients with chronic diseases such as diabetes and chronic renal failure. In the food industry, lactate level can serve as an indicator of the freshness, stability and storage quality. For the reasons above, it is desirable to develop a sensor capable of simple and direct measurements, rapid response, high specificity, and low cost. Recent research on lactic acid detection has mainly focused on amperometric sensors with lactic acid specific enzymes attached to an electrode with a mediator. Examples of materials used as mediators include carbon paste, conducting copolymer, nano-structured Si_3N_4 and silica materials. Other methods of detecting lactic acid include utilizing semiconductors and electro-chemiluminescent materials.

A ZnO nanorod array, which was used to immobilize lactate oxidase (LOx), was selectively grown on the gate area using low temperature hydrothermal decomposition (Fig. 10.10, top). The array of one-dimensional ZnO nanorods provided a large effective surface area with high surface-to-volume ratio and a favorable environment for the immobilization of LOx. The AlGaN/GaN HEMT drain-source current showed a rapid response when various concentrations of lactic acid solutions were introduced to the gate area of the HEMT sensor. The HEMT could detect lactic acid concentrations from 167 nM to 139 µM. Figure 10.10 (bottom) shows a real time detection of lactic acid by measuring the HEMT drain current at a constant drain-source bias voltage of 500 mV during exposure of HEMT sensor to solutions with different concentrations of lactic acid. The sensor was first exposed to 20 µl of 10 mM PBS and no current change could be detected with the addition of 10 µl of PBS at approximately 40 s, showing the specificity and stability of the device. By contrast, a rapid increase in the drain current was observed when target lactic acid was introduced to the device surface. The sensor was continuously exposed to lactic acid concentrations from 167 nM to 139 µM. The response was reversible.

As compared with the amperometric measurement based lactic acid sensors, our HEMT sensors do not require a fixed reference electrode in the solution to measure the potential applied between the nano-materials and the reference electrode. The lactic acid sensing with the HEMT sensor was measured through the drain current of HEMT with a change of the charges on the ZnO nanorods and the detection signal was amplified through the HEMT. Although the time response of the HEMT sensors is similar to that of electrochemical based sensors, a significant change of drain current was observed for exposing the HEMT to the lactic acid at a low concentration of 167 nM due to this amplification effect. This is about a factor of 3 lower than electrochemical sensors. In addition, the amount of sample, which is dependent on the area of gate dimension, can be minimized for the HEMT sensor due to the fact that no reference electrode is required. Thus, measuring lactic acid in the exhaled breath condensate (EBC) can be achieved as a noninvasive method.

10.7 Wireless Sensors

With many of the sensor applications, it is desirable to have the detected signal transmitted wirelessly to a central location. This could be part of an unmanned system for biotoxin detection or part of a personal medical monitoring system in which a patient could breathe into a hand-held device that then transmits the

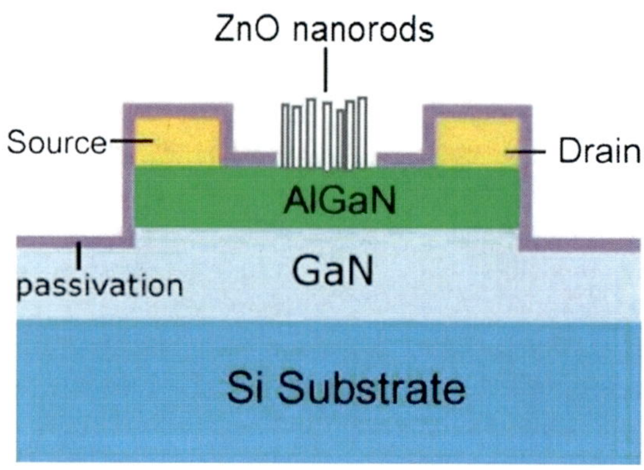
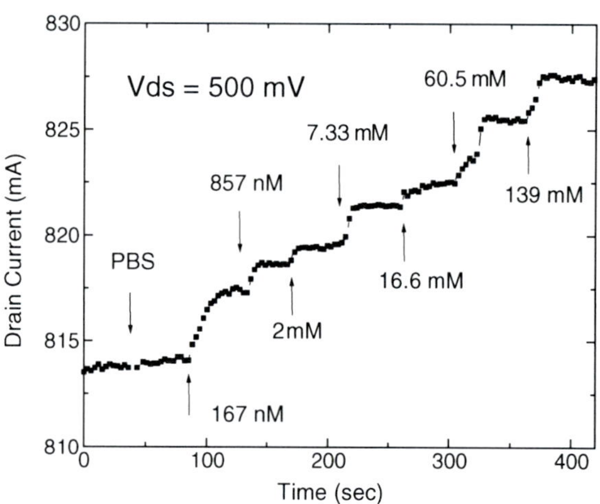

Fig. 10.10 Schematic cross sectional view of the ZnO nanorod gated HEMT for lactic acid detection (*top*) and plot of drain current versus time with successive exposure to lactic acid from 167 nM to 139 µM (*bottom*). (Reprinted with permission from Ref. 31 Chu BH, Kang BS, Ren F, Chang CY, Wang YL, Pearton SJ, Glushakov AV, Dennis DM, Johnson JW, Rajagopal P, Roberts JC, Piner EL and Linthicum KJ (2008) Appl Phys Lett 93:042114 Copyright American Institute of Physics, 2008)

encrypted signal to a doctor's office. This would allow for fewer visits to doctor's offices and less problems with false positive tests because data could be accumulated over an extended period and a more reliable baseline established.

As shown in the Fig. 10.11, we have also developed a pen-sized portable, reconfigurable wireless transceiver integrated with a pH sensor. The wireless transmitter and receiver pair was designed to acquire EBC data and transmit it wirelessly. This system is able to interface multiple different sensors and consists of a transmitter and receiver pair. The transmitter was designed such that it is the size of a marker-pen so that it could be used as an ultra-portable lightweight handheld device. The transmitter is designed to be operated on an ultra-low-power mode. The transmitter is also equipped with an on-board recharging circuit, which can be powered by using a standard mini-USB cable. The transmitter consumes on average 80 µA. The transmitter and receiver pair is designed to operate at 2.4 GHz with a range of up to 20ft line-of-sight. The receiver has USB 2.0 connectivity, which relays EBC data from the transmitter to a PC while powering the receiver.

The transmitter is designed to integrate with various different sensors through a connector. The transmitter can be reset for the required input signal range to trigger the alarm through the bi-directional wireless communication for a different sensing application. Thus this system is reconfigurable over-the-air. The wireless circuits only consume a power level around 1 μW. If the sensor consumes a similar power level, the battery installed on the transmitter package can last more than one month. This EBC sensing pair of devices can be mass-produced cost effectively well below \$100 each pair. The sensor occupies the tip of the pen-shaped layout in Fig. 10.11 and runs off a 75 mA Li ion polymer rechargeable battery. Figure 10.11 illustrates the package sensors mounted on a circuit board containing the detection circuit and microcontroller and the wireless transmitter for data collection. The sensor module is fully integrated on an FR4 PC board and is packaged with battery. The dimension of the sensor module package is: $4.5'' \times 2.9'' \times 2''$. The maximum line of sight range between the sensor module and the base station is 150 m. The base station of the wireless sensor network server is also integrated in a single module ($3.0'' \times 2.7'' \times 1.1''$) and is ready to be connected to a laptop by a USB cable. The base station draws its power from the laptop's USB interface, thus do not require any battery or wall AC transformer, which reduces its form factor. The PC is used to record the sensing data, send the data to internet, and take actions when the hydrogen is detected.

10.8 Summary and Outlook

Nitride and oxide semiconductor nanostructures are ideal materials for chemical sensing. Material properties including high chemical and thermal stability coupled with a high surface to volume ratio give these nanostructured sensors high sensitivity with detection capabilities down to the low ppm levels for gases such as hydrogen. While hydrogen sensing devices have been demonstrated on single ZnO nanowires, there are real advantages in the development of wide bandgap sensors involving contacting multiple nanowire sheets. Although the power requirement for using multiple nanowires is higher than that for single nanowire devices, the simplicity of fabrication including minimal processing steps for multiple nanowire sensors is highly favorable for future mass production of these sensors.

Selective sensing of hydrogen in nitrogen ambients has been shown using nanostructured ZnO down to 50 ppm at room temperature. Sensors were functionalized using Pt or Pd in order to facilitate dissociation of molecular hydrogen and improve sensing efficiency. All sensors displayed excellent response characteristics and decent recovery upon exposure to air or pure oxygen. Sensor recovery was slower for devices kept in pure nitrogen ambient upon removal of hydrogen. Non-functionalized devices showed little to no change upon gas exposure. In the case of ZnO nanowires, Pt functionalization produced a larger relative response to hydrogen exposure over Pd functionalization.

Fig. 10.11 Photographs of integrated pH sensor (*top*) and receiver/transmitter pair (*center*) and connection to computer (bottom)

There is also great promise for using nanowires in conjunction with HEMT sensors to enhance the detection sensitivity for glucose and lactic acid. Once again, the high surface area of nanowires provides an ideal approach for enzymatic detection of biochemically important substances.

Acknowledgments The work at UF was partially supported by ONR Grant N000140710982 monitored by Igor Vodyanoy, and the State of Florida, Center of Excellence in Nano-Bio Sensors.

References

1. Lu Y, Li J, Ng HT, Binder C, Partridge C, Meyyapan M (2004) Room temperature methane detection using palladium loaded single-walled carbon nanotube sensors. Chem Phys Lett 391:344
2. Hunter GW, Liu CC, Makel D (2001) Hak MG (ed) MEMS handbook. CRC Press, Boca Raton
3. Huang MH, Mao S, Feick H, Yan HQ, Wu YY, Kind H, Weber E, Russo R, Yang PD (2001) Room-temperature ultraviolet nanowire nanolasers. Science 292:1897
4. Lan ZH, Wang WM, Sun CL, Shi SC, Hsu CW, Chen TT, Chen KH, Chen CC, Chen YF, Chen LC (2004) Growth mechanism, structure and IR photoluminescence studies of indium nitride nanorods. J Cryst Growth 269:87
5. Chang C, Chi GC, Wang W, Chen L, Chen KH, Ren F, Pearton SJ (2006) Electrical transport properties of single GaN and InN nanowires nanowires. J Electron Mater 35:738
6. Chang C-Y, Lan T-W, Chi G-C, Chen L-C, Chen K-H, Chen J-J, Jang S, Ren F, Pearton SJ (2006) Effect of ozone cleaning and annealing on Ti/Al/Pt/Au ohmic contacts on GaN nanowires. Electrochem Solid-State Lett 9:G155
7. Mehandru R, Luo B, Kang BS, Kim J, Ren F, Pearton SJ, Fan CC, Chen GT, Chyi JI (2004) AlGaN/GaN HEMTHEMT based liquid sensors. Solid State Electron 48:351
8. Pearton SJ, Kang BS, Kim S, Ren F, Gila BP, Abernathy CR, Lin J, Chu SNG (2004) GaN-based diodes and transistors for chemical, gas, biological and pressure sensing. J Phys Condensed Matter 16:R961
9. Park WI, Yi GC, Kim MY, Pennycook SJ (2003) Quantum confinement observed in ZnOZnO/ZnMgO nanorod heterostructures. Adv Mater 15:526
10. Wang HT, Kang BS, Ren F, Tien LC, Sadik PW, Norton DP, Pearton SJ, Lin J (2005) Hydrogen-selective sensing at room temperature with ZnO nanorods. Appl Phys Lett 86:243503
11. Li C, Curreli M, Lin H, Lei B, Ishikawa FN, Datar R, Cote RJ, Thompson ME, Zhou C (2005) Complementary detection of prostate-specific antigen using In_2O_3 nanowires and carbon nanotubes. J Am Chem Soc 127:12484–12498
12. Zheng G, Patolsky F, Cui Y, Wang WU, Lieber CM (2005) Multiplexed electrical detection of cancer markers with nanowire sensor arrays. Nat Biotechnol 23:1294–1296
13. Wang B, Zhu LF, Yang YH, Xu NS, Yang GW (2008) Fabrication of a SnO_2 nanowire gas sensor and sensor performance for hydrogen. J Phys Chem C 112:6643
14. Tien LC, Sadik PW, Norton DP, Voss LF, Pearton SJ, Wang HT, Kang BS, Ren F, Jun J, Lin J (2005) Hydrogen sensing at room temperature with Pt-coated ZnO thin films and nanorods. Appl Phys Lett 87:222106
15. Tien LC, Wang HT, Kang BS, Ren F, Sadik PW, Norton DP, Pearton SJ, Lin J (2005) Room temperature hydrogen selective sensing using single Pt-coated ZnO nanowires at microwatt power levels. Electrochem Solid-State Lett 8:G239
16. Wang ZL (2006) In: Jagadish C and Pearton SJ (eds) ZnO bulk, thin films and nanostructures. Elsevier, Oxford
17. Wollenstein J, Plaza JA, Cane C, Min Y, Bottner H, Tuller HL (2003) A novel single chip thin film metal oxide array, Sensor. Actuator B 93:350
18. Wang HT, Kang BS, Ren F, Tien LC, Sadik PW, Norton DP, Pearton SJ, Lin J (2005) Detection of hydrogen at room temperature with catalyst-coated multiple ZnO nanorods. Appl Phys A Mater Sci Proc 81:1117
19. Donati S (2000) Photodetectors: devices, circuits, and applications. Prentice Hall, Upper Saddle River
20. Li YJ, Heo YW, Kwon Y, Ip K, Pearton SJ, Norton DP (2005) Transport properties of p-type phosphorus-doped (Zn, MgO) grown by pulsed-laser deposition. Appl Phys Lett 87:072101
21. Yang H, Li Y, Norton DP, Ip K, Pearton SJ, Jang S, Ren F (2005) Low resistance ohmic contacts to p-ZnMgO grown by pulsed-laser deposition. Appl Phys Lett 86:192103

22. Yang H, Li Y, Norton DP, Pearton SJ, Soohwan Jung, Ren F and Boatner LA (2005) Low resistance ohmic contacts to p-ZnMgO grown by pulsed-laser deposition. Appl Phys Lett 86:172103
23. Dong J, Osinsky A, Hertog B, Dabiran AM, Chow PP, Heo YW, Norton DP, Pearton SJ (2005) Development of MgZnO-ZnO-AlGaN heterostructures for ultraviolet light emitting applications. J Electron Mater 34:416
24. LaRoche JR, Heo YW, Kang BS, Tien L, Kwon Y, Norton DP, Gila BP, Ren F, Pearton SJ (2005) Fabrication approaches to ZnO nanowire devices. J Electron Mater 34:404
25. Wan Q, Li QH, Chen YJ, Wang TH, He XL, Li JP, Lin CL (2004) Fabrication and ethanol sensing characteristics of ZnO nanowire gas sensors. Appl Phys Lett 84:3654
26. Keem K, Kim H, Kim GT, Lee JS, Min B, Cho K, Sung MY, Kim S (2004) Photocurrent in ZnO nanowires grown from Au electrodes. Appl Phys Lett 84:4376
27. Heo YW, Varadarjan V, Kaufman M, Kim K, Norton DP, Ren F, Fleming PH (2002) Site-specific growth of Zno nanorods using catalysis-driven molecular-beam epitaxy. Appl Phys Lett 81:3046
28. Norton DP, Heo YW, Ivill MP, K.Ip, Pearton SJ, Chisholm MF and Steiner T (2004) ZnO: growth, doping and processing. Materials Today, pp. 34–40
29. Koida T, Chichibu SF, Uedono A, Tsukazaki A, Kawasaki M, Sota T, Segewa Y, Koinuma H (2003) Correlation between the photoluminescence lifetime and defect density in bulk and epitaxial ZnO. Appl Phys Lett 82:532
30. Lopatiuk O, Burdett W, Chernyak L, Ip KP, Heo YW, Norton DP, Pearton SJ, Hertog B, Chow PP, Osinsky A (2005) Minority carrier transport in p-type ZnMgO doped with phosphorus. Appl Phys Lett 86:012105
31. Shi GA, Saboktakin M, Stavola M, Pearton SJ (2004) Hidden Hydrogenn in as-grown ZnO. Appl Phys Lett 85:5601
32. Kang BS, Heo YW, Tien LC, Norton DP, Ren F, Gila BP, Pearton SJ (2005) Hydrogen and ozone gas sensing using multiple ZnO nanorods. Appl Phys A 80:1029
33. Studenikin SA, Golego N, Cocivera M (2000) Carrier mobility and density contributions to photoconductivity transients in polycrystalline ZnO films. J Appl Phys 87:2413
34. Kang BS, Ren F, Heo YW, Tien LC, Norton DP, Pearton SJ (2005) pH measurements with single ZnO nanorods integrated with a microchannel. Appl Phys Lett 86:112105
35. Heo YW, Kang BS, Tien LC, Norton DP, Ren F, LaRoche JR, Pearton SJ (2005) UV photoresponse of single ZnO nanowires. Appl Phys A 80:497
36. Heo YW, Norton DP, Tien LC, Kwon Y, Kang BS, Ren F, Pearton SJ, LaRoche JR (2004) ZnO nanowire growth and devices. Mat Sci Eng R 47:1
37. Sadik PW, Pearton SJ, Norton DP, Lambers E, Ren F (2007) Functionalizing Zn- and O-terminated ZnO with thiols. J Appl Phys 101:104514
38. Dam TV, Anh W, Olthuis, Bergveld P (2005) A hydrogen peroxide sensor for exhaled breath measurement. Sensor Actuat B, 111/112(11):494
39. Multu GM (2001) Collection and analysis of exhaled breath condensate in humans. Am J Res Crit Care Med 164:731
40. Pearton SJ, Ren F, Yu-Lin Wang, Chu BHù Chen KH, Chang CY, Wantae Lim, Jenshan Lin, Norton DP (2010) Recent advances in wide bandgap semiconductor biological and gas sensors. Progress in Materials Science, 55:1
41. Wang JX, Sun XW, Wei A, Lei Y, Cai XP, Li CM, Dong ZL (2006) Zinc oxide nanocomb biosensor for glucose detection. Appl Phys Lett 88:233106
42. Wei A, Sun XW, Wang JX, Lei Y, Cai XP, Li CM, Dong ZL, Huang W (2006) Enzymatic glucose biosensor based on ZnO nanorod array grown by hydrothermal decomposition. Appl Phys Lett 89:123902
43. Phypers B, Pierce T (2006) Continuing education in anaesthesia. Crit Care Pain 6(3):128
44. Suman S, Singhal R, Sharma A, Malthotra BD, Pundir CS (2005) Development of a lactate biosensor based on conducting copolymer bound lactate oxidase. Sens Actuators B 107:768
45. Haccoun J, Piro B, Noël V, Pham MC (2006) The development of a reagent less lactate biosensor based on a novel conducting polymer. Bioelectrochemistry 68:218

46. Di J, Cheng J, Xu Q, Zheng H, Zhuang J, Sun Y, Wang K, Mo X, Bi S (2007) Direct electrochemistry of lactate dehydrogenase immobilized on silica sol–gel modified gold electrode and its application. Biosens Bioelectron 23:682
47. Lupu A, Valsesia A, Bretagnol F, Colpo P, Rossi F (2007) Development of a potentiometric biosensor based on nanostructured surface for lactate determination. Sens and Actuators B 127:606
48. Lim W, Wright JS, Gila BP, Johnson JL, Ural A, Anderson T, Ren F, Pearton SJ (2008) Room temperature hydrogen detection using Pd-coated GaN nanowires. Appl Phys Lett 93:072110
49. Lim W, Wright JS, Gila BP, Pearton SJ, Ren F, Lai W, Chen LC, Hu M, Chen KH (2008) Selective hydrogen sensing at room temperature with Pt-coated InN nanobelts. Appl Phys Lett 93:202109
50. Wright J, Lim W, Norton DP, Ren F, Pearton SJ, Johnson J, Ural A (2010) Nitride and oxide semiconductor nanostructured hydrogen gas sensors. Semicond Sci Technol 25:024002
51. Chu BH, Kang BS, Chang CY, Ren F, Goh A, Sciullo A, Wu W, Lin J, Gila BP, Pearton SJ, Johnson JW, Piner EL, Linthicum KJ (2010) Wireless detection system for glucose and pH sensing in exhaled breath condensate using AlGaN/GaN high electron mobility transistors. IEEE Sens J 10:64
52. Anderson T, Fan Ren, Pearton SJ, Kang BS, Wang H-T, Chang C-Y and Lin J (2009) Advances in hydrogen, carbon dioxide, and hydrocarbon gas sensor technology using GaN and ZnO-based devices. Sensors 9(6):4669

Chapter 11
Metal Oxide Nanowire Sensors with Complex Morphologies and Compositions

Qiuhong Li, Lin Mei, Ming Zhuo, Ming Zhang
and Taihong Wang

Abstract Metal oxide nanowire sensors with complex morphologies and compositions have shown promising properties due to their high surface-to-volume ratio and stable structures against agglomeration. In this chapter, a series of metal oxide nanostructures modified via surface coating, morphology variation, doping and appropriate energy band engineering have been investigated, and the sensing mechanism is discussed. By using nanostructures with complex morphologies and compositions in simple material synthesis routes, the structure of the sensitive material is modified, the electronic transport of the sensor is regulated and the sensing properties can be greatly improved, including enhancing the sensitivity and selectivity, lowering the working temperatures, reducing the response time and achieving long-term stability.

11.1 Introduction

Metal oxide nanowire sensors have been investigated extensively in recent years [1–3]. They are promising in a wide range of applications including monitoring atmosphere quality, detecting hazardous and poisonous gases in mining or at home, diagnosing human health, among others. The metal oxide nanowires are usually n- or p-type semiconductors and sensitive to oxidizing or reducing ambient conditions. The chemical information about the ambient gas and its concentration can be turned into an electric one such as resistance, which can be easily detected [4].

Q. Li · L. Mei · M. Zhuo · M. Zhang · T. Wang (✉)
Key Laboratory for Micro-Nano Optoelectronic Devices of Ministry of Education,
and State Key Laboratory for Chemo/Biosensing and Chemometrics, Hunan University,
410082 Changsha, People's Republic of China
e-mail: thwang@aphy.iphy.ac.cn

M. A. Carpenter et al. (eds.), *Metal Oxide Nanomaterials for Chemical Sensors,*
Integrated Analytical Systems, DOI: 10.1007/978-1-4614-5395-6_11,
© Springer Science+Business Media New York 2013

In 2004 ZnO nanowire sensors were fabricated by a MEMS technology [5]. They had shown a sensitivity of 1.9 and 47 to 1 and 200 ppm ethanol at 300 °C, respectively. The response time was about 5 s and the recovery time was about 10 s to 1 ppm ethanol. The properties were much better than thin or thick film metal oxide sensors. This high performance has attracted much attention and more different types of nanowire sensors have since been developed. Metal oxide nanowires are usually crystalline structures with well defined chemical compositions. Sensors made of these nanowires exhibit superior properties in sensitivity, selectivity, and stability due to their high surface-to-volume ratio, high degree of crystallinity and stability against agglomeration [2].

The sensors usually consist of a large number of sensitive nanostructures (when arranged in nanowire mats) that contact each other to form a thin conduction layer. As nano-sized structures, surface effects and contact barriers dominate their sensing properties, and oxygen plays an important role in their sensing mechanism. When the sensor is in air, oxygen molecules are adsorbed on the surface of the nanostructure and deprive it of electrons, forming oxygen ions (O^-, O^{2-} or O_2^-). If the sensing material is n-type, a depletion layer forms on the surface of the nanowire's conduction channel, and the potential barriers are built up between the nanowires, thus reducing the sensor conductance. On the other hand, for a p-type material, the extraction of electrons from the nanowire increases the major carrier concentration and hence increases the conductance of the sensor. Therefore, when the n-type sensor is put in an oxidizing gas, more electrons are deprived, the thickness of the surface depletion layer increases and the potential barriers are raised. Thus the resistance of the sensor increases. While in a reducing gas, electrons are transferred from the preadsorbed oxygen ions back to the metal oxide, the depletion layer becomes narrower, the contact barrier is reduced and the conductance of the sensor increases. With a similar mechanism, the conductance of a p-type sensor increases in an oxidizing gas and decreases in a reducing gas. The contact between the sensitive nanowires and electrodes may play a role as well in the transport and sensing mechanism, but it is less important than that of a single nanowire sensor since the electrodes are in parallel connection with a large number of nanostructures.

The surface depletion and contact barriers all contribute to the sensor resistance [6–8] that can be enhanced or reduced in adsorption and desorption processes, whereas which one dominates the sensing mechanism mainly depends on the size of the nanowires. If the diameter of the nanowire is much larger than the depletion width, surface depletion does not greatly affect the density and mobility of the carriers of the nanowires but significantly modifies the potential barrier of the contacts between the nanowires. When the size of nanowires is comparable to the space charge layer $2L_d$ (L_d: Debye length), surface depletion greatly affects the density and mobility of the carriers of the nanowires and these effects becomes much more dominant in the corresponding sensing mechanism. Metal oxide nanowire sensors have shown ultra high responses as the size of the nanowire is close to or smaller than $2L_d$ [8–10].

With rapidly growing demands for better sensors, sensors from simple metal oxide nanowires need to be improved and better functional sensitive materials are under development to further increase the sensitivity, quicken the response, lower the working temperature, and enhance the selectivity. Therefore, materials with complex morphologies and composition are receiving more and more attention. The addition of noble metals has been reported to improve the selectivity and stability of thin film sensors which has been adopted in nanowire sensors in the current research [11–15]. Coating or doping is proven to be an effective method to improve the sensing performance. By using coating and doping technology or fabricating sensors from hierarchical-structure nanomaterials, the contact between sensitive materials can be modified, the adsorption and desorption process can be accelerated and the sensor properties are greatly improved. The sensing properties for some nanostructures are summarized in Table 11.1.

In this chapter, a series of representative metal oxide nanostructure sensors with complex morphologies and composition are discussed, including nanoparticle and nanowire systems, hierarchical nanostructures, metal doping in nanowires, and multiple-composition oxide nanowires. We demonstrate that the sensor properties such as sensitivity, selectivity, response time and stability can be greatly improved via the aforementioned methods.

11.2 Sensors Packaging and Measurement Setup

The sensors in this chapter are fabricated in a commercial style as shown in Fig. 11.1a. After material synthesis they are usually dispersed in a solution such as water or ethanol to get homogenous paste for coating on a ceramic tube. As shown in Fig. 11.1a, a ceramic tube is welded on a six-electrode basic stand to connect with an outside electronic circuit. The tube has a diameter of 1 mm and length of 5 mm. The platinum electrodes 1–4 on the surface of the ceramic tube work as measurement electrodes connecting with sensitive materials. Sensitive materials are coated on the surface of the tube to form the conductive route. Electrodes 5 and 6 are for heating and are connected with a platinum resistor located in the hollow ceramic tube center. By monitoring the voltage on the platinum resistor, the working temperature of the sensor is controlled. Before measurements the sensors are usually aging at 300 °C for 1–48 h by applying a heating voltage on the resistor.

The sensor is then connected in series with a reference resistor R_r as shown in Fig. 11.1b. As a voltage V_c (usually 5 V) is applied on the sensor and the reference resistor, the voltage on the reference resistor V_{out} can be measured and the resistance of the sensor R_s is then calculated as $R_s = (V_c - V_{out})R_r/V_{out}$. The sensor is first put in air with the resistance defined as R_s^a, and then put into a chamber in which a certain concentration of target gas mixed with air is already prepared. Then the sensor is taken out of the chamber and put in the air atmosphere again. The resistance in the detected gas is defined as R_s^g. The sensitivity (S) of the sensor is defined as $S = R_s^g/R_s^a$ as $R_s^g > R_s^a$ or $S = R_s^a/R_s^g$ as $R_s^g < R_s^a$. The real-time

Table 11.1 Sensing properties of various nanostructures to ethanol

Material	Size[a]	Sensitivity (at ethanol concentration, ppm)	Response time (s)	Working temperature (°C)	Reference
SnO_2 nanorods	4–15 nm	13.9 (100)	1	300	[39]
SnO_2 nanorods	3–12 nm	30.7 (100)	–	300	[40]
ZnO nanowires	25 ± 5 nm	32 (100) 1.9 (1)	∼5	300	[5]
Flower like ZnO nanorod assembles	150 nm	14.6 (100) 2.2 (1) 1.6 (0.5)	–	300	[6]
ZnO nanorods	15 nm	29.7 (100) 4.1 (1)	–	300	[8]
Highly oriented ZnO nanorods	50 nm	100 (100)	10–100	300	[41]
In_2O_3 nanobricks	Several tens to 200 nm	∼190 (100) ∼10 (1)	∼2	300	[7]
Branched SnO_2 Nanowires	Branch nanowires 20 nm, backbones 80 nm	50.6 (100) 2.3 (0.5)	2 4	300	[23]
SnO_2 nanobelts coated with CdS nanoparticles	Nanoparticles 10–20 nm, nanobelts 30–200 nm in width and 10–50 nm in thickness	90 (100)	∼5	400	[28]
ZnO nanorods coated with CdS nanoparticles	Nanoparticles 5–12 nm, nanorods 20–40 nm	33 (100)	–	300	[42]
ZnO nanorods coated with Au nanoparticles	Nanoparticles 4 nm, nanorods 15 ± 5 nm	89.5 (100)[b] 11.3 (100)[c]	2[b] ∼10[c]	300 150	[24]
Al-doped ZnO nanotetrapods	30 nm	40 (100) 5.7 (1)	–	300	[43]
In-doped ZnO nanowires	60–150 nm	27 (100) 3 (1)	2	300	[44]
In-doped SnO_2 (ITO) nanowires	70–150 nm	40 (200)	2	400	[35]
$ZnSnO_3$ nanowires	20–90 nm	18 (100)	1	300	[36]
SnO_2 nanorods loaded with La_2O_3	5–20 nm	213 (100)	–	200	[30]

[a] Size here represents diameter if not particularly pointed out
[b] at 300 °C
[c] at 150 °C

voltage on the reference resistor V_{out} while varying the ambient is recorded and the sensor resistance is calculated. The equipment for the sensor measurement is a highly precise sensor analyzer NS-4003 series made by China Zhong-Ke Micronano IOT Ltd. R_r will be chosen automatically by the equipment according to the sensor resistance. A measurement of 256 sensors can be carried out simultaneously with a time resolution of 10^{-3} s and the measurable resistance range 10^0–10^9 Ω, current range 10^{-12}–10^{-3} A, voltage range 10^{-9}–10^0 V, and capacitance range 10^{-12}–10^{-3} F. The equipment also provides accurate heating on the platinum resistor, including constant current and voltage power supply. Due to the accurate (with error less than 0.09 %) and dynamic measurement across the whole range the electric signals can be recorded precisely.

11.3 Tin Oxide Coated Multi-Walled Carbon Nanotube (MWCNT) Sensors

The sensing materials were synthesized by coating of MWCNTs with SnO_2 nanograins using an ultrasonic method [16]. The MWCNTs were prepared by an arc discharge method. The collected material was purified by reflux in H_2O_2 and then in a mixture of sulfuric and nitric acids to remove particles in the MWCNT materials. The process facilitated the formation of functional groups (mainly carboxylic acid groups) on the MWCNTs to act as sites for SnO_2 coating [17]. Such a material was sonicated in tin chloride solution with hydrochloric acid, and the resultant solution was then stirred for 30 min. Products were collected after filtration and rinsed in distilled water. High resolution transmission electron microscopy (HRTEM) studies revealed that the SnO_2 nanograins were uniformly distributed on MWCNTs with the size about 2–6 nm as shown in Fig. 11.2. Then the material was pasted on the surface of the ceramic tube for gas sensing measurement as described previously.

By changing the gas categories including NO, NO_2, C_2H_2, and ethanol with certain concentrations, the resistance was measured in real-time. The standard working temperature was about 300 °C and the resistance in air was about 130 kΩ. In oxidizing gases including NO and NO_2, the resistance increased, while in reducing gases, the sensor resistance decreased, which was in accordance with the above discussion. The sensitivity to different gases and concentrations is shown in Fig. 11.3a, b. The resistance increased to about 2.8 MΩ and 1.4 MΩ in 50 ppm of NO_2 and NO, respectively. And the resistance decreased to about 46.3 kΩ and 62.8 kΩ in 50 ppm of ethanol and C_2H_2, respectively.

The sensing performance of the SnO_2-MWCNT material can be understood in terms of the aforementioned receptor-transduction mechanism. Namely, when the sensor is in air, oxygen molecules are adsorbed on the tin oxide grains and extract electrons from them, leaving with depletion layers between the grains and thus forming barriers for electron transport. When the sensor is put in oxidizing gases

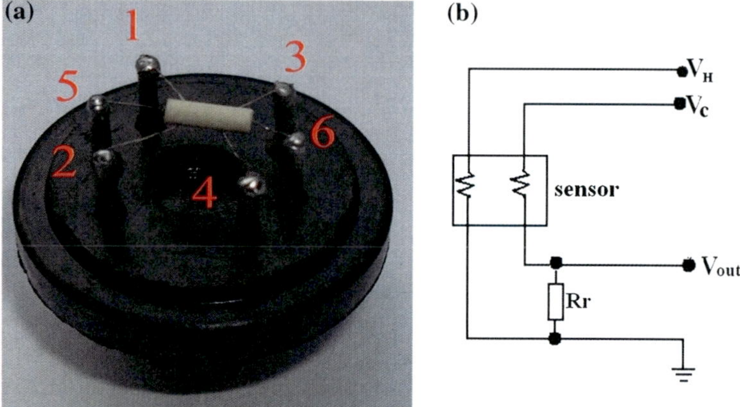

Fig. 11.1 **a** Typical photo of the sensor. *1–4* working electrodes, *5–6* heating resistor. *1* and *2* are both ends of one platinum wire, and *3* and *4* are both ends of one platinum wire. **b** Measurement circuit configuration for the sensor. V_H heating voltage, V_c voltage applied on the sensor and the reference resistor, V_{out} voltage on the reference resistor, R_r reference resistor

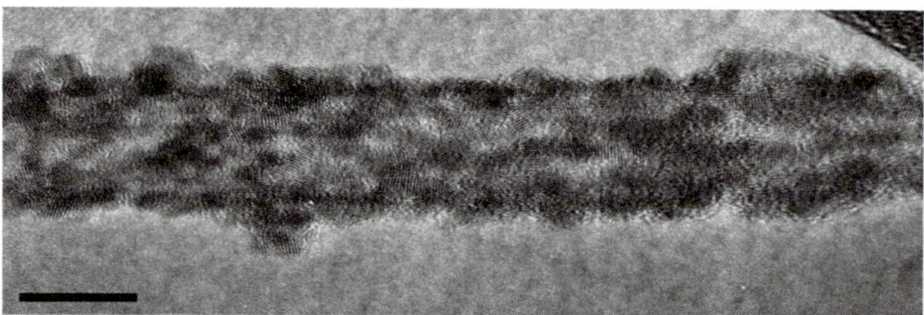

Fig. 11.2 High resolution transmission electron microscopy (*HRTEM*) image of the sensitive material for SnO_2 nanoparticle coated MWCNTs. The tin oxide nanocrystal grains can be seen clearly. *Scale bar* 10 nm

such as NO and NO_2, the molecules are directly adsorbed on SnO_2 grains and further extract electrons, so higher barriers are formed between them and the sensor resistance rises. The oxidizing response can follow the reaction path (11.1). When the sensor is taken out of the target gas and set in air again, NO_x (x: 1 or 2) gas molecules are desorbed from SnO_2 nanograins, and electrons are released. The sensor resistance then recovers. On the other hand, when the sensor is put in reducing gas such as ethanol or C_2H_2, these molecules react with preadsorbed oxygen ions and release the trapped electrons, so the barriers are lowered and the sensor resistance decreases. A representative reaction can be described as (11.2):

$$NO + e^- \rightarrow NO^- \tag{11.1}$$

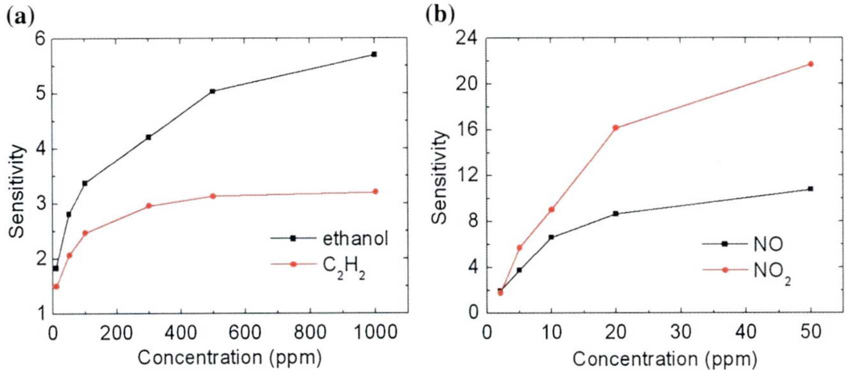

Fig. 11.3 The sensitivity in different gas concentrations for the sensors in **a** reducing gases of ethanol and C_2H_2, **b** oxidizing gases of NO and NO_2

$$C_2H_5OH + 6O^- \rightarrow 2CO_2 + 3H_2O + 6e^- \tag{11.2}$$

The sensors exhibit high selectivity to NO_x gases. SnO_2 is a typical *n*-type sensitive material and MWCNTs are highly conductive. The sensor resistance is dominated by the barriers between SnO_2 nanograins on the MWCNTs, and the barrier height is controlled by the adsorption and desorption of gas molecules, which extract or release electrons. Because the work function of the MWCNTs is approximately equal to that of SnO_2, the Schottky barrier between them is very low [18–20]. Electrons can travel through the tin oxide grains into the MWCNTs, and then conduct in MWCNTs with low resistance. The sensor exhibits rather good stability. After 3 months, the fluctuation of the sensitivity is less than ± 3 %, which indicates that the network structure of the sensor is very stable. It is assumed that MWCNTs provide a support for SnO_2 nanograins to avoid their aggregation that is a serious problem for traditional gas sensors composed of SnO_2 nanograins. The metal oxide sensors usually work at a high temperature (200–500 °C). In SnO_2 nanoparticles sensors, the nanoparticles tend to aggregate at elevated temperatures with aging that leads to structure instability [21, 22]. When the particles become larger and denser, gas diffusion into the inner part of the sensing materials becomes much more difficult. Under this configuration, a high sensitivity cannot be achieved because the resistance change occurs mostly near the surface region and the inner part remains almost inactive. Moreover, the sluggish gas diffusion through the aggregated nanostructures slows down the sensor response time.

There is another advantage of SnO_2-MWCNT system. The sensor resistance increases upon oxidizing gases. In many sensors including SnO_2 nanograins and SnO_2 thin films, the resistance is high in air. When exposed to oxidizing gases, the resistance becomes even higher (on the orders of tens or even thousands of $M\Omega$), which is hard to measure with common circuitry. The sensors with SnO_2 nanograins

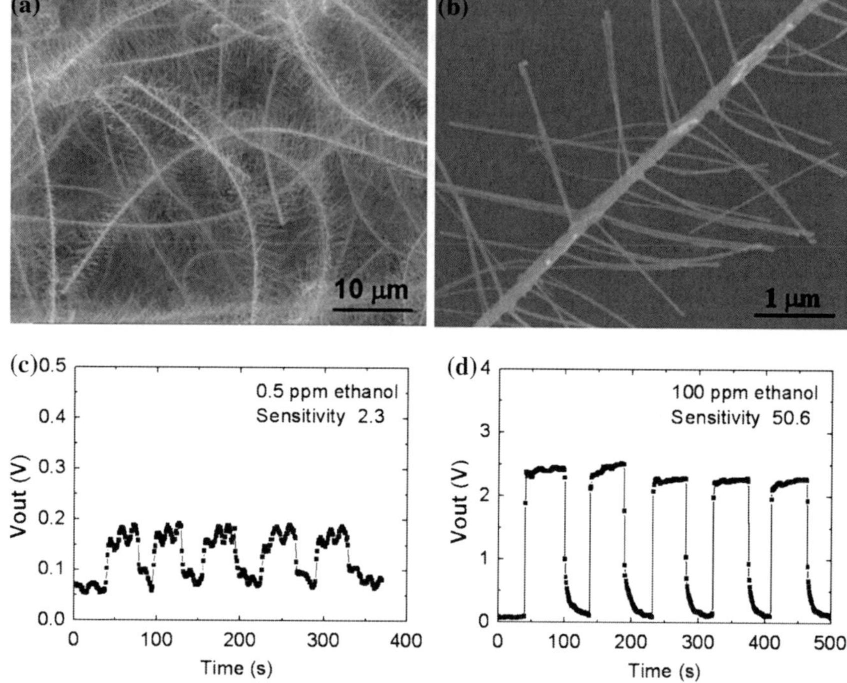

Fig. 11.4 a and **b** Scanning electronic microscope (*SEM*) images of the hierarchical nanostructures. Real-time response of the measured voltage on the reference resistor V_{out} as the ambient was switched between air, **c** 0.5 ppm ethanol, **d** 100 ppm ethanol (Reprinted with permission from [23], Copyright 2008, American Institute of Physics)

coated on MWCNT are in a suitable resistance range, which is favorable for the matching amplifying electronic circuit commonly available in industry.

11.4 Hierarchical Structures

With highly conductive Sb-doped SnO_2 nanowires as the backbone and SnO_2 nanowires as the branches, three dimensional hierarchical nanostructures were synthesized in a two-step vapor-liquid-solid (VLS) process [23]. Sb doped SnO_2 nanowires were first grown at 800 °C for 1 h in a furnace with Sn and Sb powders (20:1 in weight) as source materials and Au as catalyst. Then 5 nm-Au was deposited on the Sb-SnO_2 nanowires as catalyst. Similar processes were used to grow hierarchical structures with branched SnO_2 nanowires on Sb-SnO_2 nanowire backbones. The scanning electronic microscope (SEM) image of the materials is shown in Fig. 11.4a, b.

The sensors based on hierarchical nanostructures demonstrated good sensing properties as given in Fig. 11.4c, d. They could detect ethanol with the concentration as low as 0.5 ppm and the corresponding sensitivity was about 2.3. The sensitivity was 50.6–100 ppm ethanol with the response time about 2 s. The sensitivity vs. concentration showed an almost linear behavior with the concentration between 0.5 ppm and 500 ppm, which was favorable for a wide ethanol detection range.

Sensors based on hierarchical nanostructures had stable three-dimensional (3D) morphologies while maintaining the advantages of small size and high crystallinity of the nanowire structures without scarifying large surface-to-volume ratio. Owing to a large amount of porous space between the hierarchical structures in which gas molecules could diffuse quickly, the sensor had short response and recover times. Many paths for electrical conduction were offered by the branched nanowires structure with metallic backbones and high sensitivity could be achieved. The ability to detect ethanol down to 0.5 ppm level was probably related to the small size and large surface of the active SnO_2 nanowire branches. Branched SnO_2 nanowires were expected to be much more stable against agglomeration.

11.5 Nanoparticles Coated on Quasi-One Dimensional Materials

Quasi-one dimensional metal oxides, such as ZnO and SnO_2, demonstrate promising sensing properties. However, the development of high quality sensors based on the quasi-one dimensional materials still needs to be improved, including the sensitivity, response time and selectivity of the sensors. The sensors based on nanoparticles have very high surface-to-volume ratio and high sensitivity. However, since the sensor usually operates at high temperatures, nanoparticles become easily aggregated into larger size grains, which deteriorate the reliability and stability of the sensors. By combining large surface-to-volume ratio nanoparticles deposited on to the surface of stable quasi-one-dimensional crystalline nanostructures, one can expect to have better sensing performance.

11.5.1 Gold Nanoparticles Coated ZnO Nanorods

The noble metals are well known to be good catalysts. There are many reports where noble metal catalysts, including Au, Pt, Pd, and Ag have been used to improve the gas sensing properties of nanowire sensors [14, 15]. For example, Au nanoparticles are well known as active catalysts for e.g. CO oxidation, hydrogen peroxide synthesis from H_2 and O_2 and hydrocarbon oxidation. Therefore, coating nanowires with Au nanoparticles is an effective way to improve the gas sensing properties of metal oxide nanostructures.

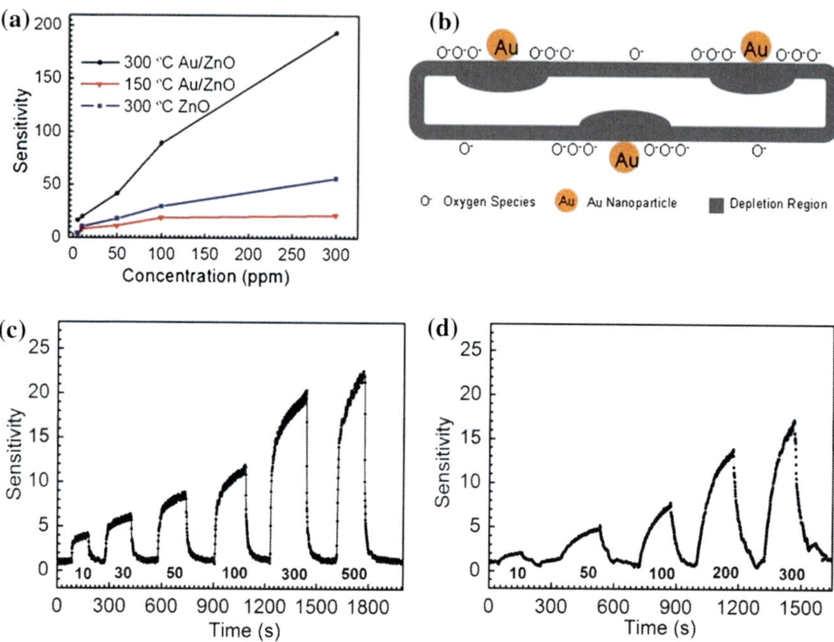

Fig. 11.5 a Sensitivity of the sensor to different ethanol concentrations with/without Au nanoparticles coating at 300 and 150 °C, **b** illustrations of chemical sensitization of the Au coated nanorods, and real-time response of the nanorod, **c** with and **d** without Au nanoparticle coating to different ethanol concentrations at 150 °C. The concentration of ethanol is indicated in **c** and **d**

ZnO nanorods were first pasted on the ceramic tubes and then Au nanoparticles were deposited on the nanorods by sputtering [24]. The sensors were annealed at 700 °C in a muffle furnace for 1 h before measurement. The average diameter and length of the nanorods were about 15 nm and 1 μm, respectively, and the diameter of Au nanoparticles was about 4 nm. Figure 11.5a, c, and d show comparatively the performance of the sensor with respect to various ethanol concentrations at 150 and 300 °C with or without Au nanoparticle coating. The sensitivity of ZnO nanorods functionalized with Au nanoparticles could reach 89.5 to 100 ppm ethanol with a response time of less than 2 s at 300 °C. Moreover, the sensor with Au nanoparticle coating could work at low temperatures, specifically the sensitivity to 100 ppm ethanol was 11.3 even as the working temperature was lowered down to 150 °C, and the response and recovery times of the sensors at 150 °C were only about 10 and 20 s, respectively. On the contrary, for the sensors without Au nanoparticle coating, the sensitivity to 100 ppm ethanol was 7, and the response and recovery times were about 80 and 55 s, respectively, which are longer than the sensors with Au nanoparticle coating.

Compared with bare ZnO nanorods, the sensing properties were greatly improved not only in sensitivity but also in response time, which could be related to the increasing rate of oxygen adsorption and desorption due to the catalytic

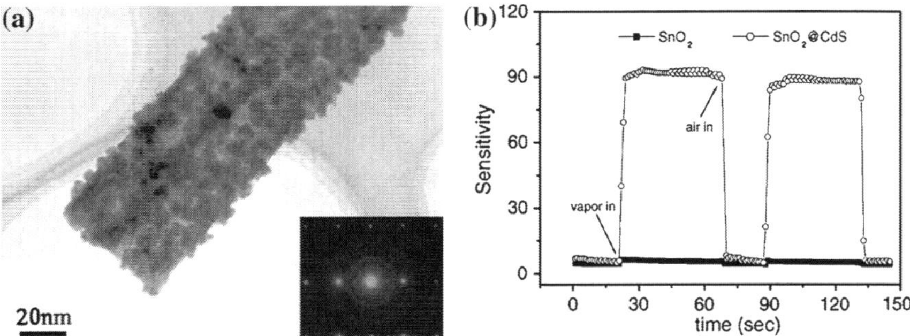

Fig. 11.6 **a** Typical transmission electron microscopy (*TEM*) image of the SnO₂ nanobelt coated with CdS, and the inset shows the corresponding selected area electron diffraction (*SAED*) patterns, **b** Response curves of the SnO₂-nanobelt and the CdS-nanoparticle-coated-SnO₂-nanobelt sensors to 100 ppm ethanolethanol at 400 °C (Reprinted with permission from [28], Copyright 2004, The Royal Society of Chemistry)

action of Au nanoparticles. As shown in Fig. 11.5b, Au nanoparticles activate the oxygen molecules, which could be more easily adsorbed on the surface of ZnO nanorods. This adsorbed oxygen could diffuse faster to surface vacancies and capture electrons from the conduction band of ZnO nanorods to become oxygen ions (O^-, O^{2-} or O_2^-). This catalyst decreases the working temperature of the sensor and increases the quantity of adsorbed oxygen. The latter results in greater and faster degree of ZnO nanorods electron depletion, which in turn defines its higher sensitivity and faster response. With reduction of the size of Au particles, the chemical sensitization effect of nanomaterials becomes much more remarkable and the working temperature could be lowered further [25–27]. These sensors demonstrated promising characteristics, especially on their low working temperatures, which could be related to Au activation and small size effect. The method provides an easy way to enhance the sensor performance.

11.5.2 SnO₂ Nanobelts Functionalized with CdS Nanoparticles

The sensing materials composed of CdS nanoparticles decorating SnO₂ nanobelts were prepared by a sonochemical synthesis method [28]. First, single crystalline SnO₂ nanobelts with 30–200 nm in width, 10–50 nm in thickness, and several hundreds of microns in length were prepared by a thermal evaporation of Sn powders at 800 °C. Then they were mixed in water with cadmium chloride and thiourea in a sonication cell. The mixture was subjected to high-intensity ultra-sonication for 1–3 h. After sonication the excess CdS nanoparticles were separated from the rest of the mixture by centrifugation. Then the products were washed, centrifuged and dried. The CdS nanoparticles were nearly spherical and had typical diameters in the range of 10–20 nm. The typical transmission electron microscopy (TEM) image of the coated SnO_2 nanobelt was shown in Fig. 11.6a.

The SnO_2 nanobelt/CdS nanoparticle sensor had a high sensitivity to 100 ppm ethanol (about 90) in air at 400 °C. As shown in Fig. 11.6b, the sensitivity of the coated structure was much higher than that of uncoated nanobelts.

As we know, SnO_2 is a wide bandgap semiconductor with Eg = 3.6 eV, and CdS is a narrow one with Eg = 2.4 eV. The energy band diagram of SnO_2 and CdS heterostructures indicates that electrons would be transported from the coated CdS nanoparticles to the SnO_2 nanobelts in a thermal equilibrium. An accumulation layer of electrons is formed in the interface of SnO_2/CdS. Thus, compared with pristine SnO_2 nanobelts, the heterojunction offers an additional source for electrons, that is, the CdS nanoparticles. Compared with general SnO_2 nanoparticles and thin films, the sensor has a lower resistance about 16 kΩ. In combination with the large surface-to-volume ratio of CdS nanoparticles and stable nature of SnO_2 nanobelts, the sensors are expected to have high sensitivity and to be more stable. Other than the noble metal addition, CdS coating is a good way to improve gas sensing properties toward specific gases.

11.5.3 p–n Junctions at the Interface Between CuO Nanoparticles and SnO₂ Nanorods

Band engineering is a very effective approach not only for microelectronic devices but also for semiconductor sensors. As an example, by using the p–n junction in CuO-SnO_2 nanostructures, we show that such sensors demonstrate ultra high sensitivity and selectivity when exposed to H_2S even at room temperature [29]. CuO and SnO_2 are known to be a p-type and n- type semiconductors, respectively. Nanoscale p–n junctions were formed by coating SnO_2 nanorods with CuO nanoparticles. The SnO_2 nanorods were synthesized by a hydrothermal method, with a typical diameter and length of 10 and 100 nm, respectively. They were added in ethanol solution with $Cu(NO_3)_2 \cdot 3H_2O$. After ultrasonic and stirring treatment, green products were collected, filtered and dried at 75 °C. Then the products were annealed with a slowly rising temperature to 800 °C and the final materials with CuO nanoparticles coated on SnO_2 nanorods were finally collected. TEM inspection revealed that CuO nanoparticles had an average diameter about 4 nm.

The gas sensing properties of this material were tested at 18, 60, 95, 180, and 300 °C and the sensitivity to 10 ppm H_2S was shown in Fig. 11.7a. It could be as high as 9.4×10^4 at 60 °C. The response time was about 30 s and the recovery time lasted several hours at 60 °C. Raising the working temperature could shorten both the recovery and response time in a large degree, but the sensitivity also decreased with elevating temperatures. When the temperature rose above 103 °C, CuS converted into Cu_2S with lower conductivity, resulting in a decrease of sensitivity with increasing temperatures. Also, the sensing properties to different gases including

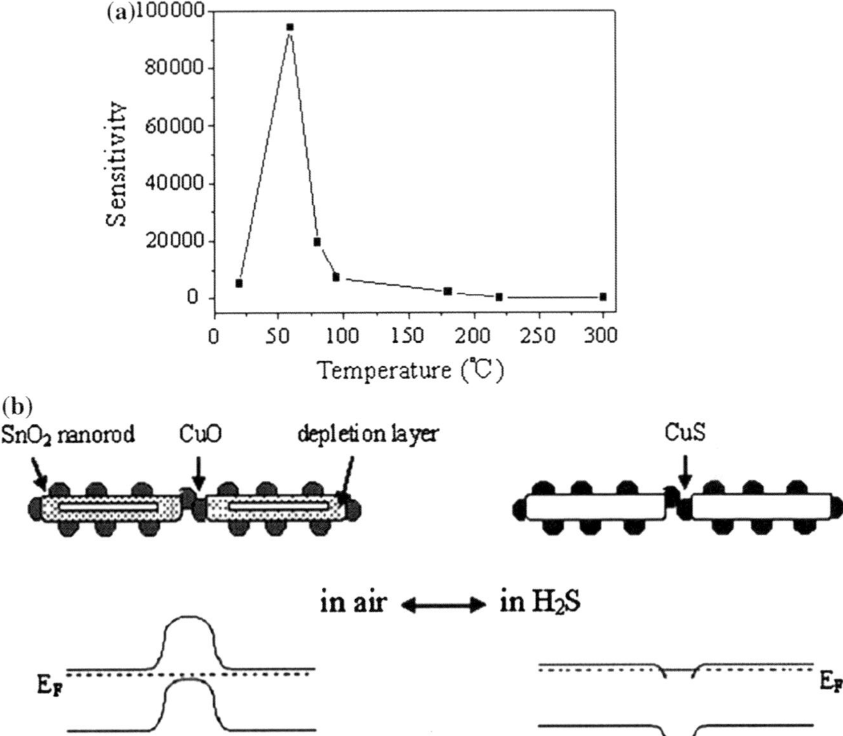

Fig. 11.7 **a** Sensitivity of the sensor at different temperatures to 10 ppm H$_2$S, **b** energy band diagram illustration of the CuO-SnO$_2$ PN junction sensor to H$_2$S gas (Reprinted with permission from [29], Copyright 2008, American Chemical Society)

ethanol, SO$_2$, and H$_2$ had been investigated, and the sensitivity toward these gases was found to be more than 1 order of magnitude smaller at room temperature. Such good selectivity toward H$_2$S can be used to avoid the interference of other gases.

The good performance in sensitivity and selectivity could be explained by the presence of the p–n junction and small size effect. *P-type* CuO nanoparticles were coated on *n*- type SnO$_2$ nanorods uniformly along the rods, forming the network of *p–n* junctions at the surface of the nanorod as shown in Fig. 11.7b [29]. When the sensor was in air, the barriers at the junctions effectively blocked the electrons flowing through the nanorods, resulting in a very low conductance. On the other hand, when the sensor was exposed to H$_2$S gas, the CuO nanoparticles reacted with H$_2$S following the chemical reaction:

$$CuO + H_2S \rightarrow CuS + H_2O \tag{11.3}$$

CuS was reported to have relatively high metal-like conductivity. Upon the reaction (11.3) *p–n* barriers disappeared. Since the work function of CuS is lower than that of SnO$_2$ as shown in Fig. 11.7b, the conductance of the sensing material become greatly

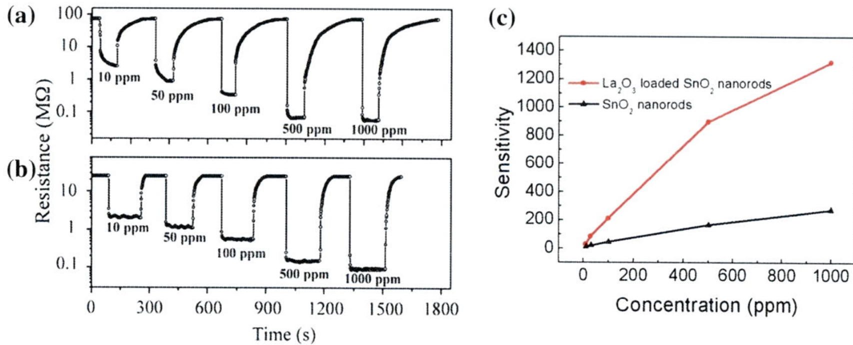

Fig. 11.8 Real-time resistance of the SnO_2 nanorod sensors **a** with and **b** without La_2O_3 loading to different concentrations of ethanolethanol (Reprinted with permission from [30], Copyright 2009, Elsevier). **c** Sensitivity of the sensor at different ethanol concentrations with and without La_2O_3 loading

enhanced. In addition to the reduction reaction with CuO, H_2S being a reducing agent, also reduced pre-adsorbed oxygen that increased the conductance further. Taking into consideration the comparability of the diameter of SnO_2 nanorods to the depletion width one can assume the conduction channel inside the nanorods could be very narrow and the resistance in air could be very large. As a result, a large resistance change could be observed upon exposure of the sensor to H_2S, which implied a super high sensitivity as large as 10^4. Moreover, the reaction in (11.3) was effective for H_2S gas but not for H_2, ethanol or SO_2, therefore a high selectivity to H_2S was achieved.

11.5.4 La$_2$O$_3$ Loaded SnO$_2$ Nanorods

Another example of the surface functionalization of nanowire sensors is SnO_2 nanorods loaded with La_2O_3. SnO_2 nanorods with diameters about 5–20 nm and length about 100–200 nm were synthesized by a hydrothermal method [30]. After dropping $La(NO_3)_3 \cdot 6H_2O$ into SnO_2 nanorod solutions, stirring for 4 h, and annealing the materials at 800 °C for 3 h, $La(NO_3)_3$ was converted into La_2O_3, and the La_2O_3 loaded SnO_2 nanorods were obtained.

The response of SnO_2 nanorod sensors with/without 5 % weight La_2O_3 towards ethanol gas at a working temperature of 200 °C is shown in Fig. 11.8a, b, respectively, and the sensitivity vs. concentration was shown in Fig. 11.8c. The sensitivity was 213 to 100 ppm ethanol. As the temperature increased, the sensitivity became smaller due to the decrease of resistance in air with increasing temperature, closely related to a thermally activation mechanism.

The high performance of the sensor can be interpreted by taking into account the surface chemical properties of these two oxides. Generally the sensing mechanism is based on the change of resistance when exposed to ethanol that reacts with the adsorbed oxygen species followed reaction (11.2) discussed above. The ethanol molecules become oxidized to CO_2 and H_2O. The reaction transfers electrons back to the conduction band, and as a result the resistance of the sensors decreases. Interactions of ethanol molecules with the oxide surfaces are quite complicated and the sensor signal depends both on the density and nature of surface active centers. Depending on the surface chemical conditions of oxides, the ethanol molecules can convert to CO_2 and H_2O through two kinds of reactions: (1) dehydration into C_2H_4 in the presence of acid surface groups as shown in (11.4), and (2) dehydrogenation into acetaldehyde (CH_3CHO) in the presence of basic surface groups as given in (11.5). The intermediates (C_2H_4 and CH_3CHO) then react with adsorbed oxygen ions and are turned into CO_2 and H_2O followed as (11.6) and (11.7), respectively:

$$C_2H_5OH \xrightarrow{acid-surface} C_2H_4 + H_2O \tag{11.4}$$

$$C_2H_5OH \xrightarrow{basic-surface} CH_3CHO + H_2 \tag{11.5}$$

$$C_2H_4 + 6O^- \rightarrow 2CO_2 + 2H_2O + 6e^- \tag{11.6}$$

$$CH_3CHO + 3O^- \rightarrow 2CO_2 + 2H_2O + 3e^- \tag{11.7}$$

The surface of the SnO_2 nanorods contains many acidic centers such as Brønsted and Lewis type [31], and ethanol molecules mainly decompose to C_2H_4 intermediates. On the other hand, La_2O_3 is a typical basic oxide [31]. Therefore, the presence of La_2O_3 at the surface of SnO_2 will reduce the amount of the acidic sites. The latter will result in the formation of Lewis acid–basic pairs, which lead to a preferred dehydrogenation process, and more ethanol molecules will convert to CH_3CHO. From the thermodynamic point of view, reaction (11.5) is more favorable than reaction (11.4) under the same conditions [32, 33], and this is the reason why SnO_2 shows a higher response to CH_3CHO than C_2H_4 with the same concentration and the La_2O_3 loading enhances the sensitivity of the SnO_2 nanorod sensor.

11.6 Doping in Oxide Nanowires and Nanowires with Multi-Compositions

Noble metal doping has a similar effect on the sensing properties to surface modification by noble metal coating that has been discussed before. The conductivity of the nanowires can be modulated by controlling the level of the doping metal, and the surface reaction path could be also modified. It is very important to control the doping level in the metal oxide nanowires during the process of their synthesis to avoid a second phase formation. To enhance the certain property of the sensor, doping with appropriate elements and levels is very promising strategy.

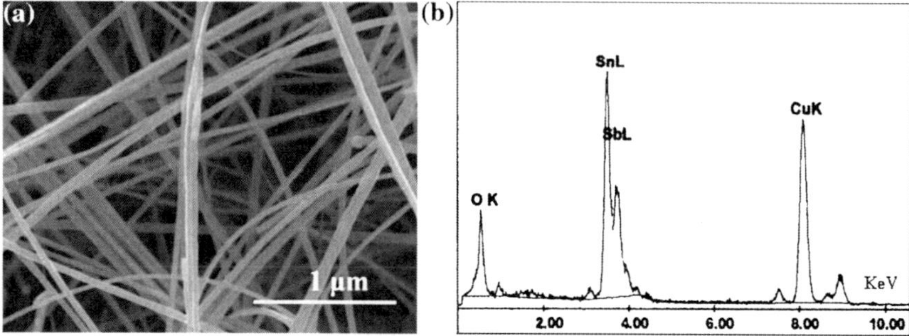

Fig. 11.9 **a** SEM image of the Sb doped SnO_2 nanowires, **b** EDS of the Sb doped SnO_2 nanowires, which inidicates the sample is composed of Sn, O and Sb (the peaks of Cu come from the Cu grids) (Reprinted with permission from [34], Copyright 2005, The Royal Society of Chemistry)

11.6.1 Sb Doped SnO₂ Nanowires

Sb doped SnO_2 nanowires were synthesized by thermal evaporation of Sn and Sb powders (the weight ratio 10:1) at 900 °C in a constant flow of 1 % oxygen and 99 % nitrogen at a rate of 5 l/min [34]. Energy dispersive X-ray spectra (EDS) revealed that the synthesized nanowires were composed of Sn, O and Sb and the atomic percentage Sb in the synthesized sample was about 3.5 %. The SEM and EDS images were shown in Fig. 11.9a, b, respectively.

The sensitivity of Sb-doped SnO_2 nanowire sensors was about 1.76 upon 10 ppm ethanol gas. The response and recovery time to 10 ppm ethanol were about 1 s and 5 s, respectively.

Sb metal doping in SnO_2 nanowires could decrease the sensor resistance in air and shorten the response and recovery time of the sensor. Presumably Sb doping accelerates the ionosorption of oxygen on the surface of the SnO_2 nanowires, which has a great significance to reduce the recovery times of the sensor.

11.6.2 Indium-Doped Tin Oxide (ITO) Nanowires and ZnSnO₃ Nanowires

Besides modifying the materials from accustomed binary metal oxide nanowires, new compounds composed of multiple elements such as ternary oxides have been also explored to test their sensing properties. Indium-doped tin oxide (ITO) and ZnO-SnO_2 compounds have a great potential for flexible transparent electronics, solar cells, light emitting diodes and etc. Here we briefly review their gas sensing properties.

ITO nanowires were synthesized by a thermal evaporation of In_2O_3, SnO, and graphite powders (weight ratio of 4:1:4) at 930 °C for 2 h in argon and oxygen mixed

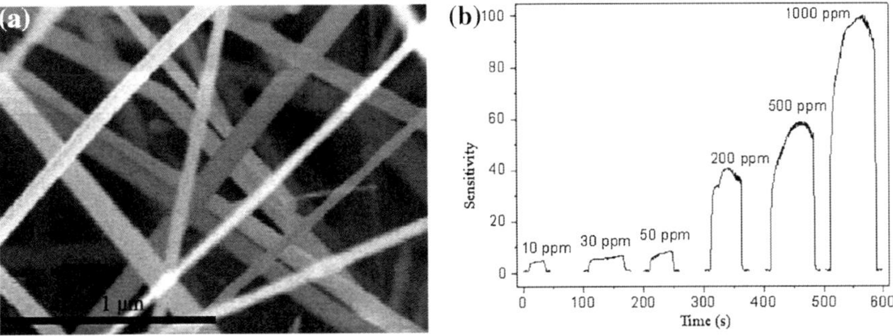

Fig. 11.10 **a** SEM image of the ITO nanowires, **b** Response of the sensor to different ethanol concentrations, insert: sketch of the nanowire in air and in ethanol (Reprinted with permission from [35], Copyright 2006, American Institute of Physics)

gases [35]. The ITO nanowires had diameters ranging from 70–150 nm and lengths of several tens of micrometers as shown in Fig. 11.10a. EDS spectrum indicated atomic ratio (In, Sn, and O) of the ITO nanowires was about 5.5: 31.6: 62.9.

The gas sensors based on ITO nanowires were very sensitive to ethanol gas as shown in Fig. 11.10b, and the sensitivity was up to 40 upon 200 ppm ethanol at 400 °C. Both the response and recovery time were less than 2 s.

$ZnSnO_3$ nanowires were synthesized by a thermal evaporation of ZnO, SnO and graphite mixture powders (weight ratio of 1:5:1) [36]. The mixture was heated to 700 °C as the pressure was kept below 4 Pa. Then a mixed gas of oxygen and argon was introduced into the tube furnace till the pressure became 1,000 Pa and the tube furnace was sealed. The mixture was heated at 990 °C for 2 h. The nanowire product was collected and analysis showed that they had a composition of $ZnSnO_3$. These nanowires had diameters ranging from 20 to 90 nm and lengths up to several tens of micrometers. EDS spectrum showed that three elements (Zn, Sn, and O) uniformly distribute over the whole nanowires, respectively.

Gas sensors based on $ZnSnO_3$ nanowires had a quick response to ethanol. The real-time response to different concentrations of ethanol at 300 °C was given in Fig. 11.11. Both response and the recovery time were about 1 s to 1 ppm ethanol. The sensitivity was about 2.7 to 1 ppm ethanol and 42 to 500 ppm ethanol, respectively.

$ZnSnO_3$ nanowires and ITO nanowires had been suggested to have higher conductivity than SnO_2 and higher sensitivity to ethanol gas than bulk ZnO, SnO_2 and Zn_2SnO_4 [37, 38]. They were potential candidates for high performance sensors and the sensing properties could also be improved by coating or doping methods in the future.

11.7 Summary

A series of metal oxide nanostructures with surface functionalization, bulk doping, and with complex morphologies or compositions are discussed and their sensing

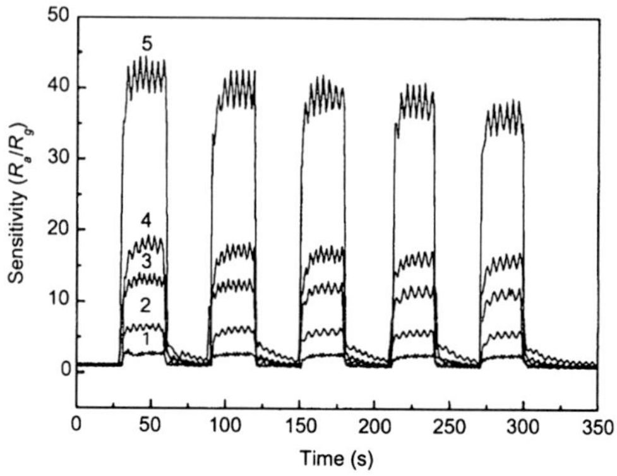

Fig. 11.11 The response of ZnSnO$_3$ nanowire sensors to ethanol at 300 °C, the curves labeled by *1–5* indicating the ethanol concentrations: 1, 10, 50, 100 and 500 ppm, respectively (Reprinted with permission from [36], Copyright 2006, American Institute of Physics)

properties along with sensing mechanisms are briefly reviewed. The nanostructure synthesis method is very simple and easy to be reproduced. It has been shown that combining the merits of functionalization, doping and morphology tuning with energy band engineering, the sensing performance of nanowires can be greatly enhanced. Among the reviewed sensors we demonstrate that the response time can be less than 1 s. The sensors exhibit high sensitivity, especially in case of SnO$_2$ nanorods loaded with La$_2$O$_3$ which demonstrate a sensitivity of 213 to 100 ppm ethanol even at a low working temperature of 200 °C. Alternatively very high sensitivity to H$_2$S has been observed for CuO-SnO$_2$ nanostructures which form *p–n* junctions. The selectivity of such sensors is also largely improved. The results indicate a number of guidelines to improve the sensor properties.

1. The hierarchical nanostructures possess a stable 3D morphology, and therefore provide a large amount of open space for facile gas diffusion, adsorption and desorption. In addition, the small dimension of nanostructured branches is favorable for high sensitivity. Therefore a stable, quick response and highly sensitive sensor can be achieved.
2. Nanoparticle coatings on the surfaces of quasi-one dimensional materials, nanowires/nanotubes/nanorods, provide a stable support for nanoparticles to prevent agglomeration that is usually the main reason for sensor deterioration with aging. In addition, the small nanoparticle size offers large surface-to-volume ratio and enhanced catalytic action which accounts for the high sensitivity and fast time response. Besides, noble metal doping can also performs as an effective catalyst and improve the sensor performance.
3. By choosing proper material combinations, the energy band, e.g. p–n junction, heterojunction can be engineered appropriately. All these can be used to enhance the sensitivity and selectivity of the sensors.
4. Rare earth oxides can "navigate" the intermediate reaction to a thermodynamically favorable direction that can enhance the rate of conversion of reducing gas and enhance the sensitivity of the gas sensor.

Acknowledgments The authors gratefully acknowledge the support from "973" National Key Basic Research Program of China (Grant No. 2007CB310500), National Natural Science Foundation of China (Grant No. 21003041), and Hunan Provincial Natural Science Foundation of China (Grant No.10JJ1011).

References

1. Huang XJ, Choi YK (2007) Chemical sensors based on nanostructured materials. Sens Actuators B 122:659–671
2. Comini E, Baratto C, Faglia G, Ferroni M, Vomiero A, Serveglieri G (2008) Quasi-one dimensional metal oxide semiconductors: preparation, characterization and application as chemical sensors. Prog Mater Sci 54:1–67
3. Comini E, Sberveglieri G (2010) Metal oxide nano wires as chemical sensors. Mater Today 13:36–44
4. Fraden J (2004) Handbook of modern sensors: physics, designs, and applications. Springer, New York
5. Wan Q, Li QH, Chen YJ, Wang TH (2004) Fabrication and ethanol sensing characteristics of ZnO nano wire gas sensors. Appl Phys Lett 84:3654–3656
6. Feng P, Wan Q, Wang TH (2005) Contact-controlled sensing properties of flowerlike ZnO nanostructures. Appl Phys Lett 87:213111
7. Feng P, Xue XY, Liu YG, Wang TH (2006) Highly sensitive ethanol sensors based on {100}-bounded In_2O_3 nanocrystals due to face contact. Appl Phys Lett 89:243514
8. Li CC, Du ZF, Li LM, Yu HC, Wan Q, Wang TH (2007) Surface-depletion controlled gas sensing of ZnO nano rods grown at room temperature. Appl Phys Lett 91:032101
9. Xu CN, Tamaki J, Miura N, Yamazoe N (1991) Grain size effects on gas sensitivity of porous SnO_2-based elements. Sens Actuators B 3:147–155
10. Hongsith N, Wongrat E, Kerdcharoen T, Choopun S (2010) Sensor response formula for sensor based on ZnO nanostructures. Sens Actuators B 144:67–72
11. Göpel W, Schierbaum KD (1995) SnO_2 sensors: current status and future prospects. Sens Actuators B 26:1–12
12. Yamazoe N (2005) Toward innovations of gas sensor technology. Sens Actuators B 108:2–14
13. Tien LC, Sadik PW, Norton DP, Voss LF, Pearton SJ, Wang HT, Kang BS, Ren F, Jun J, Lin J (2005) Hydrogen sensing at room temperature with Pt-coated ZnO thin films and nanorods. Appl Phys Lett 87:222106
14. Kolmakov A, Klenov DO, Lilach Y, Stemmer S, Moskovits M (2005) Enhanced gas sensing by individual SnO_2 nanowires and nanobelts functionalized with Pd catalyst particles. Nano Lett 5(4):667–673
15. Shen YB, Yamazaki T, Liu ZF, Meng D, Kikuta T (2009) Hydrogen sensors made of undoped and Pt-doped SnO_2 nanowires. J Alloy Compd 488:21–25
16. Liang YX, Chen YJ, Wang TH (2004) Low-resistance gas sensors fabricated from multiwalled carbon nanotubes coated with a thin tin oxide layer. Appl Phys Lett 85:666–668
17. Ago H, Kugler T, Cacialli F, Salaneck WR, Shaffer MSP, Windle AH, Friend RH (1999) Work functions and surface functional groups of multiwall carbon nanotubes. J Phys Chem B 103:8116
18. Sze SM (1981) Physics of semiconductor devices. Wiley, New York
19. Ago H, Kugler T, Cacialli F, Salaneck WR, Shaffer MSP, Windle AH, Friend RH (1999) Work function and surface functional groups of multiwall carbon nanotubes. J Phys Chem B 103:8116–8121
20. Sinner-Hettenbach M, Gothelid M, Wei T, Barsan N, Weimar U, Schenck HV, Giovanelli L, Lay GL (2002) Electronic structure of $SnO_2(110)$-4 × 1 and sputtered $SnO_2(110)$ revealed by resonant photoemission. Surf Sci 499:85–93

21. Korotchenkov G (2005) Gas response control through structural and chemical modification of metal oxide films: state of the art and approaches. Sens Actuators B 107:209–232
22. Lee JH (2009) Gas sensors using hierarchical and hollow oxide nanostructures overview. Sens. Actuators B 140:319–336
23. Wan Q, Huang J, Xie Z, Wang TH, Dattoli EN, Lu W (2008) Branched SnO_2 nanowires on metallic nanowire backbones for ethanol sensors application. Appl Phys Lett 92:102101
24. Li CC, Li LM, Du ZF, Yu HC, Xiang YY, Li Y, Cai Y, Wang TH (2008) Rapid and ultrahigh ethanol sensing based on Au-coated ZnO nanorods. Nanotechnology 19:035501
25. Qian LH, Wang K, Li Y, Fang HT, Lu QH, Ma XL (2006) CO sensor based on Au-decorated SnO_2 nanobelt. Mater Chem Phys 100:82–84
26. Hongsith N, Viriyaworasakul C, Mangkorntong P, Mangkorntong N, Choopun S (2008) Ethanol sensor based on ZnO and Au-doped ZnO nanowires. Ceram Int 34:823–826
27. Liu XH, Zhang J, Wang LW, Yang TL, Guo XZ, Wu SH, Wang SR (2011) 3D hierarchically porous ZnO structures and their functionalization by Au nanoparticles for gas sensorsgas sensors. J Mater Chem 21:349–356
28. Gao T, Wang TH (2004) Sonochemical synthesis of SnO_2 nanobelt/CdS nanoparticle core/shell heterostructures. Chem Commun 22:2558–2559
29. Xue XY, Xing LL, Chen YJ, Shi SL, Wang YG, Wang TH (2008) Synthesis and H_2S sensing properties of CuO-SnO_2 core/shell PN-junction nanorods. J Phys Chem C 112:12157–12160
30. Shi SL, Liu YG, Chen YG, Zhang JY, Wang YG, Wang TH (2009) Ultrahigh ethanol response of SnO_2 nanorods at low working temperature arising from La_2O_3 loading. Sens Actuators B 140:426–431
31. Kovalenko VV, Zhukova AA, Rumyantseva MN, Gaskov AM, Yushchenko VV, Ivanova II, Pagnier T (2007) Surface chemistry of nanocrystalline SnO_2 effect of thermal treatment and additives. Sens Actuators B 126:52–55
32. Seiyama T, Shiokawa J, Suzuki S, Fueki K (1982) Kagaku sensa. Kodansha, Tokyo
33. Rumyantseva M, Kovalenko V, Gaskov A, Makshina E, Yuschenko V, Ivanova I, Ponzoni A, Faglia G, Comini E (2006) Nanocomposites SnO_2/Fe_2O_3: sensor and catalytic properties. Sens Actuators B 118:208–214
34. Wan Q, Wang TH (2005) Single-crystalline Sb-doped SnO_2 nanowires: synthesis and gas sensor application. Chem Commun 30:3841–3843
35. Xue XY, Chen YJ, Liu YG, Shi SL, Wang YG, Wang TH (2006) Synthesis and ethanol sensing properties of indium-doped tin oxide nanowires. Appl Phys Lett 88:201907
36. Xue XY, Chen YJ, Wang YG, Wang TH (2005) Synthesis and ethanol sensing properties of $ZnSnO_3$ nanowires. Appl Phys Lett 86:233101
37. Shen YS, Zhang TS (1993) Preparation, structure and gas-sensing properties of ultramicro $ZnSnO_3$. Sens Actuators B 12:5–9
38. Wu XH, Wang YD, Tian ZH, Liu HL, Zhou ZL, Li YF (2002) Study on $ZnSnO_3$ sensitive material based on combustible gases. Solid-State Electron 46:715–719
39. Chen YJ, Xue XY, Wang YG, Wang TH (2005) Synthesis and ethanol sensing characteristics of single crystal single crystalline SnO_2 nanorods. Appl Phys Lett 87:233503
40. Chen YJ, Nie L, Xue XY, Wang YG, Wang TH (2006) Linear ethanol sensing of SnO_2 nanorods with extremely high sensitivity. Appl Phys Lett 88:083105
41. Yang Z, Li LM, Wan Q, Liu QH, Wang TH (2008) High-performance ethanol sensing based on an aligned assembly of ZnO nanorods. Sens Actuators B 135:57–60
42. Gao T, Li QH, Wang TH (2005) Sonochemical synthesis, optical propertiesoptical properties, and electrical properties of core/shell-type ZnO nanorod/CdSCdS nanoparticle composites. Chem Mater 17:887–892
43. Li LM, Du ZF, Wang TH (2010) Enhanced sensing properties of defect-controlled ZnO nanotetrapods arising from aluminum dopingdoping. Sens Actuators B 147:165–169
44. Li LM, Li CC, Zhang J, Du ZF, Zou BS, Yu HC, Wang YG, Wang TH (2007) Bandgap narrowing and ethanol sensing properties of In-doped ZnO nanowires. Nanotechnology 18:225504

Chapter 12
Optical Sensing Methods for Metal Oxide Nanomaterials

Nicholas A. Joy and Michael A. Carpenter

Abstract Optical analysis of metal oxides as a means of transduction in sensing devices provides an alternative method for the detection of target analytes. The optical properties of metal oxides are rich with opportunity as luminescence, dielectric function changes and optically active dopants are all sensitive to environmental changes and can be utilized in a sensing device. A detailed description of the latest work on the photoluminescence of metal oxides will be provided, with a focus on ZnO, SnO_2 and TiO_2 nanoscale thin films as well as nanorods. Changes in the dielectric function of perovskite metal oxide coated long period fiber gratings has been shown to be a convenient sensing device and results pertaining to these metal oxides containing alkaline earth metals and their use in the detection of CO_2 will be highlighted. Optically active dopants, pertaining specifically to metal nanoparticles such as Au, Cu and Ag, can serve as beacons for reactions and environmental changes within metal oxide nanocomposite thin films through the interrogation of changes in their plasmonic properties. In particular sensing applications within harsh conditions, including elevated temperature and either oxidizing or reduction gas environments will be detailed. Each of these optical techniques will be reviewed and any key environmental, film deposition methods or reaction conditions will be highlighted in light of their potential effects on sensor detection limits, sensitivity or selectivity characteristics.

N. A. Joy · M. A. Carpenter (✉)
College of Nanoscale Science and Engineering, University at Albany-SUNY, Albany,
NY 12203, USA
e-mail: mcarpenter@albany.edu

M. A. Carpenter et al. (eds.), *Metal Oxide Nanomaterials for Chemical Sensors*,
Integrated Analytical Systems, DOI: 10.1007/978-1-4614-5395-6_12,
© Springer Science+Business Media New York 2013

12.1 Introduction

The development of chemical sensors has been an active field of research for several decades. In terms of the choice of transduction methods there are electrical, acoustical and optical based chemical sensors. Optical transduction has several advantages over the other classes of sensing techniques. For one, the chemically sensitive component of the sensor is simply an optically active material that needs to be compatible with the sensing environment conditions. Without the need for electrical contacts, maintaining reliability is in many ways easier in a variety of operating conditions which can range from ambient temperatures to temperatures as high as 1,000 °C with either a reducing or an oxidizing environment. Maintaining an intrinsically safe sensing environment is also simpler as there are no electrically active components within the sensing environment that would require specially designed packaging or controlled operating conditions to ensure that an electrical short or a spark discharge is not possible in an explosive environment. While there are many advantages to an optical based sensing device, the use of metal oxide based sensing materials provide a wealth of chemistry, shape, and surface structure which can be tailored to optimize the detection limits, sensitivity and selectivity characteristics. In this chapter optical sensing methods based on photoluminescence, refractive index changes, and plasmonic based techniques will be highlighted. A variety of metal oxides are compatible with these optical techniques, however, the focus of this chapter will be on ZnO, TiO_2, SnO_2, and alkaline earth metal doped perovskites. The plasmonic based sensing schemes require noble metal dopants such as Au, Ag or Cu nanoparticles. This section of the chapter will focus on Cu and Au nanoparticle dopants. In particular these nanoparticles within yttia-stabilized zirconia (YSZ), TiO_2, and CeO_2 will be detailed. With such a rich chemistry and microstructure available for use as individual sensing elements it is natural to construct sensing arrays to allow for selective measurements to be made in complex sensing environments. A brief outline of the use of principal component analysis for the selective characterization and detection of gases will be provided.

12.2 Photoluminescence

Among the optical methods for probing the sensing characteristics of metal oxide nanomaterials, one is grounded in the photoluminescence properties of metal oxides. Whether wide band gap semiconducting materials like SnO_2, ZnO, TiO_2, or CeO_2 or an insulating metal oxide like ZrO_2, the photoluminescence properties of these materials is accessible with excitation photons ranging between the Uv and visible range.

Photoluminescence within bulk metal oxides is attributed to transitions between the conduction band and the valence band. Defects and impurities within the metal oxide lead to a number of states that are below the conduction band. Thus, the characteristic emission properties of the metal oxide have both intrinsic and extrinsic features. The intrinsic emission spectra are typically structure-less and arise from transitions between the conduction and valence bands. Extrinsic emission bands arise due to transitions between defect states (*i.e.* oxygen vacancy states) below the conduction band and the valence band and are thus red shifted in wavelength from the intrinsic bands [1–3]. For nanoscale metal oxides the photoluminescence properties can have similar emission characteristics as the bulk material, explained by intrinsic and extrinsic bands, or emission governed by quantum confinement. Quantum confined photoluminescence requires the metal oxide to have dimensions that are of similar magnitude or smaller than the exciton's Bohr radius. This size dependence is characteristic of the metal oxide but is typically on the order of several nanometers, with metal oxides such as ZnO and SnO_2 having Bohr radii of 3 and 2.7 nm respectively. At sizes larger than this a number of surface and defect modified emission characteristics can be realized until the metal oxide begins to show bulk emission properties. In the text that follows, utilization of the photoluminescence of metal oxides, specifically for ZnO and SnO_2 will be detailed. Specific attention will be given with respect to a characterization of the metal oxide emission properties and its corresponding sensing properties.

12.3 ZnO

ZnO is a wide bandgap, 3.4 eV, semiconductor with an exciton binding energy of 60 meV at room temperature. This has led to its recent interest as a Uv optical device. The emission properties of ZnO are characterized by superposition of a strong near band-edge photoluminescence peak as well as a broad band of photoluminescence in the visible range, typically around 2.5 eV (500 nm). The visible photoluminescence is generally attributed to defects, while the Uv emission is characterized by the crystalline regularity, thus these processes are competing phenomena. While the Uv emission has been sought to be opitimized for Uv optical devices, the visible fluorescence has been primarily explored for sensing applications as well as for visible light phosphors. Lattice defects such as oxygen vacancies and interstitial zinc cause ZnO to have an intrinsic n-doping. A deeper understanding of these lattice defects and their role in the generation of visible photoluminescence is an active area of study both for the development of optical based chemical sensors as well as the optimization of visible phosphors [2, 3].

Upon exposure to oxidizing gases that bind on the ZnO surface, near surface electron traps are formed which decreases the number of charge carriers for radiative recombination, thus leading to a decrease in the visible photoluminescence [2, 3]. It is interesting to note that the detailed characteristics of the ZnO green emission is dependent on the preparation and type of nanomaterial. Specifically, a series of films grown by pulsed laser deposition (PLD) with the substrate temperature varied between 500 and 700 °C as well as the background oxygen pressure ranging between 1 and 100 Pa resulted in both smooth and rough films as well as nanostructured thin films of ZnO [2]. Such a change in surface structure and a subsequent increase in surface area is generally beneficial from a sensing standpoint as there are more sites available for interaction. Upon exposure to NO_2 a series of surface reactions can take place which include:

$$NO_2(gas) + e^- \rightarrow NO_2^-(ads.) \tag{12.1}$$

$$NO_2(gas) + e^- \rightarrow NO(gas) + O^-(ads.) \tag{12.2}$$

$$NO(gas) + e^- \rightarrow NO^-(ads.) \tag{12.3}$$

$$NO(gas) + e^- \rightarrow \frac{1}{2}N(gas) + O^-(ads.) \tag{12.4}$$

Each of the above reactions would involve the trapping of a ZnO electron, thus creating a decrease in the number of charge carriers, resulting in a decrease in the observed photoluminescence upon exposure to NO_2. While each of the films produced a response towards room temperature exposure to NO_2 at concentrations ranging from 11 to 114 ppm in a dry air mixture, as can be seen in Fig. 12.1, the sensing response decreased with an increase in concentration. This was attributed to the irreversible binding of NO_2 related adsorbed species producing a saturated response. The microstructure of each of the films was characterized, using XRD, as being oriented along the c-axis of the Wurtzite structure. The film used for Fig. 12.1e, generally characterized with nanopillars (or pencils), also had significant contributions from other crystallographic planes, including (0 0 2) and (0 0 4). Further inspection of Fig. 12.1e shows that this film has an overall larger response towards NO_2 which may be attributed to it having a larger number of active adsorption sites, however this film also showed a steady saturation in response towards increasing $[NO_2]$ which again is attributed to irreversible adsorption. Heating each of these films to 50 °C was not enough to restore the original sensing characteristics and is evidence of binding chemistries resulting from a strong interaction upon surface adsorption.

Nanowires of ZnO can also be produced by means of vapor transport processes in which the precursor materials are vaporized into a carrier gas stream and deposited on a substrate located in a cooler portion of a furnace. Commonly a catalyst material, such as gold, is deposited on the substrate and a vapor liquid solid growth mechanism leads to nanowire growth. The recent work of Maddalena et al. have characterized both the optical properties and the sensing properties of ZnO nanowires produced in this fashion [3]. Typical SEM images of ZnO nanowires and their PL spectrum can be seen in Fig. 12.2a and b. The Uv and visible peaks are assigned to the near band edge and defect state emissions respectively. Exposure to NO_2 in a room temperature dry air mixture with concentrations ranging from 0.1 to 0.5 ppm, as shown in Fig. 12.3, shows a repeatable and reversible response with response and recovery times on the order of 10 and 8 min

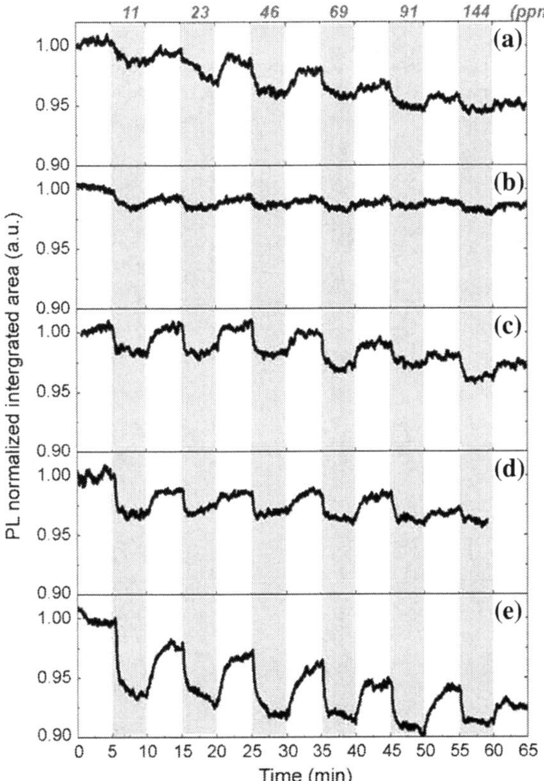

Fig. 12.1 Room temperature NO$_2$ sensing measurements using PLD grown ZnO samples. Reprinted with permission from [2]

respectively. The calibration curve of the ZnO nanowire response towards NO$_2$ followed a Langmuir isotherm with no signs of saturation for exposures as high as 2 ppm. Responses of the nanowires towards interfering gases such as ethanol and humidity levels of 20, 30 and 70 % were also studied. While NO$_2$ in dry air causes a characteristic decrease in PL yield due to electron trapping, both ethanol and the added water vapors caused an increase in PL, indicative of an effective quenching of the trapping states. Time resolved PL measurements of the visible PL band, in both dry air and 30 ppm NO$_2$ showed no change in the PL decay time constant when normalized for the overall lower PL yield. This is indicative of NO$_2$ serving as a static PL quencher leading to an overall reduction in the number of radiative sites which give rise to the visible emission. As the ZnO nanowires have both the intrinsic band edge and the extrinsic defect related emission peaks, an interesting comparison to be made is how the relative intensity of these respective bands varies with the morphological characteristics of ZnO nanowires. Such a size dependent study has been completed by Narayanamurti et al. where nanowires of diameters ranging between 50 and 250 nm in radii were grown in a single vapor deposition growth process to maintain the material integrity of the various nanowire samples [4]. As seen in Fig. 12.4 the SEM images and the corresponding PL

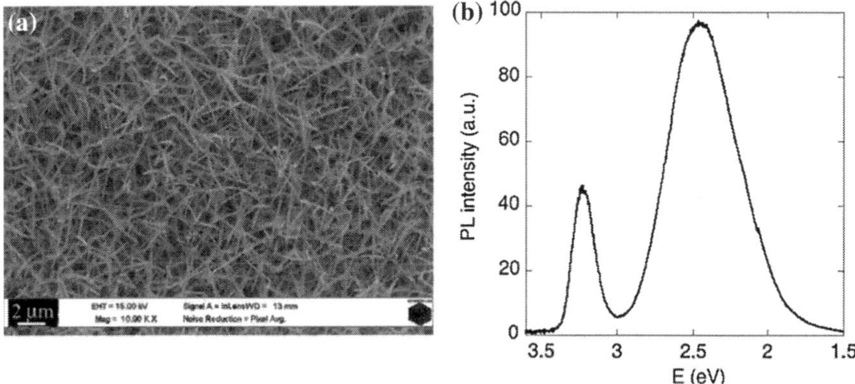

Fig. 12.2 **a** SEM image of ZnO nanowires deposited on a silicon substrate, **b** Photoluminescence spectrum of ZnO nanowires. Reprinted with permission from [3]

Fig. 12.3 Dynamic PL quenching versus time upon exposure to sub ppm concentrations of NO_2 in a dry air carrier gas. Reprinted with permission from [3]

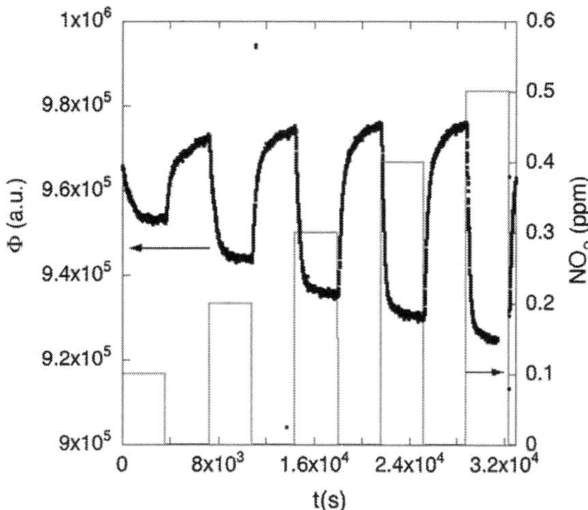

spectra show that the ratio of the band edge PL at 3.25 eV to the visible PL peaking at ~ 2.25 eV decreases with a decrease in the nanowire radius. This trend parallels the expectation that as the surface to volume ratio of the nanowire increases with a decrease in nanowire radius, the number of defects within the nanowire will also increase, leading to an increase in the visible emission peak. Figure 12.5 shows this trend more completely with wire radii decreasing from 250 to 50 nm, producing a linear decrease in the band edge to defect peak PL ratio. Since the emission peak in the visible range can be correlated with the number of defects, when dealing with nanowires of a reducing radius, an increase in the number of surface defect states leads to an increase in the visible PL. This is an interesting result as the work of Maddalena et al. [2], showed that NO_2 exposures

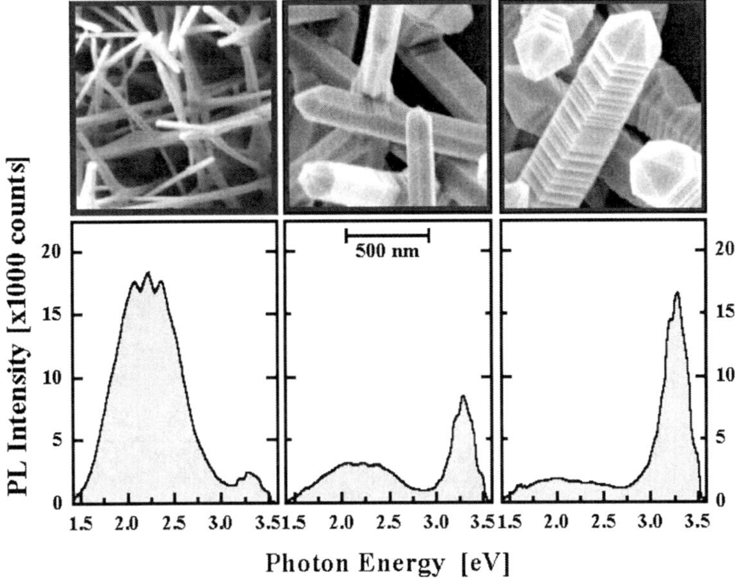

Fig. 12.4 Photoluminescence spectra obtained from ZnO wires of three different sizes as displayed in the corresponding SEM images. Reprinted with permission from [4]

decreased both the band edge and the defect emission PL intensities. Thus, a dual wavelength channel detection scheme is possible for enhanced sensitivity, however this would need to be balanced with a potential increase in cost due to hardware/software integration issues. As an alternative, if one could tailor the nanowires for production of a strong defect PL intensity within the visible range, then the NO_2 detection limits could presumably be optimized accordingly.

12.4 SnO_2

Tin oxide nanowires are routinely grown using chemical vapor deposition methods. Typically SnO is the precursor material and alumina or silica substrates are used in a tube furnace processing chamber. The precursor material is placed in the hotter zone of the furnace while the substrates are placed in cooler zones downstream from the precursor. The vaporized SnO condenses on the substrate and background oxygen levels, or added oxygen levels, lead to nanowire growth. Zappettini et al. studied the PL dependence of nanowires grown in this fashion towards exposure to NO_2, humidity, CO and NH_3 [5]. The PL spectra of the nanowires excited using a HeCd light source at 326 nm can be characterized by a broad peak centered at around 600 nm. The bandgap of SnO_2 is 3.6 eV (344 nm) and both band edge emission ~ 340 nm and defect related emission in the visible

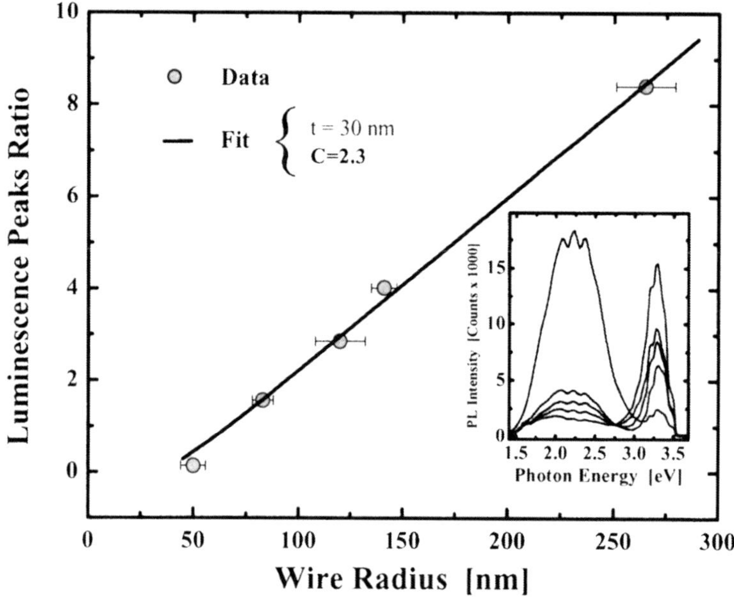

Fig. 12.5 Photoluminescence peak intensity ratios as a function of the average wire radii. Reprinted with permission from [3]

wavelengths are commonly observed. As with the case of ZnO, the visible PL emission is dominated by deep defect states and upon exposure to oxidizing gases it is expected that these will induce electron trapping states leading to a reduction in the observed visible PL. The exposure of these SnO_2 nanowires to NO_2 in dry air as well as humidified air (30 and 70 % RH) while the nanowires are held at 120 °C, produced a reversible response with regards to the change in the PL emission centered at ~ 600 nm as seen in Fig. 12.6. Also noted from this figure is that the added humidity had no apparent effect on the PL response and furthermore the response times were on the order of 30 s with a recovery time of ~ 600 s. The calibration curve representing the PL response as a function of NO_2 concentration for these nanowires was logarithmic and in this case is similar to that observed for the conductivity based SnO_2 sensors. Further experiments were done with regards to the PL response towards 1,000 ppm of CO and 50 ppm of NH_3 in dry air and there were no appreciable responses to either of these gases. As an increase in surface to volume ratio is typically beneficial to sensing applications, the synthesis of SnO_2 nanowires of various sizes as well as geometries is of particular interest for PL based sensors. Golberg et al. has synthesized SnO_2 fishbone like nano-ribbons through the use of Sn as well as $Fe(NO_3)_3$ precursors in a vapor deposition system [6]. The iron (III) nitrate serves to induce both oxidation of the Sn precursor via Fe_2O_3 which forms at the elevated temperatures, as well as inducing a variety of nucleation sites along the ribbon during the growth process, which results in the fishbone structure that is observed in Fig. 12.7. Microstructural

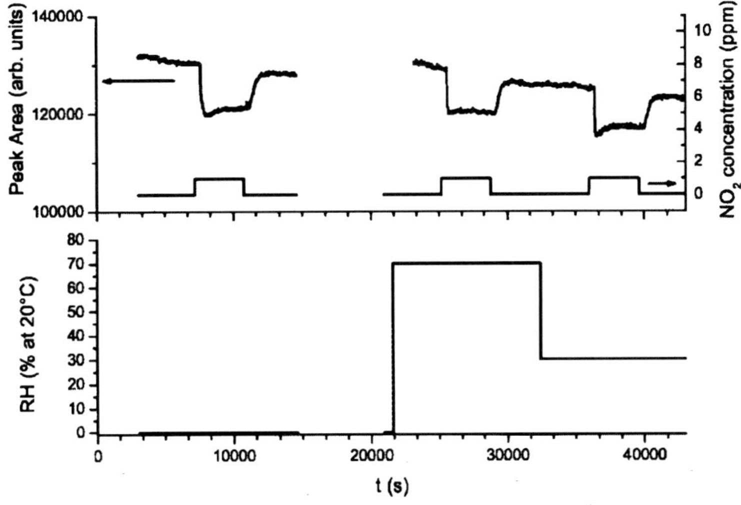

Fig. 12.6 Peak area intensity vs. time upon exposure to 1 ppm NO$_2$. Temperature is 120 °C and humidity is varied between dry air, 70 and 30 %. Reprinted with permission from [5]

analysis done on these samples indicated that the diffraction peaks could be attributed to tetragonal rutile SnO$_2$. PL emission studies of these samples have been done and a broad green emission peak is observed centered at 500 nm. This broad emission peak has been attributed to defects and while sensing studies have not been performed on this type of structure, the high surface area may prove beneficial for the detection of gases.

12.5 TiO$_2$

Titania is a wide bandgap semiconductor and has been used in a number of applications including pigments, photocatalysis, self-cleaning windows, oxidation of organic material, and is a biocompatible interface for medical implants. It has also been used in a number of sensing applications including electrochemical detectors. Titania nanopowders (grain sizes of 30–60 nm) have also been studied as PL based chemical sensors by Martinelli et al. [7]. The TiO$_2$ powders were synthesized using a sol–gel process, were screen printed onto silicon substrates and then fired at 650 °C for 1 h. PL analysis was done using a HeCd laser in the presence of dry air and NO$_2$ in dry air. Both rutile and anatase samples were produced during the process steps and the PL spectra showed a broad peak centered at ~ 2.3 eV for anatase, while a narrower peak centered in the NIR at 1.2 eV was observed for the rutile phase. While the rutile PL peak has been attributed to the ionization of oxygen vacancies, the PL peak for anatase is attributed to self trapped excitons. In this study, only the anatase phase was studied for its PL

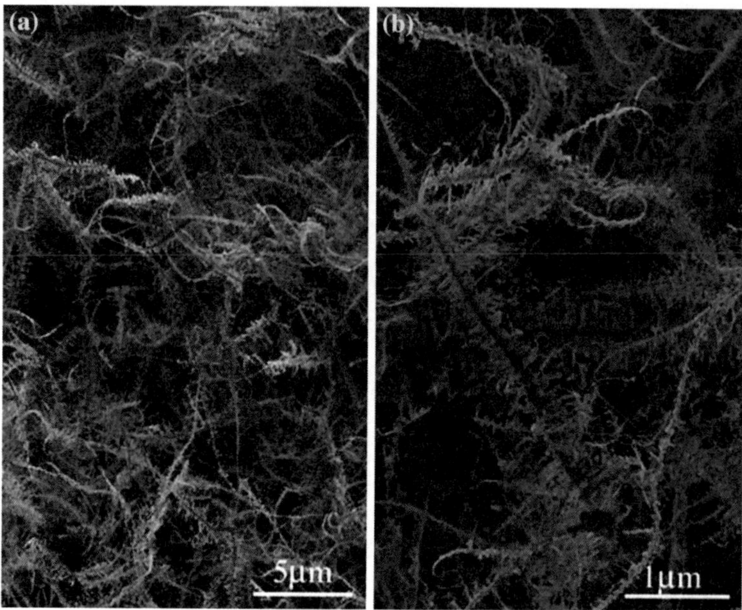

Fig. 12.7 SnO_2 fishbone-like nanoribbons **a** low magnification, **b** high magnification SEM images. Reprinted with permission from [6]

response towards exposure to NO_2. As a comparison, sol–gel synthesized SnO_2 nanopowders of a comparable grain size were also produced and studied against exposure to NO_2. As expected the SnO_2 sample showed the typical reduction in PL upon NO_2 exposure due to the enhanced trapping of charge carriers. However, the TiO_2 sample shows an increase in PL intensity upon NO_2 exposure, and thus serves as an enhancer of PL rather than a quencher. Given the nature of self trapped exciton (STE) based PL emission, the increase in electron density in the conduction band upon surface adsorption of gases increases the probability of self trapped exciton emission and thus produces a PL increase. These characteristics are evident in Fig. 12.8a which shows that upon increasing [NO_2] the peak PL intensity of anatase increases. An analysis of the PL peak position and FWHM as a function of [NO_2] ranging from 0.5 to 50 ppm, noted in Fig. 12.8b, shows that these are both essentially constant and therefore are not perturbing the nature of the STE emission properties and the sensing mechanism can be assumed stable. A Langmuir analysis was performed for the response of both the anatase and the SnO_2 samples, which matched the calibration curves reasonably well. From this a comparison can be made between these two samples if the absolute value of the relative variation in PL intensity is plotted vs. NO_2 coverage, and as can be seen from Fig. 12.9, the anatase sample is a better NO_2 sensor than the SnO_2 sample.

In conclusion the PL sensing characteristics of ZnO, SnO_2 and TiO_2 have been detailed with respect to their detection of NO_2 at concentrations as low as 0.1 ppm

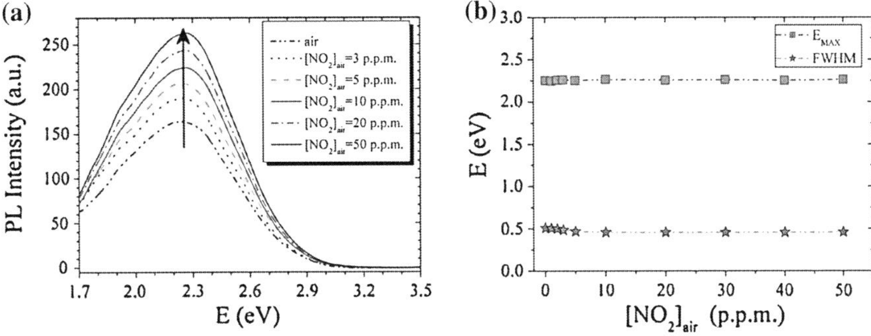

Fig. 12.8 **a** Anatase photoluminescence spectra as a function of [NO$_2$], **b** peak PL intensity and FWHM as a function of [NO$_2$]. Reprinted with permission from [7]

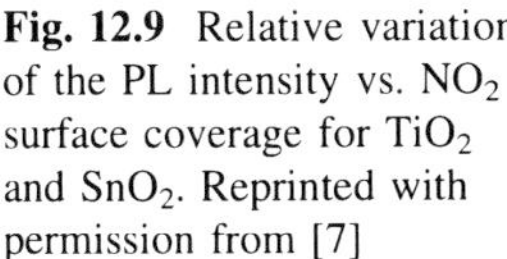

Fig. 12.9 Relative variation of the PL intensity vs. NO$_2$ surface coverage for TiO$_2$ and SnO$_2$. Reprinted with permission from [7]

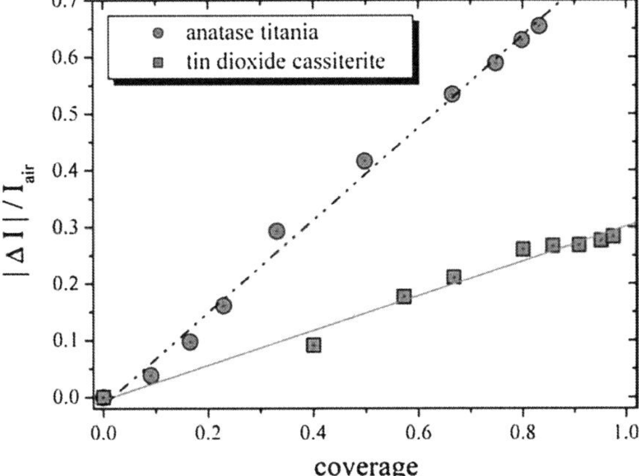

and up to 100 ppm. While the NO$_2$ sensing characteristics are quite good, especially with regards to the detection limits, less work has been done with other types of emission gases such as CO and unburnt fuel as well as the cross sensitivity with common gases such as humidity and CO$_2$. The work by Zappettini et al. [5]. did show little effects with respect to humidity, CO and NH$_3$ however more detailed work for reliability studies would be needed to completely rule these out as interfering species. Likewise the work of Maddalena et al. [1] also performed studies with ethanol and water as interferrants for the ZnO based PL detection of NO$_2$ and these initial studies showed that an increase in PL is observed, which being the opposite response with respect to NO$_2$, will require careful handling for data interpretation with mixed species. For each of these materials the response and recovery times are both somewhat limiting; on the order of tens of seconds and sometimes as high as 600 s. These are both thought to be high for many gas sensing applications and would need improvement. While the benefit of PL based sensing using metal oxide nanomaterials is that it can be used for room temperature applications, if the end-application involves their exposure to organics or particulate matter, these materials will likely become de-sensitized as the materials might become poisoned. This might be alleviated through a periodic cleaning step by heating the sample to high enough temperatures in the presence of background air and oxidizing the contaminating material. The benefits of a PL based sensing system are quite promising as demonstrated by the initial work in this field. However, further work is required on many areas to optimize this detection technique.

12.6 Long Period Fiber Gratings

Another technique which amends itself well to the optical detection of emission gases, especially at elevated temperatures, uses specialized metal oxide nano-powders which are coated onto the cladding of a long period fiber grating (LPFG). As the chemicals of interest adsorb into the metal oxide coating, the refractive index change of this sensing material induces a shift in the resonance wavelength of the LPFG indicative of the target gas concentration. Y. S. Lin et al. have made recent measurements using a Pd coated LPFG for the detection of hydrogen [8] as well as the use of perovskite metal oxides for the detection of CO_2 [9]. Perovskite metal oxides containing alkaline earth metals are susceptible to the adsorption of CO_2 and subsequent formation of carbonate species at elevated temperatures. Their recent work on perovskites has included the study of $Ba_{0.5}Sr_{0.5}Co_{0.8}Fe_{0.2}O_{3-\delta}$ (BSCF) which has shown a reversible response to CO_2 at elevated temperatures of 600 and 700 °C. The BSCF metal oxide is produced by combining the metal precursors in a combined citrate and EDTA complexing method with a final sintering step of 950 °C to form the perovskite structure [10]. The LPFG was formed using a focused CO_2 laser line and the cladding was then coated by dipping the fiber into a suspension of the BSCF powder multiple times. From SEM images of the coated fiber and XRD analysis of the powder, the metal oxide has an individual grain size which is <200 nm, with a sub monolayer coverage of 0.5–1 μm agglomerates coating the fiber. The experiments performed in this BSCF study showed that the perovskite coated LPFG is able to reversibly detect CO_2 in an air carrier gas at levels down to 1 % and up to 10 %. The response times are on the order of 8–10 min, and the BSCF coating does have a dependence on the oxygen partial pressure. A variation in O_2 levels from 0–20 % decreased the resonance wavelength shift for 0.1 atm of CO_2 by 60 pm. More work would need to be performed to determine the cross sensitivity to humidity and other common background gases, but these alkaline earth metal perovskite coated LPFGs offer another optical sensing method which is amenable to rather challenging harsh environmental conditions.

12.7 Metal Nanoparticle Doped Metal Oxides

While PL is one method to optically probe metal oxides, another technique is to perform transmission/absorption experiments. Since most metal oxides are transparent in the visible range one needs to add an "optical beacon" to the metal oxide which will provide a signal that is dependent on changes within the metal oxide that are relevant to changes in the surrounding environment. An interesting choice of beacon is one which has a strong absorption cross section in the visible range and likewise can serve as either a spectator in the sensing process or an active participant by inducing specific chemical reactions to take place. Metal nanoparticles embedded in oxide matrices have been a subject of considerable interest over the past few decades. These nanocomposite systems exhibit unique and customizable optical properties that could be exploited for a plethora of applications, such as high density information storage devices, ultrafast optical switches, and optical gas sensors [11–15]. However, the successful design of such optical devices requires the ability to customize the optical response of the nanocomposite systems.

In the case of gold (Au), the optical response of Au nanoparticles is dominated by a strong surface plasmon resonance (SPR) band (or sometimes referred as the localized surface plasmon band (LSPR)). This resonant behavior is due to the light-induced collective displacement of conduction electrons with respect to the positive ionic background in the nanoparticle, which results in a restoring force due to surface polarization [16, 17]. This collective, oscillatory motion of the conduction electrons is observed in optical extinction spectra as a selective enhancement of the absorption cross section of the nanoparticle, and is typically probed by standard absorption measurements. The frequency, width, and intensity of the surface plasmon resonance band depends on the size, size distribution, concentration, and shape of the metal nanoparticle, as well as the dielectric properties of the embedding medium. Therefore, in order to tailor the optical properties of a nanocomposite film, one needs to ensure precise control over film microstructure and identify a straightforward method for its assessment.

The optical properties of metal nanoparticles have also been studied exhaustively, yet active research is still being pursued in the development of new applications and theoretical models making use of and probing their optical properties [18–21]. One area of active research taking advantage of the optical properties of metal nanoparticles is optical sensing [22–25]. Most noble metal nanoparticles exhibit a surface plasmon resonance band in the visible spectrum and this LSPR band can be very sensitive to changes in the surrounding medium, which makes metal nanoparticles ideal optical beacons for transduction events. Likewise they can serve as either a spectator in the sensing process or an active participant by inducing specific chemical reactions to take place. Furthermore, by shrinking the size of the metal nanoparticle it is also likely to change not only the nanoparticle's optical properties but also its catalytic properties. The increase in the surface to volume ratio allows one to use less material to receive the same, if not better, catalytic behavior as bulk samples. Rodríguez et al. [26] have recently published

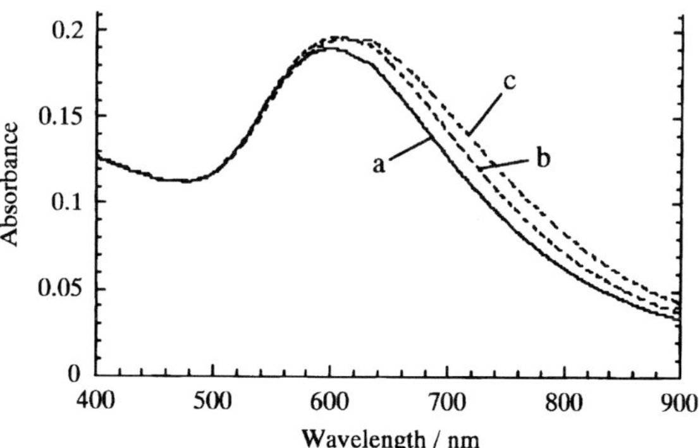

Fig. 12.10 Absorption spectrum of Au-CuO composite film at 300 °C in **a** air **b** air and 1,000 ppm CO and **c** air and 1 vol. % CO. Reprinted with permission from [25]

results showing a difference in catalytic behavior between Cu and Au nanoparticles supported on molybdenum oxides where Au nanoparticles out-perform Cu nanoparticles in the water–gas shift catalysis ($CO + H_2O \rightarrow CO_2 + H_2$). It is well known that the size of the metal nanoparticle can affect the catalytic behaviors towards certain molecules: i.e. as shown by Haruta et al., the catalytic reaction of CO on AuNPs is more efficient on smaller AuNPs [27]. In summary it is clear that the size, chemical nature and metal oxide host material will all enable characteristic catalytic properties of the composite material.

Haruta et al. [25] not only have demonstrated the catalytic behavior of gold nanoparticles (AuNPs) but also that the changes in the AuNPs LSPR band can serve as a sensing signal for carbon monoxide. The AuNPs were sputter deposited on glass substrates and then coated with copper naphthenate. The resulting sample was then fired in air at 380 °C producing a 35 nm thick film which from XRD analysis, showed that it consisted of Au and CuO. The Au particles were on average 20 nm in diameter as shown by TEM analysis. The absorbance spectrum of the Au-CuO sample was investigated as a function of CO concentration in an air carrier gas while at 300 °C. As shown in Fig. 12.10 the SPR band peaks at ∼600 nm during the air exposure and then appears to redshift and broaden upon exposure to 1,000 ppm and 1 % CO in air. Absorbance changes at a wavelength of 700 nm within the SPR absorption band showed a near linear increase with the logarithm of CO concentrations ranging between 50 ppm and 1 % CO. The response was reversible at all CO concentrations with response times on the order of 1–3 min. The sensing mechanism was attributed to the CO partial reduction reaction of CuO, forming $Cu_{2-x}O$ which is known to cause an increase in the metal oxide refractive index and thus inducing a redshift and broadening of the LSPR band.

12.8 Gold Nanoparticle Doped TiO$_2$

While copper oxide in the above example was characterized as participating in the sensing response via a refractive index change reported by the AuNPs, there are a variety of other metal oxide nanocomposites that will also serve in this manner.

TiO_2 has been the subject of research for a variety of applications including its use as a chemical sensor. Martucci et al. has been actively studying the chemical sensing properties of sol–gel processed Au-TiO_2 nanocomposites [28] for chemical sensors [29]. Two of these Au-TiO_2 films were studied, both of which contained AuNPs up to 8 wt. %. One of the films was annealed to 400 °C producing an amorphous TiO_2 matrix, and the other was annealed to 500 °C which produced the anatase TiO_2 microstructure. The AuNPs in both films were ~ 10 nm in diameter with larger polycrystalline domains comprised of these 10 nm particles. The sol gel film thicknesses deposited on silica glass slides were 50 nm. Upon exposure to either H_2 or CO in an air carrier gas at an operating temperature of 360 °C the LSPR peak of the AuNPs within the 500 °C annealed film shifts to bluer wavelengths and upon re-exposure to the air carrier gas the peak reversibly shifts to its original peak position. Exposure to CO induced a larger response than exposure to similar levels of H_2. However, for the sample annealed at 400 °C, exposure to CO in an air carrier gas at 360 °C produced no appreciable change in the LSPR peak position, while H_2 induced a measurable blue shift. The differences in the LSPR band for these two films upon exposure are portrayed in Fig. 12.11 through the use of a gas induced optical change ratio (OCR). The OCR spectrum is determined by normalization of the difference spectrum obtained by subtracting the spectrum with the exposure gas on, A_{gas}, from the spectrum with the target gas turned off, A_{air}, and thus OCR = $[A_{air}-A_{gas}]/A_{air}$. It is apparent from Fig. 12.11a that the 400 °C film is less sensitive to CO while having a large response to H_2, and that the film annealed at 500 °C, Fig. 12.11b, has a similar response to both H_2 and CO, although it is overall less sensitive than the film annealed at 400 °C. Dynamic response curves were also measured as both a function of time and exposure concentration, and these films showed a reversible response to both CO and H_2. The change in the detection characteristics for CO as a function of the annealing temperature were characterized with respect to the enhanced catalytic oxidation properties of anatase towards CO. Reaction mechanisms were proposed which could lead to the observed changes in the LSPR spectrum and these mechanisms focused on the reactive characteristics of both CO and H_2 on these Au-TiO_2 nanocomposites. Specifically, from the Drude model the LSPR peak wavelength position is dependent on both the free electron density of the metal particle as well as the dielectric function of the nanocomposite material. Reactions involving CO and H_2 may include oxidation reactions with adsorbed oxygen ion species leading to electron donation back to the AuNP, causing the observed blue shift in the LSPR peak. However, these reactions could also include adsorption within the metal oxide matrix which would cause a local change in the refractive index which could also cause a shift in the plasmon peak position. The differing response characteristics of the 400 and 500 °C annealed films towards CO was attributed to the enhanced CO oxidation properties of the anatase metal oxide. While for H_2 it has been shown to oxidize upon adsorption in the presence of Au and O_2 through formation of peroxide species, it has also been speculated to dissociatively chemisorb as a fully ionized species thus increasing the electron density.

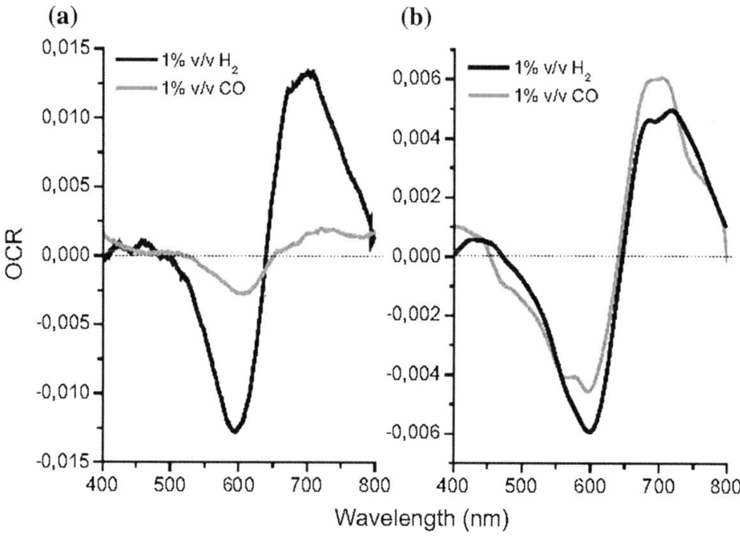

Fig. 12.11 OCR of films annealed at **a** 400 °C and **b** 500 °C after exposure to 1 % H$_2$ and 1 % CO. Reprinted with permission from [29]

12.9 Mixed Metal Oxide Matrices for Gold Nanoparticles

As the metal oxide can play a role in both changing the LSPR peak position of AuNPs as well as affect the catalytic properties of the nanocomposite and its subsequent optical response towards the target gases, it is interesting to explore mixed metal oxides as the embedding material for the AuNP optical beacons. Post et al. [30, 31] has recently performed studies resulting in a cookie like structure containing AuNPs and NiO nanoparticles embedded in SiO$_2$. The nanocomposite films were produced using sol–gel methods and the appropriate pre-cursor materials. The cookie structure producing AuNPs in contact with NiO nanoparticles was strongly dependent on the annealing temperature. Annealing temperatures of 500 °C produced a film with AuNPs and NiO nanoparticles separated within the SiO$_2$ matrix while at 600 °C there were indications that Au and NiO cookie structures were beginning to form. At 700 °C HRTEM analysis confirmed that the Au and NiO nanoparticles exhibit a twofold structure with the (1 1 1) planes of Au parallel to the (2 0 0) planes of the NiO phase. As expected, and shown in Fig. 12.12, the difference in the local surroundings of the AuNP in these three films has a pronounced affect on its resulting LSPR band. In particular for the 700 °C film portions of the AuNP is surrounded by both the SiO$_2$ as well as the NiO material, which leads to the bimodal LSPR spectrum with the red shoulder peaking at 615 nm attributed to the Au-NiO interaction and the lower intensity shoulder at 530 nm attributed to the Au-SiO$_2$ interaction. Sensing tests performed on these films showed strong responses to both H$_2$ and CO. Repeated CO exposures on the 700 °C annealed film showed a reversible response in LSPR intensity

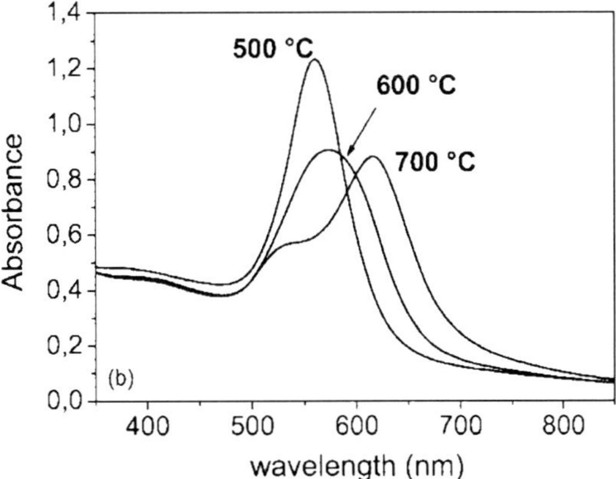

Fig. 12.12 Absorption spectra of Au-NiO-SiO$_2$ films annealed at 500, 600 and 700 °C. Reprinted with permission from [30]

as measured at 690 nm with CO concentrations ranging between 10 and 10000 ppm in an air carrier gas and operating temperature of 330 °C. It is interesting to consider this materials fabrication procedure in a broader sense. By tailoring the metal oxide matrix chemistry, it may be possible to achieve a material with multiple LSPR features whose selective response towards target gases of interest would lead to an intrinsic sensing array without having to simultaneously monitor several different spatial locations. Martucci et al. [32] has also studied AuNPs within TiO$_2$/NiO mixed metal oxide nanocomposite films. The films were made using sol–gel methods. Films containing a 50:50 or a 70:30 ratio of TiO$_2$:NiO as well as a 0.05 molar ratio of Au had characteristically better sensing properties. These films were on silica glass and tested against CO, H$_2$ and H$_2$S in an air carrier gas and elevated operating temperatures. Neither of these films had any appreciable response to CO or H$_2$ while they both responded reversibly to H$_2$S. Films exposed to H$_2$S at elevated temperatures and then cooled in the presence of H$_2$S to room temperature were analyzed using ellipsometric as well as XPS techniques. The dielectric function for each of these nanocomposite films were only different from samples unexposed to H$_2$S in wavelength regions pertaining to the LSPR band. XPS analysis showed no evidence for reactive changes to the metal oxide matrix. It was therefore surmised that H$_2$S catalytically reacts with the surface of the AuNP thus inducing a change in the LSPR band that can be reversibly monitored. Examples of this data can be seen in Fig. 12.13 for the 70:30 film annealed at 500 °C and exposed to both 0.01 % H$_2$S in air as well as 1 % H$_2$ in air at an operating temperature of 350 °C. The absorption intensity change at 605 nm as a function of time shows a strong reversible response to H$_2$S as well as no measurable response to H$_2$. A key development for chemical sensors is the production of both sensitive and selective sensing arrays for the target gases of interest. While more work is required to determine the cross sensitivity dependence of these materials against other common target gases, it is clear that the novel use of mixed metal oxides as a matrix material for AuNP optical beacons for the detection of target gases through monitoring its LSPR band offers promise for a

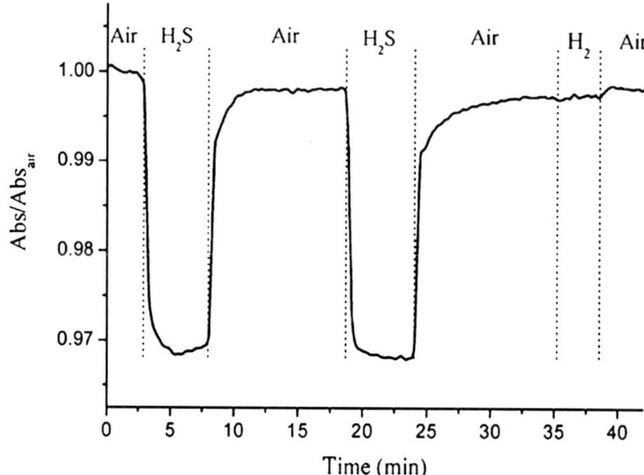

Fig. 12.13 Dynamic response of Au-70TiO$_2$-30NiO nanocomposite film annealed to 500 °C upon exposure to multiple cycles of air, 0.01 % H$_2$S in air and 1 % H$_2$ in air. Operating temperature is 350 °C. Reprinted with permission from [32]

high level of selectivity as well as sensitivity. Given the wide range of metal oxide materials, this appears to be an interesting development that will require a tailored coupling of theoretical as well as experimental work to develop arrays of materials in a logical fashion for testing against a series of target gases.

12.10 Au Nanoparticles Embedded in Oxygen Ion Conducting Metal Oxides

The combined study of metal nanoparticles embedded in metal oxide matrices is a field that has generated a great deal of interest for its potential towards development of nanodevices and for its ability to tune the optical properties of metal nanoparticles [33–39]. Yttria stabilized zirconia (YSZ) is one of the most widely used oxygen ion conductors with applications in solid oxide fuel cells, oxygen pumps and chemical sensors [40–42]. In all of these applications YSZ acts as a solid electrolyte where oxygen ions diffuse through the YSZ matrix as long as the operation temperature is above that required for oxygen ion transport (T > 300 °C) [43]. Metals are normally used as the catalytically active surface layer on porous anodes and cathodes in the above-mentioned applications for YSZ, and it has been shown that by using catalytically active metals the flux of oxygen ions through YSZ electrolytes can increase significantly [44]. By adding Au nanoparticles (AuNPs) to YSZ thin films, diffusion kinetics of oxygen ions into the oxygen vacancies in the YSZ matrix is expected to increase for the same reason metals increase electro-chemical reaction kinetics at the surfaces of YSZ electrolytes.

Metal oxides such as YSZ and SnO_2 have been used in gas sensors for about as long as commercially available gas sensors have been on the market. The sensing mechanism of these metal oxides rely on oxygen ion diffusion through, or into oxygen vacancies in the material (via a hopping mechanism) to produce a current, in the case of YSZ based O_2 sensors, or a change in film resistivity in the case of SnO_2 gas sensors. With the advent of the semiconductor industry, metal oxides such as these and others were deposited on the semiconducting portion of field effect transistors developing what is known as the ion selective field effect transistor (ISFET) [45]. The ISFET also uses the diffusion of ions in and out of the oxide, but the transduction mechanism of the sensor comes in the form of a capacitance change in the oxide material, which is observed in the I–V characteristics of the device. Nevertheless metal oxides have always been used in sensors for the very simple fact that ions can diffuse in and out of them, producing an electronic response. The use of YSZ in nanocomposite films for use in optical based chemical sensors is for this same reason; it acts as an oxygen ion sponge at elevated temperatures (>300 °C). Along with its ability to allow for the diffusion of oxygen ions, YSZ also has the added benefit of being optically transparent in the visible range and has a high refractive index (n = 2.1) [46].

12.11 Au-YSZ Nanocomposites for the Detection of CO, H_2 and NO_2

The work of Carpenter et al. has been focused on the development of plasmonic based optical chemical sensors which are able to detect emission gases under harsh environmental conditions. Films used for the detection of CO were comprised of Au-YSZ nanocomposites synthesized on sapphire substrates by RF magnetron co-sputtering followed by an annealing treatment at 1000 °C for 2 h [47]. Films with a Au content of ~ 10 at. % and a thickness of ~ 30 nm were used for this study. The microstructure of the post-annealed Au-YSZ films was examined by XRD. The XRD pattern indicated the presence of two polycrystalline phases; one corresponding to the tetragonal phase of YSZ and the other corresponding to the face centered cubic Au phase. The average YSZ and Au crystallite sizes were calculated from the Scherrer formula [48] using the YSZ (101) and the Au (111) reflections, and an average crystallite size of ~ 19 nm was obtained for both phases.

The sensing properties of the films at atmospheric pressure and elevated temperature were tested in a custom-designed quartz transmission cell housed within a tube furnace. White light from a CW lamp was collimated and transmitted through the sample held centered in the quartz cell using a Macor sample holder. The transmitted light was dispersed and detected using a spectrometer equipped with a Charged Coupled Diode (CCD) detection system. Upon exposure to an air gas cycle at temperatures above ~ 350 °C the LSPR band of the gold nanoparticles showed a red shift and a broadening of the FWHM. It is expected that YSZ will react readily with O_2 and at temperatures above $\sim$350 °C, oxygen anions species will form and O^{2-} will diffuse through the matrix and occupy the oxygen vacancies. Given that the Au-YSZ nanocomposite has a Au wt % of approximately 10 %, a likely source of electrons during the formation of O^{2-} is the AuNP. Given the Drude model for AuNPs, the peak of the LSPR band, Ω (rad/sec), will depend on

$$\Omega = \sqrt{\frac{N_0 e^2}{(1 + 2\varepsilon_m)m_e \varepsilon_0}}$$

both the free electron density, N_o as well as the dielectric function of the matrix, ε_m. A reduction in N_o will cause a reduction in Ω and the observed red shift in the LSPR peak wavelength. Furthermore, addition of O^{2-} ions into the YSZ matrix will likely increase the polarizability of the system and in doing so the dielectric function will increase, also causing a red shift in the LSRP band. Absorption spectra obtained with the CO/air gas pulse on showed LSPR peak positions that were blue shifted from the LSPR peak position during the air alone gas cycle. This blue shift during CO exposure is indicative of an interfacial charge transfer reaction of CO with the O^{2-} ions forming CO_2 with the electrons being donated back to the AuNP [49, 50]. The absorption spectra obtained with the CO/air gas pulse were subtracted from the absorption spectrum during the air gas cycle to obtain a difference spectrum. The peak to peak difference in this spectrum was labeled the sensing signal and it is this signal that was monitored as a function of time during the various gas cycles. Figure 12.14 displays the resulting CO sensing signal as a function of time for the Au-YSZ film upon exposure to 1, 0.75, 0.5, 0.25 and 0.1 vol. % CO concentrations in air at 500 °C. The change in the absorption spectra upon exposure to CO was reversible, and the sensing signal increased with increasing CO concentration. The response time, i.e. the time required for the sensing signal to obtain its maximum value upon exposure to CO was $\sim$40 s at all CO concentrations, with recovery in the subsequent air pulse displaying a two stage mechanism comprised of a fast, $\sim$60 s, initial stage followed by a slower, $\sim$1,000 s, stage. The tests performed for this experiment showed clear reversibility and likewise a dependence on the background O_2 concentration, as the LSPR band was unresponsive to CO in the presence of just a N_2 carrier gas.

12.12 Hydrogen Titration Studies

Within these studies we have performed a detailed titration study of the optical properties of Au-YSZ films as a function of H_2 concentrations ranging from 0.05 to 1 vol. % and O_2 concentrations ranging from 0.1 to 20 vol. % at an operation temperature of 500 °C. As can be seen in Fig. 12.15a and b both the LSPR band peak position and the corresponding FWHM reversibly blue shift and narrow, respectively, at all O_2 concentrations upon exposure to increasing H_2 levels. In

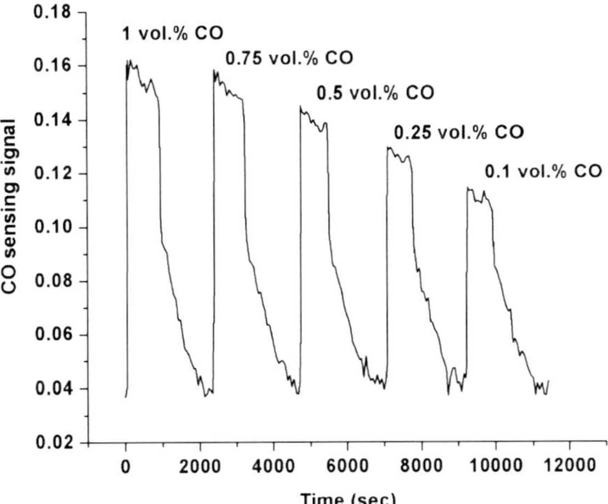

Fig. 12.14 Sensing signal response curve of the Au-YSZ nanocomposite film upon exposure to 1, 0.75, 0.5, 0.25 and 0.1 vol. % CO in air at 500 °C. Reprinted with permission from [47]

addition, through a detailed analysis of the optical response of the AuNPs to a change in their surrounding environment, this research developed and verified an electrochemical model for charge transfer between YSZ—bound oxygen ions formed through the dissociative adsorption of oxygen molecules on YSZ at high temperatures and the AuNPs. The degree of charge transfer is dictated by the equilibrium ratio, $pH_2^{1/4}\big/pO_2^{1/8}$, contributing to oxygen ion diffusion into and out of the YSZ matrix, which results in a characteristic change in the square of the LSPR band peak position [51].

12.13 NO$_2$ Sensing Studies

The primary explanation for shifts in the peak position of the LSPR band is charge exchange from the AuNPs in the YSZ matrix to diffusing oxygen ions from catalytic reactions occurring within the Au-YSZ nanocomposites, as outlined in the previous section. Exposure studies for NO$_2$, however, will lead to NO$_2$ catalytically decomposing on the hot gold surface producing an O ion for incorporation into the YSZ matrix. Evidence for this theory can be seen in Fig. 12.16. Even in the presence of a matrix oxidizing 20 vol. % O$_2$ background (air), exposure to NO$_2$ results in a red-shift of the LSPR band peak position due to further removal of charge from the AuNP by the O ion produced upon dissociation of NO$_2$ on the surface. It is interesting to note that while the optical response of the AuNP towards O$_2$ in N$_2$ is nearly saturated at levels above 10 vol. %, the addition of the oxidizing species, NO$_2$, causes a further red shift in the LSPR band at concentrations ranging from 5 to 100 ppm in an air carrier gas. A similar 5 ppm increase in O$_2$ concentration, in the air carrier gas, would not be measurable due to the matrix being saturated with the oxygen baseline reactions. While it is not fully understood why a

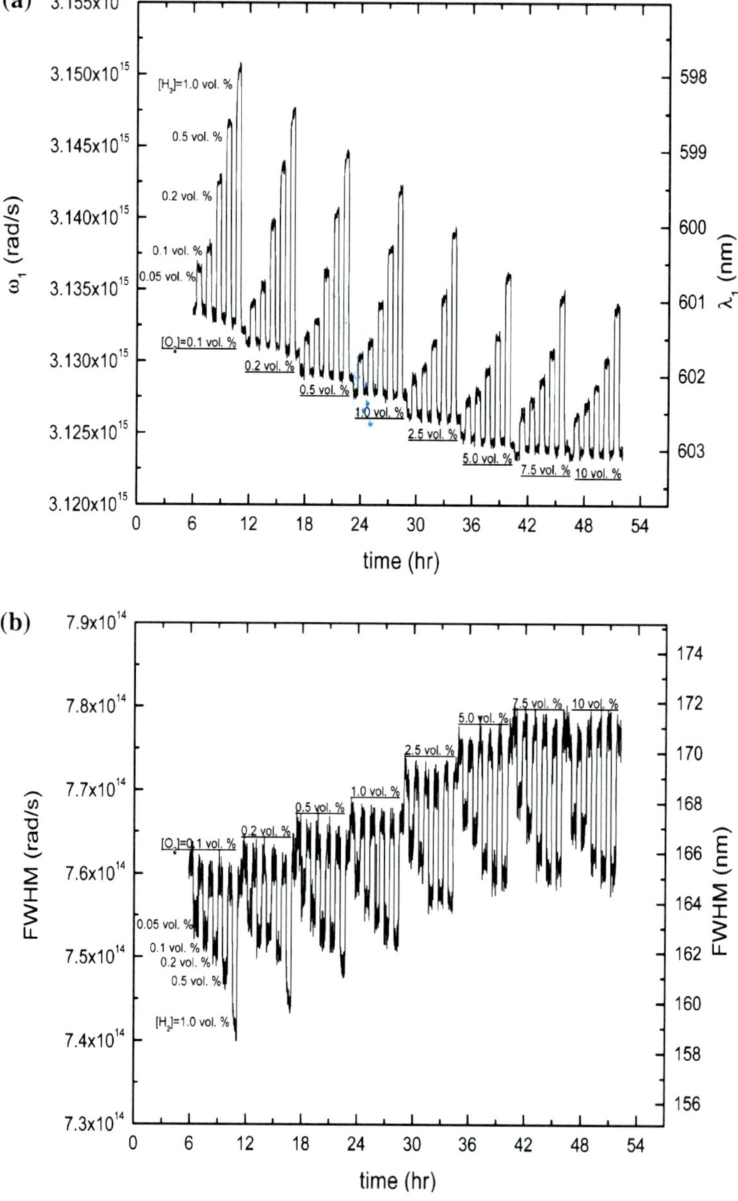

Fig. 12.15 Peak position and FWHM of 25 nm AuNP embedded in YSZ as a function of both H_2 and O_2 concentrations at a 500 °C operating temperature. Reprinted with permission from [51]

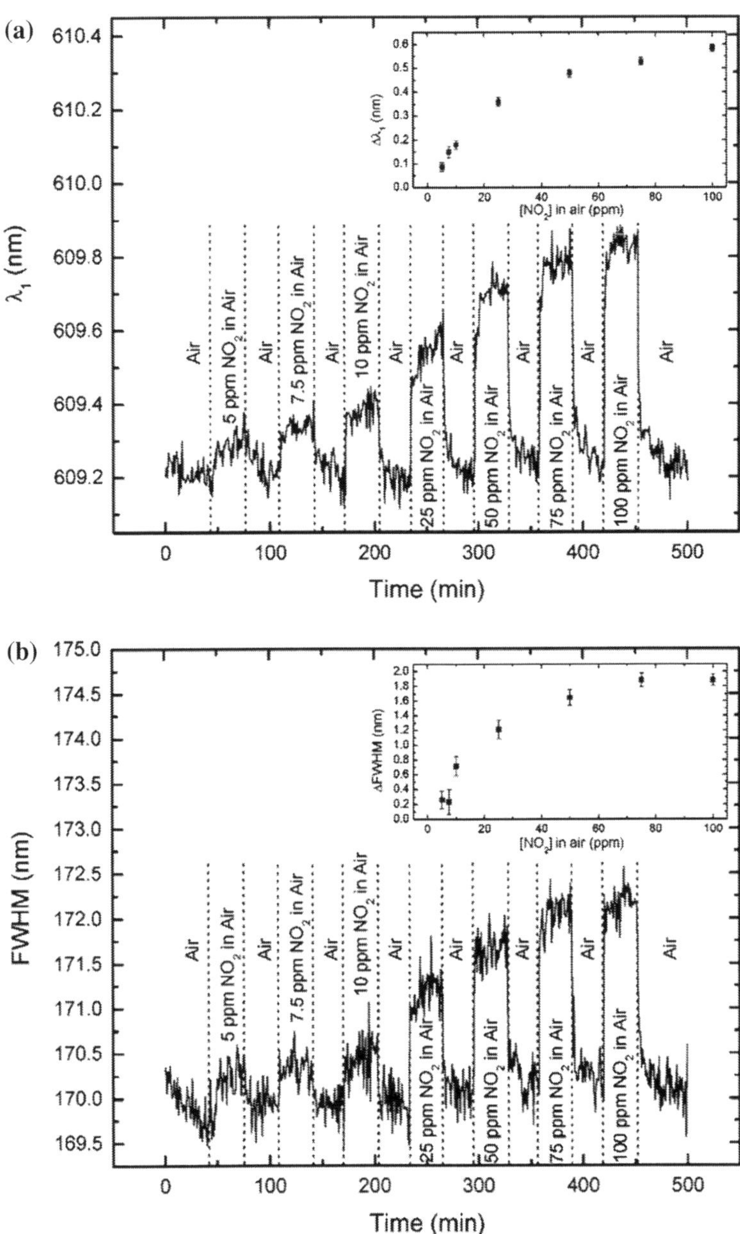

Fig. 12.16 Peak position and FWHM as a function of both NO_2 concentrations at a 500 °C operating temperature. Reprinted with permission from [52]

Au-YSZ nanocomposite is more susceptible to the addition of oxygen ions by NO_2 rather than O_2, it is likely due to differences in the interfacial reaction sites available to the two species. These may include the specific oxygen vacancy type, Y-V_O-Y, Y-V_O-Zr and Zr-V_O-Zr that is available for stabilization of the generated oxygen ion species and the reactivity differences between the two species [52].

12.14 Selectivity and Sensor Arrays

It is clear from both the work described above that both photoluminescence properties of metal oxides as well as absorption spectral properties of Au—metal oxide nanocomposite materials are sensitive to gaseous species in a variety of sensing environments. However, as with any chemical sensing application there is a need for both sensitivity and selectivity of the target gases amongst the background gases. As noted above Martucci et al. have designed a Au-mixed metal oxide material (Au-TiO_2-NiO) which appears to be sensitive towards H_2S while being insensitive towards H_2 exposures. However, in the work of Carpenter et al. both CO and H_2 cause a blue shift in the plasmon band while NO_2 induces a red shift. Evaluation of the CO and H_2 calibration curves indicates that with similar gas concentrations one apparently cannot distinguish between these two individual gases let alone a gas mixture containing both reducing and oxidizing gases. While developing a detailed understanding of the sensing mechanism will aid the design of other highly selective Au-metal oxide nanocomposite materials, it is likely that the materials design will have to be utilized in combination with statistical analysis algorithms to analyze patterns within the sensing data across a sensing array. A variety of methods are available for the statistical analysis of sensor arrays. These range from supervised learning packages such as linear discriminant analysis (LDA) or neural networks, and unsupervised learning packages such as principal component analysis (PCA). The literature is rich with references on each of these algorithms and their use in electronic nose based sensor applications, and the reader is encouraged to seek these out [53–58]. Details on these algorithms and resulting analysis of sensing data is also provided within Chaps. 14 and 15 of this text. However, specific to the statistical analysis of chemiluminescence from an array of catalytically active nanomaterials is work by Wu et al. [59]. This work used an array of 21 nanomaterials including metal oxides, metal oxides deposited on CNTs, AuNPs on metal oxides and carbonates. A chemiluminescent (CL) imaging system analyzed the 21 spots deposited on a thermally controlled ceramic chip. Through the use of temperature control during the exposure experiments a quantitative analysis of a series of organics commonly found in cigarette smoke could be determined. Using an LDA algorithm, identification of six brands of cigarettes could be accomplished. Carpenter et al. has also begun to utilize PCA analysis for the interpretation of sensing data across individual sensing elements [60] as well as an array of Au-metal oxide thin films. The sensor array consisted of 3 elements; 1) 30 nm thick Au-YSZ films containing 10 at. % Au, 2) 30 nm thick

Table 12.1 Target gas concentration exposures in ppm across a 3 element Au-metal oxide sensor array

	H_2	CO	NO_2
Exposure 1	200	200	2
Exposure 2	500	300	5
Exposure 3	1,000	500	10
Exposure 4	5,000	1,000	20
Exposure 5	10,000	2,000	98

Au-TiO$_2$ films containing 10 at. % Au and 3) 200 nm thick MBE grown ceria films with ~ 8 at. % Au implanted to a depth of ~ 75 nm. XRD analysis showed the average AuNP size to be ~ 20 nm for the Au-TiO$_2$ film, and closer to 10nm for both the Au-YSZ and Au-CeO$_2$ films. CO, H$_2$ and NO$_2$ gas exposures in a background of 5 and 10 % O$_2$ balanced by N$_2$, and in a background of dry air, were performed across this array with target gas concentrations listed in Table 12.1 at an operating temperature of 500 °C. A sample of the type of sensor response is displayed in Fig. 12.17 for the hydrogen in an air carrier gas exposure experiment. Each of the elements in the array shows a strong response towards H$_2$ with both reversibility and stability over the entire (>50 hr) experiment. A series of calibration curves have been determined for each of the target gases, however the problem is that as noted in Table 12.1, the H$_2$ and CO gas concentrations overlap at concentrations below 1,000 ppm, so a selective response determination will be difficult to ascertain without the use of a statistical algorithm. PCA analysis was used with 45 observables (5 gas concentrations, 3 target gases, and 3 O$_2$ concentrations), and for each of the three sensing elements, 175 wavelengths across the LSPR band were used as variables for input into the algorithm. Analysis of this large dataset using PCA reduces the multidimensionality of the sensing data into a few principal components while retaining the maximum variance within the dataset. The results of this analysis can be summarized by the PCA scores plots shown in Fig. 12.18. Both PC 2 vs. PC 1 and PC 3 vs. PC 2 are shown and these first 3 principal components (otherwise known as eigenvectors) represent 90 % of the variance in the data as determined from their eigenvalues. It is clear from these plots that there is good separability amongst the three target gases. While much work needs to be done for optimization of these types of Au-metal oxide sensing arrays, especially with respect to other common background gases such as water (humidity), CO$_2$, and hydrocarbons among others, these results are promising with respect to selective measurements under harsh conditions. Further work is in progress to characterize these types of background gas dependencies as well as to begin work with mixtures of the target gases.

12.15 Summary and Perspective

This chapter has provided an overview of how the optical properties of metal oxide nanomaterials can be used as a transduction mechanism for chemical

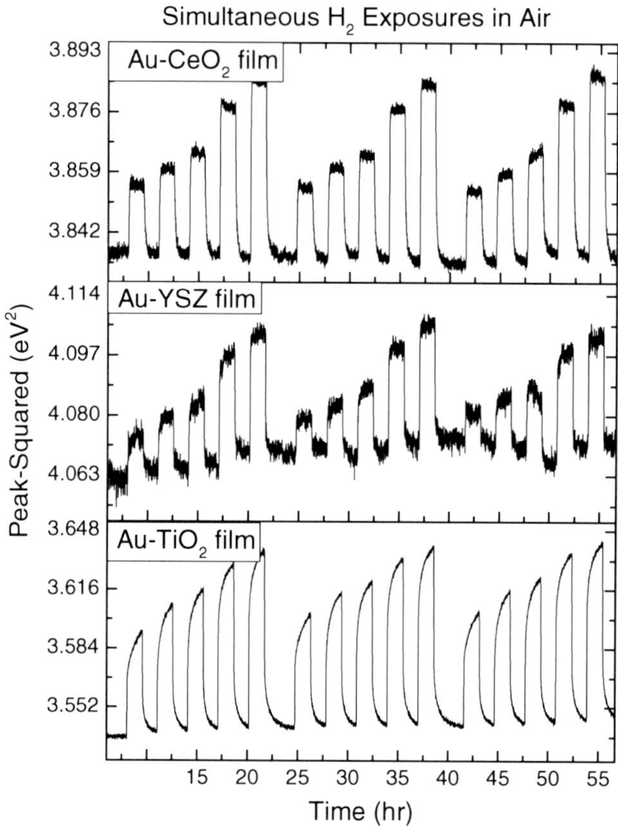

Fig. 12.17 Time dependence of the SPR peak position (eV^2) for a three element sensor array exposed to 200, 500, 1000, 5000 and 10,000 ppm H$_2$ in an air carrier gas at 500 °C

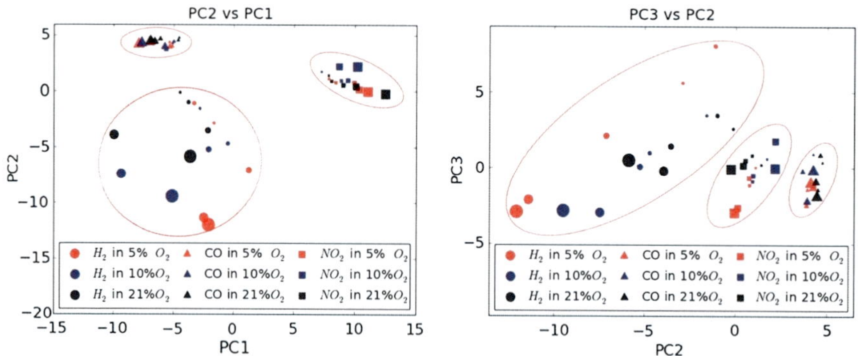

Fig. 12.18 Principal component scores plots from PCA analysis of 3 element sensing array

sensors. The photoluminescence of metal oxides have characteristic bands indicative of the transition between the valence and conduction band as well as from the valence band to defect states, or surface states. These transitions have been shown to be sensitive to changes in the surrounding gas environment and by tuning the chemistry of the metal oxide as well as its nanomaterial shape (sphere, rod or fishbone) one can change the surface area. As the target gases will interact directly with the surfaces of these metal oxides, enhancing the optical signature of the defect or surface states through design of the nanomaterial shape can likely offer advantages for sensing applications. The addition of noble metal dopants into metal oxides for use in a plasmonic based sensor has shown promise for the detection of a variety of target gases as well as operation environments. While the bulk of this work has used Au nanoparticles as the optical beacon, Ag and Cu nanoparticles also have strong plasmon bands in the visible wavelength region which may make them an attractive alternative. However, depending on the operation environment, oxidation of these nanoparticles may cause problems with reliability and reversibility. A variety of metal oxides have been used as the embedding material for these plasmonic based sensors. These were shown to include SiO_2, TiO_2, and the mixed metal oxide TiO_2–NiO, as well as oxygen ion conducting metal oxides such as YSZ and CeO_2. The addition of the oxygen ion conduction properties enhances the interfacial reaction characteristics as oxygen anions can more readily populate the entire metal oxide thin film due to their enhanced diffusion properties. This allows for sensing events to include both surface and bulk reactions. It is clear that the studies outlined above hold much promise for the use of optical based transduction methods in chemical sensors. However, as with the majority of sensing research, work is still needed to improve detection limits, reliability, and selectivity both with respect to the variety of target gases and also the background gases which are characteristic of the particular sensing application. Achievement of selectivity will most likely require the use of sensing arrays with a detailed statistical analysis to ensure the proper interpretation of the sensing "fingerprint" across the array elements. Another primary issue will entail integration of the metal oxide sensing layer into a portable optical detection scheme for future applications. While photoluminescence of materials as well as reflection spectroscopy (for monitoring plasmonic and/or refractive index changes) can be performed via a fiber optic coupling, a fiber optic based grating coated with metal oxides was demonstrated as a means for the detection of CO_2 using alkaline earth doped perovskite based metal oxides. The use of a fiber based grating for a variety of target gases will just require the tailoring of the metal oxide coating for detection. It is clear that as the optical interrogation of metal oxide based materials continues to provide the required sensing characteristics, more efforts aimed at the integration of these materials as well as their ancillary components (sources, detectors, optics) will be required for the development of cost effective sensing devices. Fortunately, the miniaturization of many of these ancillary components has been developing independently and there are a range of options to choose from for devices which function from the Uv into the NIR regions and with that it is apparent that the use of metal oxide nanomaterials in optical based chemical sensors will have a long and bright future.

Acknowledgments The author's work was supported in part by the United States Department of Energy National Energy Technology Laboratory under contract number DE-NT0007918. Any opinions, findings, and conclusions or recommendations expressed in this publication are those of the authors and do not necessarily reflect the views of the United States Department of Energy National Energy Technology Laboratory. Funding for support of this work is also acknowledged from the National Science Foundation (1006399) and the NYSTAR Environmental Quality Systems Research Center.

References

1. Littieri S, Setaro A, Baratto C, Comini E, Faglia G, Sberveglieri G, Maddalena P (2008) On the mechanism of photoluminescence quenching in tin dioxide nanowires by NO_2 adsorption. New J Phys 10:043013
2. Valerini D, Creti A, Caricato AP, Lomascolo M, Rella R, Martino M (2010) Optical gas sensing through nanostructured ZnO films with different morphologies. Sens Act B 145:167
3. Barrato C, Todros S, Faglia G, Comini E, Sberveglieri G, Lettieri S, Santamaria L, Maddalena P (2009) Luminescence response of ZnO nanowires to gas adsorption. Sens Act B 140:461
4. Shalish I, Temkin H, Narayanamurti V (2004) Size dependent surface luminescence in ZnO nanowires. Phys Rev B 69:245401
5. Faglia G, Barrato C, Sberveglieri G, Zha M, Zappettini A (2005) Adsorption effects of NO_2 at ppm level on visible response of SnO_2 nanobelts. Appl Phys Lett 86:011923
6. Hu JQ, Bando Y, Goldberg D (2003) Self catalyst growth and optical properties of novel SnO2 fishbone-like nanoribbons. Chem Phys Lett 372:758
7. Setaro A, Lettieri S, Diamare D, Maddalena P, Malagu C, Carotta MC, Martinelli G (2008) Nanograined anatase titania-based optochemical gas detection. New J Phys 10:053030
8. Wei X, Wei T, Xiao H, Lin YS (2008) Nanostructured Pd long period fiber gratings integrated optical sensor for hydrogen detection. Sens Act B 134:687
9. Wei X, Wei T, Li J, Lan X, Xiao H, Lin YS (2010) Strontium cobaltite coated optical sensors for high temperature carbon dioxide detection. Sens Act B 144:260
10. Shao Z, Yang W, Cong Y, Dong H, Tong J, Xiong G (2000) Investigation of the permeation behavior and stability of a Ba0.5Sr0.5Co0.8Fe0.2O3 − δ oxygen membrane. J Membr Sci 172:177
11. Tanahashi I, Manabe Y, Tohda T, Sasaki S, Nakamura A (1996) Optical nonlinearities of Au/ SiO_2 composite thin films prepared by a sputtering method. J Appl Phys 79:1244
12. Liao HB, Xiao RF, Wong GKL (1997) Large third-order nonlinear optical susceptibility of Au-Al_2O_3 composite films near the resonant frequency. Appl Phys B: Lasers Opt 65:673
13. Liao H, Xiao RF, Wong H, Wong KS, Wong GKL (1998) Large third-order optical nonlinearity in Au: TiO_2 composite films measured on a femtosecond time scale. Appl Phys Lett 72:1717
14. MacFarland AD, Van Duyne RP (2003) Single silver nanoparticles as real-time optical sensors with zeptomole sensitivity. Nano Lett 3:1057
15. Ando M, Kobayashi T, Iijima S, Haruta M (2003) Optical CO sensitivity of Au–CuO composite film by use of the plasmon absorption change. Sens Act B 96:589
16. Kreibig U, Vollmer M (1995) Optical properties of metal clusters. Springer, New York, p 23
17. Maier SA (2010) Plasmonics: fundamentals and applications. Springer, New York

18. Sosa IO, Noguez C, Barrera RG (2003) Optical properties of metal nanoparticles with arbitrary shapes. J Phys Chem B 107:6269
19. Jain PK, Huang X, El-Sayed IH, El-Sayed MA (2007) Plasmonics 2:107
20. Daniel M-C, Astruc D (2004) Gold nanoparticles: assembly, supramolecular chemistry, quantum-size-related properties, and applications toward biology, catalysis, and nanotechnology. Chem Rev 104:293
21. Murray WA, Barnes WL (2007) Plasmonic materials. Adv Mater 19:3771
22. Jain PK, El-Sayed MA (2007) Surface plasmon resonance sensitivity of metal nanostructures: physical basis and universal scaling in metal nanoshells. J Phys Chem C Lett 111:17451
23. Haick H (2007) Chemical sensors based on molecularly modified metallic nanoparticles. J Phys D Appl Phys 40:7173
24. Nath N, Chilkoti A (2004) Label-free biosensing by surface plasmon resonance of nanoparticles on glass: optimization of nanoparticle size. Anal Chem 76:5370
25. Ando M, Kobayashi T, Iijima S, Haruta M (2003) Optical CO sensitivity of Au-CuO composite film by use of the plasmon absorption change. Sens Act B 96:589
26. Rodriguez JA, Liu P, Hrbek J, Perez M, Evans J (2008) Water–gas shift activity of Au and Cu nanoparticles supported on molybdenum oxides. J Molec Catal A 281:59
27. Haruta M (1997) Size- and support-dependency in the catalysis of gold. Catal Today 36:153–166
28. Buso D, Pacifico J, Martucci A, Mulvaney P (2007) Gold-nanoparticle-doped TiO_2 semiconductor thin films: optical characterization. Adv Funct Mater 17:347
29. Buso D, Post M, Cantalini C, Mulvaney P, Martucci A (2008) Gold nanoparticle-doped TiO_2 semiconductor thin films: gas sensing properties. Adv Funct Mater 18:3843
30. Buso D, Guglielmi M, Martucci A, Mattei G, Mazzoldi P, Sada C, Post ML (2008) Growth of cookie-like Au/NiO nanoparticles in SiO_2 Sol-gel films and their optical gas sensing properties. Cryst Growth Des 8:744
31. Buso D, Busato G, Gugliemi M, Martucci A, Bello V, Mattei G, Mazzoldi P, Post ML (2007) Selective optical detection of H_2 and CO with SiO_2 sol-gel films containing NiO and Au particles. Nanotechnology 18:475505
32. Gaspera ED, Guglielmi M, Agnoli S, Granozzi G, Post ML, Bello V, Mattei G, Martucci A (2010) Au Nanoparticles in nanocrystalline TiO_2-NiO films for SPR based selective H_2S gas sensing. Chem Mater 22:3407
33. MacFarland AD, Van Duyne RP (2003) Single silver nanoparticles as real-time optical sensors with zeptomole sensitivity. Nano Lett 3:1057
34. Liao H, Xiao RF, Wong H, Wong KS, Wong GKL (1998) Large third-order optical nonlinearity in $Au:TiO_2$ composite films measured on a femtosecond time scale. Appl Phys Lett 72:1817
35. Yang Y, Hori M, Hayakawa T, Nogami M (2005) Self-assembled 3-dimensional arrays of $Au@SiO_2$ core-shell nanoparticles for enhanced optical nonlinearities. Surf Sci 579:215
36. Kreibig U, Vollmer M (1995) Optical properties of metal clusters. Springer, Berlin
37. Nowotny J, Bak T, Nowotny MK, Sorrell CC (2005) Charge transfer at oxygen/zirconia interface at elevated temperatures. Adv Appl Ceram 104:147–231
38. Henrich VE, Cox PA (1994) The surface science of metal oxides. Cambridge University Press, United Kingdom
39. Korotcenkov G (2007) Metal oxides for solid-state gas sensors: what determines our choice? Mater Sci Eng, B 139:1
40. Suna C, Stimming U (2007) Recent anode advances in solid oxide fuel cells. J Power Sources 171:247
41. Fergus JW (2007) Materials for high température electrochemical NOx gas sensors. Sens Act B 121:652
42. Fergus JW (2007) Solid électrolyte based sensors for the measurement of CO and hydrocarbon gases. Sens Act B 122:683
43. Docquier N, Candel S (2002) Combustion control and sensors: a Review. Prog Energy Combust Sci 28:107

44. Takeguchi T, Kikuchi R, Yano T, Eguchi K, Murata K (2003) Effect of precious metal addition to Ni-YSZ cermet on reforming of CH_4 and electrochemical activity as SOFC anode. Catal Today 84:217
45. Bergveld P (2003) Thirty years of ISFETOLOGY: what happened in the past 30 years and what may happen in the next 30 years. Sens Act B 88:1–20
46. Sirinakis G, Siddique R, Dunn KA, Efstathiadis H, Carpenter MA, Kaloyeros AE, Sun L (2005) Spectro-ellipsometric characterization of $Au-Y_2O_3$-stabilized ZrO_2 nanocomposite Films. J Mater Res 20:3320
47. Sirinakis G, Siddique R, Manning I, Rogers PH, Carpenter MA (2006) Development and characterization of Au − YSZ surface plasmon resonance based sensing materials: high temperature detection of CO. J Phys Chem B 110:13508
48. Cullity BD, Stock SR (2001) Elements of X-ray diffraction, 3rd edn. Prentice-Hall, USA
49. Hirakawa T, Kamat PV (2005) Charge separation and catalytic activity of $Ag@TiO_2$ Core − shell composite clusters under UV − Irradiation. J Am Chem Soc 127:3928
50. Oldfield G, Ung T, Mulvaney P (2000) $Au@SnO_2$ core–shell nanocapacitors. Adv Mater 12:1519
51. Rogers PH, Sirinakis G, Carpenter MA (2008) Direct observations of electrochemical reactions within Au-YSZ thin films via absorption shifts in the Au nanoparticle surface plasmon resonance. J Phys Chem C 112:6749–6757
52. Rogers PH, Carpenter MA (2008) Plasmonic based detection of NO_2 in a harsh environment. J Phys Chem C 112:8784
53. Martinez AM, Kak AC (2001) PCA vs. LDA. IEEE Trans Patt Analy Mach Intell 23:228
54. Sysoev VV, Goschnick J, Schneider T, Strelcov E, Kolmakov A (2007) A gradient microarray electronic nose based on percholating SnO_2 nanowire sensing elements. Nanoletters 7:3182
55. Hai Z, Wang J (2006) Electronic nose and data analysis for detection of maize oil adulteration in sesame oil. Sens Act B 119:449
56. Musatov VY, Sysoev VV, Sommer M, Kiselev I (2010) Assessment of meat freshness with metal oxide sensor microarray electronic nose: a practical approach. Sens Act B 144:99
57. Lerma-Garcia MJ, Cerretani L, Cevoli C, Simo-Alfonso EF, Bendini A, Gallina Toschi T (2010) Use of electronic nose to determine defect percentage in oils. comparison with sensory panel results. Sens Act B 147:283
58. Srivastava JK, Pandey P, Jha SK, Mishra VN, Dwivedi R (2011) Chemical vapor identification by plasma treated thick film tin oxide gas sensor array and pattern recognition. Sens Transd 125:42
59. Wu Y, Na N, Zhang S, Wang X, Liu D, Zhang X (2009) Discrimination and identification of flavors with catalytic nanomaterial-based optical chemosensor array. Analy Chem 81:961
60. Joy NA, Nandasiri MI, Rogers PH, Jiang W, Varga T, Kuchibhatla SVNT, Thevuthasan S, Carpenter MA (2012) Selective plasmonic gas sensing: H_2, NO_2 and CO spectral discrimination by a single $Au-CeO_2$ nanocomposite film. Analy Chem 84:5025

Part III
New Device Architectures and Integration Challenges

Chapter 13
Metal Oxide Nano-architectures and Heterostructures for Chemical Sensors

Thomas Fischer, Aadesh P. Singh, Trilok Singh,
Francisco Hernández-Ramírez, Daniel Prades
and Sanjay Mathur

Abstract Metal oxide nanostructures with hetero-contacts and phase boundaries offer a unique platform for designing materials architectures for sensing applications. Besides the size and surface effects, the modulation of electronic behaviour due to junction properties leads to modified surface states that promote selective detection of analytes. The growing possibilities of engineering nanostructures in various compositions (pure, doped, composites, heterostructures) and forms (particles, tubes, wires, films) has intensified the research on the integration of different functional material units in a single architecture to obtain new sensing materials. In addition, new concepts of enhancing charge transduction by surface functionalization and use of pre-concentrator systems are promising strategies to promote specific chemical interactions, however the challenge related to reproducible synthesis and device integration of nanomaterials persist.

13.1 Introduction

Integration of chemical sensor technology and engineered metal oxides embodies an optimal synergic platform, where the wide variety of functional properties (e.g., electrical, optical, magnetic, piezo-electric) of metal oxides and strong electronic correlations, particularly in transition metal oxides, can be adapted to the

T. Fischer · A. P. Singh · T. Singh · S. Mathur (✉)
Institute of Inorganic Chemistry, University of Cologne, 50939 Cologne, Germany
e-mail: sanjay.mathur@uni-koeln.de

F. Hernández-Ramírez
The Catalonia Institute for Energy Research, E-08930 Barcelona, Spain

D. Prades
Department of Electronics, University of Barcelona, E-08028 Barcelona, Spain

M. A. Carpenter et al. (eds.), *Metal Oxide Nanomaterials for Chemical Sensors,*
Integrated Analytical Systems, DOI: 10.1007/978-1-4614-5395-6_13,
© Springer Science+Business Media New York 2013

requirements of sensor materials. In addition to their unique electronic characteristics, metal oxides also display rich phase diagrams and abilities to form multi-cationic stoichiometries, which in conjunction with the possible modulation of morphological features and dimensions (particles, fibers, films and composites) provide new degrees of freedom to meet the requirement profile (transduction-detection-acquisition) of chemical sensor systems [1, 2]. Owing to the smaller grain sizes in nanomaterials, even small external perturbation results in a huge change in intrinsic material properties that is manifested as "response" of the material. In the last few years, tremendous efforts have been made to design sensors that show improved key features such as selectivity, low detection limits, reversibility, robustness, portability and easy handling, whereby the application of nanostructured materials has played an important role [3, 4].

For a targeted material design and development, it is essential to understand the interplay of composition, materials architecture and properties since a number of radically new phenomena (e.g., catalytic and transport properties) have been observed, when dimensions or compositions were changed in metal oxides and heterostructures thereof. Among heterostructures, the descriptors A@B, A-B and A/B denote multimodal junctions with coherent and non-coherent interfaces as well as differential work functions. Nevertheless, these terms are used interchangeably in the published literature. For instance, planar stacking of two insulating materials ($LaAlO_3$ and $SrTiO_3$) can lead to a metallic or superconducting interface [5]. The influence of size, strain and interface effects has also been a subject of significant attention in the recent past [6]. The size-effects become predominant, when long range order is perturbed, whereas the increasing surface-to-volume ratio is known to enhance the transduction behavior of nanosized materials [7]. Several reports have confirmed the intrinsic advantages of materials and interface engineering on sensor performance [8, 9]. For example, Yamazoe et al. demonstrated enhanced response to carbon monoxide (CO), when the grain size was reduced from 30 to 5 nm (6-fold increase in sensitivity) [10] and nano-porous film of SnO_2 were produced. A significant enhancement in the change in resistance was observed in sensors based on one dimensional nanostructures [11], when compared to sensors constituted by a micro-grained films. The sensor response was shown to be further improved, when the gas detection was performed on sensor devices based on individual nanowires [12]. In addition, the interfacial effects have shown to impart synergistic effects leading to improved sensing behavior.

In Field Effect Transistor (FET) sensing devices modifications of single n-type SnO_2 nanowires with n-type (ZnO) and p-type (NiO) nanoparticles resulted in enhanced selectivites towards H_2S for $ZnO@SnO_2$ or CH_4 when $NiO@SnO_2$was used [13]. In both the cases, formation of n–n or p–n heterojunctions led to modified surface states which promoted the selective surface reaction of gaseous analyte species, as no compound formation $MSnO_x$ (M = Zn, Ni) was observed under the synthesis and operating conditions. Using H_2, as the analyte gas, $ZnO@SnO_2$ n–n-heterocontacts exhibited an even more complex sensing behavior, whereby a concentration-dependent n–p–n transition was observed leading to an n-ZnO/p-Zn–O-Sn/n-SnO_2 interface which is dependent on the H_2

concentration. This mechanism is believed to be caused by shallow donors induced by H_2 and was therefore not observed with other reducing gases like CO and CH_4 as their size hinders an effective diffusion into the bulk material [14]. Not only oxide/oxide- but more often metal/oxide-heterocontacts play a crucial role in tuning the sensitivity and selectivity of gas sensors [15]. This effect not only accounts for desired doping of the active sensing material, but it needs to be considered that all electrical contacts needed to connect the metal oxide sensing element with the readout electronics essentially form Schottky barriers and catalytic sites where analyte molecules can adsorb and react [16, 17].

Among various options available for the detection of gaseous analytes such as electrochemical (solid electrolyte and amperometric), catalytic combustion (hotwire) and semiconductor (modulation in conduction), metal oxides remain the most widely used choice for sensor technology due to their low cost and the possibility to alter their structure and composition to meet the analytical requirements [18]. Numerous metal oxides both in simple (e.g., SnO_2, TiO_2, ZnO, Fe_2O_3) and multi-component (e.g., $SrTiO_3$, $BaCeO_3$, $MgAl_2O_4$, $BiFeO_3$) compositions have been used in developing metal oxide sensors [19]. The sensing mechanism principally involves charge transfer at the gas–solid interface, which can be monitored as the resistance change in the material, whereby the magnitude of change in electrical resistance is proportional to the concentration of the target gas present [18, 20]. In the case of n-type oxides, there is either donation (for reducing gases such as CO, CH_4, C_2H_5OH) or subtraction (for oxidizing gases such as NO_x, O_3) of electrons from the conduction band, whereas converse is true for p-type oxides [2, 4, 21, 22].

Solid-state gas sensors, based on conductometric metal oxide–semiconductor structures represent the most promising approach towards the development of commercial chemical sensing devices due to the simplicity of their use and the reversibility of the surface reactions [4, 23]. In most of the sensors, the chemical reactions occurring on the sensor surface are monitored by measuring the change in capacitance, work function, mass or optical characteristics upon gas–solid interactions [24]. In addition, the competitive edge given by the intrinsic advantages of using nanostructured materials in solid-state sensing has provided new impulse to the field, largely due to the possibilities of synthesizing them in various engineered compositions (pure, doped, composites, heterostructures) and forms (particles, tubes, wires, films) that provenly result in high sensitivity and faster response/recovery times and can lead to extensive miniaturization of the device structures leading to low power consumption [25]. Furthermore, innovative strategies of surface functionalization and pre-concentrators has helped in circumventing the major bottleneck of solid-state gas sensors namely cross-sensitivity (lack of chemically specific adsorption) [26, 27]. Nevertheless, the quest for new sensor materials with optimized '5S' parameters (**5 S of Sensor Technology:** Simplicity, Sensitivity, Selectivity, Signal Readout, System) continue to act as the research motor in the field of chemical sensing.

Chemical sensors based on semiconducting metal oxides such as SnO_2 ($E_g = 3.6$ eV), TiO_2 ($E_g = 3.3$ eV), WO_3 ($E_g = 2.8$), ZnO ($E_g = 3.37$ eV),

Fe_2O_3 ($E_g = 2.3$ eV) and In_2O_3 ($E_g = 2.6$ eV) [28] have been extensively studied and the effects of lattice- and surface-doping in enhancing the sensor parameters have been demonstrated [29–31]. The need for specific detection and monitoring of potentially hazardous gases (e.g., NO_x, NH_3, O_3, CO, CH_4, H_2, SO_2, etc.) and polluting volatile organic compounds (VOCs) in the urban environment has significantly increased in the past two decades with field of applications including fire detectors, detection of threshold concentrations of toxic gases, regulation of food quality, indoor air quality in residential buildings and vehicles. The major challenge of the sensor technology is to design sensor systems or multi-sensor platforms to precisely analyse complex environmental fluctuations amidst well-known cross-sensitivity problems [32]. Moreover, the development of next generation sensors demands sensing devices with better measurement precision and longevity. In addition, sensors can be multifunctional with an energy harvesting unit coupled to the sensing element in order to produce self-sustained autonomous units [33, 34]. Further improved integration technologies allow embedding multiple sensors in a single sensor unit to create more autonomous and intelligent sensor systems.

The development of new generation of sensors with advanced capabilities and constituated by advanced and adaptive materials will underpin common high-volume consumer products and higher-performance, higher-value automotive and medical products in the future. Therefore the interesting challenge would be the design, development and fabrication of new sensors devices based on materials heterostructure for detection of multiple analytes simultaneously using sensor arrays or multifunctional sensors. Among new materials, hetero-nanostructures have shown exciting opportunities to tune sensor properties emerging from the coupling effects at the interface and the heterojunction barrier characteristics (Fig. 13.1). Subject to the nature of the components, heterojunctions can be made up of metal and oxide, oxide–oxide, metal or metal oxide incorporated in an inorganic matrix as well as organic polymer-based blends. For example, metal nanoparticles dispersed in an oxide matrix are interesting systems for improving gas sensing properties due to increased surface area and altered electrical properties due to the formation of Schottky contacts.

In addition, the metal-oxide interfaces have shown to be catalytically active (e.g., TiO_2/Au) [35] which can substantiate the analyte-oxide interactions and leverage the sensing process. Further, the possible exchange of electrons between the metallic nanoparticles and metal oxide grains can change the width of the depletion layer by influencing the operation temperature (reduction) and overall sensor response. For example, ZnO films with embedded noble-metal nanoparticles (Pt, Au and Ag) were shown to exhibit enhanced sensitivity and fast initial recovery behavior, when compared to pure ZnO films [36]. The presence of electronic interaction between Au and TiO_2 causes an increase in the electrical resistance of the system, which enhances the magnitude of transduction behavior. At the heterojunction (M/MO_x) the Fermi level of metal oxide is shifted to correspond to the work function of metal. Thus, metal nanoparticles attracts electrons from the oxide to produce electron depletion (in addition to the one caused by adsorbed oxygen), which ultimately alters the Debye length.

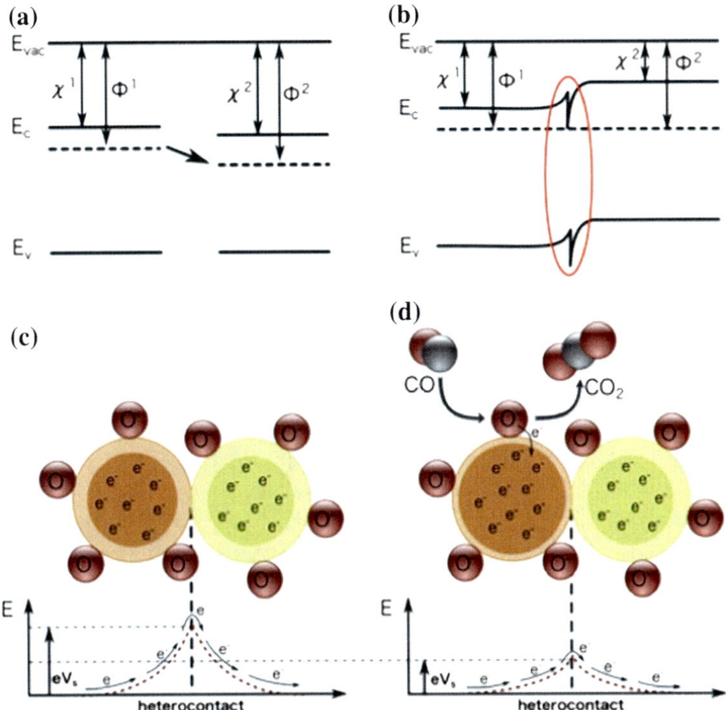

Fig. 13.1 Band bending at the interface of semiconductors heterostructures (Fe_2O_3/ZnO) a before and b after contact and barrier formation on heterocontacts without (c) and with reducing gas (d)

The hetero-structures of metal oxide nano-architecture can create intertwined nanoscopic conduction channels that enhance the charge transport kinetics responsible, for instance, for shorter response times. Furthermore, the fabrication of core–shell structures and use of porous structures (e.g., nanotubes) offers higher gas adsorption capacity, which positively influences the sensing behavior. For example, the variation of shell thickness in core–shell nanowires is reported to have a dramatic change in the ethanol sensing properties of α-Fe_2O_3/ZnO hetero-nanostructures [37]. Figure 13.1 depicts the expected band-bending phenomenon upon formation of the Fe_2O_3/ZnO interface in the semiconductor heterostructure. Similarly, the incorporation of metal or metal oxide nanostructures in polymer matrices enables a facile fabrication of planar sensor devices and improvement in gas sensing features due to the formation of metal oxide-conductive polymer junction and catalytic behavior of metal nanoparticles as shown in SnO_2-polyaniline (PANI) nanocomposites [38].

13.2 Role of Micro- and Nano-materials in Sensors

13.2.1 Nanomaterials for Chemical Sensors

Metal oxide semiconductors possess unique properties finding applications in several technological domains such as sensors, photocatalysis and varistors [39–41]. The working principle of these materials as chemometric sensors is based on the electrical transduction reactions which take place at the interface between molecular adsorbents and the metal oxide's surfaces [42–44]. However, their final performance is governed by manifold experimental parameters such as the concentration of oxygen vacancies, which are originated during the materials synthesis; and the intrinsic carrier concentration [45, 46]. Commercial sensors are commonly based on thin layers of metal oxide semiconductors deposited onto hotplates, which are used to maintain the temperature at the optimal values in order to activate the surface transduction mechanisms necessary to detect the chemical analytes[17] (Fig. 13.5a). Although this design has been successful from a commercial point of view[1], it faces two major challenges: (1) high power consumption which hinders its use in portable and autonomous systems and, (2) poor stability derived from multiple causes such as the grain boundary contribution among nanoparticles inside the sensing layers [47]. Consequently, making low-power consuming, stable and device-quality metal oxide chemical sensors remains at the forefront of this technology and represent one of major roadblocks of this approach against other more favoured alternatives (e.g., electrochemical cells).

The unique fundamental properties of metal oxide nanomaterials and related advantages have given rise to a fast and significant progress in the synthesis and characterization of nanowires and other nanostructures; and their final integration in proof-of-concept devices [48–57]. However, their manipulation and characterization is not a straightforward process due to the intrinsic limitations, when working at the nanoscale [58–60]. To accomplish electrical measurements, free of parasitic effects and to develop competitive sensors, different fabrication and characterization strategies have been successfully evaluated so far [61]. Thus, it can be asserted that the present state of development of nanomaterial-based technologies guarantees a complete and well-controlled characterization of proof-of-concept devices, overcoming most of the abovementioned matters which were considered insurmountable obstacles in the past. However, the present state of the technological development is mostly limited to the fabrication of single-nanowire prototypes (Fig. 13.2b) and the integration at a larger scale and using low-cost techniques represent a yet-to-be-solved challenge.

---

[1] There are many companies offering this type of sensors, such as Figaro, FIS, MICS, UST, CityTech, Applied-Sensors, NewCosmos, etc. See information provided by the gas sensors manufacturers on their homepages: (i) http://www.figarosens.com; (ii) http://www.fisinc.co.jp; (iii) http://www.appliedsensors.com; (iv) http://www.citytech.com; (v) http://www.microchem.com

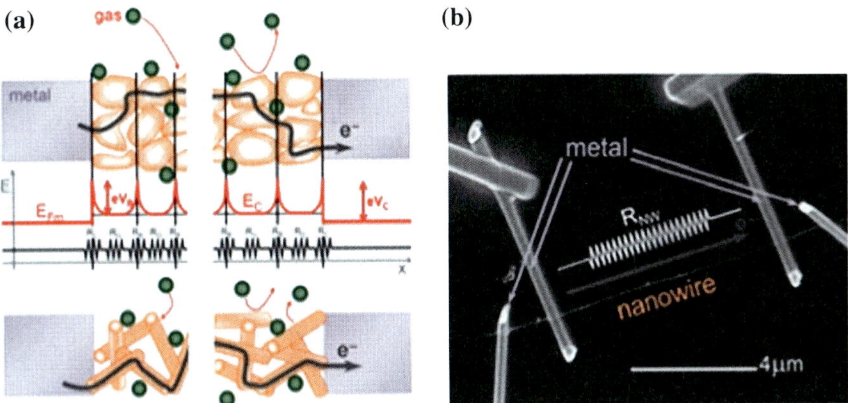

Fig. 13.2 Schematic diagrams of different types of conductometric gas sensors based on metal oxides. **a** Commercial thin-film sensor formed by a layer of nanoparticles. Here, electrons must go through a network of nanocrystals with different size and shape. From an energy point of view, electrons are to overcome potential barriers [(1) metal–semiconductor barriers (eV_C) and (2) intergrain boundary barriers (eV_B)]. The overall influence of gas on the height of the barriers determines the final response of the sensor. This is equivalent to a network of resistors [(1) metal–semiconductor contacts (R_C), (2) grain boundary interfaces (R_B) and (3) metal-oxide grains (R_G)]. **b** Single-nanowire sensor. If a nanowire is measured in 4-probe DC configuration, the conductometric response is basically determined by changes of the conduction channel along the nanowire (R_{NW}). On the contrary, contacts effects are overcome. Reproduced with permission from [53]

One of the main advantages of using individual nanostructures as building-blocks in sensor prototypes is their potential to provide a deeper comprehension of the fundamental adsorption mechanisms of analytes onto metal oxides than the one obtained upon using thin film sensors constituted by randomly oriented nanoparticles (Fig. 13.3). The low mass of individual nanowires facilitates heating them up enabling functional sensors based on individual nanowires and requiring a power consumption of only few tens of microwatts compared to thin film sensors mounted onto micro-hotplates, which usually require milliwatts to be operated in continuous mode [62, 63]. Figure 13.2a shows configuration typically used in commercial metal oxide sensors working with a thin ceramic film that is heated in order to activate adsorption/desorption of molecules at the surface, where by the modulation of conductance is measured as the function of the concentration of exposed gas. In a simplistic approximation, grain boundaries are described by highly resistive barriers, which represent the main contribution to the overall device resistance (Fig. 13.2a).

Consequently, the modulation of barriers among nanoparticles constitutes the most important transduction mechanism in porous-film sensors. However, the random aggregation of nanoparticles in the film and the size-distribution make an accurate study of the transduction phenomena rather difficult. The particular geometry of each neck and boundary, and the unique orientation of the adsorbate-modulated electrical field (E_{bar}) regarding the externally applied bias field (E_{bias})

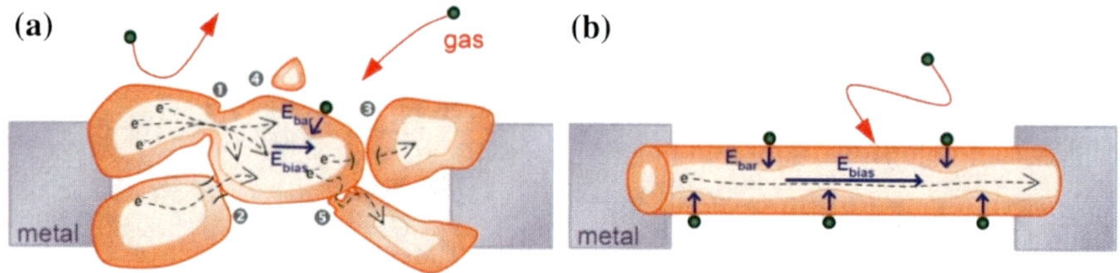

Fig. 13.3 **a** Diagram illustrating different types of necks in polycrystalline metal oxide matrices derived from the random nature of the layer. *1* and *2* Thick and thin necks between grains: the thicker the neck, the easier the electron transfer. *3* and *4* Intergrain boundary without material continuity: if the intergrain distances is short enough, electron transfer will take place by tunnel-assisted mechanisms [63]. 5 Non homogeneous intergrain interface. In this scenario the electric field that causes the band bending near the surface due to the gas interaction (E_{bar}) and the electrical field externally applied to perform the conductometric measurements (E_{bias}) are not necessarily orthogonal. **b** Diagram illustrating the gas interaction in individual nanowires: any intergrain necks or boundaries are considered. Moreover, E_{bar} and E_{bias} fields are always orthogonal and independent. Reproduced with permission from [53]

present in conductometric measurements (Fig. 13.3a) make necessary to describe thin-film sensors composed of individual nanoparticles with simplified models, and to analyze their responses as the convolution of diverse contributions. The same conclusions are found if bundles of nanowires are used instead of nanoparticles, since randomly oriented boundary effects are reduced but not totally eliminated (Fig. 13.2b). Since uncontrolled characteristics of necks and boundaries among nanoparticles has a determining influence in the gas response [64, 65]. These uncontrolled contributions are circumvented, if individual nanowire-based sensors are studied (Fig. 13.2b). In this scenario, E_{bar} is always orthogonal to E_{bias} (Fig. 13.2a), and direct responses towards gases are monitored, because the analyte diffusion among nanograins/nanowires is eliminated. Thus, the sensing principle of individual metal oxide nanowires can be theoretically described by pure surface effects. Adsorption of molecules at the nanowire surface modulates the width of a depleted region and the intensity of the associated electrical field E_{bar} (Fig. 13.3) close to the external shell. This effect based on the capture and release of electrical charges at the surface modifies the conduction channel through the nanowire and as a consequence the electrical resistance R_{NW} (Figs. 13.2 and 13.4), providing direct and fundamental information of the electrical charge exchange and the role of surface states n_v in the sensing process. According to this model, R_{NW} under exposure to a target gas is given by:

$$R_{NW} = \frac{\rho L}{\pi (r - \lambda)^2} \tag{13.1}$$

where ρ is the nanowire resistivity, L the nanowire's length, r the nanowire's radius and λ the width of the depletion layer created by adsorbed molecules (Fig. 13.4). Equation 13.1 establishes a direct connection between the nanowire's radius and the gas response: the thinner the nanowire is, the higher the resistance

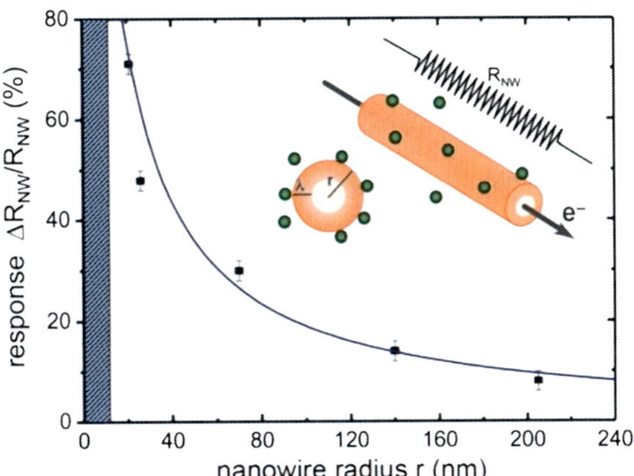

Fig. 13.4 Response of SnO_2 nanowires to synthetic air/nitrogen pulses as function of their radii measured at $T \sim 300\ °C$. Higher responses are clearly observed with shrinking dimensions in correspondence with Eq. (13.1) (*solid line*). Reproduced with permission from [53]

modulation will be observed (Figs. 13.3b and 13.4). Thus, the highest responses are always measured with ultrathin nanowires (radius below 40 nm) [7, 25].

It is noteworthy that the reduction of the sensing area of nanowire-based devices does not mean a decrease of their responses towards gas, since the modulation of the electrical resistance R_{NW} is directly proportional to the number of surface sites n_v per unit area interacting with gas molecules, and is not a function of the total effective surface. In other words, the same relative gas response is monitored with a single nanowire prototype than that obtained with multiple nanowire-based devices contacted in parallel, provided that the radii of the nanowires' are the same in both cases.

To guarantee the validity of the former assertion, gas concentration in air must be high enough to ensure the interaction of gas molecules with most of the surface sites (n_v) in a time scale shorter than the typical response times towards gas species. With regard to this point, it can be demonstrated that the interval collision between molecules and nanowires (even for the thinnest ones; $r = 20$ nm) at typical gas concentrations (from 0.1 ppm up to 1.000 ppm) is always orders of magnitude shorter than the typical dynamic response of these sensors. Finally, it should be highlighted that the absence of nooks and crannies in nanowire-based devices facilitates direct adsorption/desorption of gas molecules, improving the dynamic behaviour of these prototypes in comparison to those observed with porous-film sensors, in which gas diffusion dominates their dynamic response [66–71]. In this case, interactions between molecules and metal oxides nanowires occurs in a simplified scenario, which facilitates an easier comparison between experimental data and the results obtained in simulation results in which mostly ideal metal oxide surfaces are taken into account [30, 72]. The good agreement between the experiments and simulated systems found up to now paves the way to gain a deeper comprehension of the fundamental sensing mechanisms ruling the behaviour of metal oxides and heterostructures may become an extremely helpful tool to engineer new and better devices in the future.

13.2.2 Functional Characteristics of Nanomaterials

The synthesis of nanomaterials can be generally divided into liquid phase and gas phase methods, both strategies involving the controlled decomposition of molecular precursors like metal salts or metal–organic compounds to form the desired nanostructured material in various dimensionalities (0D: spherical; 1D: anisotropic; 2D: planar; 3D: networks and composites), morphologies (e.g., particles, wires, rods, ribbons, tubes, layers, branched structures, etc.) and compositions, for example metals, metal oxides, nitrides, sulfides, etc. [73, 74].

In the liquid phase processing a precursor solution undergoes a thermal treatment to form the desired solid phase by decomposition of the precursor or during a chemical reaction with additional chemicals the solid phase precipitates out of the solution; additional capping agents (surfactants, stabilizers) are used for controlling the shape and size of the material formed [75]. Various synthetic methods have been developed besides the simple convective heating of a precursor solution like sol–gel processing [76], hydro- and solvothermal synthesis [77] hot-injection methods, [78] microwave assisted synthesis [79], electro spinning [80] and the combinations thereof, which enable the formation of simple particular shapes, elongated and tubular structures up to complex branched morphologies through sequential precursor decomposition.

In the same way gas phase synthesis enables the synthesis of all kinds of different shapes and material combinations like liquid phase processing, starting from volatile precursors or evaporating solid material using high vacuum and temperature or sputtering techniques, respectively. Depending on the energy source applied or physical properties of the process, one distinguishes between physical vapor deposition (PVD) [43], chemical vapor deposition (CVD) [81], plasma-enhanced chemical vapor deposition (PE-CVD) [82], atomic layer deposition (ALD) [83], magnetron sputtering [84], laser ablation [85] and flame spray pyrolysis [86]. As the gas sensing properties of each material depend on the reactions taking place at the gas/solid interface, the same material produced in the same macroscopic morphology via different methods or different precursors will show varying responses to the same analytes, due to the different surface roughness, variation in lattice defects and oxygen vacancies as well as different crystallinity and stress/strain present in the material which influence drastically the interactions (adsorption, desorption, diffusion, decomposition) of gas molecules with the sensor material [87].

Synthesis methods for growing nanomaterials (nanowires, nanorods, nano ribbon, quantum dots etc.) have made significant progress in the last years, enabling the production of nanostructures with controlled size and excellent crystalline quality [88]. Moreover, recent works are demonstrating the possibility to obtain complex structural designs such as coaxial and lineal heterostructured-nanostructures [89, 90]. All these steps forwards pave the way to develop multifunctional applications and devices based on semiconductor nanowires, which make use of their high anisotropy in morphology and excellent crystalline quality.

These two intrinsic characteristics lead to new physical and electrical properties distinguished from their bulk counterparts. Present efforts focus on experimentally demonstrate these assertions; it was shown that nanostructures exhibit unique size-dependent electrical transport properties, high optical efficiency [91], enhanced thermoelectric properties compared to conventional Peltier elements [92], and good mechanical [93, 94] and electromechanical behaviors [95]. In this section, the most significant properties which encourage researchers to develop new prototypes for sensing with nanowires are briefly reviewed.

13.3 Material Nano-architectures for Sensing Application

The demand for more efficient metal oxide gas sensors requires an enhanced sensitivity, selectivity and stability, as well as reduced power consumption and further miniaturisation of devices. Therefore, the fabrication of material nano-architecture of various metal oxides is being strongly investigated in the last few years in order to exploit the complementaryjunctionbehaviors (e.g., p- and n-type semiconducting oxides) or different electronic properties (e.g., differential work function), which result in interfacial phenomenon not known in individual phases. The appropriate combination of material properties can produce heterostructures and composite materials with enhanced charge transduction properties, catalytic effects or modulated potential barriers at the grain boundaries, which are advantageous for improved sensing applications. Various material heterostructures (in the form of 0D, 1D and 2D and nano-composite architectures) with numerous binary heterostructured oxide semiconductors such as ZnO-TiO_2[96] MoO_3-TiO_2 [97–99], Fe_2O_3-SnO_2 [100], Fe_2O_3–In_2O_3 [101] and mesoporous ZnO–SnO_2 [102] etc. have been studied for their chemical sensors activity underscore the unique advantages of heterocontacts.

13.3.1 Quasi-Zero-Dimensional Metal Oxide

Spherical nanoparticles, so called zero dimensional (0D) nanostructures with a particle diameter less than 100 nm cannot be used as single sensing devices as the dimensions of the electrical contacts for the readout electronic are in the same order of magnitude thus hindering the measurements of individual particles (in contrast to 1D nanostructures). Therefore screen printed gas sensors on interdigitated electrodes based on thick-film technology using nanoparticle inks have been established as an industrial standard. Nanoparticle derived heterostructures can be produced as doped materials, core–shell structures, mixtures of different metaloxides or as decorated particles using quantum dots.

Noble metals, such as Ag, Au, Pd and Pt are the most studied additives used in gas sensing applications. These metal clusters sensitize the metal oxide as

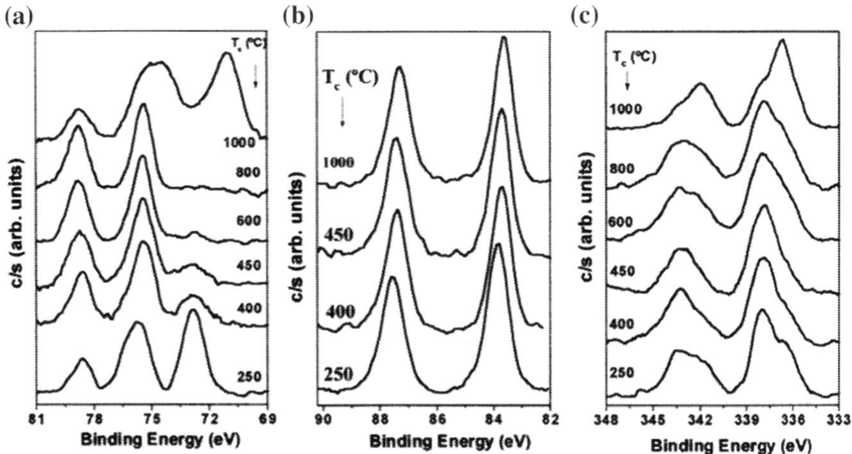

Fig. 13.5 XPS spectra of SnO_2 nanoparticles doped with 2 % noble metals after different annealing temperatures showing the Pt_{4f} (**a**), Au_{4f} (**b**) and Pd_{3d} (**c**) binding energies [104]

additional analyte-additive interactions change the overall surface chemistry and therefore the sensor's sensitivity and selectivity [103]. The evolution of XPS (X-Ray Photoelectron Spectroscopy) spectra with varying calcination temperatures (Fig. 13.5) proves that different chemical species are present on the surface of the metal oxide after calcination. Whereas Au doped SnO_2 showed no change in the signals over a calcination temperature range (Fig. 13.5b), different species of Pt and Pd on SnO_2 are present ranging from the corresponding noble metal oxides in the +2 and +4 oxidation states to metallic state (Fig. 13.5a, c).

Different chemical species present on the same metal oxide leads to an enhanced selectivity towards certain target gases, as the interactions between the gaseous analytes and the solid metal oxides differ significantly. Gaskaov et. al. demonstrated the effect of various modifiers on the selectivity of SnO_2 towards reducing and oxidizing gases [105].Although both analytes shown in Fig. 13.6 are reducing gases, thus giving an unspecific sensor response on pure SnO_2, the formation of hetero-contacts using different metal oxides and noble metals tune the selectivity dramatically.

The modifications of SnO_2 with Au (Fig. 13.6a) drastically enhance the response towards CO whereas a modification of SnO_2 with Fe_2O_3 or CuO (Fig. 13.6b) enhances the response towards H_2S. The enhanced response in the latter case can be explained by the chemical affinity of iron and copper towards sulfur, thus forming the corresponding sulfides, which are not stable under the operating conditions of the sensor and are therefore reoxidized to the starting metal oxide modifier. The effect of heterostructures on the sensitivity towards ethanol was shown by mixing Fe_2O_3 and SnO_2 nanoparticles in different stochiomteries ranging from pure Fe_2O_3 ($x = 1$) to pure SnO_2 particles ($x = 0$) resulting in the following heterostructures $(Fe_2O_3)_x(SnO_2)_{1-x}$ (Fig. 13.7).

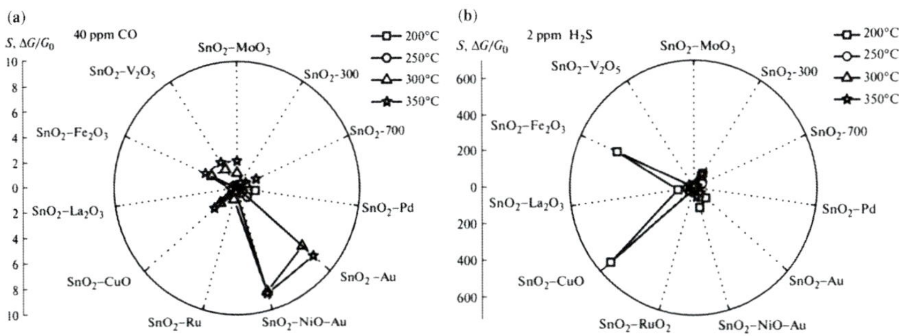

Fig. 13.6 Gas sensor response of different reducing gases towards modified SnO_2 using **a** CO and **b** H_2S [105]

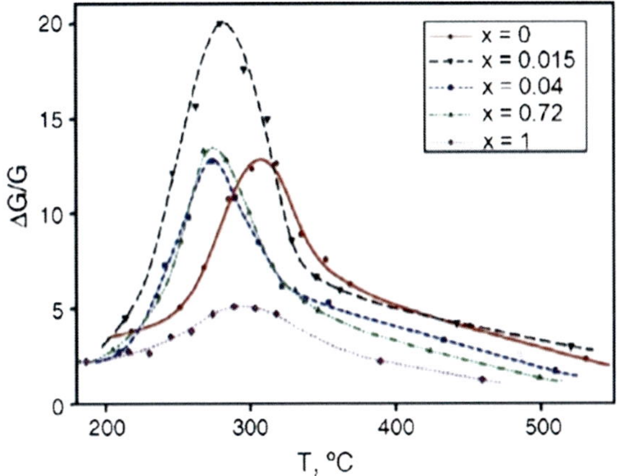

Fig. 13.7 Sensor response towards ethanol at different operating temperatures of $(Fe_2O_3)_x(SnO_2)_{1-x}$ heterostructures [106]

It could be clearly shown that the pure metal oxides exhibit a poor to decent sensor response of 5 (Fe_2O_3) and 13 (SnO_2), whereas the heterostructure $(Fe_2O_3)_{0.015}(SnO_2)_{0.985}$ yielded a superior response of 20 at a similar working temperature of 300 °C. In this case the major crystallographic phase was determined to be SnO_2 where Fe^{3+} substitutes partially the Sn^{4+} sites in the base tin oxide.

13.3.2 Quasi-One Dimensional Nanostructures and Anisotropic Heterostructures

Quasi one-dimensional (1D) semiconductor structures such as inorganic nanowires or nanotubes are interesting components for bottom-up fabrication of nanoscale devices, since electrical, optical and chemical behaviours in individual nano-structures could be potentially tuned exploiting quantum confinement of conduction electrons in reduced dimensions. A number of applications ranging across sensors, lasers, waveguides and bio-diagnostics have recently been demonstrated for 1D nanostructures, which have triggered tremendous interest in the science and technology of these tiny structures especially as functional elements in the future nanodevices. The structural hybrids (e.g., core–shell, hyperbranched structures) represent electronically coupled systems, in which the charge-transfer is defined through spatially defined functional material units. Metal decorated semiconductor nanowires have been demonstrated to promote charge separation and to exhibit charge retention (in case of metal–semiconductor interface) [107]. In addition, the nanoscale tailoring of heterostructures induce a number of size/shape-related effects making the physic-chemical behaviour of the individual material component domains strongly deviate from that of their bulk counterparts. Alterations in electronic state densities, band gap width, plasmon resonance conditions, interfacial band bending, dynamics of charge carrier recombination/delocalization, as well as modulation in reactivity arising from enhanced surface area and facet-dependent reactant adsorption and/or electron transfer capabilities offer further parameters for optimizing gas sensing capabilities of hetero-junctioned nanomaterials [108]. Therefore, the basic materials engineering platforms use anisotropic architectures, such as rod-like, wire-like, and branched morphologies and free-standing colloidal suspension on one side, and of substrate-supported nanostructures with controlled spatial organization on the other to optimize the sensing properties (Fig. 13.8). One challenge in the production of anisotropic nano-heterostructures is the diversity and complexity of the types of materials available. On the other hand inorganic nanowires-based heterostructures exhibit unique multi-functionality and can be rationally designed both axially and radially for integrating materials of different chemical compositions, sizes/shapes, morphologies and phases. This allows tuning the band offsets for photocatalytic applications.

Anisotropic nanocrystals encompass a rather broad range of fiber-like structures, which can be rationally and predictably synthesized, subject to the synthesis technique, in single or poly-crystalline forms with controllable chemical composition, diameter, length, and doping levels [114]. The synthesis of nano-heterostructures has been successfully achieved following several methodologies (Table 13.1). For instance, novel hierarchical heterostructures of TiO_2 nanofibers decorated with hematite (α-Fe_2O_3) or magnetite (Fe_3O_4) were prepared by combining the electro-spinning technique and the hydrothermal method [115]. The resulting hierarchical heterostructures revealed that the secondary α-Fe_2O_3 or Fe_3O_4 nanostructures successfully grew on the surface of the primary TiO_2 nanofibers substrates, thus integrating the magnetic and photocatalytic properties into the α-Fe_2O_3/TiO_2 and Fe_3O_4/TiO_2 heterostructures (Fig. 13.9). Nevertheless, the electronic interactions between the semiconductors phases were found to be unfavorable for the photoactivity, which was reduced due to the lower band gap of iron oxides (Fe_2O_3: 2.2 and Fe_3O_4: 0.1 eV) against that of TiO_2 (3.2 eV) mainly due to faster electron–hole recombination facilitated by the heterojunction [116].

Fig. 13.8 Overview of different 1D derived morphologies consisting of **a** OD@1D (CdS@TiO$_2$) [109], **b** 2D@1D (Fe$_2$O$_3$@SnO$_2$) [110] or **c** core shell structures (TiO$_2$@MoO$_3$) [111] and **d** 1D@1D heterostructures (Fe$_2$O$_3$@SnO$_2$) [112] Also pseudo-1D structures **e** like SnO$_2$–ZnO hollow nanofibers consisting of individual 0D nanoparticles belong to this structural family of heterojunctions [113]

Recent reports demonstrate that appropriate alignment of the work functions between metal oxides and smaller band gap semiconductors (e.g., metal chalcogenides) significantly extend their absorption cross-section and consequently enhances photocurrent efficiencies through charge injection from visible light harvester unit to the broad-band semiconductors, which can be used to generate multiple excitons with a single photon leading to high absorption cross section, and high extinction coefficient, which increases the overall efficiency of photoelectrical devices [117]. Arrays of metal oxides (MO$_x$) nanowires such as ZnO and TiO$_2$, decorated with quantum dots (QDs) have been fabricated on various conductive substrates, however they suffer from poor photo-conversion efficiency due to limited chemical stability and scattering of charge carriers due to grain boundaries. Lately the groups of Grimes [118], Aydil [119] have found a simple hydrothermal route to grow single crystalline TiO$_2$nanorods on FTO substrates. Kamat [120] and Guo's groups [121] have coated CdS QD's on such type of materials and investigated their photoelectrochemical properties to demonstrate higher photocurrent in such nanocomposite when compared to simple metal oxides.

Compared to single-phase materials, the hierarchical structures or integrated multisemiconductor systems possess significant advantages, such as promoting the separation of excess carriers and improving the optoelectronic properties of the host material. The modification of CdS nanowires by attachment of ZnO nanospheres on the surface of nanowires produced a visible light activated sensor with

Table 13.1 Selected examples of nano-heterostructures used in sensing applications

Selectivity	Hetero-structure	Sensitivity	Response/recovery time	Comment	Reference
H_2	ZnO@SnO_2 nanorods	18.4@100 ppm (350 °C)	N/A	n-p-n-type response towards H_2 and n-type response towards CO, NH_3, CH_4	[14]
EtOH	TiO_2@α-MoO_3 core/shell nanorods	4.8@10 ppm (180 °C)	<40 s/< 40 s		[111]
	SnO_2@α-Fe_2O_3 hierarchical hetero-nanostructures	4.7@10 ppm (350 °C)	N/A	No cross sensitivity towards H_2, CH_4, C_4H_{10}	[126]
	SnO2@α-MoO3 nanobelts	12@100 ppm (250 °C)	N/A		[127]
	SnO_2@CNT core shell nanotube	11.1@10 ppm (300 °C)	1 s/10 s		[128]
	TiO_2/$Ti_{1-x}Sn_xO_2$ segmented nanowires	3.54@300 ppm (160 °C)	N/A		[129]
	TiO_2@CeO_2 core/shell nanorods	4.5@1,000 ppm (270 °C)	<45 s/<45 s	Negligible cross sensitivity towards NH_3, CH_4, H_2, CO at 1000 ppm	[130]
H_2S	ZnO(0D)@SnO_2(1D) heterostructure	2.1@500 ppm (250 °C)	N/A	Low cross sensitivity towards CO, CH_4 at 500 ppm	[13]
NH_4	CdS@ZnO core shell nanocables	3800@30 ppm (20 °C)	<30 s/<30 s		[131]
NO_2	V_2O_4@CNT core shell nanotubes	18@6.5 ppm (150 °C)	N/A	Negligible cross sensitivity towards CO, ethanol, NH_3 up to 100 ppm	[132]
Toluene	SnO_2–ZnO hollow nanofibers	26.5@100 ppm (190 °C)	6–11 s/12–23 s		[113]
Cl_2	$W_{18}O_{49}$@SnO_2 nanowire hierarchical heterostructure	11@6 ppm (20 °C)	4.6 min/17 min	Negligible response towards H_2S and no cross sensitivity towards CO, NO_x, NH_3	[133]

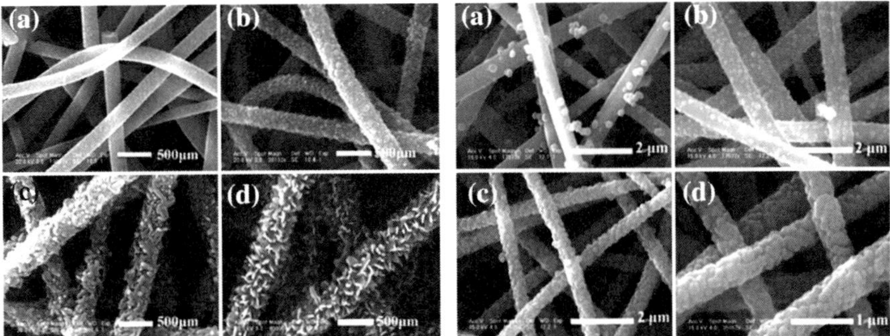

Fig. 13.9 SEM images of α-Fe$_2$O$_3$/TiO$_2$ (**a–d**, *left*) and Fe$_3$O$_4$/TiO$_2$ (**a–d**, *right*) hierarchical heterostructures

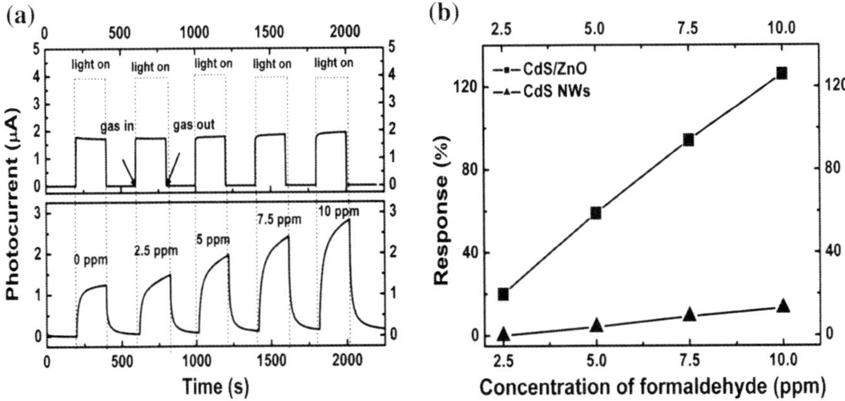

Fig. 13.10 **a** Photocurrent versus time plots of CdS nanowires and ZnO@CdS heterostructures **b** Response of CdS nanowires and ZnO@CdS heterostructures as a function of the detected analyte, formaldehyde (under visible light irradiation)

room temperature sensing properties substantiated by the photoinduced electron transfer between the ZnO nanospheres and the CdS NWs, which enhanced one order of magnitude in response to formaldehyde compared with that of CdS NWs (Fig. 13.10) [122].

The gas sensing performance of nanostructures can be further improved either by surface modification, for instance, by chemical treatments leading to surface etching, coarsening and creation of surface defects. Alternatively, the surface properties can be modified by anchoring catalytically active metal nanocrystals (e.g., Au, Ni, Pd, Pt) onto nanostructures, which modulate the surface energy band structure as well as transport and recombination dynamics of charge carriers and can improve the photoelectronic and sensing properties of metal oxidenano-heterostructures. For example, TiO$_2$ nanobelts prepared by a hydrothermal synthesis and coarsened by an acid treatment were shown to exhibit better sensing properties

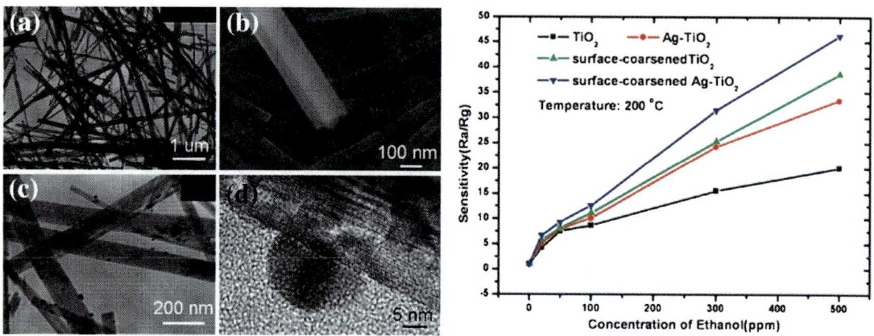

Fig. 13.11 Representative **a** TEM and **b** SEM micrographs (*left*) of TiO₂nanobelts, **c** Ag-TiO₂nanobelts, and (HR)TEM micrographs of **d** Ag-TiO₂ nanobelts Sensitivity profiles (*right*) of ethanol vapor sensors based on TiO₂ nanobelts, Ag-TiO₂ nanobelts, surface-coarsened TiO₂ nanobelts, and surface-coarsened Ag-TiO₂ nanobelts upon exposure to different concentrations of ethanol vapor at 200 °C

than as-synthesized nanobelts [123]. Similarly, the sensing properties of Ag-TiO₂ nanostructures, formed by depositing Ag nanocrystals on the surface of nanobelts, were better than the heterostructures formed by depositing Ag nanocrystals on the surface of coarsened nanobelts (Fig. 13.11). It was demonstrated that the sensor performance increases in the order of TiO₂ nanobelts < Ag-TiO₂nano-belts ≈ surface-coarsened TiO₂nanobelts < surface-coarsened Ag-TiO₂ nano-belts suggesting the impacts of surface coarsening and formation of metal-TiO₂ heterostructures on the sensor conductanceand detection performance (Fig. 13.11).

A number of heterostructures based on metal oxide nanostructures (Table 13.1) have been tested for gas sensing applications, whereby the general motivation is an expected enhancement of sensor response due to the absorption of target gas molecules at the hetero-contact, which induces charge transfer processes and modulates the barrier height to produce the conductometric change.

Detection of small- and large-sized volatile organic compounds (VOCs) has attracted considerable attention for indoor air quality and human breath analysis, respectively and demands development of new materials. Au-loaded TiO₂ nano-tubes were developed by photochemical deposition of Au nanoparticles (10 to 20 nm) on titania nanotubes for the detection of large-sized VOCs [124]. Among larger VOCs, propofol that is used as an intravenous hypnotic agent, is an important analyte because the determination of propofol (2,6-diisopropylphenol, $C_{12}H_{18}O$) concentrations in exhaled breath makes itpossible to monitor the depth of anesthesia [125]. The loading of Au nanoparticles and formation of Au-TiO₂ heterocontacts (Fig. 13.12) were verified by an increase in the electrical resistance of the Au-TiO₂ structures, when compared to neat TiO₂ nanotubes. At Au-TiO₂ heterocontacts, the Fermi level of TiO₂ shifts to correspond to the work function of Au, whereby Au attracts electrons from TiO₂ to produce enhanced electron depletion on the surface of the nanotubes increasing the electrical resistance of the composite. Upon introduction of propofol, the resistance of the devices decreased

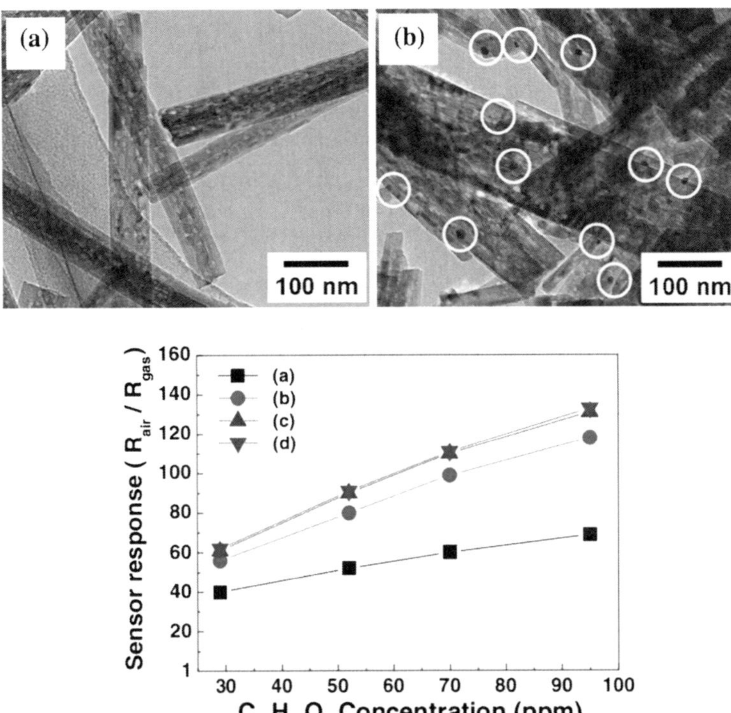

Fig. 13.12 *Top* TEM images of pure (**a**) and Au-loaded TiO$_2$ nanotubes **b** *Below* Dependence of sensor responses on propofol concentration (30, 55, 70, and 95 ppm) in nanotubes, **b** loaded with Au at 0.25 wt %, **c** 0.42 wt %, and **d** 0.92 wt %

rapidly due to the reaction of surface-adsorbed oxygen with propofol thus revealing a typicalbehavior of resistive sensors based on n-type semiconductors (Fig. 13.12).

In addition to the alteration of electrical properties, hyperbranched heterostructures can be used to modify the adsorption and wetting behaviors of the sensing materials. For instance, sequential chemical vapor deposition (CVD) of SnO$_2$ and SiO$_x$ allowed fabricating brush-type SnO$_2$@SnO$_2$ and SnO$_2$@SnO$_2$@-SiO$_x$ architectures, which exhibited hydrophobic and super-hydrophobic behaviors, respectively. Interestingly, unbranched SnO$_2$ nanowires were found to be super hydrophilic due to intrinsic porosity of nanowire mesh, whereas the decreasing contact area and hyper-branched architectures rendered the doubly- and triply-branched structures hydrophobic.

The randomly grown SnO$_2$ nanowires exhibited a contact angle of 3° attributed to the super hydrophilicity of the surface, the contact angle in SnO$_2$@SnO$_2$ heterostructures synthesized by a two-step CVD process increased up to 133°. The corresponding contact angle of SnO$_2$@SnO$_2$@SiO$_x$ heterostructures was further enhanced to 155.8° by a hydrophobic SiO$_x$ coating (Fig. 13.13). Besides

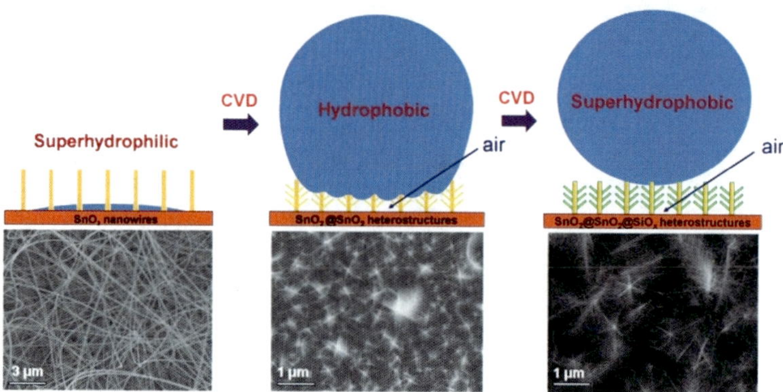

Fig. 13.13 Modification of the surface properties in tin oxide nanostructures by intentional branching ($SnO_2@SnO_2$ heterostructures) and surface coating ($SnO_2@SnO_2@SiO_x$ heterostructures)

enhancing the overall surface area, this concept of materials engineering allows to circumvent the well-known challenge of cross-sensitivity of metal oxide sensors against atmospheric moisture.

The controlled fabrication of arrays and composite architectures based on nanowires and nanotubes offers a unique platform to design and synthesize new materials based on fundamentally new effects originating from the site-selective variation of junction properties and electronic effects. For example, the photo-conductance of the SnO_2/V_2O_5 hierarchical heterostructures (Fig. 13.14), prepared by growing vanadium oxide nanorods on SnO_2 nanowire arrays, exhibited a red shift when compared to the photoresponse of the pure SnO_2 NWs [134].

The sensitivity and specificity of a sensor response can be further tuned by the addition of dopants as shown for a non-equilibrium TiO_2-SnO_2 solid-solution prepared by the sol–gel processing method. The incorporation of cadmium ions as dopants in the TiO_2-SnO_2 composite reportedly exhibits exclusive selectivity as well as high sensitivity (Fig. 13.15) towards formaldehyde [135].

13.3.3 Quasi-Two-Dimensional Metal Oxide Heterostructures

The thin film technology offers the possibility of reducing the amount of active material needed and therefore reducing the power consumption. Furthermore, in contrast to thick film metal oxide sensors, where the grain size of screen printed sensing layers is controlled by the synthetic procedure and thermal treatment of the raw materialand not the film thickness, the grain size of thin film devices directly correlates with the thickness of the film produce [47]. This enables a more precise control over the sensing capabilities of the applied material, as the grain size is one crucial factor in determining the physical response towards adsorbed analyte on

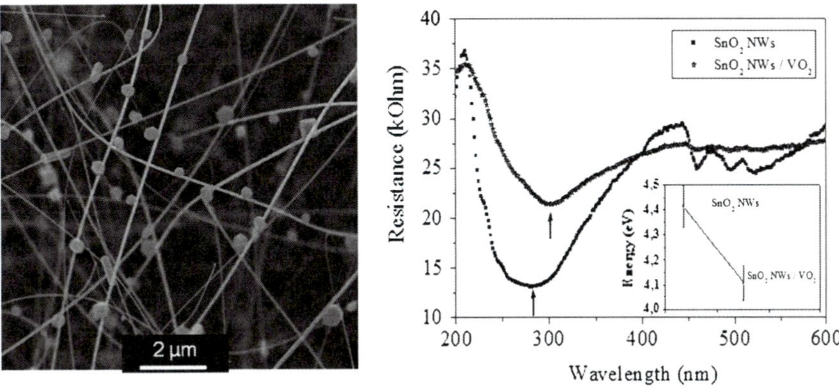

Fig. 13.14 SEM image of SnO_2/V_2O_5 hierarchical heterostructures (*left*) and the photoresponse of pure SnO_2 and SnO_2/V_2O_5 composite material (*right*)

Fig. 13.15 Response-recovery property of as-synthesized (*undoped*) and Cd-doped TiO_2-SnO_2 nanocomposite materials

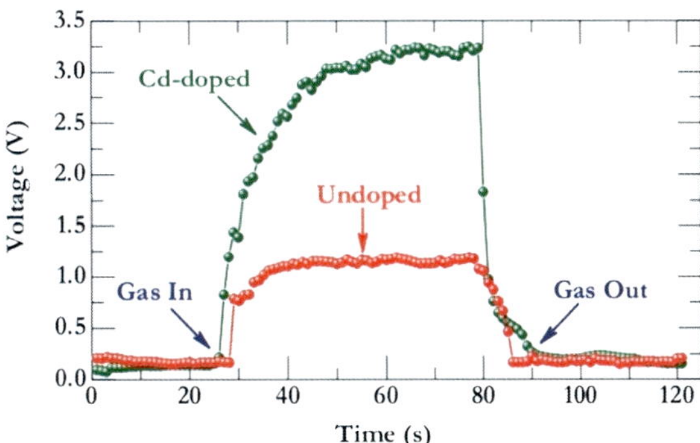

semiconductor surfaces. The grain size influence on the gas sensing properties of metal oxide films is discussed in depth in various publications [136–138]. It is widely accepted that the electro-physical properties of polycrystalline metal-oxides are strongly dependent on their microstructure, which can be described either by a "grain" or "neck" approach, where (depending on the synthetic route of producing the active material) the size of grains or width of necks are the predominant parameters in describing the gas-sensing properties of the films. Thin film devices are produced via different methods like spray pyrolysis [139], sputtering [140], chemical vapour deposition [141] or sol–gel methods [142], which yield homogeneous films with low or little porosity (see Sect. 13.2). As dense sensing layers hinder the diffusion of analytes and reduce the interface between active sensing material and target gas and hence reduce the sensitivity of the as produced device, nanostructured and/or porous thin films are needed to overcome these limitations.

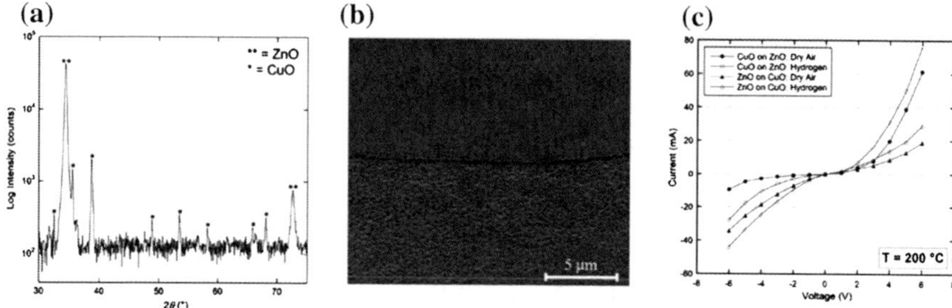

Fig. 13.16 Planar 2D thin film CuO@ZnO heterocontact: **a** XRD-Pattern showing the coexistence of pure metal oxides, **b** SEM micrograph of planar structure and **c** I–V characteristics of resulting p–n junction in dry air and 4,000 ppm hydrogen [144]

The use of oxide–oxide heterostructures in thin film technology for sensor follows different motivations:

1. New physico-chemical effects arise when combining different materials and
2. One metal-oxide acts as a template for structuring the active material which is deposited on top.

One example how porous oxide–oxide junctions enhance the sensing abilities was shown in ceramic humidity sensors, where porous Ta_2O_5/Al_2O_3 layers acted as active material [143]. Despite other sensing mechanisms used in humidity detection porous ceramics, like Al_2O_3, offer high stability especially in harsh environments and can react on adsorbed water by changing their resistance, conductance or capacitance, which is used as sensor readout. In a n-channel MOSFET (metal oxide semiconductor field effect transistor) device setup a gate electrode consisting of a Ta and Al layer is oxidised to yield a porous $Ta/Ta_2O_5/Al_2O_3$ sandwich structure which is far more sensitive and shows a much faster response, especially towards low relative humidity levels, compared to devices using pure Al_2O_3 gate materials. The origin of the observed superior performance of this heterostructure compared to pure sensing material still remains unclear, but it demonstrated that heterocontacts enhance crucial device parameters such as sensitivity, response and recovery time. Another planar thin film p-n heterojunction, consisting of CuO decorated ZnO, sensitive towards hydrogen is shown in Fig. 13.16a. The hetero-contact formed by two individual metal oxides (p-type CuO and n-type ZnO, XRD pattern in Fig. 13.16a) exhibits and I–V characteristic typical for p-n junctions, which is altered by adding reducing gases, like H_2 (Fig. 13.16c)

Despite the physico-chemical impact of oxide–oxide heterostructures simple structural features can be transferred from an oxidic substrate to the active sensing material. Using anodized alumina (AAO) membranes coated with TiO_2, efficient H_2 sensors can be produced, as the porosity of the material directly correlates with its sensitivity. These TiO_2@AAO heterostructures do not only transfer the porous structure from the template (Al_2O_3) to the active material (TiO_2), but stabilize the more sensitive anatase phase of TiO_2. The state of the art metal-oxide H_2 sensors are based on TiO_2 which tubular structure is produced via anodisation of Ti foils. This method yields the metastable anatase crystal structure and limits the operating or environment temperature to 400 °C, as a phase transformation to the thermodynamically favored rutile phase takes place and dramatically reduces the sensitivity of the device. While using a porous template of AAO, the more stable TiO_2 phase rutile can be deposited as a stable porous thin film extending the temperature regime for these sensor configurations above 600 °C. A reaction between Al_2O_3 and TiO_2 is not observed at temperatures below 1300 °C, but higher temperatures enable the formation of the mixed phase Al_2TiO_5, which is not a suitable sensing material [145].

In another example, incorporation of transition metal oxide nanoparticles NiO and Co_3O_4 in a porous silica glass matrix offered a high selectivity towards the detection of hydrogen (H_2) in a mixture of carbon monoxide (CO) and H_2. Both the NiO and Co_3O_4 doped films exhibited a conductometric p-type response, with a resistance increase upon exposure to the reducing gas. SiO_2-NiO films have shown the highest response to H_2 at 300 °C operating temperature and good selectivity to H_2 in the presence of CO as an interfering gas. Selectivity tests (Fig. 13.17) comprised of three exposure cycles including 200 ppm H_2 target gas in dry air (first section), followed by 200 ppm H_2 and to 200 ppm CO (second section) and 200 ppm CO alone (third section). The lack of CO interfering effects was explained in terms of a stronger affinity of the H_2 molecules to adsorb on the metal oxide surface (Gas sensing properties of nanocrystalline NiO and Co_3O_4 in porous silica sol–gel films [146]. Similarly, a higher sensitivity towards H_2 in a mixture of CO and H_2 was found at different operating temperatures (25–350 °C) and gas concentrations (10–1,000 ppm) in SiO_2-NiO and SiO_2-SnO_2 nanocomposite films based on p- and n-type oxides, respectively [147, 148].

The gas-sensitive properties of hetrostructures of complex structure based on both γ-Fe_2O_3 and α-Fe_2O_3 have been extensively studied by the Ivanovskaya et al. [101]. They have investigation the gas-sensitive properties of thin film sensors based on the double-layers Fe_2O_3/In_2O_3 and Fe_2O_3–In_2O_3/In_2O_3 towards gases of different chemical nature (C_2H_5OH, CH_4, CO, NH_3, NO_2, O_3). Itwas found the γ-Fe_2O_3–In_2O_3 composite (Fe: In = 9:1, mol) is more sensitive to O_3; on the contrary, the α-Fe_2O_3–In_2O_3 (9:1) system, possesses a higher sensitivity to NO_2. Figure 13.18a shows the dependence of the response values to NO_2 on the operating temperature for sensors with differentcomposition of the sensitive layer. The sensors based on heterostructure of Fe_2O_3–In_2O_3 have not only the greatestsignals, but can also operate properly at relatively lowtemperatures. α-Fe_2O_3–In_2O_3/In_2O_3 and α-Fe_2O_3–In_2O_3 sensors possesspoor responses to low concentration of 50 ppm CO (Fig. 13.18b) and almost insensitive for both CH_4 and NH_3. It was obtained that double-layer sensors exhibited the much more sensitivity towards alcohol (C_2H_5OH, CH_3OH) vapors than single-layer In_2O_3 and Fe_2O_3 samples and the maximum response was shown by the γ-Fe_2O_3–In_2O_3 heterostructure.

However, the increased sensing response of 845 for 1,000 ppm of ethanolwas obtained by the addition of α-Fe_2O_3 to SnO_2 thick films as studied obtained by Tan

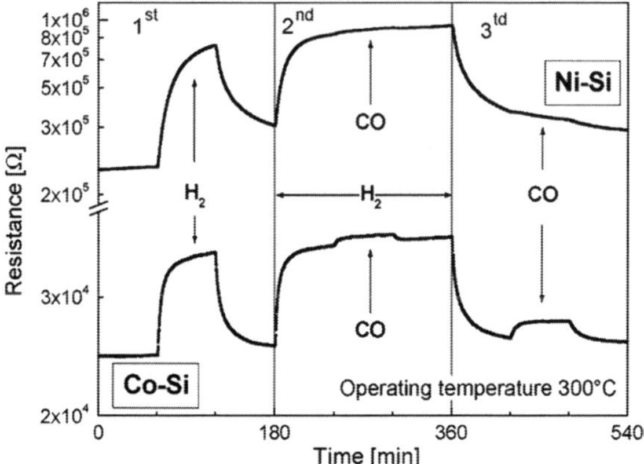

Fig. 13.17 Cross sensitivity test at 300 °C operating temperature of the SiO$_2$-NiO and the SiO$_2$-Co$_3$O$_4$ films to: (*first section*) 200 ppm H$_2$ in dry air; (*second section*) 200 ppm H$_2$ in dry air and 200 ppm CO; (*third section*) 200 ppm CO in dry air [146]

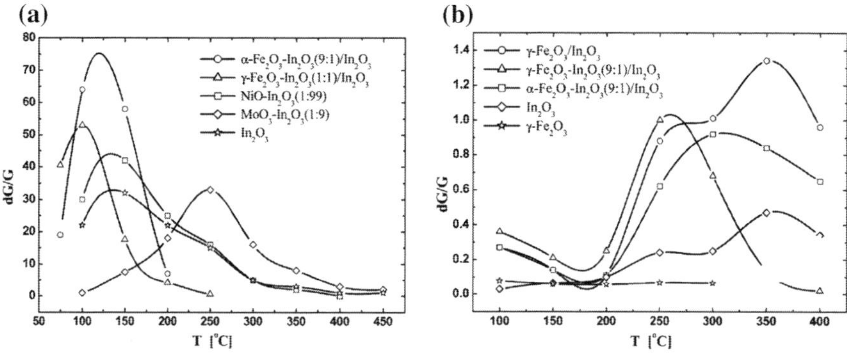

Fig. 13.18 Comparison of temperature-dependent sensitivity of In$_2$O$_3$ based double-layer sensors of different metals oxide to **a** 1 ppm NO$_2$ **b** 50 ppm CO

et al. [100], Jiang et al. [149] also studied the nano-sized SnO$_2$-α-Fe$_2$O$_3$-based powders with the grain size down to 8 nm for gas sensing and found that Sn^{4+} ions in α-Fe$_2$O$_3$ plays an important role in the gas sensitivity and the excellent sensitivity due to the nano-sized particle grains which exhibit enormous oxygen dangling bonds at their particle surfaces.

13.3.4 Nanocomposite Metal Oxides

Metal oxide based nanocomposite heterostructures are other promising directions in the development of advanced materials for sensing application. These materials have very specific properties like optical, electronic, catalytic, mechanical, and chemical properties which can be obtained by advanced nanocomposites and makes metal oxides nanocomposite very attractive material for gas sensing devices, for example, TiO_2 based mixed oxide nanocomposite films, including TiO_2-WO_3, TiO_2-MoO_3, TiO_2-NiO_x, and TiO_2-ZnO, have been characterized for gas-sensing applications by Wisitsoraat et al. [150]. NiO_x addition to TiO_2 with sufficiently high concentration has produced a p-type semiconducting thin film while WO_3, MoO_3, and ZnO inclusions result in typical n-type metal oxide semiconductors. The gas-sensing sensitivity, selectivity, and minimum detectable concentration toward different gases including acetone, CO, and NO_2 can be effectively controlled by different dopants and doping concentrations. The p-type NiO_x doped TiO_2 showed high sensitivity towards ethanol and acetone with distinct behaviors compared to other n-type TiO_2 thin films. The ZnO-TiO_2 nanocomposites have also received considerable interest, due to the synergistic combination of the peculiar component characteristics, such as the high reactivity of TiO_2 and the large exciton binding energy of ZnO (60 meV) [151, 152]. The drawbacks with ZnO systems solid-state gas sensors are relatively low selectivity and high operating temperatures [153, 154]. Such disadvantages can be overcome by the introduction of TiO_2 in ZnO matrices, resulting in improved sensor sensitivity and selectivity, as the titanium dioxide behaves as a catalytic promoter of the involved chemical processes [155, 156]. By adequateing the different choice of synthesis technique and processing parameters, properties of ZnO-TiO_2com- posite can be control. Barreca et al. [96] have also synthesized TiO_2 based ZnO nanocomposites by thermal CVD technique to improved functional performances in sensing devices for volatile organic compounds (VOCs). In this study the main attention was to develop the porous ZnO nanoplatelets (NPTs) which can be act as host matrices for the subsequent dispersion of TiO_2 nanoparticles, resulting in a suitable morphology for the development of efficient sensing devices. The obtained device as a function of the synthesis conditions, their gas sensing properties have been tested in the detection of VOCs, in particular, CH_3COCH_3, CH_3CH_2OH, and CO.

The dynamic response of ZnO-TiO_2 nanocomposite with pure ZnO NPTsfor CH_3COCH_3 detection has been shown in Fig. 13.19. In the case of ZnO-TiO_2 nanocomposites, the occurrence of reversible interactions between the analytes and the sensor elements has been obtained and the composite systems displayed both response and recovery times which is lower than pure ZnO NPTs. In the composite system, the observed current variations were systematically higher as compare to the pristine ZnO matrices with the higher TiO_2 amount, indicating a significant beneficial effect of titania addiction on the sensor performances [154, 155]. In TiO_2-ZnO composites, the higher oxygen defect concentration was cre- ated by titania dispersion, which are capable of adsorbing oxygen more efficiently than pure ZnO systems [155], providing superior functional performances. In addition, ZnO-TiO_2 coupling minimizes electron–hole recombination phenomena, thus resulting in a higher carrier concentration [156]. It was also observed that TiO_2 doping in ZnO matrices acts itself as a catalytic promoter, favoring surface

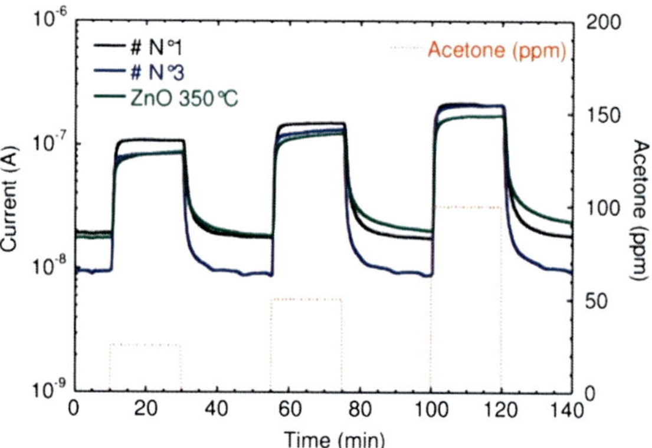

Fig. 13.19 Dynamic response of selected ZnO-TiO$_2$ nanocomposites upon exposure to concentration square pulses of acetone at a working temperature of 400 °C. The curve pertaining to a pure ZnO nanosystem is displayed for comparison

reactions between the target gases and the oxygen species adsorbed on the sensor surface [155, 156]. The reversibility and repeatability of these sensors were also satisfactory, as observed by repeating the tests many times without detecting any significant variation in the functional response.

Some reports are also available on the effect of Fe$_2$O$_3$ additives on the properties of In$_2$O$_3$ based sensors; like sputtering of Fe$_2$O$_3$ on In$_2$O$_3$ thin film increases its sensitivity to O$_3$ and reduces the optimal operating temperature [157]. O$_3$ sensitivity was considerably improved by the doping of γ-Fe$_2$O$_3$ in In$_2$O$_3$ thin film sensors as reported by Gutman et al. [158]. The high activity of γ-Fe$_2$O$_3$–In$_2$O$_3$ composite for O$_3$ detection is due to the specific features of γ-Fe$_2$O$_3$ structure, like the metal cation vacancies are create within the crystal lattice and the readiness of Fe^{2+} to Fe^{3+} transformation under exposure by gaseous species. As it is known, γ-Fe$_2$O$_3$ is characterized by a comparatively low thermal stability and the transformation of γ-Fe$_2$O$_3$ to α-Fe$_2$O$_3$ phase normally occurred at 485 °C. This peculiarity of γ-Fe$_2$O$_3$ limits its use as long-term stable gas-sensitive material.

Molybdenum trioxide (MoO$_3$) is also a promising candidate for application as a catalyst in oxidation of hydrocarbons and reduction of NO$_2$ gas [159]. It has physical and chemical properties similar to those of WO$_3$ [160]. MoO$_3$–SnO$_2$ nanocomposite is gas sensitive materials which can be incorporated as thick film, thin film and compressed powders depending on the deposition technique and the parameters used, which in turn are responsible for sensing properties towards a particular gas. MoO$_3$ doped SnO$_2$ mixed metal oxide (MoO$_3$–SnO$_2$) thin films have been prepared for chemical sensor by Galatsis et al. [161]. and Ansari et al. [162]. towards NO$_2$ and H$_2$ sensing. The addition of MoO$_3$ reduces the electrical conductivity of SnO$_2$ by two orders of magnitude in air, which may be due to the transfer of electrons trapped at oxygen vacancy sites to Mo^{6+} [163]. Kaur et al. [164]. studied the sensing behavior of MoO$_3$-doped SnO$_2$ thin film prepared by sol–gel spin coating. The gas selectivity study showed that 10 wt % MoO$_3$ added SnO$_2$ films because of its high sensitivity at low operating temperature. This high sensitivity is due to the smaller particle size and increase in the acidic nature of the films by the presence of MoO$_3$. Selectivity with CO$_2$, SO$_2$, C$_2$H$_5$OH, NH$_3$ and NO$_2$

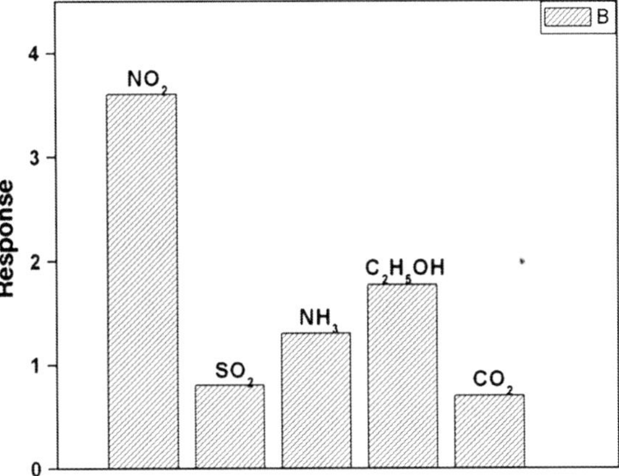

Fig. 13.20 Sensor response to 500 ppm of NO$_2$, SO$_2$, NH$_3$, C$_2$H$_5$OH and CO$_2$ measured at 170 °C of 10 wt % MoO$_3$-doped SnO$_2$ thin films

as shown in Fig. 13.20 indicated that that 10 wt % MoO$_3$-doped SnO$_2$ films show increased selectivity for NO$_2$ gas at a working temperature of 170 °C as compared to all other tested gases.

13.4 Device Integration of Sensor Materials

An easy, scalable and over all reproducible integration of sensor materials into commercial devices is the key step in the transfer of promising novel sensor materials to real life applications. Screen printing of metal oxide particles was and still is the most successful integration technique used commercially in large scale. But today's application demand more selective, sensitive and energy efficient devices, this cannot be met using simple thick film technology. By reducing the feature size using lithographically patterned substrates and more sophisticated thin film techniques, like sputtering, spray pyrolysis, sol–gel coating or chemical vapor deposition, multifunctional sensors integrating several individual sensors, can be realized on the same footprint used by thick-film-devices. In recent years aniso-tropic nanomaterials (nanowires, nanoubes, nanorods, etc.) and heterostructures thereof ((hyper-)branched, core–shell, etc.) exhibit superior material properties suitable for well-advanced sensing devices, but existing technology lacks sophisticated integration methods suitable for large scale applications, which preserve the unique properties of anisotropic morphologies.

Especially in the field of chemical sensors based on metal oxide nanowires, most of these fundamental studies make use of expensive and large lab instruments for characterization and operation tasks, and thus, the effective transfer of this nanotechnology to our everyday life remains as an unfeasible goal. For this reason, there is an increasing interest in demonstrating that low-cost portable devices with integrated nanowires can be developed to operate as sensing elements [61, 165].

In some of those preliminary works, both bottom-up and top-down fabrication techniques have been successfully combined. Nanowires are electrically contacted

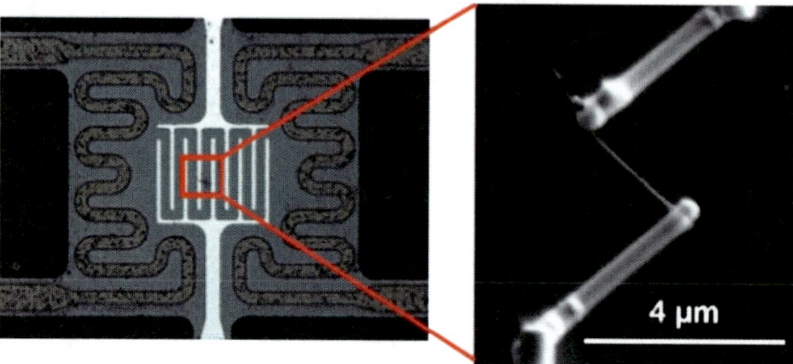

Fig. 13.21 (*Left*) Micro-hot-plate with integrated heater where the nanowire is contacted to pre-patterned microelectrodes. (*Right*) SnO2 nanowire electrically contacted to the microelectrodes. Dimensions: L = 2.1 μm (length) and R = 35 ± 5 nm (radius)

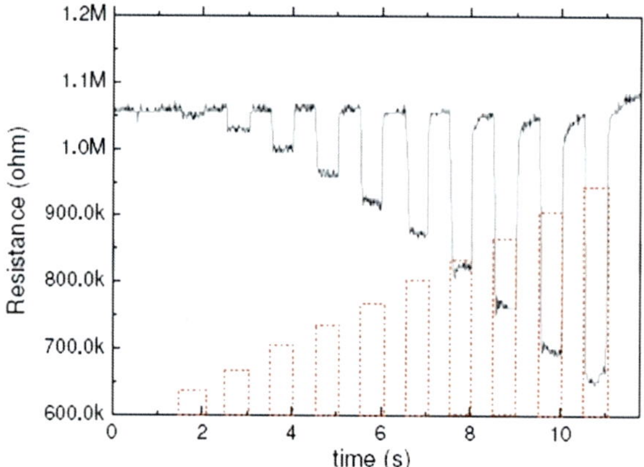

Fig. 13.22 Sensitivity of a single SnO_2 nanowire to increasing dissipated power by the heater. This experiment was performed in synthetic air. Power (0–90 mW), directly related to temperature, leads to decreasing nanowire resistance, demonstrating semiconductor characteristics. At the highest value, a nominal temperature of T = 210 °C is obtained in the centre of the micro-membrane. [Reprinted with permission from [181], Copyright @ IOP—Institute of Physics (2007)]

to a micro-hotplate with an integrated heater (Fig. 13.21), which allows modulating their effective temperature as function of the dissipated power. This process is fast and completely reproducible (Fig. 13.22).

It is well-known that metal oxide materials need to be heated at a specific temperature to maximize their response to a specific target analyte [17]. Therefore, the use of this heater becomes an extremely useful tool to modulate the final performances of these prototypes as chemical sensors (Fig. 13.23), and integrate them in standard electronics.

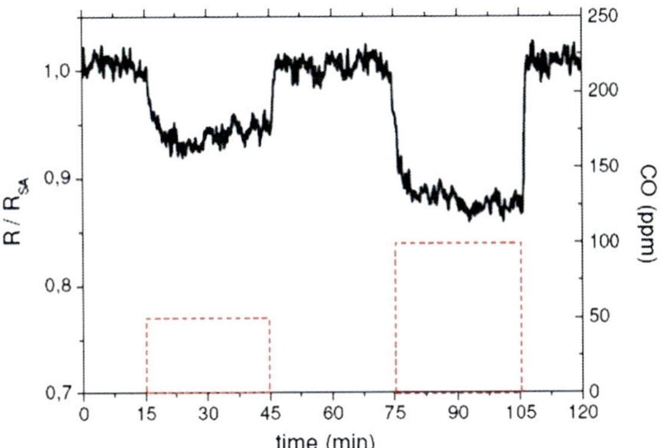

Fig. 13.23 Response of a SnO$_2$ nanowire to different CO concentrations (50 and 100 ppm), when the membrane heater is switched on (P = 56 mW, T = 120 °C). Dashed lines indicates the CO pulses. [Reprinted with permission from [181], Copyright @ IOP—Institute of Physics (2007)]

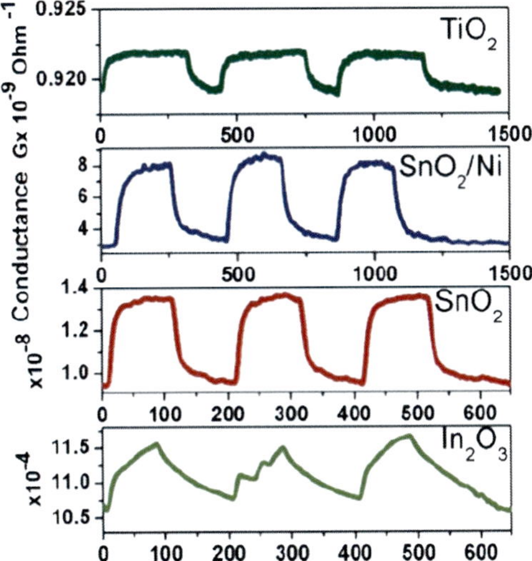

Fig. 13.24 The response of the array of metal oxide nanowires to three consecutive hydrogen pulses with partial pressure of 6.4 × 10^{-2} Pa. The background conductance is measured under the constant oxygen pressure of 1.3 × 10^{-2} Pa. [Reprinted with permission from [166], Copyright @ American Chemical Society (2006)]

The second major issue of working with metal oxide sensors from a perspective of real devices is the lack of selectivity towards analyte blends. For this reason, brand-new studies are attempting to develop electronic nose (e-nose) systems based on arrays of different metal oxide nanowires [166]. According to this approach, their responses to gases are monitored in parallel, and the specific sensing characteristics of each one are determined and electronically recorded (Fig. 13.24). Later, this knowledge is applied to perform a pattern recognition analysis of the experimental data obtained when the same nanowires are exposed to a mixture of gases, enabling to determine its composition (Fig. 13.25) [167]. Although these studies are still ongoing, they are the most promising solution to overcome the lack of selectivity, which is characteristic of metal oxide nanowires.

Finally, the need to obtain good electrical contacts in a controlled and reproducible process has forced to look for innovative nanofabrication approaches.

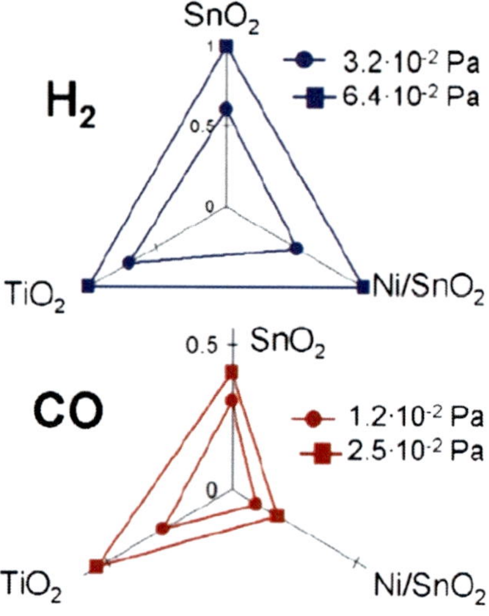

Fig. 13.25 The response of a three-chemiresistor array to H2 (*top*) and CO (*bottom*) inputs, normalized by maximum value. [Reprinted with permission from [166], Copyright @ American Chemical Society (2006)]

Nowadays, metal stripes with well-defined shapes in the nanometer range and high electrical quality are easily fabricated with different techniques such as Focused Ion Beam (FIB-), e-beam-[168] or UV- and shadow-mask-lithography [169], enabling the fast engineering of advanced proof-of-concept devices. Nevertheless, most of these techniques and particularly the nanofabrication protocols based on them evaluated so far are only suitable for research prototyping, since they are not scalable, and therefore do not fulfil the requirements to take a leap in industry.

For this reason, new solutions to solve the problem of scalability of nanowire-based devices and to edit the first complex circuit with them are currently under evaluation, such as self-assembly approaches [170, 171], and the use of other techniques like electro-spinning and microcontact/ink-jet printing [172–176]. In the former approach, metal oxide nanowires are directly deposited by a high-voltage-driven injection nozzle onto the electrodes. In the latter one, the metal electrodes are precisely printed on top of the nanowires. It should be pointed out that other alternatives such as the roll-transfer printing of devices, a method with the potential to fabricate large-area nanowire sensors and compatible with the roll-to-roll (R2R) process have been successfully evaluated to align nanowires with a good potential for the high-speed fabrication of nano-devices [177].

On the other hand and regarding the so-called self-powered sensors, Kolmakov et al. demonstrated that self-heating effect in individual metal-oxide nanowires produced by bias current was enough to activate their surface reactivity and thus, operate them as chemical sensors. This concept was further explored by Prades, validating that the probing current (I_m) applied to individual metal oxide nanowires in conductometric operation dissipates enough electrical power to self-heat their tiny mass and thus, to reach the optimum working conditions for the detection of chemicals. This experimental approach overcomes the need of external heaters and entails a dramatic reduction of the total heating volume, and as a result, the

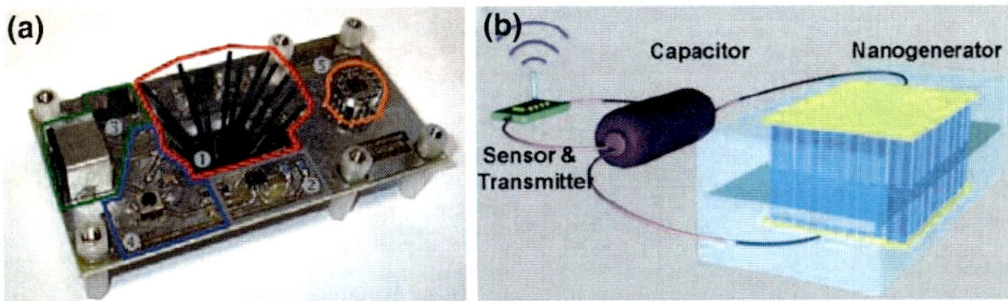

Fig. 13.26 **a** Image of the gas sensor system (Reproduced with permission from [178]) **b** The prototype of an integrated self-powered system by using a nanogenerator as the energy harvester (Reproduced with permission from [179]). The thermoelectric generator supplies regulated power to the sensor control, conditioning and output electronics. The signal conditioning circuit interfaces with the nanowires-base sensor applying a bias current (I_m) to produce self-heating

power requirements. According to preliminary measurements, metal oxide gas sensors based on individual nanowires can operate with less than 30 µW to both bias and heat them. This power value is more than three orders of magnitude lower than the typical power consumption of microheater-based devices, and envisages the development of self-powered chemical sensor systems capable to be integrated in energy scavenging and harvesting technologies.

Therefore, the possibility to use state-of-the-art energy harvesting technologies to power metal oxide gas sensors based on self-heated individual nanowires. A proof-of-concept sensor system that comes into operation harvesting energy from the ambient with a thermoelectric power unit was presented by Prades, and its response towards small concentrations of gases of interesting air quality applications, such as CO and NO_2, was shown.

Integrating an energy harvesting into a sensing device or better producing an energy generating sensing material, which can power the readout electronics will be a huge step towards autonomous devices as no separate generator would be needed anymore. Recent studies suggest that appropriate alignment of the work functions between metal oxides and smaller band gap semiconductors (e.g., metal chalcogenides) significantly extend their absorption cross-section and consequently enhance photocurrent efficiencies through charge injection from visible light harvesting unit for the broad-band semiconductors. Semiconducting metal chalcogenide nanoparticles are interesting absorbers due to their size-dependent (tunable) band gap, possibility to generate multiple excitons with a single photon, high absorption cross section, and high extinction coefficient, which knowingly reduces the dark current thereby increasing the overall efficiency of photoelectrical devices. Controlled synthesis of solar-sensitized photoelectrical material architectures and the understanding of associated photo-physical and photo-chemical processes are opening new avenues for novel sensing concepts, which are based on combination of material units offering both "powering" and "sensing" unit in one single device architecture (Fig. 13.26).

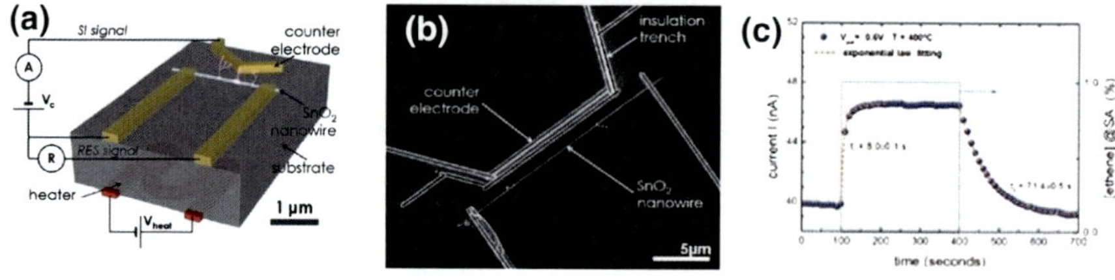

Fig. 13.27 Single nanowire surface ionization sensor setup: **a** Schematic view, **b** SEM micrograph of working prototype and **c** gas sensor response towards ethane

Fig. 13.28 Sensor grid consisting of several individual sensor nodes, interacting via wireless data links. Such sensor grids can provide real-time images of the air contamination inside buildings, air cabins and/or heavily populated city environments

13.5 Conclusions and Perspectives

The recent developments in the field of nanomaterials synthesis and engineering have opened up new avenues for next generation sensors based on new materials and combinations thereof as well as radically new concepts towards sensing principles and energy autonomy. A major advance in the direction of selectivity could be made by employing an innovative surface ionization (SI) readout, which is based on the selective thermal ionization of analytes (Fig. 13.27) [180]. It was demonstrated that a single SnO_2 nanowire which acts as an unspecific conducto-metric gas sensor can be converted into a surface ionization sensor by adding a Pt counter electrode, which detects the ions generated by the metal oxide upon heating. As the surface ionization phenomena is directly related to the work function of the metal oxide and the analyte gas, a second more specific sensing mechanism is integrated orthogonally to the conductometic sensing pathway.

By using single nanowires the temperature and power needed to generate surface ions can be reduced by many orders of magnitude compared to macro-scopic surface ionization devices. This example shows that a deep integration of different sensing principles is possible on single nanostructures which can be significantly improved by using heterostructures and junctions, in order to tune the surface ionization and conductometric response much further.

As size of sensing devices, readout electronics and power consumption is reduced using multifunctional nanostructured materials, autonomous, self sustaining sensor networks will be created, which can monitor large areas, for example detailed air quality measurements in rural areas, with a high spatial resolution. Future deploy-and-forget devices will harvest energy from the environment (photovoltaic, thermoelectric, ambient radiation, etc.) and distribute measurements over self sustained sensor networks (Fig. 13.28). Moreover, ongoing miniaturization will lead to sensor integration in most everyday products, for example clothing or packaging material. On the road to "self-aware" and "smart-products", the integration of new nanostructured materials into functional devices is the main challenge in order to utilize the unique "nano-features" of these architectures. Especially technologies for bridging the micro-nano gap have to be standardized and brought from the lab-scale to production scale.

Although heterophased sensor materials demonstrate enormous potential for chemical sensing applications, several technical challenges remain to be addressed in particular reproducible synthesis of heterostructures and their integration into devices. Subject to the materials preparation techniques, several intrinsic properties such as chemical composition, grain size and contacts as well as interfacial properties can significantly differ, which demands synthetic protocols allowing controlled and scaled-up production of nanocomposites. The same holds for novel concepts towards direct and post-synthesis integration of nanomaterials in sensing devices. Nevertheless, the increasing interests and intensive research efforts in the integration of different functional material units in a single architecture is expected to have a promising future for hetero-junction materials.

Acknowledgments Authors are thankful to the University of Cologne and the BMBF initiativeLIB-2015 (Project KoLIWin) for providing the financial assistance. The personnel support obtained by the DFG in the frame of the "Schwerpunktprogramm SPP-1166" and from the EuropeanCommission (Projects "NANOMMUNE"and "S3") in the framework of FP7 activities is gratefully acknowledged.

References

1. Franke ME, Koplin TJ, Simon U (2006) Metal and metal oxide nanoparticles in chemiresitors: Does the nanoscale matter? Small 2:36–50
2. Comini E (2006) Metal oxide nano-crystals for gas sensing. Anal Chim Acta 568:28–40
3. Zhang Y, Kolmakov A, Chretien S, Metiu H, Moskovits M (2004) Control of catalytic reactions at the surface of a metal oxide nanowire by manipulating electron density inside it. Nano Lett 4:403–407
4. Kolmakov A, Moskovits M (2004) Chemical sensing and catalysis by one- dimensional metal-oxide nanostructures. Annu Rev Mater Res 34:151–180
5. Hwang HY, Ohtomo A (2004) A high-mobility electron gas at the LaAlO$_3$/SrTiO$_3$ heterointerface. Nature 427:423–426
6. Li M, Li R, Li CM, Wu N (2011) Electrochemical and optical biosensors based on nanomaterials and nanostructures: a review. Front Biosci (Schol Ed) 3:1308–1331

7. Kummer AM, Hierlemann A, Baltes H (2004) Tuning sensitivity and selectivity of CMOS-based capacitive chemical sensor. Anal Chem 76:2470–2477

8. Traversa E (1995) Design of ceramic materials for chemical sensors with novel properties. J Am Chem Soc 78:2625–2632

9. Kuang Q, Jiang ZY, Xie ZX, Lin SC, Lin ZW, Xie SY, Huang RB, Zheng LS (2005) Tailoring the optical property by a three- dimensionally epitaxial heterostructure: A case of ZnO/SnO$_2$. J Am Chem Soc 127:11777–11784

10. Xu C, Tamaki J, Miura N, Yamazoe N (1991) Grain size effects on gas sensitivity of porous SnO$_2$-based elements. Sens Actuators B Chem 3:147–155

11. Kolmakov A, Zhang YX, Cheng GS, Moskovits M (2003) Detection of CO and O$_2$ using tin oxide nanowires sensor. Adv Mat 15:997–1000

12. Hernandez-Ramirez F, Tarancon A, Casals O, Rodriguez J, Romano-Rodriguez A, Morante JR, Barth S, Mathur S, Choi TY, Poulikakos D, Callegari V, Nellen PM (2006) Four-probe measurements and impedance spectroscopy of individual SnO$_2$ nanowires. Nanotechnology 17:5577–5583

13. Kuang Q, Lao C-S, Li Z, Liu Y-Z, Xie Z-X, Zheng L-S, Wang ZL (2008) Enhancing the photon- and gas-sensing properties of a single SnO$_2$ nanowire based nanodevice by nanoparticle surface functionalization. J Phys Chem C 112:11539–11544

14. Huang H, Gong H, Chow CL, Guo J, White TJ, Tse MS, Tan OK (2011) Low-temperature growth of SnO$_2$ nanorod arrays and tunable n–p–n sensing response of a ZnO/SnO$_2$ heterojunction for exclusive hydrogen sensors. Adv Funct Mater 21:2680–2686

15. Buso D, Post M, Cantalini C, Mulvaney P, Martucci A (2008) Gold nanoparticle-doped TiO$_2$ semiconductor thin films: gas sensing properties. Adv Funct Mater 18:3843–3849

16. Traversa E, Bearzotti A, Miyayama M, Yanagida H (1998) Influence of the electrode materials on the electrical response of ZnO-based contact sensors. J Europ Ceram Soc 18:621–631

17. Barsan N, Koziej D, Weimar U (2007) Metal oxide-based gas sensor research: How to? Sens Actuat B 121:18–35

18. Comini E, Fagliaand G, Sberveglieri G (2008) Solid state gas sensors—Industrial application. Springer, Berlin. ISBN: 978-0387096643

19. Göpel W, Hesse J, Zemel JN (1991) Functionalized polypyrrole. A new material for the construction of biosensors. Wiley-VCH, New York. ISBN-13: 978-3527267682

20. Eranna G (2011) Metal oxide nanostructures as gas sensing devices (Series in Sensors). Taylor & Francis, London. ISBN-13: 978-1439863404

21. Barsan N, Weimar U (2001) Conduction model of metal oxide gas sensors. J. Electroceramics 7:143–167

22. Barsan N, Weimar U (2003) Understanding the fundamental principles of metal oxide based gas sensors; the example of CO sensing with SnO$_2$ sensors in the presence of humidity. J Phys Cond Mat 15:R813–R839

23. Gopel W (1994) Sensor arrays calibration with enhanced neural networks. Sens Actuators B 18–19:1–21

24. Göpel W, Hesse J, Zemel JN (1991) Chemical and biochemical sensors. Wiley-VCH, New York. ISBN-13: 978-3527267682

25. Dmitriev S, Lilach Y, Button B, Moskovits M, Kolmakov A (2007) Nanoengineered chemiresistors: the interplay between electron transport and chemisorption properties of morphologically encoded SnO$_2$ nanowires. Nanotechnology 18:055707

26. Legin A, Mikhailov SM, Goryacheva O, Kirsanov D, Vlasov Y (2002) Cross-sensitive chemical sensors based on tetraphenylporphyrin and phthalocyanine. Analy Chim Acta 457:297–303

27. Albert KJ, Lewis NS, Schauer CL, Sotzing GA, Stitzel SE, Vaid TP, Walt DR (2000) Cross-reactive chemical sensor arrays. Chem Rev 100:2595–2626

28. Strehlow WH, Cook EL (1973) Compilation of energy band gaps in elemental and binary compound semiconductors and insulators. J Phys Chem Rev Data 2:163–199

29. Yamazoe N (1991) New approaches for improving semiconductor gas sensors. Sens Actuators B 5:7–19
30. Batzill M, Diebold U (2005) The surface and materials science of tin oxide. Prog Surf Sci 79:47–154
31. Rumyantseva MN, Safonova OV, Boulova MN (2003) Dopants in nanocrystalline tin dioxide. Russ Chem Bull 52:1217–1238
32. Comini E, Faglia G, Sberveglieri G, Pan Z, Wang ZL (2002) Stable and highly sensitive gas sensors based on semiconducting oxide nanobelts. Appl Phys Lett 81:1686
33. Roeck F, Barsan N, Weimar U (2008) Electronic nose: current status and future trends. Chem Reviews 108:705–725
34. Xu S, Qin Y, Xu C, Wei Y, Yang R, Wang ZL (2010) Self-powered nanowire devices. Nat Nanotech 5:366–373
35. Matthey D, Wang JG, Wendt S, Matthiesen J, Schaub R, Laegsgaard E, Hammer B, Besenbacher F (2007) Enhanced bonding of gold nanoparticles on oxidized $TiO_2(110)$. Science 315:1692–1696
36. Mishra R, Rajanna K (2005) Metal-oxide thin film with Pt, Au and Ag nano-particles for gas sensing applications. Sens Mater 17:433–440
37. Zhu CL, Chen YJ, Wang RX, Wang LJ, Cao MS, Shi XL (2009) Synthesis and enhanced ethanol sensing properties of α-Fe_2O_3/ZnO heteronanostructures. Sens Actuators B 140:185–189
38. Deshpande NG, Gudage YG, Sharma R, Vyas JC, Ki JB, Lee YP (2009) Studies on tin oxide-intercalated polyaniline nanocomposite for ammonia gas sensing applications. Sens Actuators B Chem 138:76–84
39. Law M, Goldberger J, Yang P (2004) Semiconductor nanowires and nanotubes. Annu Rev Mater Res 34:83–122
40. Kuchibhatla SVNT, Karaoti AS, Bera D, Seal S (2007) One dimensional nanostructured materials. Prog Mater Sci 52:699–913
41. Eranna G, Joshi BC, Runthala DP, Gupta RP (2004) Oxide materials for development of integrated gas sensorsa comprehensive review. Critic Rev Sol State Mater Sci 29:111–188
42. Korotcenkov G (2005) Gas response control through structural and chemical modification of metal oxides: State of the art and approaches. Sens Actuators B Chem 107:209–232
43. Comini E, Baratto C, Faglia G, Ferroni M, Vomiero A, Sberveglieri G (2009) Quasi-one dimensional metal oxide semiconductors: Preparation, characterization and application as chemical sensors. Prog Mater Sci 54:1–67
44. Kolmakov A, Klenov DO, Lilach Y, Stemmer S, Moskovits M (2005) Enhanced gas sensing by individual SnO_2 nanowires and nanobelts functionalized with pd catalyst particles. Nano Lett 5:667–673
45. Hernandez-Ramirez F, Prades JD, Tarancon A, Barth S, Casals O, Jimenez-Diaz R, Pellicer E, Rodriguez J, Morante JR, Juli MA, Mathur S, Romano-Rodriguez A (2008) An insight into the role of oxygen diffusion in the sensing mechanisms of SnO2 nanowires. Adv Funct Mater 18: 2990–2994
46. Zhang Y, Kolmakov A, Lilach Y, Moskovits M (2005) Electronic control of chemistry and catalysis at the surface of an individual tin oxide nanowire. J Phys Chem B 109:1923–1929
47. Korotcenkov G (2008) The role of morphology and crystallographic structure of metal oxides in response of conductometric-type gas sensors. Mater Sci Engin R 61:1–39
48. Mathur S, Ganesan R, Grobelsek I, Shen H, Ruegamer T, Barth S (2007) Plasma-assisted modulation of morphology and composition in tin oxide nanostructures for sensing applications. Adv Eng Mater 9:658–663
49. Pan ZW, Dai ZR, Wang ZL (2001) Nanobelts of semiconducting oxides. Science 291:1947–1949
50. Wang ZL (2004) Zinc oxide nanostructures: growth, properties and applications. J Phys Condens Matter 16:R829–R858
51. Wang ZL (2004) Nanostructures of zinc oxide. Mater Today 7:26–33

52. Wang XD, Zhou J, Song JH, Liu J, Xu NS, Wang ZL (2006) Piezoelectric field effect transistor and nanoforce sensor based on a single ZnO nanowire. Nano Lett 6:2768–2772

53. Hernandez-Ramirez F, Prades JD, Jimenez-Diaz R, Fischer T, Romano-Rodriguez A, Mathur S, Morante JR (2009) On the role of individual metal oxide nanowires in the scaling down of chemical sensors. Phys Chem Chem Phys 11:7105

54. Wang ZL, Song JH (2006) Piezoelectric nanogenerators based on zinc oxide nanowire arrays. Science 14:242–246

55. He JH, Hsin CL, Liu J, Chen LJ, Wang ZL (2007) Piezoelectric gated diode of a single ZnO nanowire. Adv Mater 19:781–784

56. Fort A, Rocchi S, Serrano-Santos MB, Mugnaini M, Vignoli V, Atrei A, Spinicci R (2006) CO sensing with SnO_2-based thick film sensors: state model for conductance responses during thermal-modulation. Sens Actuactors B Chem 116:43–48

57. Hernández-Ramírez F, Tarancón A, Casals O, Arbiol J, Romano-Rodríguez A, Morante JR (2007) High response and stability in CO and humidity measures using a single SnO_2 nanowire. Sens Actuators B Chem 121:3–17

58. Nam CY, Tham D, Fischer JE (2005) Disorder effects in focused-ion-beam-deposited Pt contacts on GaN nanowires. Nano Lett 5:2029–2033

59. Lin YF, Jian WB (2008) The impact of nanocontact on nanowire based nanoelectronics. Nano Lett 8:3146–3150

60. Hernandez-Ramirez F, Tarancon A, Casals O, Pellicer E, Rodríguez J, Romano-Rodriguez A, Morante JR, Barth S, Mathur S (2007) Electrical properties of individual tin oxide nanowires contacted to platinum electrodes. Phys Rev B 76:085429

61. Hernandez-Ramirez F, Prades JD, Tarancon A, Barth S, Casals O, Jimenez-Diaz R, Pellicer E, Rodríguez J, Juli MA, Romano-Rodriguez A, Morante JR, Mathur S, Helwig A, Spannhake J, Mueller G (2007) Portable microsensors based on individual SnO_2 nanowires Nanotechnology 18:495501

62. Prades JD, Jimenez-Diaz R, Hernandez-Ramirez F, Barth S, Cirera A, Romano-Rodriguez A, Mathur S, Morante JR (2008) Ultralow power consumption gas sensors based on self-heated individual nanowires. Appl Phys Lett 93:123110–123113

63. Strelcov E, Dmitriev S, Button B, Cothren J, Sysoev V, Kolmakov A (2008) Evidence of the self-heating effect on surface reactivity and gas sensing of metal oxide nanowire chemiresistors. Nanotechnology 19:355502

64. Scott RWJ, Yang SM, Chabanis G, Coombs N, Williams DE, Ozin GA (2001) Tin dioxide opals and inverted opals: near-ideal microstructures for gas sensors. Adv Mater 13:1468–1472

65. Scott RWJ, Yang SM, Coombs N, Ozin GA, Williams DE (2003) Engineered sensitivity of structured tin dioxide chemical sensors: opaline architectures with controlled necking. Adv Funct Mater 13:225–231

66. Maier J (2000) Festkörper-fehler und funktion: prinzipien der physikalischen festkörperchemie. Teubner studienbücher chemie, Teubner

67. Min Y (2003) Ph.D. Thesis, Massachusetts Institute of Technology

68. Korotcenkov G (2007) Practical aspects in design of one-electrode semiconductor gas sensors: Status report. Sens Actuators B Chem 121:664–678

69. Korotcenkov G, Brinzari V, Stetter JR, Blinov I, Blaja V (2007) The nature of processes controlling the kinetics of indium oxide-based thin film gas sensor response. Sens Actuators B Chem 128:51–63

70. Kamp B, Merkle R, Lauck R, Maier J (2005) Chemical diffusion of oxygen in tin dioxide: Effects of dopants and oxygen partial pressure. J Sol State Chem 178:3027–3039

71. Ponce MA, Aldao CM, Castro MS (2003) Influence of particle size on the conductance of SnO_2 thick films. J Eur Ceramic Soc 23:2105–2111

72. Prades JD, Cireraand A, Morante JR (2008) Applications of DFT calculations to chemical gas sensors: Designing and understanding, quantum chemical calculations of surface and interfaces of materials, Basiuk VA, Ugliengo P (ed), Am. Sci. Pub. (ASP). ISBN: 1-58883-138

73. Niederberger M, Pinna N (2009) Metal oxide nanoparticles in organic solvents. Springer, London. ISBN-13: 978-1848826700
74. Altavilla C, Ciliberto E (eds) (2010) Inorganic nanoparticles: synthesis, applications, and perspectives (Nanomaterial's and their applications). CrcPrInc. ISBN-13: 978-1439817612
75. Tao AR, Habas S, Yang P (2008) Shape control of colloidal metal nanocrystals. Small 4:310–325
76. Pinna N, Neri G, Antoniettiand M, Niederberger M (2004) Nonaqueous synthesis of nanocrystalline semiconducting metal oxides for gas sensing. Angew Chem Int Ed 43:4345–4349
77. Satyanarayana L, Reddy KM, Manorama SV (2003) Nanosized spinel $NiFe_2O_4$: a novel material for the detection of liquefied petroleum gas in air. Mater Chem Phys 82:21–26
78. Donega CD, Liljeroth P, Vanmaekelbergh D (2005) Physicochemical evaluation of the hot-injection method, a synthesis route for monodisperse nanocrystals. Small 1:1152–1162
79. Qurashi A, Tabet N, Faiz M, Yamzaki T (2009) Ultra-fast microwave synthesis of ZnO nanowires and their dynamic response toward hydrogen gas. Nanoscal Res Lett 4:948–954
80. Li Z, Zhang H, Zheng W, Wang W, Huang H, Wang C, MacDiarmid AG, Wei Y (2008) Highly sensitive and stable humidity nanosensors based on licl doped TiO_2 electrospun nanofibers. J Am Chem Soc 130:5036–5037
81. Mathur S, Barth S (2008) One-dimensional semiconductor nanostructures: growth, characterization and device applications. Z Phys Chem 222:307–317
82. Pan J, Ganesan R, Shen H, Mathur S (2010) Plasma-modified SnO_2 nanowires for enhanced gas sensing. J Phys Chem C 114:8245–8250
83. Du X, George SM (2008) Thickness dependence of sensor response for CO gas sensing by tin oxide films grown using atomic layer deposition. Sens Actua B Chem 135:152–160
84. Jin CJ, Yamazaki T, Shirai Y, Yoshizawa T, Kikuta T, Nakatani N, Takeda H (2005) Dependence of NO_2 gas sensitivity of WO_3 sputtered films on film density. Thin Solid Films 474:255–260
85. Williams G, Coles GSV (1998) Gas sensing properties of nanocrystalline metal oxide powders produced by a laser evaporation technique. J Mat Chem 8:1657–1664
86. Tricoli A, Righettoni M, Pratsinis SE (2009) Minimal cross-sensitivity to humidity during ethanol detection by SnO_2–TiO_2 solid solutions. Nanotechnology 20:315502
87. Kanan SM, Tripp CP (2007) Synthesis, FTIR studies and sensor properties of WO_3 powders. Curr Opi Solid State Mater Sci 11:19–27
88. Fan HJ, Werner P, Zacharias M (2006) Semiconductor nanowires: from self-organization to growth control. Small 2:700–717
89. Moore D, Morber JR, Snyder RL, Wang ZL (2008) Growth of ultralong ZnS/SiO_2 core-shell nanowires by volume and surface diffusion VLS process. J Phys Chem C 112:2895–2903
90. Liang G, Xiang J, Kharche N, Klimeck G, Lieber CM, Lundstrom M (2007) Performance analysis of a Ge/Si core/shell nanowire field-effect transistor. Nano Lett 7:642–646
91. Sham TK, Naftel SJ, Kim PSG, Sammynaiken R, Tang YH, Coulthard I, Moewes A, Freeland JW, Hu YF, Lee ST (2004) Electronic structure and optical properties of silicon nanowires: a study using x-ray excited optical luminescence and x-ray emission spectroscopy. Phys Rev B 70:045313
92. Lin YM, Sun X, Dresselhaus MS (2000) Theoretical investigation of thermoelectric transport properties of cylindrical bi nanowires. Phys Rev B 62:4610–4623
93. Rubio-Bollinger G, Bahn SR, Agrait N, Jacobsen KW, Vieira S (2001) Mechanical properties and formation mechanisms of a wire of single gold atoms. Phys Rev Lett 87:026101
94. Cuenot S, Fretigny C, Champagne SD, Nysten B (2004) Surface tension effect on the mechanical properties of nanomaterials measured by atomic force microscopy. Phys Rev B 69:65410
95. Zhao MH, Wang ZL, Mao SX (2004) Piezoelectric characterization of individual zinc oxide nanobelt probed by piezoresponse force microscope. Nano Lett 4:587–590

96. Barreca D, Comini E, Ferrucci AP, Gasparotto A, Maccato C, Maragno C, Sberveglieri G, Tondello E (2007) First example of ZnO – TiO2 nanocomposites by chemical vapor deposition: structure, morphology, composition, and gas sensing performances. Chem Mater 19:5642–5649

97. Li YX, Galatsisa K, Wlodarski W, Passacantando M, Santucci S, Siciliano P, Catalano M (2001) Microstructural characterization of MoO_3–TiO_2 nanocomposite thin films for gas sensing. Sens Actuators B 77:27–34

98. Li YX, Galatsis K, Wlodarski W, Ghantasala M, Russo S, Gorman J, Santucci S, Passacantando M (2001) Microstructure characterization of sol-gel prepared MoO_3–TiO_2 thin films for oxygen gas sensors. J Vac Sci Technol A 19:904

99. Li YX, Galatsis K, Wlodarski W, Ghantasala MK, Comini E, Faglia G, Sberveglieri G, Siciliano P, Rella R, Passacantando M, Santucci S, Cantalini C (2000) In: Proceedings of the fifth italian conference on sensors and microsystems. World Scientific, Singapore, p 139

100. Tan OK, Zhu W, Yan Q, Kong LB (2000) Size effect and gas sensing characteristics of nanocrystalline $xSnO_2$-$(1 - x)\alpha$-Fe_2O_3 ethanol sensors. Sens Actuators B 65:361–365

101. Ivanovskaya M, Kotsikau D, Faglia G, Nelli P, Irkaev S (2003) Gas-sensitive properties of thin film heterojunction structures based on Fe_2O_3–In_2O_3 nanocomposites. Sens Actuators B 93:422–430

102. Song X, Wang Z, Liu Y, Wangand C, Li L (2009) A highly sensitive ethanol sensor based on mesoporous ZnO–SnO_2 nanofibers. Nanotechnology 20:075501

103. Cabo A, Arbiol J, Morante JR, Weimar U, Barsan N, Göpel W (2000) Analysis of the noble metal catalytic additives introduced by impregnation of as obtained SnO_2 sol–gel nanocrystals for gas sensors. Sens Actuators B 70:87–100

104. Cabo A, Diegueza A, Romano-Rodrígueza A, Morante JR, Barsan N (2001) Influence of the catalytic introduction procedure on the nano-SnO_2 gas sensor performances: where and how stay the catalytic atoms? Sens Actuators B 79:98–106

105. Krivetskiy VV, Ponzoni A, Comini E, Badalyan SM, Rumyantseva MN, Gaskov AM (2010) Materials based on modified SnO_2 for selective gas sensors. Inorg Mater 46:1100–1105

106. Rumyantsevaa M, Kovalenkoa V, Gaskova A, Makshinaa E, Yuschenkoa V, Ivanovaa I, Ponzonib A, Fagliab G, Cominib E (2006) Nanocomposites SnO_2/Fe_2O_3: sensor and catalytic properties. Sens Actuators B 118:208–214

107. Mori Y, Kohno H (2009) Resistance switching in a SiC nanowire/Au nanoparticle network. Nanotechnology 20:285705

108. Pearton SJ, Kang BS, Kim SK, Ren F, Gila BP, Abernathy CR, Lin JS, Chu SNG (2004) GaN-based diodes and transistors for chemical, gas, biological and pressure sensing. J Phys Cond Matt 16:R961–R994

109. Gonzalez YM, Xu WZ, Chen B, Farhanghi N, Charpentier PA (2011) CdS and CdTeS quantum dot decorated TiO_2 nanowires. Synthesis and photoefficiency. Nanotechnology 22:065603

110. Kang J, Kuang Q, Xie ZX, Zheng LS (2011) Fabrication of the SnO2/α-Fe2O3 hierarchical heterostructure and its enhanced photocatalytic property. J Phys Chem C 115:7874–7879

111. Chen YJ, Xiao G, Wang TS, Zhang F, Ma Y, Gao P, Zhu CL, Zhang E, Xucand Z, Li QH (2011) α-MoO_3/TiO_2 core/shell nanorods: controlled-synthesis and low-temperature gas sensing properties. Sens Actuators B 155:270–277

112. Zhou W, Cheng C, Liu J, Tay YY, Jiang J, Jia X, Zhang J, Gong H, Hng HH, Yu T, Fan HJ (2011) Epitaxial growth of branched α-Fe_2O_3/SnO_2 nano-heterostructures with improved lithium-ion battery performance. Adv Funct Mater 21:2439–2445

113. Wei S, Zhang Y, Zhou M (2011) Toluene sensing properties of SnO_2–ZnO hollow nanofibers fabricated from single capillary electrospinning. Solid State Commun 151:895–899

114. Mathur S, Singh AP, Müller R, Leuning T, Lehnen T, Shen H (2011) Metal organic chemical vapor deposition of metal oxide films and nanostructures published by Wiley-VCH Verlag GmbH and Co. KGaA in the book ceramics science and technology, vol 3, synthesis and processing Ralf Riedel, I-Wei Chen (eds), pp 291–336

115. Wang H, Fei X, Wang L, Li Y, Xu S, Sun M, Sun L, Zhang C, Li Y, Yang Q, Wei Y (2011) Magnetically separable iron oxide nanostructures-TiO$_2$ nanofibers hierarchical heterostructures: controlled fabrication and photocatalytic activity. New J Chem 35:1795–1802
116. Bickley RI, Gonzalez-Carreno T, Gonzalez-Elipe AR, Munuera G, Palmisano L (1994) Characterisation of iron/titanium oxide photocatalysts. Part 2—Surface studies. J Chem Soc Faraday Trans 90:2257–2264
117. Leschkies KS, Divakar R, Basu J, Enache-Pommer E, Boercker JE, Carter CB, Kortshagen UR, Norris DJ, Aydil ES (2007) Photosensitization of ZnO nanowires with cdse quantum dots for photovoltaic devices. Nano Lett 7:1793–1798
118. Feng X, Shankar K, Varghese OK, Paulose M, Latempa TJ, Grimes CA (2008) Vertically aligned single crystal TiO$_2$ nanowire arrays grown directly on transparent conducting oxide coated glass: synthesis details and applications. Nano Lett 8:3781–3786
119. Liu B, Aydil ES (2009) Growth of oriented single-crystalline rutile TiO$_2$ nanorods on transparent conducting substrates for dye-sensitized solar cells. J Am Chem Soc 131:3985–3990
120. Bang JH, Kamat PV (2010) Solar cells by design: photoelectrochemistry of TiO$_2$ nanorod arrays decorated with CdSe. Adv Funct Mater 20:1970–1976
121. Wang H, Bai Y, Zhang H, Zhang Z, Li J, Guo L (2010) CdS quantum dots-sensitized TiO$_2$ nanorod array on transparent conductive glass photoelectrodes. J Phys Chem C 114:16451–16455
122. Zhai J, Wang L, Wang D, Li H, Zhang Y, He D, Xie T (2011) Enhancement of gas sensing properties of CdS nanowire/ZnO nanosphere composite materials at room temperature by visible-light activation. ACS Appl Mater Interfaces 3:2253–2258
123. Hu P, Du G, Zhou W, Cui J, Lin J, Liu H, Liu D, Wang J, Chen S (2010) Enhancement of ethanol vapor sensing of TiO$_2$ nanobelts by surface engineering. Appl Mater Inter 2:3263–3269
124. Seo M, Yuasa M, Kida T, Kanmura Y, Huh J, Yamazoe N, Shimanoe K (2011) Gas sensor using noble metal-loaded TiO$_2$ nanotubes for detection of large-sized volatile organic compounds. J Ceram Soc Japan 119:884–889
125. Wang P, Zhang H, Stewart JT, Bartlett MG (1998) Simultaneous detection of cisatracurium, its degradation products and propofol using positive ion detection followed by negative ion detection in a single LC/MS run. J Pharm Biomed Anal 17:547–553
126. Chen Y, Zhu C, Shi X, Cao M, Jin H (2008) The synthesis and selective gas sensing characteristics of SnO$_2$/α-Fe$_2$O$_3$ hierarchical nanostructures. Nanotechnology 19:205603
127. Xing LL, Yuan S, Chen ZH, Chen YJ, Xue1 XY (2011) Enhanced gas sensing performance of SnO$_2$/α-MoO$_3$ heterostructure nanobelts. Nanotechnology 22:225502
128. Chen Y, Zhu C, Wang T (2006) One-pot synthesis of crystalline SnO$_2$ nanoparticles and their low-temperature ethanol sensing characteristics. Nanotechnology 17:3012–3017
129. Wu JM (2011) TiO$_2$/Ti$_{1-x}$Sn$_x$O$_2$ heterojunction nanowires: characterization, formation, and gas sensing performance. J Mater Chem 11:14048–14055
130. Chen YJ, Xiao G, Wang TS, Zhang F, Ma Y, Gao P, Zhu CL, Zhang E, Xuand Z, Li QH (2011) Synthesis and enhanced gas sensing properties of crystalline CeO$_2$/TiO$_2$ core/shellnanorods. Sens Actuators B 156:867–874
131. Du N, Zhang H, Chen B, Wu J, Yang D (2007) Low-temperature chemical solution route for ZnO based sulfide coaxial nanocables: general synthesis and gas sensor application. Nanotechnology 18:115619
132. Willinger MG, Neri G, Rauwel E, Bonavita A, Micali G, Pinna N (2008) Vanadium oxide sensing layer grown on carbon nanotubes by a new atomic layer deposition process. Nano Lett 8:4201–4204
133. Sen S, Kanitkar P, Sharma A, Muthe KP, Rath A, Deshpande SK, Kaur M, Aiyer RC, Gupta SK, Yakhmi JV (2010) Growth of SnO$_2$/W$_{18}$O$_{49}$ nanowire hierarchical heterostructure and their application as chemical sensor. Sens Actuators B 147:453–460

134. Mathur S, Barth S (2007) Molecule-based chemical vapor growth of aligned SnO_2 nanowires and branched SnO_2/V_2O_5 eterostructures. Small 3:2070–2075
135. Zeng W, Liu T, Wang Z, Tsukimoto S, Saito M, Ikuhara Y (2009) Selective detection of formaldehyde gas using a Cd-doped TiO_2-SnO_2 sensor. Sensors 9:9029–9038
136. Comini E (2005) Metal oxide nano-crystals for gas sensing. Anal Chim Acta 568:28–40
137. Tiemann M (2007) Porous metal oxides as gas sensors. Chem Eur J 13: 8376–8388
138. Rothschild A, Komem Y (2004) The effect of grain size on the sensitivity of nanocrystalline metal-oxide gas sensors. J Appl Phys 95:6374–6380
139. Perednis D, Gauckler LJ (2005) Thin film deposition using spray pyrolysis. J Electroceramics 14:103–111
140. Cantalini C, Wlodarski W, Li Y, Passacantando M, Santucci S, Comini E, Faglia G, Sberveglieri G (2000) Investigation on the O_3 sensitivity properties of WO_3 thin films prepared by sol–gel, thermal evaporation and r.f. sputtering techniques. Sens Actuators B 64:182–188
141. Bruno L, Pijolat C, Lalauze R (1994) Tin dioxide thin-film gas sensor prepared by chemical vapour deposition Influence of grain size and thickness on the electrical properties. Sens Actuators B 18–19:195–199
142. Chakraborty S, Nemoto K, Hara K, Lai PT (1999) Moisture sensitive field effect transistors using $SiO_2/Si_3N_4/Al_2O_3$ gate structure. Smart Mater Struct 8:274–277
143. Lu C, Chen Z (2009) High-temperature resistive hydrogen sensor based on thin nanoporous rutile TiO_2 film on anodic aluminum oxide. Sens Actuators B 140:109–115
144. Dandeneau CS, Jeon YH, Shelton CT, Plant TK, Cann DP, Gibbons BJ (2009) Thin film chemical sensors based on p-CuO/n-ZnO heterocontacts. Thin Solid Films 517:4448–4454
145. Low IM, Oo Z, O'Connor BH (2006) Effect of atmospheres on the thermal stability of aluminium titanate. Phys B 385–386:502–504
146. Cantalini C, Post M, Buso D, Guglielmi A, Martucci A (2005) Gas sensing properties of nanocrystalline NiO and Co_3O_4 in porous silica sol–gel films. Sens Actuators B Chem 108:184–192
147. Martucci A, Pasquale M, Martucci A, Pasquale M, Guglielmi M, Post M, Pivin JC (2003) Nanostructured silicon oxide-nickel oxide sol-gel films with enhanced optical carbon monoxide gas sensitivity. J Am Ceram Soc 86:1638–1640
148. Martucci A, Buso D, Monte MD, Guglielmi M, Cantaliniand C, Sada C (2004) Nanostructured sol–gel silica thin films doped with NiO and SnO_2 for gas sensing applications. J Mat Chem 14:2889–2895
149. Jiang JZ, Lin R, Mørup S, Nielsen K, Poulsen FW, Berry FJ, Clasen R (1997) Mechanical alloying of an immiscible α-Fe_2O_3-SnO_2 ceramic. Phys Rev B 55:11–14
150. Wisitsoraat A, Tuantranont A, Comini E, Sberveglieri G, Wlodarski W (2009) Characterization of n-type and p-type semiconductor gas sensors based on NiO_x doped TiO_2 thin films. Thin Solid Films 517:2775–2780
151. Wang X, Summers CJ, Wang ZL (2004) Large-Scale Hexagonal-Patterned Growth of Aligned ZnO Nanorods for Nano-optoelectronics and Nanosensor Arrays. Nano Lett 4:423–426
152. Kim SW, Fujita S, Park HK, Yang B, Kim HK, Yoon DH (2006) Growth of ZnO nanostructures in a chemical vapor deposition process. J Cryst Growth 292:306–310
153. Baratto C, Sberveglieri G, Onischuk A, Caruso B, di Stasio S (2004) Low temperature selective NO_2 sensors by nanostructured fibres of ZnO. Sens Actuators B 100:261–265
154. Xu H, Liu X, Cui D, Liand M, Jiang M (2006) A novel method for improving the performance of ZnO gas sensors. Sens Actuators B 114:301–307
155. Zhu BL, Xie CS, Wang WY, Huang KJ, Hu JH (2004) Improvement in gas sensitivity of ZnO thick film to volatile organiccompounds (VOCs) by adding TiO_2. Mater Lett 58:624–629
156. Kanan SM, El-Kadri OM, Abu-Yousef IA, Kanan MC (2009) Semiconducting metal oxide based sensors for selective gas pollutant detection. Sensors 9:8158–8196

157. Takada T, Suzuki K, Nakane M (1993) Highly sensitive ozone sensor. Sens Actuators B 13–14:404–407
158. Chibirova FH, Gutman EE (2000) Structural defects and gas-sensitive properties of some semiconductor metal oxides. Rus J Phys Chem 74:1555–1561
159. Sunu SS, Prabhu E, Jayaraman V, Gnanasekar KI, Seshagiri TK, Gnanasekaran T (2004) Electrical conductivity and gas sensing properties of MoO_3. Sens Actuators B 101:161–174
160. Barazzouk S, Tandon RP, Hotchandani S (2006) MoO_3-based sensor for NO, NO_2 and CH_4 detection. Sens Actuators B 119:691–694
161. Galatsis K, Li YX, Wlodarski W, Comini E, Sberveglieri G, Cantalini C, Santucci S, Passacantando M (2002) Comparison of single and binary oxide MoO_3, TiO_2 and WO_3 sol-gel gas sensors. Sens Actuators B 83:276–280
162. Ansari ZA, Ansari SG, Ko T, Oh JH (2002) Effect of MoO_3 doping and grain size on SnO_2–enhancementof sensitivity and selectivity for CO and H_2 gas sensing. Sens Actuators B 87:105–114
163. Ivanovskaya M, Lutynskaya E, Bogdanov P (1998) The influence of molybdenum on the properties of SnO_2 ceramic sensors. Sens Actuators B 48:387–391
164. Kaur J, Vankar VD, Bhatnagar MC (2010) Role of surface properties of MoO_3-doped SnO_2 thin films on NO_2 gas sensing. Thin Solid Films 518:3982–3987
165. Meier DC, Semancik S, Button B, Strelcov E, Kolmakov A (2007) Coupling nanowire chemiresistors with MEMS gas sensing platforms microhotplate. Appl Phys Lett 91:063118
166. Sysoev VV, Button BK, Wepsiec K, Dmitriev S, Kolmakov A (2006) Towards the Nanoscopic "Electronic Nose": Hydrogen versus carbon monoxide discrimination with an array of individual metal oxide nano- and meso-wire sensors. Nano Lett 6:1584–1588
167. Sysoev VV, Goschnick J, Schneider T, Strelcov E, Kolmakov A (2007) A gradient microarray electronic nose based on percolating SnO_2 nanowire sensing elements. Nano Lett 7:3182–3188
168. Kim DDW, Hwang IS, Kwon SJ, Kang HY, Park KS, Choi YJ, Choi KJ, Park JG (2007) Highly conductive coaxial SnO_2-In_2O_3 heterostructured nanowires for Li-ion battery electrodes. Nano Lett 7:3041–3045
169. Kalinin SV, Shin J, Jesse S, Geohegan D, Baddorf P, Lilach Y, Moskovits M, Kolmakov A (2005) Electronic transport imaging in a multiwire SnO_2 chemical field-effect transistor device. J Appl Phys 98:044503
170. Kumar S, Rajaraman S, Gerhardt RA, Wang ZL, Hesketh PJ (2005) Tin oxide nanosensor fabrication using AC dielectrophoretic manipulation of nanobelts. Electrochim Acta 51:943–951
171. Wang D, Zhu R, Zhaoying Z, Ye X (2007) Controlled assembly of zinc oxide nanowires using dielectrophoresis. Appl Phys Lett 90:103110
172. Lin D, Wu H, Zhang R, Pan W (2007) Preparation and electrical properties of electrospun tin-doped indium oxide nanowires. Nanotechnology 18:465301
173. Sawicka KM, Prasad AK, Gouma PI (2005) Metal oxide nanowires for use in chemical sensing applications. Sens Lett 3:31–35
174. van Osch THJ, Perelaer J, de Laat AWM, Schubert US (2008) Inkjet printing of narrow conductive tracks on untreated polymeric substrates. Adv Mater 20:343–345
175. Kim YK, Park SJ, Koo JP, Oh D-J, Kim GT, Hong S, Ha JS (2006) Controlled direct patterning of V_2O_5 nanowires onto SiO_2 substrates by a microcontact printing technique. Nanotechnology 17:1375
176. Song JW, Kim J, Yoon YH, Choi BS, Kim JH, Han C-S (2008) Inkjet printing of single-walled carbon nanotubes and electrical characterization of the line pattern. Nanotechnology 19:095702
177. Chang YK, Hong FCN (2009) The fabrication of ZnO nanowire field-effect transistors by roll-transfer printing. Nanotechnology 20:195302
178. Prades JD, Jimenez-Diaz R, Hernandez-Ramirez F, Cirera A, Romano-Rodriguez A, Morante JR (2010) Harnessing self-heating in nanowires for energy efficient, fully autonomous and ultra-fast gas sensors. Sens Actuators B Chem 144:1–5

179. Hu Y, Zhang Y, Xu C, Lin L, Snyder RL, Wang ZL (2011) Self-powered system with wireless data transmission. Nano Lett 11:2572–2577
180. Hernandez-Ramirez F, Prades JD, Hackner A, Fischer T, Mueller G, Mathur S, Morante JR (2011) Miniaturized ionization gas sensors from single metal oxide nanowires. Nanoscale 3:630–634
181. Hernández-Ramírez F et al (2007) Portable microsensors based on individual SnO$_2$ nanowires. Nanotechnology 18:495501

Chapter 14
Evaluation of Metal Oxide Nanowire Materials With Temperature-Controlled Microsensor Substrates

Kurt D. Benkstein, Baranidharan Raman, David L. Lahr and Steven Semancik

Abstract Nanomaterials are becoming increasingly important for next-generation chemical sensing devices. In particular, quasi-one-dimensional materials, such as nanowires, are attracting a great deal of interest. While early examples have demonstrated the promise offered by these nanoscale materials, challenges still remain for integration, systematic characterization and evaluation of such materials in operational devices. Here, a means to assess the performance of nanowire-based materials as chemical microsensors is illustrated with two examples. Polycrystalline nanowire sensing materials are integrated with microsensor substrates that feature an embedded heater, facilitating the use of temperature to interrogate the response characteristics of sensing materials. By changing the operating temperature, different effects are observed as a function of nanowire loading density (aligned tin oxide nanowires) or overall material morphology (tungsten oxide materials, including a thin film). Further, by using conventional signal processing and data analysis approaches, the sensitivity and selectivity of these materials as a function of material scale and morphology are characterized.

14.1 Introduction

Nanomaterials are of significant interest for use as the active material in chemical sensors, because they are expected to provide benefits with respect to sensor performance versus their more macro-scaled counterparts. Nanowires in particular

K. D. Benkstein (✉) · B. Raman · D. L. Lahr · S. Semancik
Department of Biomedical Engineering, Washington University,
2007, Brauer Hall One Brookings Drive, St. Louis,
MO Campus Box 1097, USA
e-mail: kurt.benkstein@nist.gov

M. A. Carpenter et al. (eds.), *Metal Oxide Nanomaterials for Chemical Sensors*,
Integrated Analytical Systems, DOI: 10.1007/978-1-4614-5395-6_14,
© Springer Science+Business Media New York 2013

are attracting a great deal of attention, and may be incorporated into numerous sensing device architectures, from chemiresistors to chemically sensitive field-effect transistors (chemFETs) to AC impedance devices [1–3]. While there have been demonstrations of enhanced performance from nano-structured materials, questions still remain regarding the overall effectiveness of nanomaterials for chemical sensing applications [3–11]. In part, to take full advantage of the benefits of these materials, methods to integrate nanomaterials into sensor devices are of critical importance. Furthermore, it is necessary to develop approaches and metrics to evaluate the performance of the nano-structured sensor materials to determine whether they are, in fact, superior performers.

By integrating nanomaterials into a microscale platform featuring an array format, the materials may be efficiently and directly evaluated and compared under identical sensing conditions. Furthermore, the use of varied sensor operating temperatures directly impacts the evaluation of the nanomaterials, as it induces differences in sensor performance. The efficacy of the multiple elements and varied temperatures has been demonstrated before for exploring materials preparation [12, 13], nanoparticle-based sensing materials [14–16] and, as illustrated in this chapter, nanowire-based sensing materials [17–19].

14.2 Nanowire Preparation and Characterization

Quasi-one-dimensional metal oxide nanomaterials have been prepared in a variety of ways, including etching, electrospinning, chemical vapor deposition (Vapor–Liquid–Solid or Vapor–Solid) and template-directed deposition [2, 20–22]. For the studies illustrated here, polycrystalline nanowires were prepared via template-directed electrodeposition. Polycrystalline tungsten oxide nanowires were electrodeposited in track-etched polycarbonate membranes [17]. To serve as a working electrode, one side of a commercial polycarbonate filtration membrane was coated with 200 nm of silver using a thermal evaporator. The tungsten oxide was deposited in 15 min increments from a peroxytungstate solution in the pores at constant potential, monitoring the current with a potentiostat (Fig. 14.1). The solution was prepared by dissolving tungsten shot in 30 % hydrogen peroxide, followed by dilution to 25 mmol/L (by tungsten) with water:2-propanol (volume ratio of 7:3, respectively) [23]. The working electrode and template were removed with nitric acid and chloroform, respectively, and after several washings in chloroform and 2-propanol, the nanowires were ultimately dispersed in 2-propanol using ultrasonic agitation. Sensing films of tungsten oxide were prepared from the same peroxytungstate solution by thermal decomposition using the microhotplates [24]. For characterization of thermally processed nanowires, a droplet of the nanowire dispersion was deposited in a glass tube, followed by sintering in air in a furnace at 480 °C for one hour. For transmission electron microscopy (TEM), the nanowires (before and after themal processing) were dispersed in 2-propanol and

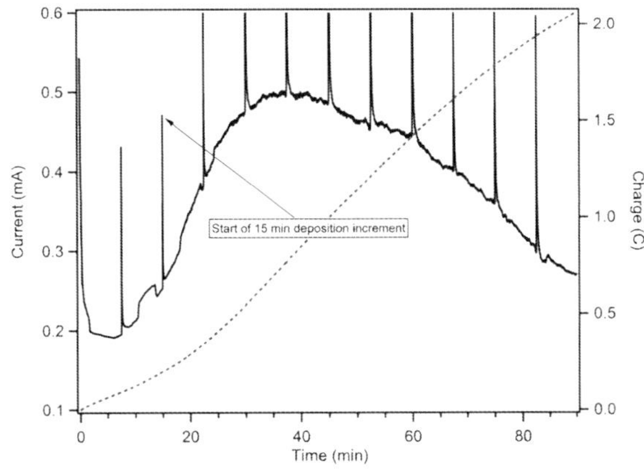

Fig. 14.1 A representative current versus time plot for potentiostatic deposition of tungsten oxide (-0.5 V versus Ag/AgCl) in a polycarbonate template with $d_{pore} = 100$ nm. The charge is shown with the *gray dashed line* (right axis)

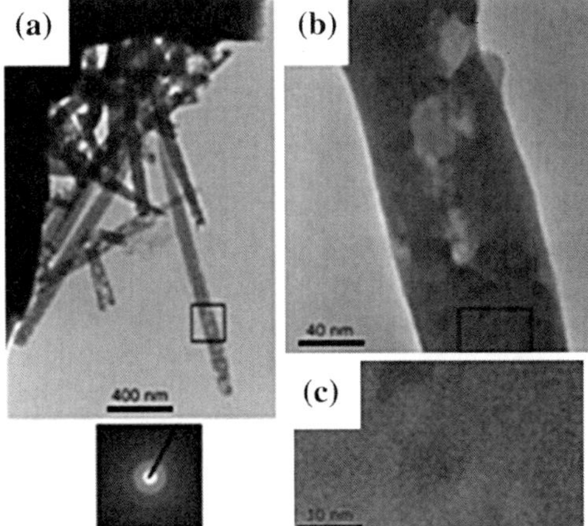

Fig. 14.2 TEM micrographs of tungsten oxide nanowires as prepared (no calcination) (**a**) Low magnification of several nanowire segments. The isolated nanocrystalline domains are shown embedded in the amorphous wire (**b**) and at high magnification (**c**). Reprinted from [17], Copyright (2009) with permission from Elsevier

deposited onto TEM grids. The WO_3 materials were also characterized by optical and scanning electron microscopy (SEM).

The analysis of as-deposited WO_3 nanowires by TEM showed that the nanowires were largely amorphous with small (≈ 10 nm) crystalline domains embedded in the amorphous material (Fig. 14.2). After thermal processing at 480 °C in air, the nanowires became poly-crystalline (Fig. 14.3), while retaining the nanocrystalline scale of 10 nm to 15 nm, as estimated from TEM (Fig. 14.3c, d). The nanowires were also examined by SEM after their deposition on an aluminum substrate, which determined that the nanowire diameter was larger than the template pore diameter. In the case of the 50 nm pores, the nanowires showed a diameter of ≈ 100 nm at the middle of the nanowire and a diameter of ≈ 60 nm at the ends. Similarly, the nanowires grown in the 100 nm pores showed diameters of ≈ 175 nm in the middle and ≈ 125 nm at the ends. The change to discrete particles was also observed in SEM, albeit in a larger scale and longer timeframe (Fig. 14.4).

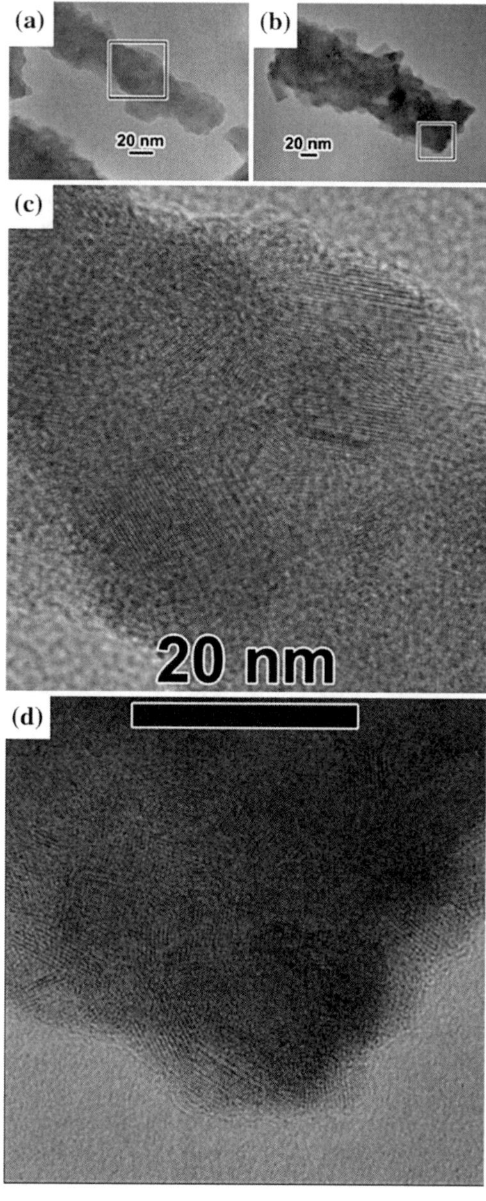

Fig. 14.3 TEM micrographs of tungsten oxide nanowires after calcination in air at 480 °C for one hour. The high-magnification micrographs (**c**, **d**) shows the crystalline structure and corrospond with the boxes in (**a**) and (**b**), respectively. Reprinted from [17], Copyright (2009), with permission from Elsevier

The oxidation state of the tungsten oxide materials was probed using X-ray Photoelectron Spectroscopy (XPS, Fig. 14.5). The XPS measurements were made using a Kratos Axis Ultra DLD spectrometer [25] with a base pressure of 1×10^{-7} Pa. Monochromatic Al K-α radiation was used. For materials deposited onto microhotplate substrates (see below), small spot spectroscopy using a 55 μm aperture was employed for all spectra. For the other samples, materials were

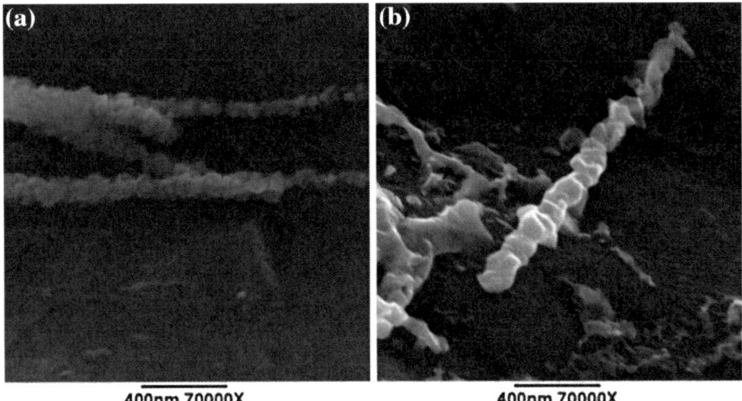

Fig. 14.4 Scanning electron micrographs of WO_3 nanowires ($d_{pore} = 100$ nm). **a** Several nanowires after 1 h of calcination at 480 °C, and **b** a single nanowire after 20 h of aging at 480 °C. The calcination and aging were done in a furnace in laboratory air, with the nanowires deposited upon an aluminum substrate

Fig. 14.5 XPS analyses of the tungsten oxide materials. The main plot is a survey scan acquired from tungsten oxide nanowires dispersed on an aluminum substrate. Inset: spectral component analysis of the W 4f peaks from nanowires (*bottom*) and from a film (*top*). The *dotted lines* are Voigt fits of the data. Reprinted from [17], Copyright (2009), with permission from Elsevier

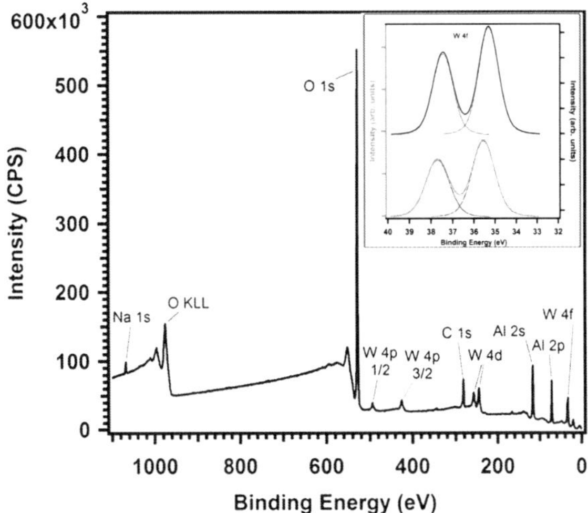

deposited on an aluminum substrate, and large-area spectroscopy was employed. A Shirley background was applied to the W 4f spectra, which were then fit using Voigt profiles. The spectra were referenced to the surface oxide of the aluminum substrate at 74.4 eV (Al 2p). Also mounted on the spectrometer is a Mini-Nova IV sputter gun [25], used for depth profiling experiments. Ar gas was used to sputter the sample. With the extractor current set to 50 µA, a 4 kV beam of Ar ions was allowed to impinge on the sample, producing 1–2 µA of current at the sample. As seen in the spectrum, tungsten peaks (4s, 4p 1/2, 4p 3/2, 4d 5/2, 4f 7/2 peaks at 596.4, 493.9, 425.4, 244.5, 35.4 eV, respectively) are observed from probing of the

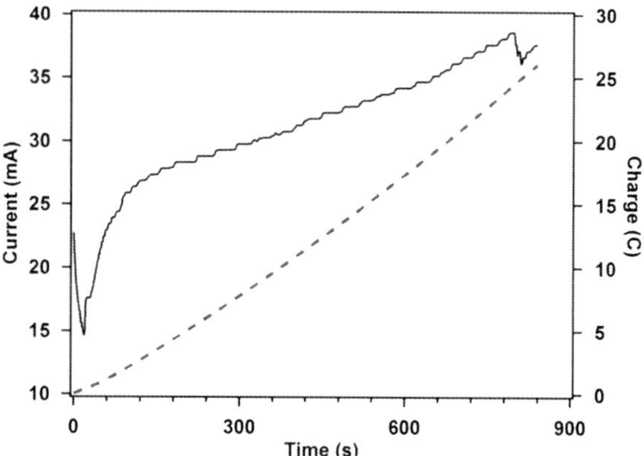

Fig. 14.6 A representative current versus time plot for potentiostatic deposition of tin (-0.8 V versus tin wire) in a polycarbonate template with $d_{pore} = 100$ nm. The charge is shown with the *gray dashed line* (right axis)

nanowire materials. The tungsten oxide film showed a similar pattern. Higher resolution scans of the W 4f region (See Fig. 14.5 inset) were fit to Voigt profiles. The peaks from each material were well fit, indicating that a single chemical species is present. Furthermore, the peak position at 35.5 eV is in good agreement with literature reports for tungsten(VI) oxide (WO_3), indicating that both are fully oxidized after processing and have similar surface stoichiometry [26].

Similarly, tin nanowires were prepared by template-directed electrodeposition, as adapted from literature reports [27]. A gold contact was deposited by thermal evaporation onto polycarbonate membranes ($l \approx 6$ μm, $d_{pore} \approx 50$ nm). Tin nanowires were electrodeposited in the membrane at constant potential -0.8 V versus a tin wire from a $SnSO_4$ solution with gelatin. In contrast to the relatively slow growth of metal oxide nanowires, the tin metal nanowires grew quickly and showed a generally monotonic increase in current as the nanowires grew closer to the bulk solution (Fig. 14.6). Rather than removing the growth electrode, it was found that the nanowires were easily separated from the gold electrode after dissolving the template by gentle rinsing with water. They were subsequently washed and finally dispersed in 2-propanol for deposition. The nanowires were characterized with SEM, XPS and electrical measurements.

A single tin nanowire is shown in the SEM image in Fig. 14.7. The as-grown nanowires are 150–190 nm in diameter and are up to 5 μm in length. The one shown in Fig. 14.7a is ≈ 4.5 μm long with a diameter of ≈ 150 nm. The associated Energy Dispersive X-ray Spectrscopy (EDS) micrograph (Fig. 14.7c) confirms the tin content of the nanowire under examination.

The chemical states of the tin nanowires were further examined using XPS. Of particular interest was the conversion of the as-deposited tin metal nanowires to tin(II) oxide and finally tin(IV) oxide nanowires by thermal processing in air [28, 29]. The conversion was first studied on an aluminum substrate, but XPS was also used to follow the oxidation of tin on the microsensor substrates, using the embedded microheater. Figure 14.8 shows the spectra from the nanowires during the course of the thermal processing on an aluminum substrate (Because the spectra were referenced to the energy of the aluminum oxide 2p peak, there is

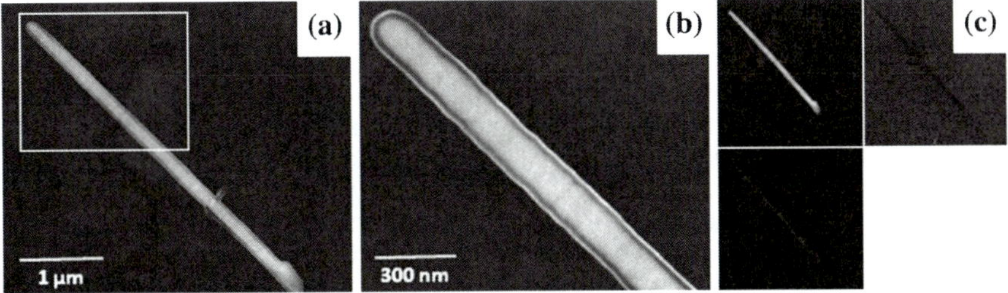

Fig. 14.7 Scanning electron micrographs of a single tin nanowire (**a**, **b**). The box in **a** corresponds roughly to the area shown in **b**. The EDS map is shown in **c** for aluminum (*upper right*) and tin (*lower left*). The *upper left* image is an SEM micrograph for reference

Fig. 14.8 XPS analyses in the Sn 3d$_{5/2}$ region of the conversion the nanowires from tin to tin oxide. **a** Shows the region for nanowires as prepared, **b** shows the region after the first thermal processing step at 200 °C for 1 h in air, and **c** shows the region after the second thermal processing at 480 °C for 1 h in air

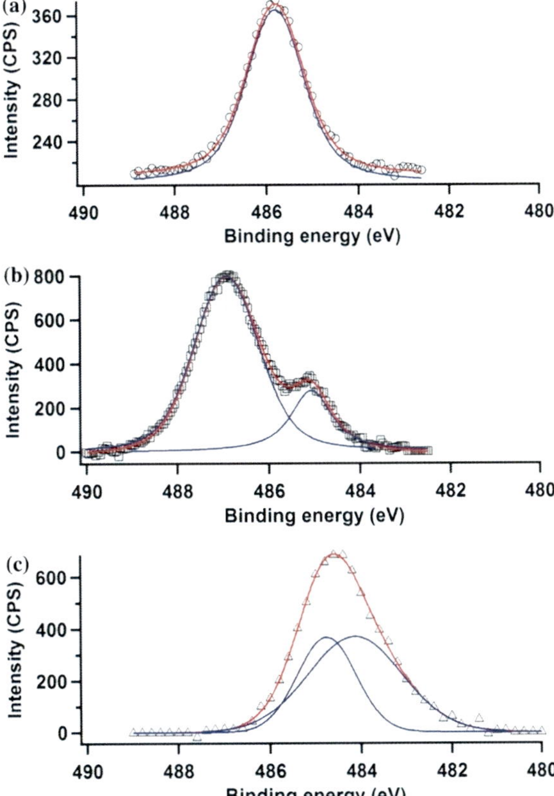

some variability in the energies for the tin 3d$_{5/2}$ peaks). In order to maintain the structure of the nanowires, a two-step processing sequence was used. Figure 14.8a shows the tin 3d 5/2 peak at ≈ 485 eV for the as-prepared tin metal nanowires. The first step of the oxidation process used a hold temperature of 200 °C for one hour to form an initial oxide shell around the nanowires, which prevented the melting of the nanowires at ≈ 232 °C. The spectrum for the nanowires after that step (Fig. 14.8b) shows two distinct tin $3d_{5/2}$ peaks with a separation of ≈ 1.8 eV. The separation indicates that the two forms are tin metal at lower energy and tin(II) oxide at higher energy [29, 30] The nanowires were further processed at 480 °C at one hour to complete the oxidation. The spectrum for the resultant nanowires is shown in Fig. 14.8c, after the nanowires have been sputtered to provide spectroscopic access to the interior material. As shown, there is one distinct peak, although it features some asymmetry, which indicates the presence of more than one chemical species. Fitting the spectrum with two Voigt peaks yields the fit shown, with a peak separation of ≈ 0.6 eV. The separation distance suggests that the two species are tin(II) oxide at lower energy and tin(IV) oxide at higher energy [29, 30]. It should be noted that ion sputtering can alter the oxygen-to-metal ratio of a material by preferentially removing the oxygen [31–34]. This may contribute to the presence of tin(II) oxide in the sample exposed to air at 480 °C. The lack of a third peak at lower energy corresponding to tin metal indicates that the nanowires have been converted to the oxide using this two-step thermal processing approach.

14.3 The Microsensor and Nanowire Integration

The microsensor device technology used in these studies has been developed and refined at the National Institute of Standards and Technology (NIST) over the course of the last two decades [32–34]. The current generation of platforms, manufactured in wafer runs at MIT Lincoln Laboratories [25] and micromachined at NIST, features a 100×100 µm silica structure suspended over a pit etched into the silicon beneath by tetramethyl ammonium hydroxide (Fig. 14.9a). A polysilicon resistive heater with an initial resistance of ≈ 2.5 kΩ is embedded in the elements. The resultant elements have a mass of ≈ 200 ng, which enables temperature control from ambient to ≈ 500 °C with millisecond time constants [16]. To power a heater to 500 °C typically takes less that 25 mW under ambient conditions. On the surface of the microhotplate elements are platinum electrical contacts of varying geometries. For the studies described in this chapter, two electrode geometries were used: (1) interdigitated electrodes with nominal width and spacing of 2 µm, covering 78×78 µm (Fig. 14.9b), and (2) in-line rectangular electrodes, 20×5.5 µm, with spacing between the long sides of the electrodes of 5 or 10 µm (Fig. 14.9c). Before deposition of sensing materials, the etched array devices were mounted into ceramic, 40-pin dual-in-line packages, surface cleaned by low-power ion-beam milling, and wire bonded. The milling utilized an argon ion beam at 300 V with a working distance of 10 cm and a total sputter time of ≈ 1.5 min [16].

Tungsten oxide nanowires were randomly deposited onto microhotplate platforms that featured interdigitated platinum electrodes. Before deposition, the

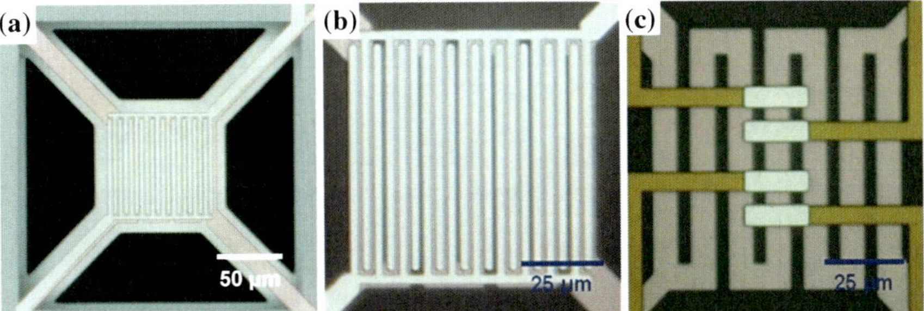

Fig. 14.9 Optical micrograph images showing a full microhotplate (**a**) and the two electrode configurations used in the example nanowire studies. **b** Shows an interdigitated platinum array with nominal spacing of 2 μm between the 2 μm wide fingers. **c** Shows a 4-point configuration with two 5 μm gaps and one 10 μm gap between electrodes

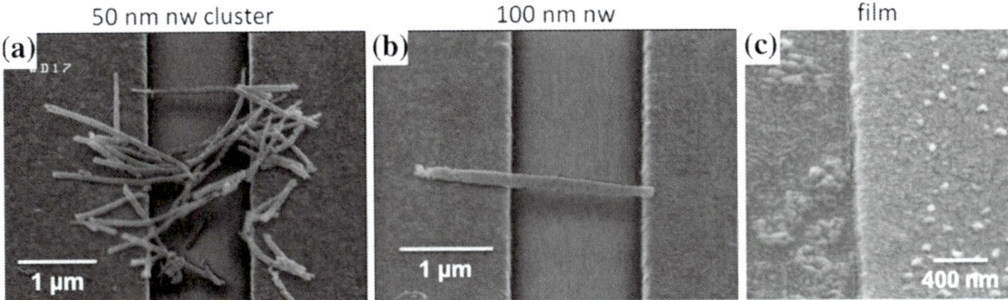

Fig. 14.10 Scanning electrode micrographs of **a** a cluster of tungsten oxide nanowires grown in a 50 nm pore diameter template, **b** a single tungsten oxide nanowire grown in a 100 nm pore diameter template, and **c** a nanostructure film of tungsten oxide prepared by the thermal decomposition of the peroxytungstate solution. All images were taken on microsensor array elements with a configuration identical to that shown in Fig. 14.9a

nanowires were dispersed in 2-propanol via sonication. To achieve localized deposition onto selected microhotplates, droplets were ejected from a microcapillary pipette (polycarbonate, capillary inner diameter, $id_{cap} = 75$ μm). For comparison with film-based sensors, a tungsten oxide film was prepared from a peroxytungstate solution: a droplet of the solution was placed on a microhotplate platform, allowed to dry, and was then thermally converted to tungsten oxide using the embedded heater. As noted in the previous section, this process yields a tungsten oxide film with similar surface stoichiometry as the thermally processed tungsten oxide nanowires. Examples of the resultant materials on the microsensor elements are shown in Fig. 14.10. While we have seen that the nanowires are composed of discreet nanocrystals by TEM, SEM analysis of the tungsten oxide film shows that it, too, has nanostructured particles, suggesting that it has a viable form for comparison with the nanowire materials.

The microhotplates are also amenable to directed alignment/deposition of nanowires, particularly so for the platform shown in Fig. 14.9c. For directed deposition, tin nanowires were integrated with the microsensor platform using micro-scale dielectrophoresis [8, 35, 36] by applying a 20 V peak-to-peak potential at 1 kHz. A capillary micropipette containing the tin metal nanowire dispersion in 2-propanol was brought in close proximity to the microhotplate substrate. The AC electric field was applied between the electrodes of the substrate, and a small droplet was extruded from the microcapillary pipette. The microcapillary was kept in contact with the substrate to minimize solvent evaporation and to provide a reservoir to compensate for any evaporation that did occur. Nanowire loading between the electrodes was controlled by the relative concentration of the nanowire dispersion (Fig. 14.11).

The tin nanowires were thermally converted to the oxide in air using the microhotplate. A stepwise temperature ramp to 200 °C, followed by an anneal at this temperature for 60 min was used to convert the nanowires to oxide before ramping past the tin metal melting point (≈ 232 °C) to 480 °C. The conversion to the oxide was confirmed by small-spot and scanning XPS.

14.4 Nanowires on Microhotplate Substrates for Chemical Sensors

The first challenge in using nanowires as sensor materials is to integrate the materials with a sensing platform. As a first approach, dispersing the nanowires in a liquid medium, followed by drop-casting onto a substrate with pre-patterned electrical contacts has been shown to be effective [19, 20, 37–39] and other sensor preparation methods yield similar structures [18, 40]. This method was used to deposit tungsten oxide nanowires onto microhotplates with interdigitated electrodes. Processing of the nanowires was achieved using the embedded heater of the microhotplate to convert the amorphous nanowires at 480 °C to polycrystalline forms with a nanocrystalline particle size of 10–15 nm. To produce chemical microsensors based upon the tin nanowires, however, a dielectrophoretic approach was used to align and deposit nanowires between the electrodes of platforms shown in Fig. 14.9c. Again, the embedded heater was employed to convert the tin metal nanowires to tin oxide for chemical sensor evaluations.

A second challenge for developing nanowires as sensing materials is assessing their performance, particularly in reference to established film materials. For this challenge, the capability to easily change the sensor operating temperature can greatly enhance the information obtained from the microsensors, facilitating evaluations of materials. Several approaches may be considered, from empirical analysis of sensor performance to modeling of the nanowire sensor-analyte interactions, as shown in a recent paper, for example [18]. The modeling of the tin oxide nanowires as a sensor (as schematically shown in Fig. 14.12) can provide useful insights into the system, predicting the responses of the sensor to changes in analyte concentration and operating temperature. In this case, the conductive

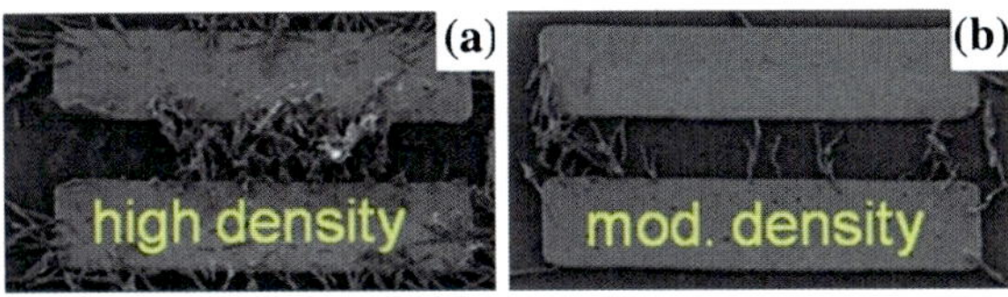

Fig. 14.11 Scanning electrode micrographs of **a** tin nanowires deposited and aligned from a high-concentration dispersion, and **b** tin nanowires deposited and aligned from a low-concentration dispersion. All images were taken on microsensor array elements with a configuration identical to that shown in Fig. 14.9b

Fig. 14.12 A model used to illustrate the resistance contributions in the case of a non-depleted nanowire bundle. Reprinted from [18], Copyright (2010), with permission from Elsevier

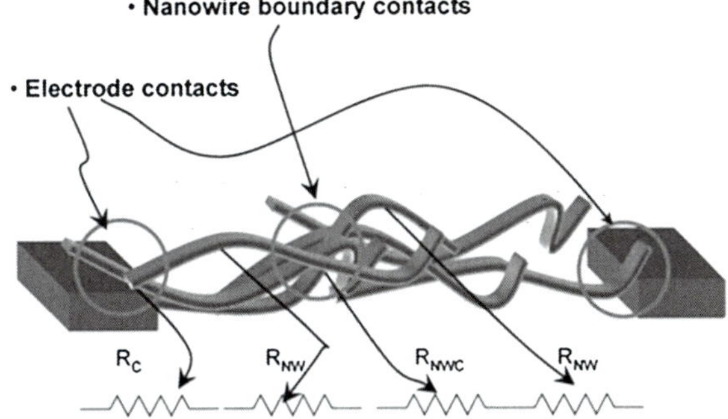

pathway is modeled as a series of junctions, in addition to the nanowire itself. The resistances of the individual nanowires are designated R_{NW}, the resistance for the nanowire to electrode junction is R_C and the resistance at the junctions between nanowires is R_{NWC}. Sensor array-based approaches with an integrated means for varying temperature offer an efficient means for making the evaluation, as multiple materials and morphologies may be studied in parallel, largely eliminating run-to-run variability in the sensing experiments (owing to, for example, analyte delivery and dilution uncertainties, changes in ambient pressure and temperature, background differences, etc.). Due to their beneficial thermal properties, microsensor arrays employed in these studies also allow for dynamic control of the sensor operating temperature. This is a critical sensing design parameter as the sensor operating temperature influences the interaction of the sensing material with an analyte of interest, potentially providing unique information across a range of temperatures as the analyte adsorbs to the surface, breaks down and reacts [34, 41–45]. Alternative strategies for inducing analytical orthogonality into sensing materials using temperature is demonstrated using a temperature-gradient approach (Fig. 14.13) [19]. In this case, a mat of nanowires was deposited onto a substrate, so that there were both bridging nanowires and a percolating network between electrodes (Fig. 14.13c). The density of the nanowires was varied across the device. Back-side platinum heaters were used to generate the thermal gradient along the device (Fig. 14.13b). The authors found that the combination of the thermal gradient and the varied nanowire densities yield sufficient analytical power to detect and discriminate between several organic compounds.

Fig. 14.13 A nanowire bundle sensor based upon tin oxide nanowires (**a**) The KAMINA microarray chip. (**b**) The temperature gradient across the sensing elements is shown in the IR image: 247 °C (*green, bottom half*) to 327 °C (*red, top area*). (**c**) The SEM micrograph shows tin oxide nanowires between Pt electrodes on the chip. Reprinted from [19], Copyright 2009 American chemical Society

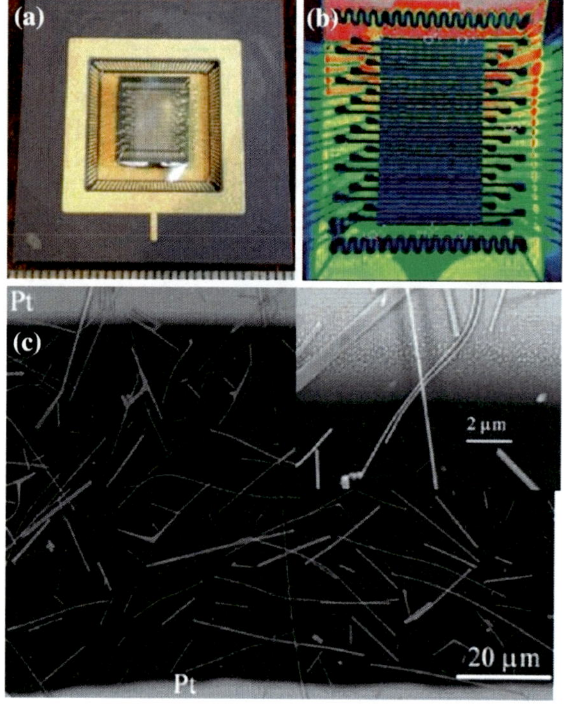

One approach employing operating temperature is illustrated using the tin oxide nanowire sensing devices similar to those depicted in Fig. 14.11. Via an operating mode termed Fixed-Temperature Sensing (FTS), the temperature of the sensors was "fixed" at select values to establish isothermal responses of the sensors to analytes. Temperature changes were effected on timescales of minutes or hours, using custom-built electronics to control the operating temperature and to record the resistance or conductance of the sensing materials. The conductance of the sensing films was monitored, with changes in conductance indicating a response to a change in analyte or a change in the sensor operating temperature. The magnitude of the change in conductance can be an indication of the degree of perturbation to the equilibrium sensor state. That is, a high concentration of analyte or a bigger change in sensor temperature would result in a bigger change in conductance than smaller concentrations/changes. In addition to the computer-controlled sensing, the experiments were performed with an automated gas-delivery system, which facilitated the matching of timing between the sensor operating temperature changes and the analyte cycles. For the experiments discussed here, the background was zero-grade dry air, generated on-site. The analyte being studied, methanol, was delivered from a gas cylinder with certified dilution in dry air. Analytes were delivered to the microsensors at concentrations less than 200 µmol/mol (moles of analyte per mole of air).

Fig. 14.14 Conductometric plots of the two tin oxide nanowire sensors with responses to temperature and exposure to varied concentrations of methanol (*gray bars* corresponding to the right axis)

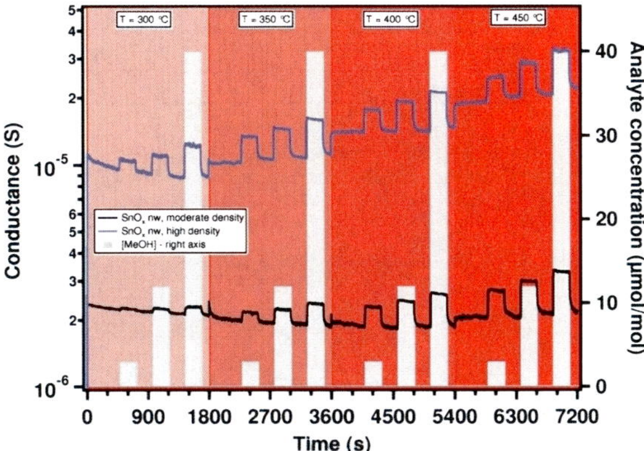

Figure 14.14 presents the responses of the two tin oxide microsensors (micrographs shown in Fig. 14.11) to varied operating temperature and methanol concentrations. One parameter that can be explored by the experiment is the role of nanowire density, or loading on the sensor platform, in affecting the sensor performance. As would be expected based on the number of conductive pathways, the high-density nanowire sensor (Fig. 14.11a) shows a higher baseline conductance at all sensor operating temperatures than the moderate density nanowire sensor (Fig. 14.11b). To determine response strength (S) for each analyte introduction, a baseline conductance was calculated by averaging 100 conductance measurements from the immediately preceding baseline (G_0) and the response was calculated by averaging 100 points from the end of the response plateau (G). The sensor response strength is then defined as $S = (G - G_0)/G_0$. The response strengths are summarized in Table 14.1. Here we see the importance of sensor operating temperature in examining the performance of the nanowire sensors. In looking at the response-strength ratio (S_{HD}/S_{MD}) between the two sensors (HD ≡ High Density, MD ≡ Moderate Density), the ratio decreases as the temperature increases. That is, the moderate-density sensor is becoming more sensitive relative to the high-density sensor as the operating temperature increases. Also of interest is that the concentration of the analyte has an impact on the change in response-strength ratios. While the high-density sensor maintains its advantage in S for the 40 μmol/mol exposures to methanol, at the lower concentrations (3 and 12 μmol/mol), the moderate-density sensor shows a larger S at the highest operating temperature (450 °C) studied for this example. This may be attributed to changes in reaction rates as the temperature increases combined with the exposed surface area of the sensor. While the moderate-density sensor shows discrete nanowires with little overlap, the high-density sensor has clustering and stacking of nanowires. At higher temperatures, as the methanol reacts at the surface of the sensor, the high-density sensor apparently loses its advantage, owing to the inaccessible interior surfaces of nanowires.

The use of temperature to evaluate nanomaterials may be taken even further, as demonstrated through an example with tungsten oxide nanowires[17]. For this

Table 14.1 Summary of responses of tin oxide nanowire sensors to methanol

Sensor operating temperature (°C)	[MeOH] (μmol/mol)	S ($\Delta G/G_0$) Moderate density nanowires	S ($\Delta G/G_0$) High-density nanowires	Response ratio (S_{HD}/S_{MD})
300	3	SNR < 3	0.09	NA
	12	0.03	0.21	6.8
	40	0.07	0.39	5.5
350	3	0.08	0.30	3.9
	12	0.14	0.37	2.6
	40	0.23	0.47	2.0
400	3	0.21	0.24	1.1
	12	0.31	0.35	1.1
	40	0.40	0.49	1.2
450	3	0.32	0.30	0.93
	12	0.45	0.41	0.91
	40	0.54	0.57	1.1

illustration, three materials are considered: a tungsten oxide film as might be typically used in a microsensor array (film), randomly dispersed clusters of tungsten oxide nanowires grown in a template with a nominal pore diameter of 50 nm (nw50), and randomly dispersed clusters of tungsten oxide nanowires grown in a template with a nominal pore diameter of 100 nm (nw100). The sensors under consideration are three elements in an array, based upon microsensor elements such as the one shown in Fig. 14.9b. Again, the comparison between the materials starts with FTS studies, if only to establish the viability of the randomly dispersed nanowires as chemical microsensors. A plot of the conductometric responses of the three tungsten oxide sensor materials to varied concentrations of methanol (similar tests were also performed to examine the responses of the sensors to carbon monoxide and nitrogen dioxide) is shown in Fig. 14.15. After an initial drift down in conductance, the sensors are generally stable, demonstrating their viability. While numerical analyses of these data are difficult owing to the slow response times of the film-based sensor and the high noise level of the nw100-based sensor, some qualitative conclusions may be drawn: the nanowire sensors show faster responses to analyte introduction/removal as compared to the film-based sensor, and the response strengths and signal-to-noise ratios are generally comparable to the film-based sensor.

While the first evaluations for novel sensing materials generally determine the raw sensing parameters (sensitivity, speed, stability and selectivity), an additional question that arises concerns the information content of the sensing responses of these materials as their scale is systematically changed. One approach to elucidate whether different sensors provide orthogonal analytical information is to perturb the sensor by changing the operating temperature over a wide range. This approach, called Temperature-Programmed Sensing (TPS) is illustrated using the three tungsten oxide sensing materials [17].

To cover a wide range of sensor-analyte interactions, a temperature ramp program (Fig. 14.16) was generated that pulsed the sensors from 60 °C to 480 °C

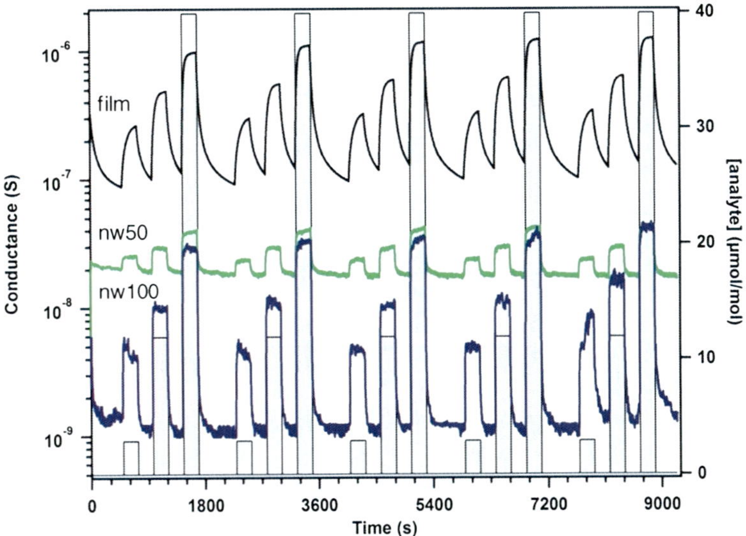

Fig. 14.15 Conductometric plots of the three tungsten oxide sensors operating at 425 °C with responses to exposure to varied concentrations of methanol (*gray bars* corresponding to the right axis)

Fig. 14.16 A temperature program used to operate the tungsten oxide sensors. Conductance measurements were made at each ramp (*circle*) and base (*diamond*) step

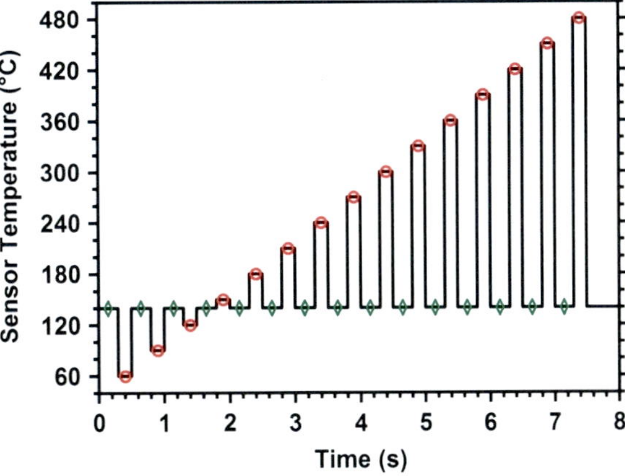

with 30 °C steps [17]. A baseline temperature was established at 140 °C to provide a base state for the sensors to equilibrate to before being pulsed to the next temperature in the ramp. The dwell time for each temperature, both ramp and base, was 250 ms. This fast temperature cycling is possible owing to the small thermal mass of the microhotplates (Fig. 14.9a). Conductance measurements are made at each temperature step to provide 30 data points per cycle for each sensor. A full sweep through the temperature cycle lasts 7.5 s. A second ramp program was generated for follow-up studies on the influence of thermal history that used a 1 °C increment between ramp temperatures, while maintaining a base temperature

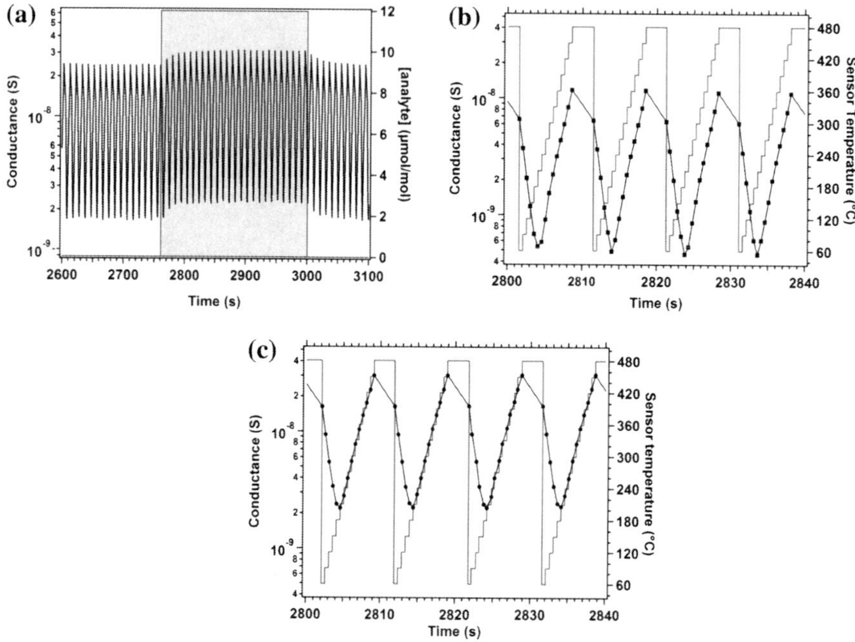

Fig. 14.17 The TPS response patterns of a tungsten oxide nanowire sensor (nw50) to presentations of (**a, b**) methanol, and **c** nitrogen dioxide in a dry air background. **a** Shows the change in sensor TPS response from air to methanol (*gray bar*, right axis). The methanol concentration is 12 μmol/mol and the nitrogen dioxide concentration is 0.3 μmol/mol. Note the pattern shapes for the two analytes as a function of temperature (*gray line*, right axis)

of 140 °C. For the subsequent analyses, only the subset of data corresponding to the 15 ramp-temperature steps of the 30 °C-increment program were used, for better comparability. The use of the second program was limited owing to the long cycle time with the additional ramp steps (≈ 3.5 min per cycle). An example of the data generated using the temperature program shown in Fig. 14.16 is seen in Fig. 14.17 (conductance measurements from the ramp temperatures only). Note that the conductance data no longer reflect equilibrium interactions between the sensor and analyte as shown in Fig. 14.15; rather it is a dynamic interaction dependent upon the current sensor operating temperature as well as the thermal history of that sensor. In Fig. 14.17a, b, a small segment of the TPS data are shown for the responses of a tungsten oxide nanowire sensor to methanol. In comparison with the FTS data, the data density per unit time is appreciably higher for the TPS data (Fig. 14.17), but the overall trends are the same. That is, the conductance increases when the sensor is exposed to methanol vapor in μmol/mol concentrations (Fig. 14.17a). Figure 14.17c shows a similar segment of the raw data upon exposure to nitrogen dioxide. The general shapes of the patterns are similar for the two analytes, but there appear to be subtle differences in the overall pattern as a function of sensor operating temperature (gray steps, right axes).

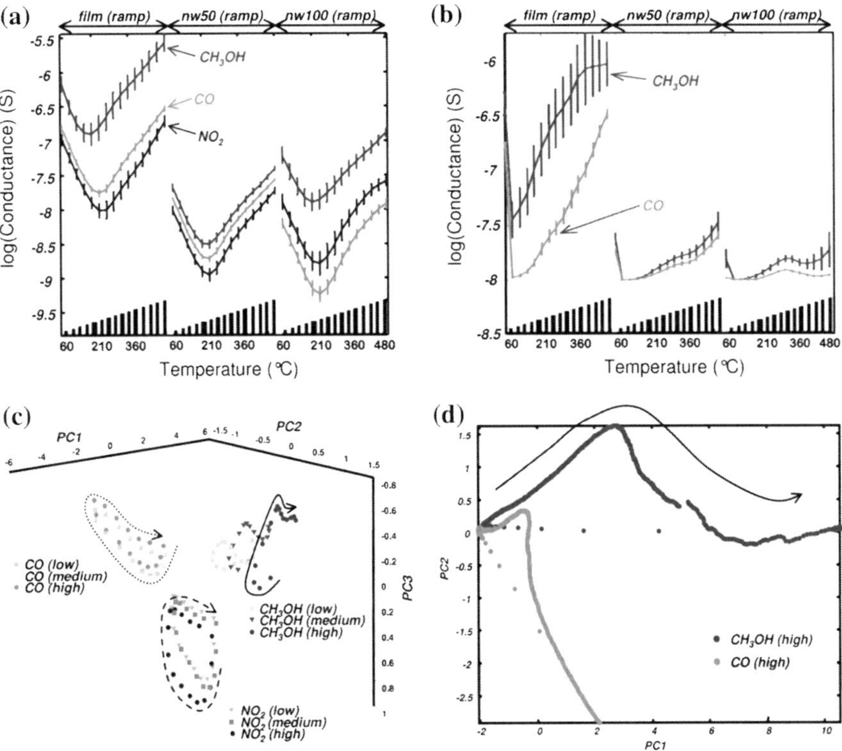

Fig. 14.18 The average conductometric measurements are shown here for the ramp measurements using 30 °C steps (**a**) and 1 °C steps (**b**). The error bars show the standard deviation. The three analytes in (**a**) are methanol (*top*), carbon monoxide (*middle*) and nitrogen dioxide (*bottom*). For (**b**), the two analytes are methanol (*top*) and carbon monoxide (*bottom*). The 6-dimensional sensor response (**c**), now including the base measurements, for the data shown in Fig. 14.18a is projected along the first three principal components (variance captured = 97.12%). The three analytes show distinct clusters, and the effect of increasing ramp temperatures is shown by the arrows. A similar projection along the first two principal components (variance captured = 93.64%) is shown in (**d**) for the data in Fig. 14.18b, now including the base-temperature measurements. Again, the arrow shows the trend with increasing ramp temperature. Reprinted from [17], Copyright (2009), with permission from Elsevier

To better visualize the data obtained from the TPS measurements and to tease out the differences in responses between the sensors, principal component analysis, a linear dimensionality reduction technique [46], was employed. The six-dimensional array responses (three sensors considering 1 ramp and 1 base temperature each) were projected onto the first three dimensions for the 30 °C step experiments and onto the first two dimensions for the 1 °C step experiments, as defined by the eigenvectors with the largest eigenvalues of the response covariance matrix that captures the most variance in the dataset. The averaged ramp data collected from the 30 °C-step TPS experiments for three analytes are shown in Fig. 14.18a. Similar plots in Fig. 14.18b show the averaged data collected from the 1 °C-step

TPS experiments for two analytes (nitrogen dioxide was excluded from this study owing to the difficulty in measuring the low conductances that result from exposure of the nanowire sensors to the gas). For clarity, only the data that correspond to the 30 °C steps are shown. Each plot shows the averaged conductance patterns for the three sensors exposed to the analytes at the specified concentrations. For the 30 °C-step experiment, the sensors were exposed to varied concentrations of analytes, but only for methanol was it possible to distinguish visually a difference in response from the raw data. The thermal history does affect the overall raw data TPS pattern, as a different pattern is seen for each material between Fig. 14.18a and b.

Figure 14.18c shows the six-dimensional sensor responses (including the base data, which were not shown in Fig. 14.18a) after linear projection along the first three principal components, which account for 97.41 % of the variance in the data. Figure 14.18d shows similar plots for the six-dimensional sensor responses from the 421 ramp data points obtained from the 1 °C-step experiment (Fig. 14.18b), albeit along the first two principal components, accounting for 93.64 % of the variance. Several noteworthy trends are observed from the PCA plot in Fig. 14.18c: (1) The identity of the analyte (labeled in the plot) introduces variance along the axis. (2) The concentration of the analyte causes variance within the analyte cluster, even for the two analytes that showed minimal separation upon visual inspection of the raw data. (3) The sensor operating temperature induces variance along a different direction, as indicated by the arrows showing the sensor ramp temperature increasing from 60 °C to 480 °C. Similar trends are observed in the 2-dimensional plot of Fig. 14.18d, excluding the analyte concentration trends. A final point to be noted is that the separability between the analytes is influenced by the sensor operating temperature, indicating the importance of the operating temperature and the thermal history of the sensor in the sensing responses. This observation illustrates that the selection of a temperature program may accentuate the differences between sensing materials, leading to a more thorough understanding of how the sensors perform differently.

To quantify the analytical orthogonality of the sensors as a function of operating temperature and of the sensing material, a measure based upon the correlation coefficients was used to compare the information content of the sensors' conductometric patterns. Suppose x is a one-dimensional vector of conductance measurements made at each of the different training conditions using a material $M1$ at temperature $T1$, and y is a one-dimensional vector of conductance measurements made using sensing material $M2$ at temperature $T2$, then the correlation between responses of $M1$ at $T1$ and $M2$ at $T2$ is calculated as

$$Corr(x, y) = \frac{Cov(x, y)}{\sigma_x \cdot \sigma_y}$$

where $Cov(x,y)$ is the covariance of x and y, σ_x and σ_y are their respective standard deviations. The greater the correlation, the greater the relation or redundancy between the selected measurements (as a function of materials and temperatures),

Fig. 14.19 The results from
the statistical analysis of the
sensor measurements made
using temperature programs
(60 °C to 480 °C) with 30 °C
increments are shown as
correlation plots. The
correlation coefficients (p <
0.05) between the three
tungsten oxide sensors at the
15 ramp- and-temperature
measurements (Fig. 14.16)
for air (*background*) are
shown. Reprinted from [17],
Copyright (2009), with
permission from Elsevier

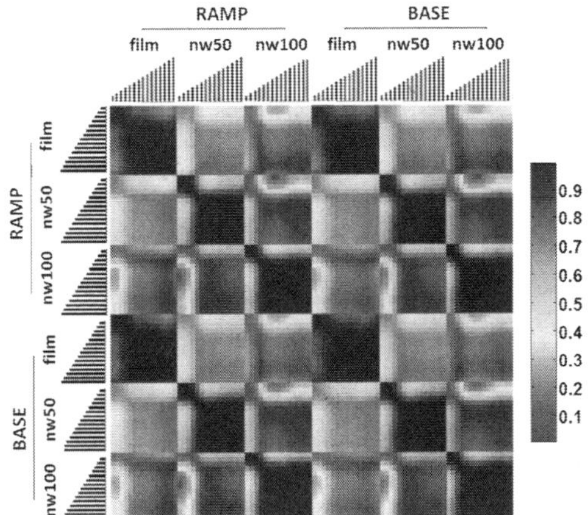

and, conversely, a low correlation typically indicates that non-redundant (analytically orthogonal) information is generated from the sensor material/temperature pair under consideration. To better visualize the correlation coefficients, the calculated correlation numbers are plotted in Fig. 14.19. Each pixel shows the absolute value of significant correlation (significant as defined by the p value, $p < 0.05$; computed by transforming the correlation to create a t-statistic having $n - 2$ degrees of freedom:

$$t = Corr(x, y) \cdot \sqrt{\frac{n - 2}{1 - Corr(x, y)^2}}$$

where n is the number of measurements) between responses of a tungsten oxide morphology at one temperature, T_1, to the background condition, dry air, versus its response at a second temperature, T_2. The darker pixels indicate higher correlations and the lighter pixels indicate lower correlations. The diagonal blocks provide self-correlation patterns and the off-diagonal patterns provide cross-correlation between tungsten oxide morphologies. This simplified figure considers only the ramp temperatures in the upper left quadrant, which is 45 × 45 pixels: three blocks of 15 × 15 pixels (one pixel for each ramp temperature conductance measurement) for each of the three materials under examinations. As expected, the diagonal blocks show large amounts of dark pixels, indicating good self-correlation. Note, however, that the correlation is somewhat lower along the edges of the diagonal blocks, indicating lower correlation between the high- and low-temperature conductance measurements. The correlation between materials is also lower than the self-correlation, especially between the film and the two nanowires materials.

The correlation analyses are expanded in Fig. 14.20, using the three analytes with the correlations from the base-temperature conductance measurements. The bands between high- and low-temperature measurements seen in Fig. 14.19 are

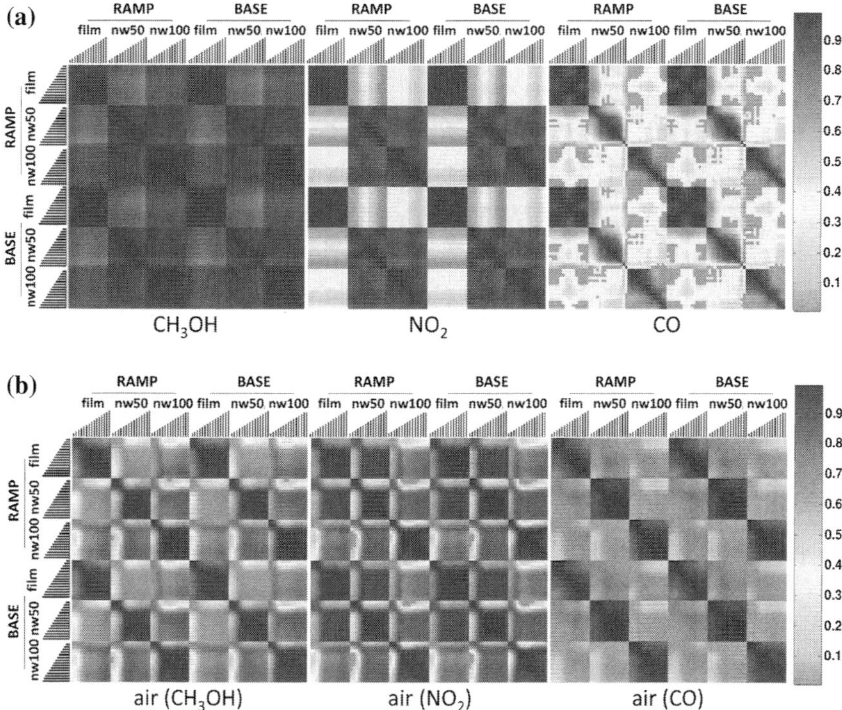

Fig. 14.20 The correlation coefficients (p < 0.05) are shown here for the ramp and base conductance measurements using 30 °C step sizes. (**a**) The three tungsten oxide sensors are compared for the three analytes studied. (**b**) Correlation plots from background exposures as a function of the analyte. Reprinted from [17], Copyright (2009), with permission from Elsevier

preserved in the expanded correlation maps for air, which now includes the ramp and base-temperature measurements for all conditions. The high correlation and preservation of patterns between the ramp-temperature and base-temperature measurements follows through to the three analytes as well, which suggests, again, that the thermal history of the sensor plays an important role in its performance. Overall, between the four conditions (the three analytes in Fig. 14.20a and the background in Fig. 14.20b), the difference in correlation patterns is striking. Examining each condition in detail, it is seen that in the presence of air, there is moderate correlation between the materials and, as noted above for Fig. 14.19, only moderate correlation between the lowest temperatures and the mid- and high-range temperatures. For methanol, the case is different as there is generally high correlation, not only within the temperature program for a select material, but also between all of the materials. Upon exposure to nitrogen dioxide, the pattern changes again to high correlation between the nanowire materials that, in turn, show low correlation with the measurements from the film. Temperature, for the case of nitrogen dioxide, does not greatly lower the correlation between measurements of a material. Finally, for carbon monoxide, there is little correlation between the materials, even for the nanowires, and distinct banding is observed for the self-correlation plots of the nanowires as a function of temperature. Thus, the three sensors provide only redundant information regarding the presence of methanol vapor, the nanowire sensors provide non-redundant information with respect to the film sensor regarding the presence of nitrogen dioxide, and each sensor provides unique information regarding carbon monoxide. Furthermore, by cycling through the range of temperatures, dissimilar information is generated for certain chemical species.

To confirm that changes in patterns for each of the analytes were not due to changes in the sensors themselves over the course of the three experiments, the correlation analysis for the background was broken down into the three separate backgrounds for each analyte experiment, as shown in Fig. 14.20b. While there are some changes in the magnitude of the correlation coefficients from experiment to experiment, overall each of the air backgrounds in Fig. 14.20b most resembles the correlation patterns seen in the air plot of Fig. 14.19. That is, each one shows high self-correlation with moderate correlation between temperature bands, and moderate correlation between materials. These results indicate that the additional particle growth seen in Fig. 14.4 after long times at high temperature are not changing the response characteristics of the sensors over the course of the experiments. Indeed, because the sensors are operated in a pulsed-temperature mode, the materials are spending much more time at more moderate operating temperatures than at the highest operating temperature of 480 °C (Fig. 14.16).

As noted above, the different temperatures do induce some analytical orthogonality in a single sensor and between sensors. The capability for rapid thermal excursions from a base temperature enables efficient (with respect to time) elucidation of those differences. This is most clearly seen in the banding patterns between the materials for nitrogen dioxide in Fig. 14.20a. While self-correlation blocks show little temperature dependence, the comparisons between materials show several distinct, low-correlation bands. Because this banding in the correlation plot is observed between the nanowire materials and the film material, it indicates that the different temperature regimes are inducing analytical orthogonality of varied degrees. This is in agreement with the expectation that different physical and chemical phenomena dominate the sensor-analyte interactions as a function of temperature. For example, one would expect adsorption and binding to be dominant at low temperatures, and desorption and reaction to be dominant at higher temperatures.

Returning to the raw TPS conductance data in Fig. 14.18a and b to expand upon the role of the temperature and thermal history in sensor performance, it is seen that the conductance does not change monotonically with increasing ramp temperature. Rather, there is an inflection point that changes with a change in the thermal history of the sensor (i.e., the size of the ramp steps). For the 30 °C step experiments, the inflection point is at ≈ 210 °C, while for those using 1 °C steps, the inflection point is at ≈ 100 °C. To determine whether different thermal histories (i.e., temperature ramps) provide different information, a correlation analysis was done between the sensor data obtained from experiments using 30 and 1 °C

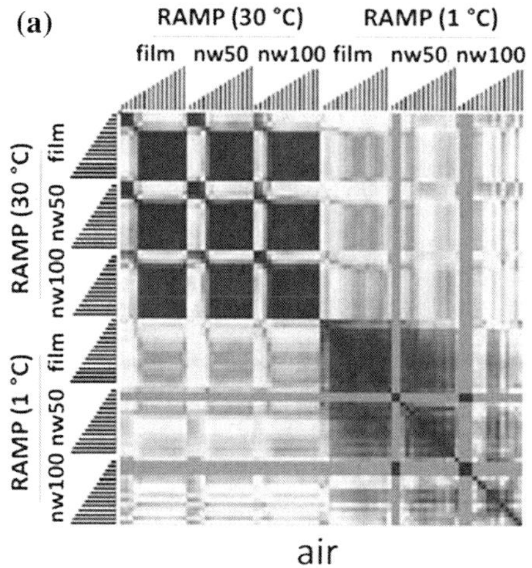

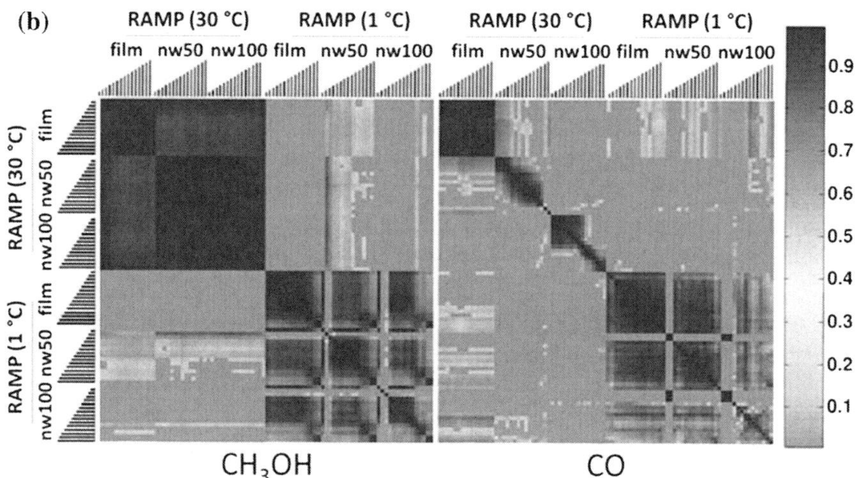

Fig. 14.21 The correlation coefficients ($p < 0.05$) are shown here for the ramp conductance measurements using both 30 °C and 1 °C step sizes. The 1 °C temperatures selected for comparison match those from the 30 °C steps. The plots for (**a**) air (*background*) and for (**b**) methanol and carbon monoxide are shown. (**b**) Reprinted from [17], Copyright (2009), with permission from Elsevier

increments. These plots are shown in Fig. 14.21 for air, methanol and carbon monoxide. The upper left quadrant for the analytes corresponds to the upper left quadrant of the appropriate analyte in Fig. 14.20a, although scaling differences may lead to somewhat different shading of the pixels. As expected from our previous discussion above, there is little to only moderate correlation between the information coming from the sensors using the two different temperature programs to produce different thermal histories on the sensors. That is, the different thermal histories for the sensors provide non-redundant information about the analytes, and even about the background air. It is of interest to note, as well, that the different temperature increments yield different patterns in the correlation plots comparing the three sensor morphologies at a specific temperature increment (upper left and lower right quadrants). This is especially apparent for carbon monoxide, where the correlation between materials generally increases with the smaller temperature steps, indicating that the larger temperature steps yield less redundant information between the different sensor morphologies.

Ultimately, the use of dynamic temperature programming enables the elucidation and accentuation of differences between chemical microsensors. Changing operating temperatures may be used to induce analytical orthogonality as seen with the temperature gradient nanowire sensor and the correlation analyses here, and they may be used as part of a model of the sensor to better understand the performance characteristics of the device. Two examples of the use of sensor temperature to accentuate differences between stoichiometrically similar but morphologically distinct materials have been expanded here with two nanowire-based microhotplate sensors: one featuring aligned tin oxide nanowires and one featuring randomly dispersed tungsten oxide nanowires. In each case, important comparisons were made. For the tin oxide sensors, the use of varied sensor operating temperatures showed differences in sensor performance owing to differences in the loading density of the nanowires between the electrodes. As the sensor operating temperature increased, the differences in performance between the sensors decreased, likely due to the increased reactivity between the sensor surface and the analyte. In the example of the tungsten oxide nanowires, differences were seen in performance using advanced temperature programming and data analysis methods. Here, it was demonstrated that non-redundant information can be obtained from microsensors with identical surface stoichiometry but different morphologies, especially with respect to a thin film-based microsensor. This capability was made possible by the changes in sensor performance over a range of operating temperatures and thermal histories. Because of the dynamic nature of the sensor-analyte interactions in TPS mode, the thermal history can have an influence on the information obtained from a sensor. This has been exploited to optimize the sensor performance, e.g., to identify regions of the program that are robust against drift or particularly relevant to discriminating between varied target analytes [47, 48].

14.5 Future Considerations

The use of temperature to probe the differences in microsensor performance has been noted before, and is a generally applicable technique. Indeed, in an array format, it makes for an efficient method for tuning the sensing materials of a chemical sensor for specific applications that range from homeland security, industrial process control, medical and health diagnostics, environmental monitoring and even space and planetary exploration. For nanowires and nanomaterials in general, which are expected to form the basis of next-generation chemical sensors, methods to evaluate and compare the different materials and morphologies are crucial. Because integration of nanowires into micro- (and eventually nano-) electronic devices is a challenge, methods like those described here may be used to not only assess the performance, but also the reproducibility from device to device. Another interesting comparison that builds off of these illustrative studies here would consider the differences in performance between poly-crystalline nanowires and single-crystal nanowires, which are expected to offer benefits for stability and reproducibility. Recent studies have shown that self-heating, single nanowire chemical sensors are also feasible [49]. Analysis approaches as described here may be useful for appreciating the role of the self-heating in the performance of the sensors, and in the differences between sensors of nominally similar construction (i.e., device reproducibility).

Acknowledgments We acknowledge C. S. Mungle for assistance in the dielectrophoretic alignment of the tin nanowires and the technical assistance of C.B. Montgomery in preparing the microsensor platforms and packaging. B. Raman was supported by a NIH–NIST Joint Postdoctoral Associateship Award and D. L. Lahr was supported by a NIST Postdoctoral Associateship Award, both administered through the National Research Council.

References

1. Choi KJ, Jang HW (2010) One-dimensional oxide nanostructures as gas-sensing materials: review and issues. Sensors 10(4):4083–4099
2. Kolmakov A, Moskovits M (2004) Chemical sensing and catalysis by one-dimensional metal-oxide nanostructures. Annu Rev Mater Res 34:151–180
3. Comini E, Sberveglieri G (2010) Metal oxide nanowires as chemical sensors. Mater Today 13(7–8):28–36
4. Franke ME, Koplin TJ, Simon U (2006) Metal and metal oxide nanoparticles in chemiresistors: does the nanoscale matter? Small 2:36–50
5. Benkstein KD, Martinez CJ, Li G, Meier DC, Montgomery CB, Semancik S (2006) Integration of nanostructured materials with MEMS microhotplate platforms to enhance chemical sensor performance. J Nanopart Res 8:809–822
6. Comini E (2006) Metal oxide nano-crystals for gas sensing. Anal Chim Acta 568(1–2):28–40
7. Donthu S, Alem N, Pan Z, Li S-Y, Shekhawat G, Dravid V, Benkstein KD, Semancik S (2008) Directed fabrication of ceramic nanostructures on fragile substrates using soft-electron beam lithography (soft-eBL). IEEE Trans Nanotech 7(3):338–343
8. Evoy S, DiLello N, Deshpande V, Narayanan A, Liu H, Riegelman M, Martin BR, Hailer B, Bradley JC, Weiss W, Mayer TS, Gogotsi Y, Bau HH, Mallouk TE, Raman S (2004) Dielectrophoretic assembly and integration of nanowire devices with functional CMOS operating circuitry. Microelectron Eng 75(1):31–42
9. Fan ZY, Ho JC, Takahashi T, Yerushalmi R, Takei K, Ford AC, Chueh YL, Javey A (2009) Toward the development of printable nanowire electronics and sensors. Adv Mater 21(37):3730–3743
10. Li XP, Chin E, Sun HW, Kurup P, Gu ZY (2010) Fabrication and integration of metal oxide nanowire sensors using dielectrophoretic assembly and improved post-assembly processing. Sens Actuators B 148(2):404–412
11. Chen PC, Shen GZ, Zhou CW (2008) Chemical sensors and electronic noses based on 1-D metal oxide nanostructures. IEEE Trans Nanotech 7(6):668–682

12. Semancik S, Xiang X-D, Takeuchi I (2003) Temperature-dependent materials research with micromachined array platforms. In combinatorial materials synthesis. Marcel Dekker, Inc., New York, pp 263–295
13. Taylor CJ, Semancik S (2002) Use of microhotplate arrays as microdeposition substrates for materials exploration. Chem Mater 14:1671–1677
14. Benkstein KD, Semancik S (2006) Mesoporous nanoparticle TiO_2 thin films for conductometric gas sensing on microhotplate platforms. Sens Actuators B 113(1):445–453
15. Cavicchi RE, Walton RM, Aquino-Class M, Allen JD, Panchapakesan B (2001) Spin-on nanoparticle tin oxide for microhotplate gas sensors. Sens Actuators B Chem 77:145–154
16. Martinez CJ, Hockey B, Montgomery CB, Semancik S (2005) Porous tin oxide nanostructured microspheres for sensor applications. Langmuir 21:7937–7944
17. Benkstein KD, Raman B, Lahr DL, Bonevich JE, Semancik S (2009) Inducing analytical orthogonality in tungsten oxide-based microsensors using materials structure and dynamic temperature control. Sens Actuators B 137(1):48–55
18. Fort A, Mugnaini M, Rocchi S, Vignoli V, Comini E, Faglia G, Ponzoni A (2010) Metal-oxide nanowire sensors for CO detection: characterization and modeling. Sens Actuators B 148(1):283–291
19. Sysoev VV, Goschnick J, Schneider T, Strelcov E, Kolmakov A (2007) A gradient microarray electronic nose based on percolating SnO_2 nanowire sensing elements. Nano Lett 7:3182–3188
20. Barth S, Hernandez-Ramirez F, Holmes JD, Romano-Rodriguez A (2010) Synthesis and applications of one-dimensional semiconductors. Prog Mater Sci 55(6):563–627
21. Chun JY, Lee JW (2010) Various synthetic methods for one-dimensional semiconductor nanowires/nanorods and their applications in photovoltaic devices. Eur J Inorg Chem 27:4251–4263
22. Wu XJ, Zhu F, Mu C, Liang YQ, Xu LF, Chen QW, Chen RZ, Xu DS (2010) Electrochemical synthesis and applications of oriented and hierarchically quasi-1D semiconducting nanostructures. Coord Chem Rev (9-10):1135–1150
23. Meulenkamp EA (1997) Mechanism of WO_3 electrodeposition from peroxy-tungstate solution. J Electrochem Soc 144(5):1664–1671
24. Yamanaka K, Oakamoto H, Kidou H, Kudo T (1986) Peroxotungstic acid coated films for electrochromic display devices. Jpn J Appl Phys 25(9):1420–1426
25. Certain commercial equipment, instruments, or materials are identified in this document. Such identification does not imply recommendation or endorsement by the National Institute of Standards and Technology, nor does it imply that the products identified are necessarily the best available for the purpose
26. Colton RJ, Rabalais JW (1976) Electronic structure to tungsten and some of its borides, carbides, nitrides, and oxides by x-ray electron spectroscopy. Inorg Chem 15(1):236–238
27. Tian ML, Wang JG, Snyder J, Kurtz J, Liu Y, Schiffer P, Mallouk TE, Chan MHW (2003) Synthesis and characterization of superconducting single-crystal Sn nanowires. Appl Phys Lett 83(8):1620–1622
28. Kolmakov A, Zhang YX, Moskovits M (2003) Topotactic thermal oxidation of Sn nanowires: intermediate suboxides and core-shell metastable structures. Nano Lett 3(8):1125–1129
29. Lee AF, Lambert RM (1998) Oxidation of Sn overlayers and the structure and stability of Sn oxide films on Pd(111). Phys Rev B 58(7):4156
30. Wang D, Miller AC, Notis MR (1996) XPS study of the oxidation behavior of the Cu3Sn intermetallic compound at low temperatures. Surf Interface Anal 24(2):127–132
31. Cox DF, Fryberger TB, Semancik S (1988) Oxygen vacancies and defect electronic states on the SnO_2(110)-1 × 1 surface. Phys Rev B 38(3):2072
32. Suehle JS, Cavicchi RE, Gaitan M, Semancik S (1993) Tin oxide gas sensor fabricated using CMOS micro-hotplates and in situ processing. IEEE Electron Device Lett 14(3):118–120
33. Semancik S, Cavicchi RE, Gaitan M, Suehle JS (1994) Temperature-controlled micromachined arrays for chemical sensor fabrication and operation. US Patent 5 345 213, 6 Sept 1994

34. Semancik S, Cavicchi RE (1998) Kinetically controlled chemical sensing using micromachined structures. Acc Chem Res 31:279–287
35. Lucci M, Regoliosi R, Reale A, Di Carlo A, Orlanducci S, Tamburri E, Terranova ML, Lugli P, Di Natale C, D'Amico A, Paolesse R (2005) Gas sensing using single wall carbon nanotubes ordered with dielectrophoresis. Sens Actuators B 111:181–186
36. Shi L, Yu CH, Zhou JH (2005) Thermal characterization and sensor applications of one-dimensional nanostructures employing microelectromechanical systems. J Phys Chem B 109(47):22102–22111
37. Sysoev VV, Schneider T, Goschnick J, Kiselev I, Habicht W, Hahn H, Strelcov E, Kolmakov A (2009) Percolating SnO$_2$ nanowire network as a stable gas sensor: direct comparison of long-term performance versus SnO$_2$ nanoparticle films. Sens Actuators B 139(2):699–703
38. Berven CA, Dobrokhotov V, McIlroy DN, Chava S, Abdelrahaman R, Heieren A, Dick J, Barredo W (2008) Gas sensing with mats of gold-nanoparticle decorated GaN nanowires. IEEE Sens J 8(5–6):930–935
39. Deb B, Desai S, Sumanasekera GU, Sunkara MK (2007) Gas sensing behaviour of mat-like networked tungsten oxide nanowire thin films. Nanotechnol 18(28):285501
40. Kim ID, Jeon EK, Choi SH, Choi DK, Tuller HL (2010) Electrospun SnO$_2$ nanofiber mats with thermo-compression step for gas sensing applications. J Electroceram 25(2–4):159–167
41. Kunt TA, McAvoy TJ, Cavicchi RE, Semancik S (1998) Optimization of temperature programmed sensing for gas identification using micro-hotplate sensors. Sens Actuators B 53(1-2):24–43
42. Barsan N, Tomescu A (1995) The temperature-dependence of the response of SnO$_2$-based gas-sensing layers to O-2, Ch4, and Co. Sens Actuators B 26(1-3):45–48
43. Cavicchi RE, Suehle JS, Kreider KG, Gaitan M, Chaparala P (1996) Optimized temperature-pulse sequences for the enhancement of chemically specific response patterns from micro-hotplate gas sensors. Sens Actuators B 33(1–3):142–146
44. Gaggiotti G, Galdikas A, Kaciulis S, Mattogno G, Setkus A (1995) Temperature dependencies of sensitivity and surface chemical-composition of Snox gas sensors. Sens Actuators B 25(1-3):516–519
45. Gutierrez-Osuna R, Gutierrez-Galvez A, Powar N (2003) Transient response analysis for temperature-modulated chemoresistors. Sens Actuators B 93(1–3):57–66
46. Duda R, Hart PE, Stork DG (2000) Pattern classification. Wiley-Interscience, New York
47. Raman B, Hertz JL, Benkstein KD, Semancik S (2008) Bioinspired methodology for artificial olfaction. Anal Chem 80(22):8364–8371
48. Rogers PH, Semancik S (2011) Feedback-enabled discrimination enhancement for temperature-programmed chemiresistive microsensors. Sens Actuators B 158(1):111–116
49. Prades JD, Jimenez-Diaz R, Hernandez-Ramirez F, Cirera A, Romano-Rodriguez A, Morante JR (2010) Harnessing self-heating in nanowires for energy efficient, fully autonomous and ultra-fast gas sensors. Sens Actuators B 144(1):1–5

Chapter 15
Multisensor Micro-Arrays Based on Metal Oxide Nanowires for Electronic Nose Applications

Victor V. Sysoev, Evgheni Strelcov
and Andrei Kolmakov

Abstract During the last decade, quasi-1D metal oxide nanostructures were proven to be a promising material platform to design new gas sensing elements. This chapter surveys the recent developments of the analytical devices based on multisensor arrays made of metal oxide nanowires. We briefly discuss the advantages and challenges of electronic noses and the major milestones of their development. We show that evolution of the nanowire based electronic noses follows the same trends: from fabrication of the devices based on discrete nanowires to creation of integrated systems made of nanowire mats and finally realization of a monolithic sensor array made from a single nanowire. The parameters and performance of such analytical systems is reviewed and fabrication protocols are discussed.

15.1 Introduction

The development of applied sciences and technologies is driven by the need for a variety of devices, which can be substituted for or used to enhance mammalian capabilities. Currently, excellently performing electronic analogs for nearly all basic human sense organs except for olfaction have become available. One of the reasons for this shortage was that until recently deep understanding of mammalian olfaction principles was lacking (see the 2004 Nobel Prize work by Axel and Buck in ref [1]. and references therein). From a technological perspective, most of the effort in gas analysis has been dedicated to the design of analytical instruments

A. Kolmakov (✉)
Department of Physics, Southern Illinois University,
1245 Lincoln Drive, Carbondale, IL 62901, USA
e-mail: akolmakov@physics.siu.edu

M. A. Carpenter et al. (eds.), *Metal Oxide Nanomaterials for Chemical Sensors*, 465
Integrated Analytical Systems, DOI: 10.1007/978-1-4614-5395-6_15,
© Springer Science+Business Media New York 2013

Fig. 15.1 Two of the first instrumental versions of e-nose systems based on sensor arrays designed: by K. Persaud (*left*); and Argonne National Laboratory team (*right picture*); from *left to right* are: Dr. S. Zaromb, Mel Findlay, Dr. W. R. Penrose and Dr. J. R. Stetter ($\sim$ 1982); Adapted from the "Electrochemical Science and Technology Information Resource" (ESTIR) http://electrochem.cwru.edu/estir http://electrochem.cwru.edu/encycl/art-n01-nose.htm

capable to sensitively detect and evaluate the molecular content of complex odors. In spite of the sensitivity and precision, these instruments are rather costly, too narrow in their functionality, and in many cases, are out-performed by the mammalian nose, under complex gaseous environments (stimuli). Driven by environmental and automotive needs in the mid-twentieth century, a variety of relatively simple and inexpensive gas sensors have been developed which detect the presence of one of the target gases and convert this information into electrical or optical signals. Compared to common analytical devices such as mass spectrometers or chromatographs, a major drawback of these devices was their poor selectivity. Despite the multiple attempts to enhance the selectivity of such sensors via fine tuning the sensor chemical composition and/or its functionalization there is not so far, a universal solution to discriminate analytes in complex gas mixtures relying solely on the development of gas-specific materials.

Alternatively, by mimicking the operating principles of the mammalian olfactory system, one may not need very selective sensing elements [2, 3]. As we now know, the mammalian (as well as insectival) olfactory system is based on an ensemble of a few hundred receptor types, each of which is not particularly selective but responds a bit differently to the specific analyte [4–6]. The ensemble of these receptors generates an odor specific set of signals while the brain extracts information about the specific gas (or their mixture) by processing the distribution of receptor signals over the receptor array [7, 8]. Following this analogy [9], the first-generation "electronic nose" devices appeared in the mid-1980s, which consisted of an array of not very selective separate gas sensors whose signals were processed using pattern recognition algorithms. To enhance the diversity in the sensor signals, the individual sensing elements were made of different materials and/or based on dissimilar operation principles [10–13] (Fig. 15.1).

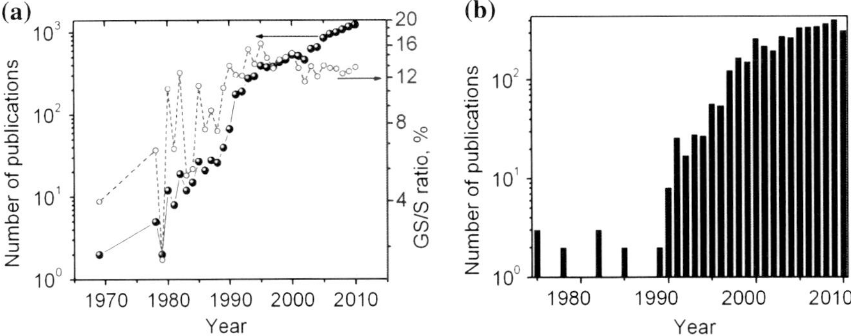

Fig. 15.2 Statistical data of the ISI web of knowledge database (http://www.isiknowledge.com) on the number of publications indexing with, **a** "gas sensor" keyword; the right y axis indicate the ratio of number of publications with "gas sensor" keyword to that with "sensor" as a keyword; **b** « electronic nose and gas sensor array » keywords

It is worth noting that the appearance of the "electronic nose" concept stimulated the development of different kinds of gas sensors. According to data from the ISI Web of Knowledge, the fraction of academic publications having "gas sensor" as a keyword was approx. 5 % of the total number of publications with "sensor" as a keyword (Fig. 15.2a). Since 1992 this fraction has increased twofold and currently fluctuates around $10 \div 14$ %.

The number of "e-nose" publications is increasing and now-a-days about 300 papers are published annually on this issue (Fig 15.2b). However, despite a successful demonstration in gas recognition, first-generation electronic nose devices have been found to have apparent limitations: (1) the cost of the serial device was quite high and comparable to that of other conventional analytical instruments; (2) signals of different type from different sensors required complicated electrical or optical interfaces; (3) arrangement of individual sensor units and signal processing circuit boards led to large dimensions and masses of the devices; (4) sensors of different type have individual drift characteristics which requires frequent and time consuming re-calibration of the instrument. All these reasons hampered the development of the e-nose market, which called for a search of alternative methods of multisensor array technology.

A technological breakthrough took place with the advances of CMOS technology and with the development of miniature field effect transistors (i.e. Chem-FETs) as gas sensing elements, in particular in the beginning of the 1990s. The integration of the ChemFET sensor array on a single wafer manifested the appearance of second generation of e-noses [14]. Since then a significant body of published papers have been dedicated to the fabrication and testing of different kinds of sensor arrays mounted on a single chip (see, for instance, Ref. [15–17]). One of the major advantages of the single-chip sensor arrays is that their production cost is comparable to that of single sensors.

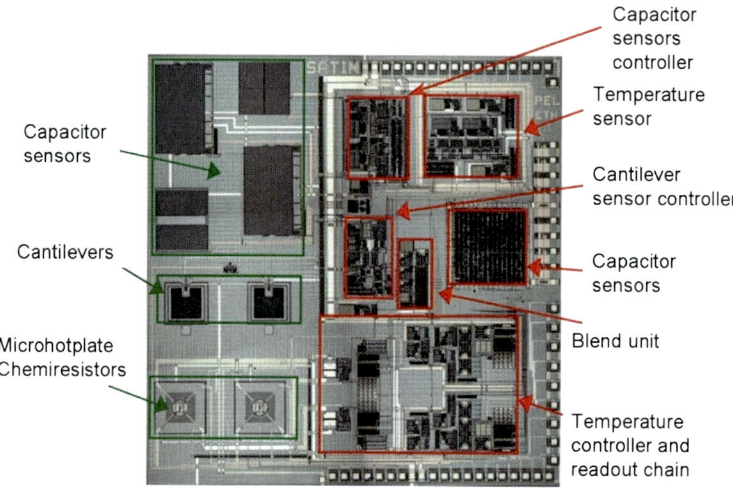

Fig. 15.3 A micrograph of a hybrid multisensor chip (7 × 7 mm^2) consisting of several sensors of different types. Reprinted with permission from [22] © Elsevier

The common requirements for the assembly of multisensor arrays are the following [18]: (1) the transduction of the sensor's signal must be carried out with the minimal expenditure of time and energy; (2) sensor's signal must be reversible and reproducible; (3) sensor's signal should have low drift; (4) sensors have to be sensitive to a wide range of analytes and their concentrations; (5) the manufacturing protocols of the sensing elements have to be compatible with those of the other electrical components; (6) the geometry sizes of the sensors must be small enough so as to allow their assemblage (often of high densities) in quite small chips; (7) sensors must tolerate many toxic environments; (8) it is desirable to have linear sensor signal, or one that could be simply linearized using known mathematical methods. The number of sensors in the array is chosen considering the specific application, possibility and means of the signal processing as well as the possible cost of such an array [18].

Currently, the development of multisensor arrays proceeds along two main directions. The first one is the fabrication of so-called high-order devices, which are based on the assembly of dissimilar sensors on a single chip using microelectronics technology [19]. These devices are probing the different (orthogonal) but complementing features of the analyte. A typical example of such hybrid multisensor micro-array is represented in a work of Balte's group [16, 20–22] (Fig. 15.3). The developed micro-array consists of microsensors of capacitive, mass-sensitive and calorimetric type fabricated on a single chip. This approach undoubtedly offers a high discrimination power and has reduced the mass and array dimensions, and, to some extent, the cost of the final instrument, however, it has not yet eliminated the physico-chemical complexity of sensors and their related limitations as described earlier.

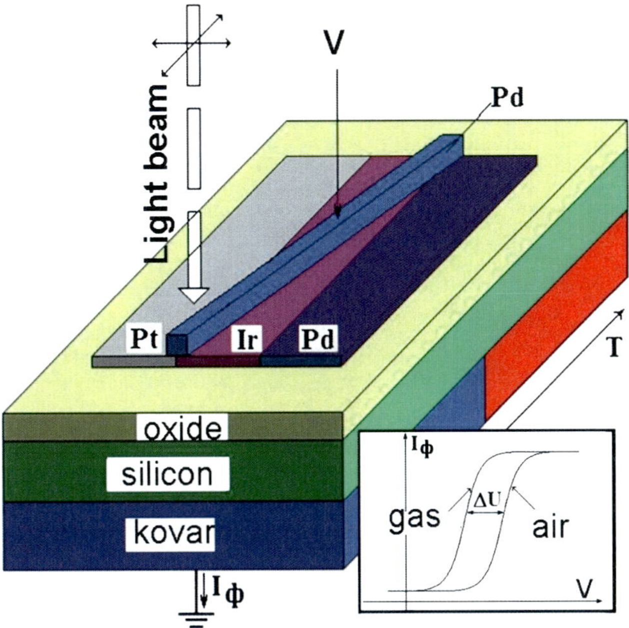

Fig. 15.4 Generation scheme of a "chemical image" via reading a gate potential of a gas-sensitive transistor. The substrate size is 4×6 mm^2. Inset: dependence of photocurrent on applied bias and the I–V curve shift in the presence of gas (adapted from [14])

The second direction constitutes the development of so-called first-order multisensor arrays made of similar sensors assembled on a single platform. In this case, sensors produce the same kind of signal (i.e. resistance or color change), whereas the gaining of the discrimination power is achieved via variation of properties and/or operating conditions along the sensor array. Being much simpler compared to higher-order devices, these inexpensive sensors have a better cost-effectiveness ratio for specific applications where the amount of possible analytes is known in advance. As a result, many laboratories are currently developing first-order multisensor arrays, where the sensing elements rely on the same receptor-transduction principle [23, 24]. A few representative examples of such systems are briefly described below.

An interesting multisensor array was proposed by Lundstrom and colleagues [14]. In this study a large working area ChemFET with variable composition and catalytic gate thickness, combined with a temperature gradient along its length was used as a 2D sensing element (Fig. 15.4) [25]. The scanning light beam was employed to generate a photo capacitive current whose local value depends on the local gate potential and gas environment. By stabilizing the photo capacitive current through adjustment of the gate potential a 2D topographic map of the surface potential distribution can be generated. The chemical image of the particular analyte can be obtained through the comparison of the surface potential distribution in a reference atmosphere (e.g. air) and in the presence of the target gas.

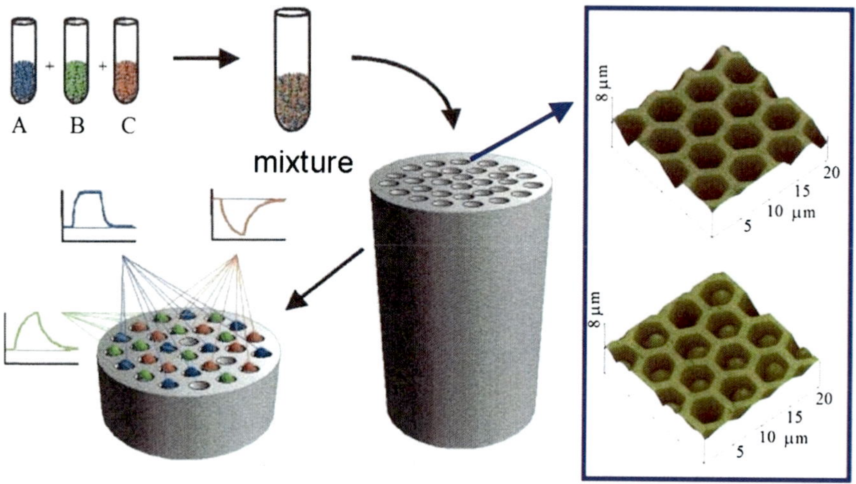

Fig. 15.5 Fabrication scheme of the matrix of the gas-sensitive fluorescent polymer micro-spheres at the end of an optical waveguide. Insets: micrographs of the "honeycomb" structure of the butt-end surface of the waveguide before and after application of polymeric microspheres Reprinted with permission from [26, 28] © American Chemical Society

Another optical multisensor arrays based on optical fibers were proposed by the Walt group [26–28]. To collect the multisensor signal, the authors employed bundles of optical fibers with terminal parts covered with fluorescent polymeric dyes (Fig. 15.5). Upon interaction with many gases these polymers change their structure that leads to shifts in the wavelength of emitted light. Prior to deposition of polymeric microspheres, the terminal part of the optical fiber is chemically etched until micro-cellular structure with pore depth of several microns appears [29, 30]. This multisensor array resembles mammalian olfactory system as it comprises of a multitude of identical sensors, yet of several different types. Modulation of intensity and wavelength of the emitted light in the presence of analytes acts as sensor signal. Currently, this product is being commercialized [31].

An alternative approach of optical reading from the gas responsive multisensor array was proposed by Suslick group employing metal-porphyrin dyes as gas sensing elements [32–34]. These substances change their colors upon exposure to reactive gases. Chelating different metal ions with porphyrin it is possible to change response (color or peak of optical spectrum intensity) to the analyte. Dyes with coordinated ions such as Sn^{4+}, Co^{3+}, Cr^{3+}, Mn^{3+}, Fe^{3+}, Co^{2+}, Cu^{2+}, Ru^{2+}, Zn^{2+}, Ag^+ are deposited from their solution onto substrate in a form of pixel matrix with subsequent vacuum drying. As a result, a colorimetric multisensor array was designed where every pixel color could alter under the influence of gas ambient. Subsequent processing of the matrix of the colorimetric signals is carried out using a scanner read-out system: the color gamut is recorded both before and after exposure to gas. Thus, it is possible to construct a map of color differences, which is later digitized using optical signal processing techniques. Notably, humidity

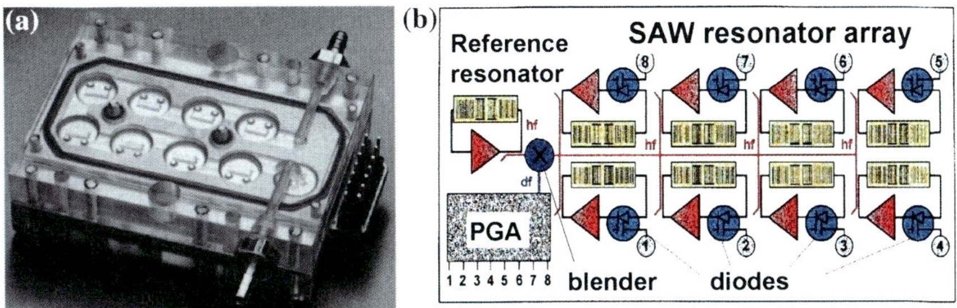

Fig. 15.6 Multisensor array based on SAW resonators: **a** photograph; **b** general scheme of the device; size is $60 \times 46 \times 17$ mm^3. Reprinted with permission from [35] © Elsevier

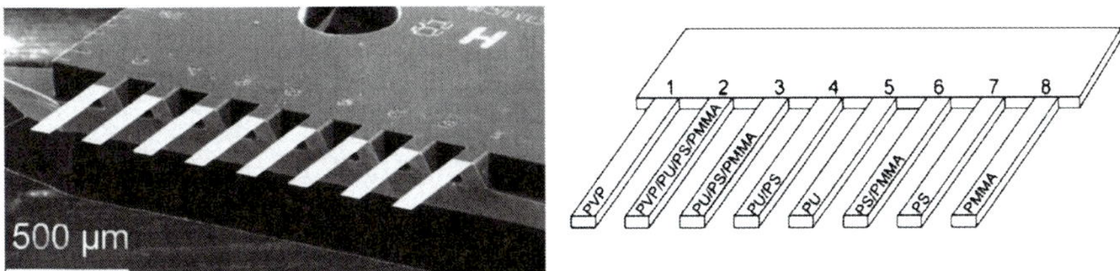

Fig. 15.7 *Left* SEM image of the micro-cantilever array, each sensor is of $500 \times 100 \times 1$ µm size; *right* cantilever tips coated with different materials Reprinted with permission from [37] © Elsevier

level does not affect the performance of the colorimetric matrix. The significant difference between the gas sensing properties of the dyes is stipulated by chemical interactions and allows one discriminating between chemically related gases at low (ppm) concentrations level.

Among the most successful multisensor arrays based on piezoelectric SAW-resonators the prototypes developed by Wohltjen group [35] and by Rapp et al. [36, 37] could be noted. An example of such a device is shown in the Fig. 15.6.

The individual SAW-resonators of the multisensor array are coated with polymers of selected composition and thickness. Exposure of such an array to test gases leads to the sorption of the analyte at the active polymer coating, which modulates the propagation parameters of the acoustic wave, frequency and amplitude. The shift of the resonant frequency can be sensitively measured. The sensitivity at the level of tens of ppm has been demonstrated, which is acceptable for many applications.

Significantly more sensitive multisensor array based on micro-cantilevers has been demonstrated by IBM lab [38, 39] (Fig. 15.7). Cantilever mechanical oscillations (or deflections from initial position) can be read out from the individual cantilever with independent lasers. The cantilever's surfaces are coated by a thin layer of gold topped with a homogeneous layer of sensing polymers of different type. Notable advantages of the cantilever-based multisensor arrays include:

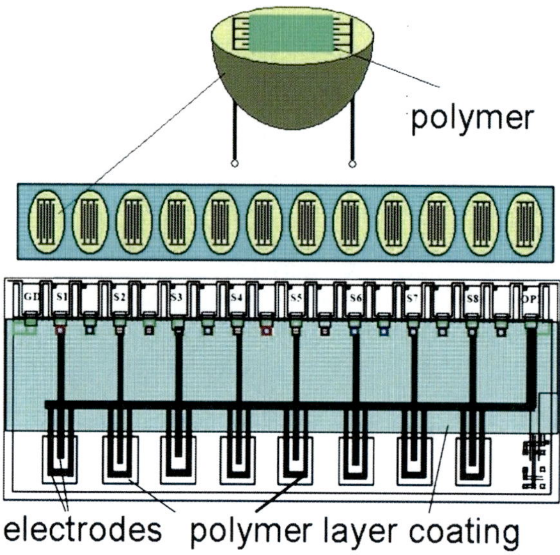

Fig. 15.8 The multisensor array of carbon-polymer chemi-resistors: the original design of the individual sensor (*top*, adapted from [41]); sensor array used in JPL E-nose device (*bottom*, adapted from [40])

(1) ability to work not only in gaseous but also in a liquid environment[1]; (2) ability to operate without external heater, at least, in case of sensing polymer coatings usage. The second feature allows one to analyze gas interaction with chemi-sensitive proteins [39, 40].

In spite of significant progress in optical and electromechanical e-nose systems, the multisensor arrays based on chemi-resistors are still popular due to their simplicity and cost effectiveness. In 1995, Lewis group has proposed carbon black filled polymer matrices as gas sensitive media for chemi-resistive arrays [41, 42] (Fig. 15.8). Precursor solutions were deposited onto interdigitated electrodes of the individual sensor of the array and got polymerized. All the sensors were mounted at a single bus, and their resistance could be read sequentially as a function of the gas exposure. The later modifications of this approach included the deposition of the compound layer onto the micro-fabricated electrode arrays [43, 44]. The simplicity and reliability of such multisensor array designed of carbon-polymer gas swellable composite combined with its good sensitivity [45] encouraged NASA to use them for air quality monitoring on board of the spacecraft.

Another single-chip multisensor array employing chemi-resistive metal oxide layers deposited over micro-hotplate platforms has been proposed by Semancik and colleagues at NIST [46–48]. This array is assembled of micro-sensor elements of ca. $100 \times 100~\mu m^2$ in size (Fig. 15.9a) each of which has individual heating element.

The architecture of the individual micro-hotplate-based sensing element is shown in the Fig. 15.9b. The heater, which also serves as a thermistor, is

---

[1] in other words, to generate "electronic tongue" or biosensor matrices signal.

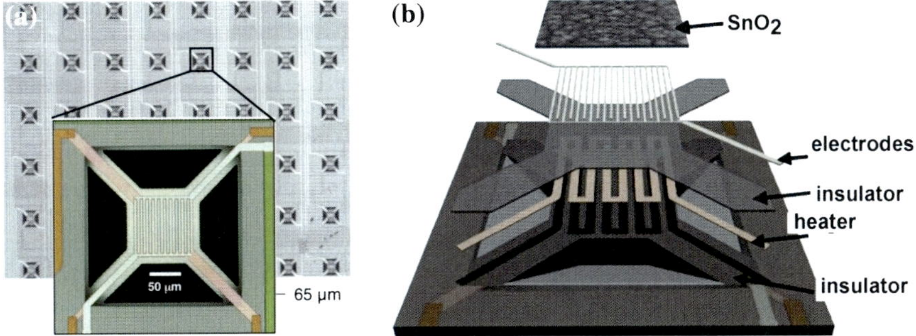

Fig. 15.9 **a** Multisensor array assembled of 48 chemiresistive metal oxide thin film elements deposited over microhotplates; **b** the structure of a microhotplate sensing element. Reprinted with permission from [46–48] © Elsevier

insulated with a silica coating on both sides. The resistance of the metal oxide sensing layer is measured with the help of four electrodes. The range of the operating temperatures which can be achieved within 2–5 ms is 20–1,000 °C. Notably, the power consumption of the device was only about 30 mW under conventional operating conditions. The important advantage of this chip is that the temperature of each sensor element in the array is individually controlled. Moreover, the individual sensing elements of the array may contain catalytic additives. All these advantages allow one, on one hand, achieving higher gas-recognition capability of the e-nose instrument, and, on the other hand, decreasing the cost of the chip fabrication down to that of hybrid chips.

One of the simplest and yet nicely performing multisensor arrays (based on chemi-resistive oxide) was proposed by Goschnick group [49, 50]. The device is known as KAMINA (Karlsruhe Micro NAse). The KAMINA chip is fabricated by deposition of monolithic gas-sensing film over (or under) a set of strip-like coplanar electrodes. Each segment of the film between pair of electrodes represents an individual sensing element of the array. A variation in sensing properties of the segments is achieved by application of temperature gradient across the array or by employing a gas-permeable coating of different thickness a top of the film. Thus thin oxide film that could have operated as one sensor works as a whole multi-sensor array. The sensor segment resistances are measured by a standard multiplexer circuitry. The reading out the resistances of the whole array might be performed at 1 Hz rate that gives an option to sample the odor evolution in a real time. The advantages of the chip are its simplicity and low manufacture cost.

An alternative to the acquisition of vector signals (of the same nature) from several sensors might be the monitoring of a set of different parameters[2] of the same sensor material [51–53]. Such structures are sometimes called as "smart" ones [54]. The segmentation of this single sensor is not physical or spatial, as in

---

[2] for instance, resistance, capacitance and potential.

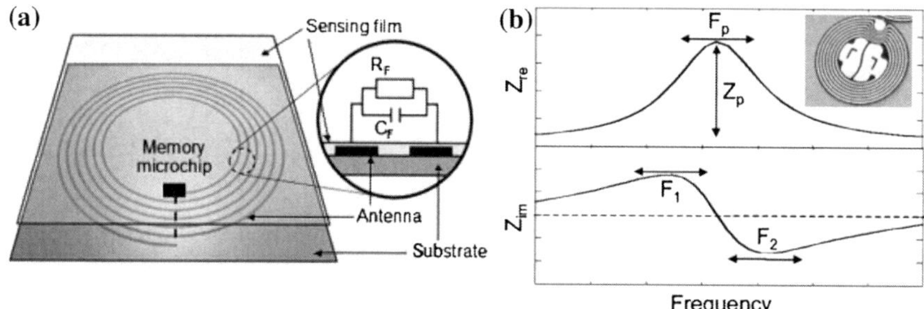

Fig. 15.10 **a** The scheme of the RFID tag sensor made of gas-sensing film characterized by resistance RF and capacitance CF; **b** the frequency spectra of real and imaginary parts of the film impedance Z indicating four parameters utilized to selectively identify the gases being in contact with the film. Inset: photo of the RFID sensor. Reprinted with permission from [51] © John Wiley & sons

the case of KAMINA device, but rather parametrical or informational. It is the information acquired from the sensor that is split up into orthogonal segments, not the sensor itself. An example of this kind of system which, in addition can be read remotely, was developed by Potyrailo group [55]. The device is a 13.56 MHz RFID tag coated by a gas-sensitive film, for example, polyaniline (Fig. 15.10). Upon exposure to analyte the film changes its resistance and a capacitance what results in modulation of the tag impedance. Readout of the impedance is carried out remotely and the sensor itself requires no power unit being a completely passive device operable at room temperature. The gas recognition is performed via the multi-variance analysis of four impedance parameters: the maximum frequency and magnitude of the impedance real part, and resonant and anti-resonant frequencies of the imaginary parts. These four parameters being orthogonal serve as four sensor signals which generate a selective vector signal to gases.

Nearly all of the described designs of integral multisensor chips have been commercialized under such brands as: Vapor Lab (Micro-sensors Systems, USA) based on SAW sensors, Cyranose (Cyrano Sciences, USA) and JPL Enose (NASA, USA) based on conductive polymer chemi-resistive sensors, Sam Detect (Daimler Chrysler Aerospace, USA-Germany) based on SAW sensors, i-PEN (WMA Airsense Analysentechnik, Germany) and KAMINA/ArtiNose (Karlsruhe Institute of Technology/SYSCA, Germany) based on semiconducting oxide chemi-resistors [13, 44, 56–59]. More details could be found in recent comprehensive reviews in the field (see, for instance Ref. [59]).

In parallel to the progress in the device architecture, new material platforms employable in e-noses have also been recently developed. The major goal of the materials science applied towards the development of active sensor elements for such multisensor arrays is the search for sensitive materials with reproducible and stable gas sensing properties, which could be controllably modified (for example via doping) to gain the required diversity in selectivity. Since the interaction between analyte and sensor material takes place at the surface of the sensing element, the use of nanostructures and nanostructured materials, which inherently possess a high surface-to-volume ratio was inevitable. The rational selection and/or tuning of chemical composition of sensing materials are often done via a combinatorial approach, common in analytical chemistry, biotechnology and pharmaceutical industry [60]. The multitude of chemically related materials with sequentially tuned parameters responsible for sensing are called "combinatorial libraries" [61]. These sets of materials with known sensitivities and selectivity toward target gases are often used as sensing elements within sensing arrays. The progress also took place in the fabrication technology of the sensing elements. In conventional microelectronics the active elements of gas sensors are produced by "top-down" methods such as sputtering, deposition, lithography, etching etc. When the characteristic dimensions of the sensing elements are on the order of microns, these technologies provide sufficient control over functional properties of the fabricated devices. However, thin film technology experiences challenges when reduction of the sensing elements to submicron range is required because the size of the sensing element in the latter case becomes comparable to the dimensions of individual grains in the sensing layer. Therefore, the "bottom-up" approaches developed by nanotechnologies find a broad interest in the development of sensor arrays.

There have been many reports during the last decade on bottom up fabrication of new morphologies such as micro-nanospheres [62, 63], nanowalls [64, 65] and nanowires [66–78] whose dimensions (at least, along one axis) are in the nanometer scale range. It is clear that these structures have a large potential to develop a new generation of gas (nano) sensors and multisensor arrays [72–74, 76]. Among the aforementioned nanostructures quasi-one-dimensional semiconducting metal oxide nanowires (nanorods, nanobelts, nanobeams etc.) which combine small transverse dimensions (in nanometer or submicron range) and macroscopically large lengths (up to few millimeters) are particularly attractive as chemi-resistors. The width of the adsorbate modulated depletion region (DR) in these nanostructures is comparable or exceeds the nanowire radii which results in an effective transduction of the gas species interaction with the oxide surface (adsorption and/or catalytic oxidation) into a measurable conductance change. The relatively large longitudinal dimensions of the nanowires allow one to confidently manipulate it and apply conventional micro fabrication techniques for contacting and functionalization of the sensing element. As a result, they are considered as promising structures for development of novel multisensor arrays with advanced characteristics for applications in electronic noses [24, 72].

15.2 Discrete Multisensor Micro-Arrays Made of Separate Nanowires

Until recently, few reports have been published on nanowire-based electronic noses. The first of these [24] appeared in 2006 and described a rudimentary e-nose made of three separate nanowires made of different metal oxides, analogous to the

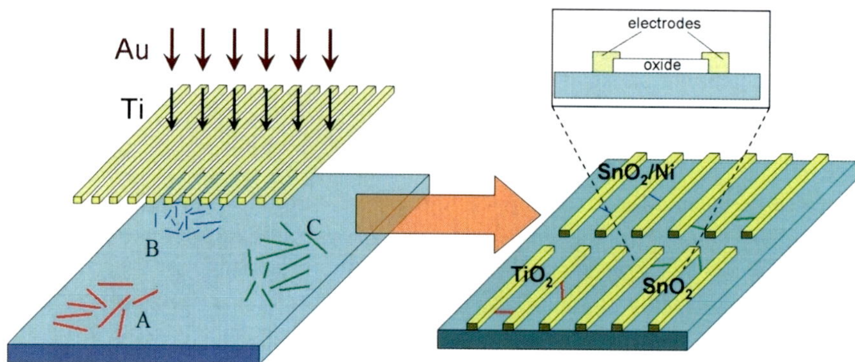

Fig. 15.11 Device fabrication: deposition of contact electrodes onto substrate with pre-deposited nanowires of different oxides (areas *A*, *B*, *C*). Inter-electrode distance is 12 μm

first-generation electronic noses based on separate gas sensors. In this design, SnO_2 and TiO_2 nanowires were deposited on a 5×5 mm^2 Si/SiO$_2$ substrate in several different places (areas A, B and C Fig. 15.11). Subsequently a set of Ti/Au electrodes (40/300 nm thick) was deposited through a shadow mask on top of the nanowires (Fig. 15.11). After the electrode deposition, the surface of selected nanowires was functionalized, in situ, with Ni nanoparticles using metal thermal evaporation through a 100 μm orifice placed above the nanowire of interest [79, 80]. Nanowire functionalization was controlled via monitoring its conductance (Fig. 15.12a) to ensure the absence of percolating conductivity between the catalytic nanoparticles. The nanowire conductance significantly decreased in the presence of these surface nickel clusters (Fig. 15.34b). The latter was explained in terms of the formation of p-NiO/n-SnO$_2$ heterojunctions on the Ni/oxide interface that leads to a localized depletion of nanowire free carriers[3].

Figure 15.13 shows SEM images of three metal oxide nanowires: SnO_2, SnO_2/Ni and TiO_2 contacted by metal pads. The nanostructures had different morphologies which along with the compositional difference, contributed to the gas-sensing selectivity of the fabricated device.

To get a deeper insight into the operation principles of the nanowire based e-nose shown in Fig. 15.13, the electrical characterization of the nanostructures was carried out at an operating temperature of 350 °C in model environments: (1) in vacuum; (2) under exposure to pure oxygen at 1.3×10^{-2} Pa; 3) under addition of controlled amounts of H_2 and CO in the oxygen background. Exposure of nanowires to oxygen led to a resistance increase which was in accordance with their n-type conductivity nature. The subsequent addition of hydrogen to the oxygen background resulted in a reproducible increase in conductance in all of the nanowires (Fig. 15.14). A similar response was also observed upon nanowire

---

[3] The influence of surface Ni doping is discussed later in the text.

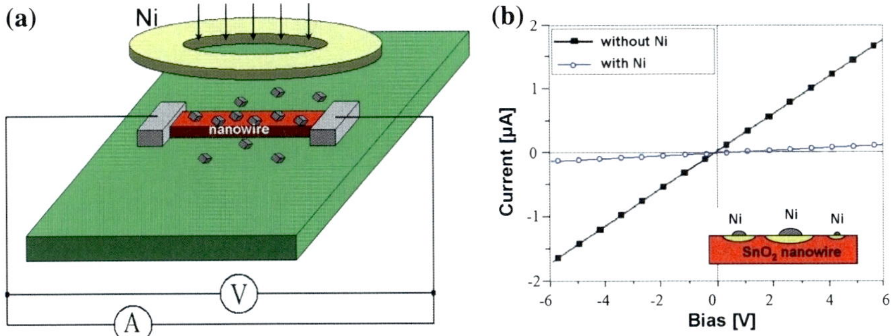

Fig. 15.12 SnO$_2$ nanowire functionalization: **a** Ni deposition through an aperture (adapted from [73]); **b** SnO$_2$ nanowire I–V curves prior and after Ni deposition, T = 250 °C. Inset in **b**: scheme of nanowire depletion by free carriers because of Ni cluster presence

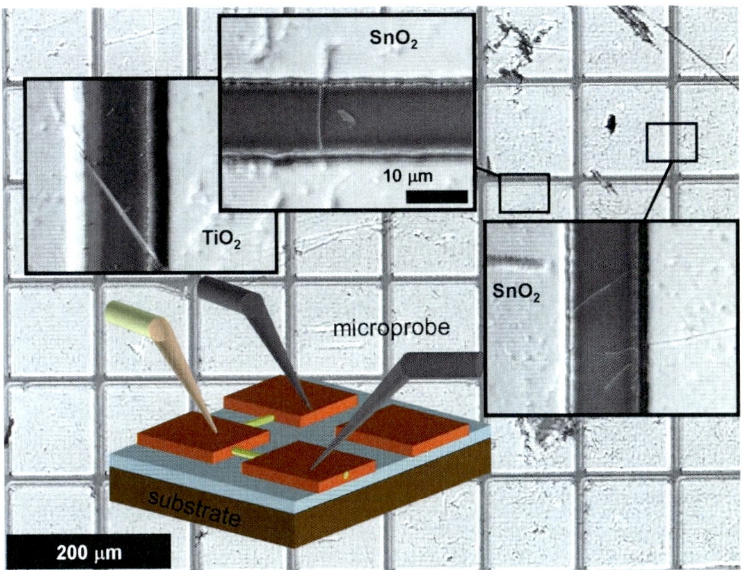

Fig. 15.13 SEM image of the studied substrate carrying few nanowires; inset shows a scheme of electrical measurements of nanowires with metal pads via press-on microprobes

exposure to CO. Assuming coaxial "core-shell" geometry[4] of the conducting channel in the nanowire the changes in the conductance are due to an adsorbate induced "shrinking" or enlargement of the conducting channel cross-section. The observed differences in gas sensitivities of the nanowires were primarily due

---

[4] The conducting channel forms a "core" while the near-surface DR represents a "shell" in the nanostructure.

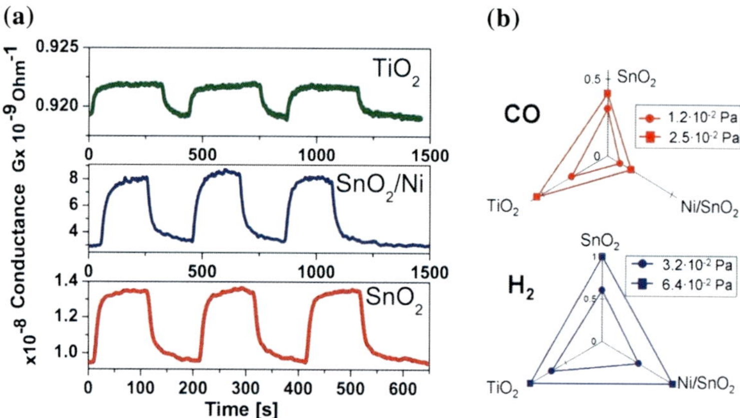

Fig. 15.14 **a** Response of chemiresistive oxide nanowires to three sequential pulses of hydrogen at partial pressure of 6.4×10^{-2} Pa. The reference conductivity of nanowires is measured at a constant oxygen pressure of 1.3×10^{-2} Pa; **b** Response of array of the three oxide nanowires to H_2 and CO. The magnitude of each of responses is normalized to the maximum value. Reprinted with permission from [24] © American Chemical Society

to the different efficiency of the increase/decrease of the depletion region (DR) width during the redox reaction at the nanowire surface. To visualize the gas selectivity, the normalized sensor signals obtained from the three nanowire sensor elements were plotted on a radar chart as shown in Fig. 15.14b.

As can be seen, two different combustible gases can be readily distinguished. Through a choice selection of array elements, it is likely that a finite increase in the number of nanostructures within the multisensor array would lead to enhanced selectivity characteristics.

Several other subsequent reports have elaborated and extended the aforementioned approach. In 2007 a multisensor array constructed of aligned Si nanowires was reported [81]. The nanowire ensembles were prepared by super lattice nanowire pattern transfer (SNAP) technique to form well-aligned arrays consisted of 400 18 nm-wide nanowires with a high aspect ratio (Fig. 15.15). Multiple Ti electrodes deposited over the nanowires sectioned the whole array into individual sensor devices. The last ones were also equipped with a gate electrode that allowed their use as a ChemFET device.

The fabricated sensors showed excellent sensitivity to NO_2 diluted in nitrogen. The reported detection limit to this gas was 20 ppb. Such a good performance was attributed to the fact that free carriers in these Si nanowires mostly reside in the near surface region which is strongly influenced by adsorbates. For e-nose applications, the nanowire surfaces were additionally functionalized with aldehyde-, amino- and alkane-silanes in order to achieve a diversity of sensors required to generate a more selective vector response. The resulting e-nose micro-array was shown to exhibit discrimination ability towards acetone and hexane vapors, of 1,000 ppm concentration.

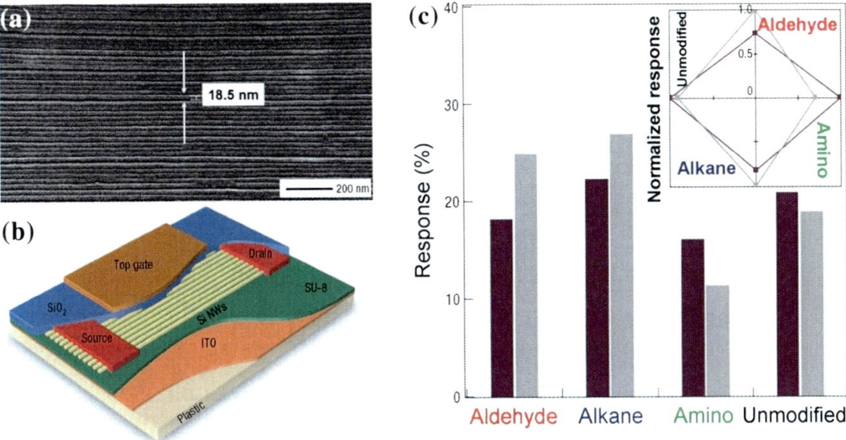

Fig. 15.15 a SEM image of Si nanowire array fabricated by SNAP process; **b** scheme of one FET segment and gas-sensing device fabricated from the Si nanowire array; **c** response magnitudes of four sensors to acetone (*purple*) and hexane (*grey*) vapors. The nanowires were functionalized with aldehyde- amino- and alkane-silanes. Reprinted with permission from [74] © Nature Publishing Group

The Zhou group [82, 83] has combined micro-hotplate technology with metal oxide nanowires and carbon nanotubes as sensing elements (Fig. 15.16). The advantages of this multisensor chip are small dimensions, 10×10 mm^2, and low power consumption, 60 mW[5], as well as the ensured possibility of independent and precise temperature control of each of the sensor elements.

Again, the chemical, morphological and thermal differences between the gas sensing elements (nanowires) resulted in the selectivity of chip response to gases. It was shown that such a device was able to discriminate between NO_2, hydrogen and ethanol vapors in air (Fig. 15.17).

Another approach reported recently by the Moskovits group [84] utilized an array of individual SnO_2 nanowire sensors fabricated on the same chip, with some of the nanowires decorated with Pd or Ag clusters. In addition to functionalization with catalytically active clusters, a temperature gradient was applied to increase the recognition power of the sensor array (Fig. 15.18a). To test its gas-sensing properties three reducing gases: carbon monoxide, hydrogen and ethylene were utilized in exposure experiments.

By processing the acquired gas-sensing data with linear discriminant analysis (LDA) [85–88] the authors demonstrated abilities of this array to discriminate the three gases of interest (Fig. 15.18b).

Thus, to-date several groups have successfully fabricated and tested e-noses using quasi-1D nanostructures. Following the notation used in the film-based e-noses, the nanowire-based prototypes correspond to first generation devices

---

[5] which are close to the parameters of the chip designed by Semancik group.

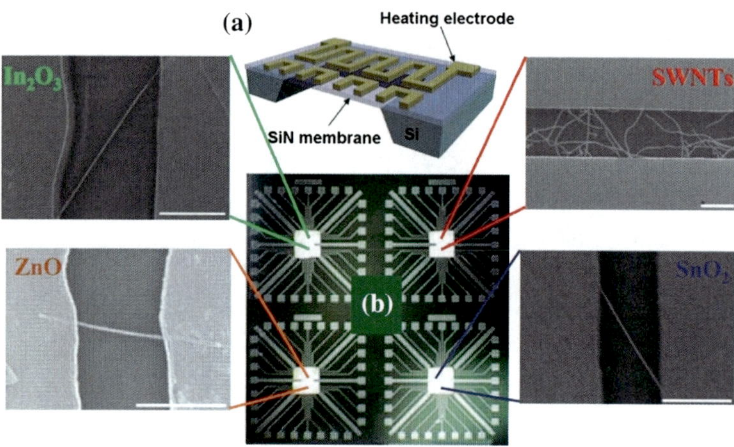

Fig. 15.16 **a** A cut-away diagram of a micromachined hotplate element **b** an array of four hotplates sensors constituting an e-nose chip. Zoomed-up areas show SEM images of indium, zinc and tin oxide nanowires, as well as single-walled carbon nanotube used as gas-sensing elements. Reprinted with permission from [75, 76] © American Institute of Physics and Institute of Physics Publishing

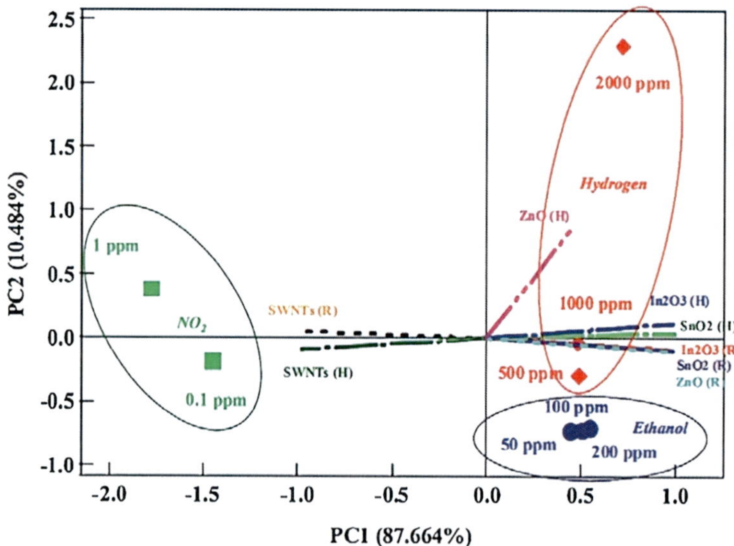

Fig. 15.17 Principle component analysis score and loading plots for gas-sensing data of all four sensors—metal oxide nanowires and carbon nanotubes to different concentrations of NO₂, hydrogen and ethanol at 25 and 200 °C. Reprinted with permission from [76] © Institute of Physics Publishing

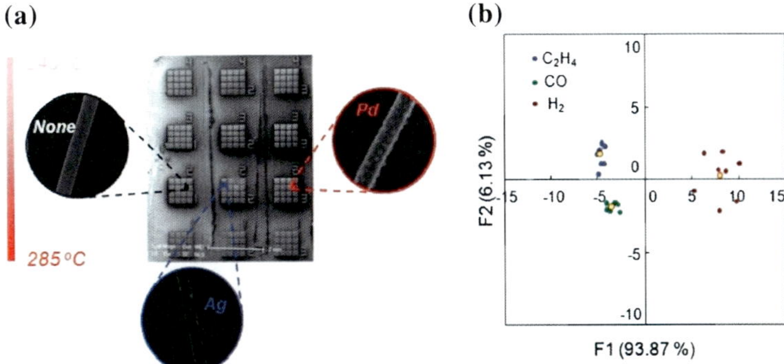

Fig. 15.18 The sensor array comprised of individual SnO$_2$ nanowires, pristine and decorated with Ag and Pd clusters, **a** the image of the array; the temperature gradient is applied perpendicular to the "functionalization axis", **b** results of the LDA processing the array signals yielding a confident discrimination between ethylene, carbon monoxide and hydrogen. Reprinted with permission from [77] © American Chemical Society

whose sensor array is formed out of separate sensing elements. Promising as they are, these designs still face the same challenges as traditional film-based electronic noses and in particular: complexity, costs in fabrication and functionalization.

15.3 Integrated Multisensor Micro-arrays Based on SnO$_2$ Nanowire Mats

To resolve the aforementioned challenges one can follow a trend similar to that taken for the fabrication of thin film based e-noses. Namely, instead of using separate sensing elements one can use a film of nanowires and introduce spatial or temporal variations of physico-chemical properties along the nanowire film to induce specificity in the sensor segments. Hereafter, the properties of such multisensor arrays fabricated on the basis of nanowire mats are discussed.

15.3.1 Fabrication and Gas-Sensing Properties of SnO$_2$ Nanowire Mats

Multisensor micro-arrays based on metal oxide nanowire monolayer (mats) were fabricated using SiO$_2$/Si/SiO$_2$ substrates equipped with co-planar multiple electrodes on the front side and with four meander-shaped heaters on the back side [89].

The fabrication of the nanowire multisensor chip was performed in two ways (Fig. 15.19): (1) the deposition of multiple Pt/Ti electrodes, thermo-resistors and

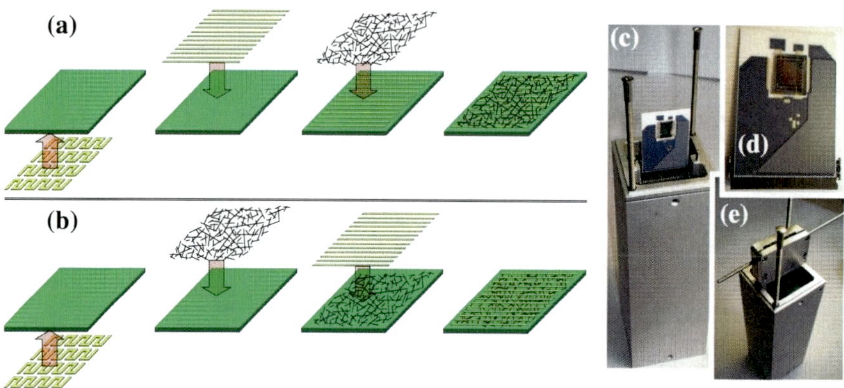

Fig. 15.19 Fabrication of the multisensor chip based on the oxide nanowire mats. **a** deposition of nanowires atop co-planar parallel electrodes, **b** deposition of the electrode arrays atop nanowire mats. The KAMINA device equipped with a 50-pin ceramic multisensor chip used in this study, **c** general view, **d** ceramic chip, **e** ceramic chip enclosed in the metallic chamber equipped with gas lines

heaters, of 1 µm thick/100 µm width, through a shadow mask by sputtering over Si/SiO_2 substrate followed by placement of the nanowire mats atop the micro-electrode array via mechanical pressing (Fig. 15.19a); (2) the placing of oxide nanowire mats by mechanical pressing against the bare $SiO_2/Si/SiO_2$ wafer followed by subsequent sputter deposition of the microelectrode array over the nanowire layer (Fig. 15.19b). The number of electrodes in the micro-array was chosen to be equal to 39 to subdivide the whole nanowire mat into 38 sensor segments. This chip design is compatible with KAMINA e-nose electronics to simplify reading and processing of the sensor signals (resistances).

Following the nanowire layer deposition, the multielectrode substrates were wire bonded to 50-pin ceramic carriers (Siegert, Germany) via ultrasonic bonding with Au wires. The photographs of the chip mounted within the KAMINA device are shown in Fig. 15.19c, d, e.

In the course of preliminary studies, there was found that enhanced gas-sensing properties could be obtained with nanowire mats of low density, where the individual nanowires form a percolating network between adjacent electrodes. The density of percolating chains, their length and number of joints (nodes) determined the resistance of the individual sensor segment between the electrodes. Segments with higher nanowire density possess, as a rule, lower resistance, whereas those of low density, close to percolation threshold, have a higher resistance (Fig. 15.20).

By tuning the density of the nanowire network either purposefully or stochastically, one can vary the resistance of the sensor segments along the array (Fig. 15.21).

In order to investigate the discrimination ability of the multisensor chips, their responses to analytes have been studied in two temperature regimes: (1) quasi-homogeneous temperature distribution (isothermal) at 300 ± 10 °C; (2) upon

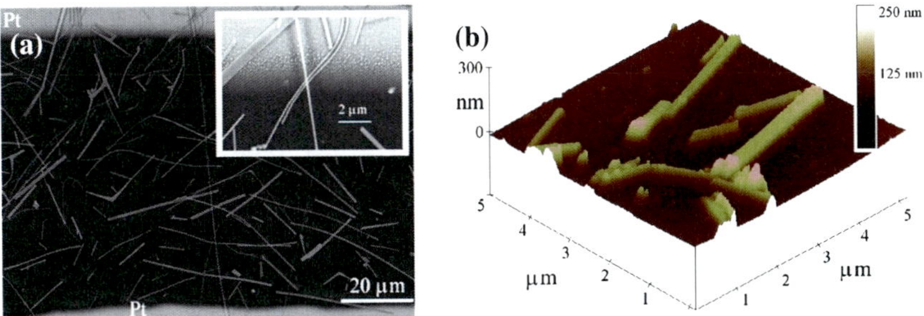

Fig. 15.20 Microphotographs of SnO$_2$ nanowires deposited onto SiO$_2$/Si/SiO$_2$ substrate of the multisensor chip between two Pt electrodes: **a** a strip of the surface of sensor segment 120 × 85 μm^2 (SEM image) Reprinted with permission from [82] © American Chemical Society; **b** a patch of the segment surface, 5 × 5 μm^2, showing crossing nanowires (AFM image)

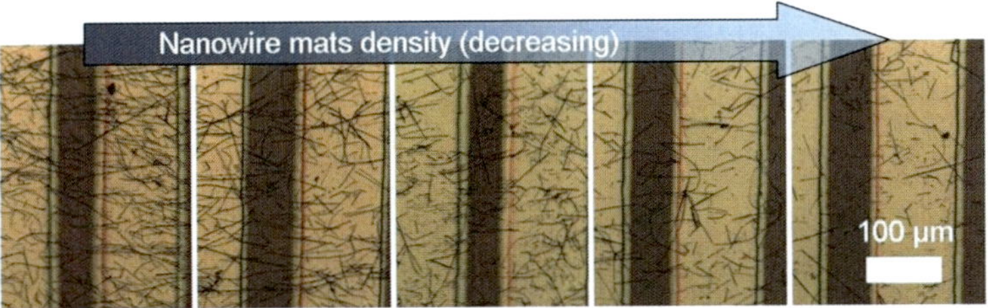

Fig. 15.21 Examples of sensor segments of different oxide nanowire mats density (*optical images, yellow strips* are gold electrodes)

application of a thermal gradient in the range 250 ÷ 320 °C along the substrate chip. The typical responses of the sensor array to trace amounts of CO in air are shown in the Fig. 15.22b. The obtained detection limits to CO were comparable to or even exceeded the data reported previously for pristine SnO$_2$ sensors [90, 91]. Defining the lower detectable limit of sensor signal as threefold excess of the noise magnitude, one can conclude that nanowire mats are capable to detect CO in air at the level of 150–200 ppb.

The response time of the nanowire mat sensor, determined by analysis of the initial slope of the resistance change, was about 30 s (Fig. 15.23) and was mainly determined by the time constant of the gas delivery system rather than the sensor itself. Such response characteristics meet the requirements of most applications.

The contribution of the morphology of nanowire mats to the mechanism of signal transduction is essential. From a fundamental knowledge, however, these effects are also not well studied. The charge transport along the chains of nanowires has a percolation nature where the conductivity between two adjacent electrodes depends on the presence of conduction "tracks" through the contacting single-crystal nanowires (Fig. 15.24). The electron transport through such a percolation chain has to be studied by numerical methods [92]. However, some basic features characterizing the gas sensitivity of such mats can be explained

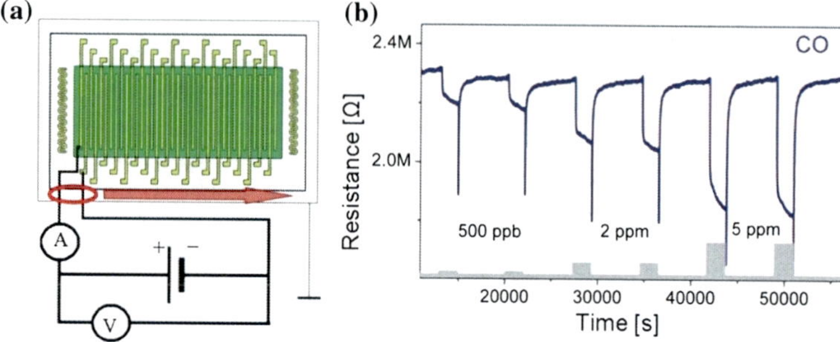

Fig. 15.22 a the circuitry for resistance measurements along the sensing array; **b** the change of the median resistance of multisensor chip based on SnO_2 nanowire mats during exposure to traces of CO in air (concentration: 0.5, 2, 5 ppm). A constant temperature, T $\sim$ 300 °C is maintained along the entire array Reprinted with permission from [82] © American Chemical Society

Fig. 15.23 Kinetics of resistance change of the "median" segment of multisensor chip upon exposure to 500 ppb of 2-propanol in air. Reprinted with permission from [82] © American Chemical Society

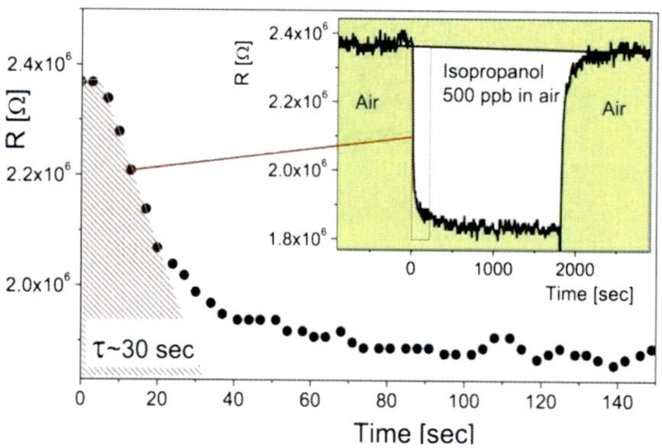

qualitatively with the help of the nanowire depletion model (diagram of Fig. 15.24). We can assume that the following mechanisms determine the transduction function of sensors based on nanowire mats: (1) adsorbate-induced change in the height of the potential barrier ΔV_S at the junction points (nodes) between nanowires effectively modulates the overall electron transport. According to theoretical models developed to calculate such a transport through the inter granular contacts [93, 94], the current is governed by the depletion zones at the contacts between nanowires. The direct visualization of the electrostatic action of the contact "node" is demonstrated in Ref. [95].

(2) Effect of adsorbate on the electron transport along the straight sections of n-type SnO_2 nanowires during exposure to gases may be described in terms of the coaxial model (Sect. 15.2), in which the central part of the semiconductor nanowire ("semiconducting core") is surrounded by a DR ("shell") of thickness W. The steady state DR depends on the Debye length λ_D of SnO_2 nanowire and the equilibrium coverage of ionosorbed acceptor groups (O^{2-}, O^-, OH^-) from the

Fig. 15.24 Receptor and transduction function of the percolating oxide nanowire network. The effect of a reducing gas in an oxidising atmosphere leads to a decrease in the potential barriers height at the contacts between nanowires and an increase in the cross-section of conducting channels in the nanowire bulk. Increasing the density of nanowire mats enhances the relative contribution of the nodes (junctions) with respect to the straight parts of the percolation paths Reprinted with permission from [82] © American Chemical Society

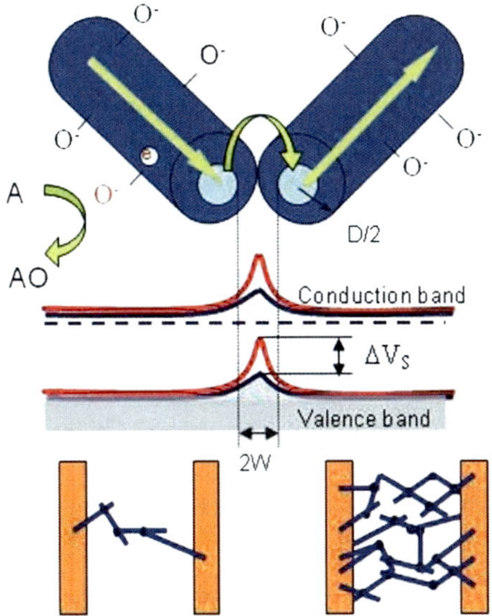

atmosphere at the given temperature. The latter is modulated by combustible analytes.

The relative contributions of these two mechanisms into the total chemiresistive response of the nanowire mats depends on the diameter of the nanowire and density of the junctions in the percolating path (Fig. 15.24, bottom). In general, unless limited by other factors (i.e. gas permeability), thinner nanowires with higher nanowire mat densities results in a larger number of nodes involved in the charge transport and likewise a higher sensitivity of the sensor can be achieved [96]. This conclusion has important implications for the gas recognition function of multisensor micro-arrays as discussed later.

The differentiation of gas-sensitive properties of similar sensor segments in the multisensor chips of electronic noses based on thin metal oxide films is achieved, for example, by applying a thermal gradient along the substrate [88, 97]. Figure 15.25a presents a response of the nanowire chip sensor segments to the vapors of 2-propanol obtained under a thermal gradient yielding a 250–330 °C temperature variation over the chip substrate.

As can be seen, at these operating temperatures the sensor segments had a significant and reproducible response to 500 ppb of 2-propanol in air. With decreasing the operating temperature the response of sensor segments to the analyte declines on average. Similar results were obtained for exposures towards carbon monoxide and ethanol vapor.

The vector response of the multisensor chip based on nanowire mats heated homogeneously and non-homogeneously was processed with LDA to estimate the effect of a thermal gradient on the gas recognition ability (Fig. 15.25b). Note that all the clusters of sensor signals corresponding to various test gases are

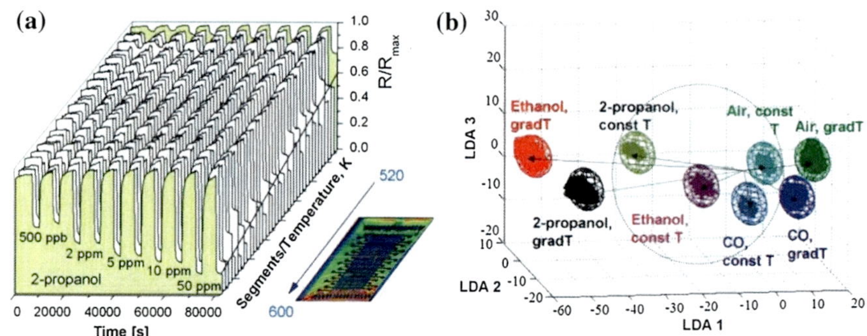

Fig. 15.25 **a** Change in the normalized resistance of the multisensor chip segments based on SnO_2 nanowire mats when exposed to vapors of 2-propanol at various concentrations in mixture with air. The temperature variation of 70 °C (250–320 °C) is applied across the chip. **b** LDA processing of the sensor array vector signals (resistances) when exposed to ethanol, 2-propanol, and CO mixed with air (range of concentrations is $2 \div 10$ ppm). The classification spheres correspond to the normal distribution of data at the level of significance of 0.9999. The micro-array has been operated under homogeneous spatial heating to T = 300 °C (const T- clusters, enclosed by the shadow ellipse) and with the T gradient application (grad T clusters). Reprinted with permission from [82] © American Chemical Society

significantly separated from each other allowing their confident recognition. Two important features of the LDA results have to be highlighted: (1) The cluster separation of the nanowire mats operated under isothermal heating is already sufficient to identify the tested gases reliably (central shaded region in Fig. 15.25b). It implies that, different from homogeneous thin films, the use of nanowire mats as sensitive elements in multielectrode micro-arrays does not necessarily require a temperature gradient. This is due to the morphology of the percolating layer of a nanowire mat, where the density of junctions and straight parts of nanowires varies stochastically from segment to segment. These variations in nanowire mat density between the electrodes ensures the specificity of the response of every single gas sensing element; (2) Upon application of an additional thermal gradient, the gas recognition power of the micro-array increases substantially in agreement with the previously reported results. The average Mahalanobis distance between clusters of data corresponding to the four reducing sample gases, increased more than twofold when compared to the one observed at isothermal measurements. This allows identification of the gases to be made with more confidence. It must be noted that even chemically similar alcohols, ethanol and 2-propanol, were easily discriminated.

An important consequence of the high gas recognition ability of nanowire mat based multisensor chips is their use for recognition of odors or complex gas mixtures composed of a large number of components present at low concentrations. Examples of such mixtures are the aromas of fruits, drinks and other consumables. To evaluate the possibility for identification of food aromas by SnO_2 nanowire mat based multisensor chips they were exposed to two classes of aromas:

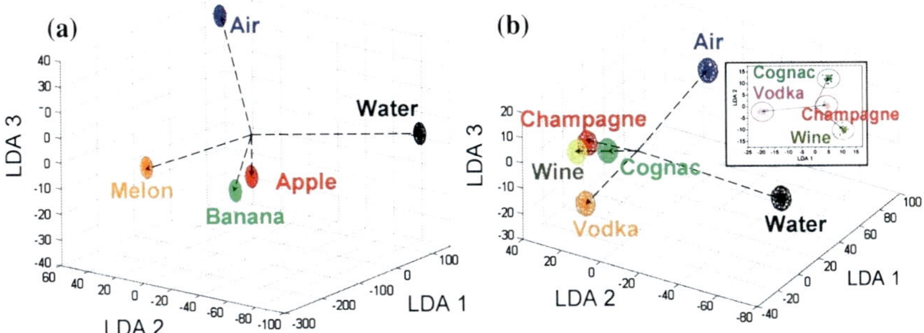

Fig. 15.26 The processing of response of SnO_2 nanowire mats based multisensor chip with LDA: **a** recognition of the fruit aroma; **b** recognition of alcoholic beverages; inset: results of processing the multisensor response to the alcoholic beverages only excluding the signals to air and water vapor. Sampling is 30 signals to each gas, the confidence level is 0.999

(1) fresh fruits (apple, banana, melon), and (2) alcohol beverages (vodka, cognac, red wine, champagne). The spirit samples were diluted with distilled water to ensure a constant alcohol concentration of 12.5 %. Given that the composition of all tested liquors included water distilled water samples were also tested in the course of these studies. A thermal gradient 250–310 °C along the nanowire mat was maintained over the chip substrate. As seen from LDA results (Fig. 15.26), aromas of fruit are confidently discriminated while the data clusters corresponding to the alcohol beverages are distanced quite close to each other. If, though, the results are analyzed without taking into account the signals of air and water, the flavors are essentially "split" (see inset in Fig. 15.26b).

Thus, the multisensor array e-nose based on percolating nanowire mats manifested a high gas discrimination ability to not only chemically various compounds, but also to complicated multi-component mixtures.

15.3.2 The Long-Term Stability of the Gas-Sensing Properties of SnO_2 Nanowire Mats

To minimize frequent repetition of the time consuming e-nose training process, one of the key requirements of these sensor elements is their long-term stability. The aging/stability studies on nanowire mats have been carried out in parallel with similar measurements of thin films consisting of SnO_2 nanoparticles [98]. The latter have been chosen because of their very high gas sensitivity observed in preliminary studies [99].

SnO_2 nanoparticles were synthesized from the vapor-gas phase by decomposition of tin-containing organic precursor. They had the crystalline structure of cassiterite with an average diameter of about 4 nm and specific surface area of about 123 m^2/g.

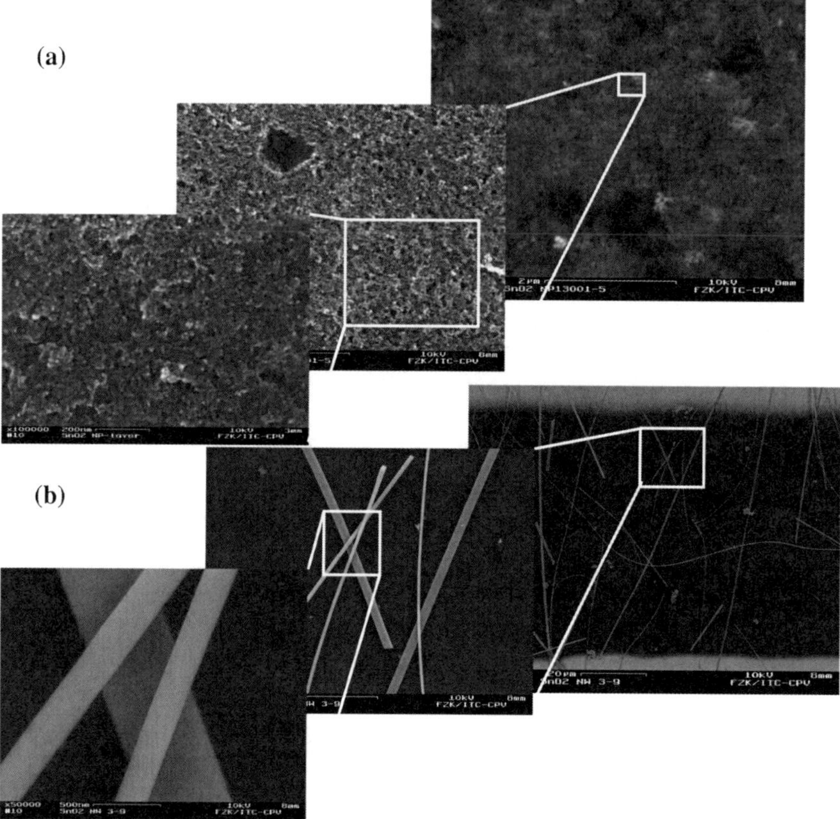

Fig. 15.27 SEM images of the surface morphology of gas-sensitive SnO_2 layers made of nanoparticles (**a**) and nanowires (**b**) Reprinted with permission from [98] © Elsevier

Figure 15.27a shows that SnO_2 nanoparticles deposited over the substrate form a mesoporous layer with most of the nanoparticles assembled in agglomerates. At the same time, the studied samples of the SnO_2 nanowires form a monolayer mat (Fig. 15.27b) with surface density of about $\rho \sim 1 \times 10^{-3} \ \mu m^{-2}$. The comparative gas sensing measurements were performed as follows.

Multisensor chips of both types (nanowires and nanoparticles) were exposed to a constant flow of synthetic air for 46 days. Dry synthetic air was applied to them during the first 25 days and in the next 21 days synthetic air was humidified to 50 rel. percent. 2-propanol vapors were added to the flow of synthetic air throughout the entire 46 days in the concentration range of 0.5–50 ppm. The operating temperature of multisensor chip was maintained to be quasi-homogeneous at the same level of ca 300 °C. The results are summarized in Fig. 15.28 for all segments of the multisensor chips based on nanowire mats and thin layer of nanoparticles of SnO_2.

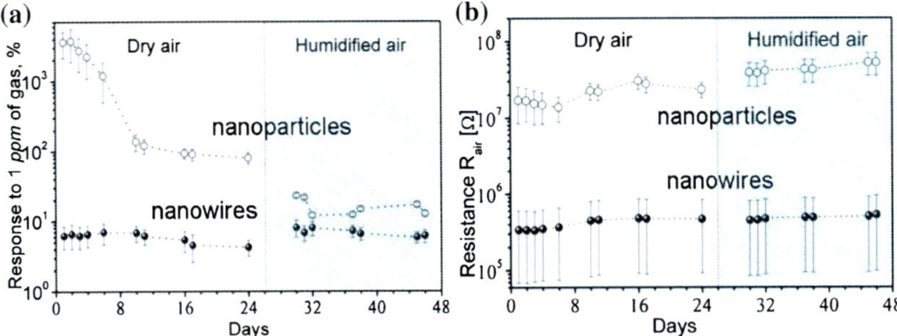

Fig. 15.28 Long-term evolution of characteristics of gas-sensitive SnO_2 layers made of nanowires and nanoparticles: **a** response to 1 ppm of 2-propanol in air; **b** the resistance in clean air. Open and shaded circles correspond to nanoparticles and nanowires, respectively. The error bars shows the standard deviation values for all segments of the multisensor chip. *Dotted curves* are drawn to indicate the trends. Reprinted with permission from [97] © Elsevier

During the long-term operation at approx. 300 °C the response of SnO_2 nanoparticles dropped down up to 2 orders of magnitude from its initial value observed in the as-prepared samples. At the same time, the response of nanowire mats to the same analyte has remained virtually unchanged. When synthetic air was humidified up to 50 rel. percent, the response of the SnO_2 nanoparticles towards 2-propanol dropped nearly another order of magnitude and then stabilized at 16 ± 5 % level of its initial value. This latter value became close to the response magnitude of the nanowire based sensor.

Both layers exhibited an immediate decrease in resistance when exposed to water, and later—a restoration of resistance to its previous value (Fig. 15.29a). This seems to confirm that the adsorbed water molecules and hydroxyl groups substitute the pre-adsorbed oxygen [100] while the total coverage of the adsorbates is defined by the concentration of surface adsorption sites and Weisz restrictions [101]. As a result, the resistance in air in over the long term, as noticed in Fig. 15.28b, for both layers is almost independent of the presence of moisture.

Since the studied nanostructured layers were made of the same material, the observed differences can be explained mainly by their difference in morphology. The long-term exposure to vapors of 2-propanol at the operating temperature around 300 °C caused, presumably, sintering of the nanoparticles, which led to their aggregation and encapsulation into larger agglomerates [102, 103] (Fig. 15.29, arrow A–B). The presence of water likely plays an important role in these processes, which is proven on the one hand, by the experimentally observed significant reduction in the response of nanoparticles to the gas when humidity appeared in the air during the measurements on days 26 through 46 (Fig. 15.28a). In addition to an increase in the effective diameter of the nanoparticles even unsintered crystallites, which may be present within the agglomerates, are excluded from gas exchange with the ambient and thus become insensitive to analytes. The number of possible percolation paths [104] in the nanoparticle gas-sensitive oxide layer also become significantly reduced with particle aggregation

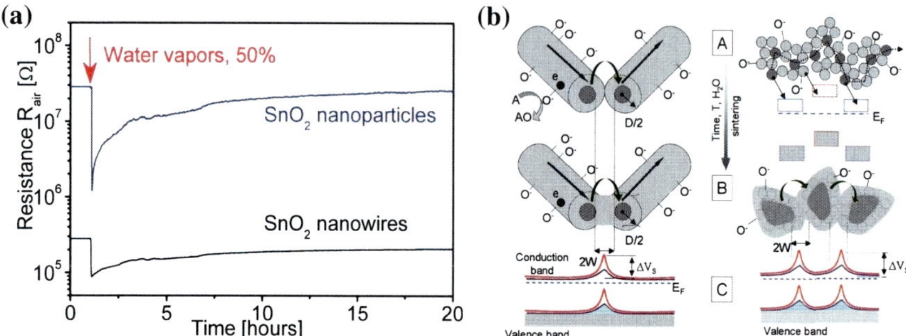

Fig. 15.29 **a** Change of resistance of the "median" sensor segment of the multisensor chips made of SnO_2 nanoparticles and nanowire mats upon exposure to water vapors, 50 rel. percent in synthetic air; **b** the diagram of electric transport mechanism in the studied nanostructures. A–C line depicts the time evolution of nanostructures morphology: from freshly-made (A) to after-long-operation (B) as well as band diagram at nanowire/nanoparticle junctions (C). Reprinted with permission from [97] © Elsevier

which leads to a decrease in gas sensitivity. Nanowire junctions can also become fused during long term operation what can deteriorate the sensor performance. However, different from nanoparticulate films, the amount of the conducting channels and open morphology of the nanowire network remained preserved, which results in better long term stability.

15.4 Monolithic Multisensor Micro-Nanoarrays Based on Single SnO_2 Meso-Nanowire: A New Generation E-nose

Different from the nanoparticles, the individual single-crystal nanowires of metal oxides have a microscopic length. Therefore, one can envision that micro processing techniques used to functionalize the sensing segments in film based multisensor arrays can be also applied to these nanostructures. This can be realized via inducing local variations of internal and/or external parameters along the length of the metal oxide nanowire, which would lead to variations of its local gas-sensing properties. Such parameters could be (Fig. 15.30): (1) surface morphology and thickness (transverse diameter) of the nanowire; (2) concentration of intrinsic (oxygen vacancies) or impurity donors/acceptors; (3) concentration of surface catalysts; (4) non-uniform heating along the nanowire. Using bottom-up methodology, the modulation of some of these parameters (e.g. thickness, doping) can be realized during the nanowire growth. After indexing of the nanowire with multiple electrodes in correlation with its functionalization profile, such a single nanowire can be considered as an array of sensor elements with diverse gas-sensing characteristics of local segments.

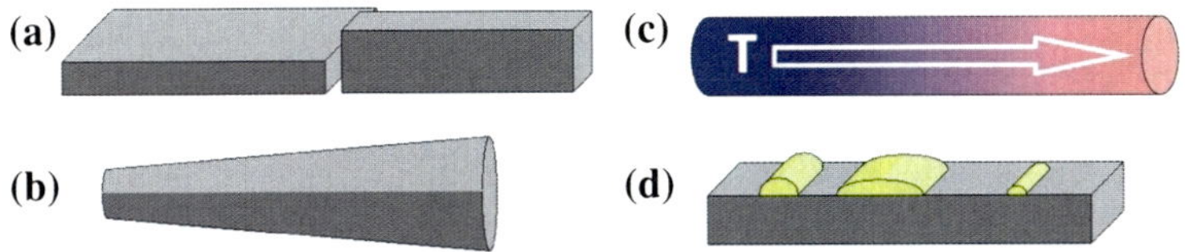

Fig. 15.30 Examples of possible variations in internal and external parameters of the oxide nanowire: **a** morphology, **b** thickness, **c** operating temperature, **d** surface or bulk doping

Since the parameters defining the gas discrimination ability in this case can be encoded into the morphology of the sensing element, this approach represents a new step in electronic nose development—an integration strategy which was not possible in traditional film-based electronic nose designs. This ultimately simple to a certain degree electronic nose can be considered as the first bottom-up engendered analytical device. Below the fabrication, performance and application of this kind of device are described.

15.4.1 Understanding the Gas Sensing Properties of the Single Nanowire Multisensor Micro-Nanoarrays

The first multisensor micro-arrays based on an individual oxide nanowire were developed using the KAMINA chip as a platform [105, 106] with an array of coplanar electrodes, planar thermistors (on the front side) and an array of heaters (on the back side of the chip); see Fig. 15.31.

To modulate the electrical properties of a nanowire along its length two approaches were used: (1) synthesis of wedge-like nanowires (nanobelts), and (2) deposition of metal catalyst (Pd) on some of the segments of the nanowire. Given that the electrodes in the KAMINA design are separated by 70–80 μm and the electrode width is approx. 100 μm, ultra-long (up to few mm) SnO_2 nanobelts were employed.

Figure 15.31 depicts the chip with a wedge-shaped SnO_2 nanobelt. The length of the nanostructure is about 2.5 mm, which allowed one to place 11 electrical contacts and have 10 sensor segments.

The widest part of the nanowire, which is labelled as the first three segments, was covered with non-percolating Pd patches of diameter less than 1 μm via thermal evaporation through the shadow mask (see Sect. 15.2 for details). The SnO_2 nanobelt was placed over the Pt electrode array. To ensure good electrical and mechanical contact of the nanobelt with the Pt electrodes they were additionally fused with Pt bridges using electron beam induced Pt deposition inside a FIB SEM. Such electrical contacts appear to be ohmic, as verified via linearity of

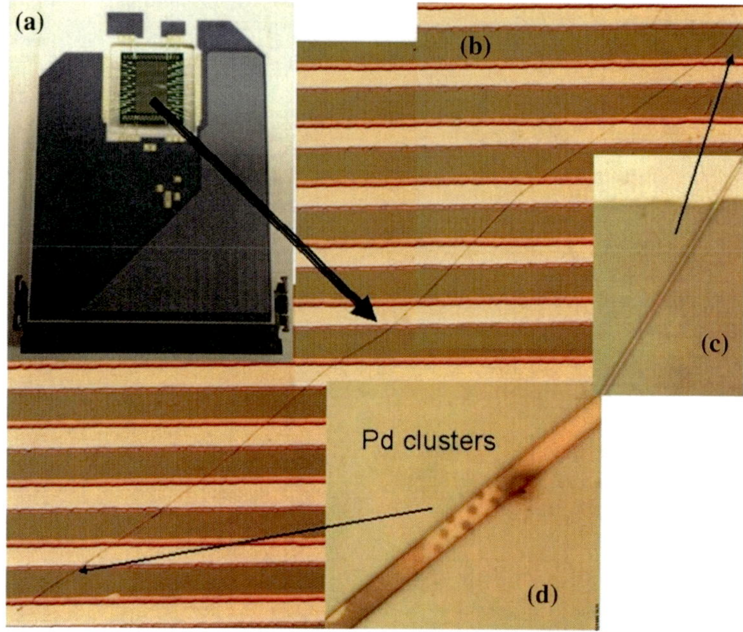

Fig. 15.31 **a** 50-pin ceramic multisensor chip; **b** wedge-shaped SnO_2 meso-nanowire that has both **c** pristine surface regions and **d** circular micro patches with Pd catalyst

I–V curves both in a vacuum (of residual pressure $P \sim 3 \times 10^{-4}$ Pa) and in synthetic air flow, over a wide temperature range (from 25 to 150 °C).

Figure 15.32 shows the distribution of resistance and extracted electrical parameters, the concentration of free carriers and the Debye length, along the segments of meso-nanowire shown in Fig. 15.31. The tests were performed in a vacuum ($P \sim 3 \times 10^{-4}$ Pa) and at an operating temperature of ca 150 °C.

As can be seen (Fig. 15.32), the resistance of the wedge-like nanobelts follows the change in its morphology. However, Pd functionalization significantly distorts the longitudinal electrical properties of the nanowire. Resistance of segments #1–3, which bear clusters of Pd, is higher compared to pristine segments. The latter presumably is due to depletion of electrons from the nanobelt by Pd, which has higher work function as compared to SnO_2. At the pristine segments the ratio of L_d/H (here L_d and H are the calculated Debye length and the thickness of the nanobelt correspondingly) varies from 0.05 to 0.1. The latter indicates that these segments will be depleted (enriched) by free electrons differently upon adsorption of the acceptor (donor)—like analytes and therefore will provide the required recognition power of an e-nose.

Figure 15.33a presents a reconstruction of the depletion profile of this SnO_2 nanobelt in air. As can be seen, by variations in the morphology and the introduction of catalyst one can significantly modulate the magnitude of the conducting

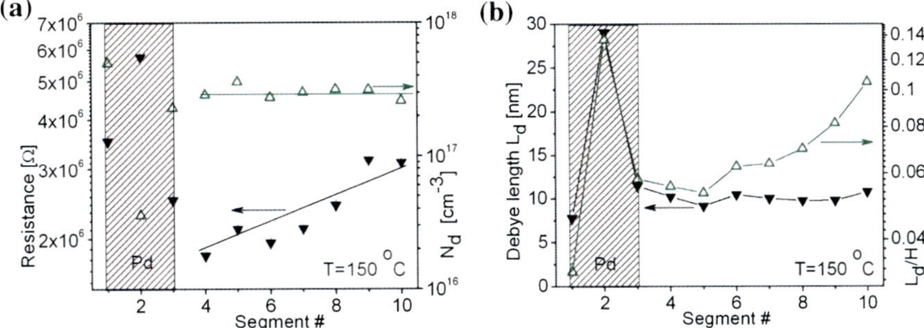

Fig. 15.32 Results of the study of electrical properties of individual SnO$_2$ meso-nanowire, segmented with electrodes, in a vacuum (P $\sim$ 3 $\times$ 10^{-4} Pa): **a** the distribution of electrical resistance and the effective donor concentration the by segments; **b** the calculated Debye length and its relation to the transverse diameter of meso-nanowire segment. Working temperature T = 150 °C

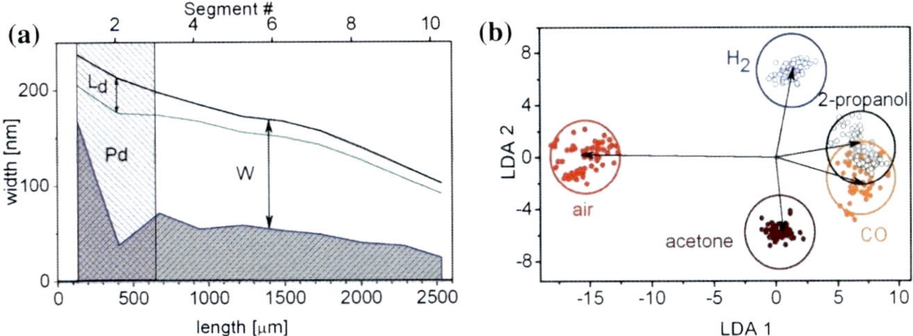

Fig. 15.33 **a** the profile of depletion at the SnO$_2$ nanobelt by free carriers; *Black and green curves* correspond to the local width and calculated Debye length in the nanobelt; *gray region* shows the conduction channel available under exposure to air; **b** LDA processing of the response of segmented SnO$_2$ nanobelt to 4 reducing gases (hydrogen, 2-propanol, CO, acetone), the concentration is 50 ppm. Signal sampling to each gas—60, confidence level—0.9

channel and, therefore, differentiate the longitudinal electric (and gas-sensing) properties of individual nanostructure.

The gas-sensing properties of the multisensor chip based on this model SnO$_2$ nanobelt were tested at quasi-homogeneous operating temperature of about 150 °C. Figure 15.33b shows the results of LDA processing of segment responses of the nanobelt when exposed to mixtures of four reducing gases with air. The feature clusters corresponding to different gases are well separated, except for CO and 2-propanol, whose clusters slightly overlap. Thus, it was proved that the modulation of the morphology and local functionalization of the individual nanostructure are sufficient factors to obtain reasonable recognition power utilizing a single nanostructure e-nose even if the latter one has mesoscopic dimensions.

15.4.2 Toward Large Scale Fabrication of Multisensor Micro-Arrays Based on Individual SnO$_2$ Nanostructures

The results presented in Sect. 15.4.1 show the possibility to differentiate the local electrical and gas-sensing properties of an individual metal oxide nanowire and to fabricate a multisensor micro-array. To benefit fully from the proposed approach one has to develop a protocol for multisensor array fabrication which would be compatible with common large scale production technologies.

We have tested the technological processes to fabricate the single nanowire micro-arrays which consisted of the following steps: (1) deposition of meander heaters at the backside of a 3 inch Si/SiO$_2$ wafer, (2) placement of individual nanowires over the front side of the wafer, (3) deposition of co-planar multi-electrodes and thermoresistors over the wafer front side on top of nanowires (Fig. 15.34). The overall technology route is nearly the same one employed to fabricate thin-film multisensor micro-arrays [107] described in Sect. 15.4.1.

Wedge-shaped SnO$_2$ nanobelts ca. 2 mm long, were used for device fabrication.

The temperature dependence of individual segments of the meso-nanobelts (and meso-nanowires) exposed to reducing gases was studied using 2-propanol vapors mixed with humidified (50 rel. percentage) synthetic air (Fig. 15.35a). The characteristic features of the response were: (1) the presence of an optimal temperature in the range where the sensitivity has a maximum; (2) for the segments with high gas sensitivity, this maximum is shifted to lower temperature range (compare segments #3 and #2 in the Fig. 15.35a). Since the pristine SnO$_2$ nanowires have their optimal performance temperature in the range of 250–300 °C one can assume that sputtered Pt electrodes induce a catalytic process at the interface with nanowire. The unintentional doping of the nanostructures with Pt during electrode deposition is well known and originates due to incomplete shadowing with the mask and/or metal diffusion (Fig. 15.35b). Apparently, different segments of the nanowire can have a dissimilar degree of doping which results in their different sensing performance.

This assumption is supported by SEM inspections and EDX analysis. Figure 15.36a shows a section of the nanowire segment in proximity to the electrode. The surface of the nanowire is covered with Pt clusters, as can be seen from EDX spectrum in the Fig. 15.36b.

This unintentional doping with metal catalyst is an effective way to differentiate the gas-sensing properties of a single-crystal nanowire sensor array. Nevertheless, it should be noted that the aforementioned doping is poorly controlled, and therefore each multisensor array has to be calibrated individually. Further development of controllable functionalization schemes for the sensor array segments is required.

Figure 15.37 shows the results of LDA processing of the multisensor response of such a segmented SnO$_2$ nanowire to four reducing test gases mixed with synthetic air at concentrations of 30 ppm measured at 250 °C (panel a) and 290 °C (panel b).

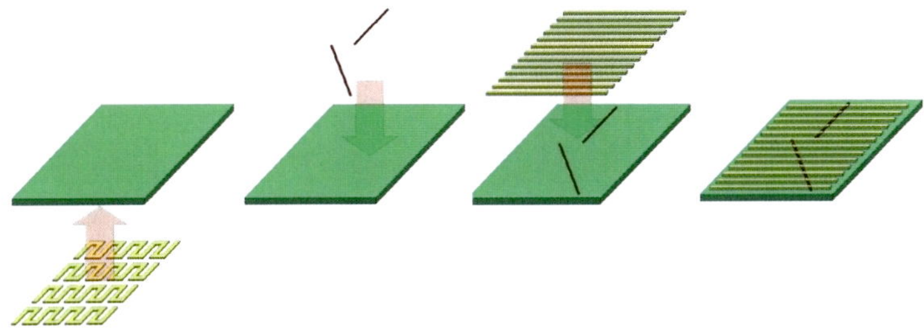

Fig. 15.34 Large scale fabrication route of micro-nanoarray based on a single nanowire. **a** deposition of heaters on the backside of the substrate; **b** deposition of nanowires on the front side of the substrate; **c** deposition of microelectrode array atop of the nanowires

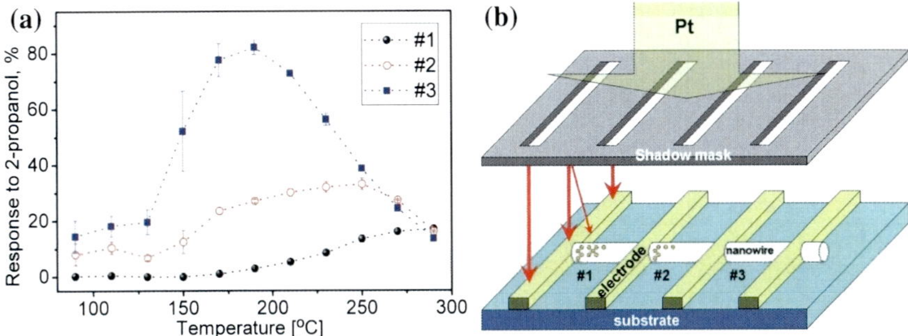

Fig. 15.35 a The temperature dependence of the response (R_{air}/R) of the three segments (# 1–3) of the nanowire to a mixture of 2-propanol with a humid (50 rel. %) air; the analyte concentration is 10 ppm; **b** unintentional surface doping of the nanowire with Pt c during the electrode deposition

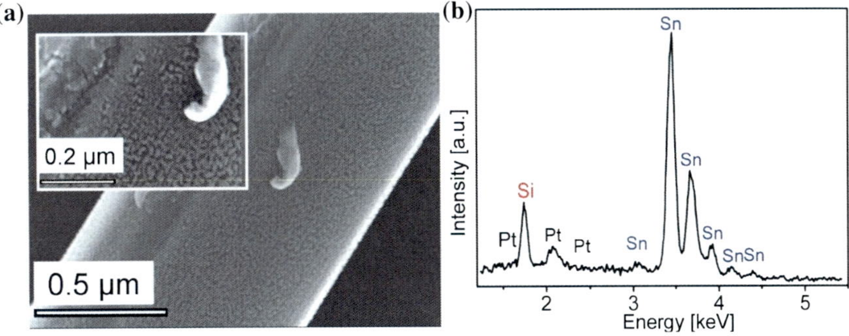

Fig. 15.36 a SEM-image of the near electrode section of nanowire. **b** EDX spectrum of this area

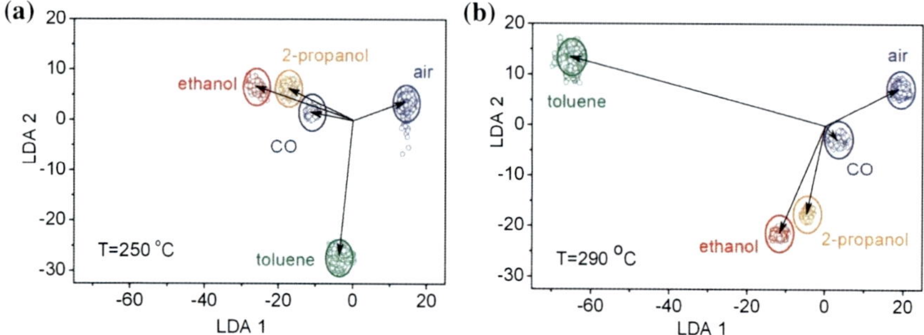

Fig. 15.37 The results of LDA from the segmented SnO_2 nanowire sensor array operating at **a** 250 °C and **b** 290 °C. The analyte concentration in synthetic air was 30 ppm. Confidence level is 0.99, sampling is 2 × 50 points. Vectors indicate the distance from the cluster centers to the origin of the LDA coordinate system

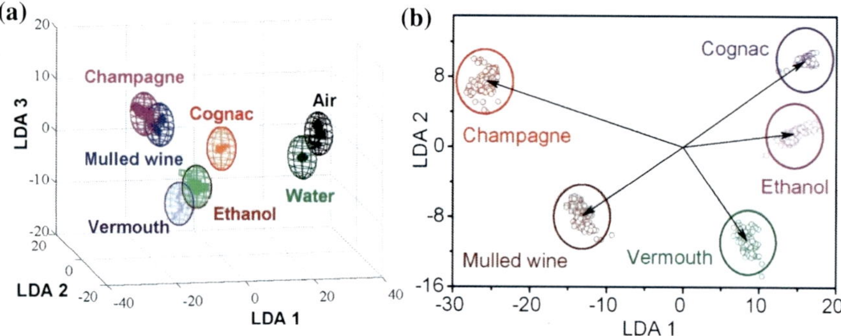

Fig. 15.38 The results of LDA processing of response of the segmented SnO_2 nanowire sensor array to aroma of some beverages: **a** recognition analysis of different drink flavors, water vapor and air; **b** recognition analysis of drink flavors only. Confidence level is 0.99; sampling is 50 × 2 points

In both cases the clusters corresponding to different analytes are confidently discriminated. The average Mahalanobis distance observed when the segmented nanowire is heated up to about 250 °C was about 19.3 units, while that at 290 °C increases to 32.1 units. This increase is primarily due to the displacement of the toluene data cluster. The obtained recognition of gases seems to mostly originate from the aforementioned differences in catalytic action of the Pt nanoparticles deposited near the electrode areas.

As noted earlier, one of the most challenging applications of e-nose devices is for food analysis in the presence of a high concentration of the disturbing agent. To test the performance of the single nanowire multisensor arrays the following analytes were used: (1) mulled wine (10 % alc., « Georg Raum » , Germany), (2) champagne (11.5 % alc., « Freixenet » , Spain), (3) Vermouth (15 %

alc., « Chinzano » , Italy), (4) Cognac (40 % alc.,). The details of the study are given in [105]. The disturbing agent in this case is the high concentration of alcohol and water vapors. Figure 15.38 presents the results of LDA processing of the segmented nanowire multisensor response to these aromas, which can be reliably distinguished from those originating from pure air and water vapors.

It may be noted that the multisensor response to Cognac is most similar to that of ethanol, which implies less influence of additives of aromatic substances in this liquor, in contrast to vermouth, mulled wine and champagne. It is possible that the concentration of aromatic substances (or their volatility) in Cognac is of minimal as compared to that in other drinks of interest.

15.5 Conclusions

Quasi-1D nanostructures possess an advantageous and controllable correlation between the Debye length and transverse dimensions, which allows development of simple chemiresistive sensor elements of high sensitivity. Single-crystal quasi-1D nanostructures, in particular, expose fewer stable facets to the ambient environment and therefore are good candidates for applications where a high density of elements, reproducibility and long-term stability are required. Utilization of metal oxide nanowires for development of discrete multisensor arrays as part of first-generation e-nose was demonstrated.

To comply with requirements of large scale production, the fabrication of the integrated multisensor array based on a layer of percolating nanowires was proposed. When compared to traditional metal oxide films the use of low-density nanowire mats have the following advantages: (1) the surface of virtually all of the nanowires in this layer is available for gas adsorption and its accessibility does not change with decreasing nanowire diameter, which is opposite to what is observed in the case of nanostructured films; (2) nanowire mats possess new transductions mechanisms which can be controlled by nanowire diameters and their density; (3) the comparative studies of the stability of the nanoparticulate film versus nanowire mats revealed that the single-crystallinity of nanowire sensor elements hampers their sintering, which minimizes the aging effects; (4) the variation of the density of the nanowire mats represent a technologically simple but effective approach for gaining diversity in sensing segments of the multisensor array.

Further integration of the sensing elements in multisensor arrays is possible via segmenting individual SnO_2 nanowires with a number of electrodes. The spatial differentiation of the functional properties of the oxide nanowire can be achieved via modulation of their morphology and doping along the length of the nanostructure. The recognition of complex mixtures was demonstrated with this monolithic multisensor array, which potentially paves the way for the bottom up fabrication of ultrasmall analytical devices.

Acknowledgments The authors thank Dr. M. Sommer, Dr. I. Kiselev, Dr. M. Bruns, Mr. W. Habicht, Mr. G. Stengel, Mr. J. Benz, Dr. D. Fuchs (KIT, Karlsruhe, Germany), Prof. S. Kar (Northeastern University, Boston, USA), Dr. S. Zemskova (CAT, Peoria, USA), Dr. L. Gregoratti, Dr. M. Kiskinova (Elettra, Trieste, Italy) for their support and assistance in the course of these studies. The research at SIUC was supported through Caterpillar Inc. research grant and at later stages through NSF ECCS-0925837 grant. V. S. thanks the financial support of his work on the project from the Fulbright postdoc scholarship, INTAS postdoc grant no. YSF 06-1000014-5877, "Michail Lomonosov" scholarship from Russ. Ministry for Education & Science and DAAD. no. A/05/58552, as well as the travel grant of Russ. Ministry for Education and Science, no. RI-111/002/012.

References

1. Buck L, Axel R (1991) A novel multigene family may encode odorant receptors: a molecular basis for odor recognition. Cell 65:175–187
2. Firestein S (2001) How the olfactory system makes sense of scents. Nature 413:211–218
3. Laurent G (2002) Olfactory network dynamics and the coding of multidimensional signals. Nat Rev Neurosci 3:884–895
4. Haddad R, Khan R, Takahashi YK, Mori K, Harel D et al (2008) A metric for odorant comparison. Nat Methods 5:425–429
5. Pearce TC, Schiffman SS, Nagle HT, Gardner JW (eds) (2003) Handbook of machine olfaction: electronic nose technology. Wiley, Weinheim, p 592
6. Turner A, Magan N (2004) Electronic noses and disease diagnostics. Nat Rev Microbiol 2:161–166
7. Kauer JS (2002) On the scents of smell in the salamander. Nature 417:336–342
8. Mori K, Nagao H, Yoshihara Y (1999) The olfactory bulb: coding and processing of odor molecule information. Science 286:711–715
9. Persaud K, Dodd G (1982) Analysis of discrimination mechanisms in the mammalian olfactory system using a model nose. Nature 299:352–355
10. Gardner JW, Bartlett PN (1996) Performance definition and standardization of electronic noses. Sens Actuators B Chem 33:60–67
11. Monkman G (1996) Bio-chemical sensors. Sens Rev 16:40–44
12. Gardner JW, Bartlett PN (1994) A brief history of electronic noses. Sens Actuators B Chem 18:210–211
13. Nagle HT, Gutierrez-Osuna R, Schiffman SS (1998) The how and why of electronic noses. IEEE Spectr 35:22–34
14. Lundstrom I, Erlandsson R, Frykman U, Hedborg E, Spetz A et al (1991) Artificial 'olfactory' images from a chemical sensor using a light-pulse technique. Nature 352:47–50
15. Goschnick J (2001) An electronic nose for intelligent consumer products based on a gas analytical gradient micro-array. Microelectron Eng 57–58:693–704
16. Hagleitner C, Hierlemann A, Lange D, Kummer A, Kerness N et al (2001) Smart single-chip gas sensor microsystem. Nature 414:293–296
17. Joo S, Brown RB (2008) Chemical sensors with integrated electronics. Chem Rev 108:638–651
18. Horner GMR, Gardner JW, Bartlett PN (1992) Odour sensors for an electronic nose. Sensors and sensory systems for and electronic nose. Kluwer Academic Publishers, Dordrecht, p 327
19. Hierlemann A, Gutierrez-Osuna R (2008) Higher-order chemical sensing. Chem Rev 108:563–613
20. Baltes H, Barrettino D, Graf D et al (2004) Microsensor and single chip integrated microsensor system. US Patent & Trademark Office, USA Patent 2004-0075140
21. Graf M, Barrettino D, Baltes HP, Hierlemann A (2007) CMOS hotplate microsensors. Springer, Berlin, p 125

22. Li Y, Vancura C, Barrettino D, Graf M, Hagleitner C et al (2007) Monolithic CMOS multi-transducer gas sensor microsystem for organic and inorganic analytes. Sens Actuators B Chem 126:431–440
23. Barsan N, Schweizer-Berberich M, Gopel W (1999) Fundamental and practical aspects in the design of nanoscaled SnO_2 gas sensors: a status report. Fresenius J Anal Chem 365:287–304
24. Sysoev VV, Button BK, Wepsiec K, Dmitriev S, Kolmakov A (2006) Toward the nanoscopic "electronic nose": hydrogen vs carbon monoxide discrimination with an array of individual metal oxide nano- and mesowire sensors. Nano Lett 6:1584–1588
25. Lundstrom I, Armgarth M, Spetz A, Winquist F (1986) Gas sensors based on catalytic metal-gate field-effect devices. Sens Actuators 10:399–421
26. Dickinson TA, White J, Kauer JS, Walt DR (1996) A chemical-detecting system based on a cross-reactive optical sensor array. Nature 382:697–700
27. Dickinson TA, Michael KL, Kauer JS, Walt DR (1999) Convergent, self-encoded bead sensor arrays in the design of an artificial nose. Anal Chem 71:2192–2198
28. Albert KJ, Walt DR, Gill DS, Pearce TC (2001) Optical multibead arrays for simple and complex odor discrimination. Anal Chem 73:2501–2508
29. Walt DR (2000) Bead-based fiber-optic arrays. Science 287:451–452
30. LaFratta CN, Walt DR (2008) Very high density sensing arrays. Chem Rev 108:614–637
31. Kermani BG, Fomenko I, Kotseroglou T, Forood B, Clark L et al (2006) Decoding beads in a randomly assembled optical nose. Sens Actuators B Chem 117:282–285
32. Rakow NA, Suslick KS (2000) A colorimetric sensor array for odour visualization. Nature 406:710–713
33. Suslick KS (2004) An optoelectronic nose: "seeing" smells by means of colorimetric sensor arrays. MRS Bull 29:720–725
34. Janzen MC, Ponder JB, Bailey DP, Ingison CK, Suslick KS (2006) Colorimetric sensor arrays for volatile organic compounds. Anal Chem 78:3591–3600
35. Snow A, Wohltjen H (2008) Materials, method and apparatus for detection and monitoring of chemical species. US patent 7,347,974, Bl, USA
36. Rapp M, Reibel J, Voigt A, Balzer M, Bülow O (2000) New miniaturized SAW-sensor array for organic gas detection driven by multiplexed oscillators. Sens Actuators B Chem 65:169–172
37. Barie N, Bucking M, Rapp M (2006) A novel electronic nose based on miniaturized SAW sensor arrays coupled with SPME enhanced headspace-analysis and its use for rapid determination of volatile organic compounds in food quality monitoring. Sens Actuators B Chem 114:482–488
38. Baller MK, Lang HP, Fritz J, Gerber C, Gimzewski JK et al (2000) A cantilever array-based artificial nose. Ultramicroscopy 82:1–9
39. Fritz J, Baller MK, Lang HP, Rothuizen H, Vettiger P et al (2000) Translating biomolecular recognition into nanomechanics. Science 288:316–318
40. Braun T, Ghatkesar MK, Backmann N, Grange W, Boulanger P et al (2009) Quantitative time-resolved measurement of membrane protein-ligand interactions using microcantilever array sensors. Nat Nanotech 4:179–185
41. Freund MS, Lewis NS (1995) A Chemically diverse conducting polymer-based electronic nose. Proc Nat Acad Sci USA 92:2652–2656
42. Lonergan MC, Severin EJ, Doleman BJ, Beaber SA, Grubbs RH et al (1996) Array-based vapor sensing using chemically sensitive, carbon black polymer resistors. Chem Mater 8:2298–2312
43. Shevade AV, Ryan MA, Homer ML, Manfreda AM, Zhou, H et al (2003) Molecular modeling of polymer composite-analyte interactions in electronic nose sensors. Sens Actuators B Chem 93:84–91
44. Ryan MA, Shevade AV, Zhou H, Homer ML (2004) Polymer-carbon black composite sensors in an electronic nose for air-quality monitoring. MRS Bull 29:714–719
45. Doleman BJ, Lewis NS (2001) Comparison of odor detection thresholds and odor discriminablities of a conducting polymer composite electronic nose versus mammalian olfaction. Sens Actuators B Chem 72:41–50

46. Semancik S, Cavicchi RE, Wheeler MC, Tiffany JE, Poirier GE et al (2001) Microhotplate platforms for chemical sensor research. Sens Actuators B Chem 77:579–591

47. Meier DC, Evju JK, Boger Z, Raman B, Benkstein KD et al (2007) The potential for and challenges of detecting chemical hazards with temperature-programmed microsensors. Sens Actuators B Chem 121:282–294

48. Suehle JS, Cavicchi RE, Gaitan M, Semancik S (1993) Tin oxide gas sensor fabricated using CMOS micro-hotplates and insitu processing. IEEE Electron Device Lett 14:118–120

49. Althainz P, Dahlke A, Frietsch-Klarhof M, Goschnick J, Ache HJ (1995) Reception tuning of gas-sensor microsystems by selective coatings. Sens Actuators B Chem 25:366–369

50. Althainz P, Goschnick J (1998) Sensor for reducing or oxidizing gases. USA patent 5,783,154, USA

51. Schierbaum KD, Weimar U, Gopel W, Kowalkowski R (1991) Conductance, work function and catalytic activity of SnO_2-Based gas sensors. Sens Actuators B Chem 3:205–214

52. Dutronc P, Lucat C, Menil F, Loesch M, Combes L (1993) A new approach to selectivity in methane sensing. Sens Actuators B Chem 15:24–31

53. Takagi T (1996) The concept and the recent research on intelligent materials. SPIE Proceedings 2779:2–15

54. Coller G (1996) Intelligent materials and systems as abasis for innovative technologies in transportation vehicles. SPIE Proceedings 2779:16–27

55. Potyrailo RA, Morris WG, Sivavec T, Tomlinson HW, Klensmeden S et al (2009) RFID sensors based on ubiquitous passive 13.56-MHz RFID tags and complex impedance detection. Wirel Commun Mob Comput 9:1318–1330

56. Young RC, Buttner WJ, Linnell BR, Ramesham R (2003) Electronic nose for space program applications. Sens Actuators B Chem 93:7–16

57. Goschnick J (2001) An electronic nose for intelligent consumer products based on a gas analytical gradient micro-array. Microelectron Eng 57(8):693–704

58. Ampuero S, Bosset J (2003) The electronic nose applied to dairy products: a review. Sens Actuators B Chem 94:1–12

59. Roeck F, Barsan N, Weimar U (2008) Electronic nose: current status and future trends. Chem Rev 108:705–725

60. Czarnic AW, DeWitt SH (1997) A practical guide to combinatorial chemistry. American Chemical Society, Washington, p 450

61. Potyrailo RA, Mirsky VM (2008) Combinatorial and high-throughput development of sensing materials: the first 10 years. Chem Rev 108:770–813

62. Scott RWJ, Yang SM, Chabanis G, Coombs N, Williams DE et al (2001) Tin dioxide opals and inverted opals: near-ideal microstructures for gas sensors. Adv Mater 13:1468–1472

63. Martinez CJ, Hockey B, Montgomery CB, Semancik S (2005) Porous tin oxide nanostructured microspheres for sensor applications. Langmuir 21:7937–7944

64. Ng HT, Li J, Smith MK, Nguyen P, Cassell A et al (2003) Growth of epitaxial nanowires at the junctions of nanowalls. Science 300:1249

65. Hong YJ, Jung HS, Yoo J, Kim Y-J, Lee C-H et al (2009) Shape-controlled nanoarchitectures using nanowalls. Adv Mater 21:222–226

66. Pan ZW, Dai ZR, Wang ZL (2001) Nanobelts of semiconducting oxides. Science 291: 1947–1949

67. Comini E, Faglia G, Sberveglieri G, Pan ZW, Wang ZL (2002) Stable and highly sensitive gas sensors based on semiconducting oxide nanobelts. Appl Phys Lett 81:1869–1871

68. Law M, Kind H, Messer B, Kim F, Yang PD (2002) Photochemical sensing of NO_2 with SnO_2 nanoribbon nanosensors at room temperature. Angewandte Chemie-Int Ed 41:2405–2408

69. Kolmakov A, Zhang YX, Cheng GS, Moskovits M (2003) Detection of CO and O_2 using tin oxide nanowire sensors. Adv Mater 15:997–1000

70. Wang YL, Jiang XC, Xia YN (2003) A solution-phase, precursor route to polycrystalline SnO_2 nanowires that can be used for gas sensing under ambient conditions. J Am Chem Soc 125:16176–16177

71. Li C, Zhang DH, Liu XL, Han S, Tang T et al (2003) In2O3 nanowires as chemical sensors. Appl Phys Lett 82:1613–1615
72. Kolmakov A, Moskovits M (2004) Chemical sensing and catalysis by one-dimensional metal-oxide nanostructures. Annu Rev Mater Res 34:151–180
73. Heo YW, Norton D, Tien L, Kwon Y, Kang B et al (2004) ZnO nanowire growth and devices. Mater Sci Eng R 47:1–47
74. Comini E (2006) Metal oxide nano-crystals for gas sensing. Anal Chim Acta 568:28–40
75. Lu JG, Chang P, Fan Z (2006) Quasi-one-dimensional metal oxide materials–synthesis, properties and applications. Mater Sci Eng R 52:49–91
76. Chen P-C, Shen G, Zhou C (2008) Chemical sensors and electronic noses based on 1-D metal oxide nanostructures. Nanotechnol IEEE Trans 7:668–682
77. Korotcenkov G (2008) The role of morphology and crystallographic structure of metal oxides in response of conductometric-type gas sensors. Mater Sci Eng R 61:1–39
78. Kolmakov A (2008) Some recent trends in the fabrication, functionalisation and characterisation of metal oxide nanowire gas sensors. Int J Nanotechnol 5:450–474
79. Kolmakov A, Klenov DO, Lilach Y, Stemmer S, Moskovits M (2005) Enhanced gas sensing by individual SnO_2 nanowires and nanobelts functionalized with Pd catalyst particles. Nano Lett 5:667–673
80. Kolmakov A, Chen XH, Moskovits M (2008) Functionalizing nanowires with catalytic nanoparticles for gas sensing application. J Nanosci Nanotechnol 8:111–121
81. McAlpine MC, Ahmad H, Wang D, Heath JR (2007) Highly ordered nanowire arrays on plastic substrates for ultrasensitive flexible chemical sensors. Nat Mater 6:379–384
82. Ryu K, Zhang D, Zhou C (2008) High-performance metal oxide nanowire chemical sensors with integrated micromachined hotplates. Appl Phys Lett 92:093111
83. Chen PC, Ishikawa FN, Chang HK, Ryu K, Zhou C (2009) A nanoelectronic nose: a hybrid nanowire/carbon nanotube sensor array with integrated micromachined hotplates for sensitive gas discrimination. Nanotechnology 20:125503-1–125503-8
84. Baik JM, Zielke M, Kim MH, Turner KL, Wodtke AM et al (2010) Tin-oxide-nanowire-based electronic nose using heterogeneous catalysis as a functionalization strategy. ACS Nano 4:3117–3122
85. Henrion R, Henrion G (1995) Multivariate datenanalyse: Methodik und Anwendung in der Chemie und verwandten Gebieten. Springer, Berlin
86. Jurs P, Bakken G, McClelland H (2000) Computational methods for the analysis of chemical sensor array data from volatile analytes. Chem Rev 100:2649–2678
87. Albert KJ, Lewis NS, Schauer CL, Sotzing GA, Stitzel SE et al (2000) Cross-reactive chemical sensor arrays. Chem Rev 100:2595–2626
88. Sysoev VV, Kiselev I, Frietsch M, Goschnick J (2004) Temperature gradient effect on gas discrimination power of a metal-oxide thin-film sensor micro-array. Sensors 4:37–46
89. Sysoev VV, Goschnick J, Schneider T, Strelcov E, Kolmakov A (2007) A gradient micro-array electronic nose based on percolating SnO_2 nanowire sensing elements. Nano Lett 7:3182–3188
90. Dolbec R, El Khakani MA (2007) Sub-ppm sensitivity towards carbon monoxide by means of pulsed laser deposited SnO_2 : Pt based sensors. Appl Phys Lett 90:173114-1–173114-3
91. Hernandez-Ramirez F, Tarancon A, Casals O, Arbiol J, Romano-Rodriguez A et al (2007) High response and stability in CO and humidity measures using a single SnO_2 nanowire. Sens Actuators B: Chem Spec Issue: 25th Anniversary Sens Actuators B Chem 121:3–17
92. Kumar S, Murthy JY, Alam MA (2005) Percolating conduction in finite nanotube networks. Phys Rev Lett 95:066802
93. Stauffer DA, Aharony A (1994) Introduction to percolation theory. CRC, London, p 192
94. Sukharev VY (1993) Percolation model of adsorption-induced response of the electrical characteristics of polycrystalline semiconductor adsorbents. J Chem Soc, Faraday Trans 89:559–572
95. Kalinin SV, Shin J, Jesse S, Geohegan D, Baddorf AP et al (2005) Electronic transport imaging in a multiwire SnO_2 chemical field-effect transistor device. J Appl Phys 98:004503-1–004503-8

96. Go J, Sysoev V, Kolmakov A, Pimparkar N, Alam M (2009) A novel model for (percolating) nanonet chemical sensors for micro-array-based E-nose applications. International Electron Devices Meeting, Baltimore, USA, art. 5424266:26.6.1–26.6.4
97. Sysoev V, Kucherenko N, Kissin V (2004) Textured tin dioxide films for gas recognition microsystems. Tech Phys Lett 30:759–761
98. Sysoev VV, Schneider T, Goschnick J, Kiselev I, Habicht W et al (2009) Percolating SnO_2 nanowire network as a stable gas sensor: Direct comparison of long-term performance versus SnO_2 nanoparticle films. Sens Actuators B Chem 139:699–703
99. Goschnick J, Hahn H, Schneider T, Shankar R (2006) Mechanism dependent detection properties of layers based on tin oxide nanoparticles prepared by chemical vapor synthesis (CVS). Proceedings of 11th International Meeting on Chemical Sensors: MP69
100. Caldararu M, Sprinceana D, Popa V, Ionescu N (1996) Surface dynamics in tin dioxide-containing catalysts II. Competition between water and oxygen adsorption on polycrystalline tin dioxide. Sens Actuators B: Chem 30:35–41
101. Weisz PB (1953) Effects of electronic charge transfer between adsorbat and solid and chemisorption and catalysis. J Chem Phys 21:1531–1538
102. Chaim R, Levin M, Shlayer A, Estournes C (2008) Sintering and densification of nanocrystalline ceramic oxide powders: a review. Adv Appl Ceram 107:159–169
103. Tielmann M (2007) Porous metal oxides as gas sensors. Chem Eur J 13:8376–8388
104. Ulrich M, Bunde A, Kohl CD (2004) Percolation and gas sensitivity in nanocrystalline metal oxide films. Appl Phys Lett 85:242–244
105. Sysoev VV, Strelcov E, Sommer M, Bruns M, Kiselev I et al (2010) Single-nanobelt electronic nose: engineering and tests of the simplest analytical element. ACS Nano 4:4487–4494
106. Sysoev V, Strelcov E, Kar S, Kolmakov A (2011) The electrical characterization of a multi-electrode odor detection sensor array based on the single SnO_2 nanowire. Thin Solid Films 520:898–903
107. Bruns M, Frietsch M, Nold E, Trouillet V, Baumann H et al (2003) Surface analytical characterization of SiO gradient membrane coatings on gas sensor micro-arrays. J Vacuum Sci Technol A: Vacuum, Surf, Films 21:1109

Chapter 16
Microhotplates and Integration with Metal-Oxide Nanomaterials

Emanuele Barborini

Abstract The current scenario of metal-oxide gas sensing shows, on one side, highly innovative silicon-based platforms, as outcomes of microelectronic and micromachining manufacturing processes, while on the other side, several techniques and methods for the synthesis of the metal-oxide active layers in the form of nano-porous-nanostructured coatings. The high specific surface area of nanoporous coatings improves the interaction with the atmosphere, while the nanostructure offers characteristic surface-dependent electrical properties. Changes in these electrical properties upon gas exposure and interfacial chemical reactions allow for the development of novel, nano-enhanced gas sensors. The base element of innovative micromachined platforms for gas sensing is the microhotplate. Although micro-hotplates have the same functional parts of traditional devices (integrated heater, electrodes for resistance readout), micromachining provides considerable improvements. These include, for example, the 2–3 orders of magnitude reduction in power consumption for heating: a feature that may disclose the possibility for remote powering through batteries or photovoltaic cells. Moreover, microhotplates originate from the manufacturing track of microelectronics, hence the concept of "system integration" turns out straightforwardly. Within this perspective, the microhotplate may be considered as just an individual component of a many-element sensing platform, including for example, other transducers, or even on-board front-end electronics. Integration concepts are also needed for optimizing the functionalization of the microhotplate with the metal-oxide nanostructured sensing layer, whose batch deposition should become one step of a device production pipeline. As two beautiful countries separated by the sea, with just few bridges in between, difficulties still exist from the point of view of the integration of metal-oxide nanomaterials on micro-hotplates and micromachined platforms in general. In fact, although many different

E. Barborini (✉)
Tethis spa, via Russoli 3, 20143 Milan, Italy
e-mail: emanuele.barborini@tethis-lab.com

M. A. Carpenter et al. (eds.), *Metal Oxide Nanomaterials for Chemical Sensors*,
Integrated Analytical Systems, DOI: 10.1007/978-1-4614-5395-6_16,
© Springer Science+Business Media New York 2013

techniques for the production of metal-oxide nanomaterials have been developed so far, each one of them suffers difficulties, at various degrees, with respect to the fundamental step of microhotplate functionalization. For example, the high temperature step required by certain techniques for stoichiometric oxide synthesis, may be incompatible with microhotplate safety, while the mechanical stress during deposition may result in microhotplate destruction and a subsequent low production yield. The chapter will describe the concepts and the technologies behind microhotplates manufacturing with respect to drawings adopted, chosen materials, and system integration approaches. Techniques and methods for metal-oxide nanomaterial production will be reviewed, highlighting weaknesses and strength points, once they would be employed for microhotplate functionalization. Recent developments on the use of nanoparticle beams to directly deposit nanoporous coatings on microhotplate batches will be included: besides providing thermal and mechanical compatibility with microhotplates, these methods also offer the possibility to synthesize a wide group of different metal-oxides, which is beneficial for an array approach to gas sensing. Relevant examples of sensing performances of microhotplates-based devices will be reported as well.

16.1 Introduction

The phenomenon of a reversible change in electrical conduction of metal-oxides at high temperature, once exposed to airborne chemical compounds, has been known for decades, and has been exploited from the very beginning to make chemical sensors, called conductometric or chemoresistive sensors. More recently, from the middle of 90s, we are experiencing an outburst of development for novel metal-oxide gas sensing solutions, driven by the progresses within each one of the technological players composing the "sensing system team": the electronics, required to deal with the electrical features of chemoresistive sensors; the software, in charge of sensor data analysis by means of "artificial intelligence" and advanced algorithms; and mostly the device platform and the active sensing material, constituting the transducer that detects the presence in air of chemical compounds. In fact, on one hand highly innovative silicon-based platforms became available from the union of microelectronic and micromachining manufacturing processes, while on the other hand novel methods for the synthesis of the metal-oxide active layers in the form of nanoporous coatings were introduced. The high specific surface area of nanoporous coatings provides an efficient interaction with the atmosphere, while the nanostructure determines the strongly surface-dependent electrical properties, and both characteristics lead to improved sensing performances.

Behind these technological efforts, there is the understanding that sensing systems may have pervasive applications if further improvements of issues such as miniaturization, reduced power consumption, portability, etc. as well as reliability of production pipeline, will be reached. Such advanced microsensors could become the basic elements for the development of complex detection systems for real-world scenarios, where the heterogeneity of operative conditions, from the point of view of the chemical composition of the detection atmosphere, and of the distribution of chemical compounds in space, is a typical monitoring situation.

An example of a crowded class of applications is atmospheric monitoring by sensor networks. This includes outdoor opportunities in metropolitan air quality monitoring or in spatially resolved early detection of environmental chemical risks (either from human activities such as airborne leakage from industrial sites or chemical spills in rivers, or from natural events such as emissions from active volcanic sites). Indoor opportunities exist as well and may include domestic applications for air quality in houses, the monitoring of atmosphere within industrial buildings where processes are executed in atmosphere-controlled conditions (such as industrial agro-food production and storage), up to the monitoring of microclimatic physical-chemical conditions within cultural heritage sites to ensure optimal preservation. In all these cases what is needed is a large number of miniaturized devices and dedicated electronics, both having low-cost and low-power consumption. These systems would provide a network of sensing nodes which would be able to detect chemicals over large areas with meaningful spatial resolution.

A second example with pervasive character is within the framework of the so called artificial olfactory systems or electronic noses, where the final goal is the recognition and quantification of different compounds (or at least their chemical family) within complex mixtures. The sensing system acting as an electronic nose should be able to conciliate unspecificity, i.e. a sensitivity extended to many different chemical compounds, with selectivity, to recognize each one of them. The approach to face this task, requiring two apparently incompatible properties, is based on microsensor arrays, having many different sensing elements whose collective response is interpreted by a trained intelligent software (principal component analysis, neural networks, etc.). In this case, the integration of a suitable number of different miniaturized sensing elements on the same platform is the issue to be addressed. An electronic nose may find important applications, for example in the food industry. Specifically, these may include the rapid characterization of product freshness, proper storage, ripening degree, and geographic origin. Other challenging applications would be within the biomedical field for assisting in disease diagnosis through breath analysis.

Due to the principle of operation of metal-oxide sensors, whose electrical resistance at high temperature changes according to the presence of reactive chemical compounds within the surrounding atmosphere, the parts constituting traditional devices include: a ceramic platform, a pair of electrodes for resistance measurements on one side of the platform, the sensing metal-oxide layer across the electrodes, and an integrated heater on ceramic platform backside, as shown in Fig. 16.1.

Although the generic term "metal-oxide" is commonly used throughout this chapter, it is worth noting that the state-of-the-art in the production of metal-oxide nanostructured layers is generally limited to a few metal-oxides, namely SnO_2, WO_3, TiO_2, among a few others. However, SnO_2 is by far the species that has been

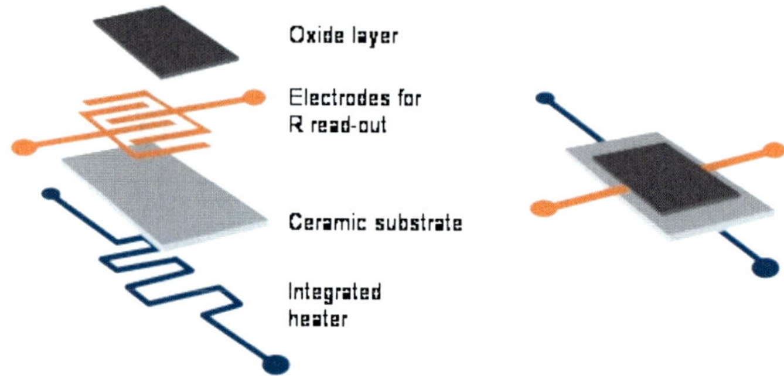

Fig. 16.1 Structure of a traditional chemical sensor based on metal-oxide sensing layer. A ceramic platform (typically alumina) has electrodes patterned on its *front-side*. The electrodes are coated with a metal oxide layer and are used for read-out of changes in the metal oxide electrical resistance. On the backside, an integrated heater provides high temperature operation

used the most. Heating power on the order of several hundreds of milliwatts is usually spent to heat up the ceramic platform to 300–400 °C, while thermal insulation is achieved by suspending the platform to bonding wires, looking for a delicate compromise between lowering thermal conduction through wires, reducing their diameter, and maintaining a sufficient wire robustness.

The base element characterizing innovative silicon-based micromachined platforms is the microhotplate: a structure constituted by a sub-millimetric sheet with a thickness of a few micrometers, hosting the electrode pair for sensor read-out and the integrated heater. Although microhotplate components implement the same functions as the parts within traditional devices (Fig. 16.1), micromachining and miniaturization provide considerable improvements. First, minimal heated area and reduced thermal contact at borders results in heating power consumption of a few tens of milliwatts (one or two order of magnitude lower than traditional platforms). This improvement opens up the potential for powering through remote sources such as batteries or photovoltaic cells. Second, the minimized thermal mass results in rapid heating and cooling times which enables modulated temperature operation mode (or temperature programmed sensing). This leads to advantages from the point of view of power consumption as well with gas selectivity exploited through the reactivity temperature dependence of the target compounds. Third, since microhotplates are manufactured using microelectronic processes, it is quite natural to address "system integration". In this framework the microhotplate is an individual component within a complex miniaturized sensing platform, having for example, other microhotplates for an array approach, or other micromachined transducers (air temperature, flow, pressure, etc.), or even on-board front-end electronics.

Unfortunately, besides these intriguing features, difficulties still exist from the point of view of the integration of active metal-oxide nanomaterials. In fact, although many different techniques for the production of metal-oxide layers have been developed so far (including either thin and thick films), each one of them, at various degrees, introduce difficulties with respect to the fundamental step of microhotplate functionalization. These difficulties may range from high temperatures required by certain techniques for stoichiometric oxide synthesis, which are incompatible with the microhotplate components, to mechanical stresses during deposition, that may result in the destruction of the microhotplates and a low production yield. Therefore, the transition from lab-scale to market-scale of sensing technology based on microhotplates is still hampered by the lack of a reliable, cost effective, and most of all a microhotplate compatible method for large-scale integration with metal-oxide nanostructured sensing layers.

Throughout this chapter we will use the term "integration" to indicate the deposition onto micromachined platforms of the active metal-oxide layer. By adopting this term we intend to underline that the functionalization of these platforms is not a merely deposition of a thin or thick film on a flat substrate: it is one task of a sequence of complex tasks leading to the final, operative sensing device. Within this meaning the functionalization has to be carried out taking care of constraints related to other tasks, and synergistically with them.

Besides the *Introduction* and the *Conclusions-Perspectives*, the contents of this Chapter are arranged into the following sections.

Microhotplates and Micromachined Platforms, where the technology behind microhotplates and micromachined platforms manufacturing will be reviewed with respect to drawings adopted, chosen materials, and system integration.

Nanomaterials Integration/Functionalization Methods, where deposition methods will be addressed, pointing out, among different deposition techniques, advantages and disadvantages once they would be employed for microhotplate functionalization. Recent developments on the use of nanoparticle beams to directly deposit nanoporous coatings on microhotplate batches will be included: besides providing thermal and mechanical compatibility with microhotplate, these methods also offer the possibility to synthesize a wide group of different oxides, which is beneficial for an array approach to chemical sensing.

Sensing Performances, that will show the most relevant characteristics of microhotplates-based chemical sensors

16.2 Microhotplates and Micromachined Platforms

16.2.1 Basic Structure Description

Many different designs of microhotplates have been developed so far, however, their general structure is the same: a thin sheet of silicon oxide or nitride, which is suspended within a bulky silicon frame (for this reason microhotplates are also called "suspended silicon membranes"). Above this sheet, all the elements required for sensor operations are stacked in a multi-layered configuration. These include the sensing layer, the electrode pair for the read-out of the sensing layer resistance, a thermometer, and a suitable heater. Passivation layers provide electrical insulation

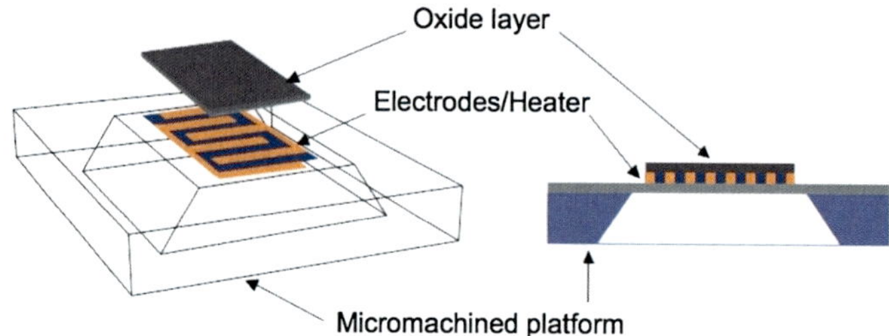

Fig. 16.2 Schematic of a microhotplate-based chemical sensor. Functional elements described in Fig. 16.1, namely the oxide layer, the electrodes and the heater, are also present within this type of devices (for simplicity, thermometer and passivation layers are omitted). Microelectronic techniques are adopted for the creation of the functional elements, while micromachining techniques are adopted to obtain the miniature base sustaining them. Typical dimensions of a microhotplate are sub-millimetric

between the heater, thermometer, and the sensing layer. Figure 16.2 schematically shows the structure of a microhotplate-based chemical sensor.

Due to the minimal thermal contact at the border along with the minimal heated area (resulting in a minimal contact, convective and radiative heat transfer) a microhotplate is able to reach temperatures up to 300–400 °C with a power consumption of the order of a few tens of milliwatts. The integrated heater typically has a serpentine-like shape, to ensure uniform heating. Microhotplate temperature may be estimated by means of the resistance of a dedicated metal film, acting as a thermometer, which is located between the heater and the sensing layer. If proper characterization of the heater is available, the measure of its resistance by suitable electronics could also be used for temperature evaluation, avoiding the use of a dedicated thermometric element (and simplifying manufacturing). A useful configuration of the electrode pair is the so called interdigitated one, where electrodes look like a pair of many-fingered hands co-penetrating each other, as visible in Fig. 16.2.

Microhotplate manufacturing exploits a microelectronic-like process and a micromachining process, using silicon wafers as the substrate. The first one, sometimes addressed as CMOS-like process (see below), regards the production of the functional elements described above—heater, thermometer, electrodes, and passivation layers. While the second one regards the chemical etching of bulk silicon in order to remove material and generate the final micrometer-thick membrane. The union of microelectronic and micromachining processes leads to the possibility of creating complex 3D structured devices generically called MEMS (micro-electro-mechanical systems), of which microhotplates are a paradigmatic example. We will see in the following some examples of microhotplate-hosting systems, where the device structure has been tailored by micromachining processes to address particular requirements. Specifically, these include having a variety of transducers on the same platform, or providing a temperature-array chemical sensor to exploit the temperature dependent chemical reactivity of different target compounds.

16.2.2 Manufacturing Route

Figure 16.3 reports the step sequence for manufacturing a microhotplate starting from a silicon wafer. Details may be found in several references such as, for example, refs. [1–3] and those therein. It is worth to note that these steps are very similar to those currently used in the microelectronic industry for microchip manufacturing.

Before describing the method and the steps to create the functional elements on the front-side of the microhotplate, it is necessary to anticipate now a concept related to the micromachining step, the last one of the manufacturing route. Bulk silicon removal is performed by chemical etching in liquid-phase by means of potassium hydroxide (KOH) or ethylenediamine and pyrocatechol in aqueous solution (EDP), at the end of front-side manufacturing. Etching is carried out on the wafer backside and removes as much bulk silicon as to preserve a micrometer-thick sheet having functional elements already created on it. The precise control of the chemical etching process is based on the knowledge that silicon oxide or silicon nitride cannot be removed by either KOH or EDP.

The first step of front-side manufacturing is therefore the oxidation (or nitridification) of a suitable layer of silicon wafer, as shown in Fig. 16.3, that will serve as the stop for chemical etching from the back-side. In addition, silicon oxide and silicon nitride possess a low thermal conductivity (1.4 and 9–30 $Wm^{-1}K^{-1}$ at 300 K, respectively [4]), a benefit for thermal insulation of the final structure. Although differences do exist, subsequent steps may be considered as standard steps of CMOS-like microelectronic production: a sequence of photolithography, thin film deposition and lift-off steps, leading to the formation of the multi-layered top structures (heater, thermometer, electrodes, and passivation layers).

One of the differences with respect to standard CMOS processes used in microelectronics is the metal to be used for heater, thermometer, and electrodes, which is usually platinum. Metallization for standard CMOS processs uses aluminum, nevertheless its limited thermal stability, causing either oxidation and atom diffusion at temperatures around 500 °C, prevents its use in microhotplates. As a noble metal, platinum offers much larger thermal stability than aluminum, it can be easily deposited, by sputtering or evaporation, and patterned by lift-off. Due to the linear behavior of resistance versus temperature, according to $R_T = R_0 \cdot (1 + TCR \cdot \Delta T)$, where R_0 is the resistance value at room temperature, R_T is the value at high temperature, TCR is the temperature coefficient of resistance, and ΔT is the temperature difference, platinum can be effectively used as material for integrated thermometric measurements. Typical TCR values of platinum are around $2 \times 10^{-3} \ K^{-1}$.

An alternative to platinum heaters is poly-crystalline silicon (polysilicon) heater, which can be introduced into the CMOS process in straightforward way, as it requires standard low-pressure chemical vapor deposition (LPCVD) as a deposition method, and standard photolithography for patterning. An advantage of polysilicon is the possibility of resistivity tuning by n-type phosphor doping, using $POCl_3$. Although much variable, TCR values of polysilicon are commonly around $0.5 \times 10^{-3} \ K^{-1}$.

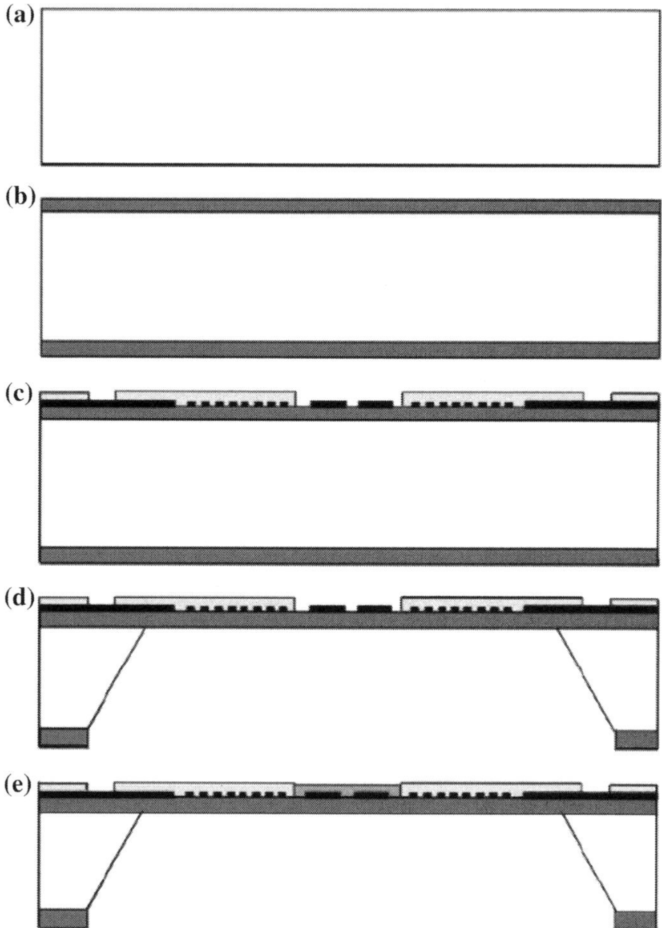

Fig. 16.3 Process steps for the production of a microhotplate starting from a silicon wafer: **a** silicon wafer with about 0.4 mm thickness; **b** dielectric layer deposition (SiO_2 or Si_3N_4); **c** deposition of metals for electrodes and heater (*black* parts), and of silicon oxide or nitride for passivation (*light grey* parts); **d** backside anisotropic etching by KOH; **e** deposition of active metal-oxide layer. Reprinted from Ref. [1] with permission from Elsevier

Electrodes for the read-out of the sensing layer resistance may be planar or interdigitated. Since metal-oxide nanostructured layers typically show very high resistance values (reaching hundreds of megaohms and more), the peculiar drawing of interdigitated electrodes has the advantage of facilitating the resistance measurements, with respect to simple planar electrodes, decreasing the read value by means of a purely geometrical argument. Referring to Ohm's 2nd law and to Fig. 16.4, the resistance of a square ideal conductor as measured by planar electrodes is $R = \rho \frac{1}{th}$ (where ρ is the material resistivity and th is the conductor

Fig. 16.4 The read-out of the resistance of metal-oxide layers may be hampered by its extremely high values. For a given area of active material, it turns out, by geometrical arguments, that interdigitated electrodes (*right*) are more effective than planar ones (*left*)

thickness) while in the case of interdigitated electrodes, resistance of the same conductor is $R \approx \rho \frac{1}{(2 \times N)^2 t h}$ (where N is the number of fingers in one electrode). The factor, $(2 \times N)^2$ results in a considerable decrease in measured resistance.

To electrically separate the three circuits for the heater, thermometer, and electrodes for sensing layer read-out, silicon oxide layers are deposited between each metallization by CVD, at a thickness typically not larger than 0.5 μm. Silicon nitride deposited by CVD (for example from ammonia and dichlorosilane) can alternatively be used for the same purpose.

When functional structures on the front-side are completed, the wafer undergoes micromachining through back-side chemical etching. Exploiting the phenomenon of anisotropic chemical etching, where the etching rate of a crystalline material depends on the crystal face exposed to the etchant, a pyramid-like pit is created, starting from the wafer backside and moving toward its front-side. Once the etching front reaches the silicon oxide (or nitride) layer, it stops. This results in the final structure having a trapezoidal cross section as shown in Figs. 16.2 and 16.3. The ratio between the etched thickness and the enlargement of the etched region with respect to final size of the suspended part is related to the anisotropy of etching rates, that is the ratio between normal-to-plane and in-plane etching rates. If structures with a higher aspect-ratio need to be produced, wet chemical etching has to be replaced by plasma etching.

Backside etching has the capability to create structures having thicknesses as small as 1–2 μm, nevertheless it has the drawback of being a double-sided process, making mask alignment more difficult than one-sided standard lithography, as usually done in microelectronics. To overcome this limitation front-side chemical etching can be introduced. An additional photolithographic step on the wafer front-side leaves four unprotected areas (etch windows) beside each of the microhotplate sides [5], as shown in Fig. 16.5. Front-side etching removes silicon from unprotected areas, starting from the front-side and moving under the central area hosting the functional structures. Once the etching step is concluded, the microhotplate will stay suspended by means of suitable beams over a closed pit etched into the bulk silicon. Backside etched microhotplates have typical dimensions between 1×1 and 2×2 mm^2, while front-side etched ones may be quite smaller (0.2×0.2 mm^2).

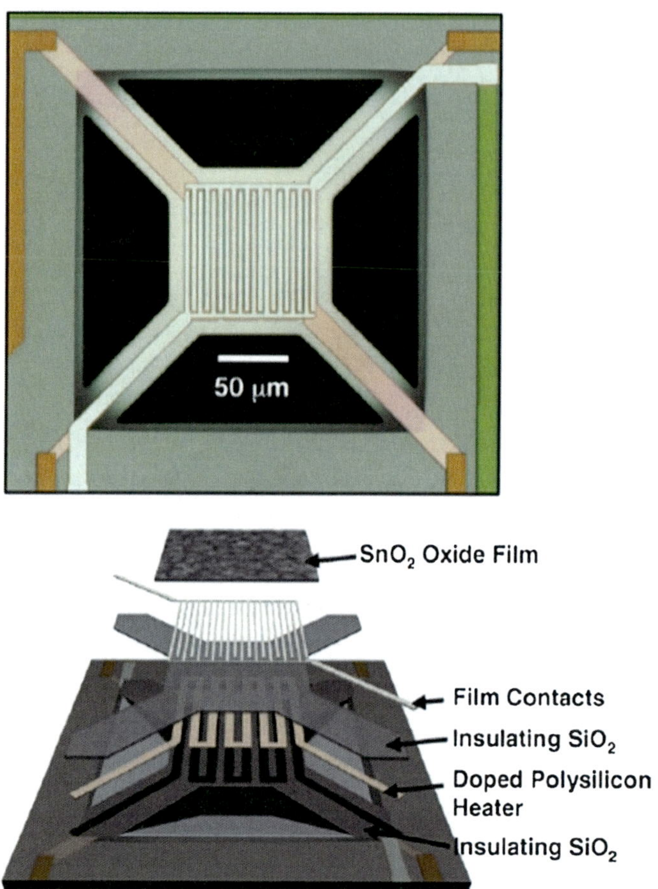

Fig. 16.5 Optical micrograph showing an example of a suspended membrane (*top*), and a lift-away scheme showing the functional elements composing the device (*bottom*). Reprinted from Ref. [5] with permission from Elsevier

Since so many details addressing each one of the manufacturing steps, from metallization to passivation, from wet etching to plasma etching, are reported in the technical literature, the general feeling a reader may have is that a large degree of freedom and space for novel experimental approaches exist on this matter. This also holds for the implemented geometries, such as the shape of the heater or that of electrodes, up to the general structure of the devices.

16.2.3 Advanced Structures

From the point of view of system integration, microhotplates could be considered as the first building block in the development of advanced platforms for sensing, based on the integration of various transducers. For example, several microhotplates could be arranged into the same platform exploiting an array approach to sensing; a microhotplate could be machined side by side with physical transducers (temperature, pressure, flow), on the same silicon platform, for hybrid physical-chemical microsensors; the silicon platform hosting the microhotplate could also host CMOS front-end electronics for impedance matching, signal pre-amplification, and data analysis.

One of the drivers for the creation of complex platforms resides in the possibility to increase the amount of simultaneously collected data, with spatial resolution, improving our capabilities to face the challenging task of physical-chemical characterization of atmospheres. An array of chemical sensors in fact generates many signals qualitatively similar each other, although not identical. By exploiting this collective output, it could be possible to associate a certain atmosphere to a certain signal fingerprint. An array-based approach to face complex scenarios requires to pass through a training step where advanced software is needed to determine the correlations among a known atmosphere composition and the array collective output. Once training is completed and correlations identified, the system should be ready not only to recognize the atmospheres seen during training, but to perform a certain degree of "extrapolation" to identify unknown atmospheres. While we do not cover this subject in this chapter, the reader is referred to other chapters within this text. Regardless, the starting point to deal with many-signals is the availability of suitable array-like platforms.

A first example of a sensor array platform is a microhotplate where the heater has been designed in a suitable way to generate a temperature gradient under a multi-electrode pattern [6], as shown in Fig. 16.6. Although this is still a single microhotplate, and will be functionalized with the same metal-oxide layer, the combination of multi-electrodes and a temperature gradient allows the read-out of portions of the active layer operating at different temperatures. This produces a sensor array which utilizes the temperature dependent reactivity of gases, to provide a multi-signal output.

A second example of multi-signal sensor array platform is a true microhotplates array, where the microhotplate structure is replicated many times on the same device [3, 7], as shown in Fig. 16.7. Operation at different temperatures can be pursued also in this case, nevertheless this setup is particularly suited to using different metal-oxides as sensing layers. Since it is not trivial at all to be able to produce and integrate on this kind of platforms many different metal-oxides, a commonly adopted approach is doping. In this case, the different sensing elements are obtained through the introduction of different noble metals into the same metal-oxide, which is typically tin oxide [8–10]. This configuration is a "material

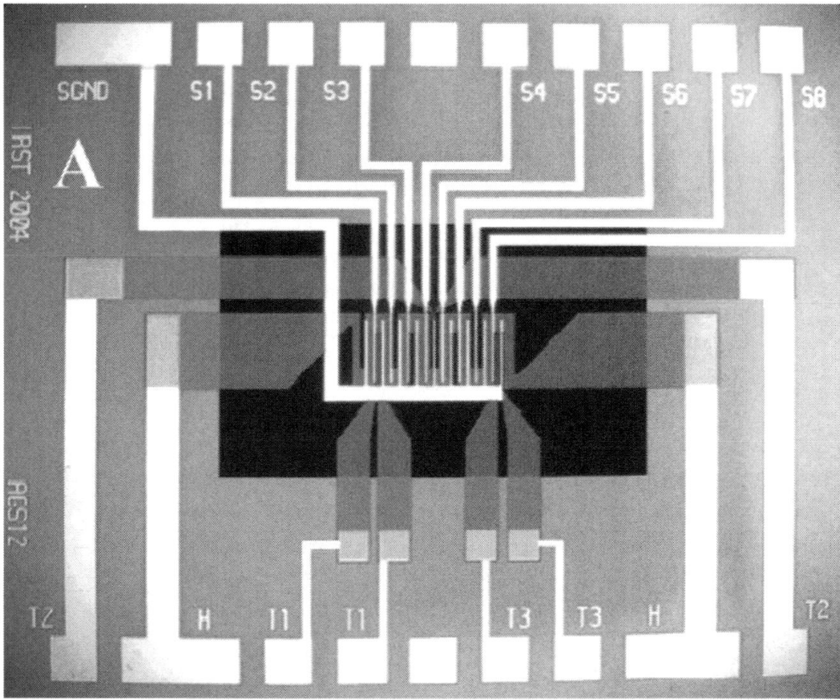

Fig. 16.6 Example of a micromachined platform where the heater (horizontal serpentine pattern in the center of the image) has been manufactured in order to generate a temperature gradient. A multi-electrodes pattern (white interdigitated structures) collects multiple signals from the same sensing material operating at different temperatures. Reprinted from Ref. [6] with permission from Elsevier

array", exploiting the different surface chemistry of gases with respect to different oxides (or different dopants), to generate a multi-signal output.

Physical sensing transducers can also be miniaturized through micromachining. Hence, beyond bare chemical sensing, micromachining techniques allow for the integration on the same platform of both chemical and physical transducers such as temperature, air flow, and pressure, to yield a multiparametric sensor. For example air flow monitoring by "hot-wire anemometry" can be performed with the micro-hotplate approach. In this case a central heater has at its sides, along the direction of the flow to be measured, two platinum thermometers; in the absence of air flow the temperature measured at the heater sides is the same; once air starts flowing a temperature mismatch between thermometers on the heater sides is revealed. Figure 16.8 (top) shows a multi-transducer platform hosting, besides the chemical sensor, a thermometer, and a micromachined hot-wire anemometer, where backside etched microhotplates are used for both the former and the latter. Figure 16.8 (bottom) shows the same kind of multiparametric platform, where micromachined

Fig. 16.7 Array of microhotplates on the same device platform. Each single element has the structure shown in Fig. 16.5, and is individually addressable and controlled in temperature. Reprinted from Ref. [7] with permission from Elsevier

structures of either the chemical sensor and the anemometer were obtained by front-side anisotropic etching, generating suspended "microbridges".

The last example of advanced sensing platforms achieved through system integration processes regards devices having on-board electronics (see Fig. 16.9 of next Section). By exploiting the affinity of micromachining methods for micro-hotplates manufacturing with CMOS methods for microelectronics production, the same silicon platform may host the transducers for chemical sensing, the ana-logical front-end electronics for signal pre-amplification, as well as the digital electronics for data pre-analysis [11, 12]. The technological driver for these efforts relies on signal handling, which gain strong benefits with respect to noise reduc-tion if connections between transducers and front-end electronics are kept as short as possible. A second technological driver deals with the possibility to distribute at least part of the data analysis, in the case of multi-device networks.

Unfortunately, with the increased platform complexity, comes an increase in the challenges for integration of metal-oxide nanomaterials on chemical sensing areas. This means for example, that some of the deposition methods may be difficult or impossible to adopt, or—more generally—that production yield could dramatically drop due to device damage during the manufacturing pipeline.

16.3 Nanomaterial Integration/Functionalization Methods

16.3.1 General Issues

The pipeline leading to operative sensing devices is a sequence of steps, namely: (1) CMOS-like processes for front-side manufacturing; (2) backside (or front-side) micromachining of suspended parts by chemical etching; (3) integration of active metal-oxide nanomaterial; (4) wafer dicing, bonding and packaging. This Section addresses the integration step, with the aim to give the reader a framework of the most important methods for metal-oxide nanomaterial deposition on microhotplates. It is worth to underline that many approaches that have adopted so far, each has their own strengths and its weaknesses. For this reason a well established and reliable route for the production of microhotplates-based gas sensors is still lacking and several approaches for functionalization are being developed and optimized. Within this competition, many elements have to be considered, from the level of compatibility with microhotplate characteristics to scalability for large-scale batch deposition, from batch-to-batch reliability to versatility in the synthesis of many different metal-oxides.

Concerning this last point, it has to be noted that the metal-oxide that is most widely used is by far SnO_2; oxides such as TiO_2, WO_3, ZnO, In_2O_3 also play a certain role in gas sensing, while others have just a marginal one. As anticipated, a commonly adopted way to extend the number of sensing materials is the use of various concentrations of noble metals (Au, Pt, Pd) as "dopants" or surface catalysts of the above oxides.

In general, the three most important features of a microhotplate to be taken into account to tailor the use of a certain deposition technique for metal-oxide layer integration are: (1) mechanical delicacy, (2) limited temperature of overall thermal treatments, and (3) the need for patterning. We already put emphasis on mechanical delicacy of MEMS, and microhotplates in particular. For this reason we are forced to exclude any deposition technique requiring mechanical contact with the suspended parts, otherwise this would frequently result in microhotplate damage, i.e. device damage, and unacceptably low production yield. Some integration techniques may require a post-deposition high temperature calcination step to complete the synthesis of the metal-oxide, and this may be incompatible with the preservation of front-side metallization. Particularly in complex, many-element platforms, the deposition of metal-oxide layer has to be spatially limited, i.e. patterned, to the region of heater and read-out electrodes, avoiding the contamination of other parts within the platform (other transducers, on-board electronics, etc.).

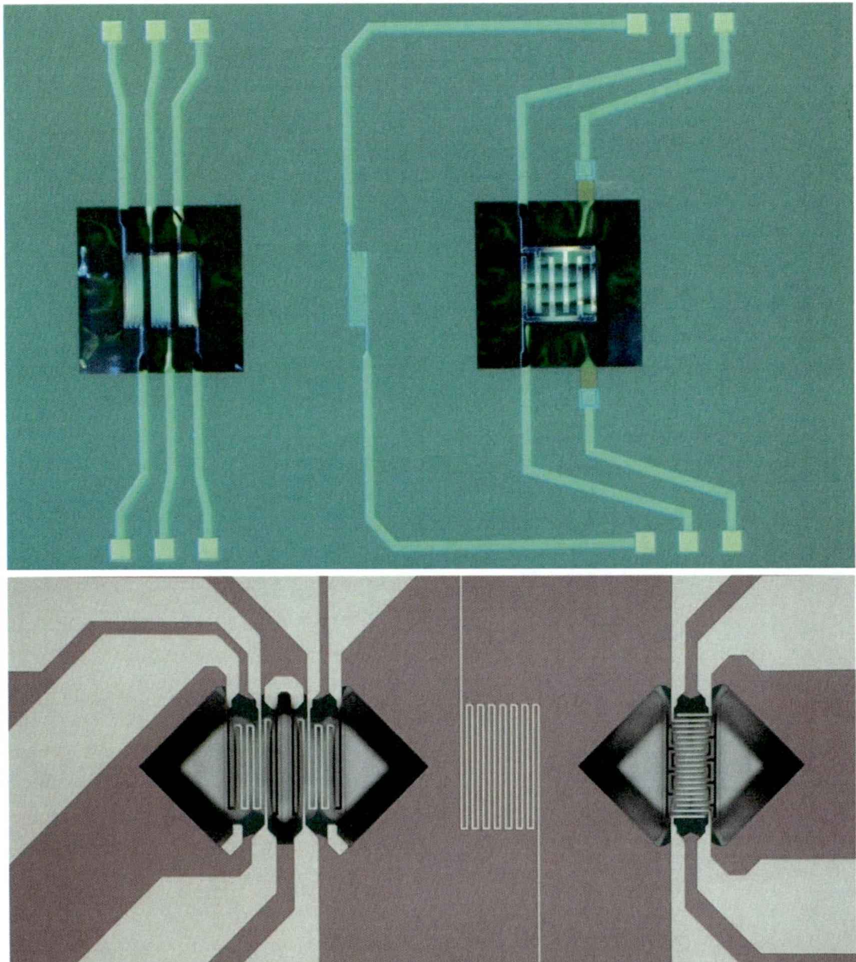

Fig. 16.8 Examples of multiparametric platforms hosting, from *left* to *right*, a micromachined hot-wire anemometer, a thermometer, and a chemical sensor. In *top platform*, backside etched microhotplates are used for anemometer and chemical sensor (courtesy from M. Lozano, CNM-Barcelona). In *bottom platform*, front-side etching was adopted, generating suspended micro-bridges (courtesy from L. Lorenzelli, FBK-Trento). In both cases the size of the entire platform is around 3×5 mm^2

Even if the step sequence reported at the beginning of this section is the most common one, alternatives do exist, where the order of the steps is modified. For example, the metal-oxide deposition step could be performed before micromachining to overcome membrane mechanical delicacy. Nevertheless, this approach imposes the need to properly protect the active layer during the subsequent etching step. As another example, last step dicing-bonding-packaging may be anticipated before metal-oxide integration. This allows for electrical connection of the microhotplate during deposition and in this way, one can exploit the selective deposition on the heated parts occurring via reactive sputtering, a precise material patterning onto the dedicated part of a microhotplate array is achievable by switching-on the desired heater [3, 13] Although this approach may be quite useful at the research level, in the case of complex arrays, it suffers scale-up limits once batch deposition is required for large-scale production, due to difficulties in making electrical connections of all the microhotplates within a given process batch.

Concerning the integration step itself, no matter where it is located within the sequence of manufacturing steps, it turns out from the literature that borders between the characteristics of various deposition techniques, methods, and approaches are often shaded, and overlapping between topics may frequently occur. Accordingly within the structure of this chapter, the following deposition techniques have been divided between those based on wet-chemistry methods and those based on gas-phase, or physical, methods. Among gas-phase methods, a novel approach to microhotplate functionalization based on the deposition of nanoparticle beam will be also presented.

As a rough overview, generally speaking gas-phase methods lead to thin films (below some hundreds of nm), while wet-chemistry methods more commonly lead to thick films (from a few to several tens of μm); the former generate compact films, the latter porous ones. Gas-phase methods offer scale-up possibility through parallel functionalization in batch depositions. Attributes of this method include the need for patterning, the compatibility with the mechanical characteristics of the microhotplates, and the possibility of high production yields. Gas-phase methods also require an expensive vacuum apparatus and hardware. Nevertheless these setups are quite similar to systems already in use in microelectronic industry, where microhotplates come from, and their introduction within the manufacturing process is therefore not so incompatible. Wet-chemistry methods such as drop coating are serial, do not require patterning since droplets containing film precursors are dispensed one by one on the desired position. On the contrary, wet-chemistry methods such as spin coating lead to parallel functionalization and do require patterning. In both cases—drop coating and spin coating—they are not mechanically safe for microhotplates, hence technological solutions providing particular care, or a manufacturing process having micromachining after deposition are required. Although wet-chemistry methods require cheap instrumentation, hardware, and chemicals, their approach is quite far from techniques commonly used within microelectronic manufacturing, raising compatibility issues with respect to the introduction of such a step into the clean-room grade manufacturing process of microhotplates.

The general issue of functional nanomaterial integration with micromachined platforms and systems—the so called Nano-on-Micro issue—is generating new visions, where system integration concepts are driven well beyond current "microelectronic meaning", up to cover the synergistic interaction of processes and methods leading to the production of complex devices, as it will be illustrated in the following.

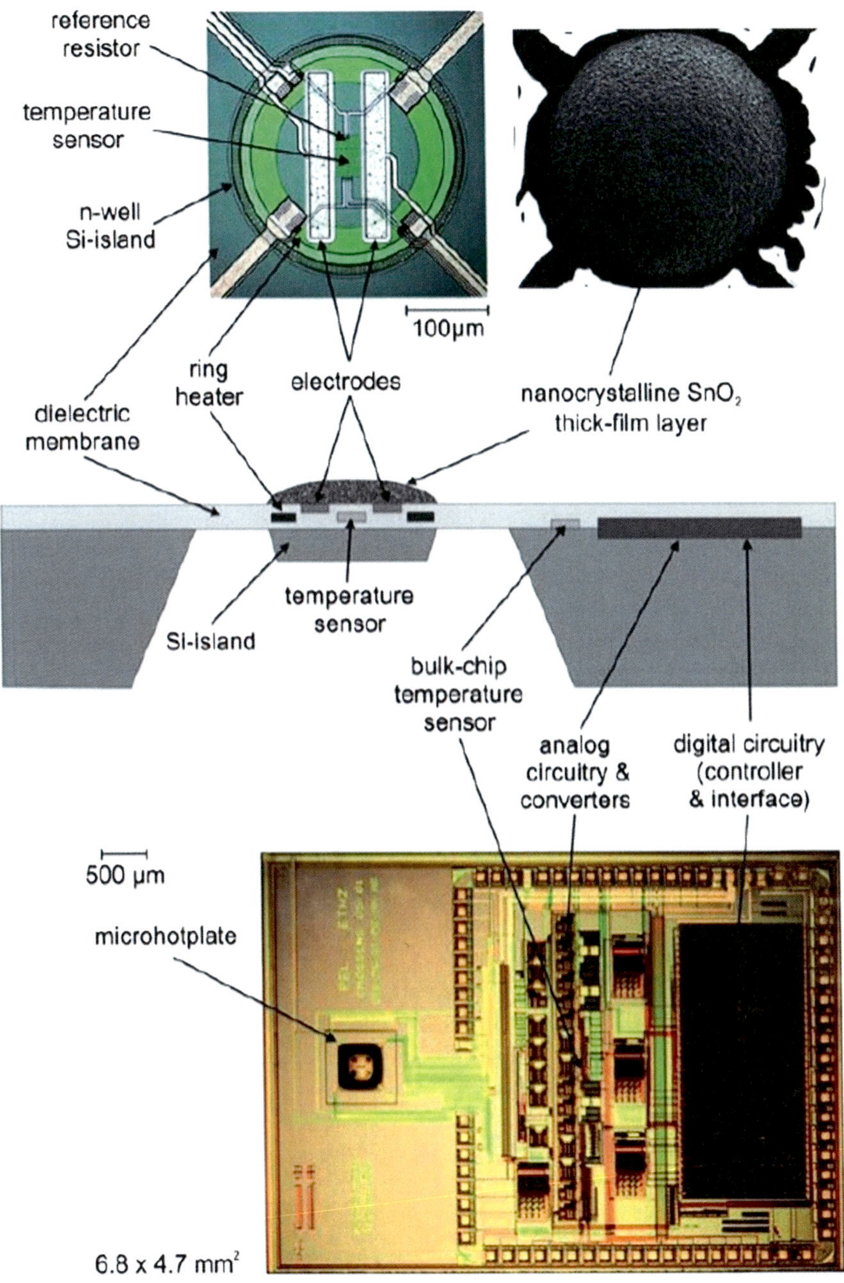

Fig. 16.9 High complexity micromachined platform for chemical sensing hosting on-board electronics. On *top*, details of the microhotplate and sensing material, which has been integrated on the microhotplate by drop coating; in the *middle*, device cross-section; on *bottom*, micrograph of the CMOS-based overall sensor system chip featuring microhotplates and circuitry. Reprinted from Ref. [14] with permission from Springer

16.3.2 Wet-Chemistry Methods

Wet-chemistry methods include various approaches to the synthesis of metal-oxide nanomaterials, namely spin coating, drop coating, dip coating, screen printing, etc., as clearly schematized in ref. [14]. In each one of these approaches, a liquid-phase or paste-form precursor of metal-oxide is added to sensor platforms in form of droplets, or in form of a continuum layer by dropping, spinning, or dipping. Advantages and drawbacks of wet-chemistry methods with respect to microhotplates integration will be highlighted below.

Starting from an initial common step consisting in the preparation of a stable suspension of a suitable precursor of the metal-oxide, two main routes can be detailed. One route passes first through the preparation of a sol-gel, then through the deposition/dropping onto the sensor platform, and finally through a post-deposition annealing step at high temperature, transforming the sol-gel into stable metal-oxide nanoparticle film (calcination). The other route, instead of creating a sol-gel from an initial suspension, anticipates the calcination step: this results in a stoichiometric and stable metal-oxide nanopowder that undergoes subsequent re-suspension into a liquid solution or into a viscous paste, and finally deposition/dropping onto the sensor platform. The difference between these routes is where the calcination step is carried out: in the latter case it belongs to paste preparation before integration onto devices, while in the former case it has to be done on microhotplates. As we already have discussed, sensing platforms may host, besides microhotplates, other components (CMOS electronic circuitry, for example) which may be incompatible with the high temperature thermal treatment required for calcination. Therefore the only way to complete nanomaterial preparation passes through the use of the integrated heater. If this is the case, not only microhotplate structure has to be as robust as it is required to survive a calcination step exploiting integrated heater temperatures of 500 °C and above, but also dicing-bonding-packaging step has to be concluded to allow the connection of the device with heater power circuitry. This "firing" step could be done at the level of manufacturing or by end users. Nevertheless, since it may deeply affect the performance of the final device, firing has to be done in well controlled and reliable conditions. Here we touch the real meaning of the "integration" concept: if a certain method for nanomaterial deposition is chosen, a certain manufacturing route in device production has to be followed.

Among wet-chemistry approaches, drop coating is the only one definitely skipping the problem of patterning: droplets with suitable size of liquid-phase precursor may be dispensed exactly onto the read-out electrodes, by means of micropipettes or microinjectors, also in the case of highly complex platforms, as shown in Fig. 16.9.

Once drop coating onto the microhotplate wafer is completed, the manufacturing route proceeds with subsequent dicing-bonding-packaging step; depending on the nature of the deposited precursor (sol-gel or calcinated paste), the additional firing step could be required or not.

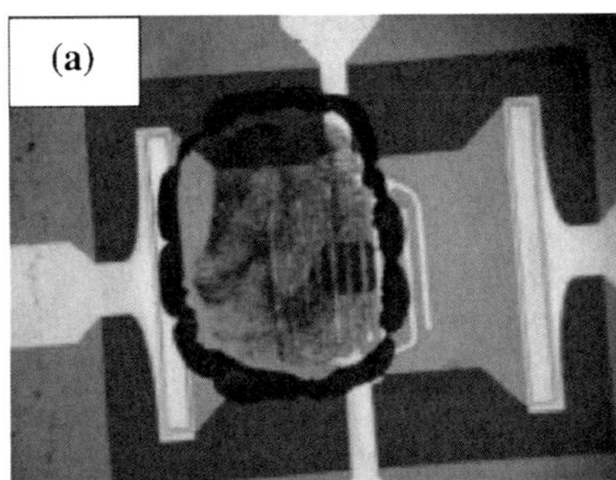

Fig. 16.10 a Difficulties in positioning and adhesion of metal-oxide layer by screen printing (SnO_2 thick film with size of $350 \times 500 \ \mu m^2$). **b** Result improvement by proper paste formulation. Reprinted from Ref. [19] with permission from Elsevier

Drawbacks of drop coating approach regard mechanical delicacy and reliability of the electrical and sensing characteristics of metal-oxide nanomaterial. The issue of mechanical delicacy concerns the way the droplets are delivered onto microhotplates: although technologically resolvable (think about the evolution of ink jet printing heads) this operation is intrinsically critical and may result in low production yield due to high rate of microhotplates damage. The issue of reliability of electrical/sensing characteristics originates from the serial nature of drop coating, where droplets are delivered one-by-one. Paste—or suspension—itself and deposition method are required to maintain stable characteristics during the whole deposition step: events such as agglomeration, precipitation, or progressive clogging of the dispensing system have to be avoided.

The nanotech evolution of drop coating is dip-pen nanolithography (DPN) [15]. It is based on the use of the tip of an atomic force microscope (AFM) to deliver a minimal amount of a suitable "ink", containing the metal-oxide precursor. Although limited so far to research labs, DPN could be used for extreme miniaturization of devices, avoiding any patterning procedures, and fulfilling safe conditions for microhotplates survival.

Even if literature reports successful examples [14, 16–19], other wet-chemistry approaches beside drop coating (from spin coating to screen printing) are still substantially unusable for microhotplate functionalization, due to the lacking of a safe and reliable method for patterning. Both contact hard mask and photolithographic soft mask methods in fact are difficult to implement, the former for mechanical reasons, the latter for chemical ones. Chemical incompatibilities arise due to the precursor solvents that may react with the polymeric resist of the soft mask. Likewise photoresist removal may result in damage or contamination of the sensing layer. For example, Fig. 16.10 shows difficulties in positioning and adhesion of metal-oxide layer by screen printing, and how results can be improved by proper paste formulation [19].

A completely different situation takes place if integration of metal-oxide nanomaterials is introduced before the chemical etching step for the formation of suspended structures, as shown for example in refs. [20, 21] Although such an approach is hampered by the necessity to have a wet-chemistry step between manufacturing of front-side structures and the micromachining step, it completely rules out any difficulty related to delicacy of suspended parts. Therefore, patterning by hard mask method is no longer an issue. On the contrary, patterning may still be a problem in the case of soft masks, for chemical incompatibilities that may exist between the photoresist and the precursor solution. Efforts trying to overcome this point are reported for example in ref. [22].

Once the nanomaterial has been deposited, backside micromachining by chemical etching has to be performed, which will require the proper protection of front-side active layer with respect to etching itself and for minimization of contamination.

16.3.3 Gas-Phase Methods

Gas-phase, or physical, methods include all standard methods for thin film deposition, namely vacuum evaporation, sputtering, pulsed laser deposition (PLD), physical as well as chemical vapor deposition (PVD-CVD). Here we do not deeply describe these techniques, whose details may be found in several publications and handbooks, such as for example refs. [23–26]. We will recall the general characteristics of the techniques and show how they could match requirements for integration of nanomaterials onto microhotplates. In contrast to wet-chemistry methods, the combination of gas-phase methods, such as vacuum evaporation or sputtering, with photolithography are standard approaches used within microelectronic manufacturing. It turns out that the introduction, somewhere along the manufacturing route, of a gas-phase step for nanomaterial integration appears to be much more natural than a wet-chemistry step. Scale-up through batch processing of wafers can also be easily addressed.

Unfortunately for gas sensing applications, gas-phase deposition methods typically produce compact thin films, where the specific surface area is by far smaller than in the case of nanomaterials. It is however, possible, for the skilled person, to properly operate deposition systems and adopt procedures to generate gas-phase nanoparticles to be used as film building blocks, or alternatively, to induce suitable growth dynamics, both leading to high-roughness high specific surface area films.

Gas-phase methods are in themselves mechanically safe with respect to microhotplate delicacy, nevertheless the issue of how to pattern the deposition has to be addressed. Gas-phase deposition methods in fact are characterized by almost isotropic diffusion of film precursors, or at least by precursor trajectories within very large solid angles (as it happens in vacuum evaporators or PLD). To effectively prevent precursor diffusion and deposition onto unwanted regions, contact masks, either hard or soft, should be used. Unfortunately this is particularly challenging, if not inapplicable, on micromachined wafers. Referring to the step sequence for device production, it turns out that the use of gas-phase methods for

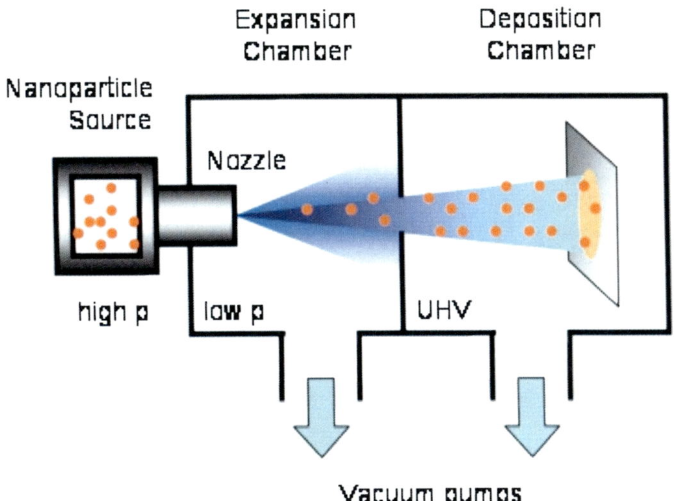

Fig. 16.11 Scheme of an apparatus for nanoparticle beam deposition. It consists of a two chamber vacuum system, where nanoparticles move from the source to deposition substrate. Once generated, for example by pulsed microplasma cluster source (*PMCS*), nanoparticles are carried by a supersonic inert gas jet expanding out of the source into the expansion chamber. The nanoparticle kinetic energy is of the order of few tenth of eV per atom: this prevents fragmentation at the impact with the substrate and allows for the soft assembly of nanoparticles, giving as a final result a nanoporous film with high specific surface area

nanomaterial integration requires some rearrangements. This could occur in two, equally challenging directions.

First, nanomaterial integration is anticipated with respect to micromachining step: as commented for wet-chemistry approaches, this definitely alleviates the microhotplate delicacy issue. Patterned functionalization of the wafer front-side may pass through standard photolithographic soft mask methods followed by lift-off, if deposition temperature is limited to photoresist-compatible values. Due to the absence of any solvent in gas-phase deposition methods, the use of photolithographic polymer soft mask is more affordable than in the case of wet-chemistry. After nanomaterial deposition, a suitable protection of the active layer against subsequent etching step has to be provided, in particular for prevention of contamination.

A second way can be followed and is based on shifting of the integration step to occur at the end on the manufacturing route, after the dicing-bonding-packaging step. By means of electrical connection to external power, integrated heaters of devices to be functionalized can be switched-on within the deposition chamber. This enables for the use of the so called self-lithographic approach, or thermally-activated CVD (it can even be applied to radiofrequency sputtering too). Self-lithography is based on the phenomenon according to which a deposition of thin film may occur from suitable precursors if the substrate is kept at a characteristic high temperature [13, 27, 28]. If the self-lithographic approach is adopted, batch

Fig. 16.12 Transmission electron microscopy (*TEM*) images of nanostructured tungsten oxide showing the structure of a film produced by nanoparticle beam deposition and the effect of thermal treatments. The as-deposited material (*top*) shows a porous structure resulting from the assembly of nanoparticles having an average size below 10 nm. The absence of lattice fringes inside the nanoparticles indicates an amorphous structure. In the 200 °C annealed material (*middle*), evolution towards a crystalline structure is evident, while the average size and porous structure are substantially identical to the as-deposited film. In the 400 °C annealed sample (*bottom*), the ordered phase fraction further increases, while maintaining the porous structure. Reprinted from Ref. [31]

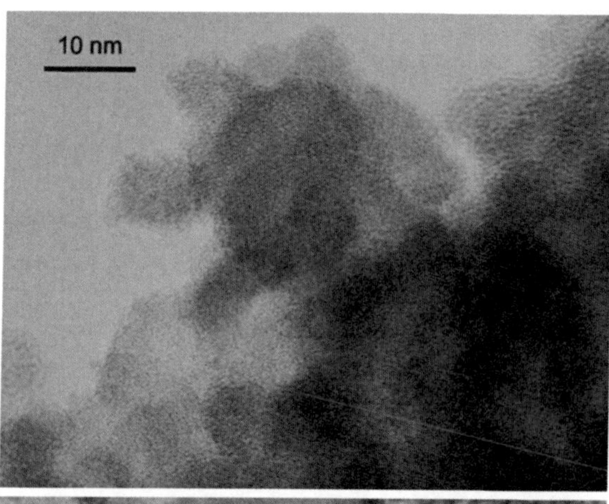

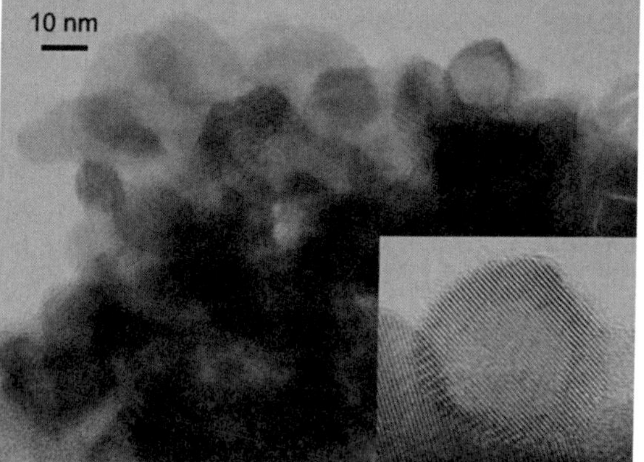

Fig. 16.13 Example of micrometric pattern of nanomaterial produced using nanoparticle beams and a non-contact hard mask. The scanning electron microscope (*SEM*) image is of a nanostructured WO_3 pattern obtained with a high-aspect-ratio micromachined silicon mask coupled to nanoparticle beam deposition system. The *dot side* is 5 μm. Reprinted from Ref. [31]

deposition on wafers is obviously no more possible since devices are separated. Even if some suitable complex wiring setup located into the deposition chamber could be imagined, the scale-up possibilities of this approach still remain quite limited.

16.3.4 Nanoparticle Beam Methods

Recently, novel gas-phase deposition techniques for the production of nanostructured films based on nanoparticle beams have been introduced [29, 30]. Their general characteristics make them particularly suited for nanomaterial integration onto micromachined substrates, since they overcome many of the limitations affecting the deposition techniques described so far. The principle at the base of nanoparticle beam deposition (also cluster beam deposition, CBD) is the use of a carrier gas undergoing a supersonic expansion, as it moves from the

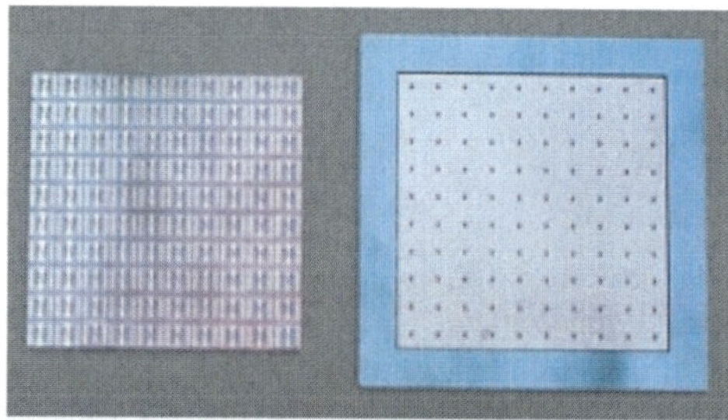

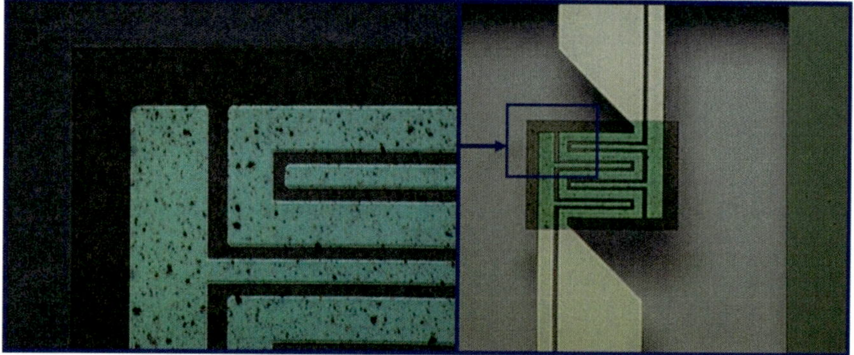

Fig. 16.14 *Top* image shows a wafer section hosting one hundred of microhotplates (*left*) and the micromachined auto-aligning silicon mask for non-contact patterning (*right*). Devices have been functionalized in batch with nanostructured WO_3 oxide, in a simple and straightforward step, by nanoparticle beam deposition, as shown in the *bottom* images. The active material is the *greenish rectangular* region on the interdigitated electrodes; it has dimensions of 0.65×0.78 mm^2. Reprinted from Ref. [31]

relatively high pressure condition of the beam source to the low pressure condition of a vacuum chamber, through a suitable nozzle, as shown in Fig. 16.11. The carrier gas, which is typically an inert gas, transports the nanoparticles, generated by bottom-up growth mechanisms and loaded into the gas stream, directly from beam source to deposition substrate, through a sequence of vacuum chambers. Once the nanoparticles reach the deposition chamber and substrate, they become the building blocks for the growth of nanostructured films. The entire process is performed at room temperature.

Although various type of nanoparticle beam sources exist, the pulsed microplasma cluster source (PMCS) appears to be one of the most promising for nanomaterial integration onto microhotplates for gas sensors. These promising attributes consist of the wide library of metal-oxides that it can produce, its stability, and the possibility of use it into large-scale production facilities.

Nanoparticle beams are characterized by low kinetic energy (fraction of eV per atom), which promotes a good adhesion as well as prevents nanoparticle destruction once they strike the substrate. Nanoparticle structure is preserved and their assembly during film growth leads to high specific surface area films which

Fig. 16.15 Atomic force microscopy (*AFM*) images showing the surface morphology of films by nanoparticle beam deposition (the side of the images corresponds to 1 μm, while the thickness of the films is a few tens of nm). Morphological features are pretty much similar for all materials: this may be ascribed to ballistic regime growth, which is characterized by nanoscale porosity, poorly-connected and non-compact structures with lower density respect to bulk and a surface roughness increasing with film thickness. Reprinted from Ref. [33]

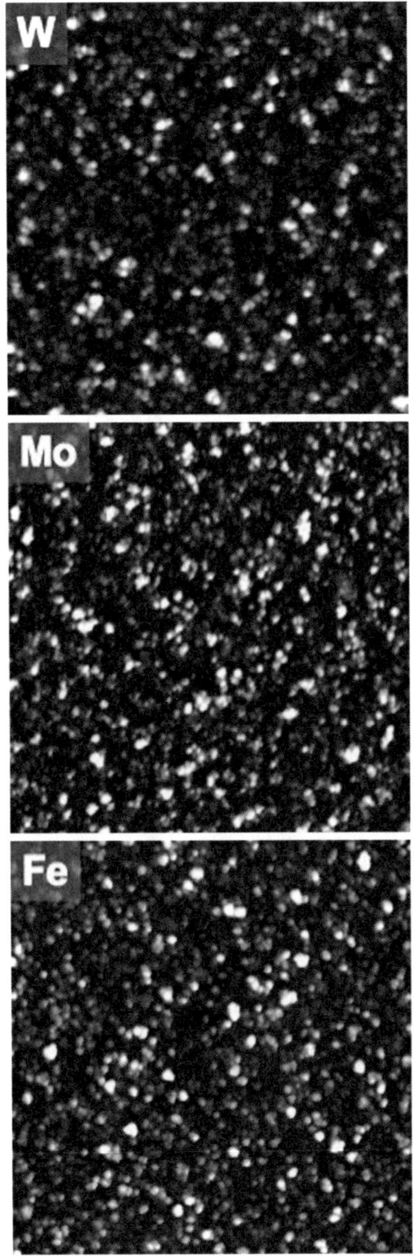

are very well suited for the interaction with the sampling atmosphere. Figure 16.12 shows an example of a film produced using nanoparticle beams.

The most relevant feature of nanoparticle beams, that strongly differentiate this technique from other gas-phase deposition techniques, is their high collimation, which is of the order of few tens of mrad. Such a narrow beam leads to various consequences. For example, the deposition region can be separated from the nanoparticle production region via the adoption of differential vacuum chambers, in order to provide cleanest, ultra-high-vacuum (UHV) growth conditions. Nevertheless the most important consequence is by far the possibility of non-contact hard mask patterning with sub-micrometric lateral resolution. Thanks to beam collimation, in fact, the drawing of the mask is exactly reproduced on the substrate, as a stencil, even if the distance between the mask and substrate is of the order of several tenth of a millimeter or more. Hard mask patterned deposition of micrometric structures spreading over an area of several cm^2 has been reported in refs. [31, 32]; and an example of them is shown in Fig. 16.13.

As with other gas-phase deposition techniques, the nanoparticle beam technique matches the delicacy requirement for microhotplate functionalization: no heating or damage of the substrate occurs. Batch deposition of hundreds of devices within a wafer can be performed—even if the beam spot size is limited to a centimeter—through substrate rastering in front of the nanoparticle beam. Non-contact hard mask methods fulfills the requirement of a mechanically safe way to pattern nanomaterials onto delicate micromachined substrates. As an example, Fig. 16.14 shows a wafer cut containing one hundred microhotplates, each of which have been functionalized in batch mode with nanostructured tungsten oxide, in a simple and straightforward step, by nanoparticle beam deposition. A micromachined silicon mask with proper sectioning provided auto-alignment with the microhotplate devices.

In addition, several nanostructured metal-oxides, such as SnO_2, TiO_2, WO_3, Fe_2O_3, MoO_3, ZrO_2, HfO_2, NbO_x, ZnO, PdO_x, can be routinely produced by PMCS. Beside those usually used in state-of-the-art metal-oxide gas sensing, many others are available, with advantages from the point of view of an array approach to chemical sensing.

A non-trivial, intriguing feature of nanostructured metal-oxides by PMCS, regards their surface morphology, which is the same for all [33], as shown in Fig. 16.15. This may be ascribed to ballistic regime growth [34], which is characterized by nanoscale porosity, poorly-connected and non-compact structures, with lower density with respect to bulk and a surface roughness which increases with an increase in film thickness.

This rules out a common drawback realized with other deposition techniques. Mainly, the surface morphology (i.e. roughness, porosity, grain interconnection, etc.) may depend on the material produced, introducing an additional complexity element that makes it difficult to ascribe functional performances, such as gas sensing, to material chemistry rather than to a "convolution" between chemistry and surface micro- and nano- structure.

In summary, among the various deposition methods, nanoparticle beam deposition offers characteristic features, matching most of the requirements to be fulfilled for the integration of nanomaterials onto microhotplates and advanced micromachined platforms for chemical sensing: delicacy of deposition in itself; mechanically safe, non-contaminating, one-step method for micrometric patterning; batch deposition of hundreds of devices; several oxides with high specific surface available.

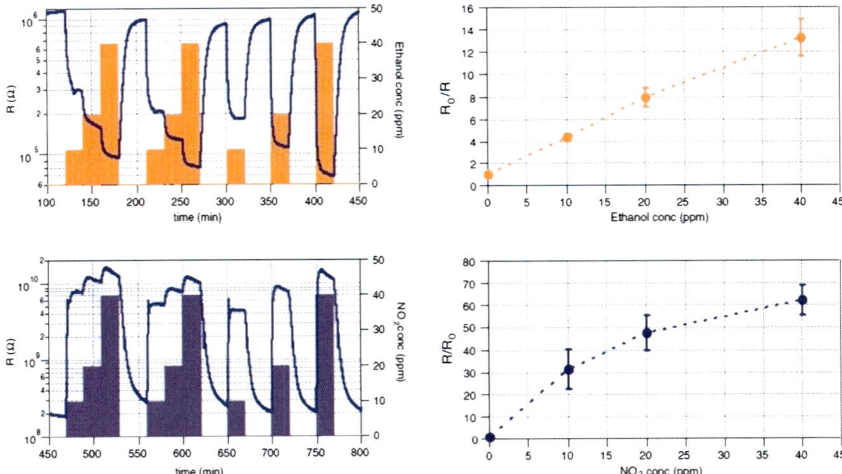

Fig. 16.16 Examples of sensing performances of microhotplates with nanoporous WO_3 by nanoparticle beam deposition, as active element. *Top* graphs show sensing results with respect to a reducing species, namely ethanol, while *bottom* graphs to an oxidizing species, namely NO_2. The operating temperatures are 300 °C for ethanol and 200 °C for NO_2, with a power of 24 and 13 mW, respectively. *Bar-like curves* indicate the nominal concentration of the test gases, referring to the *right axis*, while solid lines represent the resistance of the sensing film, referring to the *left axis*. Note that in both reducing and oxidizing cases, the use of the log scale for the *left axis* is required, due to the enormous change in sensor resistance. *Error bars* in response graphs on the *right* were calculated considering the three responses to the same gas concentration. Reprinted from Ref. [31]

16.4 Sensing Performances

16.4.1 Measurement Examples

The most important feature of microhotplate chemical sensors is by far the possibility to operate at high temperature spending a minimal amount of power. In addition, if their functionalization is done with high specific surface area porous nanomaterials, the sensitivity may be as high as to detect compound concentrations at ppb (parts per billion) levels. Figure 16.16 shows, as an example, the results of sensing characterization of microhotplates functionalized with nanostructured WO_3. In this case, the sensing material has been deposited, in the form of a nanoporous layer with a thickness of a few hundreds of nanometers, by nanoparticle beam deposition, using PMCS [31].

The Fig. 16.16 reports two examples of sensing performances relative to a reducing species (ethanol) and to an oxidizing one (nitrogen dioxide, NO_2). A decrease in resistance is observed upon exposure to a reducing gas, while an increase in resistance is observed upon exposure to an oxidizing gas. Resistance variations up to more than one order of magnitude were observed, at the injection

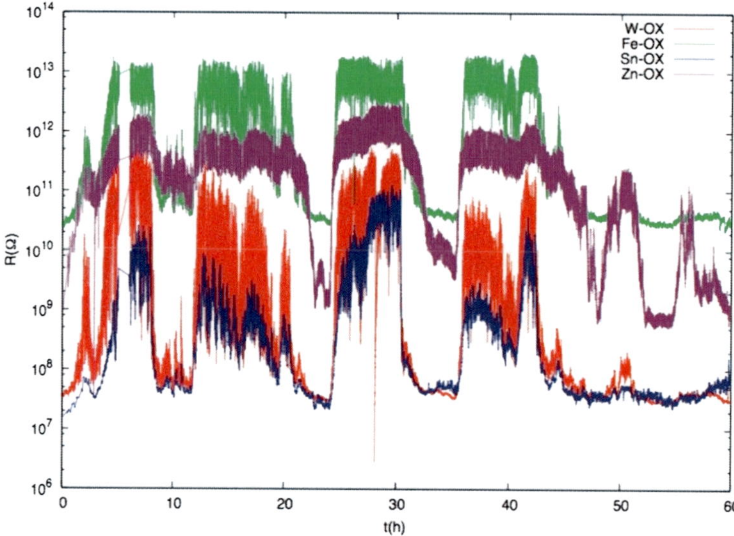

Fig. 16.17 Example of collective outputs of four nanostructured oxides (SnO_2, Fe_2O_3, WO_3 and ZnO) formed by nanoparticle beam deposition, exposed to the complex outdoor atmosphere of a car parking area. Although qualitatively similar, the sensing signals of different oxides are not identical each other: by means of advanced data analysis software with the capability of processing the collective response of the array, after suitable training it may be possible to recognize different chemical scenarios occurring in real outdoor atmosphere

of compounds with concentrations at ppm (parts per million) levels. Ethanol detection was carried out operating the microhotplate at a temperature of 300 °C, with a heating power of 24 mW, while NO_2 detection at 200 °C, with heating powers of 13 mW. Sensor responses, defined as R_0/R in the case of reducing species and R/R_0 for oxidizing species, R_0 being the sensor resistance in the presence of pure air only, and R the sensor resistance in the presence of the reactive compounds, were calculated and reported in Fig. 16.16 as well, as a function of the concentration of the test species. A saturation trend seems to appear at higher concentrations, while a good linearity is observed at lower ones, at ppm level. The strong response to NO_2 suggests a detection limit with respect to this compound in the 10–100 ppb range.

16.4.2 Microhotplate Array

A challenging problem of metal-oxide chemical sensing is the recognition of compounds, or at least their chemical family. In fact, due to the unspecific characteristics of the detection mechanism of chemoresistive sensors, different compounds at different concentrations may generate the same response in a single sensor. The strategy to try to address this issue passes through the use of sensor

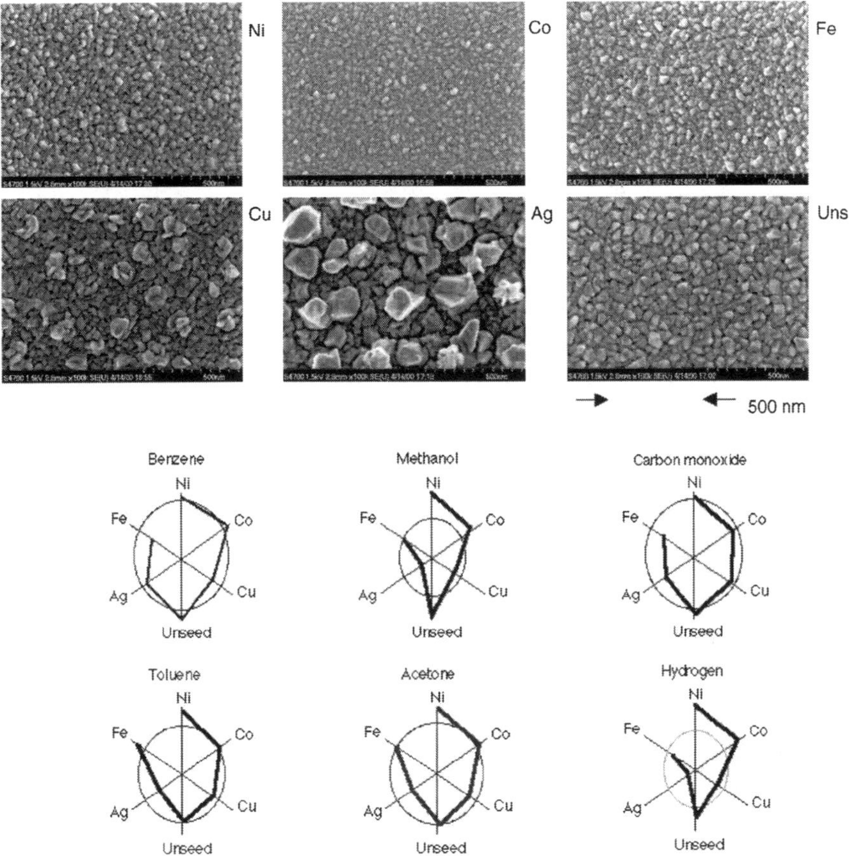

Fig. 16.18 In the *upper part* of the figure, SEM images showing the surface morphology of various sensing layers are reported. In all cases the sensing material is SnO_2 by CVD, however, the use of different metal seed layers before CVD process leads to different microstructures. *Bottom part* of the figure reports web-plots showing the relative sensitivity to six individual compounds in air of the sensing elements with an engineered microstructure. As in the case of Fig. 16.17, the use of an array of sensing materials may offer the possibility of compound recognition, although the intrinsic non-selectivity of metal-oxide chemical sensors. Reprinted from Ref. [3] with permission from Elsevier

arrays, where many sensors with different metal-oxides simultaneously react with the target chemical species. As an example of such a collective response, Fig. 16.17 shows the signals from four different nanostructured oxides deposited using a nanoparticle beam, namely SnO_2, Fe_2O_3, WO_3 and ZnO, during outdoor measurements.

Since the detection mechanism is generally the same for different oxides, sensor responses will be qualitatively similar, as clearly shown in Fig. 16.17. However, they are not identical. Hence, processing the collective response of the array, by means of advanced data analysis software (neural networks, for example),

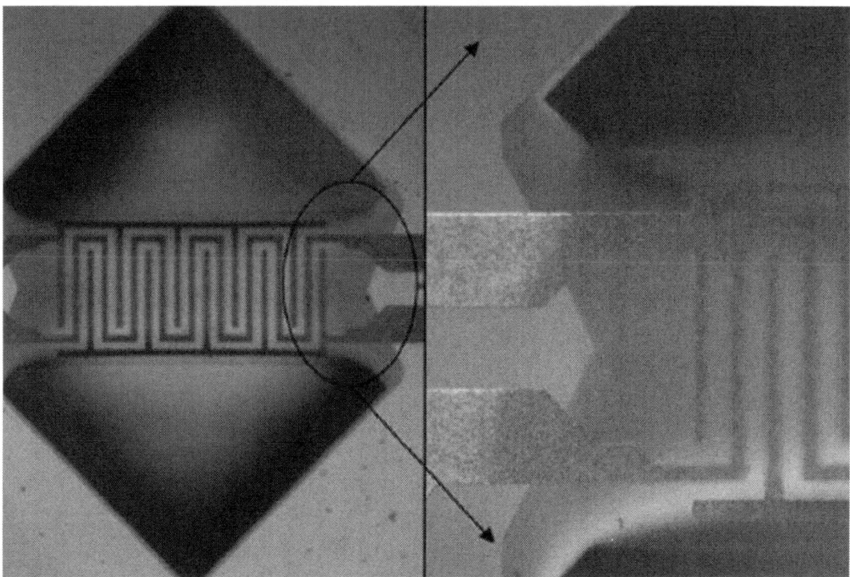

Fig. 16.19 Detail of suspended microbridge chemical sensor, within multiparametric platform. On the *left*, the entire suspended structure is visible, together with heater and interdigitated electrodes to contact the sensing layer. On the *right*, a close-up of the structure: the change of the color from *light* to *dark gray* along the *metal lines* indicates the border of the region deposited with nanostructured Fe_2O_3 by nanoparticle beam deposition. Reprinted from Ref. [36] with permission from Elsevier

precluded by a series of advanced training experiments against gases and gas mixtures of known composition, it may be possible to identify the "fingerprint" of the atmospheric gas composition.

Alternatives to arrays of different metal-oxides may be arrays of the same oxide having different microstructures, or different catalysts/dopants dispersed over the surface. In ref. [3] it is shown how the use of different metal seed layers before the deposition of SnO_2 by CVD, leads to different microstructures and to different relative sensitivities with respect to chemical compounds (Fig. 16.18).

16.4.3 Temperature Programmed Sensing

By exploiting the low thermal inertia of microhotplates, a temperature programmed sensing (TPS) mode can be adopted [35]. Analogously to array approach, the rationale at the base of TPS is the possibility to increase the data amount, by exploiting the different reactivity and different surface chemical processes in general, taking place at different temperatures, with the aim to improve

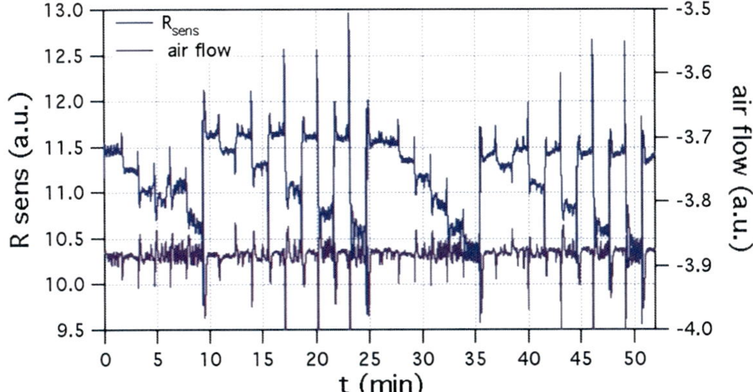

Fig. 16.20 Example of chemical sensing characterization of multiparametric device, with hydrogen in dry air at concentrations up to 30 ppm (*blue line*). The graph also reports the signal from micromachined anemometer, as a constant line with superimposed peaks (*violet line*). If the constant overall behavior of anemometric measure accounts for the constant total flow adopted—as usual—during chemical sensing experiment, superimposed peak-like features account for real fluctuations of air flow during MFCs operations, once hydrogen concentration has to be changed. Courtesy from L. Lorenzelli, FBK-Trento

the recognition capability of sensing systems with respect to complex atmospheres. The details of this methodology are highlighted in Chap. 14 in this book volume.

In an advanced approach to complex chemical sensing (as required for example to address security issues regarding the early detection of hazardous compounds such as chemical warfare agents), it is straightforward to merge together an array approach and TPS, in order to strongly increase the quantity of data characterization within a certain atmosphere [5].

16.4.4 Multiparametric Sensors

As shown in Sect. 16.2, microelectronic and micromachining methods for microhotplate production, are quite naturally extendible to include onto the same sensing platform other types of transducers, according to a common system integration approach. In particular, the example of a multiparametric platform hosting a chemical sensor, a thermometer, and an anemometer has been presented. Such a device may find applications as an in-line operating sensing system, with the capability to measure air flow and temperature, as well as to detect the presence of reactive compounds within flowing air. In a more specific example, these devices may find a role as novel miniaturized detectors, in charge of sampling the column outlet in advanced gas-chromatographic systems: feedbacks on column flow and outlet temperature, as well as, obviously, detection of chemical species, constitute the multiparametric task to be faced.

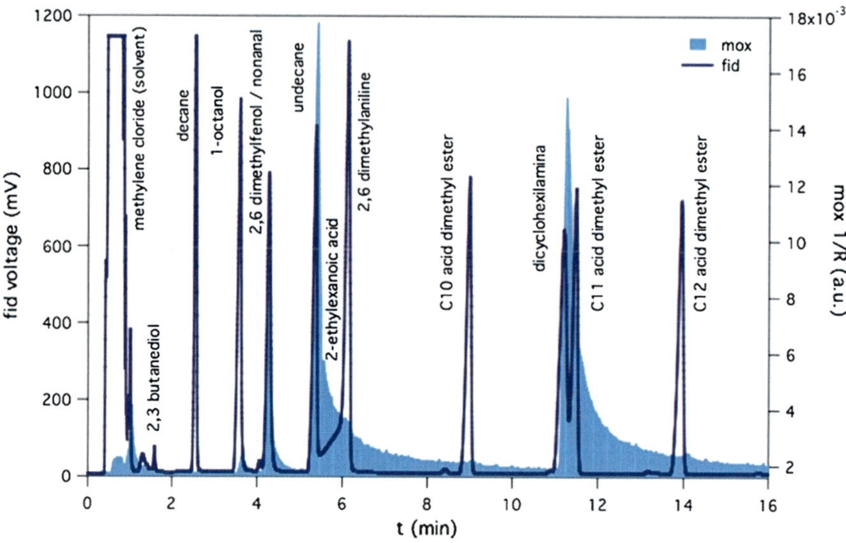

Fig. 16.21 Chromatogram acquired at the same time by a standard flame ionization detector (*FID*) and by the metal-oxide chemical sensor of a multiparametric micromachined platform. Not all the peaks visible with *FID* are detected also by metal-oxide sensor, which seems to be most sensitive to compounds containing hydroxyl group. Signal recovery after intense peaks shows long tails, probably due to inefficient atmosphere purging in the experimental setup, for gasdynamic reasons. Absence of sensitivity to the solvent (methylene chloride) could help in the detection of compounds with small retention time, or in the case of bad injections with solvent-modified baseline

Figure 16.19 shows a detail of the "microbridge" suspended structure of chemical sensor (see also bottom part of Fig. 16.8), which has been functionalized with nanostructured Fe_2O_3 by nanoparticle beam deposition [36]. Hard mask patterning limited the deposition of nanomaterial to chemical transducer only.

Platform characterization was carried out with a standard facility for gas sensor characterization, namely a test cell fluxed by means of mass flow controllers (MFCs) with the suitable atmosphere under investigation. Figure 16.20 shows an example of chemical sensing characterization with respect to hydrogen in dry air, at concentrations of 6, 12, 18, 24, and 30 ppm. Hydrogen injections correspond to descending square features of sensor resistance. On the same graph it is also reported the simultaneous signal from a micromachined anemometer: this is a constant line with superimposed upward and downward peaks. Constant behavior of anemometer signal indicates that chemical sensing characterization took place at constant total flow. Peaks are not simply signal noise: they account for real fluctuations of air flow during MFCs operations, once pure air flow is slightly decreased and hydrogen containing flow is increased to determine the proper hydrogen concentration, while maintaining constant the total flow over the device.

As an example of gas-chromatographic use of micromachined chemical sensor, Fig. 16.21 shows a chromatogram, which has been acquired at the same time from a standard FID (hydrogen-fueled flame ionization detector) and from the multi-parametric device. Although not all the compounds can be revealed by the metal-oxide chemical transducer, as well as gas sensor design needs to be improved to enhance recovery time dynamics (in order to short peaks tails), gas-chromatographic use of advanced micromachined devices seems not so far fetched, at least within portable instrumentations and "first order" analysis, skipping the needs of a hydrogen line, as imposed by FID.

16.5 Conclusions and Perspectives

Advanced chemical sensors based on micromachined platforms integrating metal-oxide nanomaterials as active elements have high potentiality to address the demand of high-sensitivity, wide-spectrum, miniaturized, low power consumption, reliable and stable devices, to bridge the gap between current solid-state sensing technology and challenging, real-world applications. These include, for example, pervasive outdoor/indoor monitoring of air conditions through wireless sensor networks, early detection of hazardous compounds for security issues, food quality monitoring, etc. up to cutting edge applications such as physiological/pathological condition identification through breath analysis, or microclimatic characterization of artistic and cultural heritages sites for conservation purposes.

Beyond technical details, to make all this real, a novel vision is developing, where the concept of system integration, i.e. the synergistic contribution of the various technological aspects playing a role in the building and in the functionality of a complex system, constitutes the background scenario. This particularly holds for the issue of microhotplate functionalization with nanomaterials, where several techniques could be used, although so far none of them seems to be the optimal one, and in any case a careful merging of microelectronic production methods, micromachining methods, and nanomaterial production and integration methods has to be reached.

From this specific point of view, many of the techniques for the production of metal-oxide nanomaterials may be refined in their features to match requirements for large-scale, reliable, safe, cost-affordable integration of active layers on microhotplates and micromachined platforms. Among those techniques, nanoparticle beam deposition may be a novel tool with particular features making it suited for nanomaterial integration purposes. Gas-phase approach, non-contact hard mask patterning, batch deposition, as well as a wide library of synthesizable oxides with nanoporous structures, are its strength points.

From a more general point of view, the feeling is that parts composing metal-oxide advanced sensing systems, namely micromachined platform, active nano-material layer, device packaging, integrated electronics, wireless communication units, and advanced software for network and data managing, are all reaching a mature development stage. However, although the richness of lab-scale experiments, where testing is carried out with respect to several different atmospheres, are claimed to reproduce real situations, a validation through long-term experiments at end-user sites, seems to be still lacking. This real-world validation is the next challenging frontier to be crossed.

References

1. Simon I, Bârsan N, Bauera M, Weimar U (2001) Micromachined metal oxide gas sensors: opportunities to improve sensor performance. Sens Actuators B 73:1–26
2. Briand D, Krauss A, van der Schoot B, Weimar U, Barsan N, Gopel W, de Rooij NF (2000) Design and fabrication of high-temperature micro-hotplates for drop-coated gas sensors. Sens Actuators B 68:223–233
3. Semancik S, Cavicchi RE, Wheeler MC, Tiffany JE, Poirier GE, Walton RM, Suehle JS, Panchapakesan B, DeVoe DL (2001) Microhotplate platforms for chemical sensor research. Sens Actuators B 77:579–591
4. Meijer G, Herwaarden A (1994) Thermal sensors, sensors series. Institute of Physics Publishing, Bristol and Philadelphia
5. Meier DC, Evju JK, Boger Z, Raman B, Benkstein KD, Martinez CJ, Montgomery CB, Semancik S (2007) The potential for and challenges of detecting chemical hazards with temperature-programmed microsensors. Sens Actuators B 121:282–294
6. Adami A, Lorenzelli L, Guarnieri V, Francioso L, Forleo A, Agnusdei G, Taurino AM, Zen M, Siciliano P (2006) A WO_3-based gas sensor array with linear temperature gradient for wine quality monitoring. Sens Actuators B 117:115–122
7. Raman B, Meier DC, Evju JK, Semancik S (2009) Designing and optimizing microsensor arrays for recognizing chemical hazards in complex environments. Sens Actuators B 137:617–629
8. Barsan N, Schweizer-Berberich M, Gopel W (1999) Fundamental and practical aspects in the design of nanoscaled SnO_2 gas sensors: a status report. J Anal Chem 365(4):287–304
9. Kohl D (2001) Function and applications of gas sensors. J Phys D Appl Phys 34:R125–R149
10. Ivanov P, Llobet E, Vergara A, Stankova M, Vilanova X, Hubalek J, Gracia I, Cané C, Correig X (2005) Towards a micro-system for monitoring ethylene in warehouses. Sens Actuators B 111–112:63–70
11. Afridi MY, Suehle JS, Zaghloul ME, Berning DW, Hefner AR, Cavicchi RE, Semancik S, Montgomery CB, Taylor CJ (2002) A monolithic CMOS microhotplate-based gas sensor system. IEEE Sens J 2(6):644–655
12. Graf M, Barrettino D, Zimmermann M, Hierlemann A, Baltes H, Hahn S, Bârsan N, Weimar U (2004) CMOS monolithic metal–oxide sensor system comprising a microhotplate and associated circuitry. IEEE Sens J 4(1):9–16
13. Suehle J, Cavicchi RE, Gaitan M, Semancik S (1993) Tin oxide gas sensor fabricated using C-MOS micro-hotplates and in situ processing. IEEE Electron Device Lett 14:118–120
14. Graf M, Gurlo A, Barsan N, Weimar U, Hierlemann A (2006) Microfabricated gas sensor systems with sensitive nanocrystalline metal-oxide films. J Nanopart Res 8:823–839
15. Su M, Li SY, Dravid VP (2003) Miniaturized chemical multiplexed sensor array. J Am Chem Soc 125:9930–9931
16. Vincenzi D, Butturi MA, Guidi V, Carotta MC, Martinelli G, Guarnieri V, Brida S, Margesin B, Giacomozzi F, Zen M, Pignatel GU, Vasiliev AA, Pisliakov AV (2001) Development of a low-power thick-film gas sensor deposited by screen-printing technique onto a micromachined hotplate. Sens Actuators B 77:95–99
17. Vincenzi D, Butturi MA, Stefancich M, Malagu C, Guidi V, Carotta MC, Martinelli G, Guarnieri V, Brida S, Margesin B, Giacomozzi F, Zen M, Vasiliev AA, Pisliakov AV (2001) Low-power thick-film gas sensor obtained by a combination of screen printing and micromachining techniques. Thin Solid Films 391:288–292

18. Riviere B, Viricelle JP, Pijolat C (2003) Development of tin oxide material by screen-printing technology for micromachined gas sensors. Sens Actuators B 93:531–537
19. Viricelle JP, Pijolat C, Riviere B, Rotureau D, Briand D, de Rooij NF (2006) Compatibility of screen-printing technology with micro-hotplate for gas sensor and solid oxide micro fuel cell development. Sens Actuators B 118:263–268
20. Jimenez I, Cirera A, Cornet A, Morante JR, Gracia I, Cané C (2002) Pulverisation method for active layer coating on microsystems. Sens Actuators B 84:78–82
21. Stankova M, Ivanov P, Llobet E, Brezmes J, Vilanova X, Gràcia I, Cané C, Hubalek J, Malysz K, Correig X (2004) Sputtered and screen-printed metal oxide-based integrated microsensor arrays for the quantitative analysis of gas mixtures. Sens Actuators B 103:23–30
22. Francioso L, Russo M, Taurino AM, Siciliano P (2006) Micrometric patterning process of sol–gel SnO_2, In_2O_3 and WO_3 thin film for gas sensing applications: towards silicon technology integration. Sens Actuators B 119:159–166
23. Smith DL (1995) Thin-film deposition: principles and practice. McGraw-Hill Professional, New York
24. Martin PM (2009) Handbook of deposition technologies for films and coatings: science, applications and technology. William Andrew—Elsevier, Oxford)
25. Chrisey DB, Hubler GK (1994) Pulsed laser deposition of thin films. Wiley-interscience, Hoboken New Jersey
26. Mattox DM (2010) Handbook of physical vapor deposition (PVD) processing. William Andrew—Elsevier, Oxford
27. Semancik S, Cavicchi RE (1991) The growth of thin, epitaxial SnO_2 films for gas sensing applications. Thin Solid Films 206:81–87
28. Cavicchi RE, Suehle JS, Kreider KG, Shomaker BL, Small JA, Gaitan M, Chaparala P (1995) Growth of SnO_2 films on micromachined hotplates. Appl Phys Lett 66:812–814
29. Milani P, Iannotta S (1999) Cluster beam synthesis of nanostructured materials. Springer, Berlin
30. Wegner K, Piseri P, Vahedi Tafreshi H, Milani P (2006) Cluster beam deposition: a tool for nanoscale science and technology. J Phys D 39:R439–R459
31. Barborini E, Vinati S, Leccardi M, Repetto P, Bertolini G, Rorato O, Lorenzelli L, Decarli M, Guarnieri V, Ducati C, Milani P (2008) Batch fabrication of metal oxide sensors on micro-hotplates. J Micromech Microeng 18:055015
32. Barborini E, Piseri P, Podestà A, Milani P (2000) Cluster beam microfabrication of patterns of three-dimensional nanostructured objects. Appl Phys Lett 77:1059–1061
33. Barborini E, Corbelli G, Bertolini G, Repetto P, Leccardi M, Vinati S, Milani P (2010) The influence of nanoscale morphology on the resistivity of cluster-assembled nanostructured metallic thin films. New J Phys 12:073001
34. Barabasi AL, Stanley HE (1995) Fractal concepts in surface growth. Cambridge University Press, Cambridge
35. Kunt TA, McAvoy TJ, Cavicchi RE, Semancik S (1998) Optimization of temperature programmed sensing for gas identification using micro-hotplate sensors. Sens Actuators B 53:24–43
36. Decarli M, Lorenzelli L, Guarnieri V, Barborini E, Vinati S, Ducati C, Milani P (2009) Integration of a technique for the deposition of nanostructured films with MEMS-based microfabrication technologies: application to micro gas sensors. Microelectron Eng 86: 1247–1249

Concluding Remarks and Outlook

The influence of nanotechnology on functional materials started about ten years ago and since then it has made a tremendous impact also on the development of metal oxide-based gas sensors. The successful demonstration of the "bottom-up" rationale as a means for creating new functionalities by designing material architectures at the nanoscale opened hitherto new possibilities to grow and engineer sensing elements with pre-defined properties. This fundamental shift in materials design approaches and miniaturization was not only a new opportunity but it also offered the conceptual basis to couple different functionalities to create multi-material units capable of doing more than what their conventional counterparts would offer. For instance, material heterostructures based on metal-oxide junctions combines advantages of both optical and microelectronic approaches to investigate gaseous media. These materials with their novel, often coupled, electrochemical properties caused the revision of the traditional receptor-transduction principles along with the sensing devices design and architecture. Furthermore, the integration of different materials characteristics into a single device has put forward the concept of self-powered sensors that can work autonomously by harvesting energy from their surroundings.

This book exemplarily reflects the decade of development of the major path-breaking directions in the field of gas sensors especially materials and systems, where nanostructured metal oxides have played a prominent role. The topics addressed present a state-of-the-art account of recent developments in reaction mechanisms, novel material deposition techniques, morphology-controlled synthesis and transduction methods as well as new device architectures and challenges pertinent to the integration of nanomaterials in device systems. While great progress has been made in each of these areas over the last decade there is still work to be done that is evident in the huge number of publications emerging in this area as noted in Fig. A.1. The general reaction mechanism of gas adsorption on metal oxide surfaces is accepted, however, the details of this mechanism and operando experiments which support this mechanism are still elusive. Meeting the sensor specifications for a variety of applications has driven the developments in

M. A. Carpenter et al. (eds.), *Metal Oxide Nanomaterials for Chemical Sensors*,
Integrated Analytical Systems, DOI: 10.1007/978-1-4614-5395-6,
© Springer Science+Business Media New York 2013

materials deposition and morphology and there now exists a broad number of chemistries and methods which will certainly lead to a number of more productive years of application driven research. Inter-related with this are the developments and need for novel device architectures and technologically viable concepts for materials integration. These studies, which utilize a variety of nano-inspired metal oxide materials and structures, are driving the state of the art in integration of enabling technologies. For chemical sensors to be as pervasive a technology as envisioned, it is these developments in integration of novel devices, which will be at the forefront of work in the next decade.

Recent advances both in growth techniques and material structures provide the opportunity to dramatically increase the response of these materials, as their performance is directly related to exposed surface volume. The recent availability of various metal oxide materials in nanoscopic form, with high surface-to-volume area, engineering of surface functionalization with the use of selected dopant application as well as implementation of newly developed fabrication techniques, offer tremendous opportunities for low detection limits. The current challenges related to the development of nanomaterials include a more thorough understanding of the underlying physical and chemical principles that can be exploited to design new and surface-modified nanostructures, for instance, chemical grafting of receptor molecules to enhance the selectivity of the sensor. Similarly, the methodologies for self-organization and self-assembly of nanostructures as well as more close interactions between the theoretical models and experimental findings will be necessary. With new materials, such as core-shell or multi-layered heterostructures, new physical principles are developing especially to assess the influence of interfacial energetics, that would require more attention and concerted efforts to exploit the benefits of these new materials.

It is legitimate at this point to assess the current status of the field and pinpoint the nucleation points for future developments. The following figure reflects the publication trends related to metal oxide based chemical sensors as determined by a search of the Scifinder^{TM} database. The blue curve depicts the evolution of the total amount of metal oxide related publications in gas sensing. As seen from the figure these trends are characteristic of a mature and steady developing field. The publication rate in the field of nanostructured metal oxide as sensing materials reflects its recent appearance in the beginning of the century and shows fast growth. However as can be seen from their ratio (white curve) the latter has become a mature field in its own and its growth currently corroborates with that of its corresponding "mother's" field.

Looking into the future, it can be anticipated that the current technology needs and microfabrication challenges will dictate the development of the nanostructured metal oxide gas sensors. The pressing need of reducing energy consumption in next-generation devices would demand materials that can deliver optimal sensing behavior at room temperature as well as development of multi-material junctions to integrate power and sensing units in a single device. The increasing safety concerns around the globe will promote the research related to faster and selective detection of explosives, drugs and other toxic materials. While nanostructured metal oxides as sensing materials demonstrated steady improvements in major 3S

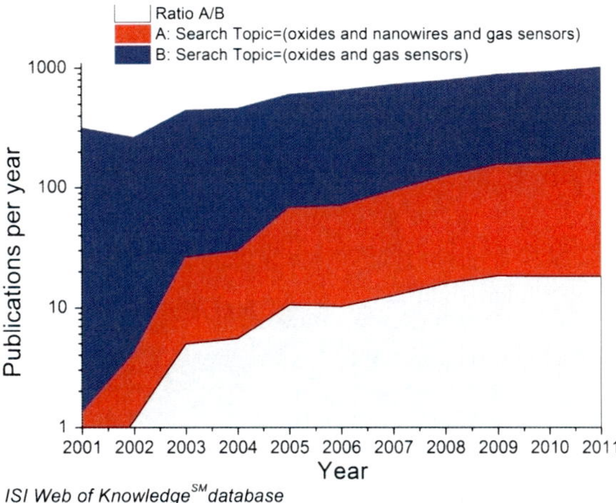

Fig. A.1 Publication trends
for metal oxide based gas sensor
research

(sensitivity, stability, selectivity) parameters in the laboratory there is a significant technological gap between research and development and industrial implementations. The major current challenges of this field lie in development of inexpensive techniques for controlled and reproducible fabrication of these novel sensing elements. In addition, since most of the proposed nanoscopic sensors rely on the same surface chemistry as their traditional counterpart the selectivity challenge remains. While demonstrations are available of the enhanced stability of the nanowire based sensors compared to their nanoparticulate traditional counterpart, further work is needed to enhance this research area.

Finally, it is interesting to observe very recent developments in quasi 1D based gas sensorics where the morphological particularities of the sensing elements has led to implementation of new receptor and transduction approaches. These are works related to reduction of the power consumption to the micro Watts level while combining sensor signals with heating function via direct Joule heating of the nanowire sensing element. Another example is enhancement of the local electric field around nanoscopic sensing elements to promote surface or field induced ionization. Both of these effects complement standard conductometric detection channels and therefore can be used as additional orthogonal signals to improve the selectivity of the sensor.

In conclusion, we strongly believe that this edition represents the cutting edge "cross-section" of the research and development in the domain of modern gas sensorics with nanostructured metal oxide sensing elements. There is a significant amount of newly discovered effects and approaches underway, which promise to act as "seeds" for future growth for many years to come.

Michael A. Carpenter
Sanjay Mathur
Andrei Kolmakov

Index

1-D, 10, 20, 130, 175, 226, 240, 246, 259, 288,
   337, 353, 362, 398, 409, 440, 475
3-D, three dimensional, 173, 195, 197, 352, 353

A
Acceptor, 84, 90, 93–96, 105, 123, 133, 276,
   304, 487, 490, 492
Acetic acid (CH_3COOH), 50, 53
Activation energy, 81, 157, 199, 204, 236, 246,
   274, 296, 310, 327
Adsorbate, 15, 35, 36, 40, 41, 43, 47, 52, 53,
   123, 193, 209, 403, 475, 477, 478, 484
Adsorption, 16, 36, 49–51, 54, 71, 78, 81, 87,
   98, 246, 260, 265, 288, 403, 404, 497
Aluminum oxide (Al_2O_3), 230, 237,
   238, 293, 444
Ammonia (NH_3), 81, 82, 96, 100,
   102, 311, 511
Anodic aluminum oxide (AAO), 293
Arrhenius Equation, 57, 204
Atomic force microscopy (AFM),
   197, 217, 527
Atomic layer deposition (ALD), 228

B
Band bending, 12, 14, 15, 36, 41, 44–47, 47,
   63, 124, 272, 401, 404, 410
Band gap (Eg), 6, 35, 41, 46, 52, 62–64,
   169, 174, 178, 180, 209, 259, 268,
   272, 295, 322, 328, 366, 410, 411
Binary metal oxides, 119, 122
Biomedical, 263, 332, 505

Bottom up, 130, 190, 227, 236, 240, 288, 409,
   423, 475, 491, 497, 529
Brønsted acids and bases, 75, 76, 83, 96, 101

C
Carbon monoxide, 95, 197, 378, 398, 419, 452,
   455, 458–461, 479, 485
Catalysis, 4, 6, 70, 132, 259, 378
Cation, 71–74, 98, 422
CdS, 294, 355, 356, 411, 413
Ceria (CeO_2), 258, 259, 388
Characterization, 5–7, 65, 119, 121, 134, 135,
   137, 139, 141, 144, 148, 149, 155, 159,
   192, 193, 196, 197, 200, 217
Charge carrier, 11, 42, 84, 92, 104, 127, 132,
   149, 169, 276, 298, 367, 368, 374, 410,
   411, 413
Chemical potential, 40, 46, 133, 310, 311
Chemical sensors, 5, 190, 226, 239, 288, 299,
   313, 366, 367, 402, 407, 423, 507, 535
Chemical vapor deposition (CVD), 173, 179,
   210, 229, 251, 290, 406, 413, 440,
   509, 522
Chemically sensitive field-effect transistors
   (chemFETs), 440
Chemiresistor, 4, 118
Chemisorption, 4, 43, 70, 84, 85
Clausius-Mozotti, 73
CMOS, 297
Colloidal, 250, 253, 254, 263, 410
Combinatorial methods
Composites, 398, 399, 406, 421
Conduction band (CB), 38

M. A. Carpenter et al. (eds.), *Metal Oxide Nanomaterials for Chemical Sensors*,
Integrated Analytical Systems, DOI: 10.1007/978-1-4614-5395-6,
© Springer Science+Business Media New York 2013

C (*cont.*)

Conductivity, 4, 8, 10, 11, 13, 17, 20, 55, 43, 45–47, 54, 70, 104, 119, 124, 273, 274, 296, 298, 329, 422, 509

Contact barrier, 346

Contact potential difference, 10, 14

Copper oxides (CuOx), 103

Correlation pattern, 457, 459

Cross sensitivity, 120, 126, 130, 131, 153, 181, 375, 376, 381, 400, 416, 420

D

Debye length, 107, 124, 226, 228, 233, 234, 239, 240, 305, 313, 346, 400, 484, 492

Defects, 272, 366

Density functional theory, 48

Depletion layer, 14, 107, 233, 346, 349, 400, 404

Desorption, 5, 16, 71, 81, 82, 90, 91, 95, 98, 104, 105, 125, 132, 210, 310, 325, 327, 346

Detection limit, 181, 201, 323, 334, 366, 371, 375, 391, 398, 478, 483, 530

Detectors, 4, 228, 327, 328, 373, 391, 400, 533

Diameter dependent, 305

Dielectric function, 379, 381, 384

Dielectrophoresis, 448

Differentiation, 6, 485, 496

Dipole, 15, 16, 37, 38, 42, 44, 46, 168, 181, 182, 246, 308

Donor, 12, 17, 19, 20, 23, 25, 79, 93, 95, 96, 98, 107, 123, 124, 195, 325

Dopants, 36, 60, 130–132, 145, 174, 178, 181, 391, 416, 421, 516

Doping, 64, 119, 121, 131, 132, 134, 140, 145, 146, 153, 168, 255, 359

Dynamic temperature programming, 461

E

Electric field, 39, 73, 84, 104, 206, 404, 448

Electrodes, 7, 119, 123, 124, 135–138, 297, 303, 308, 324, 328, 330, 347, 407, 426, 446

Electrolyte, 294, 330, 382, 399

Electronegativity, 69, 70, 72–76, 89

Electronic doping, 131–135, 143–147

Electronic nose (E-nose), 312, 313, 305

Electrons, 10–12, 17, 26, 40, 42, 52, 54, 55, 63, 72, 88, 104, 106, 123, 125, 128, 169, 193, 194, 209, 272, 273, 306, 308, 311, 330, 346, 350, 351, 356, 377, 384, 399, 401, 403, 414, 422, 492

Electrophilic, 76, 78–80, 98

Electrophysical, 8, 10, 19, 24

Electrospinning, 231, 232

Energy-dispersive X-ray spectroscopy (EDS), 360, 361, 444, 445

Ethanol, 119, 168, 175, 176, 211, 233, 235, 237, 249–251, 274, 322, 346, 350–354, 356, 358–361, 369, 375, 408, 414, 421, 479, 485, 529

Ethylene (C$_2$H$_4$), 479, 481

Explosives, 117, 118, 168, 276, 366

F

Fabrication, 61, 172, 183, 189, 226–228, 230, 232, 240, 292–294, 299, 312, 328, 381, 400–402, 409, 416, 423, 467–468, 473, 475, 476, 481, 491, 494, 495

Fermi level, 11, 12, 35, 36, 38, 41–46, 52, 54, 62, 123, 124, 126, 273, 308, 310, 311, 400, 414

Fiber grating, 376

Field effect transistor, 118, 296, 299, 398

Fixed-Temperature Sensing, 450

Fluorine-doped tin oxide (FTO), 411

Formaldehyde (HCHO), 301, 413, 416

Functionalization, 326, 327, 329, 358, 399, 465, 479, 516, 518, 521, 528, 529, 535

G

Gas adsorption, 25, 273, 325, 401, 497

Gas recognition, 467, 474, 485, 486

Gas sensing, 3, 5, 6, 8, 11, 14, 15, 18, 21, 24, 25, 35–38, 42–45, 55, 54, 64, 98, 119, 121, 122, 131, 134, 140, 275, 419

Gas sensors, 4, 6, 8, 35, 43, 64, 69, 70, 118

Geometry, 55, 130, 193, 202, 229, 233, 291, 403, 468

Gold (Au), 87, 99, 100, 101, 236, 240, 258, 290, 354, 368, 377, 380, 444, 471, 483

Gold nanoparticle (AuNP), 378

Grain boundary, 127, 210, 226, 275, 325, 403

Grain growth, 174

Grain size, 13, 84, 85, 88, 104, 125, 127, 128, 130, 132, 142, 169, 226, 228, 240, 273, 275, 374, 376, 398, 416, 417, 420, 429

Grid, 428

Group (III)-nitrides, 322

H

Hydrogen (H_2), 42, 88, 324, 384, 534

Hydrogen disulfide (H_2S), 18, 36, 43–46, 94, 96–98, 103, 110, 111, 125, 180, 191, 226, 357, 358, 362, 381, 382, 388, 398, 408, 409, 412

Hafnium oxide (HfO_2), 528

Heater, 154

HEMT, 332–338, 340

Heterocontact, 87, 418

Heterogeneous catalysis, 70, 238

Heterostructures, 356, 398, 399, 401, 405, 407–418, 420, 423, 428, 429

Hierarchical structures, 352, 353, 411

High resolution transmission electron microscopy (HRTEM), 349, 350

High throughput, 117, 120, 134, 137

High throughput experimentation, 117, 119, 121, 134

High-electron mobility transistor (HEMT), 297, 346

Hotplates, 402, 403, 480, 503, 504, 506, 515

Hydrocarbons, 93, 125, 151, 155, 226, 238, 389, 422

Hydrophilic, 249, 415

Hydrophobic, 134, 251, 415

I

Impedance spectroscopy, 6, 117, 134, 137, 148

Indium oxide (In_2O_3), 18, 47, 87, 110, 118, 119, 122, 149, 153, 154, 156, 158, 169, 191, 217, 290, 299–302, 308, 310, 312, 323, 348, 360, 400, 407, 419, 420, 422, 516

Indium tin oxide (ITO), 348, 360, 361

In-situ, 291, 313

Integrated circuit (IC), 230, 313

Integration, 311, 397, 423, 446, 503, 507

Interdigitated electrodes (IDE), 446, 472, 510, 511, 526, 532

Interfacial reactions, 388, 391

Intergrain distance, 126, 404

Interstitials, 36, 48, 60, 64

Ionic conductivity, 199, 200

Ionosorption, 8–12, 14, 15, 18, 19, 25, 226, 227, 233, 234, 275, 360

Iron oxide (FeOx), 103, 236, 263

K

Kelvin probe, 6

Kinetics, 26, 52, 80, 120, 128, 210, 234, 265, 266, 268, 270, 382, 401

L

Lanthanum oxide (La_2O_3), 81, 83, 87, 105, 110, 348, 358, 359, 362

Lattice oxygen, 11, 16, 18, 47, 50, 75–77, 87, 88, 92, 100, 236

Lattice structure, 197

Lewis acids, 94, 95

Lewis bases, 72, 81, 110

Ligands, 72, 134, 261, 265

Linear discriminant analysis (LDA), 479

Localized surface plasmon band (LSPB), 377

Low energy electron diffraction (LEED), 189, 192

M

Madelung energy, 37–40

Materials, 80, 84, 103, 199, 295, 353

MEMS, 346, 516

Mesoporous, 134, 172, 173, 175, 407, 488

Metal oxides, 110, 273, 283

Metal oxide semiconductor (MOS), 69, 399, 418

Metal oxide semiconductor field effect transistor (MOSFET), 69, 399, 418

Metastable state, 183

Methanol (CH_3OH), 119, 158, 419

Micro-arrays, 481, 485, 486, 491, 494

Microhotplate, 448, 508, 530

Microsensors, 439, 448, 450–451, 461, 468, 504, 513

Microstructure, 122–124, 126, 129, 130, 132, 146, 147, 170, 182, 191, 366, 368, 377, 379, 383, 417, 531

Microwave irradiation, 245, 246, 266, 273, 276

Microwave synthesis, 245, 249, 266, 276, 406

Mixed metal oxides, 380, 381

Molecular Beam Epitaxy (MBE), 189–193, 209, 290

Monolithic, 465, 473, 490, 497

Morphology, 36, 44, 125, 132, 194, 197, 216, 228

Multifunctional, 400, 423, 429

Multiparameteric, 567, 569

Multiparametric sensors, 514, 533

Multisensor array, 465–475, 478, 487, 489, 494, 496, 497

Multiwall carbon nanotubes (MWCNT), 233

N

Nanobelts, 190, 225, 226, 230, 235, 240, 291, 322, 324, 355, 414, 491

N (*cont.*)
Nanocomposite, 365, 377, 380–385
Nanodevices, 355, 382, 410
Nanoparticles, 362, 377, 378, 407, 525, 526
Nanoporous, 232, 238, 503, 504, 507, 523,
    529, 535
Nanorods, 353, 356, 358
Nano-scale thin films, 190, 200
Nanostructures, 230, 251–267, 324, 409, 494
Nanotubes, 172, 190, 191, 230–232,
    409, 414, 415
Nanowires, 287, 288, 321, 348, 359,
    360, 368, 439, 448, 465, 475
Neck size, 129, 130
Nitrides, 321–323, 328, 406
Nitrogen dioxide (NO$_2$), 92, 94, 95, 104,
    454, 456, 458, 459, 529
Nitrogen monoxide (NO), 6, 94, 95, 103–105,
    119, 149, 151, 153, 154, 309,
    310, 350
Non-stoichiometric, 79
n type semiconductor, 12, 84, 119, 122, 153,
    174, 180, 181, 209, 386, 415
Nucleation, 247, 248
Nucleophilic, 76–79

O
O–, 10, 14, 17, 76, 78, 105, 235
O$_2$, 9
O$_2{}^-$, 9, 10, 76, 78, 235
O^{2-}, 78, 235
OH$^-$, 75, 96, 155, 271, 335, 484
Olfaction, 465
Operando studies, 26
Optical basicity, 71, 72, 74, 75, 81, 88
Optical properties, 295, 365, 368, 377, 382,
    389
Oxidizing gases, 47, 94, 110, 125, 349, 351,
    367, 372, 388, 399, 408
Oxygen (O$_2$), 4, 7, 8, 17, 21, 60, 76, 78, 84,
    197, 272, 382
Oxygen anions, 16, 71, 72, 76, 78, 82, 87, 95,
    98, 209, 384
Oxygen vacancy, 9, 16–18, 78, 199, 204, 226,
    233, 310, 367, 388, 422
Oxygen-ion conductance, 199, 204, 382

P
Palladium (Pd), 23, 236, 240
Passivation, 41, 256, 259, 277, 297,
    507–509, 512
Pattern recognition, 310, 312, 313, 425, 466

Percolation, 124–127, 129, 130, 482,
    483, 485, 489
Perovskite, 123, 151, 176, 365, 376, 391
pH, 330–336, 338, 340, 385
Photocatalysis, 267, 269, 373, 402
Photodetector, 298, 327
Photolithography, 236, 299, 509, 522
Photoluminescence, 295, 330, 365–368,
    370–372, 375, 388, 391
Physical vapor deposition (PVD), 229, 406
Plasma enhanced atomic layer deposition, 235
Plasma enhanced chemical vapor deposition
    (PECVD), 406
Plasmon, 377, 379, 388, 391, 410
Plasmon resonance energy transfer (PRET),
    377, 410
Platinum, 87, 135, 136, 236, 238, 240, 349,
    350, 446, 447, 449, 505, 514
p–n junction, 356, 357, 362, 418
Polarizability, 26, 72, 73, 90, 384
Polycrystalline, 191, 195, 440
Porosity, 126, 129, 130, 132, 415, 417, 418,
    527, 528
Power consumption, 201, 313, 323, 399, 403,
    407, 427, 429, 473, 503–506, 535
Principle component analysis (PCA), 480
p type semiconductor, 119, 125, 151, 345
Pulsed laser ablation (PLA), 291–406

Q
Quality factors, 119, 120, 122, 140, 155
Quantum confinement, 258, 259, 367, 410
Quantum dots, 406, 407, 411
Quasi one-dimensional (Q1D), 20, 226, 287,
    288, 353, 439, 440, 475

R
Radio-frequency identification (RFID), 523
Reaction mechanism, 269, 279
Receptor, 105, 123, 349, 466, 469, 485
Recovery time, 111, 130, 155, 237, 287, 288,
    346, 361, 372, 412, 535
Redox processes, 77, 103, 269
Redox reactions, 70, 80, 169, 287, 288, 478
Reducing gases, 8, 12, 21, 25, 95, 107, 110,
    125, 409, 494
Reduction-oxidation, 21–23, 92, 169
Reflection high energy electron diffraction
    (RHEED), 189, 192, 217
Refractive index, 73, 366, 378, 379, 383, 391
Reliability, 327, 353, 366, 375, 391, 472, 504,
    516, 521

Response time, 120, 155, 234, 235, 237,
    300–302, 346, 348, 356, 376, 384, 483
Reversibility, 46, 149, 154, 389, 398, 422
Rutherford backscattering spectroscopy
    (RBS), 201–203

S
Samaria doped ceria (SDC), 193, 200–204,
    207, 208
Scanning electron microscopy (SEM),
    141–143, 231, 238, 290, 292–294,
    303, 309, 329, 331, 333, 352, 360,
    361, 477, 488
Scanning tunneling microscopy, 35, 39
Scattering, 151, 194, 271, 297, 411
Schottky-barriers, 125, 126
Schottky diode, 324, 330
Selectivity, 69, 120, 239, 388, 399, 412,
    419, 422, 475
Semiconducting oxides, 17, 45, 228, 240
Sensitive layer, 234, 419
Sensitivity, 99, 105, 120, 155, 182, 274,
    305, 347, 354, 357, 358, 360, 399,
    412, 414, 416, 424
Sensor arrays, 313, 388, 467
Sensor performance, 5, 119, 120, 130, 134,
    153, 288, 448, 459, 461
Sensor response, 23, 98, 100, 150, 153, 157,
    235, 275, 409, 416, 423, 530
Sensoric, 127
Sensors, 5, 35, 122, 170, 176, 228, 273,
    323, 324, 337, 353, 402, 448, 533
Shape control, 245, 263
Single crystal, 5, 173, 190, 201, 313, 325,
    328, 490
Single crystal surfaces, 5, 43
Sinter, 4, 125, 127, 136, 146, 148, 226,
    376, 440, 489, 497
Sintering, 136, 146, 148, 376, 440, 489, 497
Size dependent, 12, 427
Sol gel, 134, 172, 373, 374, 379, 380,
    406, 417, 423, 520
Solid-State Gas Sensors, 399, 421
Space charge layer, 11, 123, 126, 127,
    209, 210, 346
Spectroscopy, 6, 36, 39, 44, 81,
    95, 118, 443
Stability, 62, 118, 120, 135, 173, 209, 305,
    339, 346, 347, 407, 487, 509
Surface doping, 132, 133, 145, 153, 154,
    325, 400, 495
Surface functionalization, 358, 361, 399

Surface modification, 56, 69, 90, 98, 131,
    258, 413
Surface polarization, 377
Surface poisoning, 111, 133, 276, 345, 376
Surface potential barrier, 124, 125, 127, 226,
    329, 346, 403, 407, 484
Surface probes, 4, 43, 64, 72, 138, 144, 204,
    296, 377, 403
Surface structure, 39, 40, 44, 47, 56, 193, 366
Surface traps, 4
Surface-to-volume ratio, 12, 249, 313, 324,
    337, 353, 356, 398
Surfactant, 172, 246, 249
Synthesis, 80, 121, 134, 136, 140, 146–148,
    180, 249–251, 254, 258, 288, 291, 324,
    406, 410, 427, 520

T
Taguchi gas sensors (TGS), 4, 5
Temperature control, 146, 479
Temperature programmed desorption (TPD),
    16, 17, 81
Temperature programmed reduction (TPR),
    84, 88, 90–92
Temperature-programmed sensing (TPS), 452,
    454–456, 459, 532
Template based growth, 288, 292
Thermal evaporation, 328, 330, 355, 360, 361
Thermodynamics, 40
Thin film characterization, 129, 130, 141
Thin film growth, 189
Thin films, 6, 36, 47, 122, 129, 141,
    180, 190, 192, 193, 197, 212,
    215, 216, 486, 487, 518
Three dimensional (3D), 197, 353, 455
Tin oxide (SnO$_2$), 103, 234, 235, 349–351,
    360, 371, 409, 448, 450, 461, 513
Titanin (TiO$_2$, anatase, rutile, brookite), 46,
    60, 170–174, 178, 183, 255–257, 373,
    379, 419
Top down, 135, 227, 236, 240, 288,
    423, 475
Transducer, 7, 123, 127, 313, 504, 506,
    508, 513, 514, 516, 533–535
Transduction, 349, 366, 377, 383, 391, 398,
    400, 402, 407, 468, 483, 497
Transition metal, 46, 47, 69, 70, 131, 132,
    169, 237, 262, 264, 419
Transmission electron microscopy (TEM),
    141, 236, 350, 355, 440
Tungsten trioxide (WO$_3$), 176
Two dimensional (2D), 44, 80, 416, 455

U

Ultraviolet photoemission spectroscopy, 44
UV, 71, 252, 257, 259, 268–271, 276, 298, 309, 325, 332
UV-visible absorption, 252, 256, 257

V

Vacancies, 8, 17–20, 49, 52, 77, 95, 107, 200, 226, 234, 275, 384, 402, 406, 422
Valence band (VB), 38, 41, 42, 45, 46, 52, 62, 63, 124, 272, 366, 367, 391
Valence dynamics, 410, 413, 522, 535
Vanadium oxide, 232, 416
Vapor phase growth, 288, 289
Vapor-liquid-solid (VLS), 289, 291, 352
Vapor-solid, 289, 291
Volatile organic compounds (VOC), 168, 400, 414, 421
Volume doping, 119, 126, 131, 134, 152

W

Wet chemistry, 518, 520–523
Wireless, 288, 313, 322, 328, 332, 338, 339, 428

Work function, 4, 6, 8, 10, 11, 12, 14, 15, 46, 69, 70, 351, 357, 398–400, 411, 492
Wurtzite, 37, 38, 44, 170, 291

X

X-ray diffraction (XRD), 214
X-ray photoelectron spectroscopy (XPS), 215, 236
X-ray photoemission, 40, 41

Y

Yttria stabilized zirconia, 183, 382

Z

Zero dimensional (0D), 406, 407, 411, 412
Zinc oxide (ZnO), 4, 10, 17, 36–38, 41, 42, 118, 122, 236, 251, 249, 263, 276, 288, 413, 421
Zirconia (ZrO$_2$), 183, 231, 366, 382
Zinc stannate (ZnSnO$_3$), 360, 361

Printed by Publishers' Graphics LLC
BT20130206.19.28.86